D1014691

FLUID MECHANICS MEASUREMENTS

FLUID MECHANICS MEASUREMENTS

Edited by

Richard J. Goldstein

University of Minnesota

HEMISPHERE PUBLISHING CORPORATION

Washington New York London

DISTRIBUTION OUTSIDE NORTH AMERICA

SPRINGER-VERLAG

Berlin Heidelberg New York Tokyo

FLUID MECHANICS MEASUREMENTS

1 2 3 4 5 6 7 8 9 0 B R B R 8 9 8 7 6 5 4 3

This book was set in Press Roman by Hemisphere Publishing Corporation.
The editors were Brenda Brienza, Karen Zauber, and Edward Millman;
the production supervisor was Miriam Gonzalez; and the typesetter
was Sandra F. Watts.
Braun-Brumfield, Inc., was printer and binder.

Library of Congress Cataloging in Publication Data

Main entry under title:

Fluid mechanics measurements.

Includes bibliographies and index.
1. Fluid mechanics–Measurement. I. Goldstein, Richard J.
TA357.F684 1983   620.1′064′0287   83-4292
ISBN 0-89116-244-5 Hemisphere Publishing Corporation

DISTRIBUTION OUTSIDE NORTH AMERICA:
ISBN 3-540-12501-9 Springer-Verlag Berlin

CONTENTS

2 Physical Laws of Fluid Mechanics and Their Application to Measurement Techniques *E. R. G. Eckert*

3 Differential Pressure Measurement *William K. Blake*

4 Thermal Anemometers *Leroy M. Fingerson and Peter Freymuth*

9 Fluid Mechanics Measurements in Nonnewtonian Fluids

Christopher W. Macosko

10 Two-Phase Flow Measurement Techniques in Gas-Liquid Systems

Owen C. Jones, Jr.

CONTRIBUTORS

RONALD J. ADRIAN
Department of Theoretical & Applied Mechanics
University of Illinois at Urbana
Urbana, Illinois 61801

ROGER E. A. ARNDT
St. Anthony Falls Hydraulic Laboratory
University of Minnesota
2-3rd Avenue SE
Minneapolis, Minnesota 55455

WILLIAM K. BLAKE
David Taylor Naval Ship Research & Development Center
Bethesda, Maryland 20084

JAY A. CAMPBELL
Upjohn Company
Kalamazoo, Michigan 49001

E. R. G. ECKERT
Department of Mechanical Engineering
University of Minnesota
111 Church Street SE
Minneapolis, Minnesota 55455

LEROY M. FINGERSON
Thermo Systems Inc.
500 Cardigan Road
P. O. Box 43394
St. Paul, Minnesota 55112

PETER FREYMUTH
Department of Aerospace Engineering Sciences
University of Colorado
Boulder, Colorado 80302

R. J. GOLDSTEIN
Department of Mechanical Engineering
University of Minnesota
111 Church Street SE
Minneapolis, Minnesota 55455

THOMAS J. HANRATTY
Department of Mechanical Engineering
University of Illinois at Urbana
Urbana, Illinois 61801

OWEN C. JONES, JR.
Department of Nuclear Engineering
Rensselaer Polytechnic Institute
Troy, New York 12181

CHRISTOPHER W. MACOSKO
Department of Chemical Engineering & Materials Science
University of Minnesota
301 Amundsen Hall
421 Washington Avenue SE
Minneapolis, Minnesota 55455

G. E. MATTINGLY
Fluids Engineering Division
National Bureau of Standards
FM 105
Washington, D.C. 20234

THOMAS J. MUELLER
Aerospace & Mechanical Engineering Department
University of Notre Dame
Notre Dame, Indiana 46556

PREFACE

Interest in fluid mechanics by scientists and engineers continues to increase. Fluid flow is important in almost all fields of engineering, many industries, and in many studies related to oceanography, meteorology, astronomy, chemistry, geology, and physics.

The advent of sophisticated numerical methods to study and predict fluid flows has not diminished the requirement for measurements but rather has enhanced this need. The development of turbulence models requires much experimental input and eventually verification in many different flows. Flow measurements are required to increase our understanding of the physical processes in flowing systems, particularly turbulent flow and three-dimensional flows, as well as to determine flow quantities needed in a variety of industrial applications.

Many different flow parameters are often required. These include the instantaneous magnitude of velocity, the vector velocity itself, spatial and temporal correlations, turbulent shear stress, and overall volume or mass flow. There is often a need now for more complete and more accurate information than had been required recently. New variations on classical instrumentation as well as relatively new instrumentation and techniques are being used.

Chapter 1 starts this volume with a discussion of the need for different types of measurements–in particular, how the data acquired are used in turbulent flow studies. The second chapter reviews the physical laws or principles that are the bases of many flow instruments.

Chapters 3 through 5 describe techniques used in measuring local velocity. Differential pressure measurements are used in many ways. Perhaps most familiar is using the difference between total and static pressure to determine velocity or volume flow. Chapter 4 deals with hot-wire and hot-film systems. These have been used for more than 50 years in local measurements of velocity and also for a variety of spatial and temporal correlations. In the mid-1960s laser velocimetry, described in Chapter 5, was first used to make nonobtrusive measurements of velocity. Since that time a number of specialized instruments and data acquisition and processing devices have been developed for this technique.

The total volume or mass flow rate of a fluid is often required in industrial applications. Different types of instruments for flow measurement, including the general conditions under which they are best used, are examined in Chapter 6.

The next two chapters cover flow visualization for both qualitative and quantitative studies. Our understanding of a particular flow field can be greatly enhanced by a visualization system even if no quantitative information is obtained. Chapter 7 considers discrete particle and seeding systems while Chapter 8 reviews systems that make use of the variation of the index of refraction of a fluid to visualize flow.

Many flows of interest today are not those of a classical single-phase newtonian fluid. Chapter 8 examines characteristics of the flow of nonnewtonian fluids, including determination of physical properties of such fluids. Chapter 9 considers two-phase flow systems with emphasis on the different types of instrumentation to yield information on liquid-gas flows.

Chapter 11 describes techniques used in the measurement of wall shear stress. This quantity is often needed for determination of pressure drop as well as for examining the transport mechanisms in diverse flow regimes and geometries.

Much of the material in the present volume was first used at the University of Minnesota as part of a week-long course for practicing engineers and scientists reviewing the latest techniques for measurements in flowing fluids. Considerable assistance in the formation of the course was provided by R. Arndt, E. Eckert, and L. Fingerson. The final volume owes much to the efforts of K. Sikora whose patience and persistence encouraged the authors to produce a completed manuscript in a reasonable period of time.

R. J. Goldstein

FLUID MECHANICS MEASUREMENTS

CHAPTER

ONE

WHAT DO WE MEASURE, AND WHY?

Roger E. A. Arndt

1. INTRODUCTION

The purpose of this chapter is to acquaint the reader with the need for flow measurements, to provide some insight concerning various applications, and to introduce the methodology and philosophy of flow measurements. A distinction is drawn between measurements having direct application (e.g., flow rate in a pipe) and measurements that are used indirectly for the correlation of primary data, for the verification of theory, and for the tuning of mathematical models. Some emphasis is placed on the need to understand fluid mechanics for the design of experiments, for the interpretation of results, and for estimating deterministic errors due to flow modification by instrumentation placed in the flow. Examples of diagnostic techniques in research problems are drawn from various fields. The principles of fluid mechanics are reviewed to demonstrate the need for measurements, as well as to provide a basis for describing various systematic errors inherent in flow measurement devices.

Various limitations on flow measurement techniques are described in general terms. The need to understand the trade-off between temporal response and spatial resolution in a flowing system is underscored. Finally, an overview of data analysis is given to provide the reader with some insight into the interrelationship between the flow field, the measurement device, and the results.

The details of actual measurement techniques and data reduction are given in following chapters.

2. THE NEED FOR FLOW MEASUREMENTS

It would be impossible in this chapter to describe all the various types of flow measurements and their applications. However, an attempt is made to describe the overall methodology and philosophy of fluid measurement techniques and their application. One can think of several general needs for flow measurements. In some cases, the data are useful of themselves, as, for example, the rate of flow in a pipe or the mass flux of contaminants in a river. In these cases, the flow quantity desired could be measured either directly or indirectly and it would be the final result. In other cases, flow measurements are necessary for correlation of dependent variables. For example, it is well known that the lift and drag of various vehicles and structures are dependent on the density and velocity of the flow. In this case, a measurement of the velocity would be utilized to correlate force measurements. Erosion rate, heat transfer, ablation, etc., are also known to be functions of velocity, and again velocity would be an independent variable. The measurement of velocity and pressure in a flowing system can also be useful as a diagnostic for determining various quantities. For example, velocity measurements are often used in problems related to noise and vibration and as a diagnostic in heat and mass transfer research. Measurements of velocity and pressure are also used for the verification of theory and, in addition, a great deal of experimental data are necessary for calibrating mathematical models of various types. For example, there are numerous mathematical models describing the diffusion of pollutants from a point source into a large body of fluid, e.g., sewer and smokestack fallout. In these cases, a basic understanding of turbulent mixing is necessary, and appropriate mathematical models are dependent on carefully made measurements of turbulence.

Physical modeling is still very useful in many branches of engineering, ranging from wind-tunnel tests of airplanes and other aerospace vehicles, buildings, diffusion of pollutants, windmills, and even snow fences to hydraulic models of entire dams, reaches of rivers, and cooling-water intakes and outlets. Many of these applications require specialized instrumentation. An example of the detail inherent in many hydraulic models is shown in Fig. 1.

The many applications of fluid flow measurements cover a broad spectrum of activities. For example, one could think of various wind-tunnel studies related to lift and drag, vibration, and noise radiation. In engine and compressor research involving both reciprocating and rotating machinery and in the broad area of hydraulic engineering research, many different types of flow measurements are necessary. These measurements would be made in water tunnels, tow tanks, and flumes, for example. Agricultural research is now becoming more and more involved with basic fluid mechanics, and many types of fluid-mechanics measurements are related to erosion, wind effects on crops, transpiration rates for irrigation systems, and meteorological effects on buildings (including the effects of loading, both steady and unsteady, on grain storage buildings, grain elevators, and other structures).

Ocean engineering requires an extensive amount of experimentation. Measurements are required, for example, in determining the drag and seakeeping characteristics of a ship, and very detailed wake surveys are necessary for the design of the propeller. There are many other types of ocean engineering measurements, requiring a broad spectrum of fluid mechanics measurements. These include wave forces on offshore structures and air-sea interaction studies for understanding the development of waves. Flow measurements are required for fundamental research in many other fields, such as ballistics, combustion, magnetohydrodynamics (MHD), explosion studies, and heat transfer. In addition, many basic types of testing require fluid me-

Figure 1 Model of Guri Dam in Venezuela. Upon completion of the second stage of development, this will be one of the world's largest hydroelectric sites (10,000 MW). *(Courtesy St. Anthony Falls Hydraulic Laboratory, University of Minnesota.)*

chanics measurements, including cooling studies, the design of hydraulic systems, engine tests, flow calibration facilities, and various types of hydraulic machinery for handling pump flow and suspensions.

Fluid mechanics measurements are also extremely important in biomedical research related to pulmonary function, blood flow, urine flow, heart-valve models, biological cell movements, and other studies. Many flow measurements are involved in the broad area of process control measurement, such as in engines, refinery systems, semiconductor doping processes, and pilot-plant studies.

3. WHAT DO WE NEED TO KNOW?

Although it would seem to be obvious, we should have a clear picture of the requirements before undertaking any physical measurement. However, many times the application of a particular fluid flow measurement device is inappropriate, simply because the user of the instrument has not taken the time to clearly define the need. For example, we must first know the purpose of the measurement. Is the flow quantity that is measured to be used as is? Second, we must understand the fluid mechanics of the problem. Almost all types of measurement techniques depend on the nature of the flow, and this in turn governs instrument selection. We must also understand the physical principles involved in flow measurement. Almost all fluid flow measurements are indirect, in that the techniques rely on the physical interpretation of the quantity measured; e.g., a Pitot tube senses pressure. We infer the velocity from the measured pressure using the Bernoulli equation. When we measure pressure with a pressure transducer, we sense deflection of a membrane or diaphragm, usually electronically, and we interpret the deflection of this membrane in terms of pressure. Hot-wire instrumentation is dependent on the physical laws relating heat transfer with flow velocity.

Surprising as it sounds, we also need to know the answer before we begin. That is to say, we must have a reliable estimate of what the results should be. This is often overlooked, and sometimes measurements that are grossly incorrect are taken as gospel simply because the physical situation has not been carefully evaluated. It can also be said that one cannot wisely purchase flow measurement equipment without considering specific experiments. *Almost all sophisticated equipment purchased for general or unspecified use is a waste of money.*

4. EXAMPLES OF FLUID MECHANICS MEASUREMENTS

As already mentioned, there are an infinite variety of fluid mechanics measurements and applications. However, to provide some insight into the various types of measurements, a few examples are given below.

4.1 Measurement of Sediment Load in a Stream

To measure the total quantity of suspended material in a stream, both the velocity and the concentration at several depths must be measured and the results integrated:

$$G = b \int_0^d cu \, dy \tag{1}$$

The measurement equipment necessary to evaluate Eq. (1) is shown in Fig. 2. The sediment sampler shown in Fig. 2*a* is the result of extensive development by the U. S. Interagency Group on Sedimentation [1, 2]. By correctly positioning exhaust ports at a point where the pressure is the same as that in the free stream, the velocity in the intake nozzle is made always equal to the local stream velocity. Proper sizing of fins is necessary for stability, the design being an outgrowth of standard procedures for tailplane design in aerodynamics. This device is designed to automatically carry out the integration procedure in Eq. (1) during lowering and raising at a uniform rate. It is an interesting example of the application of fluid mechanical principles to a specialized measurement problem. It is typical to use a propeller meter of the type shown in Fig. 2*b* for measurements of flow velocity in large channels, streams, etc. These propeller meters are based on the simple principle that the rotational speed of the meter is

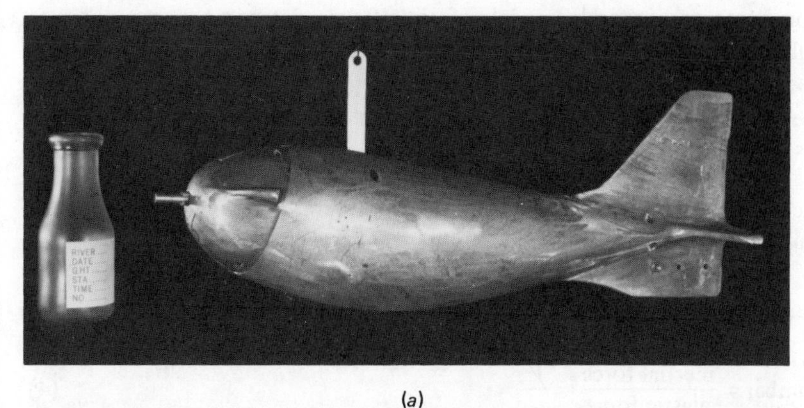

(*a*)

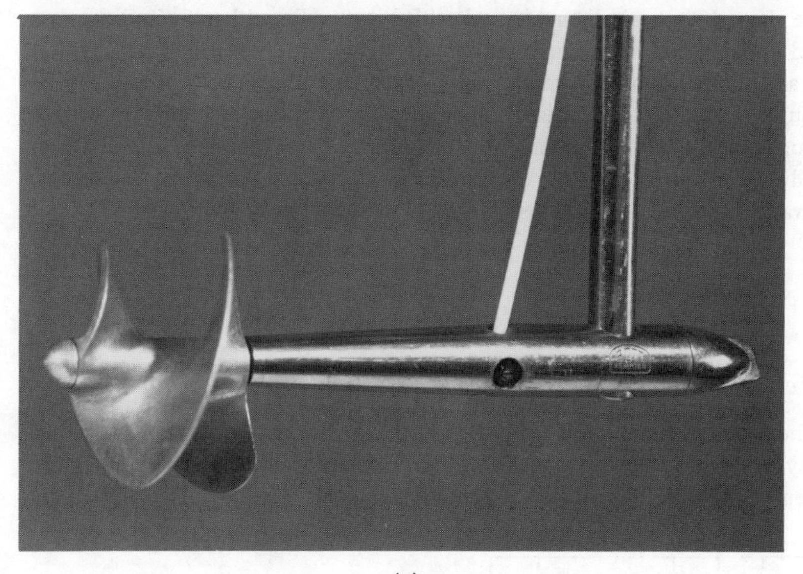

(*b*)

Figure 2 (*a*) Sediment sampler developed by United States Geological Survey. (*b*) Typical propeller meter used in hydraulic engineering studies.

proportional to the local flow velocity. The rotational speed is sensed by a counter, and the number of counts per second is displayed on a recorder. Such devices require periodic calibration, and this is usually accomplished in a laboratory flume.

4.2 Wind-Tunnel Studies

Wind tunnels are used for a myriad of investigations ranging from fundamental research to industrial aerodynamics [3, 4]. Many wind-tunnel studies involve the determination of forces on scale models of aircraft, aircraft components, automobiles, or buildings. Forces such as lift or drag are known to obey the following law of similitude:

$$F = \tfrac{1}{2}\rho C_N V^2 S \tag{2}$$

where S is a surface area or cross-sectional area, depending on the application. The force coefficient C_N is known to be a function of several nondimensional parameters. The major ones in aerodynamics are

$$\text{Reynolds number} = \frac{\text{inertial force}}{\text{viscous force}} = \frac{\rho V l}{\mu} \tag{3}$$

$$\text{Mach number} = \frac{\text{inertial force}}{\text{elastic force}} = \frac{V}{a_0} \tag{4}$$

To correlate data, V is typically measured with Pitot or Prandtl tube, and the temperature and pressure are determined with appropriate instrumentation (see Chap. 2). The forces and moments on the model are usually determined with a specially designed balance. The density is usually computed from the measured temperature and pressure. Typical wind-tunnel data are shown in Fig. 3. These data, adapted from National Advisory Committee for Aeronautics (NACA) data for an airfoil section, display the variation of section lift coefficient and moment coefficient, defined as

$$C_l = \frac{L}{\frac{1}{2}\rho U^2 c} \tag{5}$$

$$C_m = \frac{M}{\frac{1}{2}\rho U^2 c^2} \tag{6}$$

Note that the primary parameter is the angle of attack, with the influence of Reynolds number only being noted at high angles of attack where stall occurs. The maximum lift coefficient is strongly dependent on Reynolds number. This is due to the separa-

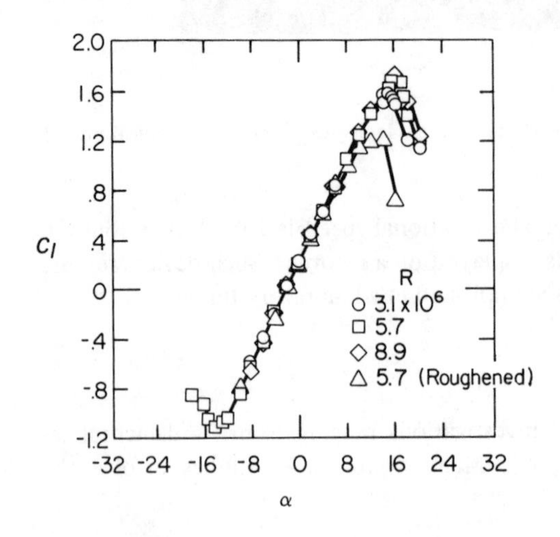

Figure 3 Typical airfoil data obtained in a wind tunnel.

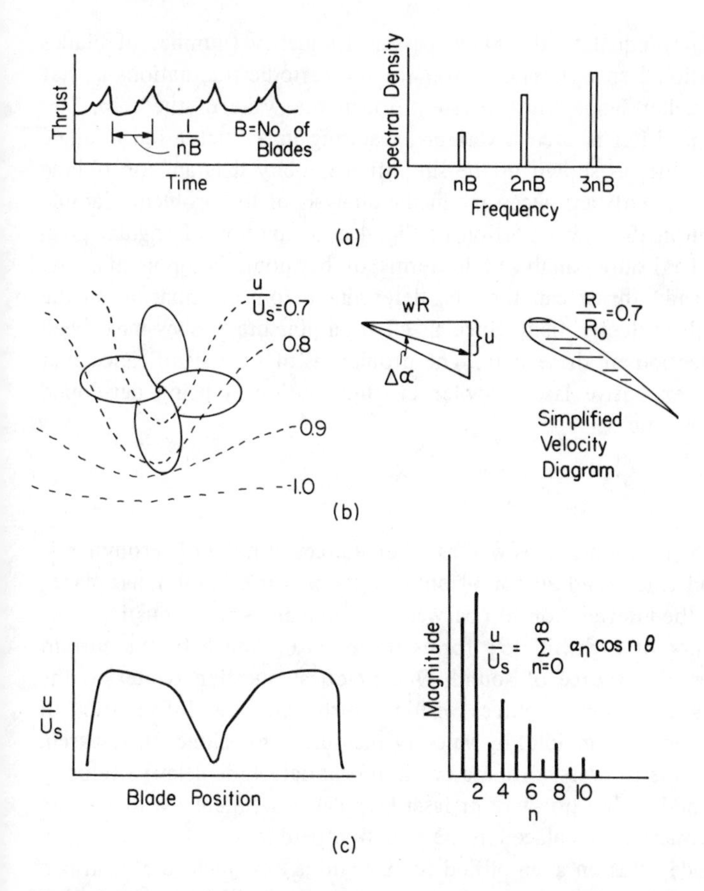

Figure 4 Use of wake velocity measurements to isolate a ship-propeller vibration problem.

tion of the boundary layer at the leading edge of the airfoil. Separation phenomena are known to be dependent on such characteristics of the boundary layer at separation as shape factor and whether it is laminar or turbulent.

4.3 Propeller Vibration

Marine propellers must operate in the confused flow in the wake of a ship. Because a propeller rotates through a nonuniform velocity field, unsteady thrust and side forces are produced. The unsteadiness of the propulsor can in some cases be extreme, producing unwanted vibration within the ship. In some cases the propeller vibration produced is so extreme that piping is shaken loose and the crew find working on the ship extremely uncomfortable. There have been cases of newly built ships that have been boycotted by the seamen's union, creating the need for an extremely expensive retrofit program. The overall solution to this problem is extremely complex. Figure 4 illustrates the problem and its analysis.

As shown, a propeller produces unsteady thrust and side forces that are periodic in time with a frequency equal to the blade passing frequency (number of blades multiplied by the rotational speed). The reason for the periodic fluctuations is that each section of a propeller blade must rotate through the wake of the ship. The axial velocity at each position is then a variable, resulting in a fluctuation in angle of attack and unsteady lift, as shown in the simplified velocity diagram for a blade section. Velocity measurements are necessary in the analysis of the problem. Sample velocity data are shown in the lower portion of Fig. 4 as a function of angular position. These data can be Fourier-analyzed in terms of harmonic components. The harmonics of the unsteady thrust can then be determined from information on the harmonic content of the velocity field. Specialized total-pressure probes have been developed for the collection of these data. The problem is of such significance that highly specialized and expensive laser-Doppler instrumentation is being developed for measurements of this type.

4.4 Aeroacoustics

Increasing concern with aircraft noise, as well as other sources of noise of aerodynamic origin (automobiles and trucks and air-conditioning systems, for example), has placed increasing emphasis on the interrelationship between aerodynamics and acoustics.

The pioneering work of Lighthill [5] forms the basis for much of the current research on turbulence as a source of sound. One practical question relates to the distribution and intensity of sound sources within a turbulent flow. Information of this type can be obtained, in principle, by velocity measurements alone. In research, a more common method is to cross-correlate (with appropriate time delay) a velocity signal (usually determined with a hot-wire or laser-Doppler velocimeter) with a sound signal determined by a microphone placed in the radiation field.

The first method depends on a simplified theory and is extremely useful in certain applications, such as the determination of sound sources in the region of a jet. The principle is based on Lighthill's theory. In simplified form, the intensity per unit volume of sound source is given by

$$\frac{I}{V_e} \sim \frac{V_e \omega^4 T_{ij}^2}{r^2 \rho_0^2 a_0^5} \tag{7}$$

In terms of near field pressure p, an alternative form of Eq. (7) is

$$\frac{I}{V_e} \sim \frac{V_e \omega^4}{r^2} \frac{\overline{p^2}}{\rho_0^2 a_0^5} \tag{8}$$

Work by Kraichnan [6], Davies et al. [7], and Lilley [8] indicate that the following simplifications can be used:

$$p \cong \rho u l \frac{\partial U}{\partial y} \tag{9}$$

$$\omega \cong \frac{\partial U}{\partial y} \cong \frac{u}{l} \tag{10}$$

$$V_e \cong l^3 \tag{11}$$

where l is a characteristic scale of turbulence usually defined by the integral scale. (More is said about integral scale in later chapters.) Hence, a simplified method for determining acoustic noise sources in a turbulent flow is based on the following equation [which follows from substitution of Eqs. (9) to (11) into Eq. (8)]:

$$\frac{I}{V_e} \sim \frac{l^5}{a_0^5 r^2} \rho \overline{u^2} \left(\frac{\partial U}{\partial y} \right)^6 \tag{12}$$

Hence, measurement of integral scale l, turbulence intensity $\overline{u^2}$, and mean velocity gradient $\partial U/\partial y$ can give an estimate of the sound-source distribution in a turbulent flow. Direct verification of Eq. (12) is difficult; however, its basis, Eq. (9), has been verified directly. An example is shown in Fig. 5. The upper portion of the figure illustrates the theoretical distribution of pressure intensity within a turbulent jet. The lower portion of the figure contains pressure data obtained with a special pressure probe developed by Arndt and Nilsen [10].

Cross-correlation techniques (discussed in detail in later chapters) are useful in aeroacoustic research. The pressure p_a in the far field is given by the volume integral

$$p_a = \frac{1}{4\pi a_0^2 r} \int_{\text{vol}} \left[\frac{\partial^2}{\partial t^2} (\rho u_r^2) \right]_{t'} dy \tag{13}$$

where $t' = t - r/a_0$ = delay time

u_r = component of velocity fluctuation in direction of observer

Since the acoustic intensity I is related to the mean square acoustic pressure, Eq. (13) can be modified to yield

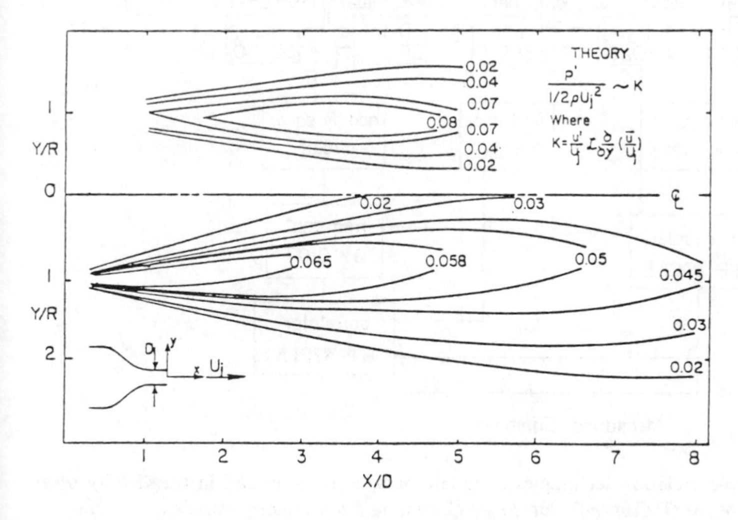

Figure 5 Measured pressure fluctuations in a turbulent jet. *(After Barefoot [9].)*

$$I = \frac{\overline{p_a^2}}{\rho_0 a_0} \simeq \frac{V_e}{4\pi\rho_0 a_0^3 r} \frac{\partial}{\partial t} \overline{(\rho u_r^2 p_a)}_{t=t'} \tag{14}$$

or, equivalently,

$$I \simeq \frac{V_e}{4\pi\rho a_0^3 r} \frac{\partial}{\partial t} \overline{(p p_a)}_{t=t'} \tag{15}$$

where p is the pressure in the flow. Hence, sound sources can be determined by cross-correlation of a near-field pressure or velocity signal and a far-field velocity signal. The principle is illustrated in Fig. 6 for the case of airfoil noise due to pressure

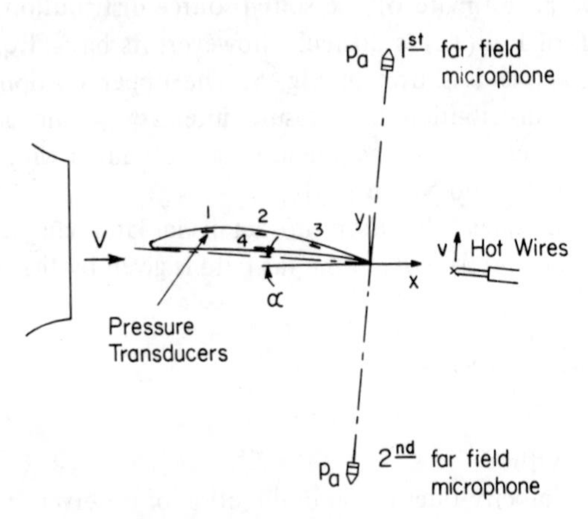

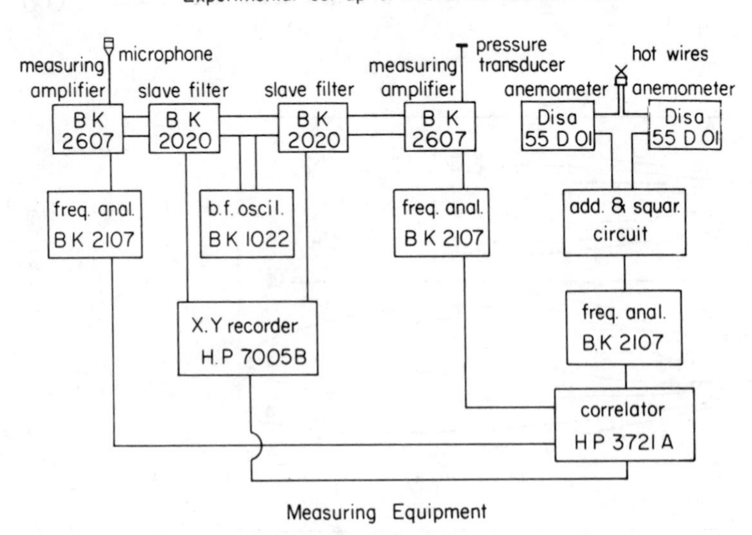

Figure 6 Use of cross-correlation techniques to isolate noise sources on and in the vicinity of an airfoil. *(Courtesy Professor G. Comte-Bellot, Ecole Central de Lyon, Ecully, France.)*

fluctuations on the surface of the foil and velocity fluctuations in the wake. A typical instrumentation package is also illustrated.

4.5 Turbulent Mixing Layer

As a final example, consider fundamental research into the properties of a turbulent mixing layer formed at the interface between two fluids of different densities and different velocities, as illustrated in Fig. 7. This is an interesting example since it illustrates how single-point measurements can be misleading. As shown in the upper portion of the figure, mean velocity and density vary uniformly across the mixing layer, with turbulence intensity peaking somewhere in the central portion of the layer. Correlation of mean velocity and density in a manner implied by classical theory results in similar profiles, as shown in the figure. Instantaneous readings of density (shown in the lower portion of the figure) indicate that sharp gradients in density exist. Further, a shadowgraph of the mixing layer indicates the existence of a quasi-orderly structure as shown. The wavelength of the structure varies with distance x and has a statistical variation as shown. It is the statistical variation (or "jitter") that has led to the misinterpretation of hot-wire data over the years. In fact, hindsight tells us that observations at one or two spatially fixed stations will include realizations from a large number of such structures at various stages in their life history. Time and space averages in the classical sense will tend to "smear out" the essential features of the quasi-orderly structure in the flow [12].

4.6 Summary

Examples of the use of flow instrumentation have been drawn from a variety of fields to illustrate the interrelated roles of flow instrumentation, the theory of fluid mechanics, and research and development. The need for a clear understanding of the problem at hand, the principles of fluid mechanics, the principles of operation of flow instrumentation, and the elements of statistical analysis is evident.

5. OUTLINE OF THE THEORY OF FLUID MECHANICS

5.1 Inviscid Flow

The reader is assumed to have a basic understanding of fluid mechanics. However, the mathematical principles are briefly reviewed here. It is sufficient for this overview to limit the discussion to incompressible fluid mechanics, to illustrate the interrelationship between theory and measurement principles.

In a eulerian frame of reference, conservation of mass and momentum yield (in standard tensor notation)

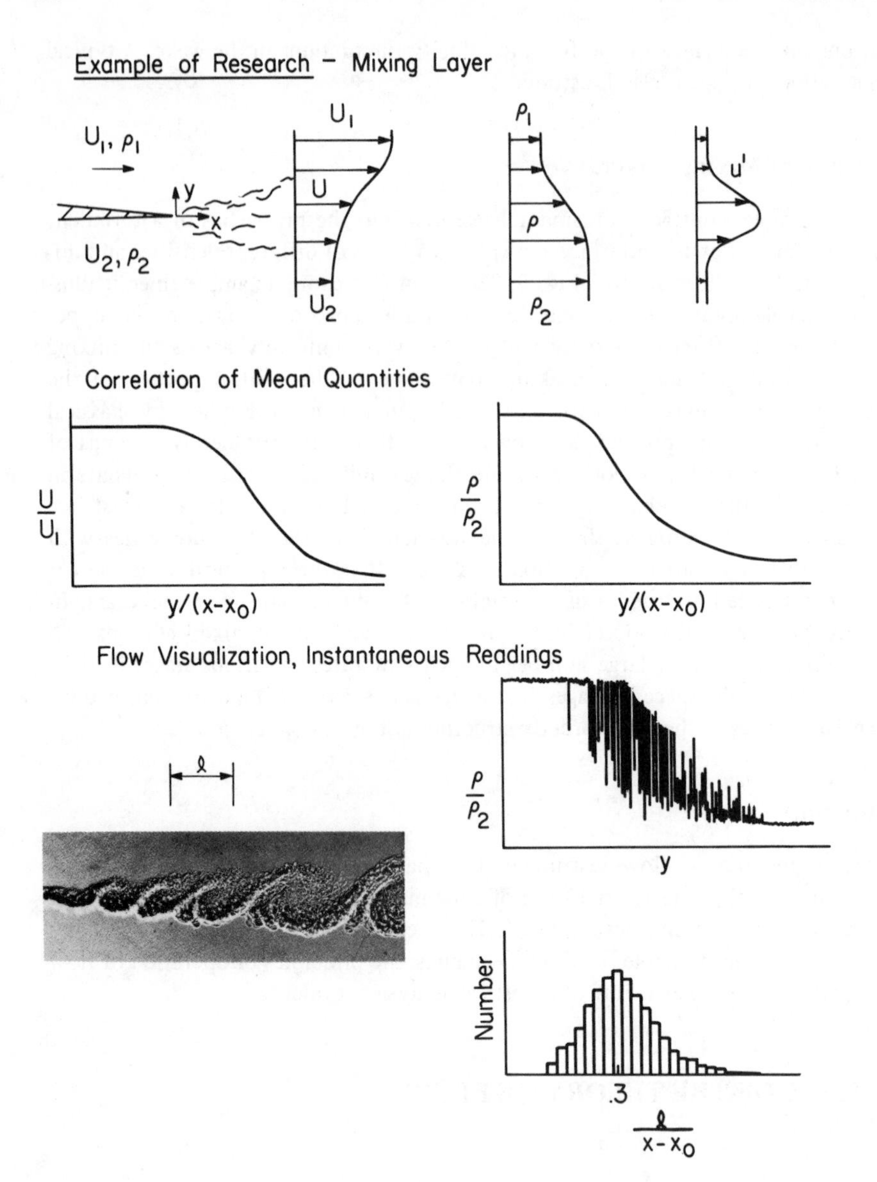

Figure 7 Example of research on the turbulence characteristics of a mixing layer. *(Adapted from [11].)*

$$\frac{\partial U_i}{\partial x_i} = 0 \tag{16}$$

$$\rho \frac{DU_i}{Dt} = \rho X_i + \frac{\partial \tau_{ji}}{\partial x_j} \tag{17}$$

$$\frac{D}{Dt} \equiv \frac{\partial}{\partial t} + U_i \frac{\partial}{\partial x_i} \tag{18}$$

where x_i are the components of a body force field. In "frictionless" flow the only term in the stress tensor τ_{ji} is the pressure p. Hence,

$$\rho \frac{DU_i}{Dt} = X_i - \frac{\partial p}{\partial x_i} \tag{19}$$

where p is positive for compression. Equation (19) can be written in vector notation as

$$\rho \left[\frac{\partial \mathbf{V}}{\partial t} + (\mathbf{V} \cdot \nabla)\mathbf{V} \right] = \mathbf{X} - \nabla p \tag{20}$$

Using the vector identity

$$\nabla(\mathbf{V} \cdot \mathbf{V}) = 2(\mathbf{V} \cdot \nabla)\mathbf{V} + 2\mathbf{V} \times (\nabla \times \mathbf{V}) \tag{21}$$

Eq. (20) can be written in the form

$$\rho \left[\frac{\partial \mathbf{V}}{\partial t} + \nabla \frac{V^2}{2} - \mathbf{V} \times (\nabla \times \mathbf{V}) \right] = \mathbf{X} - \nabla p \tag{22}$$

When the flow is irrotational ($\nabla \times \mathbf{V} = 0$), Eq. (22) can be integrated. A velocity potential ϕ exists such that

$$\mathbf{V} = -\nabla \phi \tag{23}$$

Limiting ourselves to conservation force fields, we can also write

$$\mathbf{X} = -\nabla \Omega \tag{24}$$

Thus

$$\nabla \left(-\frac{\partial \phi}{\partial t} + \frac{V^2}{2} + \frac{p}{\rho} + \Omega \right) = 0 \tag{25}$$

Equation (25) leads to

$$-\frac{\partial \phi}{\partial t} + \frac{V^2}{2} + \frac{p}{\rho} + \Omega = F(t) \tag{26}$$

where $F(t)$ is an arbitrary function of time. For steady flow,

$$\frac{V^2}{2} + \frac{p}{\rho} + \Omega = \text{const} \tag{27}$$

If the only conservative force field is that due to gravity, then

$$\Omega = gz \tag{28}$$

Equation (22) can also be integrated for inviscid rotational flow. The result is the same as Eq. (27), except that the *constant is different for each streamline.*

The velocity potential ϕ satisfies Laplace's equation,

$$\nabla^2 \phi = 0 \tag{29}$$

This follows directly from the definition of ϕ and continuity,

$$\mathbf{V} = -\nabla\phi \qquad \nabla \cdot \mathbf{V} = 0$$

The many techniques for the solution of Eq. (29) are beyond the scope of this review. In each case, the principle is the same: ϕ is determined from Eq. (29) for the given boundary conditions, the velocity field is determined from Eq. (23), and, finally, the pressure is deduced from the Bernoulli equation. An illustration of the technique is the design of a probe for measuring velocity in a uniform flow. The pressure sensed at the nose is the total pressure

$$p_t = p_0 + \tfrac{1}{2}\rho V^2 \tag{30}$$

The problem is to find a position on the probe where the pressure is p_0. The difference in the pressures at these two points can be used to determine the flow velocity

$$V = \sqrt{\frac{2(p_t - p_0)}{\rho}} \tag{31}$$

A simple probe, consisting of a blunt-nosed cylinder aligned with the flow and a vertical stem, is illustrated in Fig. 8. As illustrated, the pressure varies from p_t at the nose to values less than p_0 a short distance downstream and then increases asymptotically to p_0 further downstream. Since the stem of the probe causes an increase in pressure as shown, it is comparatively easy to find a position where the vacuum due to the nose of the cylinder is balanced by the increased pressure due to the stem. Details are given in [13].

Measurements of flow velocity are not usually made in the ideal uniform flow for which a Prandtl tube is designed. A typical situation is illustrated in Fig. 9. The measurement is being made in turbulent shear flow. Often, under these circumstances,

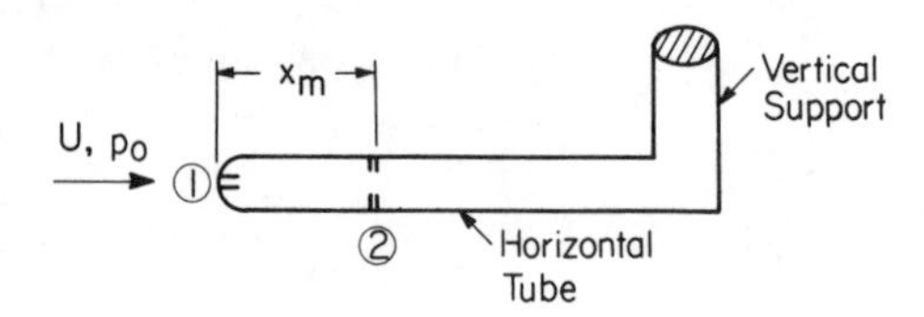

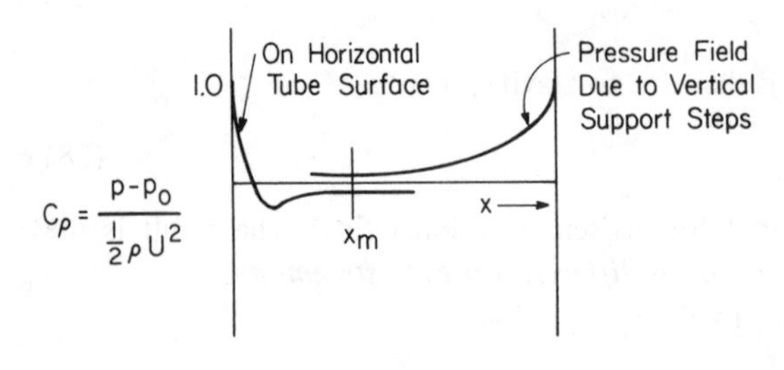

Figure 8 Fluid-dynamic considerations in the design of a Prandtl tube.

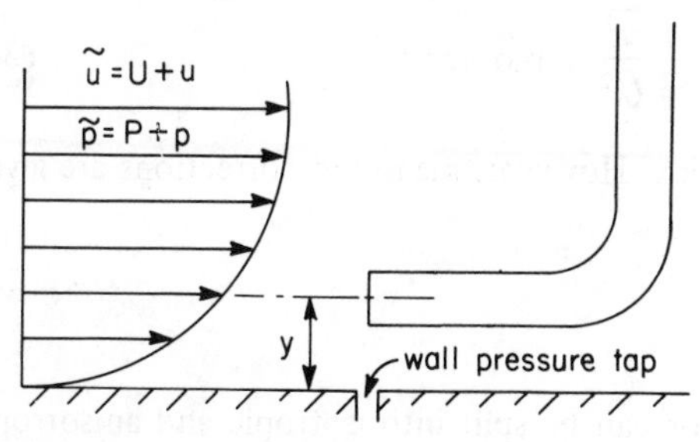

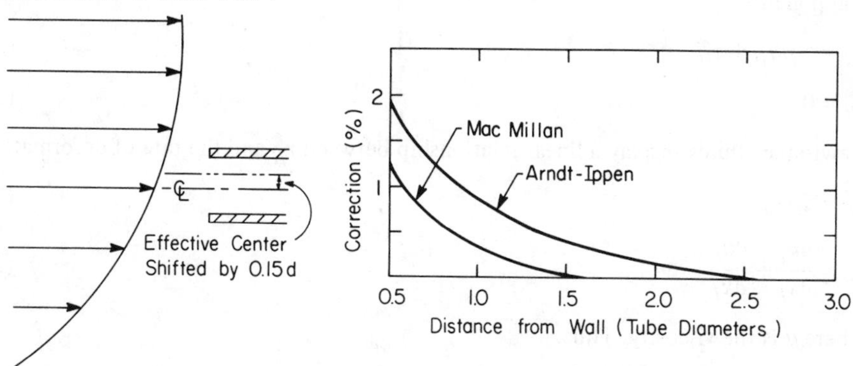

Figure 9 Effects of turbulence velocity gradients and wall proximity on measurements of total pressure.

it is convenient to measure total pressure with a total-head tube, and static pressure with a wall tap. There are three different flow effects on the total-pressure measurement: The time mean total pressure does not correspond to the time mean velocity because of the turbulence; the effective center of the probe is shifted toward the high-velocity region; and the wall creates an asymmetrical flow around the tube. The first of these effects can be estimated from the Bernoulli equation. The instantaneous total pressure is

$$\tilde{p}_t = \tfrac{1}{2}\rho(U+u)^2 + P + p \qquad (32)$$

where U and P are the time mean velocity and pressure, respectively, and u and p are the instantaneous fluctuations about the mean. Expanding Eq. (32) and taking the average yields (where bars denote time averages)

$$P_t = \frac{\rho}{2}U^2 + \frac{\rho}{2}\overline{u^2} + P \qquad (33)$$

Thus,

$$\frac{U}{[2(P_t - P)/\rho]^{1/2}} = \left(1 + \frac{\overline{u^2}}{U^2}\right)^{-1/2} = 1 - \frac{1}{2}\frac{\overline{u^2}}{U^2} + \text{h.o. terms} \tag{34}$$

The other two problems are more complex. However, measured corrections are given in Fig. 9.

5.2 Viscous Flow and Turbulence

For incompressible flow the stress tensor can be split into isotropic and anisotropic components:

$$\tau_{ji} = -p\delta_{ji} + \tau_{ji}^* \tag{35}$$

$$\tau_{ii}^* \equiv 0 \tag{36}$$

Newtonian fluids display a linear relationship between τ_{ji}^* and the rate of deformation:

$$\tau_{ij}^* = 2\mu e_{ij} \tag{37}$$

$$e_{ij} = \frac{\partial u_i}{\partial x_j} + \frac{\partial u_j}{\partial x_i} \tag{38}$$

where μ is the viscosity. Thus,

$$\tau_{ji} = -p\delta_{ji} + 2\mu\left(\frac{\partial u_i}{\partial x_j} + \frac{\partial u_j}{\partial x_i}\right) \tag{39}$$

Substitution into the momentum equation yields

$$\rho\frac{Du_i}{D_t} = \rho X_i - \frac{\partial p}{\partial x_i} + \mu\frac{\partial^2 u_i}{\partial x_j\,\partial x_j} \tag{40}$$

or

$$\frac{D\mathbf{V}}{Dt} = \mathbf{X} - \frac{1}{\rho}\nabla p + \nu\nabla^2\mathbf{V} \tag{41}$$

Equations (40) and (41) are the Navier-Stokes equations, which form the basis for theoretical fluid mechanics. They are nonlinear, and no general solution is known.

In certain cases the equations reduce to a linear form, such as for flow through a straight channel, where the convective acceleration terms are identically zero. Another example is Stokes' first problem, which considers a viscous fluid in the half-plane $y \geqslant 0$, bounded by a wall at $y = 0$, which is suddenly accelerated to a velocity U. In this case the problem reduces to

$$\frac{\partial u}{\partial t} = \nu\frac{\partial^2 u}{\partial y^2} \tag{42}$$

$$\frac{\partial u}{\partial x} = 0 \tag{43}$$

with

$$u(y,0)=0$$
$$u(0,t)=U_0 \tag{44}$$
$$u(\infty,t)=0$$

The solution is

$$\frac{u}{U_0}=\operatorname{erfc}\frac{y}{\sqrt{4\nu t}} \tag{45}$$

For $y > 4\sqrt{\nu t}$, u is negligible. This leads to the observation that viscous effects are limited to a thin layer of fluid of thickness $\sqrt{\nu t}$. To an observer moving with velocity U, the thickness of this layer is

$$\delta = 4\sqrt{\frac{\nu x}{U}} \tag{46}$$

Boundary layers, jets, and wakes all display this behavior. In such cases the Navier-Stokes equations and the continuity equation take the form

$$\frac{\partial u}{\partial t}+u\frac{\partial u}{\partial x}+v\frac{\partial u}{\partial y}=-\frac{1}{\rho}\frac{\partial p}{\partial x}+\nu\frac{\partial^2 u}{\partial y^2} \tag{47}$$

$$\frac{\partial u}{\partial x}+\frac{\partial v}{\partial y}=0 \tag{48}$$

There are two equations and three unknowns, u, v, and p. The problem is solved by first neglecting viscous effects and solving for the pressure p. Gradients in pressure normal to the boundary-layer flow are negligible, and hence p is then considered known in solving the system of equations (47) and (48). (Refer to [14] for details.)

5.3 Turbulence

Turbulent flows are highly complex. Since many measurements are made in turbulent flow, some features are reviewed here. The uninitiated should, however, refer to the several textbooks on the topic. (See, for example, [15 to 17].) Turbulence can be regarded as a highly disordered motion resulting from the growth of instabilities in an initially laminar flow. Recent information, however, indicates that the turbulence may be more orderly than has been previously thought possible [12].

It is convenient to decompose the velocity into a mean and fluctuating part. For this discussion the mean motion will be a function of position only:

$$U=\lim_{T\to\infty}\frac{1}{T}\int_0^T \tilde{u}(t)\,dt \tag{49}$$

and

$$\tilde{u}=U+u \tag{50}$$

Similarly, for pressure,

$$\tilde{p} = P + p \tag{51}$$

and so on. The turbulent velocity field u is made up of a broad range of scales. The time scale for the large eddies is proportional with l/u, where l is a characteristic length in the flow. Most of the turbulent kinetic energy is associated with large eddies. At the other end of the spectrum, viscous dissipation occurs. The time scale for viscous dissipation can be estimated from the diffusion equation

$$\frac{\partial u}{\partial t} = \nu \frac{\partial^2 u}{\partial y^2} \tag{52}$$

which can be written as

$$\frac{u}{T_m} \sim \nu \frac{u}{l^2} \tag{53}$$

or

$$T_m \sim \frac{l^2}{\nu} \tag{54}$$

Hence, time scales for turbulent and molecular diffusion have the ratio

$$\frac{T_l}{T_m} = \frac{l/U}{l^2/\nu} = \mathcal{R} \tag{55}$$

An equation for turbulent diffusion can be written in analogy with Eq. (52):

$$\frac{\partial u}{\partial t} = \nu_t \frac{\partial^2 u}{\partial y^2} \tag{56}$$

where

$$\nu_t = \frac{l^2}{T_t} = ul \tag{57}$$

The ratio of turbulent to molecular diffusivity is then

$$\frac{\nu_t}{\nu} = \frac{ul}{\nu} = \mathcal{R} \tag{58}$$

For turbulent flows the characteristic length l is proportion to δ or the width of a jet or wake, i.e., a transverse length scale. The smallest scales in a turbulent flow are controlled by viscosity. The small-scale motion has much smaller characteristic time scales than the large-scale motion and may be considered statistically independent. We may think of the small-scale motion as being controlled by the rate of supply of energy and the rate of dissipation. It is probably reasonable to assume that the two rates are equal. Thus we consider the dissipation rate per unit mass ϵ and the kinematic viscosity ν. From purely dimensional reasoning we can form length, time, and velocity scales

$$\eta = \left(\frac{\nu^3}{\epsilon}\right)^{1/4} \qquad \tau = \left(\frac{\nu}{\epsilon}\right)^{1/2} \qquad u_v = (\nu\epsilon)^{1/4} \tag{59}$$

A Reynolds number formed by these parameters is unity:

$$\frac{u_v \eta}{\nu} = 1 \tag{60}$$

which indicates that viscous effects dominate the dissipation mechanism and length scales adjust themselves to the energy supply. For equilibrium flow, we can assume that the rate of supply of energy from the large eddies is inversely proportional to their time scale. Thus,

$$\epsilon \sim \frac{u^3}{l} \tag{61}$$

This yields

$$\frac{\eta}{l} = \left(\frac{\nu^3 l}{u^3 l^4}\right)^{1/4} = \frac{1}{\mathcal{R}^{3/4}} \tag{62}$$

We can also say that

$$\frac{\tau u}{l} = \left(\frac{ul}{\nu}\right)^{-1/2} = \frac{1}{\mathcal{R}^{1/2}} \tag{63}$$

Hence the time scale of the small-eddy structure is much smaller than the large scale eddies. Since vorticity is proportional to $1/t$, we find that the vorticity is concentrated in the small scales.

It is generally accepted that turbulent flows can be described by the Navier-Stokes equations. Using the decomposition described by Eqs. (50) and (51) in Eq. (40) and taking the time mean results in

$$U_j \frac{\partial U_i}{\partial x_j} = \frac{1}{\rho}\frac{\partial}{\partial x_j}(-P\delta_{ij} + 2\mu e_{ij} - \rho\overline{u_i u_j}) \tag{64}$$

$$\frac{\partial U_i}{\partial x_i} = 0 \tag{65}$$

The important feature of Eq. (64) is the additional apparent stress tensor, $-\rho\overline{u_i u_j}$. The components of the $-\rho\overline{u_i u_j}$ are the basis for the so-called closure problem in turbulent flow computations. Subtraction of Eq. (64) from the fully expanded Navier-Stokes equations yields a system of equations for the turbulent fluctuations:

$$\frac{\partial u_i}{\partial t} + U_j\frac{\partial u_i}{\partial x_j} + u_j\frac{\partial U_i}{\partial x_j} + u_j\frac{\partial u_i}{\partial x_j} - \overline{u_j\frac{\partial u_i}{\partial x_j}} = \frac{1}{\rho}\frac{\partial p}{\partial x_i} + \nu\frac{\partial u_i}{\partial x_j\,\partial x_j} \tag{66}$$

$$\frac{\partial u_j}{\partial x_j} = 0 \tag{67}$$

$$U_j \frac{\partial}{\partial x_j}\left(\frac{1}{2} U_i U_i\right) = \frac{\partial}{\partial x_j}\left(\underbrace{-\frac{P}{\rho} U_j}_{(1)} + \underbrace{2\nu U_i e_{ij} - \overline{u_i u_j} U_i}_{(2)}\right) \underbrace{- 2\nu e_{ij} e_{ij}}_{(3)} + \underbrace{\overline{u_i u_j} e_{ij}}_{(4)} \tag{68}$$

$$U_j \frac{\partial}{\partial x_j}\left(\frac{1}{2} \overline{u_i u_i}\right) = -\frac{\partial}{\partial x_j}\left(\underbrace{\frac{\overline{p u_j}}{\rho}}_{(1)} \underbrace{- 2\nu \overline{u_i e'_{ij}} + \frac{1}{2} \overline{u_i u_i u_j}}_{(2)} \underbrace{- 2\nu \overline{e'_{ij} e'_{ij}}}_{(3)} \underbrace{- \overline{u_i u_j} e_{ij}}_{(4)}\right) \tag{69}$$

In each equation, *(1)* denotes pressure work, *(2)* denotes transport terms, *(3)* denotes dissipation terms, and *(4)* denotes production. Also, e_{ij} is the mean strain tensor, and e'_{ij} the fluctuation strain tensor. The important point is that the production term in Eq. (68) is negative, whereas in Eq. (69) it is positive. In other words, the turbulence is being produced at the expense of the mean flow.

In a steady, homogeneous, pure-shear flow in which all averaged quantities except U_i are independent of position and in which e_{ij} is constant, Eq. (69) reduces to

$$-\overline{u_i u_j} e_{ij} = 2\nu \overline{e'_{ij} e'_{ij}} \tag{70}$$

This says simply that the rate of production of turbulent energy by Reynolds stresses equals the rate of viscous dissipation. This is an idealized situation, and in most shear flows production and dissipation do not balance, although they are usually of the same order of magnitude.

5.3.1 Isotropic turbulence. The case of isotropic turbulence is of interest for two reasons: First, it is the simplest case of turbulence and is amenable to analytical treatment; second, the small-scale structure is approximately isotropic. This type of turbulence is homogeneous (the same at every point) and statistically independent of orientation and location of axis.

As a result of the definitions, we have

$$\overline{u^2} = \overline{v^2} = \overline{w^2} \tag{71}$$

$$\overline{uv} = \overline{vw} = \overline{wu} = 0 \tag{72}$$

For truly isotropic conditions the mean velocity is zero. It can be shown that this type of flow can be completely described by two functions

$$R_u(r) = \frac{\overline{u_1 u_2}}{\overline{u^2}} \tag{73}$$

and

$$R_v(r) = \frac{\overline{v_1 v_2}}{\overline{u^2}} \tag{74}$$

where $u, v = x, y$ components of velocity
$r =$ separation distance between points 1 and 2

The longitudinal and lateral velocity coordinates are not independent but are related in the form

$$R_v = R_u + \frac{r}{2}\frac{\partial R_u}{\partial r} \tag{75}$$

Based on the correlation function R_u we can define a length scale as the radius of curvature of $R_u(0)$:

$$-\frac{1}{\lambda_T^2} = \left.\frac{\partial^2 R_u}{\partial r^2}\right|_{r\to 0}$$

where λ_T is related to the radius of the smaller eddies and is referred to as the Taylor microscale. Successive differentiation of Eq. (75) yields the relation (see page 290 in [18])

$$2\overline{\left(\frac{\partial u_i}{\partial x_i}\right)^2} = \overline{\left(\frac{\partial u_j}{\partial x_i}\right)^2} \qquad \text{(no sum)} \tag{76}$$

It can be shown that

$$\frac{\overline{u^2}}{\lambda_T^2} = \overline{\left(\frac{\partial u}{\partial x}\right)^2} \tag{77}$$

In isotropic turbulence the dissipation term can be written as

$$\epsilon = 2\nu\overline{e'_{ij}e'_{ij}} = 15\nu\overline{\left(\frac{\partial u}{\partial x}\right)^2} \tag{78}$$

From Eq. (77),

$$\epsilon = 15\nu\frac{\overline{u^2}}{\lambda_T^2} \tag{79}$$

If we can approximate the fine-scale turbulence in a shear flow with this model of isotropic turbulence, it is possible to obtain an approximate energy balance in terms of the Taylor microscale. Approximating the production term by

$$-\overline{u_i u_j}e_{ij} \simeq A\frac{u^3}{l} \tag{80}$$

we obtain

$$A\frac{u^3}{l} = 15\nu\frac{u^2}{\lambda_T^2} \tag{81}$$

or

$$\frac{x}{l} = \left(\frac{15}{A}\right)^{1/2}\left(\frac{ul}{\nu}\right)^{-1/2} \tag{82}$$

Since $ul/\nu \gg 1$, we conclude that $\lambda_T/l \ll 1$.

We can reason that the time scale of the large-eddy structure is proportional to l/u, and the time scale of the strain-rate fluctuations is λ_T/u. The ratio of the two time scales is then

$$\frac{l}{\lambda_T} \sim \frac{ul}{\nu} \tag{83}$$

An important feature of turbulent flows that can be deduced from this discussion is *that production is greatest at large scales, and dissipation is greatest at small scales.*

Isotropic flow is obviously an idealization. Most practical flows, such as boundary layers, jets, and wakes, are shear flows. Hence, information concerning the Reynolds stress tensor is necessary. The many different techniques that have evolved for calculating these flows are beyond the scope of this review (cf. [17]).

6. SPATIAL AND TEMPORAL RESOLUTION IN MEASUREMENTS

In general, the following information is determined in a measurement program:

1. Physical properties such as ρ_0 and μ_0. These are not functions of the flow field.
2. Scalar fields such as temperature and pressure.
3. Vector fields such as velocity.

To select instrumentation we need to estimate the scale of the flow and the scale of any disturbances in the flow (e.g., wavelength λ and boundary-layer thickness δ). We also need to know the frequencies of any disturbances.

One of the problems we face is the distinction between spatial resolution and the temporal response of our instrumentation. This is illustrated in Fig. 10. Consider first a probe placed in a disturbance of wavelength λ, propagating with speed U_c. We have two criteria. First, the size of the probe must be small enough to resolve the spatial extent of the disturbance; i.e., the probe size must be less than $\lambda/2$. Second, since the probe is fixed in space it must have adequate frequency response; i.e., the response time must be less than λ/U_c.

Consider the more complex case of resolving a wavelike disturbance whose amplitude is decaying. The example cited is one of exponential decay, $a \sim \exp(-t/\mathfrak{I})$. We still have the spatial requirement (probe size $< \lambda/2$), but our response-time requirement is two-fold: The response time must be less than λ/U_c or $\mathfrak{I}$, whichever is less. This is a classical problem in the measurement of turbulence, where time-varying disturbances are convected by a fixed probe. The Taylor hypothesis assumes the turbulence to be frozen:

$$\left(\frac{\partial}{\partial t}\right)_{\text{measured}} \cong \left(U_c \frac{\partial}{\partial x}\right)_{\text{field}} \tag{84}$$

For this to be true,

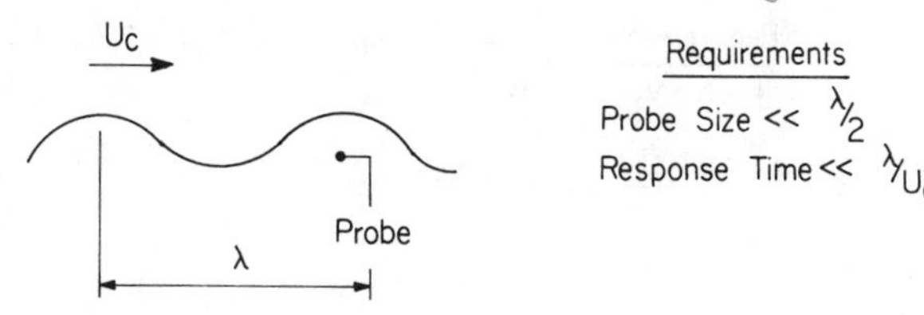

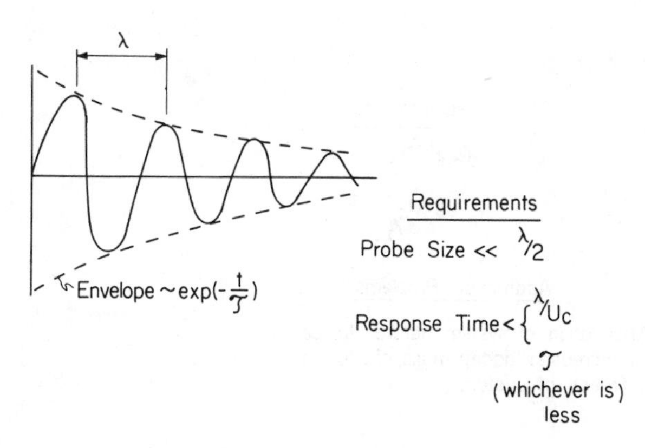

Figure 10 Spatial versus temporal resolution.

$$\frac{l}{U_c} \ll \frac{l}{u} \sim \mathfrak{I}$$

or

$$\frac{u}{U_c} \ll 1$$

Hence, Taylor's hypothesis is limited to low-intensity turbulent flows. A somewhat sophisticated way around this problem is to move the probe in a direction opposite to the flow, thereby artificially creating a high U_c. Such is the case when meteorological data is collected with flow instrumentation mounted in the wing of an airplane.

An example of the trade-off between temporal and spatial resolution is illustrated in Fig. 11. The problem is to measure unsteady wall pressure with a disturbance wavelength λ. Placing a pressure transducer flush with the wall ensures maximum frequency response. However, the diameter of the probe must be less than half a wavelength. If this is not the case, the transducer can be placed in a chamber with a pinhole leading to the surface. Here spatial resolution is achieved at the expense of temporal response. There are two methods of analysis for determining temporal response under these conditions. If the test fluid is compressible, the system can be

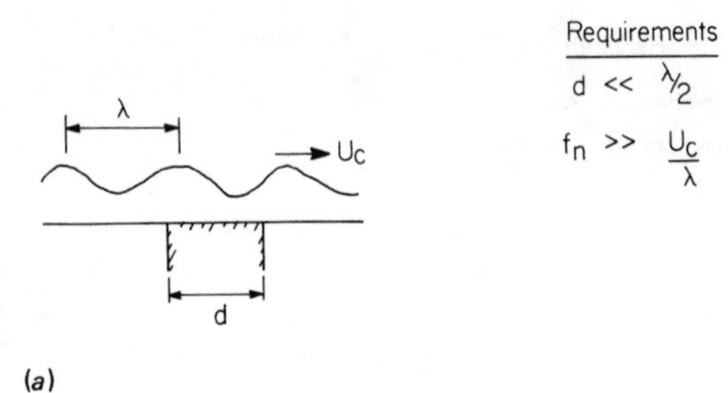

(a)

Uc

λ

d'

d

(b)

Requirements

d' << λ/2

fn >> Uc/λ

Additional Problems

Attenuation of system natural frequency by increased added mass, change in stiffness of system.

Assumptions

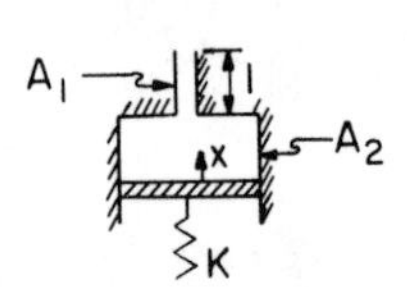

Can replace transducer diaphragm by a spring loaded piston.

Flow is one dimensional.

Viscous effects are negligible.

(c)

Figure 11 Tradeoff between spatial and temporal resolution. (*a*) Measurement of wall pressure with a flush-mounted transducer. (*b*) Same measurement using a cavity-mounted transducer. A pinhole leads from the surface to the cavity. (*c*) Idealized system for analysis.

analyzed as a Helmholtz resonator. The natural frequency f_c of the pinhole cavity system is given by

$$\frac{2\pi f_c l_T}{a_0} \tan \frac{2\pi f_c l_T}{a_0} = \frac{\pi R^2 l_T}{V_c} \tag{85}$$

where V_c = cavity volume
R = pinhole radius
a_0 = speed of sound in the medium

Equation (85) fails to predict an attenuation in incompressible fluids ($a_0 \to \infty$). A lesser-known mechanism, that of inertial damping, can result in a significant attenuation of frequency response [19]. The analysis is illustrated in the lower portion of Fig. 11. If the flow velocity in the tube is V, then $\phi = Vs$, and the unsteady Bernoulli equation (26) can be written in the form

$$\frac{1}{g}\frac{\partial V}{\partial t} = -\frac{\partial H}{\partial s} \tag{86}$$

$$H = \frac{p}{\gamma} + \frac{V^2}{2g} + z \tag{87}$$

As shown, the transducer is replaced by a spring-loaded piston. If the flow velocity in the chamber is $\dot{x}$, then V is determined from continuity:

$$V = \frac{A_2}{A_1}\dot{x} \tag{88}$$

The equation of motion for flow in the tube is then

$$\frac{1}{g}\frac{A_2}{A_1}\ddot{x} = -\frac{1}{\gamma}\frac{\partial p}{\partial s} \tag{89}$$

The force balance on the piston is

$$p_c A_2 = -Kx \tag{90}$$

If the pressure is zero at the end of the tube, then

$$-\frac{\partial p}{\partial s} = \frac{p_c}{l} = -\frac{Kx}{lA_2} \tag{91}$$

Hence,

$$\frac{1}{g}\frac{A_2}{A_1}\ddot{x} + \frac{K}{\gamma l A_2}x = 0 \tag{92}$$

$$\ddot{x} + \frac{K}{\rho l A_2}\frac{A_1}{A_2}x = 0 \tag{93}$$

The solution to Eq. (93) is

$$2\pi f_c = \sqrt{\frac{K}{\rho l A_2}\frac{A_1}{A_2}} \tag{94}$$

which should be compared with Eq. (86). Further details are given in [19]. In making unsteady pressure measurements in liquids, the problem can be further complicated by the presence of air bubbles in the cavity, which would significantly reduce the stiffness of the system.

The important point here is that we are generally limited to an eulerian frame of reference. This makes it difficult to distinguish between a truly temporal change in a flow quantity and a pseudo-temporal change due to a spatially varying disturbance convected past the fixed probe. It is essential that there be a proper balance between the spatial resolution and temporal response of the instrumentation.

7. CORRELATION OF DATA AND SIGNAL ANALYSIS

The elements of data correlation and signal analysis are presented here as an introduction to the subject. Further discussion of this important and complex topic is given in subsequent chapters and in other texts (e.g., [20]).

In a broad sense, data fall into two categories, namely, deterministic and random. The distinction between the two is somewhat nebulous since any physical process can be contaminated by unknown factors and therefore become not strictly deterministic. A physical process is deterministic if everything is known about it. We extend this definition to imply that data are deterministic if the influence of random perturbations on a physical process being measured is minimal.

7.1 Classification of Deterministic Data

The classification of deterministic data is illustrated in Fig. 12. Steady data are classified by amplitude. Sinusoidal data are classified by amplitude and frequency. Complex

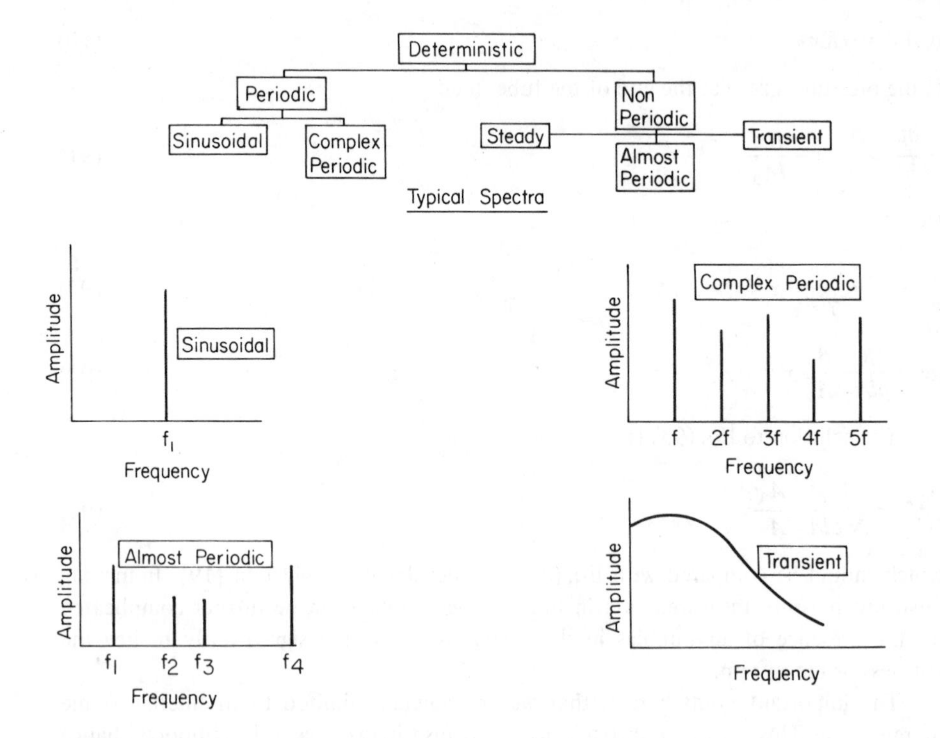

Figure 12 Classification of deterministic data.

periodic data can usually be decomposed into harmonics of the fundamental frequency, which are characterized by their amplitude and phase relationship. This is done through a Fourier series:

$$\tilde{u}(t) = \sum_{n=-\infty}^{\infty} C_n e^{in(t/T)}$$

$$C_n = \frac{1}{T} \int_0^T \tilde{u}(t) e^{-in(t/T)}\, dt$$

where T is the fundamental period.

Two or more sine waves will be periodic only when the ratios of all possible pairs of frequencies form rational numbers like 2/7 and 3/7, as opposed to $3/\sqrt{50}$. In the case of almost-periodic data, a series of the form

$$\tilde{u}(t) = \Sigma A_n \sin(2\pi f_n t + \Phi_n)$$

can usually be used to describe the function. The ratio $\tilde{u}_n/\tilde{u}_m$ is not a rational number in all cases.

7.2 Random Data and Signal Analysis

Many practical flow problems involve turbulent flow. The signal is not deterministic and is therefore classified as random. Thus, we are forced into the realm of statistics to quantify measurements. This discussion is limited to stationary random functions. Stationarity is loosely defined as the case in which the various statistical functions describing the signal are time-invariant.

7.2.1 Basic description. The simplest method of describing a function that varies randomly with time is to determine its mean value

$$U = \lim_{T\to\infty} \frac{1}{T} \int_0^T \tilde{u}(t)\, dt \tag{95}$$

The mean value does not indicate how much $\tilde{u}(t)$ is fluctuating about its mean value. A measure of the extent of deviations from the mean is the variance, or mean square value,

$$\overline{u^2} = \lim_{T\to\infty} \frac{1}{T} \int_0^T (\tilde{u} - U)^2\, dt \tag{96}$$

An estimate of the percentage of time during which $\tilde{u}(t)$ can lie within a well-defined range is given by the probability density. With reference to Fig. 13, the probability density is determined as follows.

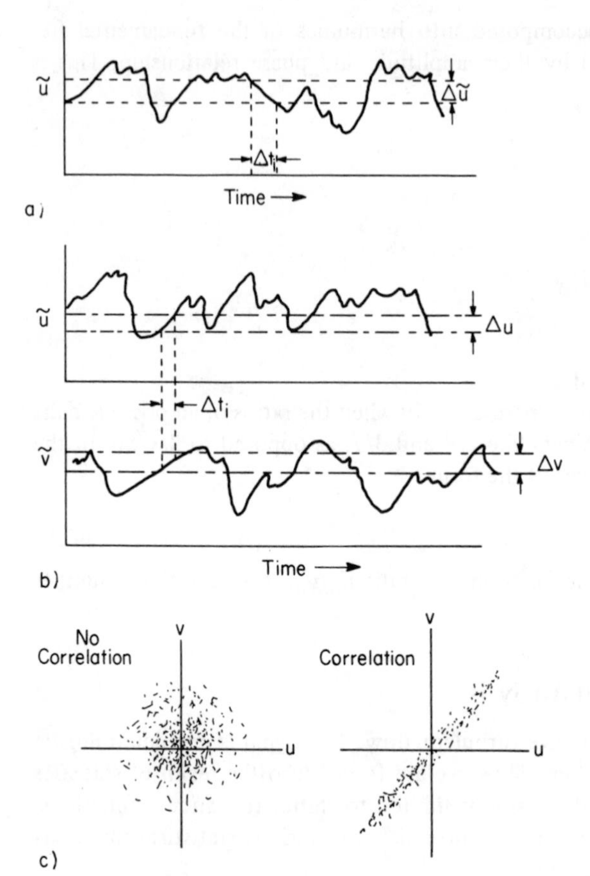

Figure 13 Determination of probability and joint probability density. Correlation of two variables.

The total time during which $\tilde{u}(t)$ lies in the range $u_1 \leqslant \tilde{u}(t) \leqslant u_2$ is given by

$$T_u = \sum_{i=1}^{\infty} \Delta t_i \tag{97}$$

where Δt_i is the length of time the function takes on values between u_1 and u_2 as it becomes larger than u_1 or less than u_2. The probability of this occurring is then

$$\text{Prob}\ \{u \leqslant \tilde{u}(t) \leqslant u + \Delta u\} = \lim_{T \to \infty} \frac{T_u}{T} \tag{98}$$

A probability density function is given by

$$p(u) = \lim_{\Delta u \to 0} \frac{\text{Prob}\ \{u \leqslant \tilde{u}(t) \leqslant u + \Delta u\}}{\Delta u} \tag{99}$$

$$p(u) = \lim_{\Delta u \to 0} \lim_{T \to \infty} \frac{1}{T} \frac{T_u}{\Delta u} \tag{100}$$

A good deal of random data follow a gaussian distribution,

$$p(u) = \frac{1}{\sqrt{2\pi}\sqrt{\sigma}} \exp\left(-\frac{u^2}{2\sigma^2}\right) \tag{101}$$

$$\sigma \equiv \sqrt{\overline{u^2}} \tag{102}$$

For this situation the time mean and variance are related to the probability density in the following way:

$$U = \int_{-\infty}^{\infty} \tilde{u} p(u)\, d\tilde{u} \tag{103}$$

$$\overline{u^2} = \int_{-\infty}^{\infty} [\tilde{u}^2 - U^2] p(u)\, du \tag{104}$$

From the definition of the probability density, Eq. (99), it is obvious that the probability that $\tilde{u}$ will fall between two values, say A and B, is given by

$$\text{Prob}\ \{A \leqslant \tilde{u} \leqslant B\} = \int_{A}^{B} p(u)\, d\tilde{u} \tag{105}$$

The cumulative probability function describes the probability that $\tilde{u}(t)$ is less than or equal to some value u. It is equal to the integral of the probability density function, taken from $-\infty$ to u:

$$P(u) = \text{Prob}\ \{\tilde{u}(t) \leqslant u\} = \int_{-\infty}^{u} p(\xi)\, d\xi \tag{106}$$

Thus, for a gaussian distribution,

$$P(u) = \frac{1}{2\pi\sigma} \int_{-\infty}^{u} e^{-\xi^2/2\sigma^2}\, d\xi \tag{107}$$

There is no closed-form solution for the integral in Eq. (107). Hence tables of $P(u)$ have been prepared through numerical integration. These are found in many references such as *Standard Mathematical Tables*, published by the Chemical Rubber Publishing Company.

7.2.2 Joint probability density. We quite often wish to investigate the interdependence of several random variables. The joint probability density function describes the probability that two functions each will assume values within certain limits at any instant of time.

In a manner similar to the computation for a single variable, we consider two continuous functions of time, say u and v. The total time during which both functions are within prescribed limits is given by

$$T_{uv} = \Sigma \, \Delta t_i \tag{108}$$

where Δt_i is the time both functions lie within a given band during each fluctuation as shown in Fig. 13. The probability of this occurring is

$$\text{Prob } \{u < \tilde{u}(t) < u + \Delta u, v < \tilde{v}(t) < v + \Delta v\} = \lim_{T \to \infty} \frac{T_{uv}}{T} \tag{109}$$

$$p(u,v) = \lim_{\substack{\Delta u \to 0 \\ \Delta v \to 0}} \frac{\Pr \{\ \}}{\Delta u \, \Delta v} \tag{110}$$

$$p(u,v) = \lim_{\substack{\Delta u \to 0 \\ \Delta u \to 0}} \lim_{T \to \infty} \frac{1}{T} \frac{T_{uv}}{\Delta u \, \Delta v} \tag{111}$$

If the two functions $\tilde{u}$ and $\tilde{v}$ are statistically independent, then

$$P(u,v) = P(u)P(v) \tag{112}$$

In this case, a three-dimensional plot is required to display the data. (See, for example, page 95 in [21].)

In a manner similar to that for a single variable, we define a joint probability function $P(u,v)$ as follows:

$$P(u,v) = \Pr \, [u(t) \leqslant u, v(t) \leqslant v] = \int_{-\infty}^{u} \int_{-\infty}^{v} p(\xi,\eta) \, d\xi \, d\eta \tag{113}$$

where ξ and η are dummy variables. If the variances in u and v are equal and u and v are statistically independent, a simplification occurs for gaussian distributed functions such that the probability density is

$$p(u,v) = \frac{1}{2\pi\sigma} \exp\left(-\frac{u^2 + v^2}{2\sigma^2}\right) = \frac{1}{2\pi\overline{(r')^2}} \exp\left[-\frac{r^2}{2\overline{(r')^2}}\right] \tag{114}$$

The joint probability function is given by

$$P(r) = \int_0^{2\pi} \left\{ \int_0^r \frac{1}{2\pi} \exp\left[-\frac{r^2}{2\overline{(r')^2}}\right] \frac{r\,dr}{\overline{(r')^2}} \right\} d\theta \tag{115}$$

or

$$P(r) = \int_0^{r^2/\overline{(r')^2}} e^{-\xi} \, d\xi = 1 - \exp\left[-\frac{r^2}{\overline{(r')^2}}\right] \tag{116}$$

$P(r)$ is, for example, the probability of being less than or equal to a distance r from a given target assuming deviations in all directions are equally likely. Note that there is a closed-form solution for $P(r)$, which was not true for a single variable.

We now want to investigate the relationship between two variables. Figure 13c can be generated by inputing two random signals $\tilde{u}$ and $\tilde{v}$ to the x and y axes of an oscilloscope. The shaded area represents the shape of the spot we would see on the oscilloscope. In the case where we consider $\tilde{v}$ and $\tilde{u}$ to be linearly related, we seek an equation of the form

$$v_c = Au + B \tag{117}$$

where A and B are constants. Their values are determined from the requirement

$$\Sigma(\tilde{v} - \tilde{v}_c)^2 = \text{minimum} \tag{118}$$

where $\tilde{v}$ represents the raw data and $\tilde{v}_c$ is computed from Eq. (117). A best-fit curve results in minimum error. This error is found to be equal to

$$E = \overline{u^2}(1 - R_{uv}^2) \tag{119}$$

with

$$R_{uv} = \frac{\overline{(\tilde{u} - U)(\tilde{v} - V)}}{\sqrt{\overline{u^2}}\sqrt{\overline{v^2}}} \tag{120}$$

where R_{uv} is a correlation function that lies in the range $-1 \leqslant R \leqslant 1$. The numerator in Eq. (120) represents the time mean of the product of the two functions, $(\tilde{u} - U)$ and $(\tilde{v} - V)$.

7.2.3 Autocorrelation. One type of correlation is that which determines the length of past history that is related to a given event. This is formed by looking at the correlation between a function of time and the same function at a time τ later:

$$R_\tau = \frac{\overline{\tilde{u}(t)\tilde{u}(t+\tau)}}{\overline{u^2}} \tag{121}$$

An important distinction between a periodic function and a random function is that as $\tau \to \infty$, $R_\tau \to 0$ for a random variable but R_τ is periodic for a periodic function. This is easily illustrated. Let

$$\tilde{u}(t) = \sin \omega t$$

Then

$$R_\tau = \lim_{T \to \infty} \frac{(1/T) \displaystyle\int_0^T \sin \omega t \sin \omega(t+\tau)\, dt}{(1/T) \displaystyle\int_0^T \sin^2 \omega t\, dt} \tag{122}$$

$$R_\tau = \cos \omega\tau \tag{123}$$

7.2.4 Spectral analysis. An important part of the description of random variables is the determination of the distribution of energy content with frequency. This is defined by the power spectral density

$$S(\omega) = \lim_{\substack{\Delta\omega \to 0 \\ T \to \infty}} \frac{1}{T\,\Delta\omega} \int_0^T u_\omega^2(t)\,dt \tag{124}$$

where u_ω is that part of the signal contained within a bandwidth of $\Delta\omega$ centered at ω. Equation (124) is not a very precise definition mathematically, but it is very clear from a physical point of view. A precise mathematical definition will follow.

The natural inclination of an engineer is to reduce a time-dependent signal into its harmonic components. If the signal is periodic, a Fourier series may be used:

$$\tilde{u}(t) = \sum_{-\infty}^{\infty} C_n e^{in\omega_0 t} \tag{125}$$

where

$$C_n = \frac{1}{T_0} \int_{-T_0/2}^{+T_0/2} \tilde{u}(t) e^{-in\omega_0 t}\,dt \tag{126}$$

$$T_0 = \frac{2\pi}{\omega_0} \tag{127}$$

If the function is *not* periodic, we have $T_0 \to \infty$, $\omega_0 \to 0$. Thus the spectrum coefficients C_n approach a continuous function of ω rather than discrete values at $n\omega_0$. Thus we let ω_0 and $\Delta\omega \to 0$, and $n\omega_0 \to \omega$ and use the Fourier-integral representation

$$F(\omega) = \frac{1}{2\pi} \int_{-\infty}^{\infty} \tilde{u}(t) e^{-i\omega t}\,dt \tag{128}$$

$$\tilde{u}(t) = \int_{-\infty}^{\infty} F(\omega) e^{i\omega t}\,d\omega \tag{129}$$

For $F(\omega)$ to exist it is necessary that

$$\int_{-\infty}^{\infty} |\tilde{u}(t)|^2\,dt = 2\pi \int_{-\infty}^{\infty} |F(\omega)|^2\,d\omega \tag{130}$$

The integrals in Eq. (130) must be finite. The spectrum density is normally defined as

$$S(\omega) = |F(\omega)|^2\,d\omega \tag{131}$$

In the case of a stationary random function the integral on the left-hand side of

(130) is infinite. Thus it would appear impossible to define a spectral density function using Eq. (131). This is overcome through an artifice. The signal is considered to be analyzed for an amount of time T and is assumed equal to zero for the rest of the time. Then

$$\int_{-\infty}^{\infty} |\tilde{u}(t)|^2 \, dt = T\overline{u^2} \tag{132}$$

Using our definitions, the following may be written:

$$\overline{u^2} = \lim_{T\to\infty} \frac{1}{T} \int_{-T/2}^{T/2} \tilde{u}^2 \, dt = \frac{1}{T} \int_{-\infty}^{\infty} \tilde{u}(t) \left[\int_{-\infty}^{\infty} F(\omega) e^{i\omega t} \, d\omega \right] dt \tag{133}$$

$$\overline{u^2} = \frac{1}{T} \int_{-\infty}^{\infty} F(\omega) \left[\int_{-\infty}^{\infty} \tilde{u}(t) e^{i\omega t} \, dt \right] d\omega \tag{134}$$

$$\overline{u^2} = \frac{1}{T} \int_{-\infty}^{\infty} F(\omega) 2\pi F^*(\omega) \, d\omega \tag{135}$$

Equation (135) can be rewritten in the form

$$\overline{u^2} = 4\pi \int_0^{\infty} \frac{|F(\omega)|^2}{T} \, d\omega \tag{136}$$

By definition of the spectrum,

$$\overline{u^2} = \int_0^{\infty} S(\omega) \, d\omega \tag{137}$$

we arrive at a modified relationship between the spectrum and the Fourier transform of $\tilde{u}(t)$:

$$S(\omega) = \frac{4\pi |F(\omega)|^2}{T} \tag{138}$$

To determine the spectrum of a pure tone, the Dirac delta function must be introduced. This is defined by

$$\int_{-\infty}^{\infty} \tilde{u}(t) \, \delta(t - t_0) \, dt = \tilde{u}(t_0) \tag{139}$$

The Fourier transform of the Dirac delta function is given by

$$F(\omega) = \frac{1}{2\pi} \int_{-\infty}^{\infty} \delta(t - t_0) e^{-i\omega t} \, dt = \frac{1}{2\pi} e^{-i\omega t_0} \tag{140}$$

Similarly, the inverse transform of the delta function is given by

$$\int_{-\infty}^{\infty} \delta(\omega - \omega_0) e^{i\omega t} \, d\omega = e^{i\omega_0 t} \tag{141}$$

Noting Eqs. (140) and (141), one obtains

$$\delta(\omega - \omega_0) = \frac{1}{2\pi} \int_{-\infty}^{\infty} e^{i\omega_0 t} \, e^{-i\omega t} \, dt \tag{142}$$

Thus, if a function can be described by a Fourier series,

$$\tilde{u}(t) = \sum_{0}^{\infty} A_n e^{i(n\omega_0 t - \phi_n)} \tag{143}$$

$$F(\omega) = \sum_{0}^{\infty} A_n e^{-i\phi_n} \, \delta(\omega - n\omega_0) \tag{144}$$

It can be inferred that the power spectral density of a periodic function is a collection of equally spaced spikes with the area under these spikes proportional to $|A_n|^2$.

7.2.5 Relationship between the autocorrelation and the spectral density function. Intuitively, one could expect a relationship between the autocorrelation and the spectral density function. By definition, the autocorrelation is given by

$$R_\tau = \lim_{T \to \infty} \frac{1}{T} \int_{-T/2}^{T/2} \tilde{u}(t) \tilde{u}(t + \tau) \, dt \tag{145}$$

We can write Eq. (145) in terms of a Fourier integral

$$R_\tau = \lim_{T \to \infty} \frac{1}{T} \int_{-T/2}^{+T/2} \tilde{u}(t) \int_{-\infty}^{\infty} F(\omega) e^{i\omega(t+\tau)} \, d\omega \, d\tau \tag{146}$$

where we replace t with $t + \tau$ in the second integration. Rearranging Eq. (146) gives

$$R_\tau = \lim_{T \to \infty} \frac{1}{T} \int_{-\infty}^{\infty} F(\omega) e^{i\omega\tau} \, dt \int_{-T/2}^{+T/2} \tilde{u}(t) e^{i\omega t} \, d\omega \tag{147}$$

or

$$R_\tau = \lim_{T\to\infty} \frac{2}{T}\int_{-\infty}^{\infty} F(\omega)F^*(\omega)e^{i\omega t}\, d\omega \tag{148}$$

where

$$F^*(\omega) = \frac{1}{2\pi}\int_{-\infty}^{\infty} \tilde{u}(t)e^{i\omega t}\, dt \tag{149}$$

$$F^*(\omega) = \frac{1}{2\pi}\int_{-T/2}^{+T/2} \tilde{u}(t)e^{i\omega t}\, dt \tag{150}$$

for $\tilde{u}(t) = 0$ except when $-T/2 \leqslant x \leqslant +T/2$. It follows that

$$R_\tau = \int_{-\infty}^{\infty} \frac{2\pi|F(\omega)|^2}{T} e^{i\omega t}\, d\omega \tag{151}$$

or

$$R_\tau = \int_{-\infty}^{\infty} \frac{1}{2} S(\omega)e^{i\omega t}\, d\omega \tag{152}$$

$$S(\omega) = \frac{1}{\pi}\int_{-\infty}^{\infty} R(\tau)e^{-i\omega\tau}\, d\tau \tag{153}$$

Since $S(\omega)$ is an even function,

$$S(\omega) = \frac{2}{\pi}\int_{0}^{\infty} R(\tau)\cos\omega\tau\, d\tau \tag{154}$$

In terms of frequency f, we obtain

$$S(f) = 2\pi S(\omega) \tag{155}$$

such that

$$\overline{u^2} = \int_{0}^{\infty} S(f)\, df \tag{156}$$

and

$$S(f) = 4\int_{0}^{\infty} R(\tau)\cos 2\pi f\tau\, d\tau \tag{157}$$

$$R(\tau) = \int_0^\infty S(f) \cos 2\pi f\tau \, df \tag{158}$$

Thus the power spectrum and the autocorrelation are Fourier-transform pairs. Functions that are transform pairs have an inverse spreading relationship. For example, if the power spectrum is wide, the energy content in the signal is spread out over a wide range of frequencies with little coherence. The autocorrelation will be nonzero only over a limited range of delay time. Conversely, if the power spectrum is very narrow, the autocorrelation function has nonzero values for very large values of τ. (One can visualize this inverse spreading by noting $\omega \sim 1/\tau$.) To be precise, the autocorrelation will oscillate with decreasing amplitude and with a frequency equal to the frequency at which the spectrum is centered. This is illustrated in Fig. 14. As an example, consider band-passed white noise:

$$S'(\omega) = \begin{cases} 0 & 0 < \omega < \omega_1 \\ 1 & \omega_1 < \omega < \omega_2 \\ 0 & \omega_2 < \omega < \infty \end{cases} \tag{159}$$

Then

$$R(\tau) = \int_{\omega_1}^{\omega_2} \cos \omega\tau \, d\omega = \left(\frac{\sin \omega\tau}{\tau}\right)_{\omega_1}^{\omega_2} = \frac{2}{\tau} \sin \frac{\omega_2 - \omega_1}{2} \tau \cos \frac{\omega_2 + \omega_1}{2} \tau \tag{160}$$

$$R(\tau) = \frac{2}{\tau} \sin \frac{\Delta\omega\tau}{2} \cos \omega_0 \tau \tag{161}$$

where

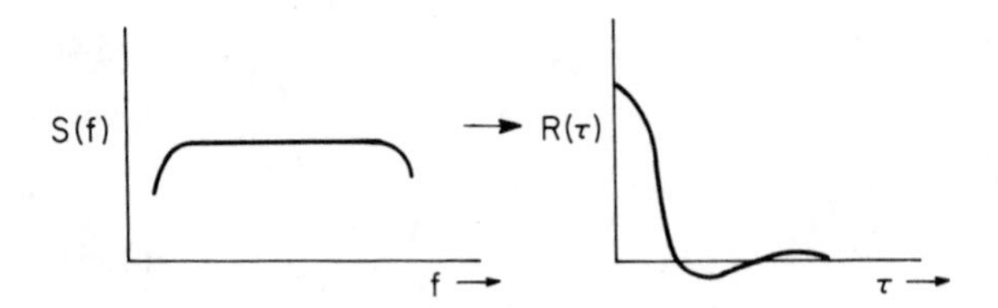

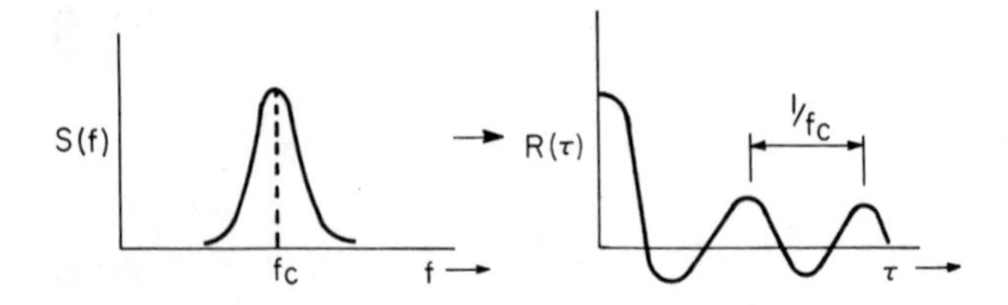

Figure 14 Relationship between spectrum and autocorrelation.

$$\omega_0 = \frac{\omega_1 + \omega_2}{2} \qquad \Delta\omega = \omega_2 - \omega_1$$

The autocorrelation depends on the bandwidth $\Delta\omega$ of the filter and the center frequency ω_0. As the bandwidth approaches that for a pure tone of frequency ω_0, $R(\tau) = \cos \omega_0 \tau$. For finite bandwidth, the autocorrelation consists of $\cos \omega_0 \tau$ with an amplitude modulation equal to $(2/\tau) \sin (\Delta\omega\tau/2)$.

7.2.6 Cross correlation and cross power spectrum. Consider two functions of time $\tilde{u}_1(t)$ and $\tilde{u}_2(t)$. The cross correlation is

$$R_{12}(\tau) = \lim_{T\to\infty} \frac{1}{T} \int_{-T/2}^{+T/2} \tilde{u}_1(t)\tilde{u}_2(t+\tau)\,dt \tag{162}$$

As an example, $\tilde{u}_1$ and $\tilde{u}_2$ could be a velocity component at two points in a flow. The cross power spectrum is the Fourier transform of the cross-correlation function. R_{12} is generally an odd function. For example, in the case of velocity, the correlation will peak at a value of delay time corresponding to some characteristic eddy size divided by some convective velocity. Hence, the Fourier transform will have a real and an imaginary part. The real part is called the cospectrum:

$$C_{12}(f) = 2 \int_{-\infty}^{\infty} R_{12}(\tau) \cos 2\pi f\tau \, d\tau \tag{163}$$

The imaginary part is denoted as the quadrature spectrum:

$$Q_{12}(f) = 2 \int_{-\infty}^{\infty} R_{12}(\tau) \sin 2\pi f\tau \, d\tau \tag{164}$$

The cospectrum represents a narrow-band correlation between $\tilde{u}_1$ and $\tilde{u}_2$. Similarly, the quadrature spectrum is a narrow-band correlation of $\tilde{u}_1$ and $\tilde{u}_2$ with a 90° phase shift.

7.2.7 Signal analysis in turbulence research. Turbulence is usually characterized by its intensity $\overline{u^2}$, its spectrum $S(f)$, and various correlation parameters as deemed appropriate. As an example, consider the measurement of turbulence in the development region of a round jet (zero to six diameters from the nozzle). The data are illustrated in Fig. 15. As shown, measurements are made in the mixing region of diameter $2R$. The width of the mixing region varies linearly with distance x from the nozzle. This is illustrated by the fact that the mean velocity profiles are similar when the distance from the center of the mixing region $(r - r_{1/2})$ is normalized with respect to x. The turbulence is a maximum in the center of the mixing region and also follows a similarity law. The measured Reynolds stress is expressed in terms of a correlation coefficient:

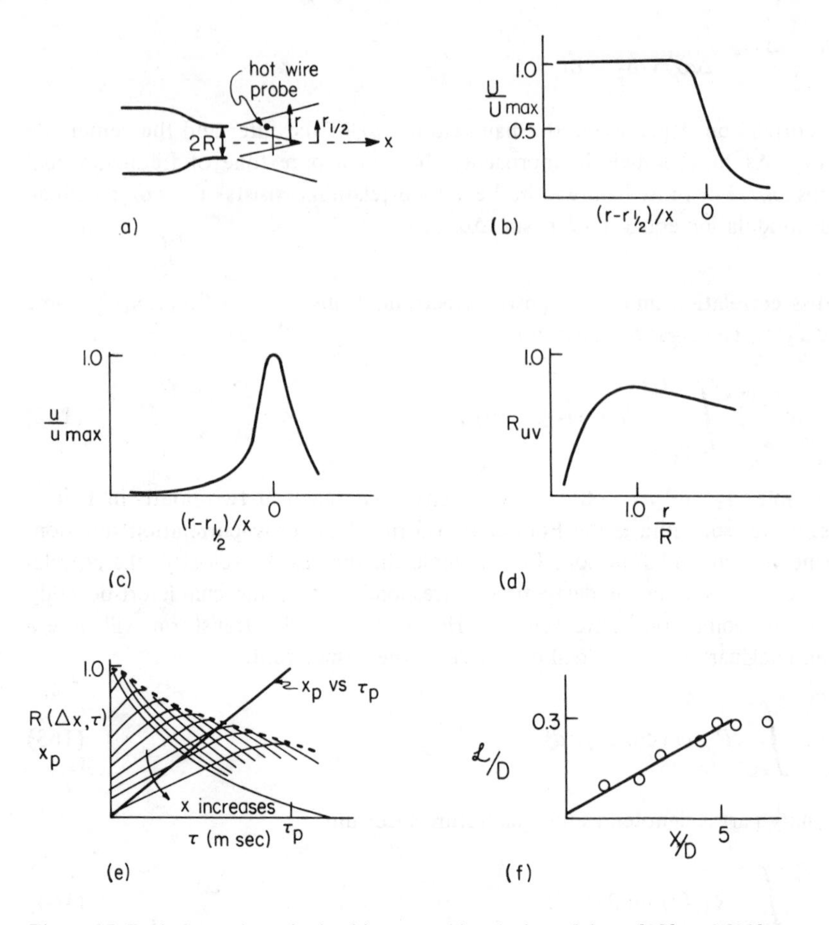

Figure 15 Turbulence data obtained in a round jet. *(Adapted from [22] and [23].)*

$$R_{uv} = -\frac{\overline{uv}}{\sqrt{\overline{u^2}}\sqrt{\overline{v^2}}} \tag{165}$$

Sample data are shown in Fig. 15*d*.

Finally, a quantitative measure of the spatial structure is obtained from the cross correlation

$$R_{12} = \frac{\overline{u_1(x,t)u_2(x+\Delta x,t+\tau)}}{\sqrt{\overline{u_1^2}}\sqrt{\overline{u_2^2}}} \tag{166}$$

where u_1 and u_2 are longitudinal velocity components in the center of the mixing region at x and Δx, respectively. As Δx is increased, the correlation peaks at a different delay time τ_p. The peak value of the correlation also decays with increasing spacing between the points x_1 and x_2. This corresponds to the true evolution of the turbulence. In fact, the envelope of the peaks is the autocorrelation that would be measured in a frame of reference moving with the speed of convection of the turbulence past a fixed point of reference. If x_p is plotted against τ_p, a linear plot results, as shown. The slope of the line is the convection velocity U_c, which, in this case, is found to be $0.54U_0$.

The value of the two-point correlation at zero time delay is defined as

$$R_{\Delta x} = \frac{\overline{u_1(x,t)u_2(x + \Delta x,t)}}{\sqrt{\overline{u_1^2}}\sqrt{\overline{u_2^2}}} \tag{167}$$

An integral length scale is defined by

$$l = \int_0^\infty R_{\Delta x}\, dx \tag{168}$$

The results are plotted in Fig. 15*f*. This shows that the integral scale is a linear function of x, in agreement with the observation that the width of the mixing region varies linearly with distance from the nozzle.

NOMENCLATURE

a_0	speed of sound in undisturbed medium
A_1, A_2	cross-sectional areas
A_n	amplitude of nth harmonic
b	breadth of river or flow passage
c	concentration, chord length
C_N	normal-force coefficient
C_l	lift coefficient
C_m	moment coefficient
C_n	Fourier coefficient
C_{12}	cospectrum
d	depth, diameter
e_{ij}	strain rate
e'_{ij}	fluctuating strain rate
E	error
f	frequency (Hz)
F	force
$F(t)$	unspecified function of time
$F(\omega)$	Fourier transform
g	acceleration due to gravity
G	mass flux of suspended material
H	total head, defined in Eq. (87)
I	acoustic intensity
K	spring constant

l	characteristic or integral length
l_T	tube length
L	lift
M	moment
p	pressure
p_a	acoustic pressure
p_t	total pressure
P	time mean pressure
P_t	time mean total pressure
$p(u)$	probability density of $\tilde{u}$
$P(u)$	cumulative probability of $\tilde{u}$
$p(u,v)$	joint probability density
Q_{12}	quadrature spectrum
r	radius, separation distance, distance to observer
r'	standard deviation in radial direction
R	pipe or jet radius
$r_{1/2}$	radial position in jet where velocity is one-half the centerline velocity
R_{12}	cross-correlation function
R_τ	autocorrelation
$R_{\Delta x}$	spatial correlation
R_u, R_v	spatial correlation functions in isotropic turbulence
s	distance along streamline
S	surface area
$S(\omega), S(f)$	power spectral density
t	time
t'	retarded time, $t - r/a_0$
T	characteristic time
$\mathfrak{T}$	characteristic decay time
T_{ij}	turbulent stress tensor
$\tilde{u}_1, \tilde{u}_2, \tilde{u}_3$	velocity in tensor notation
u_1, u_2, u_3	velocity fluctuations in tensor notation
u, v, w	velocity fluctuations in cartesian coordinates
U_1, U_2, U_3	time mean velocity in tensor notation
U, V, W	time mean velocity in cartesian coordinates
V	velocity along streamline
V_c	cavity volume
V_e	effective source volume
x_1, x_2, x_3	coordinates in tensor notation
x, y, z	cartesian coordinates
α	angle of attack
γ	specific weight
δ	boundary-layer thickness
δ_{ij}	Kronecker delta function
$\delta(t - t_0)$	Dirac delta function
Δ	incremental change

ϵ	dissipation rate
ξ	dummy variable
η	Kolomogoroff length scale
λ	wavelength
λ_T	Taylor microscale
μ	viscosity
ν	kinematic viscosity
ξ	dummy variable
ρ	density
σ	standard deviation
τ	delay time, Kolomogoroff time scale
τ_{ij}	stress tensor
τ_{ij}^*	nonisotropic stress tensor
ω	radian frequency
Ω	force potential

Subscripts

0	undisturbed medium
t	total, turbulent
a	acoustic
m	molecular
v	viscous

Other symbols

—	overbar denotes time average, $\lim_{T \to \infty} (1/T)\int(\ )\,dt$
~	vector, instantaneous value (sum of time-average and fluctuating components)

REFERENCES

1. V. A. Vanoni (ed.), *Sedimentation Engineering*, American Society of Civil Engineering, Manual 54, chap. 2, 1975.
2. M. E. Nelson and P. C. Benedict, Measurement and Analysis of Suspended Sediment Loads in Streams, *Trans. ASCE*, vol. 116, pp. 891–918, 1951.
3. A. Pope and J. J. Harper, *Low Speed Wind Tunnel Testing*, chap. 1, Wiley, New York, 1966.
4. J. E. Cermak, Aerodynamics of Buildings, in M. van Dyke, W. G. Vincenti, and J. V. Weyhausen (eds.), *Annual Review of Fluid Mechanics*, vol. 8, p. 75, Annual Reviews, Palo Alto, 1976.
5. M. J. Lighthill, On Sound Generated Aerodynamically. I, General Theory, *Proc. R. Soc. London Ser. A*, vol. 211, no. 1107, pp. 564–587, 1952.
6. R. H. Kraichnan, Pressure Field within Homogeneous, Anisotropic Turbulence, *J. Acoust. Soc. Am.*, vol. 28, no. 1, p. 64, 1956.
7. P. O. A. L. Davies, M. J. Fisher, and M. J. Barratt, Characteristics of the Turbulence in the Mixing Region of a Round Jet, *J. Fluid Mech.*, vol. 15, pt. 3, pp. 337–367, 1963.
8. G. M. Lilley, On the Noise from Air Jets, British Aerospace Research Council Rep. ARC–20, 376; N 40; FM–2724, September 1958.

9. G. Barefoot, Fluctuating Pressure Characteristics in the Mixing Region of a Perturbed and Unperturbed Round Free Jet, M.S. thesis, The Pennsylvania State University, University Park, 1972.
10. R. E. A. Arndt and A. W. Nilsen, On the Measurement of Fluctuating Pressure in the Mixing Zone of a Round Free Jet, ASME Paper 71-FE-31, May 1971.
11. G. L. Brown and A. Roshko, On Density Effects and Large Structure in Turbulent Mixing Layers, *J. Fluid Mech.*, vol. 64, pt. 4, pp. 775–816, 1974.
12. J. Laufer, New Trends in Experimental Turbulence Research, in M. van Dyke, W. G. Vincenti, and J. V. Weyhausen (eds.), *Annual Review of Fluid Mechanics*, vol. 7, p. 307, Annual Reviews, Palo Alto, 1975.
13. L. Prandtl and O. G. Tietjens, *Applied Hydro- and Aeromechanics*, p. 230, 1934 (reproduction available from Dover, New York, 1957).
14. H. Schlichting, *Boundary Layer Theory*, 7th ed., chaps. 2, 7, 8, McGraw-Hill, New York, 1979.
15. H. Tennekes and J. L. Lumley, *A First Course in Turbulence*, MIT Press, Cambridge, Mass., 1972.
16. J. O. Hinze, *Turbulence*, 2d ed., McGraw-Hill, New York, 1975.
17. P. Bradshaw (ed.), Turbulence, in *Topics in Applied Physics*, vol. 12, Springer-Verlag, Berlin, 1978.
18. H. Rouse (ed.), *Advanced Mechanics of Fluids*, chap. 6, Wiley, New York, 1959.
19. R. E. A. Arndt and A. T. Ippen, Turbulence Measurements in Liquids Using an Improved Total Pressure Probe, *J. Hydraul. Res.*, vol. 8, no. 2, pp. 131–158, 1970.
20. J. S. Bendat and A. G. Piersol, *Measurement and Analysis of Random Data*, Wiley, New York, 1966.
21. E. J. Richards and D. J. Mead (eds.), *Noise and Acoustic Fatigue in Aeronautics*, chap. 4, Wiley, New York, 1968.
22. E. D. von Frank, Turbulence Characteristics in the Mixing Region of a Perturbed and Unperturbed Round Free Jet, M.S. thesis, The Pennsylvania State University, University Park, 1970.
23. N. C. Tran, Turbulence and Acoustic Characteristics of a Screen Perturbed Jet, M.S. thesis, The Pennsylvania State University, University Park, 1973.

CHAPTER

TWO

PHYSICAL LAWS OF FLUID MECHANICS AND THEIR APPLICATION TO MEASUREMENT TECHNIQUES

E. R. G. Eckert

1. INTRODUCTION

This chapter offers a brief review of the basic laws of fluid mechanics. We are interested in the motion—or the absence of motion—of fluids, and we generally disregard the fact that fluids consist of particles that are in motion relative to each other. Rather, we are concerned with the mean motion of regions in the fluid that contain a very large number of those particles (molecules or atoms). We thus consider the fluid as an unstructured, jellylike substance which we call a *continuum.*

This idealization causes no difficulty in dealing with ordinary liquids and with gases except where they are in a rarified state at very low pressure. We have, however, to be more careful when we extend the continuum concept to two-phase or two-component fluids—for instance, to a boiling liquid in which vapor bubbles are distributed, to a fog, to a fluid containing dust, or even to a two-component mixture of liquids or gases. By the continuum concept, one considers the two phases or components again as continua, which penetrate each other (Fig. 1). Sometimes a relative movement of the two continua, a slip, has to be considered. Interdiffusion of the components of a liquid or gas mixture is an example in which the continuum concept is generally justified. Caution is, however, required in other cases. We know, for instance, that in a fluidized bed the distribution of the solid particles is often not uniform, and the liquid or gas moves in columns or large bubbles through the bed.

Many of the approaches used in the past to describe turbulence considered the turbulent structure as sufficiently small in scale so that a continuum approach to its

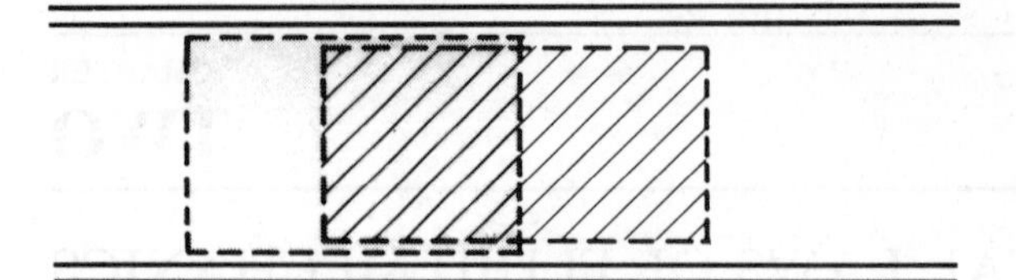

Figure 1 Interdiffusion of mass in a two-component or two-phase medium.

description could be used. However, Ludwig Prandtl observed by flow visualization that coarse structures appear frequently in a turbulent flow field. In 1903 he constructed a water tunnel with a free surface in which the fluid was circulated by a paddle wheel driven by hand (Fig. 2). The flow was made visible by scattering small particles onto the water surface. The photos reproduced in Fig. 3 were obtained with this type of device. The three lower photos were taken with a camera moving with different velocities above the water surface; but for the uppermost figure the camera was at rest. One can clearly observe that some of the vortices have diameters of the same order of magnitude as the channel width.

In this chapter we describe fluids as continua and assume that slip effects are absent or unimportant. In analyzing such fluids we have to apply the laws of

Figure 2 Ludwig Prandtl with his water tunnel (1903).

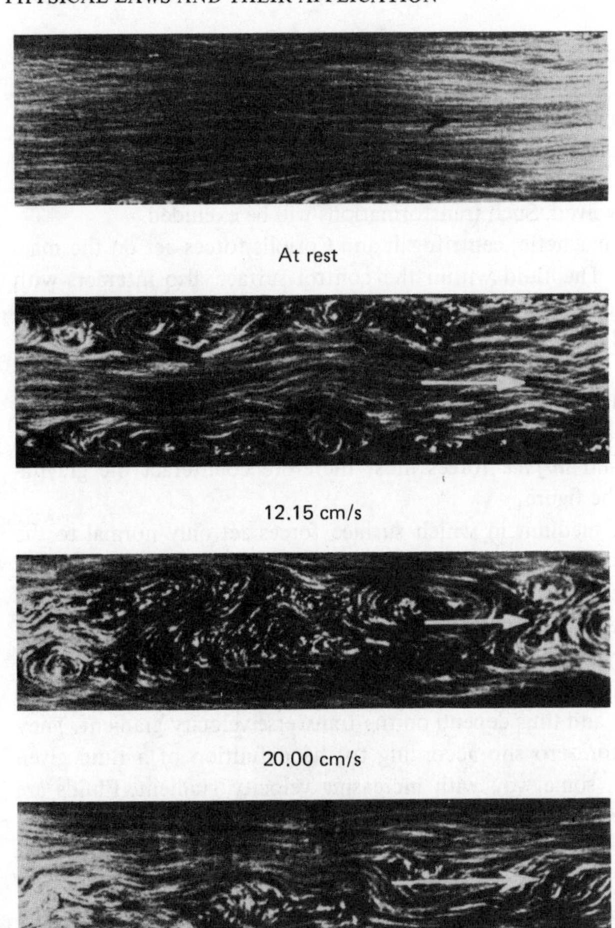

Figure 3 Visualization of turbulent flow through a channel in the tunnel of Fig. 2. The camera moved with different velocities relative to the fluid stream.

mechanics. Here we run into difficulty, because in a continuum there are no well-defined masses to which we can apply these laws. To get around this difficulty, we consider an arbitrary part of the fluid, surrounded by what we call a *control surface* and isolated from the rest of the fluid. Two different approaches are used in the formulation of the laws governing the movement of the fluid. In the lagrangian approach, the control volume is considered to move with the fluid; the history of the fluid within the control volume is studied as it moves along. In the second, or eulerian approach, the control volume is considered at rest relative to a selected coordinate system, and the fluid moves through the control surface. The second approach is used more frequently and offers advantages, especially when the flow is steady relative to the coordinate system.

The mass contained in the control volume varies in general, and this makes it necessary to introduce the *mass conservation law.* This law states that no mass is created or destroyed in nature, at least as long as nuclear processes, which transform mass into energy, are not involved. Such transformations will be excluded.

Gravitational, electric, magnetic, centrifugal, and Coriolis forces act on the mass within the control volume. The fluid within the control surface also interacts with its surroundings, and we describe this by use of forces through which the surroundings act on the control volume. This becomes very clear if we consider a liquid at rest in a gravitational field. In Fig. 4, a region of the fluid is separated from the rest of the fluid by a control surface. Gravity acts on the mass within the control surface and would cause this mass to be accelerated if no other forces were present. Actually, the fluid does not move, and *surface* forces must therefore counteract the gravitational force as indicated in the figure.

A fluid is defined as a medium in which surface forces act only normal to the surface as long as it is at rest. The force per unit area is called *pressure.*

Additional forces appear when the fluid is in motion. They are called *viscous* forces, and they may have components normal to and parallel to the surfaces of the control volume. In most situations, the components parallel to the surface dominate. They oppose the relative movement, parallel to the surface, of the fluid layers on both sides of the surface and thus depend on the transverse velocity gradient. They start with the value zero for zero slip according to the definition of a fluid given above, and they increase in some way with increasing velocity gradient. Fluids are called *newtonian* when these forces increase linearly. Some media found in nature do not follow the definition of a fluid as given above; they can support shear stresses even without movement.

According to Newton's law these forces must be in equilibrium with inertia forces. This law must be written in the form

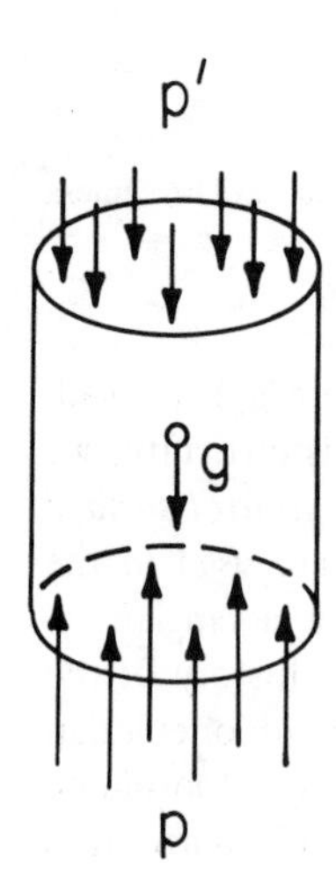

Figure 4 Gravitational and pressure forces on a fluid element at rest.

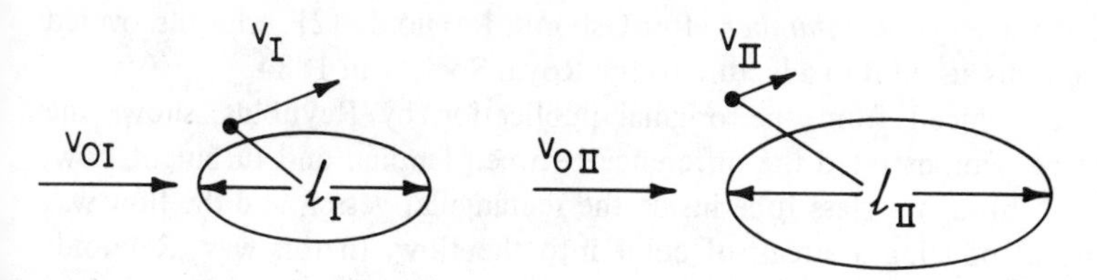

Figure 5 Two geometrically and physically similar flows.

$$\Sigma F = \frac{d}{dt}(mv) \tag{1}$$

because in general the mass in the control volume changes in time. The equation in this form is called the *momentum* equation.

Another useful law is that of *conservation of energy*. For incompressible fluids, however, this law does not provide any additional information.

2. SIMILARITY ANALYSIS

One of the most powerful tools in fluid mechanics is dimensional or similarity analysis, which permits a wide generalization of experimental results. The analysis is best performed on the equations that express the conservation laws (mass, momentum, and energy) mathematically [1]. The derivations are not discussed here, but the results obtained by dimensional analysis and their implications are discussed relative to the specific example shown in Fig. 5. The example concerns the flow of an incompressible and constant-property fluid with a uniform and constant upstream velocity v_0 over a cylinder with an elliptical cross section. Only pressure and viscous and inertia forces may be present. It is assumed that extensive experiments on the cylinder shown on the left-hand side of the figure, with major axis l_I, have resulted in information on the complete velocity and pressure field as well as on the drag exerted by the fluid on the cylinder. Our similarity analysis has the aim of extending this information to the flow over geometrically similar cylinders of different scale—for instance, to the cylinder shown on the right-hand side of the figure, with the larger major axis l_{II}. Geometric similarity requires that the ratio of the minor to the major axis have the same value for the two cylinders.

We now ask under what condition the velocity fields around the two cylinders are physically similar, in the sense that, at geometrically similar locations, the ratio of the local velocity v to the upstream velocity v_0 has the same value for both situations. Similarity analysis provides the answer that the dimensionless group $\mathrm{Re} = vl/\nu$ must have the same value for both cylinders, so that

$$\frac{v_{0,I} l_I}{\nu_I} = \frac{v_{0,II} l_{II}}{\nu_{II}} = \mathrm{Re} \tag{2}$$

This parameter expresses the ratio of the inertial forces to the viscous forces in the field. It is called the *Reynolds number* after Osborne Reynolds [2], who discovered its importance and discussed it in a lecture to the Royal Society in 1884.

Figure 6, reproduced from the original publication by Reynolds, shows the setup by which he demonstrated the difference between laminar and turbulent flow. Water was ducted through a glass tube inside the rectangular vessel, and the flow was made visible by introducing a streak of color into the flow. In this way, Reynolds demonstrated that a transition to turbulence occurs at a specific Reynolds number. He brought order and understanding to the many pressure-drop measurements that had been published prior to his experiments. A reproduction of a painting at the University of Manchester, where Reynolds was teaching, is shown in Fig. 7.

Therefore, for flow over geometrically similar objects, physical similarity of the velocity fields exists when the Reynolds number is constant. The similarity extends to the dimensionless pressure field and to the drag when it is expressed as a dimensionless drag coefficient. Experimental results can even be transferred from one fluid to another. We should expect that the generalization to objects of different scale can be made in continuum mechanics, since no inherent characteristic length is available to describe a continuum.

Figure 8 illustrates the flow around a circular cylinder for various Reynolds numbers. At Reynolds numbers smaller than 10, the flow moves around the cylinder and closes in on the downstream side without any separation from the surface. At

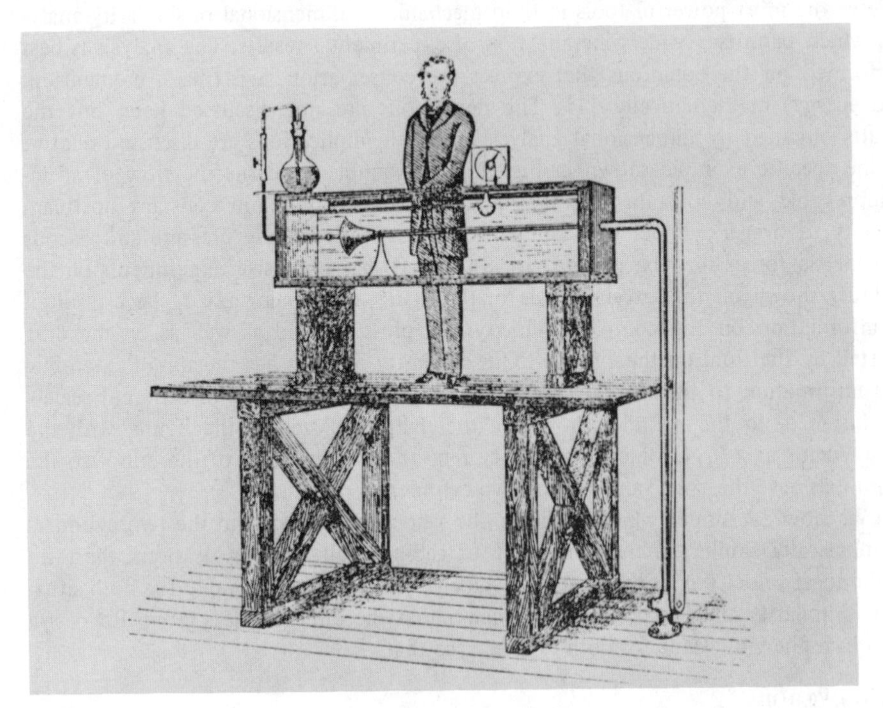

Figure 6 Demonstration of the transition to turbulence in flow through a tube by Osborne Reynolds (1880).

Figure 7 Osborne Reynolds (1904).

larger Reynolds numbers, a separation bubble occurs on the downstream side of the cylinder; the bubble increases in size with increasing Reynolds number. Inside this bubble the flow circulates slowly. Beginning at a Reynolds number of about 60, the fluid creates vortices, which periodically separate (shed) from the surface and, in moving downstream, create the *Kármán vortex street.* At Reynolds numbers above 5000, the vortex street becomes irregular and degenerates into smaller vortices, creating an appearance that is called *turbulence.*

Streamlines are shown in the two left-hand sketches of Fig. 8. They provide a clear picture of the flow because the velocity vector has, at each point, the direction of the streamline. For steady flow the streamlines also describe the path that the fluid particles follow. This, however, is not the case for unsteady flow, where the streak lines describing the fluid path are different from the streamlines that present an instantaneous picture of the direction of the velocity vectors of the fluid particles.

The periodicity of vortex shedding, when expressed as a dimensionless parameter,

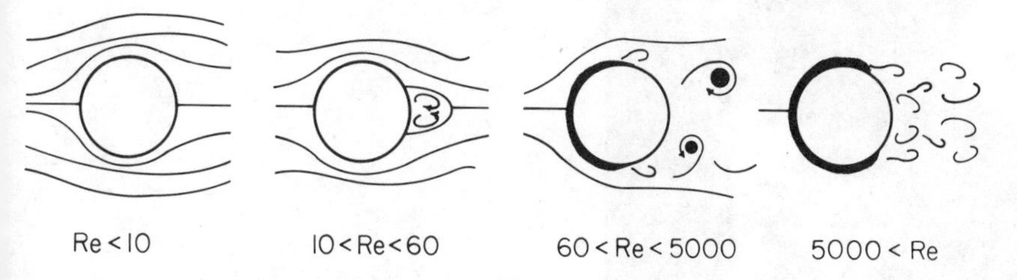

Figure 8 Flow around a circular cylinder at different Reynolds numbers.

is a function of the Reynolds number, as is the flow field in the separated region. In geometrically and physically similar situations the flow fields must be similar in all respects, including vortex shedding and the details of the turbulence. It is, however, often difficult (or even impossible) to set up situations that are exactly similar. For example, the surfaces of objects exposed to the flow may be rough. Strict geometric similarity would then require that the roughness elements be similar in shape and distribution. The flow through the boundaries may include velocity fluctuations that are not known in detail, or the pressure field may be influenced by sound waves traversing the boundaries. Experience must be used to establish the effect of such disturbances on the similarity relations. However, when used with proper caution, the similarity relations form the most powerful tool of fluid mechanics.

At low Reynolds numbers the effect of viscosity extends a considerable distance from the surface of the cylinder into the fluid stream. This distance shrinks with increasing Reynolds number and finally includes a very thin region close to the surface, called the *boundary layer*. Ludwig Prandtl, in his famous 1904 paper in which he introduced boundary-layer theory [3], unified theoretical and applied fluid mechanics. Until that time, the theoretical branch had analyzed inviscid flows in great detail but was not able to include the effects of fluid friction, and the applied branch expressed frictional effects empirically (in terms of friction factors) but could not provide any detailed understanding. Prandtl (Fig. 9) must, therefore, be considered the father of modern fluid mechanics.

The boundary layer is indicated in the right-hand sketches of Fig. 8 as a shaded area surrounding the cylinder. In the separated region, the viscosity has an important influence wherever large transverse velocity gradients exist. This is, for instance, the

Figure 9 Ludwig Prandtl (1938).

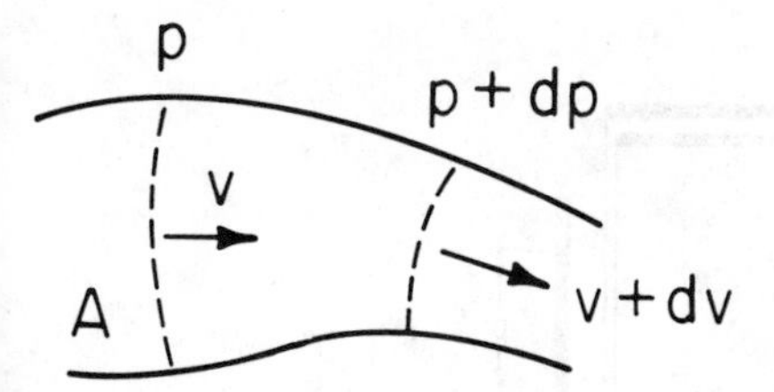

Figure 10 Inviscid flow along a stream tube.

case near the axis of the vortices. Such areas are also shaded in one of the sketches. Outside these regions, the influence of viscosity vanishes and the flow is close to the idealized flow of an inviscid fluid. In engineering applications, one usually deals with large-Reynolds-number flows, and the concept of an inviscid fluid is a useful one. The analysis in the following section deals, therefore, with the steady flow of an inviscid, incompressible (constant-density) fluid. Body forces are assumed absent or negligible. Liquids, and gases at low velocity, can be considered as incompressible.

3. INVISCID, INCOMPRESSIBLE FLUIDS

The analysis starts with the conservation of mass, momentum, and energy. For one-dimensional flow along a stream tube (Fig. 10), the mass conservation equation for steady flow has the form

$$vA = \text{const} \tag{3}$$

and the inviscid momentum equation is

$$\frac{\partial p}{\partial s} + \rho v \frac{\partial v}{\partial s} + \rho \frac{\partial v}{\partial t} = 0 \tag{4}$$

For the steady state, the last term in Eq. (4) drops out, and the equation can be integrated along the direction s of the stream tube to result in the Bernoulli equation

$$p + \rho \frac{v^2}{2} = \text{const} \tag{5}$$

The cross section of the stream tube must be small if one is to consider local values of the pressure and velocity.

The pressure p in Eq. (5) is, in principle, measured by an instrument that moves along with the fluid. This is, however, inconvenient, and the pressure is usually measured via a small hole in a wall arranged so that it does not disturb the flow. The upper part of Fig. 11 shows two ways in which this pressure, also called the *static* pressure, can be measured.

One may also define a total pressure p_t, which is obtained when the flow velocity is reduced to zero. From measurements of the total and static pressures, the velocity can be obtained as

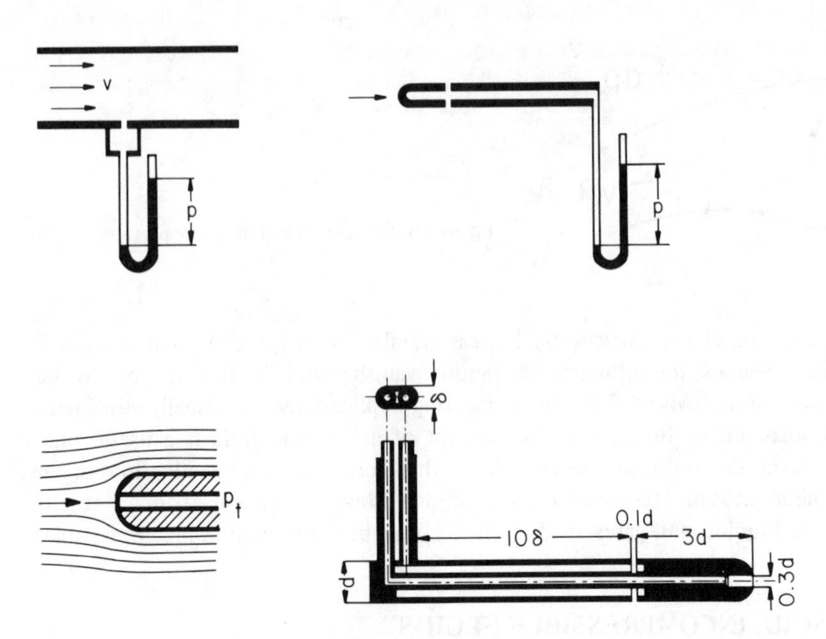

Figure 11 Measurement of the static and dynamic pressure.

$$v = \sqrt{\frac{2}{\rho}(p_t - p)} \tag{6}$$

which follows readily from Eq. (5). The lower left of Fig. 11 shows a total-pressure tube, used to measure the total pressure. The total-pressure and static-pressure tubes can be combined into one tube, as shown at the lower right of the figure. A tube with the measurements shown is called a *Prandtl tube.*

In general, a conservation equation for energy is also required for a complete description of the flow; the equation must include internal as well as mechanical energy. For an incompressible fluid, however, there is no possibility of an interchange between mechanical and internal energy. Equation (5) can thus be interpreted as a conservation equation for mechanical energy, and the conservation equation for internal energy is simply

$$T = \text{const} \tag{7}$$

along a stream tube for a fluid with constant specific heat or h (enthalpy) equals constant at variable specific heat. This is so because the viscosity and thermal conductivity of a fluid are closely connected: An inviscid fluid has, in general, a vanishing thermal conductivity.

In an experiment, a pressure measuring tube is used in a real viscous fluid and the question arises as to whether Eq. (6) applies. It was mentioned that, at high Reynolds numbers, the effect of viscosity is restricted to a very thin boundary layer along the surface of the tube; in addition, as Prandtl was the first to point out, pressure is transmitted through this thin layer without any measurable change. The use of Eq. (6) is, therefore, justified at high Reynolds numbers. However, a correction must be made when the Reynolds number (based, for instance, on the tube diameter) is small. This was established by Barker [4], who obtained the following relation:

$$\frac{p_t - p}{\rho v^2/2} = 1 + \frac{C}{\text{Re}} \qquad \text{for Re} < 30 \tag{8}$$

The value of the constant C depends on the shape of the head of the tube. Values between 3 and 5.6 are reported in the literature. A correction must also be applied where large transverse velocity gradients occur over a distance of the order of the tube diameter.

The integration of Eq. (1) in the direction n normal to the streamline leads to the equation

$$\frac{\partial p}{\partial n} = \rho \frac{v^2}{r} \tag{9}$$

This equation also states that, for a field with straight streamlines, the pressure normal to the streamlines is constant.

4. INVISCID, COMPRESSIBLE FLUIDS

Only ideal gases are considered in this section.

Pressure variations may cause considerable density variations in a compressible medium. Such a medium can give rise to pressure waves that travel with the velocity of sound:

$$a = \sqrt{\frac{dp}{d\rho}} \tag{10}$$

The change in thermodynamic state in such a pressure wave is isentropic. Since $p/\rho^\gamma =$ const, Eq. (10) becomes, for an ideal gas,

$$a = \sqrt{\gamma \frac{p}{\rho}} \tag{11}$$

Similarity analysis shows that other parameters, in addition to the Reynolds number, now determine the flow; these are the Mach number $\text{Ma} = v_0/a$, which is the ratio of a characteristic flow velocity v_0 to the sound velocity a, and the ratio of specific heats γ. One distinguishes between subsonic and supersonic flow depending on whether the Mach number is smaller or larger than 1.

The mass conservation equation for a one-dimensional treatment of steady flow along a stream tube (Fig. 10) is now

$$\rho v A = \text{const} \tag{12}$$

The energy conservation equation has the form

$$\frac{v^2}{2} + u + \frac{p}{\rho} = \text{const} \tag{13}$$

which now includes the internal energy u in addition to the mechanical energy. For an ideal gas with constant specific heat, this becomes

$$\frac{v^2}{2} + c_p T = \text{const} = T_t \tag{14}$$

The temperature T is again the quantity that, in principle, is measured by an instrument moving along with the fluid. It is shown later that this temperature cannot be measured except by a specially calibrated instrument. One can, however, measure the total temperature T_t that, according to Eq. (14), is obtained when the velocity in the fluid is decreased to zero. An instrument with which this temperature is measured to a satisfactory approximation is shown in Fig. 12. The flow approaching the instrument is slowed down at the entrance of the horizontal tube, and the temperature there is measured by a sensor–for instance, a thermocouple. An erroneous measurement would be obtained if the tube were closed at its downstream end, because the fluid at rest in the tube would lose heat to the tube walls. Therefore, a small amount of fluid is vented through an orifice at the downstream end of the tube; this flow is adjusted so that the overall error, which results both from heat loss and because the kinetic energy $v^2/2$ does not completely vanish, is minimized.

The manner in which the flow velocity is determined from pressure measurements depends on whether the flow is subsonic or supersonic. For subsonic flow, the deceleration of the flow ahead of a total-pressure tube is isentropic; integration of the momentum equation (4) and use of the relation

$$\frac{p}{\rho^\gamma} = \text{const} \tag{15}$$

result, for steady flow and subsonic velocities of the order of the sound velocity, in the equation

$$v = \sqrt{\frac{2\gamma}{\gamma - 1} R T_t \left[1 - \left(\frac{p}{p_t}\right)^{(\gamma-1)/\gamma}\right]} \tag{16}$$

For supersonic flow, the deceleration ahead of the tube occurs partially in the form of a shock, as indicated in Fig. 13. The stream tube that ends at the mouth of

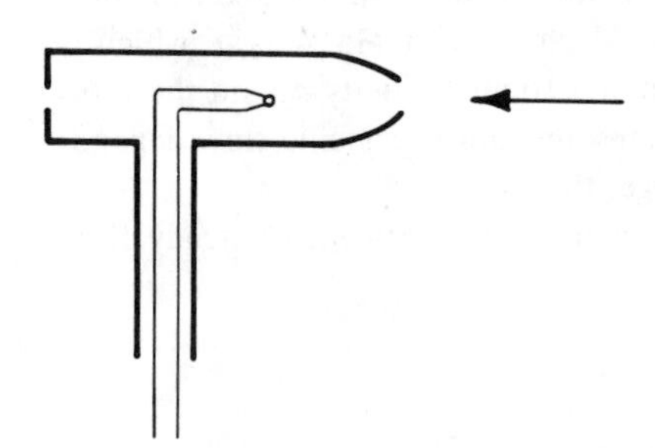

Figure 12 Total-temperature measurement.

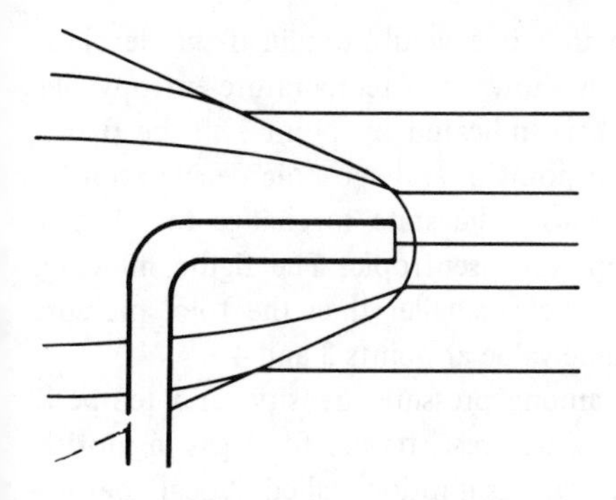

Figure 13 Supersonic flow around a total heat tube.

the measuring tube passes through a normal shock; its conservation equations, in terms of the control volume shown in Fig. 14, are as follows:

Mass: $\rho v = \rho' v'$

Momentum: $\rho v(v - v') = p' - p$ (17)

Energy: $\frac{v^2}{2} + c_p T = \frac{(v')^2}{2} + c_p T'$

These equations, together with the state equation of an ideal gas,

$$\frac{p}{\rho} = RT \tag{18}$$

are sufficient to allow calculation of the velocity, pressure, and temperature downstream of the shock, provided the values upstream of the shock are known.

The equations result in an entropy increase when the velocity decreases, and the pressure increases in passage through the shock. The deceleration in a shock is irreversible because of the effects of heat conduction and dissipation within the shock, which has a small but finite thickness. As a consequence, the pressure indicated by

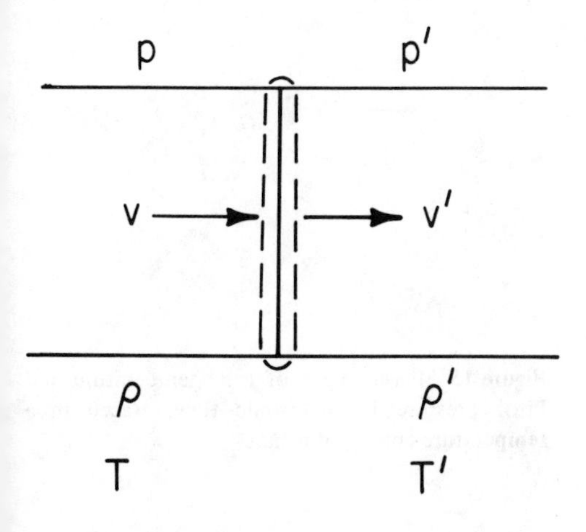

Figure 14 Compression shock in supersonic one-dimensional flow.

the tube is different from the total pressure that one would obtain if the deceleration ahead of the tube were isentropic. This is shown in a temperature-entropy diagram in Fig. 15. The state ahead of the tube is indicated by point 1 in the figure. Deceleration in the shock changes the state to point 2, and subsonic deceleration to the velocity zero downstream of the shock causes the state to change to point 3. Point 4 would be reached if the deceleration were isentropic. The figure indicates that the Pitot pressure p_p measured by the tube is smaller than the total pressure p_t, whereas the total temperature T_t has the same value at points 3 and 4.

Equation (18) describes the connection among pressure, density, and temperature for a gas at thermodynamic equilibrium, i.e., at rest. Its use for a gas in motion implies that this equilibrium persists locally, an assumption called "local thermodynamic equilibrium." This is a valid assumption in most cases, but exceptions do exist (for instance, in high-temperature plasma flow).

5. VISCOUS FLUIDS

In the discussion of Fig. 8, it was mentioned that at low Reynolds numbers the viscosity influences the entire flow field. Such flow fields (up to Re ~ 100) can today be studied by numerical computation, with the help of electronic computers. The results can be generalized using similarity analysis. At higher Reynolds numbers,

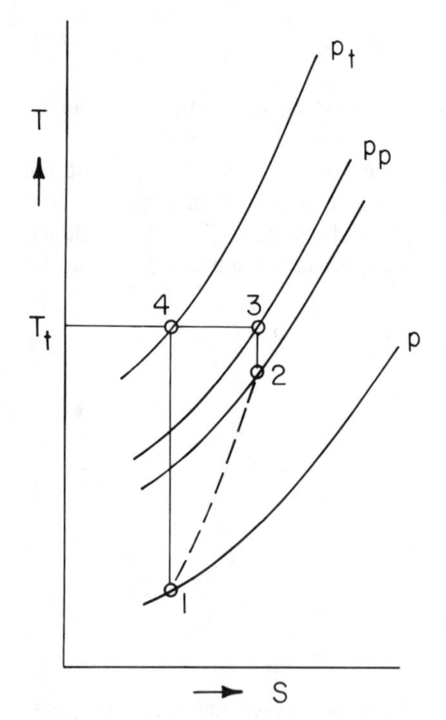

Figure 15 Measurement of total temperature and Pitot pressure in supersonic flow, traced in a temperature-entropy diagram.

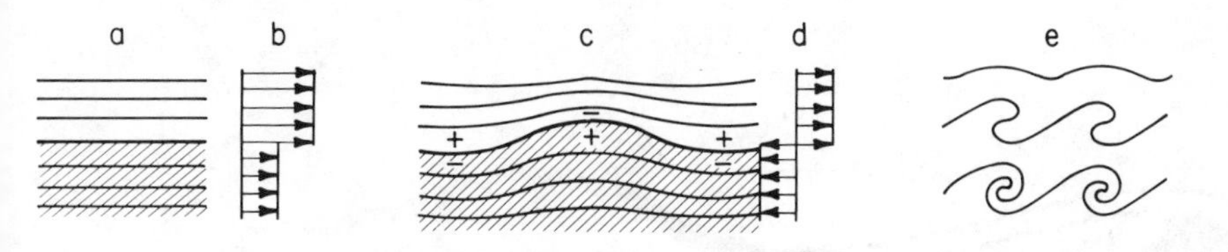

Figure 16 Wave formation in two inviscid fluid streams moving with different velocities side by side.

information on separation and on the flow in the separated regions has to be obtained primarily by experiments.

An analytical prediction of turbulent flow is the most difficult and challenging problem in fluid mechanics. In spite of intensive effort during the last hundred years, we are still far away from a complete understanding. A number of the other chapters in this book deal with turbulent flow, and only a few introductory remarks can be made here.

The fact that the flow of a fluid has the tendency to become unstable and to create vortices can readily be understood for inviscid flow. Figure 16*a* shows, following Ludwig Prandtl, two layers of an inviscid fluid; the layers move side by side with velocities that are uniform within each layer but different for the two layers (Fig. 16*b*). According to our discussion, the pressure is constant throughout the two layers. Consider now a small disturbance that deforms the separation line between the two layers into a slightly wavy shape. In Fig. 16*c* this is sketched using a coordinate system that moves along with the velocity of the separation line, which should be halfway between the velocities of the two layers (Fig. 16*d*). In this coordinate system one expects the flow to be steady; the Bernoulli equation predicts that the pressure is higher at locations indicated by a plus sign and lower at locations indicated by a minus sign. This pressure distribution obviously will increase the waviness, and the velocities in the layers into which the waves penetrate cause these layers to deform and roll up into vortices, as shown in Fig. 16*e*.

That such a development occurs in viscous flow as well was demonstrated for the first time by Tollmien [5], who succeeded in solving the laminar boundary-layer equations by superimposing on this flow a sinusoidal disturbance. His analysis resulted in the statement that the disturbance waves are damped out at low Reynolds numbers but increase in intensity for values above a certain "instability" Reynolds number. Experiments by Schubauer and Skramstad [6], performed in 1943 but published only after World War II, confirmed this result experimentally. Instability waves in a natural-convection boundary layer on a vertical surface were observed by flow visualization using a Zehnder-Mach interferometer (Fig. 17). In some flow visualization studies it has been found that the vortices created by a similar process become unstable and break up into a group of smaller vortices. The continuation of this process can be visualized to lead finally to turbulence. To date, however, no analysis has been successful in describing such a sequence.

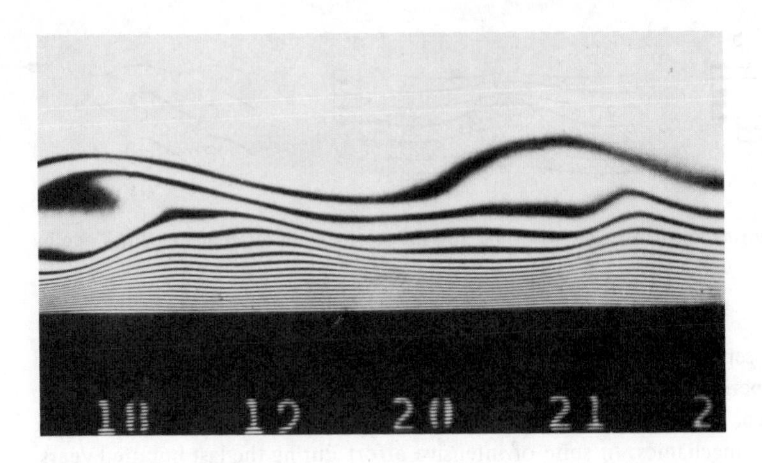

Figure 17 Interferogram showing instability waves in a natural-convection boundary layer on a vertical surface.

Osborne Reynolds, in a paper in which he derived the time-averaged continuity and momentum equations, was the first to present a method for predicting those characteristics of turbulence that are most important for engineering purposes. He demonstrated that turbulence creates turbulent shear stresses of the form

$$\tau_t = \rho \overline{u'v'} \tag{19}$$

in which u' and v' denote two instantaneous velocity components normal to each other, and the bar indicates the time average of the product of these fluctuating velocities. The time-averaged equations demonstrate the effect of turbulence, but the equations cannot be solved because the turbulent shear stresses or the fluctuating velocities appear as additional unknown parameters whereas the number of available equations is not increased. The equations can, therefore, be solved only with the use of additional experimental information. The first successful approach is again due to Prandtl, who argued that the fluctuating velocities (for instance, u') should be related to the local gradient of the time-averaged velocity through an expression of the form $l\,du/dy$, where l has the dimension of a length. He obtained the equation

$$\tau_t = \rho l^2 \left(\frac{du}{dy}\right)^2 \tag{20}$$

Through a reasonable assumption concerning the variation of the parameter l, which he called the *mixing length*, he was able to calculate the turbulent boundary-layer profile in reasonable agreement with experimental results. His approach has been refined in the years since its introduction, and it is still useful today.

In the rest of this section it is shown why the temperature T cannot be measured analogously to the measurement of the static pressure p as sketched in Fig. 11. The discussions of the preceding sections consider an ideal gas with no viscosity or heat conductivity. These assumptions apply, at high Reynolds number, to the flow

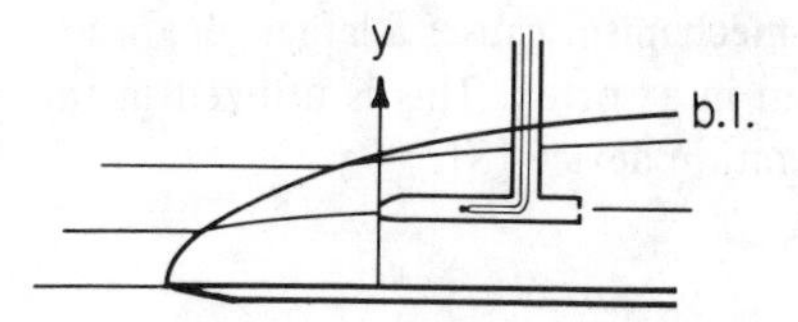

Figure 18 Total-temperature measurement in a boundary layer on an adiabatic flat plate.

in the mainstream. In a boundary layer or, generally, for low-Reynolds-number viscous flow, the situation is different. Additional terms appear in the energy equation (13), because energy can now be transferred from one streamline to another, either as heat by conduction or as mechanical energy through work when neighboring streamlines have different velocities. Near an adiabatic wall, the flux of heat and the flux of work occur in opposite directions. Whether the total temperature measured by an instrument within the boundary layer (Fig. 18) is larger or smaller than the total temperature given by Eq. (14) depends on whether the transport of work is larger or smaller than the transport of heat. In a fluid with a Prandtl number equal to 1, the two processes compensate each other everywhere within the boundary layer; the measured temperature is constant throughout the layer and equal to the total temperature outside the boundary layer. For other Prandtl numbers, the temperature measured with the instrument varies throughout the boundary layer, as shown in Fig. 19.

The total temperature at the wall itself, which is equal to the temperature that the adiabatic wall assumes, is called the *recovery temperature* T_r. Figure 19 shows that, for Prandtl numbers smaller than 1, which is in general the case for gases, the recovery temperature is smaller than the total temperature in the mainstream. C has, in a laminar boundary layer of air flowing over a flat plate, the value 0.16 in the equation

$$T_{t\infty} - T_r = C\frac{v_\infty^2}{2c_p} \tag{21}$$

where v_∞ indicates the velocity outside the boundary layer. The constant C has been measured to be 0.12 in a turbulent boundary layer over a flat plate.

The recovery temperature, therefore, is the temperature that a probe with the shape of a static pressure tube (Fig. 11) indicates, when a temperature sensory is lo-

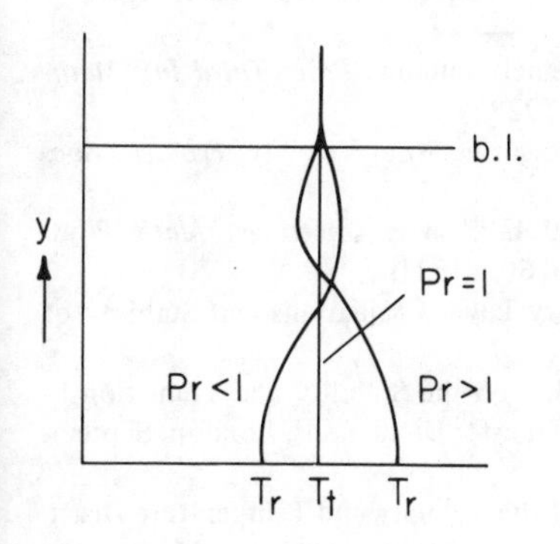

Figure 19 Profiles of the local total temperature in the boundary layer of the adiabatic flat plate of Fig. 18 for fluids with different Prandtl numbers.

cated where the pressure is measured. The same mechanism causes a larger separation of energy between a hot and a cold fluid stream in a vortex. This is utilized in the Hilsch or vortex tube to construct a simple refrigeration device [8].

NOMENCLATURE

A	area
a	sound velocity
c_p	specific heat at constant pressure
F	force
h	enthalpy
l	length
Ma	Mach number
m	mass
n	length, normal to streamline
p	pressure
R	gas constant
Re	Reynolds number
r	radius of curvature
s	length in streamline direction
T	temperature
t	time
u	internal energy
v	velocity
γ	specific heat ratio
ν	kinematic viscosity
ρ	density
τ	shear stress

REFERENCES

1. E. R. G. Eckert and Robert M. Drake, Jr., *Analysis of Heat and Mass Transfer*, pp. 389, 523, 728, McGraw-Hill, New York, 1972.
2. Osborne Reynolds, *Philos. Trans. R. Soc. London,* 1883, or *Collected Papers,* vol. II, 1901, p. 51, Cambridge University Press, Cambridge.
3. Ludwig Prandtl, Über Flüssigkeitsbewegung bei sehr kleiner Reibung, *Proc. Third Int. Math. Congr.,* Heidelberg, 1904, or *NACA Tech. Memo. p. 452,* 1928.
4. M. Barker, On the Use of Very Small Pitot-Tubes for Measuring Wind Velocity, *Proc. R. Soc. London Ser. A,* vol. 101, p. 435, 1922.
5. W. Tollmien, Über die Entstehung der Turbulenz, *Nachr. Akad. Wiss. Goettingen Math. Phys. Kl. 2A,* pp. 21–24, 1929. Translation: NACA Tech. Memo. 609, 1931.
6. G. B. Schubauer and H. K. Skramstad, Laminar Boundary Layer Oscillations and Stability of Laminar Flow, *J. Aero. Sci.,* vol. 14, pp. 69–78, 1957.
7. E. R. G. Eckert and E. Soehngen, Interferometric Studies on the Stability and Transition to Turbulence of a Free-Convection Boundary Layer, Heat Transfer Discussions, London, September 11–13, 1951.
8. E. R. G. Eckert and J. P. Hartnett, Experimental Study of the Velocity and Temperature Distribution in a High Velocity Vortex-Type Flow, *Trans. ASME,* vol. 79, pp. 751–758, 1957.

CHAPTER

THREE

DIFFERENTIAL PRESSURE MEASUREMENT

William K. Blake

1. INTRODUCTION

This chapter examines the instrumentation used to measure static and dynamic differential pressures and the uses of the measurements. The many practical and scientific uses of steady and fluctuating pressures are briefly reviewed. Multiholed Pitot tubes for vectorially decomposing complex steady-flow geometries are also described. The most generally used types of transducers for measuring either hydrodynamic or acoustic pressures in flows are described in general terms. Both condenser and piezoelectric transducers for sensing unsteady pressures are discussed in considerable detail, but the electromechanical properties of these sensors are simplified for clarity. Emphasis is instead placed on practical aspects of usage. To this end, we examine the simplified electronic circuitry and the limitations to and environmental influences on the respective sensititivities, as well as the frequency response. Physical errors in the measurement of either steady or fluctuating pressures that are distinct from statistical uncertainties are discussed. These errors include alignment uncertainties, differentials due to surface irregularities, spatial resolution, aerodynamic interference, and acoustic reflection. An introduction to the spatial filtering of unsteady pressures with arrays of transducers ends the chapter.

2. USES OF DIFFERENTIAL PRESSURE MEASUREMENTS

Differential pressures may be conveniently defined as differences between time-averaged pressures at two points in a fluid flow, or between a time-averaged and an instantaneous value of pressure evaluated at a point in the flow. In flows that are free of turbulence or other unsteadiness, pressures and velocities at a given point in space are constant in value. In these cases, differentials in the steady pressure at two points in the flow reflect differences in velocity.

The time variation or unsteadiness in the pressure sensed at a point in the flow occurs because of flow-velocity unsteadiness and because of local pressure disturbances. The pressure disturbances may be the result of hydrodynamic or acoustic fluctuations.

Differential pressure measurements provide a useful alternative to hot-wire and hot-film anemometry for determining complex flow directions and even turbulence intensity. For purposes of clarity it should be noted that pressures and velocities will each be regarded as a superposition of a mean and a fluctuating contribution. The fluctuating part of either quantity can be virtually eliminated by suitable time averaging. In the traditional use of differential pressures illustrated in Fig. 1, the difference between the time-averaged stagnation pressure and time-averaged "static" pressure obtained with a Pitot static tube is directly interpreted as the local mean velocity in the flow.

More generally, however, the properties of steady potential flow over simple bodies (e.g., cones, spheres, cylinders) may be exploited to elucidate magnitudes and directions of velocity vectors in complicated flows such as three-dimensional wakes, vortex flows, and fluid machinery. As long as the simple shape is small enough so that it does not disturb the flow and the velocity over it is uniform, yet large enough so that laminar separation does not generate any self-turbulence, the flow over it will be streamlined. Pressures vary over such surfaces because of the

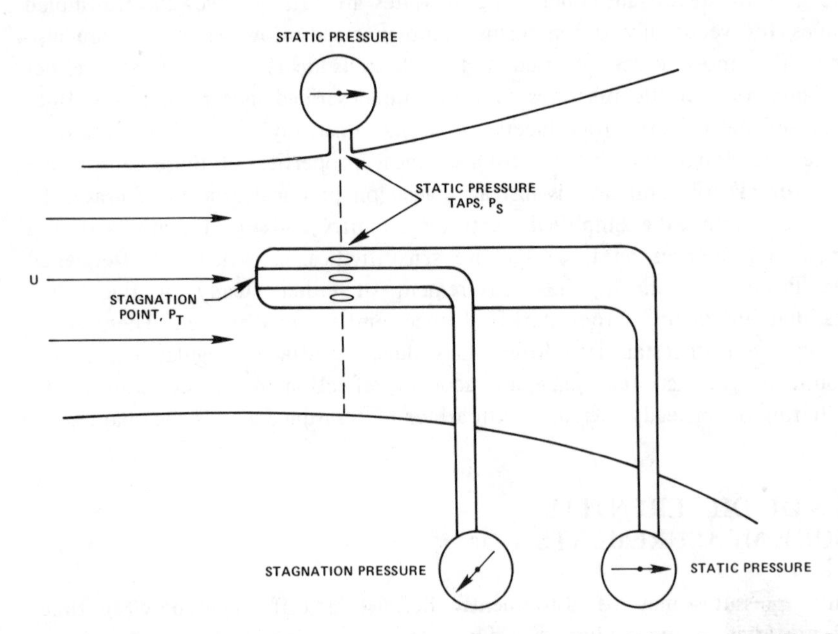

Figure 1 Typical installation of a Pitot-static tube in a ducted flow facility. The Pitot tube combines total pressure and static pressure into one instrument.

expansion of flow around them, and these pressures are functions of the probe orientation to the flow direction. Therefore, differences between pressures at suitably selected points on such surfaces are determined by both the magnitude and direction of flow relative to the body.

If the incident flow is turbulent, the total or stagnation pressure will contain both a mean and a temporally varying component. It is shown below that the temporally varying component can be used to elucidate the fluctuating velocity in the streamwise direction. Such a method of turbulence measurement is a useful alternative to hot-film anemometry, especially in liquids.

The so-called static pressure in a flow is generally measured on a surface on which streamline curvature is small. Such a surface is illustrated on the cylindrical body of the Pitot-static tube in Fig. 1. Perforations are placed around the circumference of the tube to average out any flow irregularities around the tube, as well as to minimize misalignment errors. The resultant pressure may have an instantaneously fluctuating component that is associated with any combination of acoustic disturbances and hydrodynamic pressure pulsations in the flow. The device has been used to determine both acoustic and aerodynamic pressure fluctuations in ducts and wind tunnels. Nose cones for microphones in wind streams are variations of this usage. Pressure taps in the walls of fluid machinery and lifting surfaces may also be used to determine both steady and unsteady surface pressures. Often, when only the fluctuating part of the surface pressure is desired, the sensor used is a microphone (generally in air) or a hydrophone (generally in water), either flush mounted or connected to the flow via a thin tube, or a small volume connected to the flow through a small "pinhole." Steady and unsteady differential surface pressures on airfoils and hydrofoils are measured to quantify fluid-dynamic loading on lifting surfaces.

Special forms of impact tubes, called *Preston tubes*, have been used for the measurement of wall shear stress in boundary layers over smooth walls. These are essentially hypodermic needles that respond to the mean velocity profile very near the wall.

3. PRINCIPLES INVOLVED IN MEASURING VELOCITIES WITH DIFFERENTIAL PRESSURE

The fundamentals of deducing fluid velocities, both mean and unsteady, from pressure measurements are discussed below. The discussion is divided into two sections. First we consider one-dimensional instantaneous velocity information, determined from the differential between the impact (stagnation) pressure and the static pressure. Then we consider multidimensional mean-velocity measurements in complex flows. The review article by Chue [1] should be studied by the reader interested in extensive coverage of calibration methods, errors, and special problems of high-speed flow for a variety of one-dimensional and multidimensional mean-velocity probes.

3.1 Pitot-Static and Impact Pressure Tubes

The Pitot-static tube is the most widely used and most reliable differential pressure probe for flow measurement. It is highly accurate, of simple construction, and governed by the most rudimentary of physical principles. As illustrated in Fig. 1, this probe has two sets of pressure taps included in an overall cylindrical body with a hemispherical nose. Along the barrel of the probe the flow lines are parallel, so the pressure there is equal to the static pressure in the stream. At the nose of the probe the fluid dams up so that the pressure there (the stagnation pressure) is greater than the pressure on the barrel by an amount $\rho_0 U^2/2$, where U is the local velocity of the fluid. Therefore the differential between these two pressures gives the fluid velocity. Accordingly, the device is outfitted with two conduits; one transmits the stagnation pressure, and the other the static pressure. To reduce probe alignment errors, the static pressure is transmitted to the appropriate conduit by a series of perforations around the perimeter of the barrel.

When the Pitot-static tube is placed in a laminar one-dimensional flow in which viscosity may be ignored, the pressures and velocities may be regarded along streamlines. The arrangement is idealized in Fig. 1, which illustrates pressures on the side of the probe, where the velocity is U, and at the nose of the probe, where the velocity is zero. Bernoulli's equation applies to this situation and gives

$$P_T = P_s + \tfrac{1}{2}\rho_0 U^2 \tag{1}$$

where P_T is the pressure at the nose of the probe, called the stagnation or total pressure, and P_s is the pressure on the side of the tube, called the static pressure. The difference between the static and stagnation pressures gives the local fluid velocity:

$$U = \sqrt{\frac{2(P_T - P_s)}{\rho_0}} \tag{2}$$

This pressure differential may be measured directly by connecting the separate tubes to an anemometer board or to a more sophisticated differential pressure gauge [2]. As long as the streamlines are not curved, there is no gradient of pressure perpendicular to the flow direction. Therefore the pressure at the probe wall equals the static pressure on the nearby wall of the duct.

For Eqs. (1) and (2) to apply, the probe must not disturb the flow, and its axis must be carefully aligned parallel to the stream. Although Pitot-static tubes are used as illustrated, they are generally regarded as error prone for certain applications that require miniature sensors. This is both because of limitations on size and because the static pressure on the side of the probe may be altered by self-induced disturbances on the probe. An alternative to the use of the combined Pitot-static tube, therefore, is a simple impact pressure probe that may be traversed in the flow, combined with a fixed, static, wall pressure tap, as illustrated in Fig. 2*a*. The impact probes may be miniaturized for use in boundary-layer traverses, and they may be of either circular or rectangular cross section. In special applications these probes may be used for shear-stress measurement [3, 6].

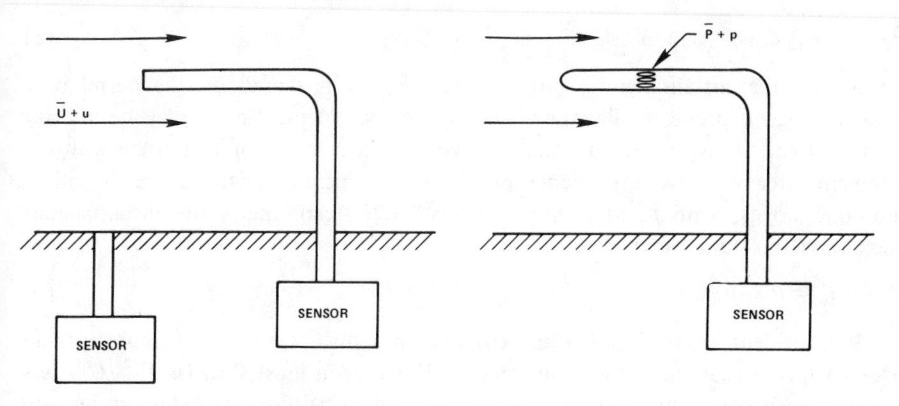

Figure 2 Illustrations of pressure probes used for the measurement of impact or stagnation pressure, or for the measurement of static pressure.

A generalization of Eq. (1) may be made to include unsteadiness in both the pressure and the velocity. If we assume a simple two-dimensional flow, then Euler's equation becomes

$$\frac{\partial(\bar{U}+u)}{\partial t}+(\bar{U}+u)\frac{\partial(\bar{U}+u)}{\partial s}=-\frac{1}{\rho_0}\frac{\partial(P+p)}{\partial s} \tag{3}$$

where $\bar{U}+u$ is the instantaneous velocity in the stream direction. For inviscid, irrotational flow (zero vorticity), Eq. (3) integrates to

$$\frac{\bar{P}+p}{\rho_0}+\frac{(\bar{U}+u)^2}{2}+\dot{\phi}=\text{const} \tag{4}$$

where the pressure and velocity are the sums of both mean and fluctuating quantities. The time average of Eq. (4) is

$$\frac{\bar{P}}{\rho_0}+\frac{(\bar{U}^2+\overline{u^2})}{2}=\text{const} \tag{5}$$

The difference between Eqs. (4) and (5) is

$$\frac{P}{\rho_0}+uU+\dot{\phi}=0$$

Now, ϕ is the fluctuating potential, which in a locally irrotational region of the flow is the same at the stagnation point in the flow and on a nearby point away from the stagnation point. Therefore, within the vicinity of the probe it can be hypothesized that the instantaneous total or stagnation pressure is

$$(P+p)_T=\bar{P}+p+\tfrac{1}{2}\rho_0(\bar{U}+u)^2 \tag{6}$$

The fluctuating part of the total pressure is given by

$$p_T=p+\rho_0\bar{U}u+\tfrac{1}{2}\rho_0(u^2-\overline{u^2}) \tag{7}$$

and the time-averaged pressure is

$$\bar{P}_T = \bar{P} + \tfrac{1}{2}\rho_0(\bar{U}^2 + \overline{u^2}) \tag{8}$$

Ideally the instantaneous static pressure $P + p_s$ is sensed on the barrel of a properly aligned probe, as illustrated in Fig. 2*b*, but in practice it may be in error by an amount p_e owing to pressures induced by interaction of the probe with the turbulent stream. These interference pressures are generally small compared with P and, one hopes, with p, as is noted in Sec. 6.2. Accordingly, the instantaneous pressure at the static pressure tap is

$$(P + p)_s = P + p \tag{9}$$

It is evident from Eqs. (2) and (6) that in a turbulent flow $\overline{u^2}$ represents an error to the measurement of mean velocity. However in most flow $(\overline{u^2})^{1/2}/\bar{U}$ is less than 0.2 so it contributes little to the average differential pressure. This can be seen from

$$\begin{aligned}\bar{U} &\simeq \bar{U}\left(1 + \frac{\overline{u^2}}{\bar{U}^2}\right)^{1/2} \\ &\simeq \bar{U}\left(1 + \frac{1}{2}\frac{\overline{u^2}}{\bar{U}^2}\right) \\ &\simeq \bar{U}(1 + 0.02)\end{aligned}$$

for 20 percent turbulence intensity.

Separate measurements of the total and static pressures can yield both the mean and fluctuating components of the velocity and the pressure. Turbulence intensities have been deduced from total-pressure fluctuations by Ippen and Raichlen [7] and more recently by Arndt and Ippen [8]; unsteady fluid pressures have been deduced from the fluctuating static pressure by Strasberg [9], Sidden [10], and Elliott [11]. To perform these measurements the fluctuating pressures must be transmitted down the tubes to a microphone or other dynamic pressure-sensing device, as is discussed at length in subsequent sections.

Just as unsteady fluid motions influence the value of the mean velocity deduced from the measurement of the differential of average pressures, the unsteadiness influences a determination of the turbulence level from the fluctuating total pressure. The mean square of the fluctuating total pressure is, from Eq. (7) (neglecting the squares of fluctuating velocities),

$$\overline{p_T^2} = \overline{p^2} + \rho_0^2\bar{U}^2\overline{u^2} + 2\rho_0\bar{U}\,\overline{pu} \tag{10}$$

The term we are interested in is $\rho_0^2\,\bar{U}^2\overline{u^2}$, but it must be established first that $\overline{p^2}$ and $\overline{pu}$ are negligible by comparison. In practical flows there is little or no information on $\overline{p^2}$ and $\overline{pu}$, but there is guidance to be taken from the measurement and theory of isotropic turbulence (see [12]). In isotropic turbulence it can be argued on kinematic grounds that $\overline{pu} = 0$ identically. Assuming that this is an indication of approximately similar behavior in most flows, we can state at least that

$$\frac{\overline{pu}}{\rho_0\bar{U}^3} \ll \frac{\overline{u^2}}{\bar{U}^2}$$

Also, in isotropic turbulence it is now well known [12] that

$$\overline{p^2} \sim \tfrac{1}{3}\rho_0(\overline{u^2})^2$$

Accordingly, it can be stated that

$$\frac{\overline{p^2}}{\rho_0^2 \bar{U}^4} \sim \frac{1}{3}\left(\frac{\overline{u^2}}{\bar{U}^2}\right)^2 \ll \frac{\overline{u^2}}{\bar{U}^2}$$

Equation (10) then gives the turbulence intensity as

$$\frac{\overline{p_T^2}}{(\bar{P}_T - \bar{P}_S)^2} = 4\,\frac{\overline{u^2}}{\bar{U}^2} \tag{11}$$

Figure 3 shows an example of the use of the total-pressure tube for the measurement of streamwise turbulence intensities.

An alternative approach to the measurement of turbulence by the use of Pitot tubes has been described by Becker and Brown [13]. In fact they report great success in determining transverse turbulence intensities from total-pressure fluctuations. To do this they note that the directional responses of a Pitot tube depends on the geometry of the nose. The differential between total pressures sensed by differently shaped probes are then used to determine the turbulence intensities using empirically deduced directionality factors.

3.2 Multidimensional Mean-Velocity Measurement

Although a variety of pressure probes have been devised for decomposing the flow velocity vector, the most well known and widely used is the five-hole pressure probe. This device is a streamlined axisymmetric body that points into the flow. The pressure distribution on the surface of the probe depends on the angle of incidence of the mean flow vector relative to the axis of the probe. To determine the magnitude and orientation of the flow vector, the surface pressure is sampled at five locations: on the axis of the probe and at four equispaced points on a line encircling this central point. The pressure differentials between selected pairs of these points may be related to the inflow velocity vector by using an appropriate calibration to deduce pitch and yaw directions. Hypothetically, theoretical relationships for the potential flow around the body may also be used, but the flow is generally not ideal, which makes such calculations impractical. Two commonly used shapes are the cylindrical tube with a hemispherical nose and a spherical ball at the end of a slender cylinder. These arrangements are sketched in Fig. 4. Less streamlined shapes, such as prismatic forms and cones, are also used, with their calibrations wholly empirical. The central pressure tap gives the conventional stagnation pressure when the flow vector is perpendicular to that point on the surface. Yaw and pitch inclinations of the flow vector with the axis of the probe result in an imbalance of pressures on pairs of holes.

In the case of the spherical probe, the theory yields a format for interpreting the differential pressures between pairs of holes as functions of angles of pitch and yaw. For calibration, the probe may be installed in a water or wind tunnel and

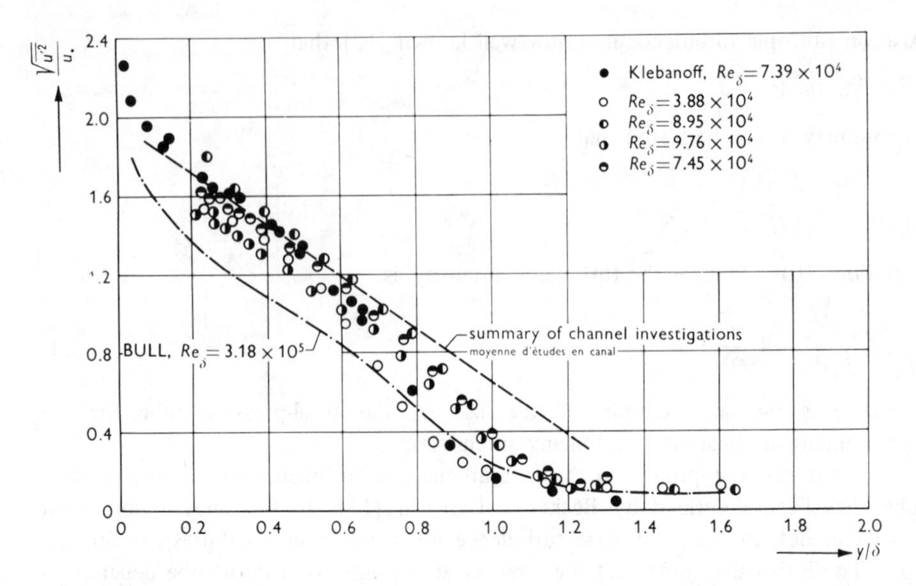

Figure 3 Turbulence intensities in boundary layers on flat walls. Points designated ○, ◐, ◑, and ◓ were determined from fluctuating impact pressure; those designated ● were determined with a hot-wire anemometer. *(From [8].)*

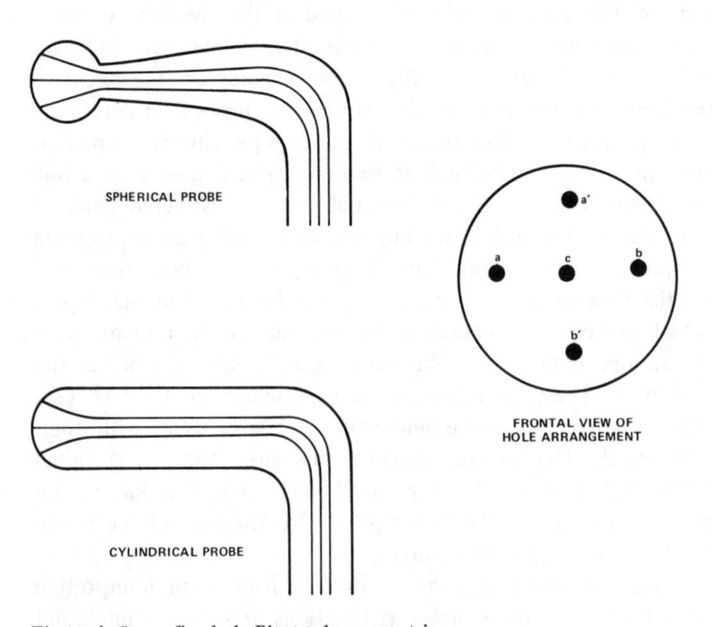

Figure 4 Some five-hole Pitot-tube geometries.

aligned with the known flow direction. Pressure differentials are then measured for selected angles of yaw and pitch placed on the probe relative to the flow direction.

As an example, consider in Fig. 5 the vector decomposition of the flow velocity U,* which is incident on the five pressure taps *a, b, c, a'*, and *b'* of a spherical probe. The center of the coordinate axes coincides with the center of the probe. The holes *a, c,* and *b* lie on a circle in the *xy* plane; *a'*, *c*, and *b'* lie on a circle in the *xz* plane. The angles between holes in their respective planes all equal α. The velocity vector U has a component U_h in the *xy* plane at an angle β_n with the *x* axis,† and a component U_v in the *xz* plane at an angle β_v with the *x* axis. The flow vector makes an angle γ with the *z* axis. Potential flow over a sphere gives rise to a surface pressure p given by

$$\frac{p - p_0}{\frac{1}{2}\rho_0 U^2} = 1 - \frac{9}{4} \sin^2 \beta$$

where β = angular position of point on sphere from flow axis
p_0 = static pressure in free stream

Pien [14] has shown that this relationship gives, for the geometry shown in Fig. 5,

$$\frac{P_a - P_b}{2P_c - P_a - P_b} = \frac{\sin 2\alpha}{1 - \cos 2\alpha} \tan 2\beta_h \tag{12}$$

and that

$$\frac{P_a - P_b}{\frac{1}{2}\rho_0 V_h^2} = \frac{9}{4} \sin 2\alpha \sin 2\beta_h \tag{13}$$

Note that the velocity component U_v in the vertical plane does not appear in these relationships. Measurements of the pressures P_a, P_b, and P_c give the angle β_h in Eq. (12); $U_h = U \sin \gamma$ is given by Eq. (13). Since Eqs. (12) and (13) do not include U_v

*In the following, the mean velocity will be denoted by a capital U with the U' bar being dropped.

†Also the probe axis.

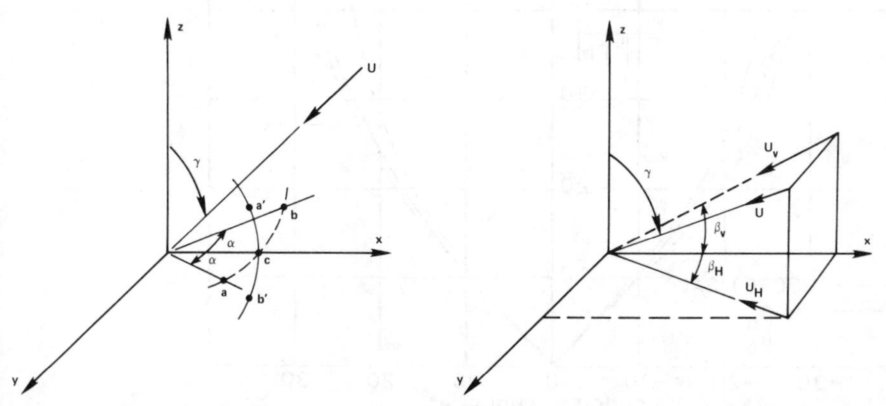

Figure 5 Vector decomposition of hole geometry and flow direction for a spherical-head Pitot tube. Points *a, c,* and *b* lie in the *xy* plane; *a', c,* and *b'*, in the *xz* plane.

or β_v, similar equations with $P_{a'}$ replacing P_a, etc., will give β_v and U_v independently of U_h and β_h.

As a practical matter, the probe can be calibrated in a wind or water tunnel, as mentioned above, so that the flow angles γ and β_h or β_v may be carefully controlled [15]. Figure 6 is a comparison of measured curves for β_v versus

$$C_{p_h} = \frac{P_{a'} - P_{b'}}{2P_c - P_{a'} - P_{b'}}$$

for β_h versus

$$C_{p_h} = \frac{P_a - P_b}{2P_c - P_a - P_b}$$

and for Eq. (12) with $\alpha = 20°$. The curves were determined by Hale and Norrie [16], who also determined a means of calibration when a controlled calibration flow is not available. Since the shape is axisymmetric, the spherical probe has equal sensitivities to yaw β_h and pitch β_v. Other probes, such as the prismatic probes used by Treaster and Yocum [17], have calibrations that are not so symmetrical.

Few direct comparisons between five-hole probes and alternative flow-measuring devices are available. One comparison by Scragg and Sandell [18] concerned the measurement of the tip vortex of a finite-aspect-ratio airfoil. Figure 7 shows measurements made with a spherical five-hole probe (HPT) of 9.5 mm diameter, a cylindrical five-hole probe with a hemispherical head of 3.4 mm diameter, and a three-component hot-wire anemometer with an overall diameter of 3 mm. The three methods give velocities that agree to within 10%, with some systematic differences

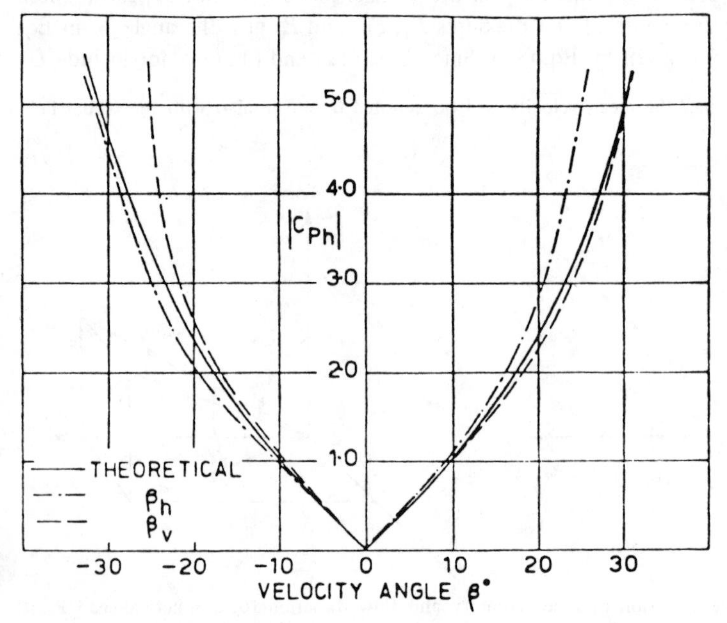

Figure 6 Typical calibration curve for a $\frac{3}{8}$-in-diameter spherical-head Pitot tube. *(From [16].)*

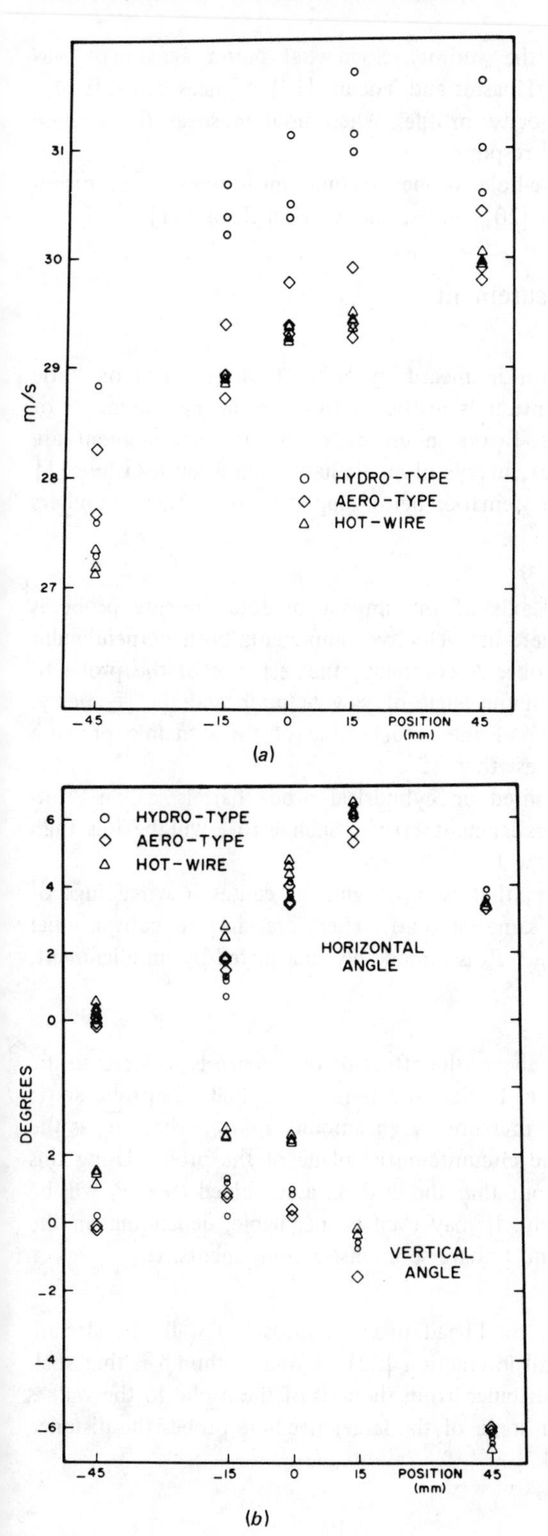

Figure 7 Measurements of (*a*) resultant velocity and (*b*) flow angles at the tip and 0.6 m downstream of an airfoil. Points designated "hydro-type" and "aero-type" pertain to cylindrical and spherical probes, respectively. *(From [18].)*

that could not be explained by the authors. Somewhat better agreement was obtained in a comparison made by Treaster and Yocum [17] of measurements of a one-dimensional boundary-layer velocity profile. When used in shear flows these probes may give spurious directional response.

Additional applications for five-hole probes include measurements in turbomachinery [19], wakes of ship hulls [20], and special vortical flows [21].

3.3 Physical Errors in the Measurement of Steady Pressures

The most common sources of error in measuring either total (impact) or static pressure are described below. Attention is restricted to errors in measurements of time-averaged, or steady, pressures. Errors in unsteady pressure measurement are discussed in Sec. 6. More exhaustive surveys of errors have been given by Chue [1] and Beckwith and Buck [2]. The remarks below apply to flow Mach numbers below 0.6.

3.3.1 Alignment errors. When the axis of the impact or total-pressure probe is misaligned with the flow vector, there are velocity components both perpendicular and parallel to the face of the probe. Accordingly, the response of the probe to misaligned flow will depend on both the angle of yaw or pitch and the geometry. Typically, errors are less than 0.05 of the actual value of the dynamic pressure ($\frac{1}{2}\rho_0 U^2$) as long as these angles are less than 15°.

When static pressures are measured on cylindrical probe barrels, as on Pitot-static tubes, there will also be misalignment errors. Such errors will be less than 0.05 of $\frac{1}{2}\rho_0 U^2$ for probe angles below 10°.

These errors are both negative; that is, misalignment causes low readings of both P_T and P_s by nearly the same amount. Therefore, in the conventional Pitot-static tube, the differential $P_T - P_s$ is remarkably unaffected by misalignment, at least for angles of less than 15°.

3.3.2 Turbulence errors. As shown above, the effect of turbulence is to increase the sensed value of mean total pressure. In Sec. 6.2 it is shown that the probe static pressure will exceed the true static pressure by an amount $\frac{1}{4}\rho_0 \overline{u_n^2}$, where $\overline{u_n^2}$ is the resultant turbulence intensity in the circumferential plane of the probe. Using this factor and Eq. (8), one can work out that the error in a combined $P_T - P_s$ will be less than in each pressure separately. It may even be negligible, depending on the relative magnitudes of streamwise and transverse intensity components.

3.3.3 Wall proximity effects. When total-head probes approach a wall, the streamlines are deflected by the probe-wall interaction [22]. A rule of thumb is that such interactions cause error when the distance from the axis of the probe to the wall is less than two probe diameters. For some of the larger five-hole probes the distance may be more like four diameters [17].

3.3.4 Hole geometry for static-pressure taps. An "ideal" tap geometry is a small circular hole of less than $\frac{1}{4}$ mm diameter drilled perpendicular to the surface on which the pressure is to be measured; the corner of the hole is perfectly sharp and squared off. Any departure from this geometry will introduce error. A 1-mm diameter should introduce an error of less than 1% of $\frac{1}{2}\rho_0 U^2$ compared with a hole diameter of $\frac{1}{4}$ mm. According to Shaw [23], errors with practical-sized holes occur because of flow in and around the hole opening. Rounding of the hole corners (up to a radius of curvature equal to the hole diameter) and nonperpendicularity of the hole with the wall (up to 45°) introduce errors of less than 1% of $\frac{1}{2}\rho_0 U^2$. Burrs on the edge of the hole, with heights of less than $\frac{1}{30}$ of a hole diameter and extending into the flow, introduce errors of less than 1% of $\frac{1}{2}\rho_0 U^2$. For probes used in boundary-layer measurements, errors of less than 3% of τ_w are encountered for $(D/\nu)\sqrt{\tau_w/\rho_0}$ less than 400, where τ_w is the wall shear stress, ν is the kinematic viscosity, and D is the hole diameter. (See [1, 23].)

3.3.5 Nose geometry for total-head and static-head probes. For angles of yaw below 15°, P_T and P_s are sensitive to nose geometry. In the case of P_s, the taps must be set far enough downstream of the nose to be uninfluenced by the pressure gradient there. A rule of thumb is three to four times the probe diameter, although a distance of five diameters has been used in transducers designed for measuring unsteady "static" pressures (see Sec. 6.2).

3.3.6 Influence of probe supports. The pressure gradients associated with the curvature of flow lines around probe supports can be avoided by proper displacement of the probe. For cylindrical probe supports perpendicular to the probe axis, this distance should be five support diameters to avoid an error amounting to 2% of $\frac{1}{2}\rho U^2$. For an aerodynamically faired strut the rule for low Mach numbers might be relaxed to three support thicknesses.

3.3.7 Sizes of probes. In the case of Pitot and total-head probes, Eq. (2) is within 2% of the calibration for Reynolds numbers based on external tube radius that are greater than 300. This applies to nose shapes that are blunt or round.

Spherical-nosed five-hole probes should be large enough to prevent flow disturbances by laminar separation. Reynolds numbers based on diameter that are greater than 10^4 generally provide steady, reliable calibrations [16]. A miniature five-hole probe with a conical shape has been used down to a Reynolds number of 2000 [21]. This probe had a physical diameter of 2 mm. Typically, however, sphere diameters of 8 to 10 mm are used; hemispherical-nosed five-hole probes can be as small as 3 mm.

4. TYPES OF TRANSDUCERS FOR MEASURING UNSTEADY PRESSURES

We turn now to the transducers that are used in the measurement of unsteady hydrodynamic or acoustic pressures. Microphones of various types have long been used in the audio industry for acoustic pressure measurements. In underwater acoustics, hydrophones too have long been used for acoustic monitoring. Until, perhaps, 20 years ago these were the principal uses of pressure transducers. The late 1950s brought about the beginnings of miniature transducer technology, and now miniature transducers of nominally 3.18 mm diameter or less are commercially available for many applications that require high spatial resolution and high-frequency response.

For the measurements of aeroacoustics and for aerodynamic measurements in subsonic flow, the use of condenser microphones is common. These devices can be either flush mounted in a wall, fitted with windscreens to be positioned in the flowing gas, or connected to the pressure leads of a total- or static-pressure tap. Hydrophones are generally more durable but less sensitive for their sizes; they find widespread use in underwater acoustics, monitoring dynamic loads imposed on structures adjacent to hydraulic machinery, as well as monitoring pressures in shock tubes, high-pressure vessels, and reactor vessels. Either microphones or hydrophones (depending on application) are used as detectors of turbulence, dynamic stall, and cavitation (hydrophones).

Primarily, pressure transducers consist of either flexible diaphragms or piezoelectric elements; the fundamental behavior of each major type is shown in Fig. 8. Of the membrane or diaphragm type, the most commonly used for aeroacoustics are

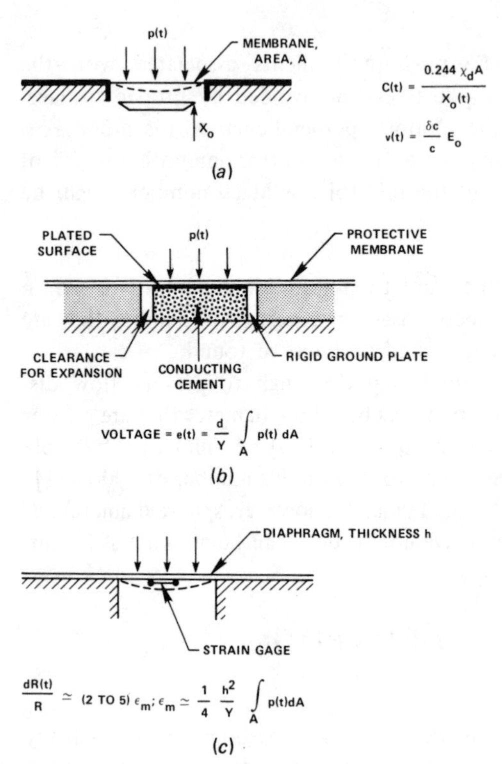

Figure 8 Schematic representation of the leading types of pressure transducers. (*a*) Capacitance device ($\chi_d = 1$ in air). (*b*) Piezoelectric device (d = piezoelectric coefficient). (*c*) Strain-gage device.

condenser microphones, which are really capacitors whose capacitance varies as the plate separation changes under the influence of applied pressure. These must be delicate transducers if they are to provide an adequate plate displacement $x_0(t)$ under the influence of small pressures (100 μbar or so). Furthermore, the dielectric constant κ_0 of the capacitor is determined by the humidity and temperature of air between the plates. The delicate nature of the device rules out the condenser microphone for general use in liquids. In-depth coverage of the characteristics of condenser microphones may be found in [24, 25]. Reference [26] has a short descriptive section.

The piezoelectric transducer has as its element a crystalline substance that generates an electric field as it is mechanically deformed. The crystal is shaped as a cylinder, wafer, or bar; it is generally plated on opposite surfaces, and the electric charge is generated across those plates. The impedance of the transducer is thus basically capacitive. The transduction of force to charge is described in terms of any of the various piezoelectric constants, say d, which relate the stress and field in either parallel or orthogonal axes. The leading materials in use are compounds of lead, zirconate, and titanate (PZT); the latter has long retention of polarity and good temperature stability. These crystals are not found in nature, but require artificial polarization. Piezoelectric transducers are reversible, in contrast to capacitive transducers; they may be electrically driven to generate mechanical thickness deformations. They therefore make ideal acoustic transmitters (loudspeakers) for liquid application. Kinsler and Frey [26] give a good summary of the electromechanical behavior of piezoelectric materials.

A third type of pressure transducer utilizes a flexible diaphragm on which is attached a strain gauge. When the diaphragm is deflected by the action of a pressure fluctuation, the strain amplitude ϵ_m is sensed by a resistance change dR in a strain gauge that forms a branch of a Wheatstone bridge. The advantage of these transducers is that they are very thin and can, therefore, be installed in such bodies as airfoils near trailing edges [28]. They are also less sensitive to vibration than piezoelectric transducers. Their main drawback is dc drift, which may cause combinations of nonlinearity of voltage output per unit fluctuating pressure and amplifier overload. In special applications it should be possible to design a strain-gauge pressure gauge for underwater applications.

A fourth type of transducer responds not to a pressure at a point, but to a difference in pressure between two points. These are called *pressure-gradient* transducers, and they are not used as widely as pressure transducers. Since the pressure gradient in an acoustic field is proportional to the fluid velocity, these transducers are used to determine acoustic particle velocity. They therefore may be used to determine wave propagation direction. A form of gradient microphone has also been used to determine unsteady differential aerodynamic loads on the rotor blades of turbomachines [27, 28]. A good summary of the electromechanical operation of pressure-gradient transducers has been given by Beranek [24] and Olsen [25]. An elementary treatment of the general use of pressure transducers has been given by Beckwith and Buck [2].

4.1 Condenser Microphones

Figure 9 is a sketch of the typical construction of a miniature condenser microphone. The capacitor consists of the membrane and back plate as opposite plates. Under the action of an applied pressure the membrane defects so that the spacing changes; accordingly, so does the capacitance. When a constant-bias-voltage potential is applied across the membrane and back plate, a stable charge is maintained in the capacitor. The change in capacitance causes a fluctuation in voltage that is proportional to the fluctuating pressure amplitude. A hole is provided to the side of the instrument so that no static pressure differential exists across the membrane, which may be made of nickel, plated mylar, or plated glass. Since the operation of the sensor depends upon the dynamic deflection of the membrane, it has a resonance frequency at which the sensitivity is controlled by applying the appropriate damping to the membrane.

The microphone is used in conjunction with a cathode follower, or microphone preamplifier (Fig. 10), which provides a constant-bias unity-gain preamplifier of very low input capacitance and high output impedance. The preamplifier unit is required because the capacitance of the transducer is small (100 down to ~4 pF) so that the addition of even a small amount of cable capacitance (~30 pF/ft) is sufficient to cover the signal.

The simple equivalent circuit is shown in Fig. 11. The open-circuit voltage $e_{oc}(t)$ created is proportional to the applied bias voltage E_0, the area of the microphone face A, and the effective mechanical stiffness or capacitance (displacement/force) of the membrane C_m. The voltage fluctuation increases as the plate separation X_0 decreases. An analysis of the circuit in Fig. 11 for the *blocked* voltage $v_0(t)$ (that is, the voltage output from the unity-gain preamplifier when no current flows) resulting from $e(t)$gives

$$\frac{e_{oc}(t)}{v_0(t)} = \frac{1 + i\omega R_p C_p + i\omega R_p C_t}{iR_p \omega C_t} \tag{14}$$

where R_p = input resistance of cathode follower
C_p = input capacitance
C_T = transducer capacitance

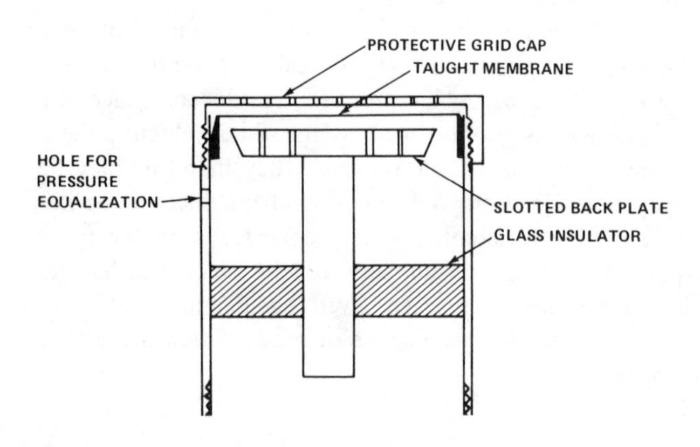

Figure 9 Typical condenser microphone construction.

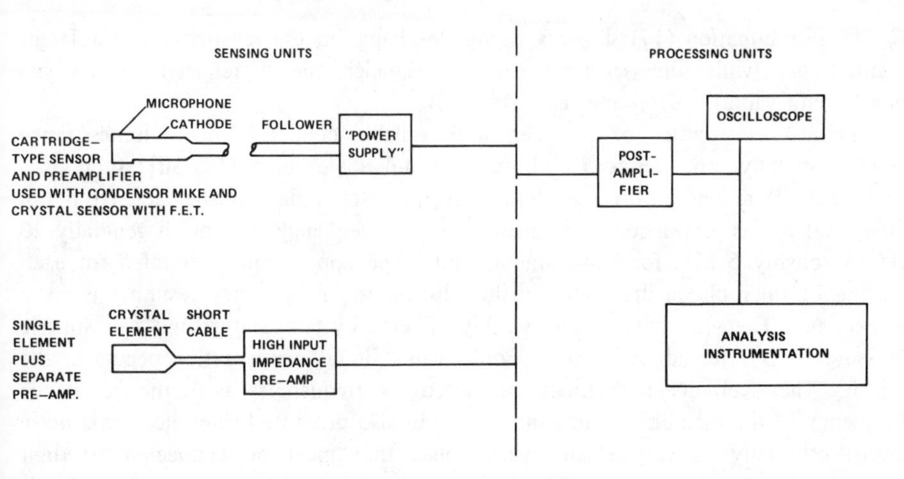

Figure 10 Pressure transducers and their typical instrumentation chains.

If the input resistance is very high (i.e., if $R_p C_p \omega \gg 1$), then Eq. (14) reduces to

$$\frac{v_0(t)}{e_{oc}(t)} = \frac{C_T}{C_T + C_p} \tag{15}$$

Equation (15) is important, for it shows how the blocked voltage is reduced by the capacitance of the preamplifier.

Since the open-circuit sensitivity S_{oc} of the microphone is

$$\frac{e_{oc}(t)}{p} = S_{oc} = \frac{C_m A E_0}{X_0} \tag{16}$$

the effective sensitivity of the microphone-cathode follower is reduced by the factor $C_T(C_T + C_p)^{-1}$; that is,

$$\frac{v_0(t)}{p} = S_{\text{cable}} = S_{oc} \frac{C_T}{C_T + C_p} \tag{17}$$

As examples we compare the sensitivity losses of two microphones: a 1-in-diameter microphone with $C_T = 100$ pF and a $\frac{1}{8}$-in diameter microphone with $C_T = 4$ pF and

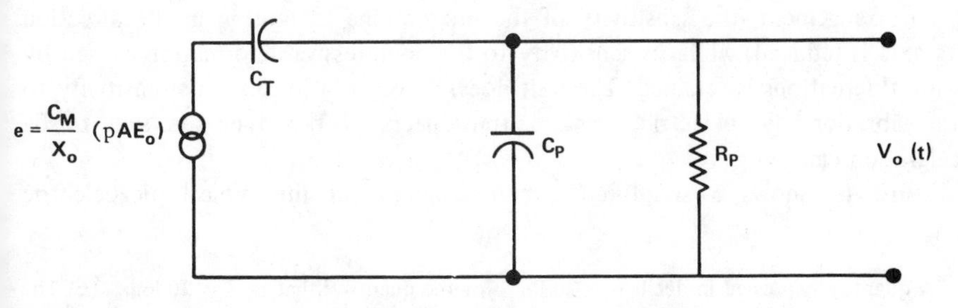

Figure 11 Equivalent circuit for capacitor microphone. *(From [30].)*

$C_p \simeq 6$ pF. Equation (17) shows a negligible change in the sensitivity of the larger microphone, while the sensitivity of the smaller one is reduced to 0.4 (its open-circuit value), a drop equivalent to 8 dB.*

Typical sensitivities of commercially available condenser microphones range from nominally −60 dB re 1V/μb for a 1-in-diameter unit [29, 30]; there is a roughly 6-dB reduction for each halfing in diameter, reflecting the fact, from Eq. (16), that S_{oc} is proportional to sensor area. The frequency response is generally 20 Hz to roughly 5 kHz for 1-in-diameter units. The upper limit is doubled for each halving of microphone diameter. Further discussion of frequency response is given in Sec. 6.3. The sensitivity is only weakly affected by temperature and atmospheric pressure; it is affected by humidity only when liquid condensation occurs in the sensor. These effects are largely restricted to frequencies near the resonance frequency of the membrane. High humidity can also promote higher electronic noise levels, especially in very small microphones that must be connected to their preamplifiers through adaption. The vibration sensitivity of microphone and cathode follower is on the order of 90 dB re 1 b of equivalent sound pressure level for 1 g of acceleration; it is more sensitive to vibration in the direction of the face of the transducer than in directions perpendicular to it.

4.2 Piezoelectric Transducers

Piezoelectric materials are crystalline substances that, when deformed, produce an electric field; the voltage generated is proportional to the force applied. Conversely, if an electric field is imposed along one of the piezoelectric axes, the crystal will be deformed in proportion to the voltage applied. The most commonly used materials are quartz, Rochelle salt, ammonium dihydrogen phosphate (ADP), barium titanate, and lead zirconate titanate (PZT). The piezoelectric effect must be induced in PZT crystals by polarizing. Barium titanate and PZT crystals are most frequently used in miniature transducers. Discussions on the electromechanical conversion properties of these materials and the elements of transducer design are given in simplified form by Beranek [24] and by Kinsler and Frey [26]. More detailed treatment is given by Mason [31].

Figure 12 shows one possible transducer configuration. Generally, the crystal is plated on two surfaces separated by the thickness of the crystal. As shown, two crystals are mounted back to back, with their piezoelectric polarities reversed. In such an arrangement, the sensitivity of the microphone to motion in the direction of its axis is reduced, while its sensitivity to the compressive deformation caused by pressure fluctuations is retained. The unit does, however, also retain its sensitivity to lateral vibration. Piezoelectric pressure transducers of this type are said to be "acceleration canceling."

Figure 13 shows a simplified circuit diagram for the typical piezoelectric

*A quantity expressed in decibels is $20 \log_{10}$ of the quantity; that is, $8 = 20 \log_{10} 2.5$. The sensitivity in decibels is $20 \log_{10} S$. Therefore, $S = 10^{-3}$ V/μb corresponds to −60 dB re 1V/μb.

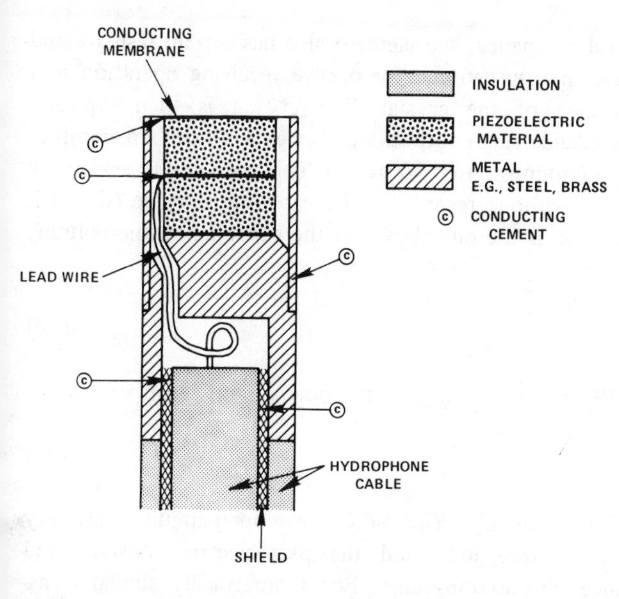

Figure 12 General arrangement of a piezoelectric pressure transducer made with disk ceramics. Two elements are used in tandem to provide for acceleration cancellation in the axial direction.

hydrophone. When a time-varying force $F(t) = p(t)A$ is applied, an open-circuit voltage $e(t)$ is produced; its magnitude is

$$e(t) = p(t)A \frac{Y}{dl} \tag{18}$$

where d = piezoelectric electromechanical coupling constant
Y = elastic modulus of material
l = a lateral dimension

We have not specified the directions of applied force or the axis of polarization; therefore, Eq. (18) simply expresses the dimensional considerations in the process. The transducer has an effective electromechanical capacitance

$$C_m = \frac{d^2 l}{Y} \tag{19}$$

and a true electrical capacitance C_0. It turns out that $C_m \ll C_0$. At high frequencies,

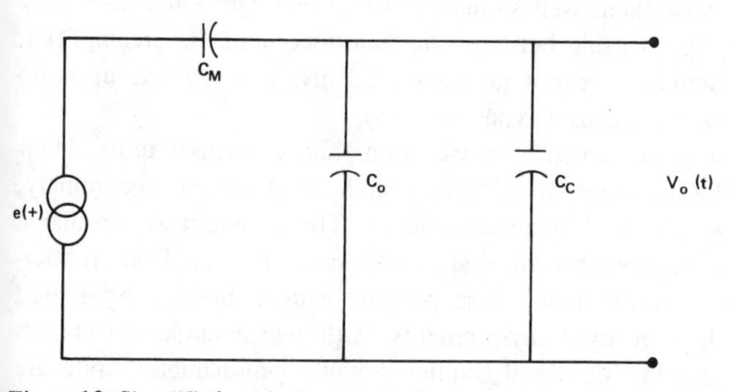

Figure 13 Simplified equivalent circuit for a piezoelectric transducer.

near and above the mechanical resonance, the element also has certain resistive and inductive properties, but these may be ignored for passive receiving operation well below the mechanical resonance of the crystal. The crystal is then typically connected to a low-input-impedance (high-output-impedance) low-noise preamplifier via a cable of capacitance C_c (amounting to ~90 pF/m). The open-circuit sensitivity of the transducer is calculated without regard for the cable capacitance ($C_c = 0$). The circuit calculation is the same as the one above for the capacitance microphone, and it gives

$$\frac{e_{oc}(t)}{e(t)} = \frac{C_m}{C_m + C_0} \tag{20}$$

Substituting Eqs. (18) and (19) into Eq. (20) gives the open-circuit sensitivity as

$$S_{oc} = \frac{e_{oc}(t)}{p_{oc}(t)} = \frac{Ad}{C_0} \tag{21}$$

where we used the approximation $C_c + C_m \simeq C_0$. The open-circuit sensitivity therefore increases with the surface area and the piezoelectric constant and decreases with the capacitance of the transducer. For geometrically similar transducers of the same material, $C_0 \propto l$; therefore S_{oc} increases in proportion to the linear dimension of the crystal.

The effect of the cable can be very important. By allowing the circuit in Fig. 13 to include the cable capacitance, we find

$$\frac{v(t)}{e(t)} = \frac{C_m}{C_m + C_0 + C_c}$$

therefore,

$$\frac{S_{\text{cable}}}{S_{oc}} = \frac{C_0 + C_m}{C_0 + C_m + C_c} \simeq \frac{C_0}{C_0 + C_c} \tag{22}$$

Capacitances of ceramic transducers range from several hundred picofarads (600 pF, say) to tens of thousands (12,000 pF), while sensitivities are on the order of −90 to −130 dB re 1V/μb, depending on the size and construction. Smaller flush-mounted units are generally less sensitive. The larger sensor capacitance of piezoelectric transducers make them well suited for situations where it is necessary to have substantial quantities of cable between the transducer and the preamplifier. (Microphone-type power supplies are not necessary, because piezoelectric units do not need the bias voltage that condenser types require.)

Piezoelectric units are more sensitive to vibration than condenser units. Their sensitivities depend on the direction of vibration and construction; accordingly, ceramic transducers are widely used as accelerometers. The commercially available units may be found in a wide variety of shapes and sizes, ranging from rubber-sheathed cylindrical units to screw-mount flush pressure gauges. Some can be used in either low-pressure or high-pressure environments. Although ceramic transducers are generally considered mainly for use in liquids, small $1\frac{1}{2}$-in-diameter units are now commercially available as microphones that are comparable in characteristics to their condenser counterparts [29].

Because of the higher acceleration sensitivity, the piezoelectric units must be installed in facilities with great care. Flow-induced vibration of the facility walls, if directly imparted to the transducer, will often mask the desired pressure-induced signals. Thus, special mounting procedures designed to vibration-isolate the sensor are required in those circumstances.

4.3 Strain-Gage Transducers

Another type of membrane transducer that utilizes a strain gage to translate a membrane displacement into an electric signal is the strain-gage transducer. As mentioned above, these transducers are useful for applications where clearances are so small that other types of transducers simply will not fit into the apparatus. As illustrated in Fig. 14, a recently used device was less than 1 mm thick, and it was installed in the surface of an airfoil [28]. Being a strain-gage device, it responds to a displacement from dc to 5000 Hz or so. It is, therefore, particularly suited to low-to-moderate frequency requirements. Strain-gage sensors have long been used on rotating devices; the signals are transmitted from the gage through either slip rings or telemeter electronics.

The electric circuitry is typically that used in standard strain-gage applications and will not be discussed here in any great detail. Resistance strain gages, whose resistance changes in response to an applied strain, are connected to one arm of a Wheatstone bridge. Electrical processing of the signal is then accomplished in a manner that is completely analogous to that used in hot-wire anemometry and in other types of resistance devices that depend on resistance fluctuations. The problems of dc drift and sensitivity changes are also analogous to similar problems with hot wire anemometer probes. The units are, however, versatile, reliable, and relatively insensitive to vibration. A more extensive treatment of strain-gage technology and circuitry in general, and strain-gage pressure gauges in particular, may be found in [2].

5. MECHANICAL TRANSDUCTION OF TIME-VARYING PRESSURE SIGNALS

In many instances, the total and static pressures associated with flow must be transmitted to a pressure sensor via a tube of some length. The steady-state average pressure at the end of such a tube presents no difficulty. Since the transducer at the termination provides a large impedance, there is no flow in the tube; therefore the pressure at the probe equals the pressure at the sensor. In the case of time-varying signals, caused by either transient or undulating pressures, the tube acts as a transmission line.

The first time-varying effect with which we are concerned is the transient response of the probe-transducer system reacting to a step change in pressure differential. Especially when the pressure gauge has a small enough stiffness (e.g., a simple fluid manometer), one wonders whether to expect a significant lag between a change in pressure differential and a steady-state differential pressure reading. The pressure gauges with the smaller stiffnesses are often those of large sensitivity. The

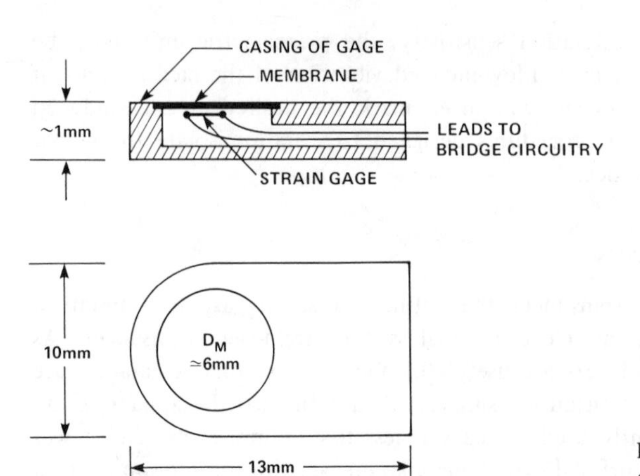

Figure 14 Schematic of a thin strain-gage microphone.

probe-gauge system resembles a simple *RC* electrical analog, where the *R* is provided in the thin probe constriction and the *C* is provided by the stiffness of the gauge.

As an example, consider the simple probe-manometer system illustrated in Fig. 15. The resistance to flow is assumed to be confined to the thin probe of cross-sectional area *A* and length *L*. Further, if it is assumed that the flow across the tube follows the familiar parabolic velocity profile of Stokes flow, then we find

$$R_m = \frac{\Delta P\, A}{U} = 8\pi\, \nu_g \rho_g L \tag{23}$$

where ρ_g is the density of the flowing fluid. The capacitance is assumed to be controlled by the manometer tube. Accordingly we define

$$C_m = \frac{\text{gauge displacement}}{\text{applied force at total-head tube}} = \frac{\Delta h}{\Delta P\, A} = \frac{1}{A\rho L g}$$

where ρ_L is the manometer fluid density. Therefore the time constant for the device is

$$\tau = R_m C_m = \frac{8\pi\, \nu_g L \rho_g}{g A \rho_L} \tag{24}$$

and the response will vary with time as $\exp(-t/\tau)$. Accordingly the system will reach 5% of its steady-state reading in 3τ. When air is the flowing medium, and alcohol the manometer medium, a 1-mm-diameter probe, 50 mm long, will respond in about 1 ms. For the same tube, if the flowing medium is water and $\rho_g/\rho_L \simeq 1$, the time increases to 0.1 s. Accordingly, the transient response time of the probe-manometer system is, practically speaking, instantaneous.

If a measurement of the fluctuating total or static pressure is desired, then such fluctuations must be transmitted through tubes to the transducer. In addition to the schemes indicated in Fig. 2, schemes such as that shown in Fig. 16*a* may be utilized. These tubes act as acoustic transmission lines that have characteristic masses, resistances, and capacitances. Such a tube behaves as a linear harmonic oscillator with a series of resonance frequencies f_m that satisfy the equation

$$f_m = \frac{C_0(2m-1)}{4L} \qquad \text{for } m = 1, 2, 3, \ldots \tag{25}$$

At each of these frequencies the pressure in the tube resonates so that its value p_m at the microphone face is much larger than the value p_0 at the open end of the tube. The equation for the ideally transmitted pressure p_m relative to the open-end pressure p_0 is

$$\frac{|P_m|^2}{P_0^2} = \frac{1 + \eta^2 (f/f_m)^2}{[1-(f/f_m)^2]^2 + \eta^2 (f/f_m)^2} \tag{26}$$

as long as the mass impedance per unit length of fluid in the tube is much less than the mechanical impedance of the transducer, i.e., as long as $2\pi \rho_g A f \ll Z_m$. This behavior is illustrated in Fig. 17, and it is normally the case, for tubes connected to condenser microphones, below the first membrane resonance of the microphone.

The loss (or damping) factor governs the response of the tube at resonance;* Eq. (26) reduces to

$$\frac{|P_m^2(f = f_m)|}{|P_0^2|} = \frac{1+\eta^2}{\eta^2} \tag{27}$$

at each of the resonances. When $\eta \ll 1$, the pressure at the transducer end of the tube will be $1/\eta^2$ larger than the pressure at the open end.

Since the loss factor is a crucial parameter in determining the pressure response

*For tube resonances near the resonance of the transducer, the response near $f = f_m$ will also depend on the damping in the transducer, which is designed to be high.

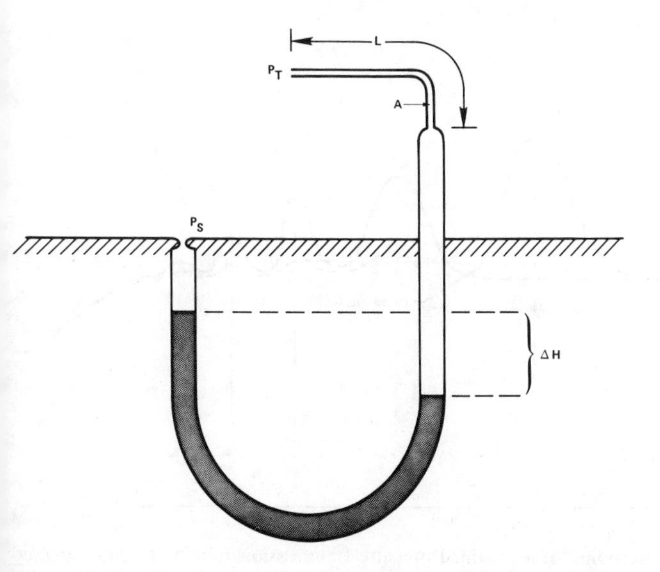

Figure 15 Total-head tube and static-pressure tap linked to a simple manometer tube.

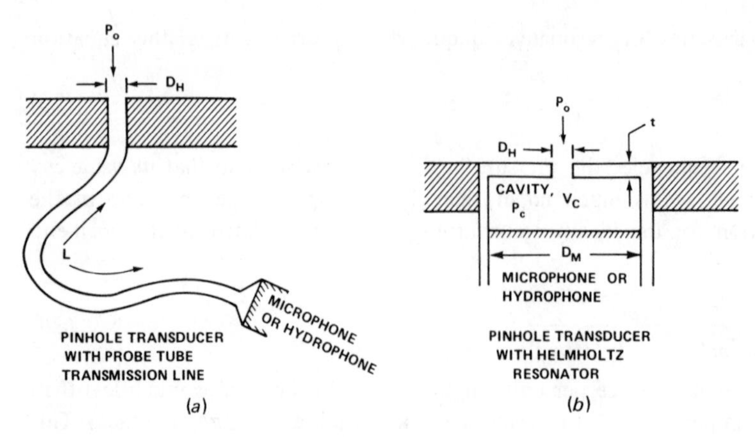

Figure 16 Methods of improving spatial resolution through the use of pinholes and acoustic resonators.

of such tubes, it is derived here in terms of fluid parameters. By definition,

$$\eta = \frac{R_m}{M\omega}$$

where R_m is the viscous fluid resistance and M is the fluid mass, both per unit length L of tube. The viscous resistance for small tubes is given by Eq. (23); the mass is $\rho_g AL$. Accordingly we find the loss factor of the tube to be

$$\eta = \frac{8\pi\nu_g}{\omega_A} \tag{28}$$

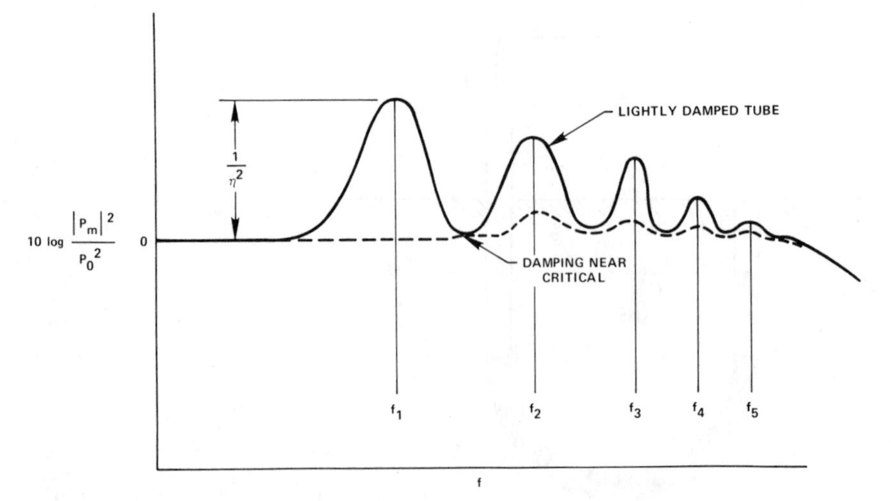

Figure 17 Typical pressure response of a gas-filled pressure transmission tube with radius greater than 1 mm, connected to a condensor microphone.

When the tubes are critically damped, i.e., when $\eta = \eta_{cr} = 2$, no resonant behavior occurs. The condition for critically damped air-filled tubes is, for inner radii R,

$$R < \frac{3}{\sqrt{f}} \quad \text{mm}$$

Practically speaking, for tubes with internal radii larger than 1 mm, resonance will occur at the frequencies given by Eq. (25). Damping in such tubes can be increased to near the critical value by loosely packing cotton or steel wool in the opening. One must be careful not to overdamp the tube, however.

Figure 16*b* is a diagram of the pinhole microphone, which consists of a connection between the transducer and the fluid through a cavity and a small hole. This device is used as a means of reducing the sensitive area of a transducer while maintaining its high-pressure response. The acoustic response of such a gas cavity as is illustrated in Fig. 16*b* also satisfies Eq. (26). For air-filled cavities this frequency is given [26] by

$$f_m = \frac{C_0 m}{2\pi} \sqrt{\frac{A_H}{l_c V_c}} \qquad m = 1, 2, 3, \ldots \tag{29}$$

where $A_H = \pi D_H^2/4$

$l_c = h + 8D_H/3\pi \simeq h + D_H$

At frequencies well below this frequency the transduction of pressure is stiffness-controlled because the cavity is acoustically very stiff; therefore, $p_0 = p_c$. In the author's experience these pressures are equal when the frequency is within one octave of the resonance frequency, i.e., within $f < f_1/2$. At resonance the amplification of the cavity pressure amounts to an enhancement of $p_c \simeq 3p_0$. The damping at the resonance frequency is probably controlled by dissipation in the membrane.

6. PHYSICAL ERRORS IN THE MEASUREMENT OF UNSTEADY PRESSURES

We consider now the three leading physical causes of errors in dynamic-pressure measurement. These errors, it may be stated, effectively act to prevent the pressure measurement from meaning what we think it means, and they are common to all types of pressure transducers. The leading sources of error that we examine here are spatial-resolution errors, aerodynamic interference by gusting, and acoustic reflections.

6.1 Spatial-Resolution Errors

This is the most important source of errors in aerohydrodynamic measurements. Hence, trade-offs must be made in experiment design between small size, with associated good spatial resolution and high-frequency response, and large size, with associated good sensitivity and lower-frequency response. In the special case of boundary-layer pressures the subject has been given extensive treatment in [32 to

43]. The problem, though, has general application to all types of measurements in traveling (convected) pressure fields.

For all three of the transducer types discussed above, the output is proportional to the integrated response to a distributed pressure over the face of the transducer. If we let $S(X)$ be the voltage output (sensitivity) per unit area of the transducer of area A, as when an elemental force

$$dF = p\, d^2x$$

is applied at X, then the net voltage output of the transducer is

$$e(t) = \iint p(\mathbf{x}, t) S(\mathbf{x})\, d^2\mathbf{x} \tag{30}$$

The temporal autocorrelation of the voltages at times t and $t + \tau$ is defined as

$$R_{ee}(\tau) = \frac{\overline{e(t)e(t+\tau)}}{[\overline{e^2(t)}\,\overline{e^2(t+\tau)}]^{1/2}} = \frac{\overline{e(t)e(t+\tau)}}{\overline{e^2}} \tag{31}$$

where $\overline{e^2}$ is the (statistically stationary) mean square voltage from the transducer.

In terms of the transducer response, the processes of time-averaging and spatial integration may be interchanged to give

$$\overline{e^2}\, R_{ee}(\tau) = \iint_A \left[\iint_A \overline{p^2}\, R_{pp}(\mathbf{x} - \mathbf{y}, \tau) S(\mathbf{x}) S(\mathbf{y})\, d^2\mathbf{x} \right] d^2\mathbf{y} \tag{32}$$

where the cross correlation of the pressure is defined as

$$\overline{p^2}\, R_{pp}(\mathbf{x} - \mathbf{y}, \tau) = \overline{p(\mathbf{x}, t) p(\mathbf{y}, t + \tau)} \tag{33}$$

and

$$\overline{p^2} = \overline{p^2(\mathbf{x}, t)} = \overline{p^2(\mathbf{y}, t)} \tag{34}$$

If we assume that the statistics of the pressures are the same at locations $\mathbf{x}$ and $\mathbf{y}$, then $R_{pp}(\mathbf{x} - \mathbf{y}, t)$ is a function only of the separation $\mathbf{r} = \mathbf{x} - \mathbf{y}$.

We introduce the space-time Fourier transform of the fluctuating pressure as

$$R_{pp}(\mathbf{r}, \tau) = \iiint_{-\infty}^{\infty} \phi_p(\mathbf{k}, \omega) e^{i(\mathbf{kr} - \omega)}\, d^2\mathbf{k}\, d\omega \tag{35}$$

where $\overline{p^2}\phi_p(\mathbf{k}, \omega)$ is the wave number-frequency spectrum of the pressures. By definition, then, the frequency spectrum of the pressures measured at a point is just

$$\overline{p^2}\phi_p(\omega) = \iint_{-\infty}^{\infty} \overline{p^2}\, \phi_p(\mathbf{k}, \omega)\, d\mathbf{k} \tag{36}$$

and the autospectrum is defined so that the mean square pressure is

$$\int_{-\infty}^{\infty} \phi_p(\omega)\, d\omega = 1$$

Continuing with the analysis, we form an expression for the voltage autocorrelation from Eqs. (32) and (35):

$$\frac{\overline{e^2}}{\overline{p^2}} R_e(\tau) = \int_A \int \left[\int_A \int \int\int_{-\infty}^{\infty} d^2\mathbf{k} \int_{-\infty}^{\infty} \phi_p(\mathbf{k}, \omega) e^{-i\omega\tau} S(\mathbf{x}) e^{i\mathbf{k}\mathbf{x}} S(\mathbf{y}) e^{-i\mathbf{k}\mathbf{y}} \, d\omega \, d^2\mathbf{x} \right] d^2\mathbf{y} \tag{37}$$

This rather complex-looking integral simplifies if we make the substitution

$$S(\mathbf{k}) = \iint_A S(\mathbf{x}) e^{i\mathbf{k}\mathbf{x}} \, d\mathbf{x} \tag{38}$$

and similarly for $S(\mathbf{y})$, which yields $S^*(\mathbf{k}) = \iint_A S(\mathbf{y})\, e^{i\mathbf{k}\mathbf{y}} d^2\mathbf{y}$, and if we note the relationship between the autocorrelation and the autospectrum:

$$R_e(\tau) = \int_{-\infty}^{\infty} \phi_e(\omega) e^{-i\omega t} \, d\omega \tag{39}$$

Making the necessary substitutions, we find

$$\frac{\overline{e^2}\phi_e(\omega)}{\overline{p^2}} = \iint_{-\infty}^{\infty} \phi_p(\mathbf{k}, \omega) |S(\mathbf{k})|^2 \, d\mathbf{k} \tag{40}$$

This is the key equation of this section, and it is the analog of the linear passive filter response of electronic frequency filtering. Here, rather than thinking of the pressure being continuously distributed over a range of frequencies (or time scales), we must also regard it as distributed over a range of wave numbers (or space scales) in two dimensions. The effect of the finite size of the transducer is to provide a wave-number band limit (low-pass filtering) to the measured pressure.

Let us say that $S(\mathbf{x})$ is uniform and that the transducer is circular; i.e.,

$$S(\mathbf{x}) = \frac{S_{oc}}{\pi R^2} \qquad 0 < r \leqslant R \tag{41}$$

$S(\mathbf{k})$ turns out to be

$$S(\mathbf{k}) = S_{oc} \frac{2J_1(kR)}{kR} \tag{42}$$

where $J_1(x)$ is the first-order Bessel function of the first kind, and S_{oc} is the open-circuit sensitivity of the transducer, so that

$$\frac{\overline{e^2}\,\phi_e(\omega)}{\overline{p^2}} = \int_0^{2\pi} \int_0^{\infty} \phi_p(\mathbf{k}, \omega) \frac{S_{oc}}{2} \left| \frac{2J_1(kR)}{kR} \right|^2 k \, dk \, d\theta \tag{43}$$

The function $J_1(x)/x$ is shown in Fig. 18 with an illustration of a plausible shape of $\phi_p(\mathbf{k}, \omega)$ at a fixed frequency for a typical convected, turbulent pressure field. The shape shown for $\phi_p(\mathbf{k}, \omega)$ is typical of that associated with the convection of

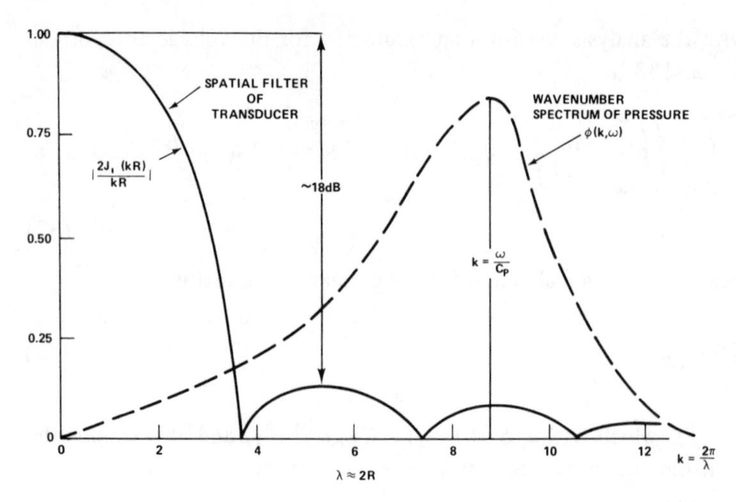

Figure 18 Illustration of wave-number (wavelength) selectivity of transducers of finite size, shown for a large circular sensor exposed to a pressure field of predominantly short-wavelength pressure.

disturbances as in turbulence measurements. The spectrum is strongly peaked at a wave number $k_x = \omega/U_c$ and $k_y = 0$, where U_c is the phase (or convection) speed of the disturbance of frequency ω past the transducer. It may be a convection velocity of eddies, or the trace velocity of an incident pressure wave. If the pressure pulses at a given frequency are the result of straight-crested waves with only one phase speed as illustrated in Fig. 19, then $\phi_p(\mathbf{k}, \omega)$ has a nonzero value only at a streamwise wave number $k_x = \omega/U_c$ and at a spanwise wave number $k_y = 0$. This type of pressure disturbance is an idealization of the turbulent flow pressure, but

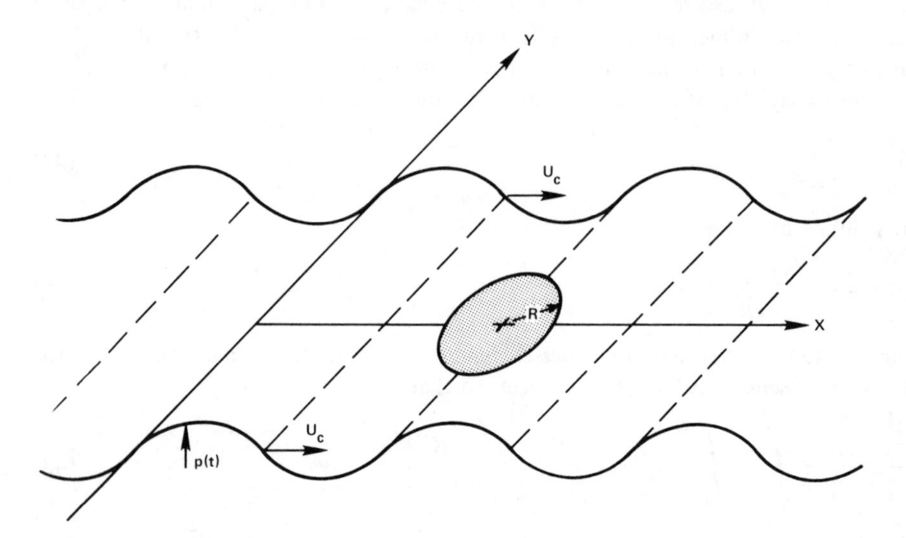

Figure 19 Straight-crested pressure waves being convected past a microphone.

fairly representative of an incident acoustic wave. Thus, we have

$$\phi_p(\mathrm{k}, \omega) \simeq \delta\left(\mathrm{k}_x - \frac{\omega}{U_c}\right) \phi_p(\omega)\, \delta(k_y) \tag{44}$$

A physical situation to which this formula applies is depicted in Fig. 20, in which the turbulence is idealized as a group of eddies, elongated in the cross-stream direction, of sizes $\Lambda = 2\pi U_c/\omega$ and passing the sensor at a speed $U_c \simeq 0.6\ U_\infty$. Equations (43) and (44) give

$$\overline{e^2}\,\phi_e(\omega) = \overline{p^2}\,\phi_p(\omega) \left| S\left(\frac{\omega R}{U_c}\right)\right|^2 = \overline{p^2}\,\phi_p(\omega) S_{oc}^2 \left|\frac{2J_1(\omega R/U_c)}{\omega R/U_c}\right|^2$$

For a transducer of vanishing size, which implies $\omega R/U_c \to 0$, we have the point pressure spectrum

$$\overline{e^2}\,[\phi_e(\omega)]_0 = \overline{p^2}\,\phi_p(\omega) S_{oc}^2$$

that we would have written down at the outset, if we had not provided for the effects of spatial averaging. The wall pressure spectrum measured with a finite-sized transducer,

$$\overline{p^2}\,[\phi_p(\omega)]_m = \frac{\overline{e^2}\,\phi_e(\omega)}{S_{oc}^2} \tag{45}$$

is therefore only a fraction

$$\frac{[\phi_p(\omega)]_m}{\phi_p(\omega)} = \left| S\left(\frac{\omega R}{U_c}\right)\right|^2 \frac{2J_1(\omega R/U_c)}{\omega R/U_c} \tag{46}$$

of the spectrum level that would be measured with a point transducer. Figure 21 shows the reduction in the measured spectrum relative to the true spectrum as a function of the frequency and the sensor radius. For turbulent

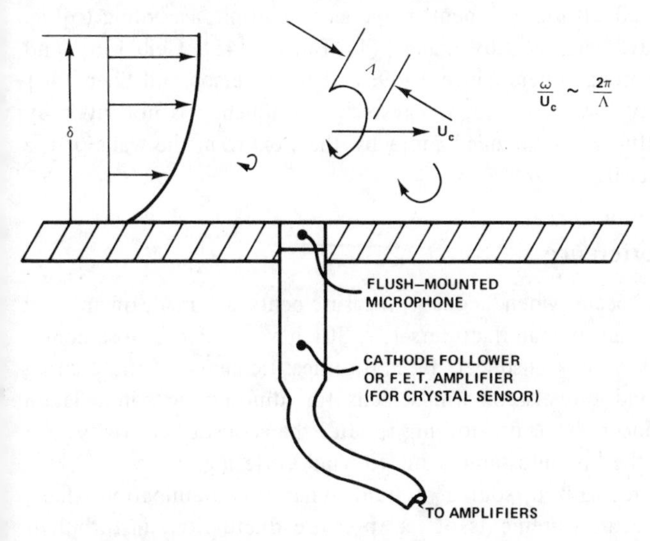

Figure 20 Example of a flush-mounted pressure transducer in a hydrodynamic application.

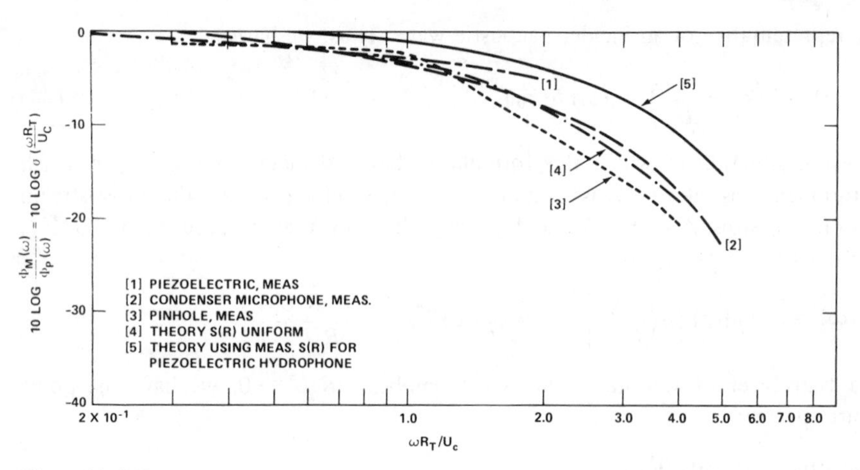

Figure 21 Effect of spatial averaging on the wall pressure spectrum observed in various facilities.

boundary layers over smooth walls, $U_c \simeq 0.6U$. The measurements shown in Fig. 21 come from a variety of sources, and the scatter reflects the combined effects of the departure of $S(\mathbf{x})$ from Eq. (41) and a departure of the true spectrum function from Eq. (44). Although the measured resolution functions in Fig. 21 are for turbulent boundary-layer pressures, the theoretical result, Eq. (46), is not sensitive to the source of the pressures as long as they may be thought of as progressive waves in the sense of Eq. (44) and Fig. 19. Therefore, Eq. (46) applies to the measurement of acoustic waves as well.

The spatial resolution of transducers can be improved by either of the two schemes shown in Fig. 16. Each of these acoustic-resonator transducers provides a resolution advantage D_M/D_H. At high frequencies, $fD_M/U_c > 1$; a ratio $D_M/D_H = 4$ may provide as much as 20 dB improvement in spatial resolution, according to Fig. 21. These microphones have been used by Blake [44], Burton [45], Geib [36], and Bull and Thomas [41], among others, since 1969, and by Peterson and Shen [46] in water. When the cavity is water-filled, the resonance frequency is not given by Eq. (29) because the stiffness is controlled more by the flexure in the walls of the cavity than by the fluid itself.

6.2 Aerodynamic Interference

These errors traditionally occur when acoustic measurements are made in moving fluids. To this end, some sensor manufacturers [29, 30] have produced nose cones, to aerodynamically shape the transducer face for measurements of fluctuating pressure in ducts and wind tunnels, or windscreens for atmospheric wind. These adaptations to the transducer shape do nothing to alter the acoustic sensitivity, but they substantially reduce the "pseudosound" due to wind buffeting.

Another, less widely recognized, source of fluid dynamic contamination occurs when it is desired to make measurements of real pressure fluctuations in turbulent shear flows, as described in Sec. 3.2. Such measurements have been made and errors analyzed in detail by Strasberg and Cooper [48], Strasberg [9], and Siddon [10] and were recently used by Armstrong [49]. The problem is to determine the fluctuating pressure in a flow field that includes gustiness due to turbulent eddies. The eddies produce velocity fluctuations u_1, u_2, u_3, so that, in a crude sense, what is possible is a signal that not only contains the required pressure information but also is contaminated by pressures induced on the probe surface by the turbulence components buffeting it. The problem is analogous to the sensing of mean static pressure in a flow. The measured static pressure P_m exceeds the true static pressure P_s by an amount

$$P_m - P_s \simeq a\rho_0(\overline{u_2^2} + \overline{u_3^2}) \tag{47}$$

where $a \simeq \frac{1}{4}$ and $\overline{u_2^2}$ and $\overline{u_3^2}$ are the two mean-square turbulence intensities directed perpendicular to the axis of the probe [50].

The devices used by Strasberg and by Siddon follow the general design shown in Fig. 22. Behind the nose cone are slits in the casing that transmit the pressure into the tube and ultimately to the microphone. The diameter d of the probe is approximately $\frac{1}{10}$ in, and the length L depends on the relationship d/D. It should be as short as possible, consistent with appropriate aerodynamic fairing and with acoustic transmission losses, as discussed in Sec. 5. Theoretically, the aerodynamically induced errors in such a measurement, according to Strasberg [9], amount to approximately

$$\overline{[p(t)_m - p(t)_s]^2} \simeq \tfrac{1}{8}[\rho_0(\overline{u_2^2} + \overline{u_3^2})]^2 \tag{48}$$

whereas Siddon's empirical results would suggest that the numerical coefficient should instead be between $\frac{1}{4}$ and $\frac{1}{2}$. A point made by Strasberg is that the magnitude of the theoretical mean square error is comparable to the mean square pressure extant in a gaussian isotropic turbulent field, that is,

$$p^2_{\text{isotropic}} \simeq \tfrac{1}{3}(\rho_0\overline{u_2^2})^2$$

because since $\overline{u_2^2} \simeq \overline{u_3^2} \simeq \overline{u_1^2}$, and u_2^2, Eq. (48) gives the error as

$$\overline{[p_m(t) - p_s(t)]^2} \simeq \tfrac{1}{2}(\rho_0\overline{u_1^2})^2 \tag{49}$$

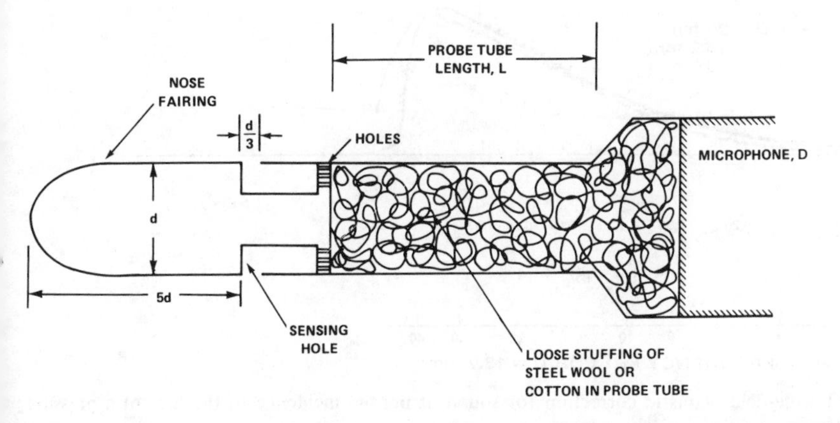

Figure 22 Schematic of a free-stream probe for measuring pressure fluctuations in flowing fluids.

6.3 Acoustic Reflection

Sound waves incident on a microphone placed in a sound field are reflected by it. The reaction pressure on the face of the microphone to reflected waves causes an enhancement of its apparent sensitivity. This enhancement is frequency-dependent; it is shown in Fig. 23 as the ratio of the perceived sound pressure P_m to the actual incident sound pressure P_0. Two abscissa girds are shown–one is the ratio of the face diameter to the acoustic wavelength, and the second is the actual frequency for the particular diameter of the Bruel and Kjaer nominal 12.7-mm microphone [30]. The curves show that the apparent pressure exceeds the free-field pressure by an amount that is a function only of the ratio of the diameter to the wavelength [51]. This effect is dependent on the angle of incidence of the sound field; Fig. 23 is drawn for the worst case, normal incidence to the microphone.

Microphones that are designed for free-field acoustic measurements are compensated for the reflection effect by making the resonance frequency of the membrane nearly coincident with the maximum of P_m/P_0 and then overdamping the membrane resonance so that the free-field acoustic response is as frequency-independent as possible. When microphones are calibrated for so-called "pressure" sensitivity, i.e., open-circuit voltage per unit pressure on the face, Fig. 23 must be used to adjust for the anomaly when the sound field is known to be normally incident; other values [30, 51] must be used when the incidence angle is other than 0°.

7. SPECIAL TECHNIQUES WITH MICROPHONE ARRAYS

It is to be noted that Eq. (41) has general application. The transducer response function $|S(k)|^2$ takes an additional meaning when we speak of the voltage output

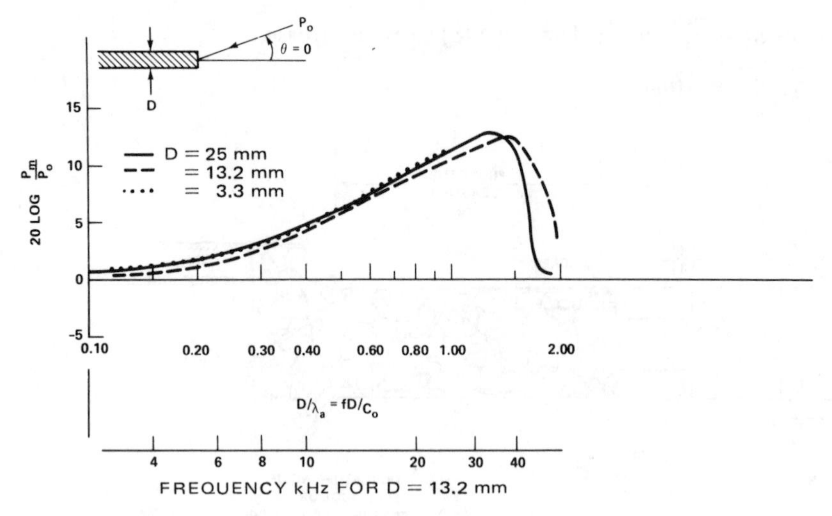

Figure 23 Free-field acoustic correction for sound at normal incidence to the face of a pressure microphone.

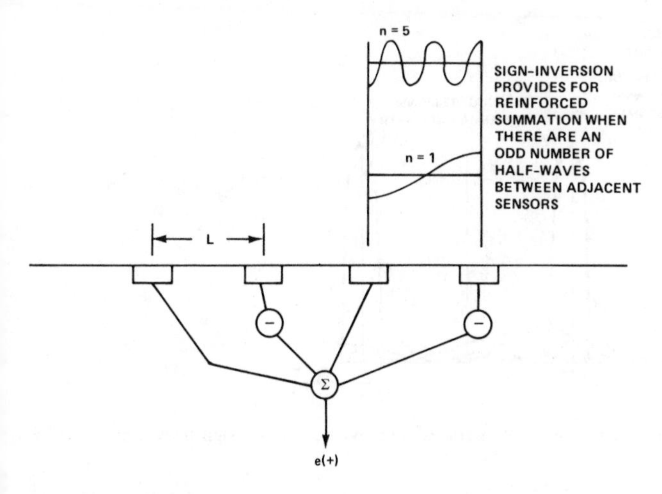

Figure 24 Spatial filter with alternate sign inversion.

from an array of transducers that are exposed to an unknown pressure field. Such arrays, well known in acoustical usage, have been examined analytically [52 to 56] and have been used to elucidate the contribution to the overall hydrodynamic pressure of components of specific wavelengths [56, 57]. Figure 24 shows one possible spatial filter in which alternate transducers are phase-inverted and the signals are processed. When a pressure field is imposed on the array so that an odd number of half-waves fit between the sensors, i.e., so that

$$L = (2n - 1)\frac{\lambda}{2} \tag{50}$$

the processed sensor signal reinforces. Alternatively, when an even number of half-waves coincides with L, then the array has a null response. This behavior is indicated in Fig. 25 as an endless series of acceptance regions at every

$$k = (2n - 1)\frac{\pi}{L}$$

However, the individual facial response of the transducer imposes a high-wave-number limit on this so that the effective wave-number response function is the product of an array function, say $|A(k)|^2$ and the individual response function $|S(k)|^2$. Therefore, in Eq. (41) we replace $|S(k)|^2$ with a function

$$|H(k)|^2 = |S(k)|^2 \, |A(k)|^2$$

which is illustrated by the solid line in Fig. 25. Since $|A(k)|^2$ is controlled by L, and $|S(k)|^2$ by R, it is possible to adjust L and R to give a moderate degree of wave vector discrimination. Furthermore, by increasing the number N of elements, the filter bandwidth

$$\Delta k = \frac{2\pi}{NL}$$

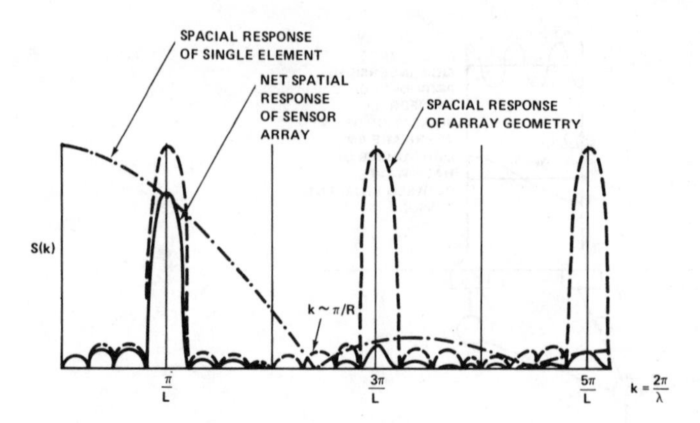

Figure 25 Spatial filtering of an array of circular transducers with alternate sign inversion.

may be made narrow. For only two elements there is a relatively small amount of spatial, or wave number, discrimination. The number of minor bumps between the major bumps is equal to $N - 1$.

NOMENCLATURE

A	area of microphone or hydrophone face
C_c	cable capacitance
C_m	effective mechanical stiffness or capacitance (displacement/force) of membrane
C_0	true electrical capacitance
C_p	input capacitance
C_T	transducer capacitance
dA	elemental area
dF	elemental force
dR	differential resistance change
$d^2\mathbf{x}$	elemental area for $\bar{x}$ coordinate
$d^2\mathbf{y}$	elemental area for y coordinate
D	transducer or hole diameter
E_0	applied bias voltage
$\overline{e^2}$	(statistically stationary) mean square voltage from the transducer
f	circular frequency (Hz)
g	gravitational constant
$\mathbf{k}$	vector wave number
l	thickness dimension of piezoelectric transducer
p	steady or time-averaged pressure
$\overline{p^2}$	mean square pressure
$p(\mathbf{x}, t)$	pressure at position $\bar{x}$, time t
$R_{ee}(\tau)$	autocorrelation function of voltage $e(t)$
R_p	input resistance of cathode follower

$R_{pp}(x-y, \tau)$	cross correlation of pressures at locations $\bar{x}$ and $\bar{y}$
R_T	transducer radius
$\mathbf{r}$	separation vector $x - y$
S_{cable}	transducer sensitivity adjusted for cable effects
S_{oc}	open-circuit sensitivity of transducer element
$S(\mathrm{x})$	voltage output (sensitivity) per unit area of pressure transducer
t	time
U	fluid velocity
x_0	plate separation, displacement
$\mathbf{y}$	spatial coordinate
$\delta(x)$	Dirac delta function
ϵ_m	strain amplitude
κ_0	dielectric constant
λ	wavelength
ν_0	blocked voltage
$\nu(t)$	blocked voltage in Eqs. (1) and (2)
ν_g	kinematic viscosity of the flowing medium
ρ_0	fluid density; also subscripted g, gas and l, liquid
τ	time delay in correlation functions
$\phi(\omega)$	autospectral density
ω	radian frequency $= 2\pi f$
$\int\int_A$	integration over area A

Subscripts

oc	open circuit
p	pressure
e	voltage
f	force
T	total or impact pressure, when used with p
s	static pressure, when used with p
c	convection
∞	free-stream

REFERENCES

1. S. H. Chue, Pressure Probes for Fluid Measurement, *Prog. Aerosp. Sci.* vol. 16, pp. 147–223, 1975.
2. T. G. Beckwith and N. L. Buck, *Mechanical Measurements*, Addison-Wesley, Reading, Mass., 1961.
3. J. H. Preston, The Determination of Turbulent Skin Friction by Means of Pitot Tubes, *J. Roy. Aeronaut. Soc.*, vol. 58, p. 109, 1954.
4. M. R. Head and I. Rechenberg, The Preston Tube as a Means of Measuring Skin Friction, *J. Fluid Mech.*, vol. 14, pp. 1–17, 1962.
5. V. C. Patel, Calibration of the Preston Tube and Limitations on Its Use in Pressure Gradients, *J. Fluid Mech.*, vol. 23, pp. 185–208, 1965.
6. L. Hwang and E. M. Laursen, Shear Measurement Technique for Rough Surfaces, *J. Hydraul. Div. Proc. ASCE*, paper 3451, pp. 19–35, 1963.

7. A. T. Ippen and F. Raichlen, Turbulence in Civil Engineering: Measurements in Free Surface Streams, *J. Hydraul. Div. Proc. ASCE,* paper 1392, 1957.
8. R. E. A. Arndt and A. T. Ippen, Turbulence Measurements in Liquids Using an Improved Total Pressure Probe, *J. Hydraul. Res.*, vol. 8, pp. 131-158, 1970.
9. M. Strasberg, Measurements of the Fluctuating Static and Total Head Pressure in a Turbulent Wake, DTMB Rep. 1779, 1963.
10. T. E. Siddon, On the Response of Pressure Measuring Instrumentation in Unsteady Flow, UTIAS Rep. 136, 1969.
11. J. A. Elliott, Microscale Pressure Fluctuations Measured within the Lower Atmosphere Boundary Layer, *J. Fluid Mech.*, vol. 53, pp. 351-383, 1972.
12. J. O. Hinze, *Turbulence,* McGraw-Hill, New York, 1959.
13. H. A. Becker and A. P. G. Brown, Response of Pitot Probes in Turbulent Streams, *J. Fluid Mech.,* vol. 62, pp. 85-114, 1974.
14. P. C. Pien, Five-Hole Spherical Pitot Tube, DTMB Rep. 1229, 1958.
15. V. Silovic, A Five Hole Spherical Pitot Tube for Three Dimensional Wake Measurement, Hydro-Og Aerodynamisk Laboratorium, Lyngby, Denmark, Rep. Hy 3, 1964.
16. M. R. Hale and D. H. Norrie, The Analysis and Calibration of the Five-Hole Spherical Pitot, presented at the ASME Winter Meeting, 1967.
17. A. L. Treaster and A. M. Yocum, The Calibration and Application of Five-Hole Probes, A.R.L. Tech. Memo. 78-10, Pennsylvania State Univ., 1978.
18. C. A. Scragg and D. A. Sandell, A Statistical Evaluation of Wake Survey Techniques, *Proc. Int. Symp. on Ship Viscous Resistance,* Goteborg, Sweden, 1978.
19. B. Lakshminarayana and P. Randstadler, Jr., Measurement Methods in Rotating Components of Turbomachinery, ASME Joint Fluids Eng. Gas Turbine Conf. and Products Show, 1980.
20. Y. Himeno and T. Okuno, Study on Cross Flow in Ship Boundary Layer, *Proc. Int. Symp. on Ship Viscous Resistance,* Goteborg, Sweden, 1978.
21. I. Tanaka and T. Suzuki, Interaction between the Boundary Layer and Longitudinal Vortices, *Proc. Int. Symp. on Ship Viscous Resistance,* Goteborg, Sweden, 1978.
22. F. A. MacMillen, Experiments on Pitot-Tubes in Shear Flow, A.R.C. Rep. Memo. 3028, 1956.
23. R. Shaw, The Influence of Hole Dimensions on Static Pressure Measurements, *J. Fluid Mech.,* vol. 7, pp. 550-569, 1960.
24. L. Beranek, *Acoustics,* McGraw-Hill, New York, 1954.
25. H. F. Olsen, *Elements of Acoustical Engineering,* Van Nostrand, Princeton, N.J., 1957.
26. L. E. Kinsler and A. R. Frey, *Fundamentals of Acoustics,* Wiley, New York, 1962.
27. H. H. Heller and S. E. Widnall, The Role of Fluctuating Forces in the Generation of Compressor Noise, NASA CR 2012, 1972.
28. R. N. Hosier, et al., Helicopter Roter Rotational Noise Predictions Based on Measured High Frequency Blade Loads, NASA TN D-7624, 1974.
29. A. P. G. Peterson and E. E. Gross, *Handbook of Noise Measurement,* 7th ed., General Radio Co., Concord, Mass.
30. Bruel and Kjaer Instruments, *Instruction and Operation Manuals for Types 4133 and 4134 Microphones,* 1962.
31. W. P. Mason, *Physical Acoustics,* vol. I, pt. A, Academic, New York, 1964.
32. G. M. Corcos, The Resolution of Pressure in Turbulence, *J. Acoust. Soc. Am.,* vol. 35, pp. 192-199, 1963.
33. J. H. Foxwell, Wall Pressure Spectrum Under a Turbulent Boundary Layer, Admiralty Underwater Weapons Establishment TN 218/66 Portland, England, 1966.
34. R. B. Gilchrist and W. A. Strawderman, Experimental Hydrophone-Size Connection Factor for Boundary Layer Pressure Fluctuations, *J. Acoust. Soc. Am.*, vol. 38, pp. 298-302, 1965.
35. D. M. Chase, Turbulent Boundary Layer Pressure Fluctuations and Wave Number Filtering by Non-Uniform Spatial Averaging, *J. Acoust. Soc. Am.,* vol. 46, pp. 1350-1365, 1969.
36. F. E. Geib, Jr., Measurements on the Effect of Transducer Size on the Resolution of Boundary Layer Pressure Fluctuations, *J. Acoust. Soc. Am.,* vol. 46, pp. 253-261, 1969.

37. P. H. White, Effect of Transducer Size, Shape and Surface Sensitivity on the Measurement of Boundary Layer Pressure, *J. Acoust. Soc. Am.,* vol. 41, pp. 1358–1363, 1967.
38. G. J. Kirby, The Effect of Transducer Size, Shape, and Orientation on the Resolution of Boundary Layer Pressure Fluctuations at a Rigid Wall, *J. Sound Vib.*, vol. 10, pp. 361–368, 1969.
39. K. L. Chandiramani, Interpretation of Wall Pressure Measurements under a Turbulent Boundary Layer, Bolt, Beranek and Newman Rep. 1310, 1965.
40. W. W. Willmarth and F. W. Roos, Resolution and Structure of the Wall Pressure Field beneath a Turbulent Boundary Layer, *J. Fluid Mech.*, vol. 22.
41. M. K. Bull and A. S. W. Thomas, High Frequency Wall Pressure Fluctuations in Turbulent Boundary Layers, *Phys. Fluids,* vol. 19, pp. 597–599, 1976.
42. G. D. Haddle and E. J. Skudrzyk, The Physics of Flow Noise, *J. Acoust. Soc. Am.,* vol. 46, pp. 130–157, 1969.
43. G. M. Corcos, The Resolution of Turbulent Pressures at the Wall of a Boundary Layer, *J. Sound Vib.,* vol. 6, pp. 59–70, 1967.
44. W. K. Blake, Turbulent Boundary Layer Wall Pressure Fluctuations on Smooth and Rough Walls, *J. Fluid Mech.,* vol. 44, pp. 637–660, 1970.
45. T. Burton, On the Generation of Wall Pressure Fluctuations for Turbulent Boundary Layers Over Rough Walls, MIT Acoustics and Vibration Laboratory Rep. 70208-4, 1971.
46. Y. T. Shen and F. B. Peterson, Unsteady Cavitation on an Oscillating Foil, *Twelfth Symp. on Naval Hydrodynamics,* Washington, D.C., 1978.
47. W. K. Blake and L. J. Maga, Near Wake Structure and Unsteady Pressures at Trailing Edges of Airfoils, *Int. Symp. on Mech. of Sound Generation in Flows,* Gottingen, Germany, 1979.
48. M. Strasberg and R. D. Cooper, Measurements of the Fluctuating Pressure and Velocity in the Wake behind a Cylinder, *Proc. Ninth Int. Congr. in Appl. Math.,* vol. 2, p. 384, 1957.
49. R. Armstrong, *Einfluss Der Machzahl Auf Die Koharente Turbulenzstruktur Eines Runden Freistrahls,* Berlin, 1977.
50. S. Goldstein, *Modern Developments in Fluid Dynamics,* vol. VI, Dover, New York, 1965.
51. V. Bruel and G. Rasmussen, Free Field Response of Condenser Microphones, *B&K Tech. Rev. No. 1,* pp. 12–17, *No. 2,* pp. 1–15, 1959.
52. G. Maidanik and D. W. Jorgensen, Boundary Wave-Vector Filters for the Study of the Pressure Field in a Turbulent Boundary Layer, *J. Acoust. Soc. Am.,* vol. 42, pp. 494–501, 1967.
53. D. W. Jorgensen and G. Maidanik, Response of a System of Point Transducers to Turbulent Boundary-Layer Pressure Field, *J. Acoust. Soc. Am.,* vol. 43, pp. 1390–1394, 1968.
54. G. Maidanik, Flush-Mounted Pressure Transducer Systems as Spatial and Spectral Filters, *J. Acoust. Soc. Am.,* vol. 42, pp. 1017–1024, 1967.
55. G. Maidanik, System of Small-Size Transducers as Elemental Unit in Sonar System, *J. Acoust. Soc. Am.,* vol. 44, pp. 488–496, 1968.
56. W. K. Blake and D. M. Chase, Wavenumber-Frequency Spectra of Turbulent Boundary Layer Pressure Measured by Microphone Arrays, *J. Acoust. Soc. Am.,* vol. 49, pp. 862–877, 1971.
57. P. W. Jameson, Measurement of Low Wavenumber Component of Turbulent Boundary Layer Wall Pressure Spectrum, BBN Rep. 1937, 1970.

CHAPTER

FOUR

THERMAL ANEMOMETERS

Leroy M. Fingerson and Peter Freymuth

1. INTRODUCTION

Thermal anemometers measure fluid velocity by sensing the changes in heat transfer from a small, electrically heated sensor exposed to the fluid motion. Their generally small size and good frequency response makes them especially suitable for studying flow details, particularly in turbulent flow.

In many applications fluid temperature, composition, and pressure are constant so the only variable affecting heat transfer is fluid velocity. When other parameters are varying, accurate velocity measurements with a thermal sensor become more difficult. At the same time, sensitivity of thermal sensors to other fluid parameters presents the possibility of measuring more than just velocity by using more than one sensor.

A simple thermal anemometer is shown in Fig. 1, along with some typical sensor and operating parameters. Assuming a linear relation between temperature and resistance (usually adequate in thermal anemometry), the resistance R of the sensor can be represented as

$$R = R_r[1 + \alpha(T_m - T_r)] \tag{1}$$

where R_r = resistance at reference temperature T_r
T_m = average sensor temperature
α = temperature coefficient of resistance

The value of α is critical, since if the sensor did not vary in resistance with temperature there would be no signal from a thermal anemometer. For convenience, the fluid temperature T_a is often used for the reference temperature T_r. The value of α depends on the reference temperature used.

In Fig. 1, if the resistance of resistor R_1 is large compared to that of the sensor

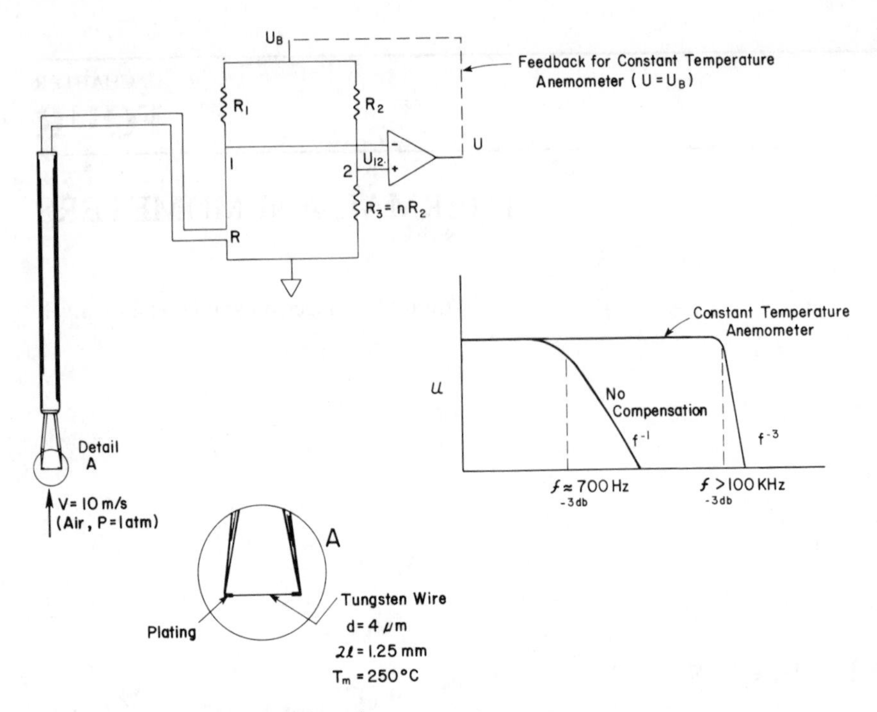

Figure 1 Basic elements of a hot-wire anemometer.

(R), then the current I is nearly constant, and any increase in heat-transfer rate from the sensor to its surroundings will cause the sensor to cool. Because of the temperature coefficient of resistance α, this cooling will cause a decrease in R, a decrease in U_{12}, and a decrease in amplifier output U. A decrease in heat transfer between sensor and fluid would have the opposite effect. When the changes in heat transfer are caused by changes in fluid velocity, the result is a thermal anemometer.

The system of Fig. 1 (without the feedback) is an uncompensated constant-current hot-wire anemometer. According to Freymuth [1] systems of this nature were first considered during the last two decades of the nineteenth century. The earliest work frequently referred to in current literature is that by King [2]. Subsequently there have been about 1300 publications relating to thermal anemometry techniques.

Advances have taken place in sensors, electronic control circuits, and data analysis. Small-diameter wires are still frequently used as sensors and, in fact, are still optimal for much research work in fluid mechanics. Film sensors, introduced by Lowell and Patton [3] and Ling [4], have been a major addition to sensor technology, especially for liquids or gases with particles. Cooled-film sensors [5] permit measurements in high-temperature gases.

Since transistors became available, the constant-current operation of thermal sensors has been largely replaced by constant-temperature operation. Stability criteria and techniques for checking response are now well understood for constant-temperature systems. In addition, constant-temperature systems work equally well for both wire and film sensors, and frequency compensation is not as sensitive to mean velocity.

Digital techniques–either in specialized instruments or with general-purpose computers–have significantly expanded capabilities for analyzing data from thermal anemometers. Correlation, spectrum, and amplitude probability density distributions are all readily obtainable. Special techniques for conditional sampling have also been developed.

Many excellent researchers have made significant contributions to thermal anemometry and should be given credit in any complete review of the subject. However, in this chapter the focus is on the problems of and procedures for getting good measurements with current thermal-anemometer and data-analysis techniques. The reader is referred to Freymuth [1] and a review paper by Comte-Bellot [6] for a more thorough review of the literature.

In the next section, a brief comparison is made between the thermal anemometer and the laser velocimeter, since the instruments are used for similar purposes. Then a rather complete discussion is given of the characteristics of hot-wire sensors operated in a constant-temperature circuit. Finally, film sensors are compared with hot-wire sensors.

2. STRENGTHS, LIMITATIONS, AND COMPARISONS WITH LASER VELOCIMETERS

To measure velocity details in a flowing fluid, the ideal instrument should

1. Have high-frequency response to accurately follow transients
2. Be small in size for an essentially point measurement
3. Measure a wide velocity range
4. Measure only velocity, and work in a wide range of temperature, density, and composition
5. Measure velocity components and detect flow reversal
6. Have high accuracy
7. Have high resolution (low noise)
8. Create minimal flow disturbance
9. Be low in cost
10. Be easy to use

For many years, only the hot-wire anemometer satisfied enough of these criteria to be used extensively in turbulence studies. Pitot probes, flow visualization, and other techniques complemented hot-wire data but generally could not get details as well as the hot wire.

For most applications, items 4, 5, 6, and 10 are perhaps the weakest areas of thermal anemometers. To be even more specific, the primary practical limitations are (1) fragility and (2) sensitivity to contamination. From a theoretical point of view, the limitations include the following:

1. Velocity is not measured directly but deduced from a measurement of convection heat transfer from the sensor.
2. Normal configurations limit the turbulence intensity that can be measured accurately.
3. Heat losses from the sensor other than by convection can cause errors.

The importance of these limitations depends on the application. Relative to item 6 (accuracy), thermal-anemometer measurements are very repeatable when the conditions are reproduced exactly. It is the effect of contamination and the effect of variables other than flow on the heat transfer that cause inaccuracies.

Since its introduction in 1964 [7], the use of the laser velocimeter to measure velocity details of flowing fluids has expanded rapidly. Since it, too, satisfies many of the "ideal instrument" criteria, a comparison between it and the thermal anemometer follows. It should be emphasized from the start that, rather than replacing thermal anemometers, the laser velocimeter complements their use. As is often the case in making measurements, it is not a question of the best instrument but rather which instrument will perform best for the specific application.

The laser velocimeter uses Doppler-shifted light scattered from particles in the flow to deduce the velocity of those particles. If the particles are small enough to follow the flow, then the flow velocity is measured. A window to the flow must be provided for both the incident light and the scattered light, but no probe needs to enter the flow field. Of course, there must be enough particles of the appropriate size and concentration so the desired statistical data can be determined.

In the following, the thermal anemometer and laser velocimeter are compared on the basis of the above criteria for the ideal instrument.

1. *Frequency response.* In current practice, the thermal anemometer is definitely superior. Measurements to several hundred kilohertz are quite easily obtained, with 1 MHz being feasible.

   Theoretically, a laser velocimeter could approach the response of a thermal anemometer. Practically, spectra up to only about 30 KHz have been measured. In many applications the problem is adequate size and concentration of the scattering centers (particles). At sufficiently high frequencies, electronic limitations also start to enter.
2. *Size.* A hot-wire type thermal sensor is typically 5 μm in diameter by about 2 mm long, although wires as small as 1 μm by 0.2 mm are feasible.

   For a laser velocimeter, measuring volumes of 50 μm by 0.25 mm are common, while a 5- by 5-μm measuring volume is achievable in very small test sections. If the distance from the focusing lens to the measuring point is long (e.g., over 400 mm), then small measuring volumes are difficult to achieve. They may also be impractical in some flows, owing to movement of the incident beams caused by refractive-index variations.
3. *Velocity range.* Both techniques have a very wide velocity range. The laser has the advantage at very low velocities because the "free convection" effects that affect hot-wire readings are usually not a problem.

   Although both measure high-speed (compressible) flows, laser data are easier to interpret because they are sensitive only to particle velocity and no calibration is required. At the same time, providing particles that follow the flow, scatter enough light, and have high enough concentration for spectral measurements can be difficult.

The thermal anemometer is sensitive to recovery temperature, Mach number, and Reynolds number in compressible flows. Measurements in transonic flows require considerable calibration while, for $M > 1.5$, Mach-number independence makes measurements more feasible.

4. *Measure only velocity over wide temperature density, and composition ranges.* The laser velocimeter measures only the velocity of the scattering center (particle), and it measures it in a known direction (pure cosine response). The hot-wire sensor measures heat transfer to the environment. This can be a plus since, for example, by using two sensors, both temperature and velocity fluctuations can be measured. Generally, though, it is preferable to be sensitive to velocity only.

   Both instruments will operate over wide temperature, density, and composition ranges. However, even though the cooled-film probe (a thermal sensor with operating principle similar to that of a hot wire) can be used at high temperatures, its application is limited by sensitivity to other variables. At low density, measurements become more difficult for both. Conduction losses become excessive, and slip flow effects complicate hot-wire anemometry, while in laser velocimetry there are problems in finding particles that both follow the flow and scatter enough light.

   The laser must "see" into the flow. In liquid metals, for example, thermal anemometers can be used but generally not laser velocimeters.

5. *Component resolution.* The hot wire can be used to resolve one, two, or all three components of a flow field by using one, two, or three sensors respectively. However, it appears limited to rather low turbulence intensities even with rather sophisticated data-reduction procedures. Film sensors give the potential for measuring at any turbulence intensity [8], but the complexity of present techniques and probe size limit the applications.

   The laser velocimeter can resolve components and, with frequency shifting, can detect flow reversals. It is difficult to obtain the third component, since a separate lens system is generally required–causing alignment problems.

6. *Accuracy.* Hot-wire results can be very repeatable, so accuracy is really a function of how closely the calibration conditions are reproduced in the flow to be measured. In practice, contamination, temperature changes, and other factors generally limit accuracy to a few percent.

   The laser velocimeter can give very high accuracy (0.1%) in carefully controlled experiments. In many practical measurements, refractive-index variations, limited accuracy on beam-crossing angle, and signal-processor limitations make a value of 1% more realistic.

7. *Resolution.* The hot wire is clearly superior, since it can have a very low noise level. Resolution of 1 part in 10,000 is easily accomplished, while with a laser velocimeter 1 part in 1000 is difficult with present technology.

8. *Flow disturbance.* Since only light needs to enter the flow, the laser is clearly better. The normally required size and concentration of particles do not measurably alter the flow field.
9. *Cost.* The hot wire is lower in cost by a factor of three to ten for most applications.
10. *Ease of use.* At present, the laser velocimeter is probably more difficult to set up and start getting valid data with. Once it is set up, the laser velocimeter may be easier to use, since there are no fragile sensors to get dirty, break, or shift calibration. A complexity in data interpretation with the laser velocimeter is the fact that discrete measurements are made (on each measurable particle), which gives a discontinuous output.

As a general rule, if other instruments (such as a Pitot tube or pressure transducer) cannot give the detailed measurements required, one should consider a hot-wire (or hot-film) anemometer. If high temperatures, moving objects in the flow, proximity to walls, dirt in the flow, or some other problem makes hot wires difficult or impossible to use, then a laser velocimeter should be considered.

In summary, thermal anemometers can theoretically be used in almost any fluid-flow situation. However, sensor fragility, calibration shifts due to contamination, or difficulties in separating out variables make many potential applications difficult. The most common and easiest measurements with thermal anemometers are in constant-temperature gases near atmospheric pressure, at relatively low turbulence intensities, and at flow velocities low enough so that the assumption of incompressibility is adequate. But when the need is sufficient, good measurements can be made over a much wider range of conditions.

3. HOT-WIRE SENSORS

Common hot-wire materials include tungsten, platinum, and platinum-iridium (80% Pt, 20% Ir). These materials are used partly because of their properties but also because of their availability in the small diameters of interest in hot-wire anemometry. Table 1 gives some of their properties. Ideally the first three items should be high and the last item low; low thermal conductivity reduces conduction losses to the supports.

Tungsten is desirable because of its high temperature coefficient of resistance and high strength. The major disadvantage is the rate at which it oxidizes, especially at temperatures above about 300°C [9]. Tungsten wires are available commercially in diameters down to 2.5 μm. They are made by first drawing and then etching to the final diameter.

Platinum is available in very small sizes (0.5 μm), has a good temperature coefficient, and does not oxidize. It would be the ideal wire if it were not so weak, especially at high temperatures. At high wire temperatures and high air velocities, aerodynamic drag alone can cause the stress on a platinum wire to exceed its limit.

Platinum-iridium is a compromise wire that does not oxidize, has better

Table 1 Some properties of common hot-wire materials

	Tungsten	Platinum	80% Platinum, 20% iridium
Temperature coefficient of resistance α, °C^{-1}	0.0045	0.0039	0.0008
Resistivity, $\Omega\cdot$cm	5.5×10^{-6}	10×10^{-6}	31×10^{-6}
Ultimate tensile strength, Kg/mm^2 (lb/in^2)	420 (60×10^4)	24.6 (3.5×10^4)	100 (14.22×10^4)
Thermal conductivity, cal/(cm·°C)	0.47	0.1664	0.042

strength than platinum, but has a low temperature coefficient of resistance. It finds application where the wire temperature must be too high for tungsten and platinum is too weak.

These wires have all been used since 1950 [9], and probably long before that. Newer materials might make superior hot wires, but no extensive study has been reported recently. In any case, the three materials in Table 1 seem to satisfy most requirements for hot-wire sensors. Sandborn [10] provides considerable details on wire properties and mounting procedures.

In the selection of wire diameter and length, many conflicting criteria come into play. For example, as regards sensor length:

1. A *short* sensor is desired to:
   a. Maximize spatial resolution
   b. Minimize aerodynamic stress
2. A *long* sensor is desired to:
   a. Minimize conduction losses to supports
   b. Provide a more uniform temperature distribution
   c. Minimize support interference

And, with regard to sensor diameter:

1. A *small* diameter is desired to:
   a. Eliminate output noise due to separated flow around the sensor (sensor-generated flow fluctuations)
   b. Maximize the time response of the wire due to lower thermal inertia and higher heat-transfer coefficient
   c. Maximize spatial resolution
   d. Improve the signal-to-noise ratio at high frequencies
2. A *large* diameter is desired to:
   a. Increase strength
   b. Reduce contamination effects due to particles in the fluid

For research work, 2.5- to 5-μm wires are the most common, with the length-to-diameter ratios $2l/d$ of from 100 to 600. Commercially mounted wires typically have $2l/d \simeq 300$. Possible errors due to the effects listed above are discussed later.

4. PROBE SUPPORTS AND MOUNTING

Figure 2*a* shows a typical probe support for a hot-wire probe. The ends of tungsten wires are generally electroplated to isolate the sensing element from the supports. With platinum wire, the Wollaston process used to make the wire leaves a silver coating on the ends, since only the silver on the sensor portion is etched away.

It is preferable to have both wire supports electrically insulated from the body and, therefore, from the test section. In locations with strong electromagnetic fields or in water flows, an outside ground shield connecting the electronics and probe, independent of the sensor leads, can often reduce background noise.

The plug-in tip shown on Fig. 2*a* is a convenience. Wires are fragile, so it is useful to have replacement wires already mounted when running tests. At the same time, in a clean gas a hot wire will last indefinitely if it is not physically damaged or burned out by a malfunction in the control circuit. This should not happen with modern transistorized constant-temperature anemometers.

A wide variety of probe supports have been designed, including miniature

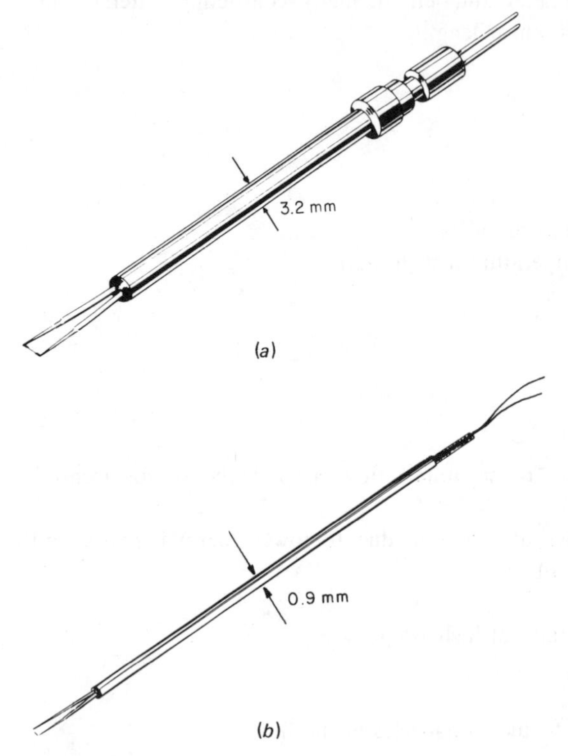

Figure 2 Examples of probes for thermal anemometers. (*a*) Typical plug-in probe. (*b*) Subminiature probe.

versions with much smaller bodies, as shown in Fig. 2*b*. Sandborn [10] discusses some special considerations in the design of supports for measurements in supersonic flow.

5. CONTROL CIRCUIT

Figure 1 shows a very simple electric circuit for heating a hot-wire probe. The next step in the evolution of hot-wire circuitry was to provide open-loop frequency compensation for the hot wire. This was easy to do electrically, since hot-wire response is similar to an *RC* circuit (first order). Until the introduction of transistorized circuitry, the compensated constant-current circuit was the most common for high-frequency hot-wire anemometry.

Since about 1960, the application of constant-temperature systems has increased, owing to both improved circuitry and improved understanding of the system's characteristics. Those, combined with its applicability to both films and wires as well as to a wide range of flows, have caused it to essentially displace the constant-current approach. Therefore, in the following, we concentrate on the constant-temperature type control circuit.

Adding a feedback line converts the bridge circuit and amplifier of Fig. 1 to a constant-temperature system. The feedback from the output of the high-gain amplifier to the top of the bridge acts to maintain the resistance of the sensor (hence, the average temperature) essentially constant. The step-by-step operation is as follows:

1. A velocity increase past the hot wire cools it, lowering the temperature, the resistance R, and the voltage at point 1.
2. The lowered voltage at the negative input to the amplifier causes the voltage U_{12} to increase.
3. The increased amplifier input voltage U_{12} increases the output voltage of the amplifier U.
4. The increased voltage on the bridge U increases the current through the sensor.
5. The increased current heats the sensor, resulting in a decrease in U_{12} until the entire system is again in equilibrium.

All this takes place almost simultaneously, so an increase in velocity is seen as an increase in the voltage U. This voltage is generally used as the anemometer output. In some cases the voltage across the fixed resistors in series with the sensor is used as a direct measure of sensor current I_p, independent of the sensor operating resistance R.

Figure 1 also shows a sketch of the amplitude response of the anemometer. By adding the feedback, the frequency response (−3-dB point) can be increased from about 700 Hz for the wire and environment shown to over 100 KHz. The latter depends on the characteristics of the bridge and amplifier as well as the hot wire and its environment.

6. CALIBRATION OF A HOT-WIRE ANEMOMETER

A typical calibration curve for a hot-wire anemometer is given in Fig. 3. The two outstanding characteristics are:

1. The output is very nonlinear with velocity.
2. The sensitivity decreases as velocity increases.

It turns out that the *sensitivity as a percent of reading stays nearly constant.* This helps make the hot-wire anemometer useful over a very wide range of flow velocities.

Figure 4 shows a typical calibration system. By measuring the total pressure upstream, the velocity in the nozzle can be precisely calculated with the simple relation

$$V = \sqrt{2gh_t}$$

where g = gravitational constant

h_t = total pressure in height of fluid flowing

This relation is exact for the centerline velocity of a properly designed nozzle and incompressible flow.

Once a calibration curve similar to Fig. 3 is obtained, good measurements in the unknown environment can be made directly, providing:

1. The fluid temperature, composition, and density in the unknown environment are the same as those during calibration or, if not, are corrected for.
2. The turbulence intensity is below 20%.
3. The flow is incompressible.

With these restrictions, the single hot-wire anemometer can be used as the basic transducer to measure mean velocity, turbulence intensity, turbulence spectrum, waveform, and flow transients. Mean flow, true waveform, and flow transients must be interpreted by converting voltage to velocity by use of the calibration curve. Spectrum and turbulence are interpreted by measuring the slope of the curve at the operating point. Electronic linearizing greatly simplifies this process and is common. It can also reduce errors due to nonlinearity effects, but this is certainly not assured [11, 12].

The above list of limiting factors applies only to this straightforward application of hot-wire anemometry. As the complexity of the instrument, calibrations, and/or data-reduction technique increase, the potential range of application of the basic thermal anemometer can become very broad.

7. HEAT TRANSFER FROM FINE WIRES

If an actual velocity calibration can be made that covers the test conditions, then an analytical relation between flow velocity and anemometer output may not

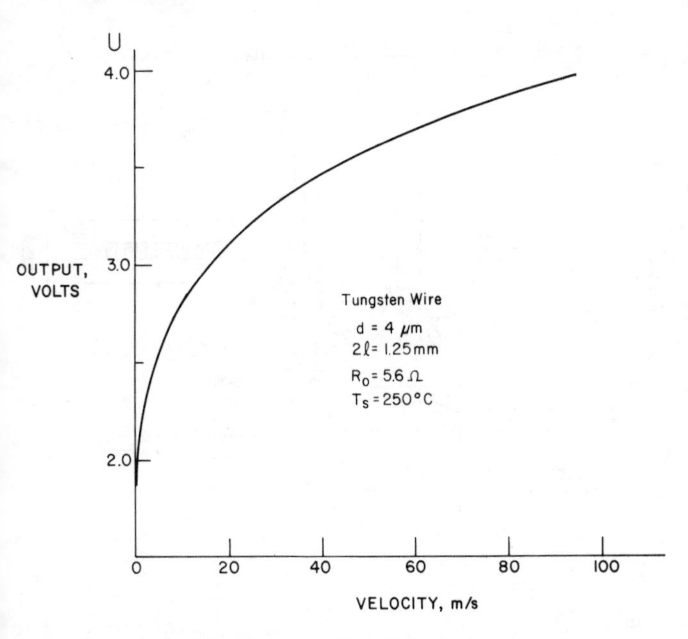

Figure 3 Typical calibration curve for a hot-wire anemometer.

be required. Situations where an equation is useful or necessary include:

1. Complex measurement situations such as compressible flow, where a complete set of calibration curves is impractical
2. Investigating or correcting for various error sources that are difficult to determine experimentally

In both cases some calibration should still be done on each sensor. The reproducibility of thermal sensors is typically not sufficient to give identical calibration curves for two different sensors of the same type.

It is useful to think of heat transfer H from the sensor to its environment as the fundamental variable. From Fig. 1, for a constant-temperature anemometer, it is related to the anemometer output voltage U by the equation

$$H = P = \frac{U^2 R}{(R + R_1)^2} = \phi + K \tag{2}$$

where P = electric power input to sensor
H = heat transfer from sensor to environment

Equation (2) is valid as long as R represents the resistance of the sensor only.

The more difficult relation is that between total heat transfer H and flow velocity V. The heat transfer H consists of two parts:

1. Convective heat transfer ϕ between the heated portion of the sensor and the flowing fluid.
2. Conduction heat transfer K between the heated portion of the sensor and its

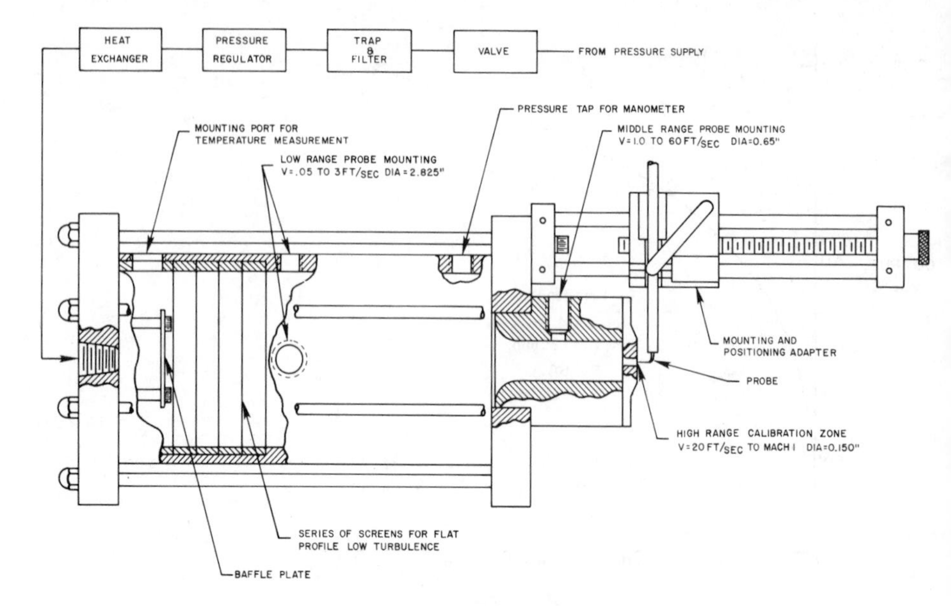

Figure 4 Calibration system–air.

supports. This heat K is generally transferred by convection from the supports to the flowing fluid.

As is shown later, K influences the dynamic response as well as the steady-state heat balance on the sensor. Radiation is usually negligible, except for measurements in rarefied gases.

The quantity of primary interest here is the convective heat transfer ϕ; conduction is treated below as a deviation from the ideal. For the hot wire,

$$\phi = \text{Nu}\ \pi\ 2lk_f(T_m - T_a) \tag{3}$$

where Nu $= (h_c d)/(k_f)$ = Nusselt number
d = sensor diameter
h_c = heat-transfer coefficient for convection
k_f = thermal conductivity of fluid at "film" temperature $T_f = (T_m + T_a)/2$
T_m = mean cylinder temperature
T_a = ambient temperature of fluid
$2l$ = length of sensitive area of hot wire

The problem is to find a representative expression for the Nusselt number in terms of the fluid and sensor parameters.

A general expression for Nu would be

$$\text{Nu} = f\left(\text{Re}, \text{Pr}, \alpha_1, \text{Gr}, \text{Ma}, \gamma_h, a_T, \frac{2l}{d}, \frac{k_f}{k_w}\right) \tag{4}$$

where Re $= \rho V d/\mu$ = Reynolds number
V = free-stream velocity

ρ = fluid density
μ = fluid viscosity
$\mathrm{Pr} = \mu C_p/k$ = Prandtl number
C_p = specific heat of fluid at constant pressure
α_1 = angle between free-stream flow direction and normal to cylinder
$\mathrm{Gr} = \rho^2 g \beta_v (T_m - T_a)\, d^3/\mu^2$ = Grashof number
$\beta_v = 1/T_a$ = volume coefficient of expansion
$\mathrm{Ma} = V/\gamma_h R_0 T_a^{1/2}$ = Mach number
R_0 = gas constant
$\gamma_h = C_p/C_v$
C_v = specific heat at constant volume
$a_T = (T_m - T_a)/T_a$ = overheat ratio or temperature loading
k_w = thermal conductivity of sensor material

Fortunately, most applications permit a significant reduction in the number of parameters that must be included. The simplest expression is often referred to as "King's law" [2]:

$$\mathrm{Nu} = A' + B'\,\mathrm{Re}^{0.5} \tag{5}$$

where A' and B' are empirical constants, usually determined by calibration.

Although Eq. (5) does not accurately represent Nu over a wide range of velocities, it is still frequently used in calculations because of its simplicity. A more accurate expression for air is that of Collis and Williams [13]:

$$\mathrm{Nu} = (A + B\,\mathrm{Re}^n)\left(1 + \frac{a_T}{2}\right)^{0.17} \tag{6}$$

where

$A = 0.24 \quad B = 0.56 \quad n = 0.45 \quad \text{for } 0.02 < \mathrm{Re} < 44$

$A = 0 \quad B = 0.48 \quad n = 0.51 \quad \text{for } 44 < \mathrm{Re} < 140$

At this point, it is useful to discuss why the remaining parameters of Eq. (4) are ignored in Eqs. (5) and (6).

Equations (5) and (6) are derived for forced convection normal to the wire ($\alpha_1 = 0$) and for incompressible flow. Since Pr depends only on fluid properties, limiting the discussion to air eliminates Pr as a variable in Eq. (4). It was found [13] that buoyancy effects (free convection) are important in air only if

$$\mathrm{Gr}^{1/3} > \mathrm{Re}$$

For the hot wire of Fig. 1, this occurs at a forced convection velocity of 5.2 cm/s. This value can be reduced further by lowering the sensor temperature (overheat). For higher velocities, Gr can be ignored.

For high-velocity or low-density flows, the Mach number Ma and specific heat C_p must be considered as variables. For low density, the most relevant parameter is the Knudsen number, which is

$$\mathrm{Kn} = \frac{\lambda}{d} = \left(\frac{\pi \gamma_h}{2}\right)^{0.5} \frac{\mathrm{Ma}}{\mathrm{Re}} \tag{7}$$

where λ is the molecular mean free path. Three ranges of Kn are usually defined:

Continuum flow:

$\mathrm{Kn} < 0.01$

Slip flow:

$0.01 < \mathrm{Kn} < 1$

Free molecular flow:

$\mathrm{Kn} > 1$

Again for the conditions of Fig. 1, $\mathrm{Kn} \simeq 0.02$, which is in the slip-flow region. While most measurements with fine wires are in slip flow, continuum-flow assumptions are usually adequate as long as density changes are small.

At high velocity, two fluid temperatures are normally defined–static temperature T_s and total temperature T_0. They are related by (assuming C_p constant)

$$T_0 - T_s = \frac{V^2}{2C_p} \simeq \frac{V^2}{2000} \tag{8}$$

for air, with T in degrees kelvin and V in meters per second. The total temperature is the temperature the fluid attains when brought to rest; it is both the most convenient reference temperature in thermal anemometry and the easiest temperature to measure in high-speed flows. However, when exposed to a high-velocity airstream, a hot wire does not always attain the total temperature but instead equilibrates at a recovery temperature T_r between T_0 and T_s. The recovery factor η is defined as:

$$T_r = \eta T_0$$

From [14], for $\mathrm{Kn} < 0.1$ and $M < 1$, the recovery factor η is greater than 0.98. It should be noted that, experimentally, when the "cold" wire temperature is measured in the flow stream it is the recovery temperature that is measured, and T_r should be substituted for T_a in heat-transfer equations.

The value of a_T appears in Eq. (4) to account for the change in fluid properties with temperature. For incompressible flow this effect is minimized by using the film temperature T_f when selecting fluid properties. However, a weak effect is still present, as shown in Eq. (6).

The aspect ratio $2l/d$ enters Eq. (4) because three-dimensional aerodynamic effects near the prongs may affect flow. Champagne [15] discusses this as it relates to Eq. (6) for $\mathrm{Re} < 44$ and a platinum wire. While n was unaffected, B decreased by 5% and A doubled when $2l/d$ was decreased from 10^3 to 10^2.

Finally, k_f/k_w may enter Eq. (4) because the temperature distribution along and around the sensor depends on it. In actual practice this effect is not generally taken into account and, in any case, it is considered as a "perturbation" or correction to relations such as Eq. (5). Of course, both $2l/d$ and k_f/k_w are very important parameters when conduction to the supports K is considered. However here we are concerned only with convective heat transfer ϕ.

Equation (6) applies for air only. A more general equation often used is that proposed by Kramer [16]:

$$\mathrm{Nu} = 0.42\ \mathrm{Pr}^{0.26} + 0.57\ \mathrm{Pr}^{0.53}\ \mathrm{Re}^{0.50} \tag{9}$$

which covers the range $0.71 \leqslant \mathrm{Pr} \leqslant 525$ and $2 \leqslant \mathrm{Nu} \leqslant 20$. Again, fluid properties are to be selected at the film temperature T_f. Equation (9) covers a wide range of fluids and is very useful for liquids and gases other than air [where the more precise Eq. (6) should be used].

7.1 High-Speed Flows

Measurements in high-speed flows present a number of special problems, and the reader is referred to publications on the subject [10, 14, 17 to 21]. Some comments may, however, be helpful.

In high speed flows, Eq. (3) should be reformulated as

$$\phi = \mathrm{Nu}\ \pi 2 l k_0 (T_m - T_r) \tag{10}$$

where k_0 is the thermal conductivity of the gas at the stagnation temperature T_0. According to Kovasznay et al. [17], it is advantageous to formulate Nusselt-number dependence in terms of the Reynolds number

$$\mathrm{Re}_0 = \frac{\rho_\infty V_\infty d}{\mu_0}$$

where the subscript ∞ refers to free-stream conditions, and the subscript 0 refers to stagnation conditions. In supersonic continuous flow in air ($\mathrm{Ma} > 1.5$, $\mathrm{Kn} < 0.01$) they found experimentally that the Nusselt number is independent of Mach number and for vertical flow incidence can be represented by a relation

$$\mathrm{Nu} = (A\ \mathrm{Re}_0^{1/2} - B)\left(1 - C\,\frac{T_m - T_r}{T_0}\right) \tag{11}$$

where $A = 0.58$, $B = 0.8$, and $C = 0.18$. However, according to [10] and [14], temperature loading effects are much smaller; that is, C may be negligible, and Nu is about 20% too high [22]. Because of Mach-number independence (Fig. 5), the anemometer in supersonic continuum flow is sensitive to $\rho_\infty V_\infty$ and to T_0, just as the incompressible-flow anemometer is sensitive to ρV and T_a.

In the low-density flow regime, as shown in Fig. 5, the Mach-number dependence of Nu sets in strongly as a parameter. In this regime Nu becomes proportional to Re rather than $\mathrm{Re}^{1/2}$, and heat transfer becomes mainly sensitive to density and rather insensitive to velocity; as a consequence, hot wires have marginal value as anemometers in the molecular flow regime. In the low-density regime, the recovery factor η is a function of Re and M [14].

The influence of aspect ratio on Nu seems to be sufficiently small to have received no particular attention, and the influence of M on directionality also remains uninvestigated.

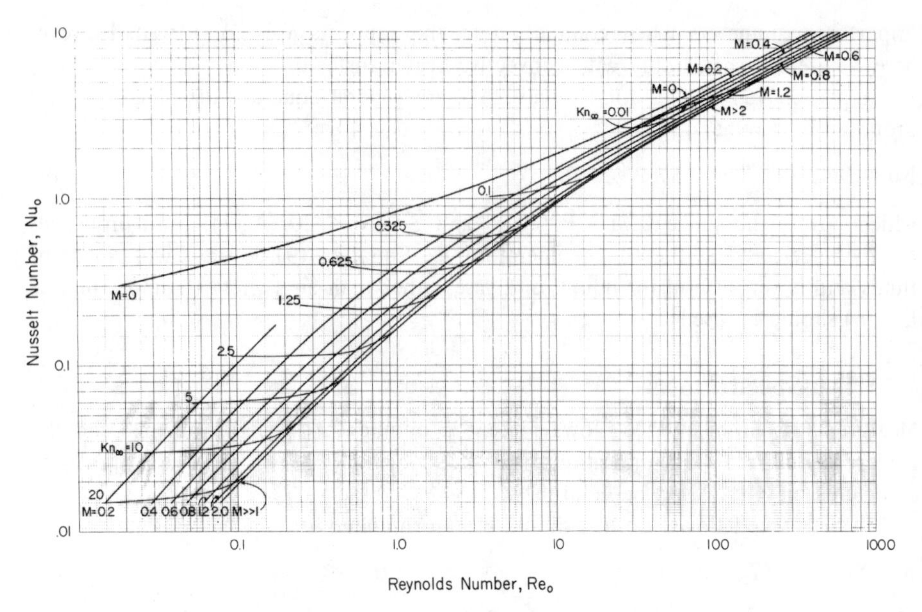

Figure 5 Empirical correlation of cylinder heat transfer at low Reynolds number. *(From [14].)*

7.2 Conduction to Walls

As a hot wire approaches a wall, the temperature and velocity field around the wire is modified, owing to the presence of the solid surface. Any modification of this boundary layer will change the relation between velocity and the output voltage of the anemometer.

According to Wills [23], the primary parameter is b/a, where b is the distance from the wall and a is the sensor radius. For b/a greater than 500, the effect is negligible (2 mm for a 4-μm-diameter wire). For the range tested ($Re < 1$), if $Re > 0.1$ the wall effect seemed to be a constant correction on the A term of Eq. (6). At $b/a \to \infty$, Wills [23] found an A of 0.26 rather than 0.24 [13]. This increased to 0.32 at $b/a = 100$, and to 0.47 at $b/a = 20$. Since only the A term was affected, the percent error decreases as velocity increases.

8. CONDUCTION TO THE SUPPORTS

Any heat transfer K from the sensor to the supports by conduction is a "loss" and a potential error source. Again, for steady-state measurements this conduction loss will be included in the calibration process. However, calibration under actual operating conditions is not always possible. Perhaps even more important, conduction losses cause dynamic effects that are difficult to measure experimentally. While in this section our concern is only with steady-state heat transfer, in Sec. 11.4 the strong relation between steady-state and dynamic effects is shown.

Figure 6 shows the temperature profiles of a hot-wire sensor for various values of the parameter $\sqrt{C_0}l$. From [24], $\sqrt{C_0}l$ can be calculated from

$$\sqrt{C_0}l = \frac{(2l/d)(k_a \text{ Nu}/k_w)^{1/2}}{\{1 + \alpha(T_m - T_a) + [\alpha(T_m - T_a)]/(\xi - 1)\}^{1/2}} \tag{12}$$

where

$$\xi = \sqrt{C_0}l \coth \sqrt{C_0}l$$

For values of $\sqrt{C_0}l > 2$, Eq. (12) quickly converges by calculating an initial value of $\sqrt{C_0}l$ using only the first two terms in the denominator for an estimate of ξ. For the sensor of Fig. 1, $\sqrt{C_0}l = 2.93$.

In the case of small overheats $\alpha(T_m - T_a) \ll 1$ and $\alpha(T_m - T_s) \ll 1$, Eq. (12) can be written

$$\sqrt{C_0}l = \frac{2l}{d}\left(\frac{k_a}{k_w}\text{ Nu}\right)^{1/2} \tag{12a}$$

Therefore, the sensor temperature profile (and resulting conduction losses) depends on the length-to-diameter ratio $2l/d$, the thermal-conductivity ratio between the fluid and the sensor, and the Nusselt number.

The question of primary interest is the amount of heat transferred to the fluid by convection, as compared with the total. Let us introduce the ratio

$$\epsilon = \frac{\phi}{\phi + K} \tag{13}$$

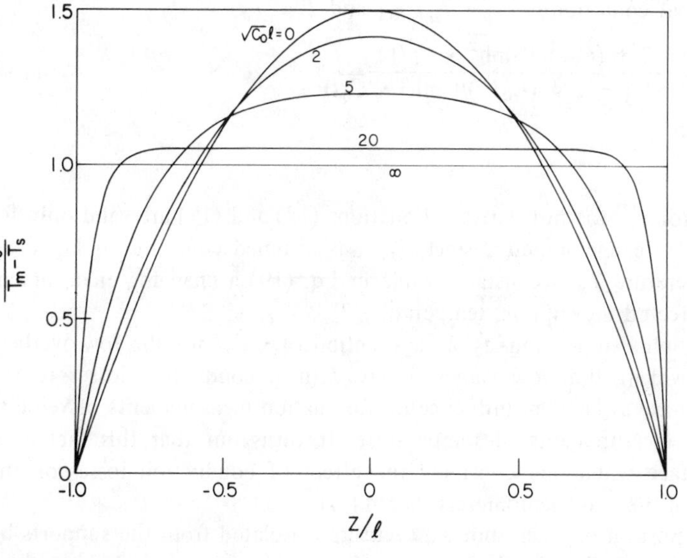

Figure 6 Temperature distribution along hot wires with constant mean temperature for various values of the parameter $\sqrt{C_0}l$.

In [24] the solution was

$$\epsilon = 1 - \frac{(T_m - T_s)/(T_m - T_a)}{\xi[1 + \alpha(T_m - T_a)] + (T_m - T_s)/(T_m - T_a) - 1 + \alpha(T_a - T_s)} \tag{14}$$

which, for small overheat and $T_s \simeq T_a$, reduces to

$$\epsilon = 1 - \frac{1}{\xi} \tag{15}$$

For the conditions of Fig. 1 (assuming $T_a = T_s$), $\epsilon = 0.826$. In other words, about 17% of the heat generated in the sensor by the electric current is conducted to the supports, while over 82% is transferred directly to the fluid by convection.

Equation (13) can be analyzed further to determine the effect of changes in the fluid parameters on ϵ. Two cases may be considered [24]:

$$\epsilon_1' = \left.\frac{d\phi}{d\phi + dK}\right|_{T_a, T_s = \text{const}} \tag{16}$$

$$\epsilon_2' = \left.\frac{d\phi}{d\phi + dK}\right|_{\text{Nu}=\text{const},\ T_a = T_s,\ \sqrt{C_0}\, l = \text{const}} \tag{17}$$

In the first case Nu (velocity) varies, while in the second fluid temperature T_a is the variable. The calculated value of ϵ_1' is, for the general case,

$$\begin{aligned} \epsilon_1' = 1 - \{[(T_m - T_s)/(T_m - T_a)]\,[(\xi - 2 + (C_0 l^2)/(\sinh^2 \sqrt{C_0} l)]\}/ \\ 2[1 + \alpha(T_m - T_a)](\xi - 1)^2 + [(T_m - T_s)/(T_m - T_a) + \alpha(T_m - T_s)] \\ \times \{[\xi - 2 + (C_0 l^2)/(\sinh^2 \sqrt{C_0} l)]\} \end{aligned} \tag{18}$$

For the low-overheat condition $\alpha(T_m - T_a) \ll 1$ and $T_s = T_a$,

$$\epsilon_1' = 1 - \frac{\xi - 2 + (C_0 l^2)/(\sinh^2 \sqrt{C_0} l)}{2(\xi - 1)^2 + \xi - 2 + (C_0 l^2)/(\sinh^2 \sqrt{C_0} l)} \tag{18a}$$

$$\epsilon_2' = \epsilon = 1 - \frac{1}{\xi} \tag{19}$$

The general case for ϵ_2' was not derived. Equations (18) and (19) are valid only for a constant-temperature anemometer, since T_m was assumed constant. In Eq. (18), the support temperature T_s is constant, while in Eq. (19) a changing environment temperature T_a affected the support temperature T_s.

Figure 7 is a plot of ϵ_1' and ϵ_2' as a function of $\sqrt{C_0}\, l$ for the low-overheat condition. It is evident that low values of $\sqrt{C_0}\, l$ (high conduction losses to the supports) do not have as large an influence on fluctuation measurements of velocity as for fluctuation measurements of temperature. It turns out that this fact is of particular significance when one considers the effect of conduction losses on the dynamic response of thermal anemometers (Sec. 11.4).

The sensitive portion on many hot-wire sensors is isolated from the supports by a plated area, as shown in Fig. 1. The support effectively starts at the plating, which provides a very small, fragile support. In this case $T_s > T_a$, where T_s is now the

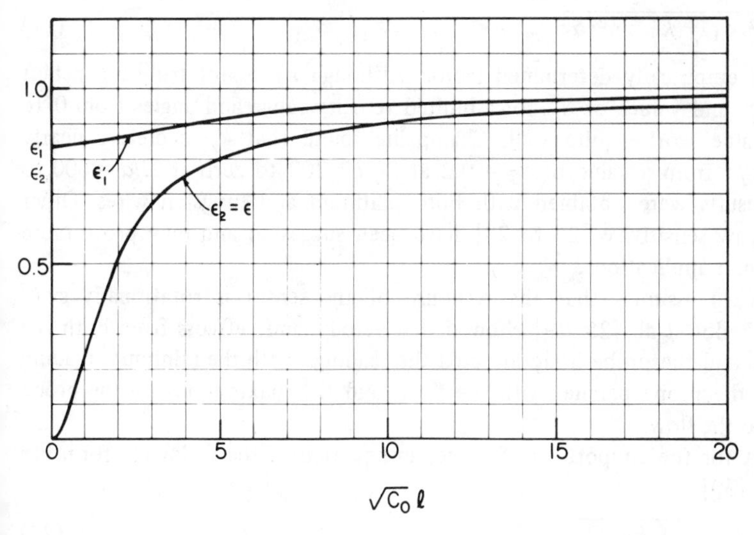

Figure 7 Heat-loss fluctuation ratios ϵ_1' and ϵ_2' in isothermal and nonisothermal constant-velocity flow as functions of the Biot number $\sqrt{C_0}l$ (small-overheat case) of the hot wire.

temperature of the junction between the plated area and the sensitive area of the sensor. This, however, can only decrease conduction losses K, resulting in ϵ and ϵ_2' being closer to 1. The effect on ϵ_1' is not so clear, and it is discussed further in Sec. 11.4.

In high-speed and low-density flows, the end-loss equations are valid as long as the recovery temperature T_r is substituted for ambient temperature T_a in calculating the heat loss from the sensor. Also, the support temperature should be calculated in the same way, although the larger size of the supports would give $\eta = 1$ and $T_s = T_0$ for most flow environments.

9. ANGLE SENSITIVITY AND SUPPORT INTERFERENCE

For an infinitely long wire, the angle sensitivity of the hot wire is expressed (Fig. 8) as

$$V_{\text{eff}} = V \cos \alpha_1 \tag{20}$$

where V_{eff} is the effective cooling velocity past the sensor. Equation (20) essentially states that the velocity along the sensor ($V_T = V \sin \alpha_1$) has no cooling effect on the sensor and that the sensor is rotationally symmetrical in both construction and response. In many calculations and experiments, Eq. (20) is adequate, maintains simplicity, and, depending on probe and sensor design, can be quite accurate [25].

Because the sensor has a finite length, there is heat transfer due to the flow parallel to the sensor (V_T). To account for this, a second term is added [22, 15] to give

$$V_{\text{eff}} = V\sqrt{\cos^2 \alpha_1 + k_T^2 \sin^2 \alpha_1} \tag{21}$$

where k_T is an empirically determined factor. Although k_T is not truly a constant for all velocities and values of α_1, for a limited velocity range and angles from 0° to 60° a fixed value works quite well. Champaine found that k_T decreases nearly linearly with $2l/d$ from a value of $k_T = 0.2$ at $2l/d = 200$ to zero at $2l/d = 600$ to 800. Similar results were obtained with both platinum and tungsten wires. Other equations for yaw sensitivity [26 to 28] have been suggested and may prove more accurate in certain applications.

Equation (21) assumes that the response of the sensor is rotationally symmetric. Comte-Bellot et al. [25, 29] showed that aerodynamic effects from both the support needles and the probe body affected the readings, with the minimum reading occurring with the probe parallel with the flow, and the maximum with the probe perpendicular to the flow.

To account for the support interference, an equation of the following form has been suggested [30]:

$$V_{\text{eff}} = \sqrt{V_N^2 + k_T^2 V_T^2 + k_N^2 V_{BN}^2} \tag{22}$$

where V_{BN} is the velocity vector perpendicular to both the sensor and the support prongs. The value of k_N can range from 1.0 to 1.2, depending on the design of the probe support and needles [28]. As is true of k_T, k_N is not constant for all angles and velocities, but careful use can improve accuracy as compared with assuming that $k_N = 1$ (theoretical value). It was found in [30] that plating the wire ends as shown in Figs. 1 and 8 reduced the values of k_T and k_N.

Equations (21) and (22) are given here to provide concrete examples. There is no intent to imply that they always represent the best functional relationships.

It should be emphasized that, in using a single calibrated sensor in turbulence intensities under 20%, good accuracy can be obtained without the above equations. It is when multisensor probes are used in large turbulence intensities, or when the sensor orientations during calibration and use are different, that the above considerations become important.

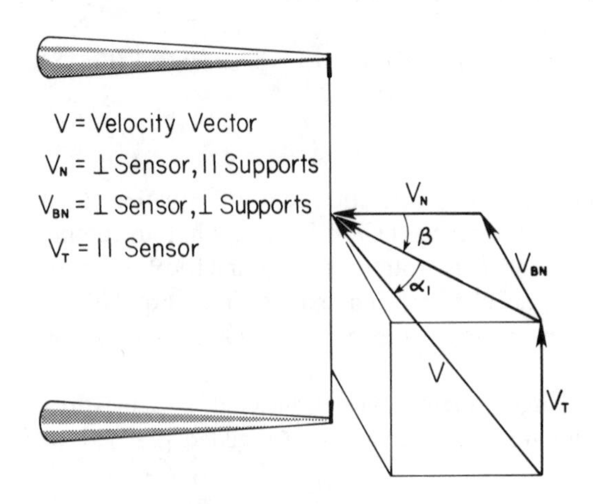

Figure 8 Velocity components at sensor.

10. MEASURING MEAN VELOCITY, VELOCITY COMPONENTS, AND TEMPERATURE

The most common measurement is the use of a single hot-wire probe, perpendicular to the flow, to measure mean velocity $\bar{V}$ and fluctuations in the mean flow direction $\overline{v_1^2}$. Two components are often measured with an X probe, while three components can be measured by adding a third sensor or by rotating the X probe. In addition, both temperature and velocity can be obtained by operating two parallel sensors at different temperatures. In all these measurements there are limitations that must be observed.

10.1 One Component Using a Single Hot Wire

In Fig. 8, assume the sensor is oriented in the flow stream so that $V_N = V_1$, $V_T = V_2$, and $V_{BN} = V_3$, where V_1, V_2, and V_3 are the desired orthogonal velocity components of the velocity vector **V**. From Eq. (22), the effective cooling velocity past the sensor is

$$V_{\text{eff}} = \sqrt{V_1^2 + k_T^2 V_2^2 + k_N^2 V_3^2} \tag{23}$$

If, further, the mean flow is in the V_1 direction, then $\bar{V}_2 = \bar{V}_3 = 0$. If the fluctuations are v_1, v_2, and v_3, then

$$V_{\text{eff}} = \sqrt{(\bar{V}_1 + v_1)^2 + k_T^2 v_2^2 + k_N^2 v_3^2} \tag{24}$$

Since k is small and $h \simeq 1$, this can be simplified to

$$V_{\text{eff}} = \sqrt{(\bar{V}_1 + v_1)^2 + v_3^2} \tag{25}$$

If we neglect v_3, then

$$\bar{V}_1 = \bar{V} = \bar{V}_{\text{eff}} \tag{26}$$

$$\sqrt{\overline{v_1^2}} = \sqrt{\overline{v^2}} \tag{27}$$

The value of $\bar{V}_1$ is obtained with a mean-value (averaging) meter, and $\sqrt{\overline{v_1^2}}$ is obtained with an ac-coupled true-rms meter.

When $\sqrt{\overline{v_1^2}}/\bar{V} = 0.2$, the error due to ignoring v_3 is about 2% for isotropic, normally distributed, and normally correlated turbulence [12]. The mean velocity error is also about 2%.

10.2 Two Components Using an X Probe

The cross-wire or X probe is frequently used to measure two velocity components (V_1 and V_2). In this case two sensors (A and B in Fig. 9) are made sensitive to V_2 by orienting them at an angle to the mean flow. With the two sensors in the $v_1 v_2$ plane and oriented at 90° to each other, Eq. (22) can be written for sensor:

$$V_{A,\text{eff}}^2 = (V_1 \cos\alpha_1 - V_2 \sin\alpha_1)^2 + k_T^2(V_1 \sin\alpha_1 + V_2 \cos\alpha_1)^2 + k_N^2 V_3^2 \tag{28}$$

$$V_{B,\text{eff}}^2 = (V_1 \sin \alpha_1 + V_2 \cos \alpha_1)^2 + k_T^2(V_1 \cos \alpha_1 - V_2 \sin \alpha_1)^2 + k_N^2 V_3^2 \quad (29)$$

where α_1 is the angle between V_1 and sensor B.

The coordinates are usually selected such that $\bar{V}_3 = 0$. If the sensors are sufficiently long so $k_T \to 0$ and $k_N \to 1$, then Eqs. (28) and (29) reduce to

$$V_{A,\text{eff}}^2 = (V_1 \cos \alpha_1 - V_2 \sin \alpha_1)^2 + v_3^2 \quad (30)$$

$$V_{B,\text{eff}}^2 = (V_1 \sin \alpha_1 + V_2 \cos \alpha_1)^2 + v_3^2 \quad (31)$$

In large turbulence intensities, Eqs. (30) and (31), like Eq. (25), cannot be further reduced. However, if v_3 is small, then the last term can be ignored. Further, orienting the sensors so that $\alpha_1 = 45^\circ$ and rearranging Eqs. (30) and (31) give

$$V_1 = 2^{-1/2}(V_{A,\text{eff}} + V_{B,\text{eff}}) \quad (32)$$

$$V_2 = 2^{-1/2}(V_{A,\text{eff}} - V_{B,\text{eff}}) \quad (33)$$

With these simplifying assumptions, summing the linearized output voltages of the two constant-temperature anemometers gives V_1, and differencing them gives V_2.

In most applications the orientation is such that the mean flow is in the V_1 direction, so that $\bar{V}_2 = 0$. The results are then

$$\bar{V} = 2^{-1/2}\,\overline{(V_{A,\text{eff}} + V_{B,\text{eff}})} \quad (34)$$

$$\overline{V_1^2} = \tfrac{1}{2}\,\overline{(v_{A,\text{eff}} + v_{B,\text{eff}})^2} \quad (35)$$

$$\overline{v_2^2} = \tfrac{1}{2}\,\overline{(v_{A,\text{eff}} - v_{B,\text{eff}})^2} \quad (36)$$

$$\overline{v_1 v_2} = \tfrac{1}{2}\,\overline{(v_{A,\text{eff}} + v_{B,\text{eff}})(v_{A,\text{eff}} - v_{B,\text{eff}})} \quad (37)$$

These are the equations used in most measurements with X probes.

Neglecting v_3 gives an error of about 8% when the turbulence intensity is 20%, with the same flow field as discussed for the single wire [12]. It should be emphasized that $k_T \neq 0$ can also significantly influence the accuracy of the results. Finally, the thermal wake from one sensor can influence the other [31]. All these considerations suggest the use of sensors with a high $2l/d$ ratio, as well as isolating the sensitive area from the supports by plating the wire ends.

10.3 Three Components

It is often desirable to measure all three components of the flow. This gives more information about the flow field and provides data on V_3 to improve the accuracy of measurement of V_1 and V_2.

In Fig. 9, adding a third sensor C (not shown) whose axis is at an angle α_3 to V_1 and in the $V_1 V_3$ plane gives the equation

$$V_{C,\text{eff}}^2 = (V_1 \sin \alpha_3 + V_3 \cos \alpha_3)^2 + k_T^2(V_1 \cos \alpha_3 - V_3 \sin \alpha_3)^2 + k_N^2 V_2^2 \quad (38)$$

Equations (28), (29), and (38) now give three equations in three unknowns that, in theory, can be solved for V_1, V_2, and V_3. Once these instantaneous components

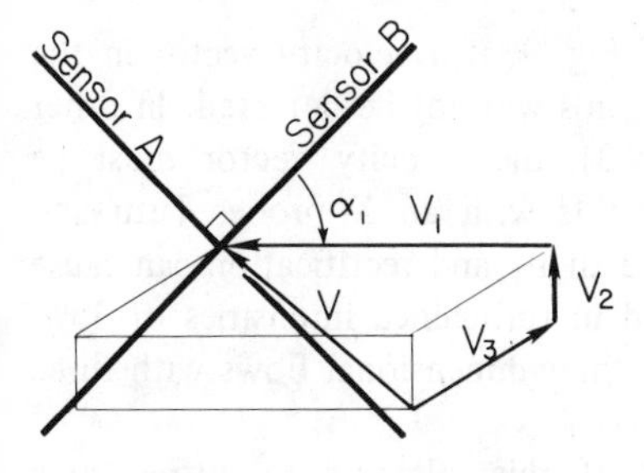

Figure 9 Configuration of X probe.

are available, all the turbulence parameters can be calculated. Again, the equations can be simplified if the sensors are long so $k_T \to 0$ and $k_N \to 1$, if the sensors are oriented so $\bar{V}_2 = \bar{V}_3 = 0$, and if α_1 and α_3 are 45°. Fabris [32] used this approach for making three-component velocity measurements.

The maximum turbulence intensity that can be measured is still limited. For example, with ideal sensors, flow reversals cannot be detected no matter how many

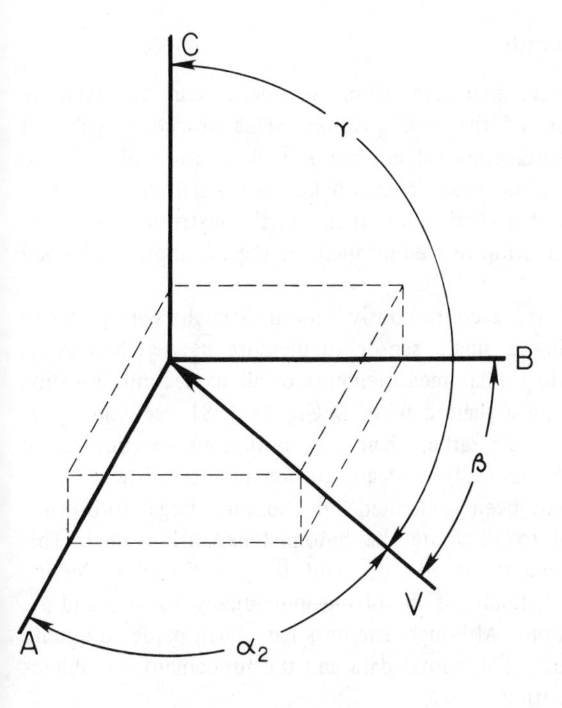

$$V_{A,\mathrm{eff}}^2 = V^2 \left(\sin^2 \alpha_2 + k^2 \cos^2 \alpha_2\right)$$

$$V_{B,\mathrm{eff}}^2 = V^2 \left(\sin^2 \beta + k^2 \cos^2 \beta\right)$$

$$V_{C,\mathrm{eff}}^2 = V^2 \left(\sin^2 \gamma + k^2 \cos^2 \gamma\right)$$

$$V^2 = \frac{V_A^2 + V_B^2 + V_C^2}{2 + k^2}$$

Figure 10 Direction sensitivity using three mutually perpendicular sensors.

sensors are used. In fact, with the X probe of Fig. 9, if a velocity vector in the plane of the sensors crosses the axis of sensor B, this will not be detected. In other words, even with two-dimensional flow ($V_3 = 0$), the velocity vector must be limited to a single quadrant for valid measurements with an X probe. Tutu and Chevray [33] report that the combined influence of v_3 and rectification can cause errors in $\overline{v_1 v_2}$ of 28% when an X probe is used in turbulence intensities of 35%. Similar restrictions hold for the measurement of three-dimensional flows with three sensors.

Olin and Kiland [8] provide a way out of this dilemma by using three orthogonal split-film cylindrical sensors. The splits were used to detect the octant, while the effective velocity readings were used to detect where the vector was within the octant. Figure 10 shows the configuration and equations [34], where $k_N \rightarrow 1$ was assumed. Atmospheric measurements have been made with this probe by Tielman et al. [35]. The size of the probe (~1 cm) restricts its use where spatial resolution is important.

10.4 Multiposition Measurements

As an alternative to the X probe, two sets of measurements can be taken by orienting a single sensor in each of the two positions. This procedure will not permit the measurement of instantaneous values, but it will measure $\overline{v_1^2}$, $\overline{v_2^2}$, and $\overline{v_1 v_2}$. Using one sensor eliminates the need to match hot-wire sensitivies but does add to the stability requirements of both the flow field and the instrumentation. As with X probes, the usual data-reduction procedure includes the assumption of small fluctuations.

Fujita and Kovasznay [36] used a continuously rotating straight wire probe to improve the accuracy when using a single sensor to measure $\overline{v_1^2}$, $\overline{v_2^2}$, and $\overline{v_1 v_2}$. Bissonnette and Mellor [37] made similar measurements of all six second moments ($\overline{v_1^2}$, $\overline{v_2^2}$, $\overline{v_3^2}$, $\overline{v_1 v_2}$, $\overline{v_1 v_3}$, $\overline{v_2 v_3}$) using a slanted wire. DeGrande [38] used the same basic technique but discrete positions rather than the continuous rotation. Kool [39] extended this technique to periodically unsteady turbomachinery flow.

Recently, the technique has been extended to measure large turbulence intensities by squaring the signal to eliminate the binomial expansion [40]. This appears to permit the measurement of $\overline{v_2^2}$, $\overline{v_3^2}$, and $\overline{v_1 v_2}$ without neglecting higher-order terms. The primary difficulty is in solving individually for $\overline{V_1^2}$ and $\overline{v_1^2}$, which does require some assumptions. Although attempts have been made to extend these results further [41], lack of experimental data and the fundamental problems of sensing flow reversal suggest caution.

10.5 Nonisothermal Flows

A thermal anemometer is sensitive to fluid temperature changes. These may be either slow changes due to mean temperature changes or high-frequency temperature changes resulting from a heat source in the flow or compressibility effects.

To examine temperature effects, it is useful to combine Eqs. (6) and (3) to examine the case for airflow:

$$\phi = 2l\pi k_f(A + B\,\mathrm{Re}^n)\left(1 + \frac{a_T}{2}\right)^{0.17}(T_m - T_a) \tag{39}$$

For velocities for which the A term is small, one can write

$$\frac{\Delta V}{V} \simeq \frac{\Delta T_a/n}{T_m - T_a} \tag{40}$$

Therefore, the velocity error ΔV due to a change in fluid temperature ΔT_a is minimized by maintaining a high overheat $T_m - T_a$. Often this precaution is sufficient to keep errors due to temperature changes within acceptable limits.

In Eq. (39), the following substitutions, based on fluid-property dependence on temperature, can be made [13]:

$$k \simeq (T_m + T_a)^{0.8} \tag{41}$$

$$\rho \simeq (T_m - T_a)^{-1} \tag{42}$$

$$\mu \simeq (T_m + T_a)^{0.76} \tag{43}$$

Substituting into Eq. (39), with $n = 0.45$, gives [42]:

$$\phi = [A_1(T_m + T_a)^{0.8} + B_1 V^{0.45}]\left(1 + \frac{a_T}{2}\right)^{0.17}(T_m - T_a) \tag{44}$$

It is interesting to compare this with the simple relation

$$\phi = H_1(V)(T_m - T_a) \tag{45}$$

With $T_m = 230°\mathrm{C}$ and $T_a = 23°\mathrm{C}$ and for a 50°C increase in T_a, the velocity calculated using Eq. (45) is within ±3% of that using Eq. (44), for the range 6 to 100 m/s. That is why the common technique of manually setting T_m to maintain $T_m - T_a = \text{const}$ works well for many measurements when the test temperature is different from that during calibration. This gives the same value of ϕ ($\simeq P$) for the same velocity.

A more convenient technique is to replace R_3 (Fig. 1) with a temperature-sensitive resistor exposed to the flow. For convenience, this is set to maintain U constant and independent of temperature, rather than ϕ. Again, this technique is used primarily to correct for low-frequency temperature changes and utilizes a rather large sensor that can approach the environment temperature. Trying to compensate in this manner for fast temperature fluctuations with a small sensor presents several problems. One of the most serious is the effect on the anemometer output U of the thermal capacity of the velocity sensor itself as it is heated and cooled to follow temperature changes.

To compensate for higher-frequency changes, an alternative technique is to measure the temperature separately and then correct the output data. For small temperature changes Eq. (45) is adequate, while for large temperature changes a more complex formula can be used with computer data reduction. The temperature sensor must be small enough so it will follow the temperature changes in the flow. For example, the wire in Fig. 1 will follow 700 Hz at 10 m/s. By using 0.625-μm-diameter wires, Fabris [32] was able to follow 4500 Hz at 6.5 m/s. For even better response, frequency compensation could be used as long as velocity changes are small.

As frequency increases, the interest is not only in correcting velocity measurements, but also in measuring statistical parameters such as the rms of temperature fluctuations and the cross correlation of temperature and velocity [43]. Of course, a single temperature sensor can also be used with X probes and other multisensor probes [32].

Perhaps the most flexible technique is the use of two sensors at different overheats. From Eq. (45),

$$\phi_1 = H_1(V)(T_{m1} - T_a) \tag{46}$$

$$\phi_2 = H_2(V)(T_{m2} - T_a) \tag{47}$$

Then, if $H_1(V) = H_2(V)$,

$$H(V) = \frac{\phi_1 - \phi_2}{T_{m1} - T_{m2}} \tag{48}$$

$$T_a = T_{m1} - \frac{\phi_1}{H(V)} \tag{49}$$

With this technique, the only frequency limitation is that of the constant-temperature anemometers themselves. For maximum sensitivity, $T_{m1} - T_{m2}$ should be large. Although more complex equations can be used with computer data reduction, Eqs. (48) and (49) should give good results for modest temperature fluctuations.

As with velocity components, second moments can be determined by using a single sensor at more than one overheat [44]. In fact, this technique has been used extensively in compressible flows. To improve accuracy, several overheats are generally used [17, 19].

11. DYNAMICS OF THE CONSTANT-TEMPERATURE HOT-WIRE ANEMOMETER

As shown in Fig. 1, adding the feedback loop to maintain T_m constant extends the frequency range from about 700 Hz to over 100 KHz. One of our concerns in this section is to establish the upper frequency limit for a given sensor, environment, amplifier, and bridge. A related problem is adjusting the system properly, so the amplitude response is as flat as possible. Finally, dynamic effects that are not compensated by the feedback system need to be considered. These include heat waves along the wire and temperature fluctuations of the probe support. In addition, the spatial resolution of the sensor, due to its finite size, and possible boundary-layer effects can limit the effective frequency response to velocity fluctuations.

All these effects could be tested experimentally if one could generate velocity and temperature fluctuations of known amplitude over a wide range of frequencies. Since this is not practical for routine measurements, an electrical test is used to optimize the system. The interpretation of the electrical test is based on a theoretical model of the anemometer. Other effects can be examined theoretically to obtain a measure of their importance.

Freymuth [45] has developed a detailed theory of electronic sine- and square-wave testing for the constant-temperature hot-wire anemometer. He uses a third-order linear equation that is consistent with the concept of two adjustable controls for optimizing the response of the anemometer. As it turns out, the optimization suggested by the linear equations also provides optimal response for large fluctuations [46].

11.1 Frequency Response of a Constant-Temperature Hot-Wire Anemometer

Figure 11 is a block diagram of a constant-temperature anemometer. It is similar to Fig. 1 except for the addition of U_t, the electronic test signal. It is this test signal that is used to optimize the frequency response by adjusting the two controls.

Freymuth [45] solved the third-order equation for three cases, but the one of most interest is the "maximally flat" response. In this case the cutoff frequency is:

$$f_{\text{cut}} = \frac{1}{2\pi}\left(\frac{G/M''}{M}\right)^{1/3} \tag{50}$$

where M'' is a second-order time constant, G is the amplifier gain, and

$$M = \frac{(n_b + 1)^2}{2n_b} \; \frac{R}{R - R_a} \; \frac{c}{H(V)} \tag{51}$$

is the time constant of the wire, where c is the thermal capacity of the wire.

Equation (51) indicates that reducing the sensor thermal capacity c, increasing the overheat $R - R_a$, or increasing the velocity $H(V)$ will increase f_{cut}. For example, since $c \simeq d^2 l$ and $H(V) \simeq d^{1/2} l$, it follows that $f_{\text{cut}} \simeq d^{-1/2}$. Therefore f_{cut} is less sensitive to wire diameter than the time constant M of the wire given above.

11.2 Optimization and Electronic Testing of the Dynamics of the Hot-Wire Anemometer

Figure 12 is a qualitative representation of the results of Freymuth's theory. For a properly adjusted hot-wire anemometer, an electrical sine-wave input (U_t in Fig. 11) will give an output signal U whose amplitude versus frequency graph is similar to curve 1*a*. Velocity fluctuations for the same anemometer will give an output whose amplitude and frequency follow curve 1*ac*. Finally, the uncompensated sensor will give the results represented by curve 1. The important point is that f_{cut} can be

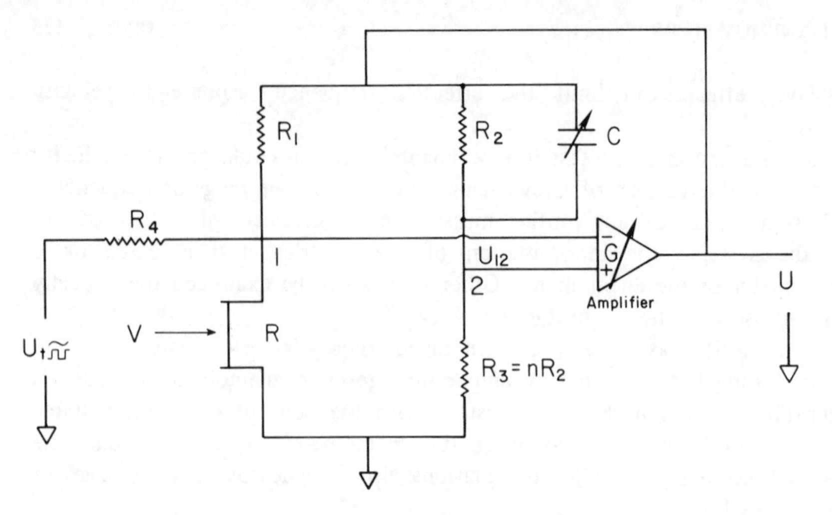

Figure 11 Constant-temperature anemometer control circuit.

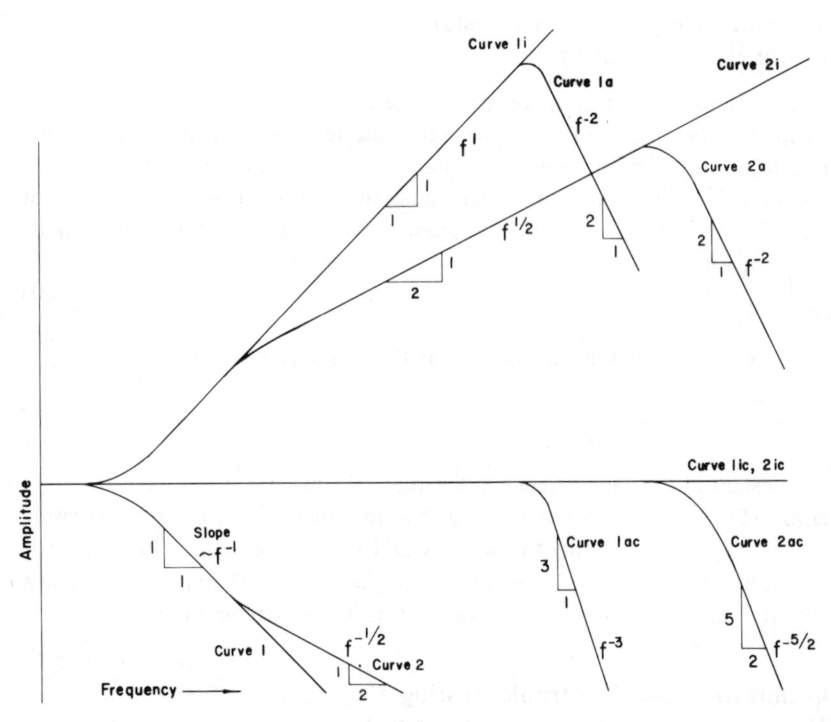

Figure 12 Relative frequency response (logarithmic scale) of hot-wire and cylindrical hot-film sensors. *Lower curves*: Response to velocity fluctuations. *Curve 1*: Hot wire with uncompensated constant-current operation. *Curve 2*: Hot film with uncompensated constant-current operation. *Curves 1ic, 2ic*: Ideal response of both hot wire and hot film with constant-temperature operation. *Curve 1ac*: Actual constant-temperature hot-wire system with optimized third-order response. *Curve 2ac*: Constant-temperature hot-film system with optimized $\frac{5}{2}$ order response. *Upper curves*: Response to sine-wave test on constant-temperature anemometer. *Curve 1i*: Ideal response with hot wire. *Curve 2i*: Ideal response with hot film. *Curve 1a*: Actual hot-wire system with optimized third-order response. *Curve 2a*: Actual hot-film system with optimized $\frac{5}{2}$ order response.

established directly from curve $1a$, so curve $1ac$ need not be established experimentally.

Figure 13 shows experimental data taken using the sine-wave test and a TSI Model 1050 anemometer [47]. It can be seen that the experimental points follow the theory closely, even to an approximately f^{-2} attenuation after the cutoff frequency. For curve 1, f_{cut} is about 98 KHz, while for curve 2, it is about 238 KHz. For the conditions of Fig. 13, M is 3.95×10^{-8} W·s/°C. Putting these values into Eq. (50) gives a value of M''/G of 3×10^{-15} s^2 for curve 1, and 2×10^{-16} s^2 for curve 2.

The response of the sensor without frequency compensation can also be determined from Fig. 13. Since curves 1 and 2 are horizontal at an abscissa of 2.38, the 3-dB point for the wire would be at $2.38 \times \sqrt{2} = 3.37$. The frequency corresponding to this amplitude is about 700 Hz. At the higher velocity (curve 3), this frequency increases to about 1200 Hz.

Although the sine-wave test is very helpful in analyzing a constant-temperature anemometer, it is not convenient for making the required two-parameter optimization. For this, Freymuth [45] has also analyzed the expected output for a step input of current. Repetitive step inputs for U_T are easily provided by a square-wave generator. For the maximally flat case, the output should be a pulse that has undershoot of 13% relative to the maximum. The inserts in Fig. 13 show the appropriate output when the system is properly adjusted. The cutoff frequency is

$$f_{cut} = \frac{1}{1.3t} \tag{52}$$

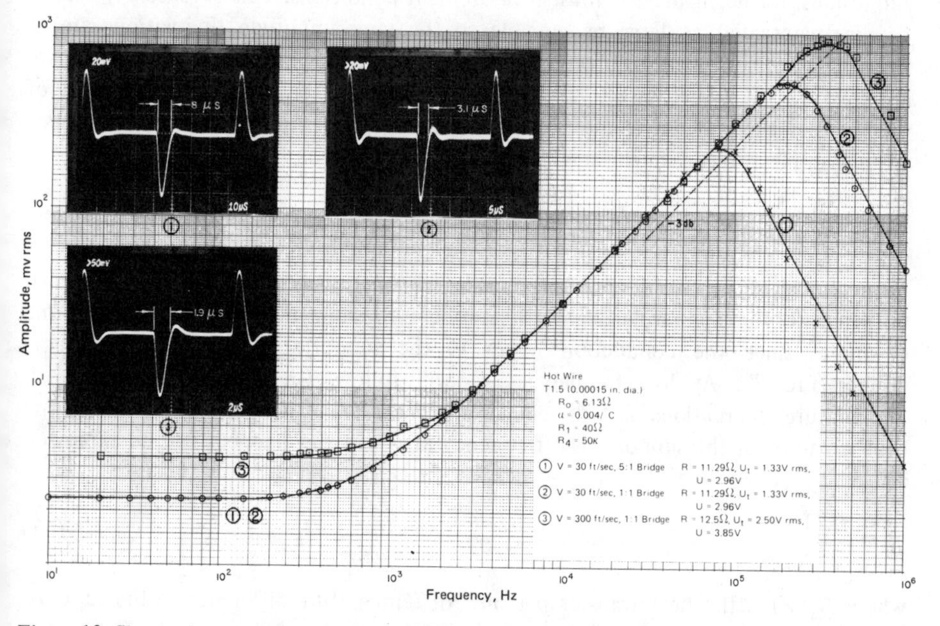

Figure 13 Sine- and square-wave tests on TSI T1.5 hot wire with air flow.

where t is measured from the start of the pulse to where it has decayed to within 3% of its initial value. For curve 1, f_{cut} is calculated to be 96 KHz, and for curve 2 it is 248 KHz. This checks well with the sine-wave results.

The optimization should be done at the mean velocity expected during measurements when the fluctuations are small. When large changes in mean velocity are expected, optimization should be done at maximum velocity. As the velocity decreases, f_{cut} will also decrease, but much more slowly ($f_{cut} \simeq V$) than the decrease in maximum frequency required ($f_{max} \simeq V^{7/4}$) in turbulent flow measurement [48, 49].

The shape of the output pulse should be adjusted carefully, using the controls. Improper adjustment will give a response curve (to velocity) that is not flat over the frequency range.

11.3 Large Velocity Fluctuations

An important characteristic of the constant-temperature anemometer is its ability to respond to large fluctuations. Freymuth [46] has analyzed the dynamic effects of large fluctuations. The results indicate that fluctuations of 50% of the mean value at a frequency of $f_{cut}/2$ will give an error in the mean of less than 0.1%, an error in the mean square of about 2.3%, and an error in the skewness of 0.3. It is this last error that is significant, since this compares with a value for isotropic turbulence of about 0.6. Alternatively, if the frequency is $f_{cut}/10$, then even the skewness error is only about 0.02, and the others are truly negligible.

To summarize, errors due to nonlinearities in the dynamic response at large amplitudes are negligible for most measurement conditions. This is especially true if f_{cut} is maintained as large as possible so the large-amplitude fluctuations are at frequencies of less than 10% of f_{cut}. This is also the range in which phase shifts are less than 12°. It is then safe to say that the "foolproof" or "safe" dynamic range of the thermal anemometer is the range below 10% of its cutoff frequency. Of course, in most applications a much larger range gives good results.

11.4 Dynamic Effects of Conduction Losses to the Supports

Even though T_m is held constant by the feedback electronics, other effects can influence response. Thermal lag of the probe support in nonisothermal flow is one of these, since the conduction losses to the supports depend on the support temperature T_s. At low frequencies T_s will follow T_a, but for high-frequency temperature fluctuations T_s will remain at some average temperature, owing to the heat capacity of the support. The frequency at which this happens can be expressed as

$$f_s = \frac{H_s(V)}{2\pi c_s} \tag{53}$$

where $H_s(V)$ is the heat transfer per unit of temperature difference to the support, and c_s is the heat capacity of the support. At frequencies well below f_s, $T_s = T_a$; at

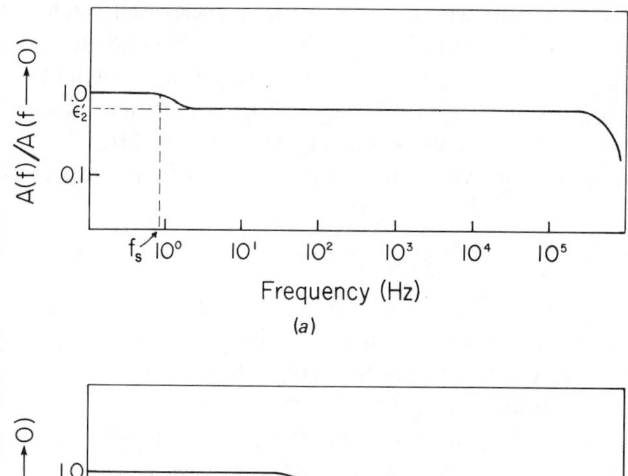

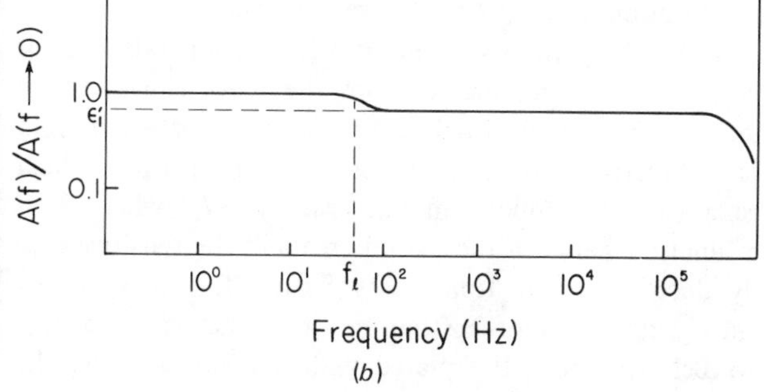

Figure 14 Dynamic effect of conduction loss to supports. (*a*) Effect of temperature changes of supports, Nu = const, T_a varying. (*b*) Effect of heat waves along sensor, T_a = constant, Nu (velocity) varying.

frequencies well above f_s, $T_s = \overline{T}_a$. This phenomenon results in an attenuation of the amplitude response above f_s (Fig. 14*a*). In the case of high-speed flows, T_a would be replaced by the recovery temperature of the support.

Since at high frequencies $dK = 0$, one can write

$$\left.\frac{dH(f \gg f_s)}{dH(f \ll f_s)}\right|_{\mathrm{Nu=const}} = \frac{d\phi + dK = 0}{d\phi + dK} = \epsilon_2' \tag{54}$$

Assuming small overheat, ϵ_2' is given by Eq. (19) and in Fig. 7. Thus, for short wires with high end losses (small $\sqrt{C_0} l$), the dynamic error at high frequencies can be significant when compared with low-frequency or steady-state conditions. For the conditions of Fig. 1, $\epsilon_2' = 0.826$, giving an amplitude error of over 17% (Fig. 7).

Heat waves along the cylinder in isothermal flow can also cause deviation from the ideal response. As velocity changes, the value of $\sqrt{C_0} l$ changes, causing a change in temperature distribution along the wire (Fig. 6). According to [24], the frequency f_l at which this occurs can be estimated as

$$f_l \simeq 6.4 \frac{D}{l^2} \tag{55}$$

where D is the thermal diffusivity of the wire material. At frequencies well below f_l the wire will have time for the temperature distribution to equilibrate. At frequencies well above f_l, the wire will have a temperature distribution represented by some average velocity $(\overline{\sqrt{C_0} l})$. Again, this results in an attenuation of the amplitude response above f_l (Fig. 14*b*). For the sensor of Fig. 1, $f_l \simeq 82$ Hz.

The ratio of the fluctuations at high frequencies to those at low frequencies can be represented as

$$\frac{dH(f \gg f_l)}{dH(f \ll f_l)} = \frac{d\phi + dK = 0}{d\phi + dK}\Bigg|_{T_a, T_m = \text{const}} = \epsilon_1' \tag{56}$$

where ϵ_1' is given by Eqs. (18) and (18*a*) and in Fig. 7. For the wire of Fig. 1, $\epsilon_1' = 0.908$. If the overheat is taken into consideration [Eq. (18)], then $\epsilon_1' = 0.953$. Therefore, the low-overheat assumption gives a conservative estimate.

This calculation assumes a relatively massive support whose temperature T_s is not significantly affected by the heat K conducted from the wire. However, the sensitive portion on most hot-wire and cylindrical hot-film sensors is isolated from the supports by a plated area. Effectively, the support starts at the plating, which can be described as a very small, fragile support. In this case $T_s > T_a$, where T_s is now the temperature of the junction between the plated area and the sensitive area of the sensor. This can only decrease conduction losses K, resulting in ϵ' and ϵ_2' being closer to 1. For ϵ_1' and ϵ_2', it adds another time constant that is between f_l and f_s. Finally, according to Beljaars [50], the plated ends actually decrease the value of ϵ_1' for short sensors, so for this particular parameter large supports, where $T_s \rightarrow T_a$, are desirable. On the other hand, plated ends are very helpful in reducing the aerodynamic influences of the support needles and probe body [28].

11.5 Attenuation of Heat Waves across the Thermal Boundary Layer of the Sensor

The thermal boundary layer surrounding the sensors of medium thickness δ_T readjusts to new flow conditions only if

$$f \leqslant f_\delta \simeq \frac{D_a}{\pi \delta_T^2} \tag{57}$$

where D_a is the thermal diffusivity of the fluid, and where dynamic flow effects have been neglected. For $f > f_\delta$ heat waves will be more and more attenuated, and thus the sensitivity of the sensor decreases. According to Clark [51], an increase in sensitivity may be observed prior to a decrease for bulky sensors in water or blood flow as a consequence of dynamic flow effects.

For thermal boundary layers,

$$\delta_T \simeq \left(\frac{D_a L}{V}\right)^{1/2} \tag{58}$$

where L is a characteristic dimension of the sensor (equal to d for cylindrical sensors). Thus, by combining Eqs. (57) and (58), one arrives at an estimate for f_δ of

$$f_\delta \simeq \frac{V}{2\pi d} \tag{59}$$

For the situation in Fig. 1, $f_\delta = 400 \times 10^3$ Hz, which is above the range of concern for $V = 10$ m/s. On the other hand, for an extremely bulky sensor as used, for instance, in water or blood flow, $d = 1$ mm and $V = 1$ m/s, and one obtains $f_\delta \simeq 160$ Hz. Such a sensor might not be acceptable for some tasks and then would have to be replaced with a smaller one.

A similar attenuation effect can occur for a coated sensor at frequencies

$$f > f_{\text{coat}} \simeq \frac{D_{\text{coat}}}{\pi\delta_{\text{coat}}^2} \tag{60}$$

where D_{coat} = thermal diffusivity of coating material
δ_{coat} = thickness of coating material

This discussion of boundary-layer lag is intended to be introductory. For a detailed study, the reader is referred to the timely thesis by Lueck [52].

11.6 Finite Resolution due to Finite Sensor Size

To accurately follow flow fluctuations, the sensor itself must be small in comparison to the wavelength of the maximum frequencies of interest. This kinematic effect was first considered by Frenkiel [53], and in more detail by Uberoi and Kovasznay [54] and Wyngaard [55, 56]. The results show that for one-dimensional spectra of turbulence, errors on the order of 5% show up for $f_5 \simeq 0.08V/2l$; at $f_{20} \simeq 0.5V/2l$, the error is about 20%. Again, in Fig. 1 the value of f_5 is 640 Hz, and f_{20} is 3200 Hz. This turns out to be one of the most restrictive conditions on the frequency response of thermal sensors.

12. NOISE IN CONSTANT-TEMPERATURE THERMAL ANEMOMETRY

The noise in any measurement system limits the minimum measurable change. This is important in hot-wire anemometry because measurements at low turbulence intensities and wide signal bandwidth are often desired. In addition to the usual increase in noise due to bandwidth, at fixed bandwidth the electronic compensation for the thermal lag of the sensor causes the noise amplitude to increase with center frequency the same as the sine-wave signal of curve 1*a* of Fig. 12.

Noise can be classified as avoidable and unavoidable. Electronic pickup from power lines, radio or television stations, and magnetic fields can usually be eliminated by proper shielding, grounding, etc. However, Johnson noise from the bridge resistors and electronic noise generated by the bridge amplifier cannot be eliminated.

An analysis of the electronic noise in thermal anemometers [57] yields, as the input noise to the amplifier within a small frequency range,

$$u_{12,n}^2 = \left[K_a^2 + 4k_b T_a R \frac{(1 + n_b)(1 + R_2/R_1) + (T_m - T_a)/T_a}{(1 + n_b)^2} \right] \Delta f \tag{61}$$

where K_a = equivalent input noise of amplifier, $\mathrm{V \cdot Hz^{-1/2}}$
k_b = Boltzmann's constant

It is assumed that all resistors are at temperature T_a, except the sensor which is at temperature T_m.

With current amplifier technology, values of K_a on the order of 1.5×10^{-9} $\mathrm{V \cdot Hz^{-1/2}}$ are attainable. The second term of Eq. (61) for the hot wire and bridge of Fig. 1 (with $n_b = 0.36$ and $R_2/R_1 = 1$) is $(0.6 \times 10^{-9}\ \mathrm{V \cdot Hz^{-1/2}})^2$. Therefore,

$$\sqrt{u_{12,n}^2} = 1.69 \times 10^{-9}\ \Delta f^{1/2}\ \mathrm{V}$$

and the sensor and bridge resistors contribute only about 12% of the total noise.

Often R_2/R_1 is increased to reduce the current drain through R_2 and R_3 (Fig. 1). However, if $R_2/R_1 = 20$, the above calculation would give $\sqrt{u_{12,n}^2} = 2.45 \times 10^{-9}$ $\Delta f^{1/2}$ V. The sensor and bridge resistors now contribute over 60% of the noise.

In the range where thermal lag dominates and where the influence of noise is most critical, the output noise of the constant-temperature anemometer is

$$u_N = \frac{\pi(n_b + 1)^2}{n_b} \frac{R}{R - R_0} \frac{c}{H_1(V)} f \sqrt{u_{12,n}^2} \tag{62}$$

The velocity signal is

$$u_v = \frac{n_b + 1}{n_b} [H_1(V)(T_m - T_a)R]^{1/2} \frac{dH_1}{2H_1\, dV} v \tag{63}$$

If we let $H_1(V) = B\sqrt{V}$, then

$$u_v = \frac{n_b + 1}{4n_b} [H_1(T_m - T_a)R]^{1/2} \frac{v}{V} \tag{64}$$

and the signal-to-noise ratio, $N_w = u_v/u_N$, of the wire becomes

$$N_W = \frac{\alpha R_a}{n_b + 1} \frac{B^{3/2} V^{3/4} (T_m - T_a)^{3/2}}{2\pi f c \qquad 2R^{1/2}} \frac{v}{V} \frac{1}{\sqrt{u_{12,n}^2}} \tag{65}$$

Therefore, for a given v/V and equivalent input noise, to maximize the signal-to-noise ratio one should

1. Operate at high overheat to maximize $T_m - T_a$.
2. Use a wire material with a high temperature coefficient of resistance.
3. Keep n_b small compared with 1.
4. Use a thin wire to minimize thermal capacity c.

The most difficult measurements are at high velocities, where high frequencies are required.

To get total output noise, Eq. (62) must be integrated over the frequency range to be observed after substitution for $\sqrt{u_{12,n}^2}$ from Eq. (61). For the wire of Fig. 1, this results in about a 4% noise contribution when measuring 0.1% turbulence with a 0- to 50-KHz bandwidth. At 10 m/s, a bandwidth of 2 KHz is more appropriate. Then less than 0.01% turbulence intensity can be measured with negligible noise contribution. Therefore, it is very important to use a low-pass filter set at the maximum frequency of interest in the flow to minimize the noise contribution, even though the anemometer should be tuned for maximum response.

13. FILM SENSORS

The concept of film sensors was introduced in 1955 [3, 4] and has become a major addition to sensor technology. Film sensors have been particularly useful for measurements in liquids or gases with particle contamination and in conducting liquids, especially water.

As shown in Fig. 15*a*, the hot wire is a homogeneous, electrically conducting material. On cylindrical film sensors (Fig. 15*b*) the substrate is an electrical insulator, with the conducting film deposited on the surface. This permits selection of the substrate for its strength and low thermal conductivity, while the conducting film can be selected for its electrical properties. In addition, since the sensitive portion is on the surface, the frequency response of a film sensor is superior to that of a hot wire of similar dimensions when operated with a constant-temperature anemometer. A thin overlayer of insulating material is often added for electrical insulation from the fluid or to protect the metal film from erosion by particles in the fluid.

The construction of film sensors permits flexibility in shape, as shown in Fig. 15*c*. While the shapes in Fig. 15*c* have advantages in terms of strength and ability to remain clean in contaminated fluids, they do give rise to some special problems in frequency response [58] not shared by the simple cylindrical configuration.

13.1 Cylindrical Film Sensors

Cylindrical film sensors (Fig. 15) share most of their measurement characteristics with hot wires. From the performance viewpoint, the primary advantage of hot-wire sensors is that, with present technology, they can be made much smaller in diameter. A typical hot wire is 4 μm in diameter, while typical film sensors are 50 μm in diameter. Using larger-diameter hot wires results in either excessive length or excessive conduction losses to the supports. The low thermal conductivity of film-sensor substrates (typically less than 1% of that of hot-wire materials) permits relatively short sensors while conduction losses to the supports are maintained at permissible levels.

The larger diameter of film sensors has these advantages for air measurements:

1. In most applications, particles in the fluid will not break and cannot strain the sensor.
2. Smaller particles will not intercept the sensor because of its larger diameter.

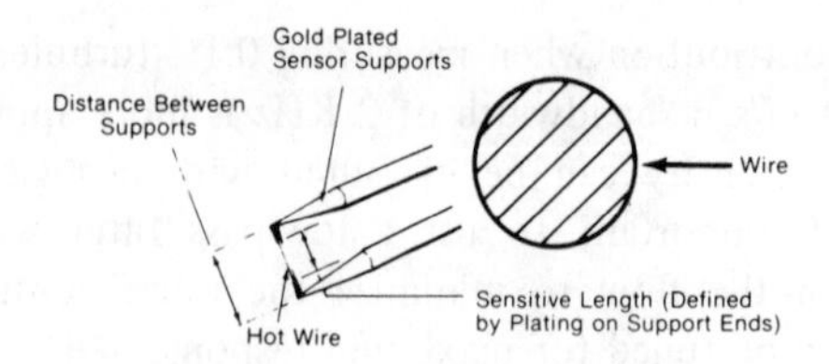

a CROSS SECTION OF HOT-WIRE SENSOR

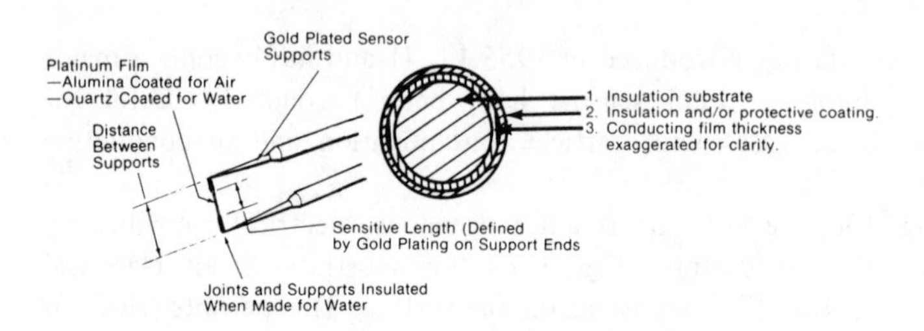

b CROSS SECTION OF HOT-FILM SENSOR

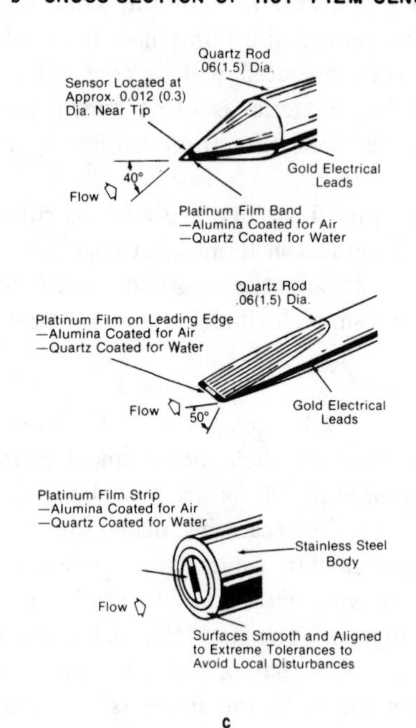

Figure 15 Thermal sensor configuration.

3. Since the sensor is rigid, it is always effectively "taut," creating the possibility of better repeatability on angular sensitivity for X probes and other multisensor measurements.

On the other hand, the primary disadvantages are:

1. At Re > 150 (V > 50 m/s for a 50-μm-diameter sensor in air), self-generated "turbulence intensity" may limit the performance of large-diameter cylindrical sensors for the measurement of low turbulence intensities.
2. The recommended operating temperature of film sensors is below 370°C, with an absolute maximum of 760°C.

Another factor is the generally greater cost of film sensors. However, if particles in the fluid are breaking wires or contamination is causing signal calibration shifts, the longer life of film sensors more than compensates for their higher initial cost.

In tap water, sea water, and other conducting liquids, film sensors are used almost to the exclusion of hot wires for velocity measurements. The ability to isolate the sensor from the fluid with a thin overlayer of insulating material is the primary reason for this. Of course, the problem of self-generated turbulence at high Re remains and will occur at much lower velocities in water than in air. In addition, at sufficiently high velocities, cavitation can occur. Even so, cylindrical films are used extensively; some of the theoretical details are treated in the following.

13.1.1 Construction and operation of cylindrical film sensors. To date, the most common substrate materials for cylindrical films have been pyrex and quartz, because they can be drawn to small diameters. Platinum and nickel are the most common conducting films. Protective layers of pyrex, quartz, and alumina have been used. Alumina is particularly effective in preventing erosion of the sensor, since it is harder than most contaminates in air.

Regarding the length-to-diameter requirements, the conflicting criteria given in Sec. 3 are valid for cylindrical films. The discussions of Sec. 4 and 5 also hold for film sensors, with one exception. Film sensors cannot be mounted on stiff supports if they are to be exposed to shocks or large transients in the flow. The needle supports must have some flex, since the films themselves are quite rigid.

The basic operation of the constant-temperature anemometer is the same for both films and wires. Constant-current compensation has not been used with films, owing to the difficulty of matching the frequency characteristics of films (curve 2 in Fig. 12). Calibration procedures are, of course, identical.

13.1.2 Heat transfer for cylindrical film sensors. The heat-transfer equations cited for hot wires are also valid for cylindrical films. The formula $Gr^{1/3} > Re$ for calculating when free convection is important is not affected by sensor diameter. However the Knudsen number for a 50-μm film, if substituted for the wire in Fig. 1, is 0.0016. This is well into the continuum flow region, whereas the wire was in slip flow. Similarly, the recovery factor for the larger film sensors is closer to 1.0.

One difference from hot wires is that the larger diameter of film sensors and their small thermal conductivity guarantee that the surface temperature of the film sensor is not uniform circumferentially. In fact, the operation of split films [8] depends on this. At the same time, the influence of this phenomenon for normal cylindrical film sensors has not been investigated.

Because of the larger diameters of film sensors, conduction to the walls will influence readings at larger distances from the wall than for fine hot wires.

13.1.3 Conduction to the supports. The equations of Sec. 8 hold for films, with the film substrate properties substituted for the wire properties. The electrically conductive film is so thin that its heat conduction along the sensor is less than 2% of the total for a 50-μm-diameter film sensor. Of course, if very small diameter film sensors were used, this would no longer be true.

Again, substituting a film sensor 50 μm in diameter and 1 mm long in Fig. 1 gives the following comparisons with the hot wire, according to the equations of Sec. 8:

Parameter	Hot wire	Film
$2l/d$	333	20
$\sqrt{C_0}\,l$	2.93	3.48
ϵ	0.826	0.820
ϵ_1'	0.953	0.931
$\epsilon_2'(T_m - T_a \ll 1)$	0.660	0.713

Therefore, the effects of conduction losses to the supports are similar for a tungsten wire and a film sensor whose $2l/d$ ratio is approximately one-sixteenth that of the tungsten wire.

13.1.4 Angle sensitivity and support interference. Work by Friehe and Schwarz [26] and Jorgenson [30] indicates that cylindrical film sensors and hot wires give very similar directional response. This is somewhat surprising, since the $2l/d$ ratio of film sensors is generally lower by a factor of 10 to 20. One advantage of the film is its rigidity, giving the potential for better repeatability of measurements. In any case, the discussion in Sec. 9 is valid for cylindrical film sensors as well as hot wires. The same is true for Sec. 10, with one exception. As is pointed out in the next section, compared to hot wires the hot films make rather slow sensors for the direct measurement of temperature (no frequency compensation), because of their larger diameters.

13.1.5 Dynamic response of cylindrical film sensors. Freymuth [59] has extended the dynamic theory of the constant-temperature hot-wire anemometer to constant-temperature cylindrical hot-film anemometers. The result gives the following value for the cutoff frequency:

$$f_{\text{cut}} = \frac{0.9}{2\pi}\left(\frac{G/M''}{M\omega_c^{1/2}}\right)^{2/5} \tag{66}$$

Above ω_c, the heat transfer of the film is dominated by the skin effect. Its definition is

$$\omega_c = \frac{4D_{su}}{d^2} \tag{67}$$

where D_{su} is the diffusivitiy of the substrate. As a consequence of the influence of ω_c, $f_{\text{cut}} \simeq d^{1/5}$. This implies a cutoff frequency nearly independent of diameter, but one that actually increases as the diameter increases. This is in contrast to the hot wire, for which $f_{\text{cut}} \simeq d^{-1/2}$.

13.1.6 Optimization and electronic testing of cylindrical film sensors. Figure 12 gives a qualitative representation of Freymuth's dynamic theory for both hot-wire and hot-film sensors. Curve 2 is the response of an uncompensated film sensor showing the expected $f^{-1/2}$ response at high frequencies. Curve $2a$ is the response to sine-wave testing when operated with a constant-temperature anemometer, and curve $2ac$ is the response of the same system to velocity changes.

Figure 16 shows experimental data taken via the sine-wave test on a TSI Model 1050 anemometer [47] with a 50-μm-diameter hot-film sensor. Again, the experimental points follow the theory closely. However, both sets of points do deviate from the $f^{1/2}$ line by a few percent. This was predicted by [59]. Although this implies some lack of flatness, even for an optimally adjusted system, the error of a few percent is generally small enough to be negligible in high-frequency measurements. If it is important, the sine-wave test can be used to determine what the error is, and appropriate corrections can be made to the data.

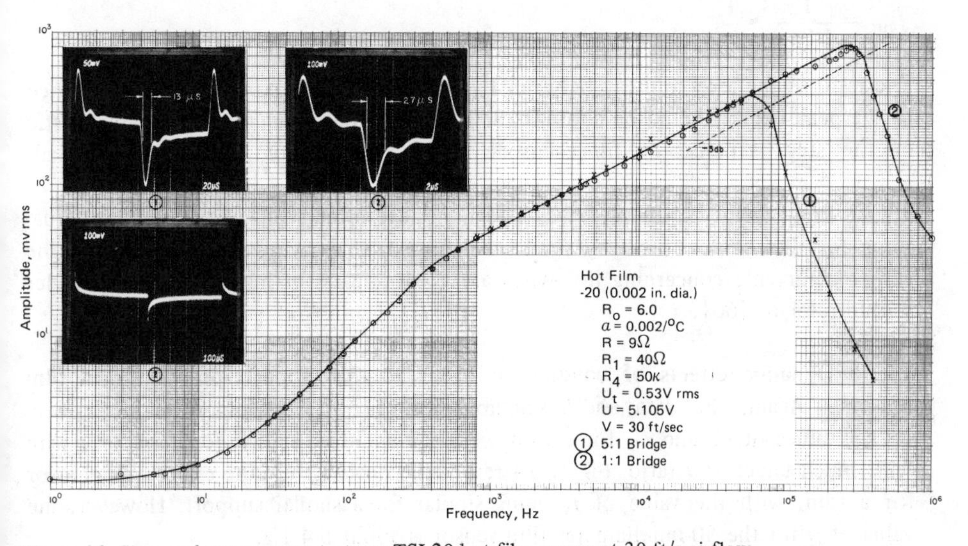

Figure 16 Sine- and square-wave tests on TSI-20 hot-film sensor at 30 ft/s airflow.

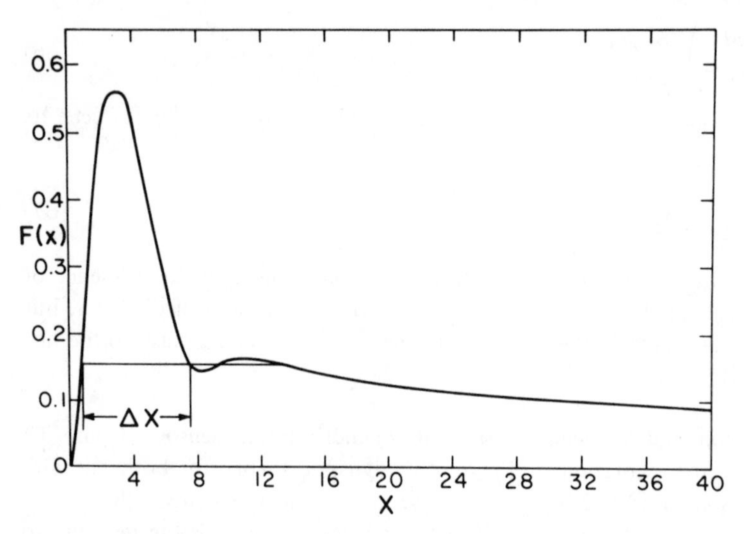

Figure 17 Optimized response to square-wave test for film sensors.

Again, the response of the sensor alone can be taken from Fig. 16; it is found to be about 13 Hz, compared with about 700 Hz for the hot wire. At the same time, f_{cut} for the film is about 77 KHz, which is comparable to the 96 KHz for the hot wire. This shows the influence of the $f^{-1/2}$ region in making the film sensor easier to compensate to high frequencies. The value of f_c from Eq. (67) is 216 Hz. This compares with the approximately 350 Hz from Fig. 16.

Recently, the theory for square-wave testing of film sensors was worked out [59]. Figure 17 shows the optimized output for a step input. With $t\ (= \Delta x)$ measured as shown in Fig. 17,

$$f_{\text{cut}} = \frac{1}{1.04t} \simeq \frac{1}{t}$$

The ratio of the level midway between the minimum and maximum of the resonance to the peak height should be about 0.28. The $1/\sqrt{x}$ "tail" is an inherent characteristic of the step response of film sensors; hot wires do not exhibit such a tail.

13.1.7 Large velocity fluctuations. The nonlinear effects of large velocity fluctuations for film sensors are very similar to those for hot wires [60]. Therefore, the earlier comments concerning hot wires are appropriate; for more details, the reader should refer to [60].

13.1.8 Dynamic effects of conduction losses to the supports of cylindrical film sensors. Again, the data and discussions for hot wires are appropriate. In Sec. 13.1.3 values of ϵ_1' and ϵ_2' for a film were shown to be similar to those for a wire with much larger $2l/d$ ratio. Figure 14 represents qualitatively the expected behavior for a film, with the value of f_s being similar for a similar support. However, the value of f_l for the 50-μm-diameter film sensor is about 5.4 Hz.

For film sensors, the larger diameter (and therefore larger total heat transfer), along with the somewhat flexible supports, makes the assumption that $T_s = T_a$ even more suspect than for wires. In addition, film sensors are isolated from the supports by heavy plating. Therefore the comments at the end of Sec. 11.4 are particularly appropriate for cylindrical film sensors.

13.1.9 Other dynamic effects on cylindrical film sensors. The ability of the thermal boundary layer to respond to new flow conditions is inversely proportional to diameter [Eq. (59)]. Therefore, while the 4-μm hot wire of Fig. 1 would have a boundary-layer response of 400 KHz, for a 50-μm film sensor exposed to the same conditions the response is 32 KHz. This is still more than adequate. However, it does point out that even in gases there may be conditions under which the thermal boundary of a large film sensor would attenuate high frequencies.

Arguments regarding spatial resolution are based on sensor length and are identical for both films and wires.

13.1.10 Noise with cylindrical film sensors. As pointed out in Sec. 12, output noise increases with frequency in the same manner as sinusoidal electronic test signals. From Fig. 12 it can be observed that the rate of increase of noise with a film sensor will be less than with a hot wire at high frequencies, owing to the $f^{-1/2}$ response region. At the same time, the large diameter of the film sensor provides significant noise amplification between 10 and 10^3 Hz. Figure 18 shows a comparison of the signal-to-noise ratio for a 4-μm-diameter wire and a 50-μm-diameter film exposed to 30 ft/s in air. The larger noise figure for the film at about 1 KHz is not a problem for almost all practical applications of film sensors.

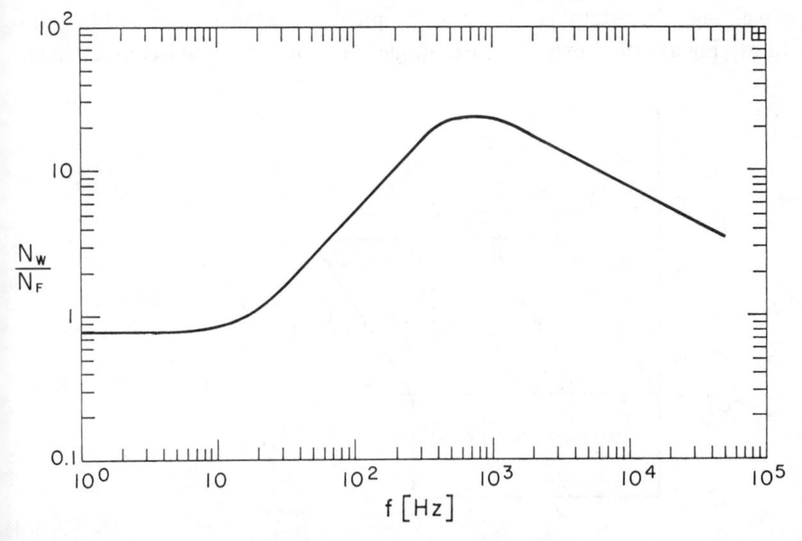

Figure 18 Frequency dependence of the signal-to-noise ratio of a 4-μm-diameter hot wire compared with that of a 50-μm-diameter film sensor for 30 ft/s air velocity. *(From [62].)*

13.2 Noncylindrical Film Sensors

Figure 15*c* shows three configurations of noncylindrical film sensors. The cone-shaped sensor is the most frequently used, with the flush mount often used to examine boundary layers and shear stress. The shapes of these sensors avoid the self-generated turbulence that causes a problem with the large cylindrical sensors. In addition, they can be used in high-speed water flows with no cavitation problems. Finally, their configuration also minimizes problems due to contaminants in the flow.

The major difficulty with noncylindrical sensors has been proper interpretation of the amplitude response as a function of frequency. At the time they were originally suggested, the dynamic effects of conduction losses were not considered. As was pointed out earlier, even with cylindrical sensors these effects can be on the order of 10% in air. The design of noncylindrical films substantially increases these effects in gases. Fortunately, the greatest need for noncylindrical sensors is in liquids, where the high convective heat transfer reduces the dynamic effect of conduction losses to the order of 10%.

Since the dynamic effect of conduction losses is the most important problem with noncylindrical sensors, this is discussed first. Then some general comparisons with other sensors are made.

13.2.1 Conduction losses on noncylindrical sensors. For cylindrical hot wires and hot films, end losses are an important part of the heat transfer. For noncylindrical hot-film sensors, heat transfer into areas not covered by the film plays a similar role and is appropriately called "side losses." An estimation of such losses is complicated by the three-dimensional nature of heat transfer through the substrate. A simple, empirical one-dimensional model devised by Bellhouse and Schultz [58] greatly helps in explaining the static and dynamic phenomena associated with noncylindrical films. Figure 19 shows the basic model, in which the fluid convects heat

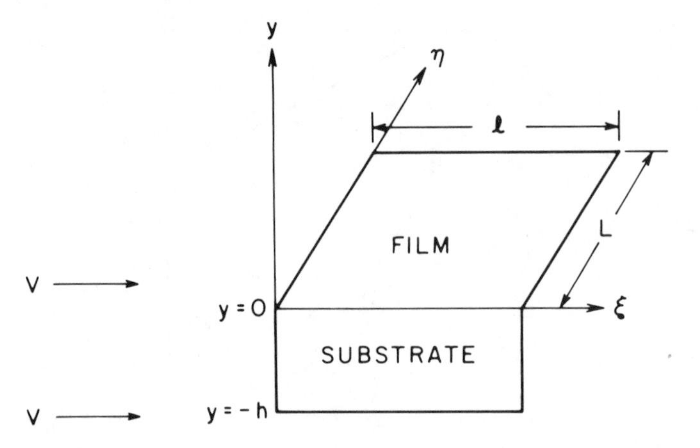

Figure 19 Heat loss from a noncylindrical hot-film sensor according to the Bellhouse-Schultz model.

from the surface of a film at constant temperature T_m and heat is conducted one-dimensionally across the substrate to the lower substrate surface not covered by the film. From this lower surface heat is again convected to the liquid, resulting in an equilibrium temperature T_{su} at the surface. The thickness of the substrate is h, and the film thickness is negligibly small.

The heat transfer from the electrically heated surface via convection is ϕ, and conduction K across the substrate is

$$K = k_{su} A_{su} \frac{\partial T}{\partial y} = k_{su} A_{su} \frac{T_m - T_{su}}{h} \tag{68}$$

where A_{su} = surface area of film
k_{su} = heat conductivity of substrate
With $\phi \simeq T - T_a$, it follows that, for the lower surface,

$$K = k_{su} A_{su} \frac{T_m - T_{su}}{h} = \phi \frac{T_{su} - T_a}{T_m - T_a} = \phi \left(1 - \frac{T_m - T_{su}}{T_m - T_a}\right) \tag{69}$$

Further evaluation of Eq. (69) yields

$$K = \phi \frac{1}{1 + x} \tag{70}$$

where the dimensionless quantity x is the Biot number defined by

$$x = \frac{h\phi}{k_{su} A_{su}(T_m - T_a)} \tag{71}$$

Introducing ϵ as previously, we have

$$\epsilon = \frac{\phi}{\phi + K} = \frac{1 + x}{2 + x} \tag{72}$$

Furthermore, for velocity fluctuations in isothermal flow,

$$\epsilon' \Big|_{T_a, T_m \text{ const}} = \frac{d\phi}{d\phi + dK} \Bigg|_{T_a, T_m = \text{const}} = \epsilon_1' = \frac{1 + 2x + x^2}{2 + 2x + x^2} \tag{73}$$

and for temperature fluctuations of T_a and constant velocity,

$$\epsilon' \Big|_{T_m, x = \text{const}} = \epsilon_2' = \frac{1 + x}{2 + x} = \epsilon \tag{74}$$

ϵ_1' and ϵ_2' for the Bellhouse-Schultz model are both represented in Fig. 20 as a function of the dimensionless thickness x of the substrate. These results take on added importance when the dynamics of noncylindrical film sensors are considered.

13.2.2 Dynamic effects of conduction losses on noncylindrical film sensors according to the Bellhouse-Schultz model. For velocity fluctuations in isothermal flow, conduction fluctuations across the substrate are attenuated at high frequencies such that

$$\frac{dH(\omega \gg \omega_h)}{dH(\omega \ll \omega_h)} = \frac{d\phi + dK = 0}{d\phi + dK} \Bigg|_{T_a, T_m = \text{const}} = \epsilon_1' = \frac{1 + 2x + x^2}{2 + 2x + x^2} \tag{75}$$

where ϵ_1' was obtained from Eq. (73) and the transition frequency ω_h is estimated as

$$\omega_h = \frac{2D_{su}}{h^2} \tag{76}$$

D_{su} is the thermal diffusivity of the substrate. For ambient temperature fluctuations dT_a, on the other hand, with velocity kept constant, we have, by the same reasoning, from Eq. (74),

$$\left.\frac{dH(\omega \gg \omega_h)}{dH(\omega \ll \omega_h)}\right|_{T_m, x = \text{const}} = \epsilon_2' = \frac{1 + x}{2 + x} \tag{77}$$

ϵ_1' and ϵ_2' are represented in Fig. 20. Equations (75) and (76) show that anemometer dynamic attenuation to velocity fluctuations differs from dynamic attenuation to temperature fluctuations such that

$$\epsilon_1' = \frac{(\epsilon_2')^2}{2(\epsilon_2')^2 - 2\epsilon_2' + 1} \tag{78}$$

with ϵ_1' and ϵ_2' varying between 0.5 and 1. Bellhouse and Schultz [58] were the first to show how static calibration leads to errors in dynamic measurements, especially in air where x is small, and Comte-Bellot [62] realized that attenuation differs for velocity and temperature fluctuations.

A difficulty with the Bellhouse-Schultz model is that the Biot number x of the model has not yet been related to sensible geometric parameters of actual sensors. Furthermore, the maximum attenuations ϵ_1' and ϵ_2' are 0.5 ($x = 0$), whereas for actual sensors attenuation may be to even lower levels (in this case the effective lower surface area would have to be modeled larger than the upper surface area of the Bellhouse-Schultz model). The Bellhouse-Schultz model nevertheless remains the theoretical mainstay in the assessment of noncylindrical sensor performance, and it can be generalized to overcome shortcomings. (Note the similarity of "end" and "side" loss effects by comparing Figs. 7 and 20.)

The use of noncylindrical films for research measurements in gases has been limited by difficulties in interpreting transient data. Although wires and cylindrical films also have conduction losses, they are small (less than 10%), can be estimated quite accurately, and can be made smaller by increasing the value of $2l/d$. For noncylindrical sensors in air, these effects are often 50% or more and are difficult to estimate.

The ideal solution is a support material whose thermal conductivity approaches zero. Thin-film gauges utilizing plastic have been tried [63]. Certainly, the construction of noncylindrical sensors opens the possibility of a wide range of materials. This flexibility is, of course, limited by the need for very high stability of the conductive film for good measurements.

13.2.3 Electronic testing of noncylindrical film sensors. Figure 21 shows experimental data from a sine-wave test on a conical sensor. As with the noncylindrical film, the features of the response at high frequencies can be determined by how

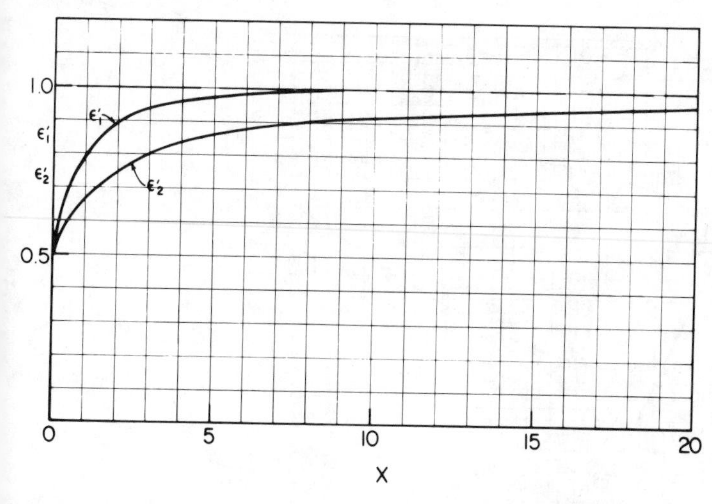

Figure 20 Heat-loss fluctuation ratios ϵ_1' and $\epsilon_2' = \epsilon$ in isothermal flow and in nonisothermal constant-velocity flow as functions of the Biot number x of the noncylindrical hot film.

well the sine-wave results follow the $f^{1/2}$ line. As can be observed on Fig. 21, they follow quite well from 200 Hz to 100 KHz, or range of about 500:1. It is below 200 Hz that attenuation due to "side losses" occurs.

It is this region below 200 Hz that is influenced by the "side losses" of the noncylindrical film. In other words, to completely interpret this part of the curve would require a detailed calculation of the transient heat transfer in the substrate.

The value of f_{cut} for the cone is about 122 KHz, compared with 77 KHz for the 50-μm-diameter cylindrical film. This is at least partial confirmation of the rather surprising prediction that f_{cut} increases with diameter ($\sim d^{1/5}$). The 3-dB point of the cone sensor by itself is, from Fig. 21, about 2.5 Hz in air at 30 ft/s.

A first theoretical attempt at sine-wave testing according to the Bellhouse-Schultz model is that of Freymuth [64]. Whether sine-wave testing can become a quantitative tool for assessing the dynamic side-loss as well as end-loss effects of heat conduction for noncylindrical films and short hot wires remains to be seen. Up to now such effects can only be assessed quantitatively by "direct calibration," in which the probe is exposed to appropriate velocity or temperature fluctuations and its frequency response is directly measured. Relevant methods of direct calibration are probe shaking [65], probe rotation of a slanted wire such that there is an oscillation of angle of attack [66], exposure to a pulsating flow [67], and exposure to a turbulent flow with known spectrum [68], originally proposed by Bellhouse and Schultz [58].

13.2.4 General comments on noncylindrical film sensors. While the flexibility in shape of noncylindrical sensors can be very promising, applications have been limited because of difficulties in interpreting transient and, in some cases, even steady-state signals. Therefore the volume of data available on noncylindrical sensors is very small compared with that for cylindrical configurations.

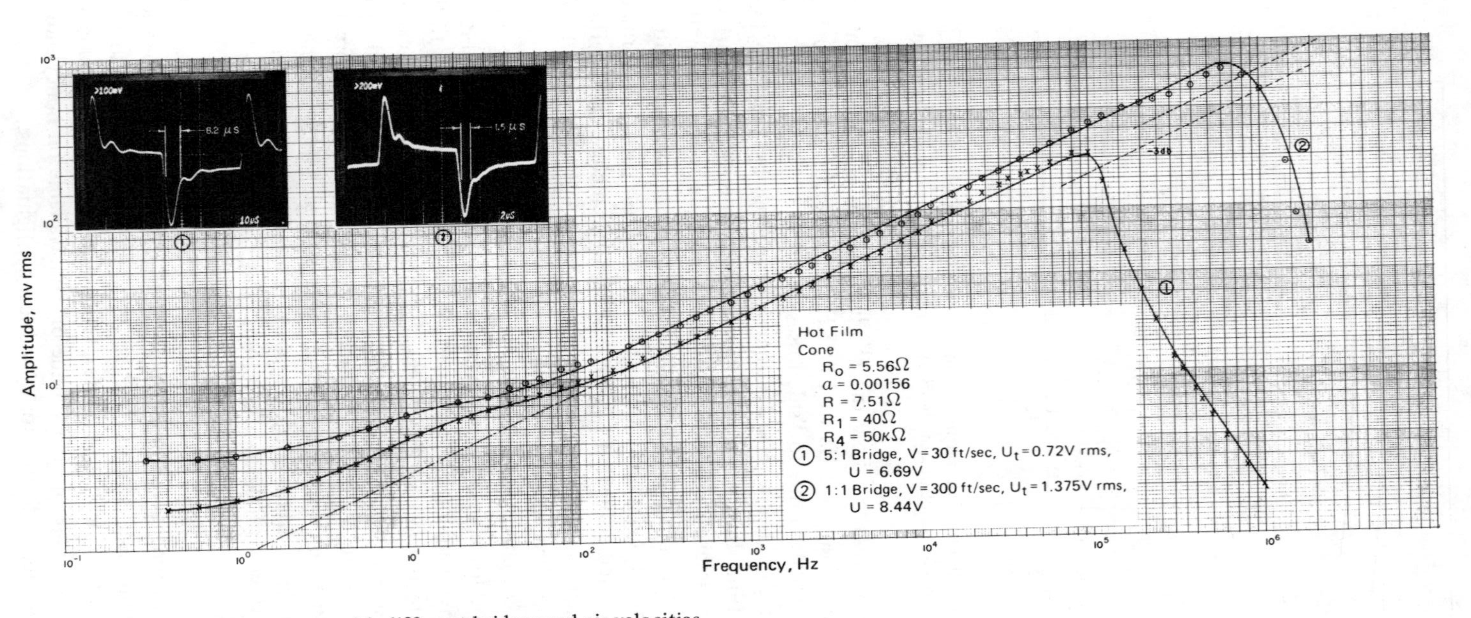

Figure 21 Tests on a cone sensor with different bridges and air velocities.

For steady-state measurements, calibrations are performed. However, the calibration cannot be extended reliably using heat-transfer relations because of inadequate data as well as, perhaps, the lack of repeatability between sensors. Similarly, the angle sensitivities of noncylindrical sensors has not been studied extensively. Ling [4] took some data on wedge sensors, and Frey [69] has some data on cones. The cone sensor is relatively insensitive to changes in flow direction.

14. CONSTANT–CURRENT OPERATION

In constant-temperature operation there must always be a flow of heat from the sensor to its environment. With the exception of the cooled-film sensor, this means that the sensor temperature must always be above the environment temperature. If the environment temperature is changing, then the maximum environment temperature determines the minimum sensor temperature.

The constant-current anemometer does not have this limitation. Therefore, in high-speed flows where several overheats are used to separate the modes, the constant-current anemometer is still often used. At very low heating currents the compensated constant-current hot-wire anemometer essentially becomes a fast thermometer.

As pointed out by Freymuth [57], the noise is identical in constant-current and constant-temperature systems if a similar amplifier and bridge are used. Since noise increases as overheat decreases [Eq. (65)], there may be a minimum overheat at which an adequate signal-to-noise ratio is obtained. If this overheat is at the point where $T_m > T_a$ at all times, then the advantage of the constant-current system would seem to vanish. Then the ability to measure large fluctuations, ease of optimizing response, ability to compensate films for frequency, and even the simplification of the equations [21] would favor constant-temperature operation.

15. OTHER MEASUREMENT TECHNIQUES AND APPLICATIONS USING THE CONSTANT-TEMPERATURE ANEMOMETER

The usual application of the thermal anemometer is to the measurement of the details of flow velocity and direction. However, the unusual sensitivity, small size, and good response have led to the adaptation of the anemometer for other measurements.

15.1 Aspirating Probe

If the flow past a thermal sensor is controlled by a sonic nozzle, then velocity past the sensor depends only on gas composition, pressure, and molecular weight. In other words, the effect of environment velocity has been removed. Velocity will influence pressure if the probe is pointed directly upstream.

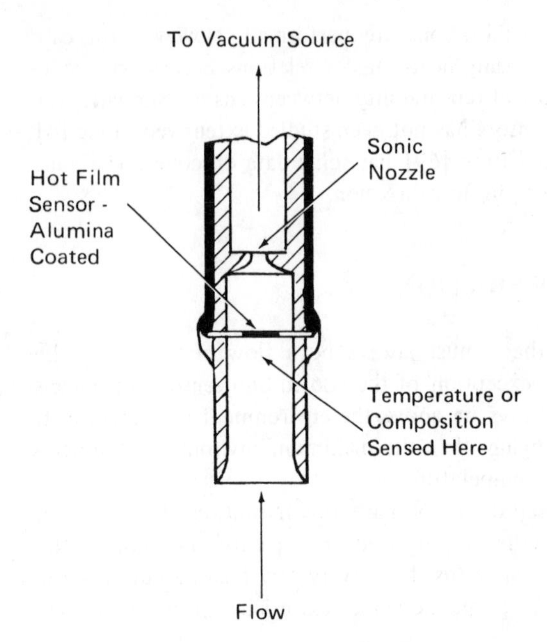

Figure 22 Aspirating probe.

Figure 22 is a sketch of an aspirating probe with the sensor in front of the sonic nozzle. Blackshear and Fingerson [70] used a probe of this type for concentration measurements in a helium jet mixing with air. Brown and Rebollo [71] used a smaller probe with the sensor downstream of the sonic nozzle for measuring binary mixtures. Recently, a similar probe has been used to detect methane in spill experiments [72]. In constant-composition flows, the same probe configuration can be used for temperature measurements.

15.2 Pressure Measurements

Pressure is another flow variable of interest. Figure 23 shows a configuration that has been used to measure pressure fluctuations. Spencer and Jones [73], Planchon [74], and Jones et al. [75] used a probe of this type to measure static-pressure fluctuations after some modifications to the tip. Remenyik and Kovasznay [76] and Wills [77] used similar systems to measure wall-pressure fluctuations using hot wires.

15.3 Total Flow

Any instrument designed to measure velocities at a point can measure total flow if it is in a nozzle. Figure 24 shows an example configuration. The justification of using thermal sensors is generally their speed of response or their wide velocity range, combined with the fact that they are primarily mass-flow sensors. In clean gases, their measurements are also highly repeatable and, for example, they are used

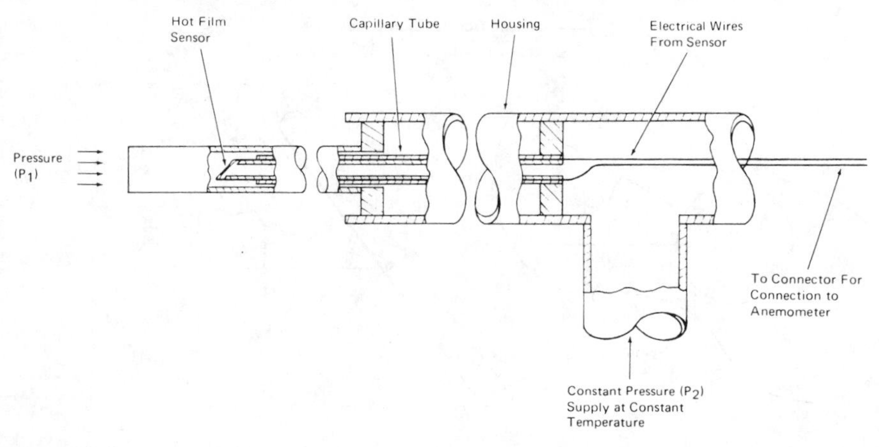

Figure 23 Pressure probe.

in production test stands to check carburetor orifice flows. The ability to sense very low flows has resulted in their use in leak detection.

15.4 Split-Film Sensors

Figure 25 shows a split-film sensor and the associated concept. These devices have found application in sensing angle of attack [78] as well as in some turbulence measurements [79 to 81]. As pointed out by Spencer and Jones [81] and Young [68], the normal 300-μm-diameter split films suffer from attenuation at higher frequencies. Spencer and Jones [81] did not find this to be true for a 50-μm split film.

16. CONCLUSION

The constant-temperature hot-wire anemometer satisfies most of the criteria for an ideal measuring tool. Its primary disadvantages are practical ones such as fragility

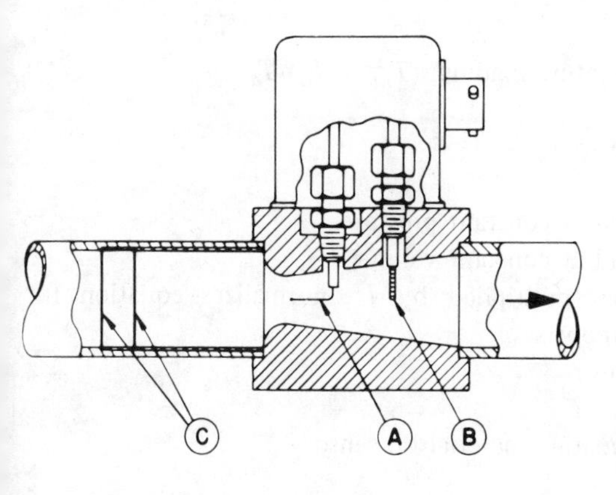

Figure 24 Mass-flow transducer. Transducer Venturi section showing flow element (A), temperature compensator (B), and inlet screens (C) in place.

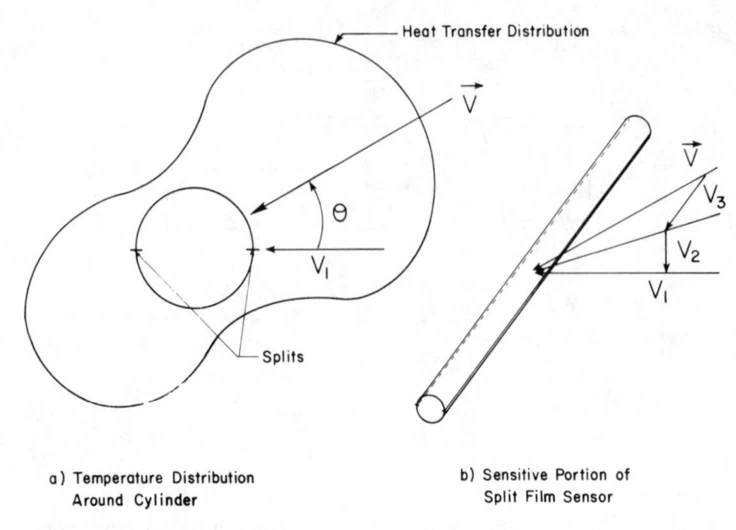

Figure 25 Split-film cylindrical sensor.

and calibration shifts due to contamination. In addition, to maintain accuracy all heat-transfer effects must be considered.

The cylindrical hot-film sensor overcomes some of the practical problems of the fine hot-wire sensor and is quite well understood. In fact, its operation is basically identical to that of a hot wire in most respects.

The noncylindrical hot-film sensor can further reduce the fragility and contamination effects, but at present the interpretation of the dynamic response is difficult–especially in gases. However, noncylindrical film sensors such as surface sensors permit measurements not readily available using hot wires.

NOMENCLATURE

A	constant
A_{su}	surface area
a	radius of sensor
a_T	overheat ratio or temperature loading, $(T_m - T_a)/T_a$
B	constant
b	distance from sensor to wall
C	constant
C_p	specific heat of the fluid at constant pressure
C_v	specific heat of the fluid at constant volume
C_0	parameter which, when multiplied by l^2, normalizes equation for conduction losses to supports
c	thermal capacity of sensor
D	thermal diffusivity
D_{coat}	thermal diffusivity of coating material on sensor

d	sensor diameter
f	frequency (Hz)
f_{coat}	frequency where attenuation due to a coating on the sensor becomes significant
f_{cut}	frequency where amplitude has decreased 3 dB (cutoff frequency)
f_l	transition frequency where longitudinal temperature distribution is no longer in equilibrium with the environment
f_s	transition frequency where sensor supports (prongs) no longer follow changes in environment temperature
f_δ	transition frequency where thermal boundary layer around sensor is no longer in equilibrium
f_5	frequency where spatial-resolution limits cause turbulence measurement error of 5%
f_{20}	frequency where spatial-resolution limits cause turbulence measurement error of 20%
G	amplifier gain
Gr	Grashof number, $\rho^2 g_c \beta(T_m - T_a)\, d^3/\mu^2$
g	gravitational constant
H	heat transfer from sensor to environment
h	substrate thickness in theoretical model for noncylindrical film sensors
h_c	convective heat-transfer coefficient
h_t	total pressure in height of fluid flowing
I	electric current in sensor
K	heat conducted to sensor supports
K_a	equivalent input noise of amplifier
Kn	Knudsen number, λ/d
k_T	factor for cosine-law deviation of flow tangential to sensor
k_N	factor for deviation of measured velocity with flow normal to supports (prongs)
k_b	Boltzmann's constant
k	thermal conductivity
L	characteristic length (equals d for cylindrical sensors)
l	half length of sensor
M	time constant of sensor
M''	second-order time constant of bridge-amplifier system
Ma	Mach number
Nu	Nusselt number, $h_c d/k$
N_w	signal-to-noise ratio
n	exponent of Reynolds number (velocity) in heat-transfer correlations
n_b	ratio of sensor resistance R to resistance R_1 in series with sensor when bridge is in balance
P	electric power to sensor
Pr	Prandtl number, $\mu C_p/k$
R_0	gas constant
R	sensor resistance at temperature T_m

R_a	sensor resistance at temperature T_a
R_1	resistor in series with sensor
R_2	bridge resistor (Fig. 1 or 10)
R_r	resistance of sensor at reference temperature T_r
Re	Reynolds number, $V\rho d/\mu$
Re_0	Reynolds number for heat-transfer correlation in supersonic flow, $\rho_\infty V_\infty d/\mu_0$
T_a	ambient temperature of fluid
T_f	film temperature, $(T_m + T_a)/2$
T_m	mean temperature of sensitive area of sensor
T_0	total temperature of fluid
T_{re}	recovery temperature
T_r	reference temperature
T_s	support temperature
T_{st}	static temperature of fluid
T_{su}	temperature of lower surface in theoretical model of noncylindrical film sensor
t	time
U	output voltage of constant-temperature anemometer
U_B	voltage impressed on bridge of anemometer (equals U for constant-temperature anemometer)
U_{12}	bridge off-balance voltage (equals amplifier input voltage)
U_T	test signal
u_v	output voltage due to velocity changes
u_n	output voltage due to noise
$u_{12,n}$	voltage due to noise at amplifier input
V	velocity vector at sensor location
V_1, V_2, V_3	orthogonal components of V relative to flow facility
V_{eff}	effective cooling velocity past sensor (equivalent value of V_N)
$V_{A,\mathrm{eff}}$	effective velocity as seen by sensors (and similarly for sensors B and C)
V_N	velocity vector normal to sensor and parallel to supports
V_{BN}	velocity vector normal to sensor and perpendicular to supports
V_T	velocity vector tangent to sensor axis
v	small fluctuations in velocity V
x	Biot number for theoretical model of noncylindrical films
α	temperature coefficient of resistance of sensor
α_1	angle between velocity vector and sensor axis
β_v	volume coefficient of expansion, $1/T_a$
γ_h	ratio of specific heats, C_p/C_v
δ_{coat}	thickness of coating material on sensor
δ_T	thermal boundary-layer thickness
ϵ	ratio of convective to total heat transfer from sensor to surroundings
ϵ_1'	change in ϵ due to changes in Nu with T_a constant
ϵ_2'	change in ϵ due to changes in T_a with Nu constant

η	recovery factor in high-speed flows
λ	molecular mean free path
μ	dynamic viscosity of fluid
ξ	$\sqrt{C_0}l \coth \sqrt{C_0}l$
ρ	density of fluid
ϕ	convective heat transfer between sensor and surrounding fluid
ω	frequency, $2\pi f$ (rad/s)
ω_c	transition frequency where skin effects start to dominate on cylindrical film sensors
ω_h	transition frequency where skin effects start to dominate on noncylindrical film sensors (theoretical model)

Subscripts

a	conditions at ambient temperature T_a
f	conditions at film temperature T_f
s	sensor support
su	substrate
w	wire
0	stagnation conditions in high-speed flow
∞	free-stream conditions in high-speed flow

REFERENCES

1. P. Freymuth, A Bibliography of Thermal Anemometry, *TSI Q.*, vol. 4, p. 2, 1978.
2. L. V. King, On the Convection of Heat from Small Cylinders in a Stream of Fluid: Determination of the Convection Constants of Small Platinum Wires, with Applications to Hot-Wire Anemometry, *Proc. R. Soc. London,* vol. 90, pp. 563–570, 1914.
3. H. H. Lowell and N. Patton, Response of Homogeneous and Two-Material Laminated Cylinders to Sinusoidal Environmental Temperature Change, with Applications to Hot-Wire Anemometry and Thermocouple Pyrometry, NACA TN 3514, 1955.
4. S. C. Ling, Measurement of Flow Characteristics by the Hot-Film Technique, Ph.D. thesis, State University of Iowa, Iowa City, 1955.
5. L. M. Fingerson and P. L. Blackshear, Heat Flux Probe for Dynamic Instantaneous Temperature and Three Velocity Components in Turbulent Flow, *Rev. Sci. Instrum.,* vol. 49, pp. 654–664, 1962.
6. G. Comte-Bellot, Hot-Wire Anemometry, *Annu. Rev. Fluid Mech.,* vol. 8, pp. 209–231, 1976.
7. Y. Yeh and H. Cummins, Localized Fluid Flow Measurement with a He Ne Laser Spectrometer, *Appl. Phys. Lett.,* vol. 4, p. 176, 1964.
8. J. G. Olin and R. B. Kiland, Split-Film Anemometer Sensors for Three-Dimensional Velocity-Vector Measurement, *Proc. Symp. on Aircraft Wake Turbulence,* pp. 57–79, Seattle, 1971.
9. H. H. Lowell, Design and Application of Hot-Wire Anemometers for Steady-State Measurements at Transonic and Supersonic Air Speeds, NACA TN 2117, 1950.
10. V. A. Sandborn, *Resistance Temperature Transducers,* Metrology Press, Fort Collins, Colo., 1972.
11. W. G. Rose, Some Corrections to the Linearized Response of a Constant-Temperature

Hot-Wire Anemometer Operated in a Low-Speed Flow, *Trans. ASME J. Appl. Mech.*, vol. 29, pp. 554–558, 1962.
12. S. P. Parthasarathy and D. J. Tritton, Impossibility of Linearizing Hot-Wire Anemometer for Turbulent Flows, *AIAA J.*, vol. 1, pp. 1210–1211, 1963.
13. D. C. Collis and M. J. Williams, Two-Dimensional Convection from Heated Wires at Low Reynolds Numbers, *J. Fluid Mech.*, vol. 6, pp. 357–389, 1959.
14. C. F. Dewey, A Correlation of Convective Heat Transfer and Recovery Temperature Data for Cylinders in Compressible Flow, *Int. J. Heat Mass Transfer*, vol. 8, pp. 245–252, 1965.
15. F. H. Champagne, *Turbulence Measurements with Inclined Hot-Wires*, Boeing Scientific Research Laboratories, Flight Science Laboratory, Rep. 103, 1965.
16. H. Kramers, Heat Transfer from Spheres to Flowing Media, *Physics*, vol. 12, pp. 61–80, 1946.
17. L. S. G. Kovasznay and S. I. A. Toernmark, *Heat Loss of Wires in Supersonic Flow*, Bumblebee Ser. Rep. 127, 1950.
18. L. V. Baldwin, V. A. Sandborn, and J. C. Lawrence, Heat Transfer from Transverse and Yawed Cylinders in Continuum, Slip and Free Molecular Air Flows, *Trans. ASME, Ser. C, J. Heat Transfer*, vol. 82, pp. 77–78, 1960.
19. M. V. Morkovin, Fluctuations and Hot-Wire Anemometry in Compressible Flows, AGARDO Graph 24, 1956.
20. M. V. Morkovin and R. E. Phinney, Extended Application of Hot-Wire Anemometry to High-Speed Turbulent Boundary Layers, AFOSR TN 58-469, Johns Hopkins University, Department of Aeronautics, 1958.
21. C. L. Ko, D. K. McLaughlin, and T. R. Troutt, Supersonic Hot-Wire Fluctuation Data Analysis with a Conduction End-Loss Correction, *J. Phys. E.*, vol. 11, pp. 488–494, 1978.
22. J. O. Hintz, *Turbulence*, 2d ed., McGraw-Hill, New York, 1975.
23. J. A. B. Wills, The Correction of Hot-Wire Readings for Proximity to a Solid Boundary, *J. Fluid Mech.*, vol. 12, pp. 388–396, 1962.
24. P. Freymuth, Engineering Estimates of Heat Conduction Loss in Constant Temperature Thermal Sensors, *TSI Q.*, vol. 5, pp. 3–8, 1979.
25. G. Comte-Bellot, A. Strohl, and E. Alcaraz, On Aerodynamic Disturbances Caused by Single Hot-Wire Probes, *Trans. ASME J. Appl. Mech.*, vol. 38, pp. 767–774, 1971.
26. C. H. Friehe and W. H. Schwarz, Deviations from the Cosine Law for Yawed Cylindrical Anemometer Sensors, *Trans. ASME J. Appl. Mech.*, vol. 35, pp. 655–662, 1968.
27. J. C. Bennet, Measurement of Periodic Flow in Rotating Machinery, *AIAA 10th Fluid and Plasmadynamic Conf.*, pp. 770–713, 1977.
28. R. E. Drubka, J. Tan-atichat and H. M. Nagib, On Temperature and Yaw Dependence of Hot-Wires, IIT Fluids and Heat Transfer Rep. R77-1, Illinois Institute of Technology, Chicago, 1977.
29. A. Strohl and G. Comte-Bellot, Aerodynamic Effects Due to Configuration of X-Wire Anemometers, *Trans. ASME J. Appl. Mech.*, vol. 40, pp. 661–666, 1973.
30. F. E. Jorgensen, Directional Sensitivity of Wire and Fiber Film Probes, DISA Inf. 11, pp. 31–37, 1971.
31. F. E. Jerome, D. E. Guitton, and R. P. Patel, Experimental Study of the Thermal Wake Interference between Closely Spaced Wires of an X-Type Hot-Wire Probe, *Aero. Q.*, vol. 22, pp. 119–126, 1971.
32. G. Fabris, Probe and Method for Simultaneous Measurement of "True" Instantaneous Temperature and Three Velocity Components in Turbulent Flow, *Rev. Sci. Instrum.*, vol. 49, pp. 654–664, 1978.
33. N. K. Tutu and R. Chevray, Cross-Wire Anemometry in High Intensity Turbulence, *J. Fluid Mech.*, vol. 71, pp. 785–800, 1975.
34. L. M. Fingerson, Practical Extensions of Anemometer Techniques, in W. L. Melnik and J. R. Weske (eds.), *Advances in Hot-Wire Anemometry*, University of Maryland, College Park, pp. 258–275, 1968.
35. H. W. Tielmann, K. P. Fewell, and H. L. Wood, An Evaluation of the Three-Dimensional Split

Film Anemometer for Measurements of Atmospheric Turbulence, College of Engineering, VPI-E-73-9, Virginia Polytechnic Institute and State University, Blacksburg, 1973.

36. H. Fujita and L. S. G. Kovasznay, Measurement of Reynolds Stress by a Single Rotated Hot-Wire Anemometer, *Rev. Sci. Instrum.*, vol. 39, pp. 1351–1355, 1968.
37. L. R. Bissonnette and S. L. Mellor, Experiments on the Behaviour of an Axisymetric Turbulent Boundary Layer with a Sudden Circumferential Strain, *J. Fluid Mech.*, vol. 63, pt. 2, 1974.
38. De Grande, Three-Dimensional Incompressible Turbulent Boundary Layers, Ph.D. thesis, Vrye Universiteit, Brussels, 1977.
39. P. Kool, Determination of the Reynolds-Stress Sensor with a Single Slanted Hot-Wire in Periodically Unsteady Turbomachinery Flow, ASME Pub. 79-GT-130, 1979.
40. W. Rodi, A New Method of Analyzing Hot-Wire Signals in Highly Turbulent Flow and Its Evaluation in a Round Jet, DISA Inf. 17, pp. 9–18, 1975.
41. M. Acrivlellis, An Improved Method for Determining the Flow Field of Multi-Dimensional Flows of Any Turbulence Intensity, DISA Inf. 23, pp. 11–16, 1978.
42. P. Freymuth, Hot-Wire Anemometer Thermal Calibration Errors, *Instrum. Control Syst.*, vol. 43, no. 10, pp. 82–83, 1970.
43. A. A. Townsend, The Diffusion of Heat Spots in Isotropic Turbulence, *Proc. R. Soc. London Ser. A*, vol. 209, pp. 418–430, 1951.
44. S. Corrsin, Extended Application of the Hot-Wire Anemometer, *Rev. Sci. Instrum.*, vol. 18, pp. 469–471, 1947.
45. P. Freymuth, Frequency Response and Electronic Testing for Constant-Temperature Hot-Wire Anemometers, *J. Phys. E*, vol. 10, pp. 705–710, 1977.
46. P. Freymuth, Further Investigation of the Non-Linear Theory for Constant-Temperature Hot-Wire Anemometers, *J. Phys. E*, vol. 10, pp. 710–713, 1977.
47. P. Freymuth and L. M. Fingerson, Electronic Testing of Frequency Response for Thermal Anemometers, *TSI Q.*, vol. 3, pp. 5–12, 1977.
48. M. S. Uberoi and P. Freymuth, Spectra of Turbulence in Wake behind Circular Cylinder, *Phys. Fluids*, vol. 12, pp. 1359–1363, 1969.
49. M. S. Uberoi and P. Freymuth, Turbulent Energy Balance and Spectra of Axisymetric Wake, *Phys. Fluids*, vol. 3, pp. 2205–2210, 1970.
50. A. C. M. Beljaars, Dynamic Behaviour of the Constant Temperature Anemometer Due to Thermal Inertia of the Wire, *Appl. Sci. Res.*, vol. 32, pp. 509–518, 1976.
51. C. Clark, Thin Film Gauges for Fluctuating Velocity Measurements in Blood, *J. Phys. E*, vol. 7, pp. 548–556, 1974.
52. R. G. Lueck, Heated Anemometry and Thermometry in Water, Ph.D. thesis, Department of Physics and Institute of Oceanography, University of British Columbia, Vancouver, B.C., 1979.
53. F. N. Frenkiel, Etude Statistique de la turbulence, 1 mesure de la turbulence avec un fil chaud non-compense, 2 Influence de la longueur d'un fil chaud compense sur la mesure de la turbulence, O.N.E.E.A. Rapp. Tech. 37, 1948.
54. M. S. Uberoi and L. S. G. Kovasznay, On Mapping and Measurement of Random Fields, *Q. Appl. Math.*, vol. 10, pp. 375–393, 1953.
55. J. C. Wyngaard, Measurement of Small-Scale Turbulence Structure with Hot Wires, *J. Phys. E*, vol. 1, pp. 1105–1108, 1968.
56. J. C. Wyngaard, Spatial Resolution of the Vorticity Meter and Other Hot-Wire Arrays, *J. Phys. E*, vol. 2, pp. 983–987, 1969.
57. P. Freymuth, Noise in Hot-Wire Anemometers, *Rev. Sci. Instrum.*, vol. 39, pp. 550–557, 1968.
58. B. J. Bellhouse and D. L. Schultz, The Determination of Fluctuating Velocity in Air with Heated Thin Film Gauges, *J. Fluid Mech.*, vol. 29, pp. 289–295, 1967.
59. P. Freymuth, Calculation of Square Wave Test for Frequency Optimized Hot-Film Anemometers, *J. Phys. E*, vol. 14, pp. 238–240, 1981.
60. P. Freymuth, Extension of the Non-Linear Theory to Constant-Temperature Hot-Film Anemometers, *TSI Q.*, vol. 4, pp. 3–6, 1978.

61. P. Freymuth, A Comparative Study of the Signal-to-Noise Ratio for Hot-Film and Hot-Wire Anemometers, *TSI Q.*, vol. 11, pp. 3–6, 1978.
62. G. Comte-Bellot, The Physical Background for Hot-Film Anemometry, *Proc. Symp. Turbulence in Liquids,* Rolla, pp. 1–13, 1977.
63. W. J. McCrosky and E. J. Durbin, Flow Angle and Shear Stress Measurements Using Heated Films and Wires, ASME Paper 71-WA/FE-17, 1971.
64. P. Freymuth, Sine Wave Testing on Noncylindrical Hot-Film Anemometers According to Bellhouse-Schultz Model, *J. Phys. E,* vol. 13, pp. 98–102, 1980.
65. C. Salter and W. G. Raymer, Direct Calibration of Compensated Hot-Wire Recording Anemometer (by Dynamic Calibration), A.R.C. R&M 1628, 1934.
66. J. R. Weske, A Hot-Wire Circuit with Very Small Time Lag, NACA Tech. Note TN 881, 1943.
67. H. H. Lowell, Early (1944–1952) Hot-Wire Anemometer Developments at NACA Lewis Research Center, in W. L. Melnik and J. R. Meske (eds.), *Advances in Hot-Wire Anemometry,* University of Maryland, College Park, pp. 29–37, 1968.
68. M. F. Young, Calibration of Hot-Wires and Hot-Films for Velocity Fluctuations, Department of Mechanical Engineering, Stanford University, Rep. TMC-3, 1976.
69. H. R. Frey, An Investigation of Instrumentation and Techniques for Observing Turbulence In and Above the Oceanic Bottom Boundary Layer, New York University, School of Engineering and Science, 1970.
70. P. L. Blackshear and L. M. Fingerson, Rapid Response Heat Flux Probe for High Temperature Gases, *ARS J.,* (Nov.) pp. 1709–1715, 1962.
71. G. L. Brown and M. R. Rebollo, A Small Fast-Response Probe to Measure Composition of a Binary Gas Mixture, *AIAA J.*, vol. 10, pp. 649–652, 1972.
72. Ronald Koopman, Data presented at a meeting arranged by Gas Research Institute, Chicago, August 27, 1979.
73. B. W. Spencer and B. G. Jones, A Bleed Type Pressure Transducer for In-Stream Measurement of Static Pressure Fluctuations, *Rev. Sci. Instrum.*, vol. 42, pp. 450–454, 1971.
74. H. P. Planchon, The Fluctuating Static Pressure Field in a Round Jet Turbulent Mixing Layer, Ph.D. thesis, University of Illinois, Urbana-Champaign, 1974.
75. B. G. Jones, R. J. Adrian, C. K. Nithianandan, and H. P. Planchon, Jr., Spectra of Turbulent Static Pressure Fluctuations in Jet Mixing Layers, *AIAA J.*, vol. 17, no. 5, pp. 449–457, 1979.
76. C. J. Remenyik and L. S. G. Kovasznay, The Orifice Hot-Wire Probe and Measurements of Wall Pressure Fluctuations, *Proc. Heat Transfer and Fluid Mech. Inst.,* Stanford University Press, 1962.
77. J. A. B. Wills, A Traversing Orifice-Hot-Wire Probe for Use in Wall Pressure Correlation Measurements, National Physical Laboratory Aero. Rep. 1155, 1965.
78. M. D. Mack, H. C. Seetharam, W. G. Kuhn, and J. T. Bright, Jr., Aerodynamics of Spoiler Control Devices, AIAA Aircraft Systems and Technology Meeting, New York, 1979.
79. W. H. Wentz, Jr., and H. C. Seetharam, Split-Film Anemometer Measurements on an Airfoil with Turbulent Separated Flow, *Proc. of the Fifth Biennial Symp. on Turbulence,* pp. 31–33, University of Missouri, Rolla, 1977.
80. C. T. Crowe, D. E. Stock, M. R. Wells, and A. Barriga, Application of Split-Film Anemometry to Low-Speed Flows with High Turbulence Intensity and Recirculation as Found in Electrostatic Precipitators, *Proc. of the Fifth Biennial Symp. on Turbulence,* pp. 117–123, University of Missouri, Rolla, 1977.
81. B. W. Spencer and B. G. Jones, Turbulence Measurements with the Split-Film Anemometer Probe, *Proc. of Symp. on Turbulence in Liquids,* pp. 7–15, University of Missouri, Rolla, 1971.

CHAPTER

FIVE

LASER VELOCIMETRY

Ronald J. Adrian

1. INTRODUCTION

Laser-Doppler velocimetry (LDV) is the measurement of fluid velocities by detecting the Doppler frequency shift of laser light that has been scattered by small particles moving with the fluid. The technique was originally discussed in a pioneering paper by Cummins, Knable, and Yeh [1] in which they measured the Brownian motion of an aqueous suspension of micron-sized particles by observing the spectrum of the scattered light. In these measurements the quantity of interest was the broadening of the laser light spectrum due to the random particle motion. However, they also observed a net shift in the frequency of the light, an effect that they attributed to small convection currents that generated mean velocities in their water cell. Hence, almost inadvertently, they performed the first measurements of fluid velocity by laser-Doppler velocimetry. Shortly thereafter, Yeh and Cummins [2] carried out an experiment intended expressly to demonstrate the measurement of fluid velocities.

The LDV concept rapidly attracted the attention of numerous experimental fluid dynamicists, and within a few years various research groups had communicated the results of successful LDV measurements of laminar water flow in square ducts [3, 4], laminar water flow in round ducts [5], laminar gas flow [6, 7], turbulent water flow in pipes [8], and wind-tunnel turbulence [9]. At this stage of development, measurements were performed using spectrum analysis of the scattered light, a technique that required relatively long averaging times and, hence, precluded velocity tracking, e.g., measurement of rapidly fluctuating velocity as a function of time. Even so, the results were very encouraging. The technique offered numerous

This work was supported by the Atmospheric Sciences Section of the National Science Foundation.

advantages: It was nonintrusive, so it could be used in flows that were hostile to material probes or that would be altered by the presence of a material probe; it did not depend on the thermophysical properties of the fluid, in contrast to thermal probes or chemical probes; it allowed the unambiguous measurement of one or more components of the velocity vector, independent of the fluctuation intensity (e.g., flow reversals could be sensed); it offered reasonably good spatial resolution; and it appeared to be capable of tracking very high frequency fluctuations of the flow velocity, provided that sufficiently fast electronics could be developed. On top of all this, the achievable accuracies were impressive: Goldstein and Kreid [4] reported 0.1% absolute accuracy for measurements of flow development in a square duct.

There were, of course, many problems to be solved. The very weak intensities of light scattered by small particles resulted in noisy signals that were difficult to analyze. The random locations of the scattering particles in the fluid created new types of data-analysis problems, since data arrived randomly and often far too infrequently to resolve the time history of the velocity. The optical equipment was sometimes sensitive to vibrations, and electronics that were developed to follow the Doppler frequency often failed to do so, tracking, instead, noise and extraneous radio-frequency signals. Even so, the natural appeal of the laser velocimeter led to an intense period of development over the ensuing decade; this effort has produced, at the time of this writing, an experimental technique that is routinely used in hundreds of research laboratories and industrial applications throughout the world, that is available through several commercial manufacturers of scientific equipment (TSI, Inc., St. Paul; DISA, Inc., Denmark; and Precision Instruments, Inc., England, for example), and that has reached a high level of maturity.

The power and versatility of the laser-Doppler velocimeter are probably best conveyed by a list (admittedly incomplete) of the types of flows in which it has been used successfully. These include supersonic flow, recirculating flow, natural free convection, flow in internal combustion engines, steam turbines, and gas turbines, chemically reacting flows including premixed flames and diffusion flames, jets, drag-reducing polymer flows, rotating flows, helicopter rotor studies, two-phase flows, atmospheric turbulence, arterial flow, capillary blood flow, both simulated and *in vivo*, ocean-bottom flows, high-temperature plasmas, and MHD channel flows. The range of flow velocities measured extends from less than 10 μm/s to 1 km/s.

In view of the widespread interest in laser-Doppler velocimetry, it is not surprising that an abundant body of scientific literature exists on the topic. In 1974 a literature survey by Durst and Zaré [10] included over 600 papers, and the total number of papers has easily doubled since then. For the newcomer to the field the task of assimilating all this information is clearly formidable. Fortunately, there are a number of books that deal in whole or in part with the subject and that are relatively comprehensive as of the time of their writing [11 to 15]. In addition, the proceedings of numerous workshops and symposia contain a wealth of information on the theory and application of laser velocimetry [16 to 24].

From this brief description of the extensive scope of laser-Doppler velocimetry, it should be clear that an exhaustive and intensive discussion of the subject here is precluded by space limitations. Instead, we attempt to survey those topics that are of most significance to the understanding and successful application of laser velocimetry as comprehensively as possible, leaving out many of the background details. Often this means that final results are presented with little detailed derivation. In these cases the reader who is interested in the details is referred to appropriate papers in the literature. However, certain topics are so fundamental to the field that a more or less complete discussion seems advisable, and these topics are presented in detail.

2. BASIC PRINCIPLES

2.1 Doppler Shift of Light Scattered by Small Particles

The fundamental phenomenon in laser-Doppler velocimetry is the Doppler shift of light that is scattered from small (typically 0.1 to 10 μm) particles. The usual situation is shown in Fig. 1, where the ith particle located at $\mathbf{x}_i(t)$ scatters the light wave with complex electric vector $\mathbf{E}_{li}$ from an incident illuminating beam $\mathbf{E}_{0l}$. Here the subscript l is used to denote the lth illuminating beam. For reasons that are explained later, the illuminating beam may be assumed to be a plane wave, linearly polarized in the spatial region where it illuminates the particle. Its frequency is ω_{0l}, its direction of propagation is $\hat{\mathbf{s}}_l$ (unit vector), its wave number is $k = 2\pi/\lambda$, its direction of linear polarization is $\hat{\mathbf{p}}_l$, and its intensity is I_{0l} (W/m^2). It may be represented by the complex wave

$$\mathbf{E}_{0l} = \sqrt{I_{0l}(\mathbf{x})}\; e^{j\Phi_{0l}(\mathbf{x})}\hat{\mathbf{p}}_l \tag{1}$$

where $\Phi_{0l}(\mathbf{x})$ is the phase evaluated at $\mathbf{x}$,

$$\Phi_{0l}(\mathbf{x}) = \omega_{0l}\, t - k\hat{\mathbf{s}}_l \cdot \mathbf{x} \tag{2}$$

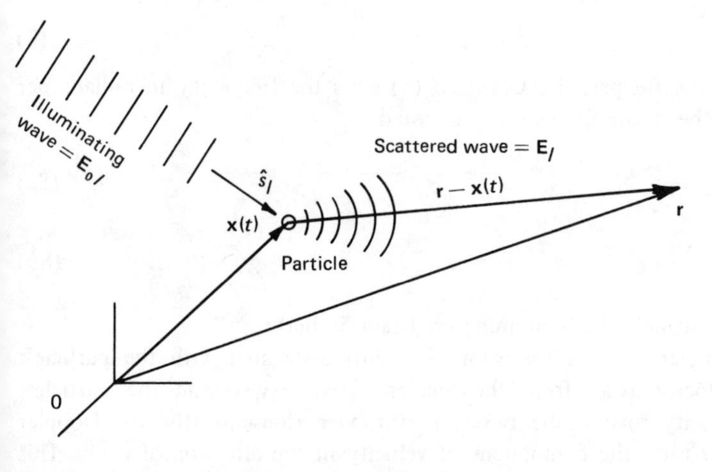

Figure 1 Light scattered by a moving particle.

and it is understood that the physical electric vector is given by the real part of $\mathbf{E}_{0l}$. The particle scatters a light wave from $\mathbf{E}_{0l}$ in all directions, and a point $\mathbf{r}$ is said to be in the *far field* of the particle if the distance $r = |r|$ is much greater than both the wavelength of the light and the mean diameter of the particle. In the far field the scattered light wave is a spherical wave, regardless of the particle's shape, given [25] by

$$\mathbf{E}_{li} = \sqrt{I_{0l}(\mathbf{x}_i)}\ \frac{\sigma_{li}}{k|\mathbf{r} - \mathbf{x}_i|}\ e^{j[\Phi_{0l}(\mathbf{x}_i) - k|\mathbf{r} - \mathbf{x}_i|]} \tag{3}$$

where σ_{li} is a scattering coefficient for the ith particle that specifies the intensity, phase shift, and polarization of the scattered wave relative to the illuminating wave $\mathbf{E}_{0l}$.

Equation (3) is simplified by noting that under normal conditions the region illuminated by the incident wave is very small because the wave has been focused. Hence, $|\mathbf{x}_i| \ll |\mathbf{r}|$, the vectors $\mathbf{r} - \mathbf{x}_i$ and $\mathbf{r}$ are nearly parallel, and by simple geometry $|\mathbf{r} - \mathbf{x}_i| \cong r - \mathbf{x}_i \cdot \hat{\mathbf{r}}$ in the far field. Then

$$\mathbf{E}_{li} = \sqrt{I_{0l}}\ \frac{\sigma_{li}}{kr}\ e^{j\Phi_{li}} \tag{4}$$

$$\Phi_{li} = \omega_{0l} t - kr + k\mathbf{x}_i \cdot (\hat{\mathbf{r}} - \hat{\mathbf{s}}_l) \tag{5}$$

where the factor $\mathbf{x}_i \cdot \hat{\mathbf{r}}$ must be retained in the phase but can be ignored safely in the denominator. Equation (4) implies that the scattered wave is approximately a spherical wave diverging from the origin (since $|\mathbf{x}_i|$ is small) whose phase depends on the particle position through the term $k\mathbf{x}_i \cdot \hat{\mathbf{s}}_l$. Now, the instantaneous frequency of a nearly sinusoidal signal is defined as the time derivative of its phase. From Eq. (5) this frequency is

$$\dot{\Phi}_{li} = \omega_{0l} + k\mathbf{v}_i(t) \cdot (\hat{\mathbf{r}} - \hat{\mathbf{s}}_l) \tag{6}$$

where the last term is the Doppler shift caused by the particle motion, and

$$\mathbf{v}_i(t) = \dot{\mathbf{x}}_i(t) \tag{7}$$

is the velocity of the ith particle. Equation (6) gives the frequency in radians per second. In units of hertz the frequency is denoted by

$$\nu_{li} = \frac{\dot{\Phi}_{li}}{2\pi} \tag{8a}$$

$$= \nu_{0l} + \frac{\mathbf{v}_i \cdot (\hat{\mathbf{r}} - \hat{\mathbf{s}}_l)}{\lambda} \tag{8b}$$

where ν_{0l} is the frequency of the illuminating beam, in hertz.

The total Doppler shift is the sum of a shift associated with the particle's component of velocity away from the incident wave, $-\mathbf{v}_i \cdot \hat{\mathbf{s}}_l$ and the particle's component of velocity toward the observer at $\mathbf{r}$, $\mathbf{v}_i \cdot \hat{\mathbf{r}}$. Consequently, the Doppler shift depends *linearly* on the component of velocity in the direction of $\hat{\mathbf{r}} - \hat{\mathbf{s}}_l$. This is one of the more desirable features of laser-Doppler velocimetry, because it gives the experimentalist the freedom to study a single component by choosing $\hat{\mathbf{s}}$ and $\hat{\mathbf{r}}$ accordingly. Note further that the Doppler shift has a sign associated with it so that flow in the positive $\hat{\mathbf{r}} - \hat{\mathbf{s}}_l$ direction can be distinguished from flow in the negative $\hat{\mathbf{r}} - \hat{\mathbf{s}}_l$ direction.

2.2 Optical Heterodyne Detection

The maximum Doppler shift occurs where $\hat{\mathbf{r}} = -\hat{\mathbf{s}}$, and its value is typically 4 MHz/(m•s^{-1}). Usually $\hat{\mathbf{r}}$ is more nearly parallel to $\hat{\mathbf{s}}$, and a typical value in practice is about 0.4 MHz/(m•s^{-1}). Thus, for example, the Doppler frequency produced by a 500-m/s gas flow would be around 200 MHz, while the shift observed in a 1-mm/s convection flow would be about 400 Hz. While these frequencies can be measured readily and accurately by modern electronics, they are, nonetheless, miniscule compared to the basic frequency of the laser light, which is of the order of 10^{14} Hz. For example, a relatively large Doppler shift of, say, 10^8 Hz, represents a change in the light frequency of only one part in a million. [Since the fractional change in the wavelength would be the same, changes in $k = 2\pi/\lambda$ were ignored, implicitly, in Eq. (3).] Hence, while direct spectroscopic detection of Doppler shift is possible in high-speed flows [26, 29], it is not used in lower-speed flows.

The method used in almost all laser-Doppler velocimetry is to subtract the ω_{0l} term from the total frequency, leaving a signal that oscillates at the Doppler shift frequency. The technique whereby this is accomplished is called *optical heterodyne detection* or *optical mixing.* The essence of this technique is revealed by the simple trigonometric identity $\sin \omega_1 t \sin \omega_2 t = \frac{1}{2} [\cos (\omega_1 + \omega_2)t + \cos (\omega_1 - \omega_2)t]$. Thus, by multiplying two light waves, a signal that oscillates at their difference frequency can be obtained. In actuality, the multiplication is performed by combining two light waves on the surface of a photodetector. Since the photodetector is a square-law device, its output will be of the form $(\sin \omega_1 t + \sin \omega_2 t)^2$, from which the cross product $\sin \omega_1 t \sin \omega_2 t$ is obtained. The output of the photodetector does not contain the sum-frequency component of this cross product because that frequency, of the order of 10^{14} Hz, is much higher than the frequency response of any available detector. Thus, the output oscillates only at frequency $\omega_1 - \omega_2$.

2.3 Basic Optical Systems

There are three distinct types of LDV optical systems, corresponding to three different methods of combining the Doppler shift phenomenon with the optical heterodyne technique to produce a flow-measuring instrument: the reference-beam system, the dual-beam system, and the dual-scatter system. Figure 2 depicts the optical geometries corresponding to these three types schematically, and Fig. 3 shows typical optical systems used in practice. Many minor variations are, of course, possible. To facilitate comparisons, each of these systems is configured to measure the same component of velocity, and the coordinates are always defined with respect to the optical system so that the origin is imbedded at the center of the

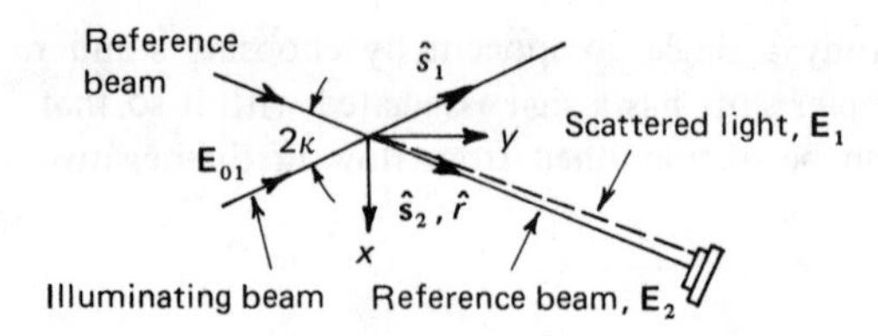

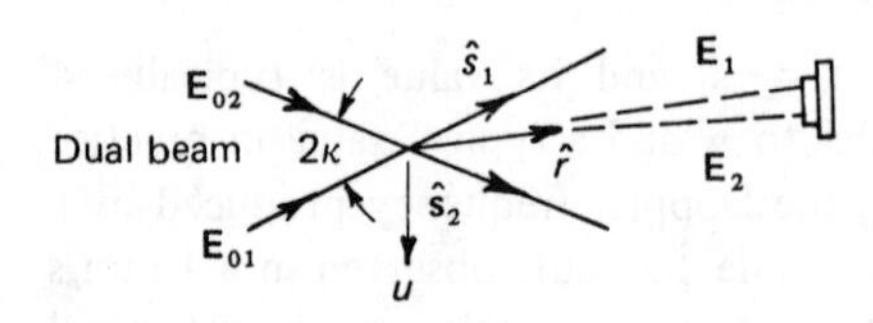

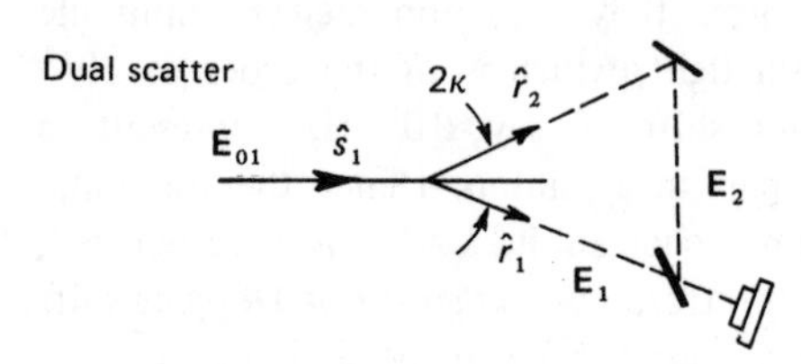

Figure 2 Three modes of heterodyne detection in laser-Doppler velocimetry. The geometry of each mode is shown such that the Doppler difference frequency $\nu_D = 2u \sin \kappa/\lambda$. *(From [47].)*

measurement volume and the system measures the x component of velocity, u. The y axis lies in the plane of the light beams and is referred to as the "axis" of the system.

The basic principle of the *dual-beam* LDV is to illuminate a scattering particle with two plane light waves, $\mathbf{E}_{01}$ and $\mathbf{E}_{02}$ propagating in two different directions, $\hat{\mathbf{s}}_1$ and $\hat{\mathbf{s}}_2$, respectively. The ith particle scatters two waves, $\mathbf{E}_{1i}$ from $\mathbf{E}_{01}$ and $\mathbf{E}_{2i}$ from $\mathbf{E}_{02}$, and the frequencies of these waves in the scattering direction $\hat{\mathbf{r}}$ are

$$\nu_{1i} = \nu_{01} + \frac{\mathbf{v}_i \cdot (\hat{\mathbf{r}} - \hat{\mathbf{s}}_1)}{\lambda} \tag{9a}$$

and

$$\nu_{2i} = \nu_{02} + \frac{\mathbf{v}_i \cdot (\hat{\mathbf{r}} - \hat{\mathbf{s}}_2)}{\lambda} \tag{9b}$$

The frequency difference is

$$\nu_{1i} - \nu_{2i} = \nu_S + \nu_{Di} \tag{10}$$

where

$$\nu_S = \nu_{01} - \nu_{02} \tag{11}$$

is a constant frequency difference determined by the illuminating-beam frequencies, and

$$\nu_{Di} = \frac{\mathbf{v}_i \cdot (\hat{\mathbf{s}}_2 - \hat{\mathbf{s}}_1)}{\lambda} \tag{12}$$

is the difference between the Doppler shifts. This difference is independent of the scattering direction $\hat{\mathbf{r}}$, so the heterodyne frequency is the same at every point on the photodetector and independent of the detector's location. It is convenient to follow [28] and write Eq. (12) in the form

$$\nu_{Di} = \frac{\mathbf{K}\cdot\mathbf{v}_i}{2\pi} \tag{13}$$

where

$$\mathbf{K} = \frac{2\pi(\hat{\mathbf{s}}_2 - \hat{\mathbf{s}}_1)}{\lambda} \tag{14}$$

is a wave vector in the direction $\hat{\mathbf{s}}_2 - \hat{\mathbf{s}}_1$. Then, if we always agree to let $u_i(t)$ be the component of $\mathbf{v}_i$ in the $\hat{\mathbf{s}}_2 - \hat{\mathbf{s}}_1$ direction, we have

$$\nu_{Di} = \frac{Ku_i(t)}{2\pi} \tag{15}$$

where

$$|\mathbf{K}| = K = \frac{4\pi \sin \kappa}{\lambda} \tag{16}$$

by simple geometry. Thus, ν_{Di} depends only upon κ, λ, and the single velocity component u_i, which lies in the plane of the illuminating beams and is perpendicular to their bisector. In terms of circular frequencies,

$$\omega_{1i} - \omega_{2i} = \omega_S + Ku_i(t) \tag{17}$$

where $\omega_S = 2\pi\nu_S$.

Ordinarily, the illuminating beams are obtained by splitting the original laser

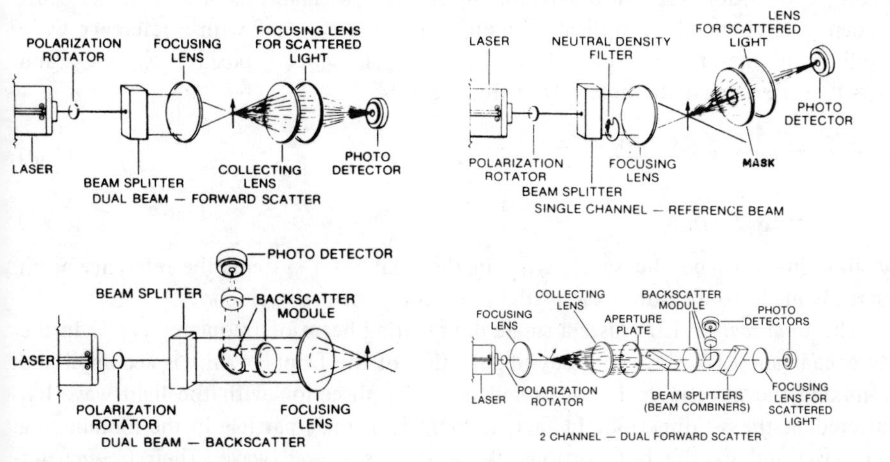

Figure 3 Practical LDV systems for the dual-beam, reference-beam, and dual-scatter modes. *(Courtesy TSI, Inc.)*

output of frequency ν_0 so that $\nu_{01} = \nu_{02} = \nu_0$. Then $\nu_{1i} - \nu_{2i} = \nu_{Di}$, and the heterodyne frequency of the photodetector output is directly proportional to u_i. In this case, however, a sign change in u_i, and hence ν_{Di}, simply corresponds to a 180° phase change in the sinusoidal heterodyne signal; this is difficult to detect electronically, particularly when the signal is contaminated by noise. Consequently, whenever bipolar velocities are anticipated, e.g., in reversing flows or in high-intensity turbulence, it is common practice to generate a frequency difference ν_S between the illuminating beams so that fluctuations about zero velocity correspond to frequency fluctuations about the shift frequency ν_S. The signal then looks much like an FM radio signal with ν_S corresponding to the carrier frequency. In addition to resolving the polarity of u_i, frequency shifting offers other advantages that are discussed later.

Frequency shifting is accomplished in practice by splitting the original laser beam and shifting the frequency of one of the beams, say beam 1, by ν_S, so that $\nu_{01} = \nu_0 + \nu_S$ and $\nu_{02} = \nu_0$. Frequency shifts can be produced by electrooptic Pockels cells and Kerr cells [29], rotating diffraction gratings [30, 31], or acoustic-optic Bragg cells [32], the latter being the most commonly used by far. Typical shifts available from glass Bragg cells are rather large, in the 10 to 80 MHz range, but they can be very accurately controlled, and when the basic shift is too large compared to the Doppler frequency, the signal from the photodetector can be electronically shifted downward to a more convenient value. For example, it is common to use a 40-MHz Bragg cell with electronic down-mixing to produce effective shift frequencies as low as a few kilohertz.

The *reference-beam* LDV, shown in Figs. 2 and 3, uses a single illuminating beam from which a light wave is scattered with frequency

$$\nu_{1i} = \nu_{01} + \frac{\mathbf{v}_i \cdot (\hat{\mathbf{r}} - \hat{\mathbf{s}}_1)}{\lambda} \tag{18}$$

in the $\hat{\mathbf{r}}$ direction. A photodetector signal that oscillates at the Doppler shift frequency is obtained by optically mixing the scattered wave with a reference wave traveling in the $\hat{\mathbf{r}}$ direction whose frequency is ν_{02}. Typically, $\nu_{01} = \nu_0$ and $\nu_{02} = \nu_0 - \nu_S$ so that the heterodyne frequency is

$$\nu_{1i} - \nu_{2i} = \nu_S + \mathbf{v}_i \cdot \frac{\hat{\mathbf{r}} - \hat{\mathbf{s}}_1}{\lambda} \tag{19a}$$

$$= \nu_S + \nu_{Di} \tag{19b}$$

By choosing $\hat{\mathbf{r}}$ to be the same as $\hat{\mathbf{s}}_2$ in the dual-beam system, the reference-beam system is made to measure exactly the same component of velocity.

The *dual-scatter* LDV uses a single illuminating beam of frequency ν_{01} as in the reference-beam LDV, but heterodyne detection of the Doppler shift is accomplished by mixing the light wave $\mathbf{E}_{1i}$ scattered in the $\hat{\mathbf{s}}_1$ direction with the light wave $\mathbf{E}_{2i}$ scattered in the $\hat{\mathbf{s}}_2$ direction. In fact, if there is a single particle in the illuminating beam, $\mathbf{E}_{1i}$ and $\mathbf{E}_{2i}$ are but portions of the same scattered wave. Their frequencies are

$$\nu_{1i} = \nu_{01} + \mathbf{v}_i \cdot \frac{\hat{\mathbf{r}}_1 - \hat{\mathbf{s}}_1}{\lambda} \tag{20a}$$

$$\nu_{2i} = \nu_{01} + \mathbf{v}_i \cdot \frac{\hat{\mathbf{r}}_2 - \hat{\mathbf{s}}_1}{\lambda} \tag{20b}$$

with heterodyne difference frequency

$$\nu_{1i} - \nu_{2i} = \mathbf{v}_i \cdot \frac{\hat{\mathbf{r}}_1 - \hat{\mathbf{r}}_2}{\lambda} \tag{21a}$$

$$= \nu_{Di} \tag{21b}$$

Hence, the heterodyne frequency is independent of the direction of the illuminating beam.

While the heterodyne frequency equations for the dual-beam, reference-beam, and dual-scatter systems are very similar, there are, in fact, significant differences among other properties of the signals produced by these systems. In particular, the strengths and signal-to-noise ratios of the signals may differ substantially in a given application. These differences arise primarily because of differences in the *efficiency* with which scattered light is collected and heterodyne mixing is accomplished. It is shown below that the size of the scattered-light collecting aperture is severely restricted for both the reference-beam system and the dual-scatter system, the former because large apertures cause poor mixing efficiency, and the latter because the Doppler frequency shift varies over the aperture [cf. Eq. (21*a*)]. Thus, only the dual-beam system is capable of effectively utilizing a large light-collecting aperture to produce strong signals, and for this reason it is the most commonly used type of LDV (although the other types can be superior in certain applications). Consequently, the ensuing discussion concentrates on the dual-beam system.

3. THE DUAL–BEAM LDV

3.1 Practical Dual-Beam Optics

Before discussing the theory of the dual-beam anemometer, it is best to look at the way these systems are constructed for practical applications. Figure 3 shows the most commonly used configurations. The intense, highly collimated light beam from a CW (continuous-wave) gas laser, usually helium-neon or argon ion, is divided into two parallel beams of equal power by a beam splitter. The parallelism of the beams is critical, so the beam splitter is either a precisely constructed single-piece unit or a multicomponent unit that provides the very fine adjustments needed to maintain parallelism. Passing the parallel beams through a good-quality focusing lens causes the beams to intersect in the focal plane of the lens and simultaneously focuses the illuminating beams to small spots coincident with the intersection. In the region of the focal spots the light beams are approximately cylinders with diameters determined ideally by the diffraction-limited spot size, and the light waves are approximately plane parallel. The spatial region from which measurements are obtained is essentially the intersection of these beams, since a scattering particle must scatter light from *both* illuminating beams to produce a heterodyne signal.

Scattered light is collected by a set of receiving lenses and focused through a pinhole aperture onto a photodetector that may be either a photomultiplier tube or a photodiode. The process of heterodyne detection takes place on the active surface of the photodetector, where the two scattered light waves are "mixed" together. The illuminating beams are normally blocked from the detector to avoid swamping the weak scattered-light signals. In principle, if only one particle were present in the fluid, one could replace the receiving lens with a photodetector of equal area and achieve the same results as with the lens-pinhole system, because the flux of light energy through the light-collecting aperture would be the same as the flux through the pinhole. However, in practice, the photodetector also receives light scattered from particles distributed throughout the illuminating beams, light from extraneous sources such as background radiation (room lights, radiation from flames, etc.) and extraneous laser light from flare and reflections at the surfaces of the transmitting optical elements and test-section windows. The function of the lens-pinhole combination is to reject most of this light by allowing only light from a small region around the beam intersection to enter the photodetector. The size of this region is essentially the diameter of the image of the pinhole in the plane of the beam intersection. It is usually chosen to be slightly smaller than the beam intersection itself, but when the beam intersection is undesirably large, the pinhole may be made even smaller so that only a portion of the intersection can be "seen" by the detector. In this case the measurement volume is the intersection of the pinhole field of view with the illuminating beam intersection.

If the collecting lens is centered on the axis of the system, e.g., the y axis in Fig. 2, the configuration is called *coaxial*. "Forward scatter" and "back scatter" refer to locations of the collecting aperture that receive light scattered forward from the illuminating beams and scattered backward, respectively. Thus, the first dual-beam system in Fig. 3 is a coaxial forward-scatter system. While coaxial systems enjoy a certain symmetry, it is often better to use "off-axis" light collection (e.g., not centered on the y axis) to reduce the amount of extraneous light seen by the detector and reduce the size of the measurement volume in the y direction.

The photodetector signal is a series of short, random bursts of Doppler-frequency sine waves caused by particles passing through the measurement volume. For setup purposes it is always advisable to observe these signals on an oscilloscope, but for the purpose of data acquisition special electronic signal processors are needed to measure the frequencies of the Doppler bursts and convert them into either a voltage proportional to frequency or a digital number proportional to frequency. Likewise, because of the random times at which particles enter the measurement volume, special data processing methods may be needed to properly extract the desired information, such as mean velocity, rms velocity, and power spectrum. Signal processing and data processing are discussed in Secs. 10 and 11.

3.2 Characteristics of the Dual-Beam Signal

The dual-beam LDV can produce two types of signals: "incoherent" and "coherent." The incoherent signal occurs when a *single* particle scatters two light waves, one from each illuminating beam, and these waves subsequently mix on the photodetector. The coherent signal occurs when *at least two* particles reside in the measurement volume simultaneously. Then each particle produces an incoherent signal but, in addition, each *pair* of particles produces a coherent signal. The coherent signal is the result of the light wave scattered from one beam by the first particle mixing with light scattered from the other beam by the second particle, and vice versa. Since the scattered light waves originate from two different points in space, they may not mix efficiently unless the particles are so close together that the waves appear to come from a small, coherently illuminated region. Hence, the term "coherent." The incoherent signal is more useful than the coherent signal in most cases, so it is examined first and in the most detail.

3.2.1 Single-particle signal. Consider the case of the *i*th particle crossing the illuminating beams with trajectory $\mathbf{x}_i(t)$ as in Fig. 4, and assume that the only light reaching the collecting aperture is that which is scattered by the particle, i.e., background light, laser flare, etc., are negligible.

In the region of the beam intersection, each focused illuminating beam is approximately a plane wave with variable intensity I_{0l}, where $l = 1, 2$; thus, $\mathbf{E}_{1i}$ and $\mathbf{E}_{2i}$, the light waves scattered from beams 1 and 2, respectively, are represented in the far field by Eq. (4) with $l = 1$ and 2. The scattering coefficients σ_{1i} and σ_{2i} will differ, in general, because each depends upon the scattering angle relative to the direction of the illuminating beam and its polarization, for example, $\sigma_{li} = \sigma_{li}(\hat{\mathbf{p}}_l, \hat{\mathbf{s}}_l, \hat{\mathbf{r}})$ [33, 34]. At the point $\mathbf{r}$ the total electric vector is

$$\mathbf{E}_i = \mathbf{E}_{1i} + \mathbf{E}_{2i} \tag{22}$$

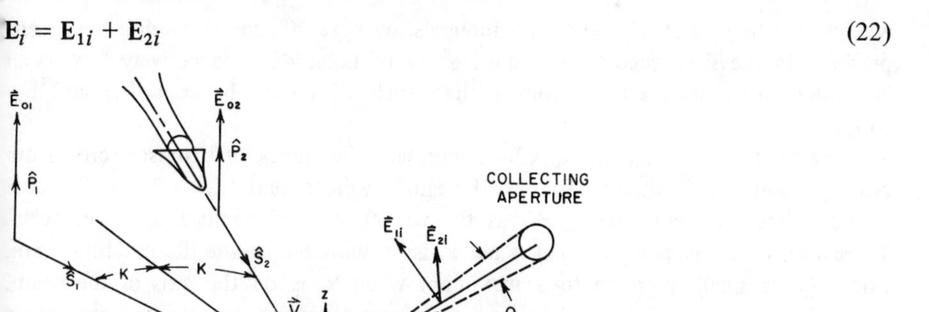

Figure 4 Dual-beam LDV geometry. *(From [47].)*

and the intensity, defined as the flux of light energy per unit area averaged over a time much longer than the light period but much shorter than the period of the heterodyne frequency, is given [32] by

$$I = \mathbf{E}_i \cdot \mathbf{E}_i^* \tag{23}$$

where * denotes complex conjugate. Equations (4), (5), (14), (22), and (23) give

$$I_i(\mathbf{r}, \mathbf{x}_i, t) = \mathbf{E}_{1i} \cdot \mathbf{E}_{1i}^* + \mathbf{E}_{2i} \cdot \mathbf{E}_{2i}^* + \mathbf{E}_{2i} \cdot \mathbf{E}_{1i}^* + \mathbf{E}_{1i} \cdot \mathbf{E}_{2i}^* \tag{24a}$$

$$= \frac{I_{01}\,\sigma_{1i} \cdot \sigma_{1i}^*}{k^2 r^2} + \frac{I_{02}\,\sigma_{2i} \cdot \sigma_{2i}^*}{k^2 r^2} + \frac{2\sqrt{I_{01} I_{02}}}{k^2 r^2} \operatorname{Re}\left(\sigma_{1i} \cdot \sigma_{2i}^*\, e^{j\Phi_i(t)}\right) \tag{24b}$$

where

$$\Phi_i(t) = \Phi_i(\mathbf{x}_i(t), t) = 2\pi\nu_S t + k\mathbf{x}_i(t) \cdot (\hat{\mathbf{s}}_2 - \hat{\mathbf{s}}_1) \tag{25a}$$

$$= \omega_S t + K x_i(t) \tag{25b}$$

and the intensities I_{01} and I_{02} are evaluated at the location of the particle $\mathbf{x}_i(t)$. The photodetector output is proportional to the light flux hitting the detector, and this in turn will be equal to the light flux J_i through the light-collecting aperture, which subtends the solid angle Ω, assuming that no pinhole blocks the light:

$$J_i(t) = \int_\Omega I(\mathbf{r}, \mathbf{x}_i, t) r^2 \, d\Omega \tag{26a}$$

After some algebraic manipulation, this becomes

$$J_i(t) = \frac{1}{k^2} \{I_{01} P_{1i} + I_{02} P_{2i} + \sqrt{I_{01} I_{02}}\, D_i \cos [\Phi_i(t) - \Psi_i]\} \tag{26b}$$

where P_{1i}, P_{2i}, and D_i represent integrals over Ω of the scattering-coefficient products in the first, second, and third terms of Eq. (24b), respectively.* Ψ_i is an integrated phase shift arising from a phase-angle difference between $\mathbf{E}_{1i}$ and $\mathbf{E}_{2i}$ [33].

The first two terms in Eq. (26b) represent the fluxes of light scattered from beams 1 and 2 individually. That is, if beam 2 were absent ($I_{02} = 0$) the detector would still see a light flux $I_{01} P_{1i}$ as the particle crossed beam 1, and vice versa. These terms will usually be small when $\mathbf{x}_i$ is at the edge of the illuminating beam, where I_{0l} is small, increase to a maximum when $\mathbf{x}_i$ is on the axis of the beam, where I_{0l} is maximum, and then decrease again. Hence, a particle will generate a pulse as it crosses each beam. Historically the sum of these pulses is called the *pedestal*. The pedestal light flux is denoted by

$$J_{P_i}(t) = k^{-2} [I_{01}(\mathbf{x}_i(t)) P_{1i} + I_{02}(\mathbf{x}_i(t)) P_{2i}] \tag{27}$$

The third term in Eq. (26b) is the heterodyne mixing term that arises from the interference of $\mathbf{E}_{1i}$ and $\mathbf{E}_{2i}$. Frequently it is called the *Doppler signal* or *Doppler*

*P_1, P_2, D, and Ψ are the same as $\bar{P}_1$, $\bar{P}_2$, $\bar{D}$, and $\bar{\Psi}$ in [33, 34].

light flux, since it oscillates at the Doppler difference frequency. It is denoted by

$$J_{D_i}(t) = a(\mathbf{x}_i(t))D_i \cos [\Phi_i(t) - \Psi_i] \tag{28}$$

where

$$a(\mathbf{x}_i) = k^{-2}\sqrt{I_{01}(\mathbf{x}_i)I_{02}(\mathbf{x}_i)} \tag{29}$$

The amplitude of $J_{D_i}(t)$ must always be less than or equal to $J_{P_i}(t)$; otherwise the total light flux would be negative when the cosine term is equal to -1, an obvious impossibility.

A useful quantity is the *visibility* of the signal, defined as the ratio of the amplitude of the Doppler signal to the amplitude of the pedestal:

$$V_i(t) = \frac{\sqrt{I_{01}I_{02}}\, D_i}{I_{01}P_{1i} + I_{02}P_{2i}} \tag{30}$$

This quantity is a function of time because I_{01} and I_{02} change as the particle crosses the intensity profiles of the beams. However, when $I_{01}(\mathbf{x}_i(t)) = I_{02}(\mathbf{x}_i(t))$ the value of the visibility is

$$V_i\big|_{I_{01}=I_{02}} \equiv \bar{V}_i = \frac{D_i}{P_{1i} + P_{2i}} \tag{31}$$

which depends only upon the geometry of the optical system and the scattering properties of the particle and is independent of time. For this reason $\bar{V}_i$ is often used to characterize the signal, i.e., the efficiency with which the scattered light waves from a single particle mix to produce the heterodyne signal. $\bar{V}_i = 1$ represents 100% efficiency and, of course, $\bar{V}_i \leqslant 1$ always. It is sometimes convenient to express $J_i(t)$ in the form

$$J_i(t) = J_{P_i}(t)\{1 + V_i(t) \cos [\Phi_i(t) - \Psi_i]\} \tag{32}$$

The spatial resolution of the dual-beam LDV is determined in part by the distribution of light intensity at the intersection of the focused laser beams. Almost universally, one uses a laser operating in the TEM_{00} mode, meaning that the laser cavity sustains a purely longitudinal standing-wave oscillation along its axis, with no transverse modes. In this case the output of the laser has an axisymmetric intensity profile that is very nearly a gaussian function of radial distance from the axis. The diameter of the beam is measured between the points where the intensity is $1/e^2$ of the peak intensity at the centerline and is denoted by $D_{e^{-2}}$. Typically $D_{e^{-2}}$ is about 1 mm. The output beam diverges such that in the far field it appears to be a spherical wave originating from a point source that is located at the front mirror of the laser, or perhaps somewhat further back, depending upon the design of the laser cavity. The full divergence angle $\theta_{e^{-2}}$ is on the order of a few milliradians, so the beam is essentially parallel by most standards.

The propagation of gaussian laser beams and their focusing characteristics are discussed in detail in an excellent review paper by Kogelnik and Li [35]. The report by Weidel and Pedrotti [36] is also a concise summary of the relevant equations. The behavior is shown in Fig. 5. The effect of an ideal thin lens with focal length f

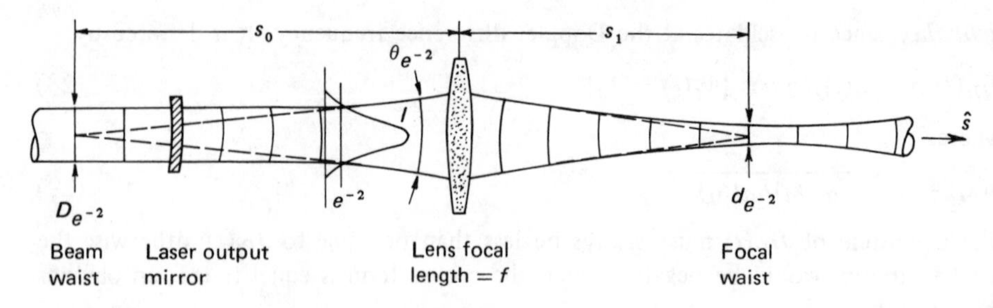

Figure 5 Properties of a focused gaussian beam.

(i.e., a lens with zero aberration) is to convert the spherical wave diverging from the laser into a converging spherical wave whose radius of curvature first decreases as though the wave were converging to a point at s_1, and then increases until it is infinite exactly at the point s_1 (i.e., the wave is planar) where the beam coincidentally has a minimum diameter $d_{e^{-2}}$. This focal point is located at

$$s_1 = f + \frac{s_0 - f}{(s_0/f - 1)^2 + (\pi D_{e^{-2}}^2/4f\lambda)^2} \tag{33}$$

and the diameter of the focal spot is given by

$$\frac{1}{d_{e^{-2}}^2} = \frac{1}{D_{e^{-2}}^2}\left(1 - \frac{s_0}{f}\right)^2 + \left(\frac{\pi D_{e^{-2}}}{4f\lambda}\right)^2 \tag{34}$$

Equation (33) states that the minimum beam diameter does not occur exactly in the focal plane of the lens unless $s_0 = f$. Since the lens in a dual-beam LDV will cause the beams to cross at f when the beams are initially parallel, errors can occur if $s_1 \neq f$ because the wave fronts will not be planar in the beam intersection. These errors have been analyzed in [37]. For typical LDV parameters the errors are negligible for focal lengths up to several hundred millimeters, and the following approximate equations are satisfactory:

$$s_1 \simeq f \tag{35}$$

$$d_{e^{-2}} \simeq \frac{4f\lambda}{\pi D_{e^{-2}}} \tag{36}$$

This last equation is especially important. It implies that small measurement volumes require either large beam diameters or short focal lengths. Practically, these equations are also valid for long-focal-length LDV systems since, to maintain spatial resolution and high focal-spot intensity, one commonly employs beam expansion before the beam splitter to increase the effective value of $D_{e^{-2}}$.

In the neighborhood of the focal spot the exact equations show [35] that the focused gaussian beam is essentially a plane wave whose diameter is nearly constant and whose intensity distribution is gaussian. It can be represented by

$$\mathbf{E}_{0l} = \sqrt{I_{0l}}\; e^{j(\omega_{0l}t - k\hat{\mathbf{s}}_l\cdot\mathbf{x})}\,\hat{\mathbf{p}}_l \tag{37}$$

wherein

$$I_{0l} = \frac{8P_{0l}}{\pi d_{e^{-2}}^2} \exp\left(\frac{-8\zeta^2}{d_{e^{-2}}^2}\right) \tag{38}$$

where P_{0l} = total beam power

ζ = radial distance from centerline of beam

The region of nearly constant diameter and nearly plane-wave behavior, called the *beam waist*, extends for approximately $(f/D_{e^{-2}})d_{e^{-2}}$ on either side of the focal spot. Hence, it is very long compared to the spot diameter whenever the lens F number, $f/D_{e^{-2}}$, is large–i.e., almost always.

Equation (37) is identical to Eq. (1), so the earlier assumption about illuminating the particle with a plane wave is justified in the region of the beam waist. We note that *the intensity distribution in the beam waist of a focused gaussian beam is gaussian.* Also, the peak intensity at the centerline is inversely proportional to $d_{e^{-2}}^2$; that is, the illuminating-beam intensity is inversely proportional to the square of the focusing-lens F number.

Conventionally, the LDV measurement volume has been defined as the region in which the amplitude of the Doppler signal is greater than $1/e^2$ of the maximum amplitude that the particle could produce, i.e., all points $\mathbf{x}$ such that

$$a(\mathbf{x}_i)D_i \geqslant e^{-2}\, a(\mathbf{0})D_i \tag{39}$$

Using Eq. (38) it can be shown [38] that

$$a(\mathbf{x}_i) = k^{-2}\sqrt{I_{01}(\mathbf{0})\,I_{02}(\mathbf{0})}\,\exp\left[-\frac{8}{d_{e^{-2}}^2}\,(x_i^2\cos^2\kappa + y_i^2\sin^2\kappa + z_i^2)\right] \tag{40}$$

so that the amplitude ratio will exceed e^{-2} whenever the magnitude of the exponential is less than 2. This defines an ellipsoidal measurement volume as shown in Fig. 6, with axes in the x, y, and z directions given by

$$d_m = \frac{d_{e^{-2}}}{\cos\kappa} \tag{41}$$

$$l_m = \frac{d_{e^{-2}}}{\sin\kappa} \tag{42}$$

$$h_m = d_{e^{-2}} \tag{43}$$

respectively. The volume enclosed by the ellipsoid is

$$V_D = \frac{\pi d_{e^{-2}}^3}{6\cos\kappa\,\sin\kappa} \tag{44}$$

Note that the volume becomes very long in the y direction when $\sin\kappa$ is small. Typical dimensions in a system that has not been designed specifically for high spatial resolution are $d_m \approx 0.1$ mm, $l_m \approx 0.8$ mm, and $h_m \approx 0.1$ mm. With special design these values can be reduced by more than one order of magnitude.

The e^{-2} definition of the measurement volume (hereafter abbreviated mv) provides a convenient basis for comparing different LDV systems; moreover, it is an appropriate definition when certain types of signal processors (e.g., spectrum analyzers or correlators) are used, or if there are many particles contributing to the signal simultaneously. However, if the signal is a series of nonoverlapping Doppler bursts from single particles and if the signal processor is a counter or a frequency tracker, it is inappropriate because these processors will not process the signal at all, unless the amplitude exceeds a fixed threshold voltage level at the input to the processor. The threshold level is a characteristic of the particular processor, and it can often be adjusted by the operator, essentially by varying the gain of the signal-processor preamplifier. It can be related to a threshold light-flux level $J_{\min}$ with knowledge of the proportionality between the light flux and the voltage at the processor input.

In this case, the mv is defined by requiring

$$a(\mathbf{x}_i)D_i \geqslant J_{\min} \tag{45}$$

This again defines an ellipsoidal volume, but the lengths of the major axes are now

$$\begin{bmatrix} l_x \\ l_y \\ l_z \end{bmatrix} = \begin{cases} \dfrac{d_{e^{-2}}}{\sqrt{2}} \left(\ln \dfrac{a(\mathbf{O})D_i}{J_{\min}} \right)^{1/2} \begin{bmatrix} \dfrac{1}{\cos \kappa} \\ \dfrac{1}{\sin \kappa} \\ 1 \end{bmatrix} & a(\mathbf{O})D_i \geqslant J_{\min} \\ \begin{bmatrix} 0 \\ 0 \\ 0 \end{bmatrix} & a(\mathbf{O})D_i < J_{\min} \end{cases} \tag{46}$$

Thus, the size of the mv depends upon the laser power and the illuminating-beam focusing [through $I_{01}(\mathbf{0})$ and $I_{02}(\mathbf{0})$], the scattering properties of the particle and the light-collecting efficiency (through D_i), and the threshold voltage-detection level, photodetector sensitivity, and postdetector amplification (through $J_{\min}$). When polydisperse suspensions of particles are used, e.g., natural aerosol or hydrosol, the scattering characteristics will vary dramatically, with larger particles usually (but not always) producing bigger signals. A large particle might produce a detectable signal when its trajectory is far from the center of the mv, while a small particle might not produce a detectable signal unless its trajectory passes through a region that is rather smaller than the nominal $1/e^2$ volume. Finally, certain particles may scatter so weakly as to produce no detectable signals, even when they pass through the center of the mv.

To summarize, if the photodetector views the entire beam intersection, and if the particles are polydisperse, then the mv is not a precisely defined region. Rather, it can be thought of as an expandable volume whose size is determined by the value of D_i for each particle. This uncertainty is eliminated if the particles are monodisperse or if the measurements are interpreted statistically.

If the particle follows the fluid motion, its trajectory $\mathbf{x}_i(t)$ is a lagrangian

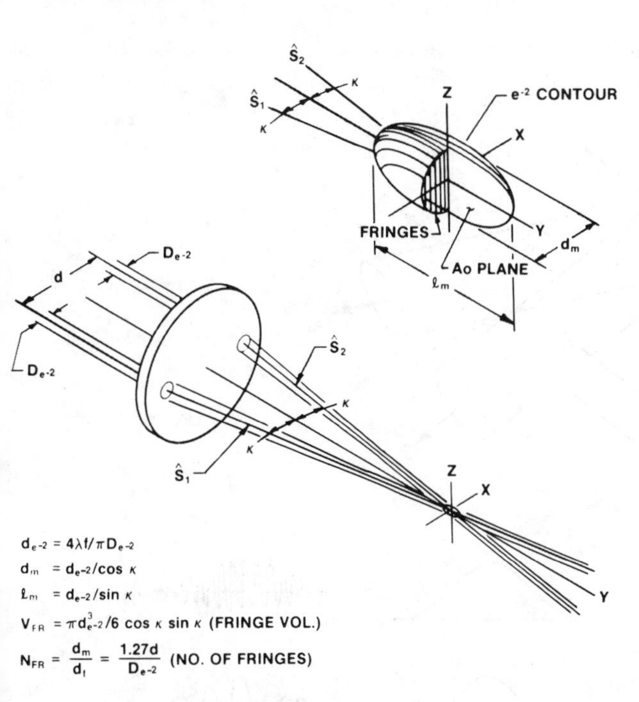

Figure 6 Geometry of the nominal LDV measurement volume. *(Courtesy TSI, Inc.)*

trajectory of the fluid, and the velocity $\mathbf{v}_i$ measured by the LDV is a lagrangian fluid velocity. However, since the LDV measures this velocity only when the particle is in a small neighborhood of a fixed point in the flow, the measurement also represents the eulerian velocity at that point. This is true so long as the mv is small compared to the size scale of the fluid motion. Thus, one can view the individual particle LDV signal as a sequence of eulerian velocity samples that are obtained at random points in time because the particles are originally located at random points in space.

The form of the signal burst produced by the ith particle can be found as a function of time and velocity by substituting the equation for its trajectory,

$$\mathbf{x}_i(t) = \mathbf{x}_i(0) + \int_0^t \mathbf{v}_i(\xi)\, d\xi \tag{47}$$

into Eq. (26b) and using the gaussian intensity distribution described by Eq. (40). When the mv is so small that the particle trajectory is essentially a straight line during its crossing of the mv, it is easy to show that $J_i(t)$ always consists of two gaussian pedestals plus a burst of Doppler sine wave at frequency ν_{Di}. *The envelope of the sine wave is always a gaussian function of time, for any particle trajectory.*

Some typical signals are shown in Fig. 7 for various trajectories and signal visibilities. Note that the individual pedestals coincide for trajectories in the xz plane, so that they appear to be a single pedestal (trajectories a and b). While this is

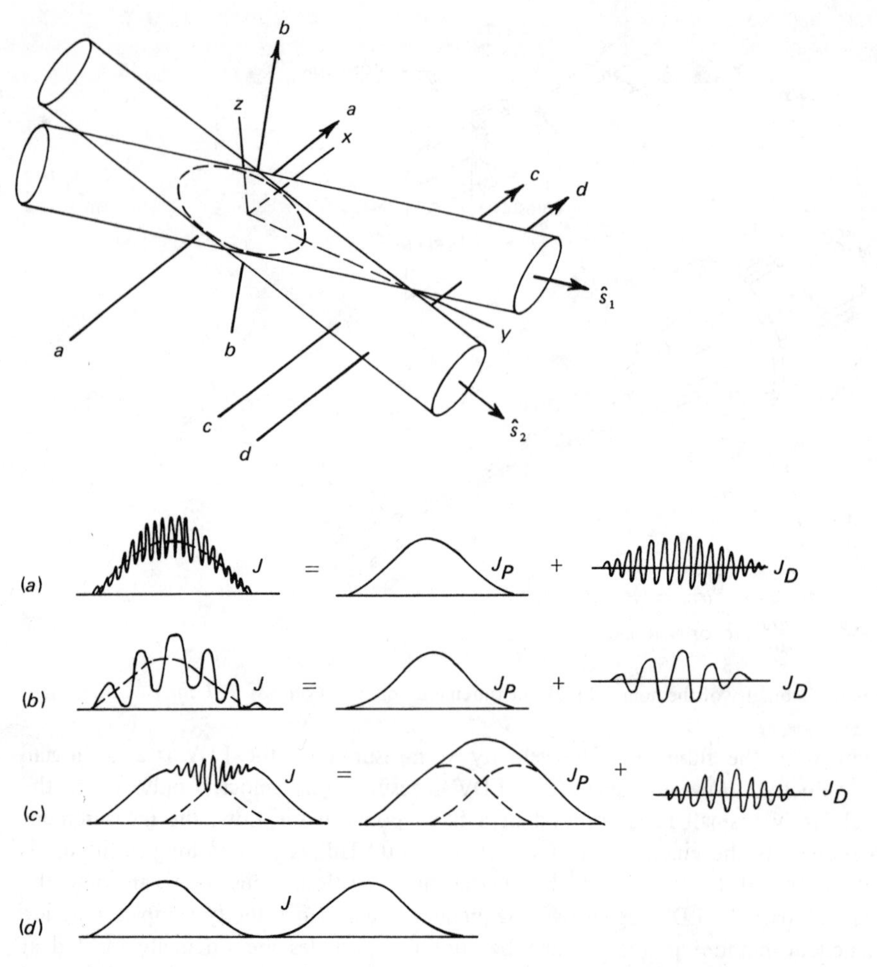

Figure 7 Typical LDV signals produced by various partial trajectories decomposed into pedestals plus Doppler bursts.

the situation that is most often sketched in the LDV literature, it should be remembered that it is a special case. The pedestal signals vary slowly relative to the rate at which the Doppler sine wave oscillates, so they consist of low-frequency components that can be removed from the total signal by high-pass filtering, leaving only the Doppler signal. Decomposing the total signals in this way yields the separate pedestal and Doppler components shown in Fig. 7.

3.2.2 Fringe Model of the Dual-Beam LDV. The fringe model of the dual-beam LDV is an alternative explanation of the operation of the dual-beam LDV that avoids reference to the Doppler shift effect. This model, first proposed by Rudd [39], is useful because it is usually much easier to visualize than the Doppler shift model. Unfortunately, it is not entirely correct, and it can predict unrealistic results, a fact that was not realized in some of the early work. Even so, it does yield correct results in many regards, and certain concepts based on the fringe model have become so ingrained in the LDV literature that a discussion of the model is warranted.

The fringe model is based on the observation that the light waves of the illuminating beams will interfere to form a set of interference fringes at the intersection of the beams. We can see this mathematically by calculating the intensity of the sum of the two illuminating waves, $I_{00}(\mathbf{x}) = (\mathbf{E}_{01} + \mathbf{E}_{02}) \cdot (\mathbf{E}_{01}^* + \mathbf{E}_{02}^*)$. The algebra is exactly the same as in Eq. (24*a*) with $\mathbf{E}_{li}$ replaced by E_{0l}. The result is

$$I_{00}(\mathbf{x}) = I_{01}(\mathbf{x}) + I_{02}(\mathbf{x}) + 2\sqrt{I_{01}(\mathbf{x})I_{02}(\mathbf{x})} \cos(\omega_S t + 2kx \sin\kappa) \tag{48}$$

This intensity distribution is drawn in Fig. 8 assuming gaussian illuminating beams with $P_{01} = P_{02}$, and $\omega_{01} = \omega_{02}$. The interference fringes, i.e., the last term in Eq. (48), lie in planes parallel to the yz plane, like a deck of cards. The spacing between fringes is given by $2k\, d_f \sin\kappa = 2\pi$, so that

$$d_f = \frac{\lambda}{2 \sin\kappa} \tag{49}$$

Now suppose that a very small particle crosses this fringe pattern and scatters light such that the intensity of the scattered light is proportional to I_{00}. Then the scattered light flux will oscillate sinusoidally as the particle crosses the fringes, and the frequency of the oscillation will be

$$\nu_D = \frac{u}{d_f} = \frac{2u \sin\kappa}{\lambda} \tag{50}$$

since d_f/u is the time for the particle to cross one fringe. The velocity components v and w are, of course, irrelevant, since they represent motion *parallel* to the fringes. This frequency is exactly the same as that obtained from Doppler shift considerations, and it will be noted that the wave number defined in Eq. (16) is just

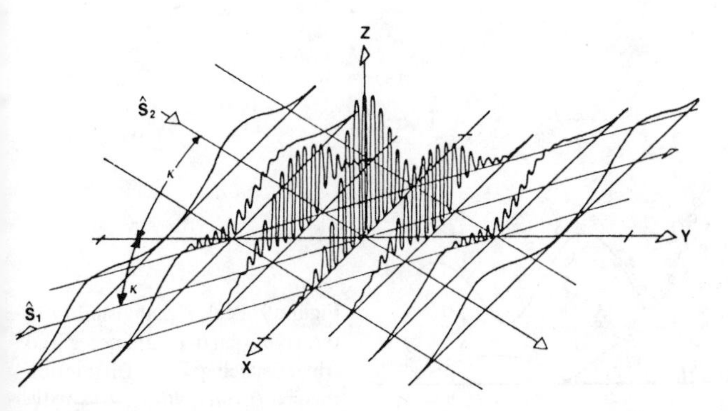

Figure 8 Intensity distribution of the fringes in the dual-beam measurement volume. *(Adapted from [39a].)*

$K = 2\pi/d_f$. Moreover, the signal characteristics are also similar. From Fig. 8 the intensity consists of two pedestals (the intensity profiles of each beam) and a gaussian amplitude-modulated sine wave (the interference term).

If $\omega_{02} - \omega_{01} \doteq 2\pi\nu_S$ the cosine term in Eq. (48) implies that the fringe pattern moves with velocity $u_f = \nu_S\, d_f$, while the envelope of the cosine, that is, the mv, remains fixed. In effect, the interference of two equal-frequency waves creates a standing-wave interference pattern, while the interference of two unequal-frequency waves creates a traveling-wave pattern. The velocity u_f is such that fringes would sweep across a stationary particle at a rate equal to the shift frequency ν_S. If the particle moves with (or against) this fringe motion, the crossing rate is lower (or higher) than the frequency difference by the amount u/d_f. Here too, the results are identical to the Doppler shift theory.

It will be seen later that the number of fringes in the mv, defined as

$$N_{FR} = \frac{d_m}{d_f} \tag{51}$$

is an important parameter that characterizes many of the LDV signal properties.

Thus far, the fringe model is entirely adequate. The deficiency is manifested when one attempts to calculate the signal strength when the particle is not small compared to d_f. In the context of the fringe model the signal should be proportional to the total light flux striking the particle, i.e., the integral of $I_{00}(\mathbf{x})$ over the particle's surface. Farmer [40] has performed this calculation by integrating I_{00} over a disk whose diameter equals the particle diameter d_p. A primary result of this calculation is that the peak visibility $\bar{V}$ of the Doppler signal depends only on the ratio d_p/d_f of the particle diameter to the fringe spacing when both illuminating beams have equal intensities and polarizations in the z direction. This dependence is shown by the dashed line in Fig. 9. For small d_p/d_f the visibility equals unity, implying that the amplitude of the Doppler signal is just proportional to the Doppler component of $I_{00}(\mathbf{x})$, as expected. However, when $d_p/d_f = 1.22$,

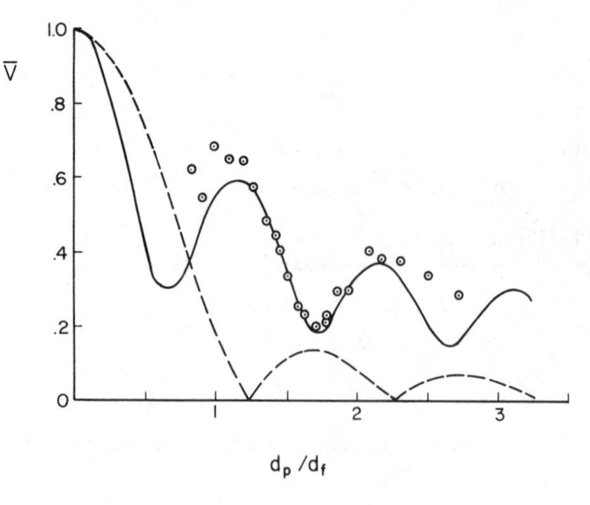

Figure 9 Peak signal visibility in terms of particle diameter and fringe spacing: - - - Diffraction theory *(from [40])*; —— Mie scattering theory *(from [34])*; ⊙ experimental data.

corresponding to the particle being covered equally by one bright fringe and one dark fringe, the visibility vanishes because the integral of I_{00} remains constant as the particle moves, implying zero Doppler signal amplitude. $\bar{V}$ also vanishes at each value of d_p/d_f that corresponds to an integer number of whole fringes.

These implications of the fringe model are not in general, correct. For example, the data points in Fig. 9, obtained by using 9.8-μm-diameter particles in back scatter and varying the fringe spacing, exhibit a maximum at about the same d_p/d_f value that the fringe model predicts as a minimum. The reason for this discrepancy is that the fringe model deals with intensity at the particle, but in reality detection of the light intensity occurs far away, at the photodetector. The scattering process is an intervening step between illumination and detection that must be accounted for in the correct analysis. Calculation of $\bar{V}$ using Mie scattering theory for spherical particles to evaluate the scattering coefficients in Eq. (24*b*) yields the solid curve in Fig. 9, and this does agree well with the data [34]. Scattering theory predicts that the signal visibility depends, in general, upon the values of $I_{02}(\mathbf{x})/I_{01}(\mathbf{x})$, $\hat{\mathbf{p}}_1 \cdot \hat{\mathbf{p}}_2$, $\hat{\mathbf{s}}_l \cdot \hat{\mathbf{p}}_m$, d_p/d_f, $2\pi d_p/\lambda$, the location and geometry of the solid angle for light collection, and the ratio of the refractive index of the particle to the refractive index of the fluid.

The relationship between the fringe model and the scattering model is discussed in [34, 41, 42], where it is shown that the fringe model is correct in the scalar diffraction-theory limit $2\pi d_p/\lambda \gg 1$, provided that light is collected in forward scatter, and the light-collecting aperture is very large. Thus, when d_p and d_f are large enough, e.g., a 20-μm particle in 10-μm fringes, the fringe model is accurate, in agreement with physical intuition. But when the particle diameter too small ($2\pi d_p/\lambda$ less than about 60), it fails, together with our intuitive understanding of scattering on such small scales.

3.2.3 Signal Coherence. Suppose there are two particles in the mv of the dual-beam system, instead of one. Call these particles *i* and *j*. Particle *i* will scatter two waves: $\mathbf{E}_{1i}$ from beam 1 and $\mathbf{E}_{2i}$ from beam 2. Likewise, particle *j* will scatter $\mathbf{E}_{1j}$ and $\mathbf{E}_{2j}$. Heterodyne mixing of $\mathbf{E}_{1i}$ with $\mathbf{E}_{2i}$ produces the single-particle signal discussed in Sec. 3.2.1, and likewise for $\mathbf{E}_{1j}$ mixing with $\mathbf{E}_{2j}$. However, there will be additional signals due to $\mathbf{E}_{1j}$ mixing with $\mathbf{E}_{2i}$ and $\mathbf{E}_{1i}$ mixing with $\mathbf{E}_{2j}$. These signals will not be as strong as the single-particle signal because the scattered waves originate from two points in space $\mathbf{x}_i$ and $\mathbf{x}_j$, which may not, in general, be coincident. This loss of signal strength is referred to as "spatial incoherence."

The easiest way to understand this effect is to consider the mixing that occurs at the detector surface when light waves from two separated sources are collected and focused onto the surface, as shown in Fig. 10. If the light waves look like spherical waves coming from point sources (such as a micron-sized particle), diffraction at the lens produces an image of the source whose diameter is on the order of $\lambda f_c/D_a$, where D_a is the diameter of the light-collecting aperture. Hence, if the images of the two sources are separated by more than the diffraction-limited spot size, they will not overlap, and no heterodyne mixing can occur. In this case

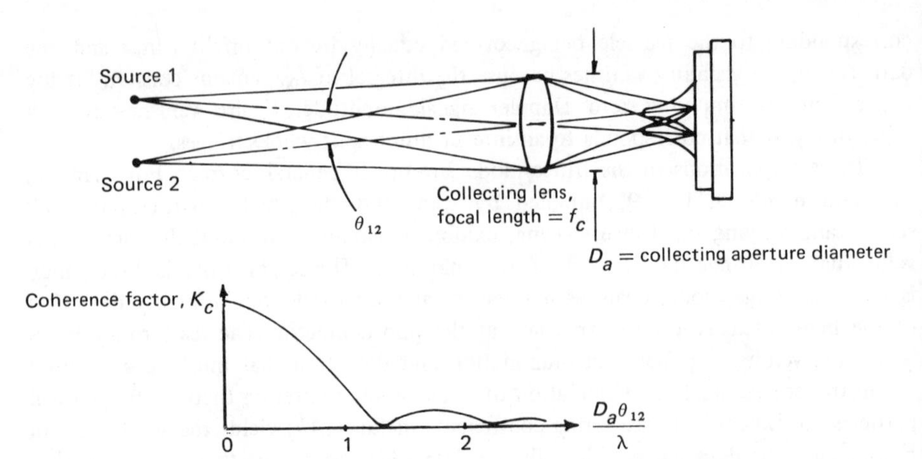

Figure 10 Source-aperture coherence function. *(From [47].)*

no Doppler signal will be observed, although there will be pedestals associated with each particle crossing each beam.

At the other extreme, if the angular separation θ_{12} is zero, the images overlap perfectly and there is perfectly coherent heterodyne mixing so long as the particles are both in focus at the detector. The efficiency of mixing for partial overlap is defined as the amplitude of the mixing signal divided by the amplitude for perfectly coherent mixing. It is denoted by K_c and is commonly called the *coherence factor* [43, 44]. K_c depends on the ratio of the image separation to the image diameter, $f_c\theta_{12}/(f_c\lambda/D_a) = D_a\theta_{12}/\lambda$, as shown qualitatively in Fig. 10. The dual source signal will have poor mixing efficiency when either D_a or θ_{12} is large. Since the two particles must be within the mv, the maximum possible value of θ_{12} is about $d_{e^{-2}}/f_c$. The maximum diameter of the light-collecting aperture that will still yield efficient mixing is $D_c \approx f_c\lambda/d_{e^{-2}} \approx (f_c/f)D_{e^{-2}}$. That is, if $f_c \approx f$, the "coherent aperture diameter" D_c is about the same size as the initial laser-beam diameter, which is only a few millimeters!

Obviously one would like to use a large light-collecting aperture to receive as much scattered light energy as possible, but from the foregoing arguments the loss of mixing efficiency associated with increasing D_a may offset the gain in collected light energy to the extent that the Doppler signal amplitude actually decreases as the aperture is increased. This result pertains only to the dual source signal which, somewhat confusingly, is called the "coherent" signal because the aperture must satisfy the spatial coherence requirements. The signal from a single particle is called "incoherent" because the waves scattered from it always appear to originate from the same point and, hence, always satisfy the coherence requirement; that is, $K_c = 1$ for the coherent signal, regardless of the collecting-aperture diameter.

Suppose that the scattering particle concentration C (number per unit volume), and hence the mean number of particles in the mv,

$$N_D = CV_D \tag{52}$$

can be adjusted. When $N_D \ll 1$, the probability of two or more particles in V_D is negligible, and the only signals will be the incoherent signals from individual particles. When $N_D \gg 1$, there will be about N_D^2 pairs of particles producing N_D^2 coherent signals plus N_D incoherent signals. Now, is it better to use a small coherent aperture and many particles so that the N_D^2 coherent signals will mix efficiently, or is it better to use a large aperture and few particles so that each particle produces a large signal? The answer depends, of course, on the specifics of the application, but in the vast majority of applications the large-aperture, incoherent signal provides the best result. In fact, the coherent signal is usually used only when the particle concentration is naturally or unavoidably large. It should be noted that the coherent signal can be reduced to negligible amplitude simply by making the collecting aperture much larger than D_c. Then, even if $N_D \gg 1$, the total signal will be the sum of many incoherent signals. Quantitative analyses of the coherent and incoherent signals are presented in [43, 44].

4. THE REFERENCE-BEAM LDV

As discussed in Sec. 2.3, the reference-beam LDV mixes light $\mathbf{E}_{1i}$, scattered from an illuminating beam, with a reference beam to detect the frequency difference. The analysis of the reference-beam signal is the same as the analysis of the dual-beam signal, except that $\mathbf{E}_2(\mathbf{r})$, the reference wave evaluated on the surface of the photodetector, is a spherical wave that diverges from the focal point of the reference beam. For best performance this focal point *appears* to be located at the center of the mv, and in practice this is usually accomplished by actually placing it there, as in Figs. 2 and 3. Thus, the reference-beam LDV may look rather similar to the dual-beam LDV, but this is not necessary. For example, the reference beam may be passed around the test section and still be made to originate apparently from the center of the mv by using a system of mirrors, lenses, and beam splitters, as in the system described in [45]. Foreman [46] has shown that when this is done the path-length difference between the reference beam and the illuminating beam-scattered light path must be an integer multiple of twice the laser cavity length to achieve efficient heterodyne mixing with lasers operating in multiple longitudinal modes. The best efficiency occurs for zero path-length difference. If the difference must be large, then the laser should be operated in a single longitudinal mode, i.e., "single frequency." The arrangement shown in Fig. 3 allows equal path lengths.

The principal drawback of the reference-beam LDV is that it must satisfy a coherent aperture condition, much like the dual particle signal of the dual-beam LDV. This is explained by letting source 1 in Fig. 10 represent the focus of the reference beam, fixed at the center of the mv, and letting source 2 be a scattering particle. The arguments presented for the coherence aperture of the dual particle signal apply equally well to this situation, with the result that the coherence aperture diameter is just the diameter attained by the illuminating beam at a distance f_c from the mv. Increasing the aperture diameter beyond this would produce increasingly inefficient mixing, but in compensation more light would be collected. In general, the mixing efficiency ultimately decreases faster than the scattered light flux increases, so that the best aperture diameter is not more than a few times D_c. Since this is relatively small, reference-beam systems are not usually used unless the aperture must subtend a small solid angle for other reasons, e.g., in a long-range system wherein D_a is constrained by cost whereas f_c is large, or unless the particle concentration is large. Drain [43] indicates that when N_D is large the reference-beam signal-to-noise ratio is twice the signal-to-noise ratio obtained from a dual-beam coherent signal.

If one considers the usual system, as shown in Fig. 3, the illuminating beam and the reference beam are both gaussian beams focused to the same spot size. Hence, the diameter of the reference beam at the light-collecting aperture matches the coherence aperture diameter and is so small that the scattering coefficient σ_1 is essentially constant over the solid angle Ω. (A simple, approximate criterion for σ_1 to be effectively constant over Ω is that the particle diameter must be less than the focal-spot diameter, which is almost always the case [47]). Analysis of this system shows that the Doppler signal is given [38] by

$$J_D = \frac{32\sqrt{P_{01}P_{02}}\ \mathrm{Re}\,(\sigma_{1i}\cdot\hat{\mathbf{p}}_2)}{d_{e^{-2}}^2 k^2} \exp\left[-8\left(\frac{x_i^2}{d_m^2} + \frac{y_i^2}{l_m^2} + \frac{z_i^2}{h_m^2}\right)\right] \cos\,[\Phi_i(t) - \Psi_i] \tag{53}$$

where Φ_i is given by Eq. (25*a*) (wherein $\hat{\mathbf{s}}_2$ is the direction of the reference beam), and Ψ_i is a phase shift determined by σ_{1i}. Comparison of Eqs. (28) and (53) shows that the single-particle reference-beam Doppler signal is identical in form to the single-particle dual-beam signal, aside from a constant factor in the amplitude. In particular, the reference-beam mv is the same as in Fig. 6, provided that the reference- and illuminating-beam focal spots are equal and the reference beam is not blocked at the collecting aperture. Of course, the total reference-beam signal also contains two pedestals, each of gaussian shape.

The amplitude of the reference-beam Doppler signal is proportional to the square root of the reference-beam power P_{02}. Hence, it can be increased almost without limit by the simple expedient of increasing P_{02}. This is a useful feature when the scattered light flux is small compared to a background radiation level (e.g., atmospheric air measurements in daylight). Even in laboratory applications the scattered light power is so weak (of the order of 10^{-12} to 10^{-6} W) that the reference beam must be attenuated substantially to reach a level that is comparable to the scattered light power. The system in Fig. 3 includes a variable neutral density filter for this purpose.

The dual-scatter system is used so infrequently (because of small aperture limitations and alignment difficulties) that it will not be discussed here.

5. MULTI-VELOCITY-COMPONENT SYSTEMS

It is possible to measure two or even three components of the velocity vector by using several single-component systems. Although this seems a simple task, it is complicated by the necessity of identifying which one-component system the scattered light signal is coming from. This is easy with the reference-beam and dual-scatter systems because the direction of the velocity component to be measured is determined by the direction in which scattered light is collected. But when using the dual-beam approach, one must supply two basic illuminating beams plus at least one additional illuminating beam for each additional velocity component. Then the scattered light waves from each beam will heterodyne with the waves from every other beam to produce several signals, and the problem is to identify the signals coming from each pair of beams. (Alternatively, one can imagine one set of fringes being formed by each beam pair, and the subsequent complexity of the signals produced by a particle crossing all the fringe systems simultaneously.)

The identification problem is solved by labeling each pair of beams using some kind of optical property. The three possible methods are: different colors, different frequencies, and different polarizations. For the purpose of measuring the u and w components of velocity perpendicular to the optic axis of the LDV system there are also two possibilities: a four-beam arrangement and a three-beam arrangement that can be oriented in two ways. The various combinations are shown schematically in Fig. 11, wherein the beams appear as they would be seen when viewed from behind the focusing lens, looking toward the mv.

The four-beam, two-color, dual-beam LDV is shown in the upper left-hand square of Fig. 11. The blue and green lines from an argon-ion laser are each split and focused to form two mutually orthogonal dual-beam systems. Scattered light is collected by a pair of receiving optics/photodetector subsystems, one accepting only

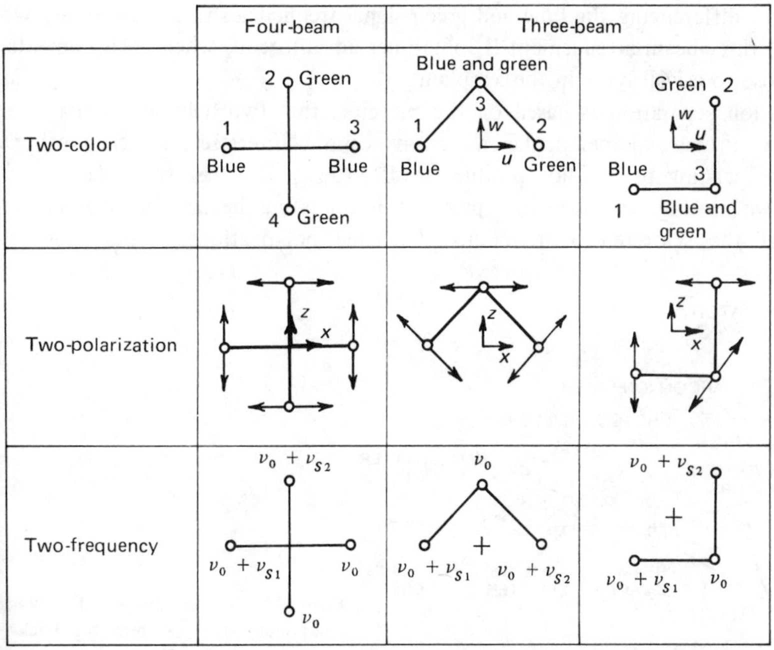

	Four-beam	Three-beam	
Two-color	2 Green 1 Blue, 3 Blue 4 Green	Blue and green 3 w, u 1 Blue, 2 Green	Green 2 w, u Blue 3 1 Blue and green
Two-polarization	z, x	z, x	z, x
Two-frequency	$\nu_0 + \nu_{S2}$ $\nu_0 + \nu_{S1}$, ν_0 ν_0	ν_0 + $\nu_0 + \nu_{S1}$, $\nu_0 + \nu_{S2}$	$\nu_0 + \nu_{S2}$ + $\nu_0 + \nu_{S1}$, ν_0

Figure 11 Illuminating-beam geometries for two-component LDV systems.

blue light and the other accepting only green light. The outputs of the blue and green detectors then correspond to the velocity components measured by the blue and green illuminating-beam pairs, u and w, respectively.

While simple in principle, the two-color system is more complicated than two one-color systems, largely because of problems associated with keeping the blue and green channels separate as the light propagates through the optical system. A typical two-color design using a prism to separate the colors at the laser head is shown in Fig. 12. Assuming that the primary focusing lens is achromatically corrected, the only condition needed for coincidence of the measurement volumes of the blue and green pairs is that the blue and green lines be parallel ahead of the beam splitters (which, presumably, maintain parallelism). This is accomplished by fine tilting adjustments of the mirrors that deflect the beams toward the beam splitters. Light scattered from both beam pairs is collected by the same lens. Thereafter a dichroic mirror transmits the blue light and reflects the green light to accomplish separation. The various polarization rotators in the system are used to align the laser polarization for optimal operation of the other optical elements (see [48] for applications details).

The two-color, three-beam systems in Fig. 11 eliminate one beam by combining blue and green light into a single beam. Light scattered from beam 1 and the blue portion of beam 3 mixes to form a blue heterodyne signal, and light scattered from beam 2 and the green portion of beam 3 forms a green heterodyne signal. In the second of the three-beam arrangements (upper right-hand corner of Fig. 11) the blue signal measures u and the green signal measures w. In the other three-beam arrangement the blue signal measures $u + w$, and the green signal measures $u - w$. Summing and differencing the blue and green signal frequencies then yield u and w. This type of three-beam arrangement is sometimes advantageous when measurements are made close to a wall lying in the xy plane.

Polarization separation is based on the principle that two light waves that are polarized in mutually perpendicular directions cannot interfere with each other. [This is the meaning of the dot product in Eq. (23).] The first two-polarization system shown in Fig. 11 uses two pairs of illuminating beams that are cross-polarized. If the scattered light retains the same polarizations as the incident

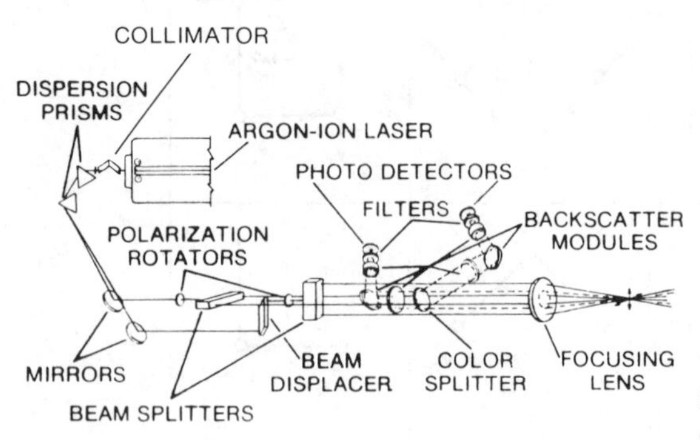

Figure 12 Typical two-color, two-component LDV system for backscatter. *(Courtesy TSI, Inc.)*

illuminating beams, the signals can be separated by passing the scattered light through a polarization cube or a similar device that reflects one polarization and transmits the other. The difficulty with this technique is that the scattering process does in fact depolarize the light, especially when the scattering particles are large (of the order of 10 μm diameter). Its advantage is simplicity. In particular, it does not require a multicolor laser. The three-beam polarization systems employ one beam whose direction of polarization has vector components in the directions of each of the other two beams.

The four-beam frequency separation system in Fig. 11 produces signals centered at $\nu_{S2} - \nu_{S1}$, ν_{S2}, 0, and ν_{S1} by heterodyne mixing of the light scattered from beams 1 and 2, 2 and 3, 3 and 4 and 1 and 4, respectively. By choosing ν_{S1} and ν_{S2} judiciously, these four carrier frequencies can be separated sufficiently to permit signal separation by band-pass filtering. For example, $\nu_{S1} = 30$ MHz and $\nu_{S2} = 40$ MHz yield carrier frequencies of 10 MHz, 30 MHz, 40 MHz and 0 MHz. Discarding the 0- and 10-MHz signals leaves the 30-MHz signal from pair 1-3, which measures u, and the 40-MHz signal from pair 2-4, which measures w. This signal can be separated as long as the Doppler shifts for u and w do not exceed 5 MHz. Thus, this technique is limited to moderate flow rates, roughly less than 20 m/s. Hornkohl et al. [49] describe a system of this type that uses a two-dimensional Bragg cell, and a three-beam system that uses a single Bragg cell driven at 30 MHz and 40 MHz is described in [50].

Three-component LDV systems present special problems because the v component of velocity (the "on-axis" component) is awkard to measure directly with a dual-beam system. One approach is to use a reference-beam system to measure v and a two-component dual-beam system to measure u and w. The v component is measured by illumination along the y axis and light collection along the negative y axis; that is, $2\kappa = 180°$. In this way all the light-collecting optics can also be kept on the same side of the test section as the transmitting optics.

When simultaneous measurements of u, v, and w are not required, one can use a single-component LDV tilted at three different angles to measure three components that are not necessarily orthogonal. The mean components U, V, and W can be determined from these data, and in certain instances the higher-order moments such as rms or Reynolds stress can also be calculated [51, 52].

6. PHOTODETECTORS

The photodetector system converts J_{tot}, the total flux of light energy striking the detector, into a voltage $e(t)$ that is the input to the signal processor. The total light flux is the sum of the background light flux J_B (e.g., room lights, radiation from the fluid, reflections, and flare), the pedestal light flux J_P, the Dopper light flux J_D, and, if appropriate, the reference-beam light flux J_R:

$$J_{tot} = J_B + J_P + J_D + J_R \tag{54}$$

Ideally, $e = S\tilde{J}_{tot}$, where S is the sensitivity of the detector system in volts per watt, and $\tilde{J}_{tot}$ denotes the value of $J_{tot}(t)$ after it has been filtered by high-pass and/or low-pass filters following the detector. However, real detector systems always contain noise. Part of the noise is generated in the electronics between the photodetector and the signal processor, and part is generated within the detector itself. The latter, called *shot noise*, is inherent in the photodetection process, and it places a fundamental limit on the signal-to-noise ratio.

6.1 Detector Characteristics

The photomultiplier tube (PMT) uses the photoelectric effect wherein photons striking a coating of photoemissive material on the photocathode cause electrons to be emitted from the material. The *quantum efficiency* of the photocathode is defined as the mean number of electrons emitted per photon. Since the mean number of photons per second is $J_{tot}/h\nu_0$, the *mean emission rate*, i.e., the mean number of photoemissions per second, is given by

$$\dot{\epsilon} = \frac{J_{tot}\eta_q}{h\nu_o} \tag{55}$$

where η_q is the quantum efficiency. In general η_q depends on the wavelength of the light. The photocurrent is amplified within the PMT by accelerating the electrons from the cathode through a high-voltage field and impacting them on a dynode that emits more than one electron, on the average, for each electron striking it. Repeating the amplification process through several stages of dynodes yields gains between 10^3 and 10^7, depending on the number of stages in the dynode chain and the applied voltage, which is usually 1 to 3 kV.

The PMT behaves as a light-controlled current source with a very high source impedance. Its output current at the anode is converted to a voltage by passing it through a "load resistor" whose value for high-frequency work is typically $R_L = 50$ Ω. The unfiltered anode current is a series of current pulses, each pulse being caused by a single cathode photoemission, but consisting of many electrons after dynode amplification. The width of these pulses is determined first by the time spread associated with electrons taking different paths through the dynode chain, and second by the anode capacitance C_a. In combination with the load resistance R_L that is connected to the anode, this capacitance forms a low-pass RC filter with time constant $R_L C_a$ and frequency response $1/2\pi R_L C_a$. With care, the pulse width may be made to approach the transit-broadened value, but normally the filtering by the anode capacitance dominates and the pulse width is of order $R_L C_a$. Tubes with frequency response up to 200 MHz are used in LDV. In applications requiring less frequency response, the output signal is filtered further to remove noise above the maximum frequency of interest. It may also be high-pass filtered to remove the signal pedestals.

The photodiode is a light-sensitive semiconductor junction whose reverse-biased conductivity changes with the incident light flux. By passing a constant current through the reverse-biased diode, the conductivity variations are converted to a voltage signal with properties much like those of the PMT signal. The chief differences between the PMT and the photodiode are (1) the photodiode works well at relatively high light levels, whereas the PMT will burn out if the light flux is too great; (2) the PMT works well at low light levels because of its high gain, whereas the ordinary photodiode has no internal gain or, in the case of avalanche photodiodes, very little gain (of the order of 10^2); (3) the quantum efficiency for good photodiodes is about 80% in the visible wavelengths, but only about 20% for good PMTs; and (4) photodiodes are much smaller and require smaller power supplies. PMTs are usually used in LDV because of their high sensitivities, but photodiodes should be considered for forward-scatter situations in which the intensity is high.

6.2 Photoemission Statistics

The photoemission process is inherently random, and this property greatly influences the approaches taken to extract the Doppler-shift information from the detector signal. For this reason, we discuss the PMT signal statistics in considerable detail. Similar considerations apply to photodiodes, with some changes in the details.

The primary source of randomness in the PMT signal is the fact that the time t_j at which the jth photoelectron is emitted is a random variable. If it is assumed for simplicity that J_{tot} is constant, the probability that the emission time t_j occurs in a small time interval $(t_1, t_1 + dt)$ is the same as the probability that it occurs in any other time interval $(t_2, t_2 + dt)$ and is proportional to dt. Hence, the probability of $n = k$ emissions in Δt seconds is given by the Poisson distribution,

$$\text{Prob } \{n = k\} = \frac{e^{-\dot{\epsilon}\,\Delta t}(\dot{\epsilon}\,\Delta t)^k}{k!} \tag{56}$$

with rate parameter $\dot{\epsilon}$. Here n is a random variable. Some well-known properties of this distribution [53] are that the mean number of emissions in Δt is $\langle n \rangle = \dot{\epsilon}\,\Delta t$; the rms fluctuations in the number of emissions in Δt is $\langle(n - \langle n \rangle)^2\rangle^{1/2} = (\dot{\epsilon}\,\Delta t)^{1/2} = \langle n \rangle^{1/2}$; and the probability density given by Eq. (56) is a maximum at $k = \dot{\epsilon}\,\Delta t$ for large $\dot{\epsilon}\,\Delta t$. Thus, while the rms fluctuation in n increases as $\langle n \rangle^{1/2}$, the *relative* rms fluctuation decreases as $\langle(n - \langle n \rangle)^2\rangle^{1/2}/\langle n \rangle = \langle n \rangle^{-1/2}$ and vanishes in the limit $\langle n \rangle \to \infty$. In this limit, the number of emissions in Δt is a deterministic value proportional to J_{tot}. However, for more modest mean emission rates the number fluctuates randomly about $\langle n \rangle$, and these fluctuations are the source of shot noise in $e(t)$.

6.3 Shot Noise

The foregoing results can be extended readily to allow for time-varying light fluxes by defining a time-dependent emission-rate parameter

$$\dot{\epsilon}(t) = \frac{\eta_q J_{tot}(t)}{h\nu_0} \tag{57}$$

and treating the statistics of the emission process in the sense of conditional averages, i.e., the mean number of emissions at time t given the value of $J_{tot}(t)$, the rms fluctuations given $J_{tot}(t)$, etc. [54].

As sketched in Fig. 13, the cathode current can be represented as a sequence of current impulses (with charge q_0) at random times t_j such that the number of impulses per second is large when $J_{tot}(t)$ is large. The dynode chain broadens the jth impulse because of travel-time spread and amplifies it by a random amount g_j so that the total anode current is represented by $\Sigma_j g_j h_p(t-t_j)$, where $h_p(t)$ is essentially the impulse response function of the dynode chain. Here again the pulse density is proportional to $J_{tot}(t)$. With the exception of photon correlation spectroscopy, which is covered later, one does not usually work directly with the raw anode current. Instead, the photocurrent is converted to a voltage and smoothed by analog filters to give $e(t)$.

Suppose that we let $g_j h(t-t_j)$ denote the filtered output of the system in response to a photoemission at t_j. The function $h(t)$ is the impulse response of the complete system consisting of the PMT, the load resistor, and any additional filters that operate on the signal, e.g., low-pass or high-pass filters applied to the PMT output. [If $h_f(t)$ represents the impulse response of the postdetector filters, $h(t)$ is

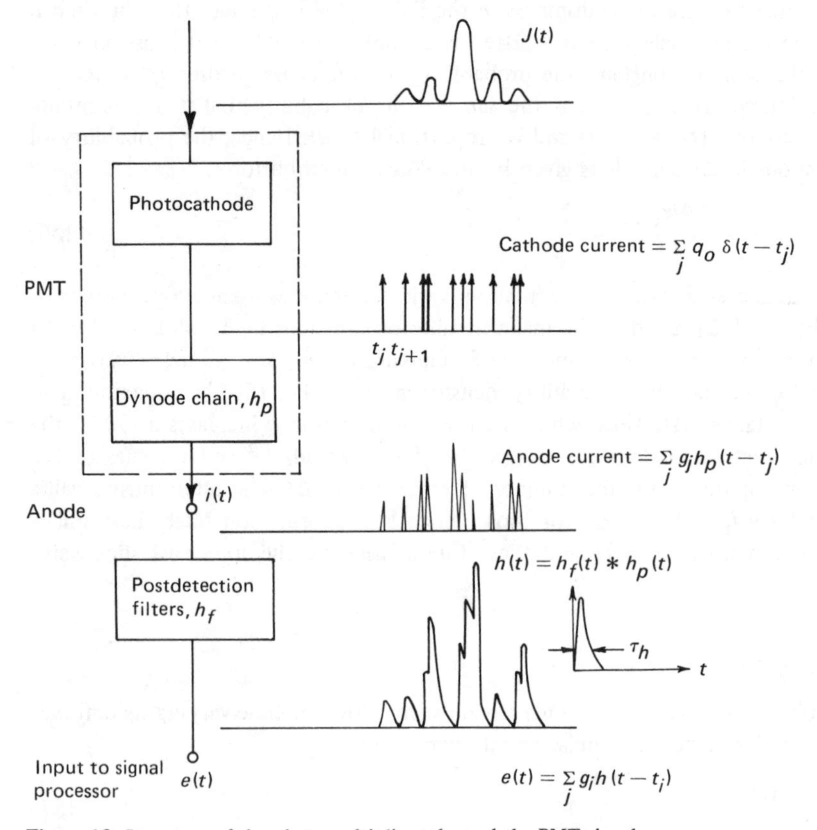

Figure 13 Structure of the photomultiplier tube and the PMT signal.

the convolution of $h_p(t)$ with $h_f(t)$.] Then the filtered voltage signal will be the sum of the responses to the individual photoemissions:

$$e(t) = \sum_{j=-\infty}^{\infty} g_j h(t - t_j) \tag{58}$$

This signal is sketched in Fig. 13.

The degree to which $e(t)$ corresponds to $J_{tot}(t)$ depends on the value of $\dot{\epsilon}\tau_h$, where τ_h is the characteristic time constant of $h(t)$, defined by

$$\tau_h = \frac{1}{2\pi\ \Delta f} \tag{59}$$

where Δf is the bandwidth of the PMT-filter system. Physically, $\dot{\epsilon}\tau_h$ represents the average number of photoemissions in one time constant. If $\dot{\epsilon}\tau_h \ll 1$, the average time between photoemissions $\dot{\epsilon}^{-1}$ is large compared to τ_h, and the filtered pulses rarely overlap. In this case, referred to as "low photon density," $e(t)$ would be a sequence of random pulses much like the anode current sketched in Fig. 13. Signal processors based on conventional frequency-demodulation techniques are developed essentially for modulated sine waves and would fail, utterly, to measure the Doppler frequency of this type of signal. Instead, one must use techniques such as photon correlation that work directly on the number of emissions per second.

In the other extreme, when $\dot{\epsilon}\tau_h \gg 1$, the pulses overlap to produce a more continuous signal, and $e(t)$ becomes a fairly good representation of $J_{tot}(t)$, although randomness in the emission times still results in some differences. These differences are called the shot noise. This case is called "high photon density," and conventional frequency demodulators (counters and trackers) work well on signals that fall into this class. The signal sketched in Fig. 13 actually corresponds to the intermediate case in which $\dot{\epsilon}\tau_h$ is of order 1. It is a recognizable, albeit ragged, replica of $J_{tot}(t)$.

It should be noted that $\dot{\epsilon}\tau_h$ is determined in equal measure by the optical system (through J_{tot}) and the bandwidth Δf. For example, the signal produced by a particular LDV system may be of low photon density when a 200-MHz bandwidth ($= \Delta f$) is used, but of high photon density when a 1-MHz bandwidth is used, quite simply because band-pass filtering rejects those components of the shot noise whose frequencies lie outside the pass band.

As in turbulent flow, it is convenient to describe the high-photon-density signal in terms of a mean value plus random fluctuations about the mean. However, time averaging must be avoided in defining the mean because a time average of $e(t)$ would not contain any information about the temporal structure (and hence the frequency) of $J_{tot}(t)$. The appropriate type of average is the conditional average of $e(t)$ given the value of J_{tot}. It can be shown that this average is given [54] by

$$\langle e(t) | J_{tot}(t) \rangle = K_1 \int_{-\infty}^{\infty} \dot{\epsilon}(\xi) h(t - \xi)\, d\xi \tag{60a}$$

or, $\langle e(t)|J_{tot}(t)\rangle = S\tilde{J}_{tot}$ (60*b*)

where K_1 is a proportionality constant, S is the sensitivity of the detector, and $\tilde{J}_{tot}$ is the filtered value of J_{tot}:

$$\tilde{J}_{tot}(t) = \int_{-\infty}^{\infty} J_{tot}(\xi)h(t-\xi)\,d\xi \tag{61}$$

For example, if the signal is high-pass filtered to remove the pedestals, $\tilde{J}_{tot}$ is just the Doppler signal, because the high-pass filter will also remove any dc components. Equation (60*a*) assumes that "dark currents," i.e., photoemissions that occur spontaneously, independent of J_{tot}, are negligible. The total signal is

$$e(t) = \langle e(t)|J_{tot}(t)\rangle + e_N(t) \tag{62}$$

where the total noise $e_N(t)$ is the sum of the shot noise in the detector e_n the electronics noise generated in the postdetector filters e_E, any RF noise picked up in the lines e_{RF}, and any spurious heterodyne signals e_H. (Including the latter in the noise term means that the spurious heterodyne light flux should not be included in $\tilde{J}_{tot}$.)

The electronics noise is independent of the light flux, and its power is usually constant in time. It is a combination of Johnson noise, whose power spectrum is "white," i.e., constant power-spectral density up to the filter cutoff frequency, and semiconductor surface noise whose power spectrum is inversely proportional to frequency. The RF pickup may include local radio and television broadcasts, but these are usually weak and can be shielded against. The primary source of RF pickup is the oscillator that drives the Bragg cell, and this can be quite troublesome because it appears to the signal processor as a very steady, continuous Doppler signal corresponding to zero velocity. Spurious heterodyne signals occur when *laser light* enters the photodetector from two or more sources such as reflections or flare from optical elements, flare from test-section windows, or scattering from particles either inside or outside the mv. When these sources are strong enough, the heterodyne signal may be large even though the coherence factors are small.

The shot noise is unique because it is an intrinsic part of the detection process. The expected value of the square shot noise given $J_{tot}(t)$ is [54, 47]

$$\langle e_n^2(t)|J_{tot}(t)\rangle = K_2\langle g_i^2\rangle \int_{-\infty}^{\infty} \dot{\epsilon}(\xi)h^2(t-\xi)\,d\xi \tag{63a}$$

$$\simeq 2\,\frac{h\nu_o}{\eta_q}\,\Delta f\,S^2 J_{tot} \tag{63b}$$

where $K_2 = \text{const}$ and Eq. (63*b*) is a satisfactory working formula.

7. SIGNAL–TO–NOISE–RATIO EFFECTS

At present there are a number of signal-processing techniques that will extract the Doppler frequency from relatively noisy signals ranging from high-photon-density signals that are only slightly contaminated by shot noise to low-photon-density signals that barely resemble $\tilde{J}_{tot}(t)$. One generally obtains the maximum amount of flow information per second when the signal-to-noise ratio (SNR) is high, and therefore it is advisable to strive for the best SNR that can be obtained within the constraints of the experiment. In this section we discuss the most important factors that affect the SNR.

The light flux J_{tot} in Eq. (63*b*) is the total flux of light energy striking the PMT, only a part of which is useful–i.e., the reference beam and/or the Doppler-shifted scattered light fluxes. The pedestal light flux is an inevitable part of the heterodyne detection method, but the background light flux J_B is nonessential. Moreover, it is thoroughly detrimental in that it contributes nothing to the signal power while increasing the shot-noise power, often substantially.

One of the first tasks in the setup of an LDV system is the minimization of extraneous background light. Even with commercial LDV systems this task falls squarely upon the shoulders of the user because extraneous light is primarily dependent on the flow apparatus. The pinhole aperture in front of the PMT rejects light that is not in the field of view of the pinhole, and narrow-band optical interference filters are often used to block light that is not close to the laser wavelength. Then, the primary sources of extraneous light are laser flare (e.g., the nonspecular reflection and diffraction of the laser light where it enters or leaves an optical surface) and reflections. These can be very large. For example, a typical 1-μm scattering particle illuminated by a 10-mW beam may scatter about 10^{-8} W through a backscatter collecting aperture. In contrast, 4% of a laser beam will be reflected at an air-glass interface, corresponding to 4×10^{-4} W from a 10-mW beam. If only $\frac{1}{4}$% of the reflected light were to be collected through the pinhole it would still be 100 times greater than the light flux from the particle, and the shot-noise power would be 100 times greater than necessary. Hence, very substantial improvements in the SNR can be achieved by minimizing laser flare and reflections.

The methods used to reduce J_B are (1) tilt the optics so that reflections miss the PMT pinhole; (2) use clean optical surfaces, including the test-section windows, or use antireflection-coated surfaces; (3) block stray light with black tape before it enters the collecting lens; and (4) place the pinhole far away from the collecting lens so that stray light that is not in focus is collected over a smaller solid angle. If background light has been reduced to a negligible level, the crossed laser beams will be seen clearly against a dark background when viewed through the receiving optics. If the background is brighter than the light scattered from the beams, extraneous light will contribute appreciably to the shot noise.

In situations such as measurements very close to a wall, the flare from the wall can be very difficult to eliminate by any of these methods because it originates so close to the mv. One proposed technique is to use fluorescent particles that would fluoresce, as they crossed the fringe pattern, at a wavelength *different* from the laser wavelength; then flare could be discriminated against by a narrow-band optical filter at the fluorescence wavelength. This technique would work well, provided that care is taken to avoid coating the wall with fluorescent dye via particle deposition. Alternatively, the wall can be painted with fluorescent paint to provide high absorption and low flare. Stevenson et al. [55] discuss the fluorescent-particle technique.

The ideal dual-beam LDV signal is obtained when all background light, spurious heterodyne signals, and electronics noise are negligible, and a single scattering particle resides at the center of the mv, where the illuminating beams, assumed to be of equal powers $P_{01} = P_{02} = P_0$, have maximum intensities. In this case the light flux on the detector is just the light flux scattered from the particle, given by Eq. (26*b*). The signal $e(t)$ still contains shot noise due to the scattered light itself, and this noise imposes a fundamental limit upon the maximum attainable signal-to-noise ratio. The signal power and the noise power are conveniently defined by time-averaging over one Doppler cycle. The period-averaged signal power is 0.5 $S^2 a^2(\mathbf{0}) D_i^2 = 0.5 \quad S^2 \bar{V}_i^2 J_{Pi}(\mathbf{x}_i = \mathbf{0})$, and the period-averaged noise power is $2(h\nu_0/\eta_q)\, \Delta f\, S^2 J_{Pi}(\mathbf{x}_i = \mathbf{0})$, yielding

$$\mathrm{SNR}_{\mathrm{peak}_i} = \frac{\eta_q}{4h\nu_0\ \Delta f}\ J_P(\mathbf{x}_i = \mathbf{0})\,\bar{V}_i^2 \tag{64a}$$

or

$$\mathrm{SNR}_{\mathrm{peak}_i} = \frac{\pi^2}{256}\,\frac{\eta_q P_0}{h\nu_0\ \Delta f}\left(\frac{D_a\, D_{e^{-2}}}{f_c f}\ \frac{d_{pi}}{\lambda}\right)^2 G_i\,\bar{V}_i^2 \tag{64b}$$

Equation (64*b*) pertains to a spherical particle with diameter d_{pi}, and the "scattering gain"

$$G_i = \frac{2(P_{1i} + P_{2i})}{k^2 d_{pi}^2\ \Omega} \tag{65}$$

is the ratio of the actual flux of light scattered through Ω, to the flux that would go through Ω if energy $[I_{01}(\mathbf{0}) + I_{02}(\mathbf{0})]\,\pi\ d_{pi}^2/4$ were scattered isotropically. The primary implication of Eq. (64*b*) is that the peak signal-to-noise ratio for the *i*th particle is proportional to the scattering power J_{Pi} of the particle and the *square of the visibility* of the heterodyne signal $\bar{V}_i^2$. Thus, doubling the signal visibility by an appropriate choice of particles is as effective as quadrupling the laser power.

The factor in parenthesis in Eq. (64*b*) contains the first-order effects of the LDV geometry, and it implies that the SNR decreases as the fourth power of the focal distance if $f_c \simeq f$, which is the case in most applications. Thus, long-range measurements usually suffer from very low signal-to-noise ratios. One remedy is to expand the unfocused illuminating-beam diameter so as to keep $D_{e^{-2}}/f$ constant. Beam expansion is generally desirable because it also produces smaller measurement volumes for the same focal distance. However, the limitations of physical size and the alignment accuracy required to cross two small focal spots at large distances usually make beam expansions greater than about 10:1 difficult.

8. SCATTERING PARTICLES

The scattering particles are the basic source of the Doppler signal, and their importance in the overall performance of an LDV system should not be underestimated. They can have more influence on the quality of the signal than any other component of the system. For example, the signal strength can be increased by a factor of 10^2 to 10^4 by increasing the particle diameter from several tenths of a micron to several microns. Similar improvements in scattered light intensity can be achieved by observing the light scattered in the forward-scatter direction rather than the back-scatter direction. Improvements of these orders of magnitude are difficult, expensive, or perhaps impossible to achieve by increasing the laser power or otherwise improving the optical system.

The most important properties of an individual scattering particle are the signal-to-noise ratio that it produces and its aerodynamic size, which is a measure of its ability to follow the flow. From the discussion in Sec. 7, good SNR requires that the particle is an effective scatterer (large J_{Pi}) and that it scatters light waves that can mix efficiently (large $\bar{V}_i$). In addition, the concentration and uniformity of the particle population play important roles. Ideally, one would like particles that have the same density as the fluid, large effective area in regard to scattering power, very uniform properties from particle to particle, and easily controlled concentration. They should also be inexpensive. In reality there are no known particles that satisfy all these requirements.

A wide variety of particles have been used in LDV applications. Naturally occurring aerosols and hydrosols are, of course, the most convenient, and often they can yield satisfactory results. However "natural" does not always imply "best," and this is certainly true in LDV. Natural particles usually have a very wide size distribution (cf Fig. 14) so that many of the particles are too small to produce measurable signals while others may be too large to follow rapidly accelerating flows. Also, their concentrations may not be controlled. Since these factors influence the design of the LDV optics, signal processing, and data reduction, it is often simplest and most satisfactory to artificially seed the flow with appropriately chosen particles.

In water flows with forward-scattered light, the natural hydrosol is ordinarily quite satisfactory, even with low-power lasers in the 5- to 25 mW range. In back scatter, high laser power and seeding with 5-μm or larger particles is needed to achieve strong signals. Plastic spheres are available in precisely sized batches with specific gravities very close to unity so that settling is not a problem [56, 57], but they are not suitable for large flow systems because of their expense. Plastic spheres with 0.5-μm diameters are produced by Dow Chemical Co. in bulk lots for paper coating. These are very inexpensive and quite suitable for forward scatter. Silicon carbide particles back scatter effectively because of their high refractive index, but they settle too quickly for use in flows below about 1 m/s. Other particles that have been used in water include milk, latex paint pigment (usually too small), and various fine powders.

Scattering power is not as low in air as it is for back scatter in water, and it is relatively easy to get good signals unless the velocity is very high or the focal length of the transmitting lens is large (more than 0.5 to 1 m without beam expansion). Figure 14 shows the number density of particles in a typical urban aerosol versus aerodynamic diameter D_P. The distribution is actually trimodal, but the sum of

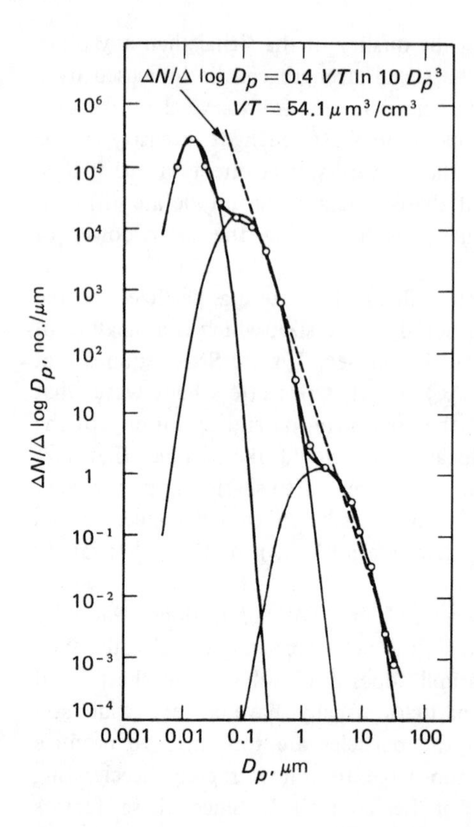

Figure 14 Urban aerosol number density distribution versus aerodynamic diameter. *(From [57a].)*

these three modes is often described by a power law. The mode in the range 0.1 to 1 μm is fairly universal and can be used for LDV estimates. When the natural aerosol is unsatisfactory, one tries to seed the flow with the largest particles that are still capable of following the flow accelerations. Typically, 1- to 2-μm particles are required for most low- to moderate-speed airflows. Particles may be liquid droplets such as water, various oils (vegetable oil, motor oil, etc.), or solid particles such as sieved grinding powder [47, 58, 59].

The sizes given above are guidelines for typical low-speed flows. For a particular flow it is prudent to make some estimate of the particle response time and the flow acceleration, to estimate the maximum allowable particle size. For gases (particle density $\gg$ gas density), a simple first-order model of the particle's response to a step change in velocity yields [58] the time constant

$$t_1 = \frac{d_p^2 \rho_p}{18\mu_f} \tag{66}$$

The 3-dB frequency at which the particle would follow a sinusoidal velocity variation with an amplitude of 0.707 of the fluid-velocity amplitude is $f_{3\text{-dB}} = 1/2\pi t_1$. This frequency should be larger than the frequency of the velocity fluctuations seen by the particle. In turbulent flows the appropriate fluid frequency is the lagrangian frequency, since the particle motion is approximately lagrangian.

Table 1 Properties of common scattering particles in air

Particles*	n_p	ρ_p	0.5 μm t_1, μs	0.5 μm $f_{3\text{-dB}}$, kHz	1 μm t_1, μs	1 μm $f_{3\text{-dB}}$, kHz	2 μm t_1, μs	2 μm $f_{3\text{-dB}}$, kHz
Silicon carbide	2.6	3.3	2.55	62	10.2	16	41	3.9
Alumina	1.76	3.8	2.93	54	11.7	14	47	3.0
Polystyrene	1.6	1.05	0.81	196	3.24	52	13	13
Peanut oil	1.47	0.91	0.70	227	2.81	57	11.2	14
Microballoons		0.23	0.18	897	0.71	224	2.84	56

*For air, settling velocity = $9.8t_1$ (m/s).

The time constants and response frequencies of particles that are commonly used in air are given in Table 1.

In liquid flows the velocities are usually so small that the primary limitation on particle size comes from the settling velocity rather than the ability to follow the flow. Settling velocities for various particles in water are given in Table 2. References [60, 61] contain useful information on natural hydrosols.

The scattering properties of small particles are very sensitive to size, refractive index, and scattering angle, so it is difficult to make simple, general statements about their behavior. The computed results for dioctyl phthalate (DOP) particles in air (shown in Fig. 15) illustrate this complexity. On average, the following inferences are valid: (1) big particles scatter more than small particles; (2) back scatter is very weak compared to forward scatter for the 0.5- to 20-μm range; (3) high ratios $m = n_p/n$ of the particle refractive index to the fluid refractive index yield better scattering; and (4) particle diameters larger than the fringe spacing do not necessarily imply low signal visibility. They often yield better signals than the smaller particles, especially in back scatter. Computer codes such as the one used to compute Fig. 15 are available for the computation of the LDV signal properties in terms of the parameters of the LDV system and the particle [33, 62, 63]; accurate evaluation of particle-size effects requires the use of such codes because of the highly complex nature of the light scattering phenomenon. For rougher estimates that yield order-of-magnitude accuracies, a procedure is outlined in [64].

Table 2 Settling velocities for common scattering particles in water

Particles	m	Settling velocity, μm/s 0.5 μm	1 μm	2 μm	5 μm
Silicon carbide	1.95	17	70	280	1700
Alumina	1.32	21	85	340	2100
Polystyrene	1.2	0.37	1.5	6	37
Microballoons		Most float			

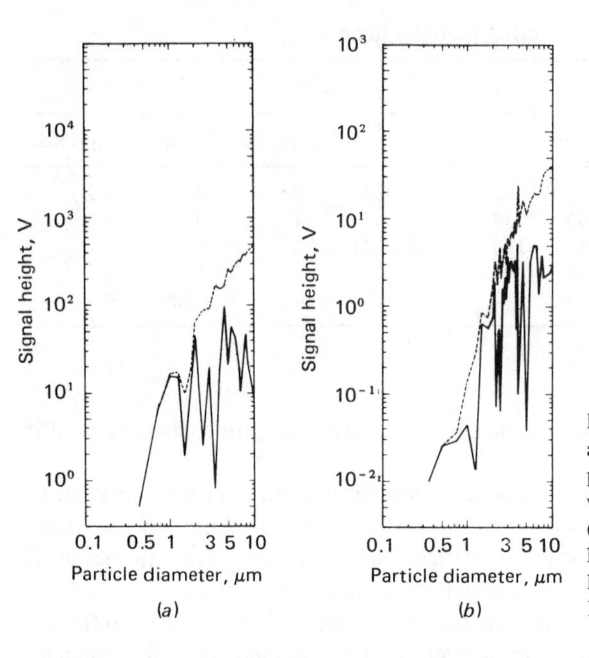

Figure 15 Strength of the Doppler and pedestal signals as a function of particle diameter for typical forward-scatter (*a*) and back-scatter (*b*) LDV systems. (——, Doppler height; - - - -, pedestal height; focal length = 120 mm, $m = 1.48$, e.g., DOP in air.) *(From [62].)*

9. PROPERTIES OF THE RANDOM LIGHT FLUX

9.1 Signal Representation

The signal $e(t)$ is random because the photodetector introduces random noise and because the scattering particles are located at random points in the flow. Further, the particles of a polydisperse suspension have random scattering coefficients, and in turbulent flows their velocities are also random. These random elements pervade all aspects of signal processing and data analysis in laser-Doppler velocimetry, and a thorough description of their statistics is essential. Fortunately, it is possible to isolate certain of these elements. It is shown in Sec. 6 that the photodetector noise is either electronics noise that is statistically independent of the light flux, or shot noise that can be described in terms of conditional statistics of $e(t)$, given the value of the light flux. Unconditional statistics of $e(t)$, corresponding to long time averages, can be obtained by averaging over all random values of the light flux. Thus, the statistical description of the signal from the photodetector-filter system can be completed by analyzing the statistics of the light flux separately from the statistics of the shot noise.

9.2 Random Doppler Light Flux

In general, the total Doppler light flux is a sum of the individual bursts caused by each particle:

$$J_D(t) = \sum_i J_{D_i}(t) \tag{67a}$$

$$= \sum_i a(\mathbf{x}_i(t)) D_i \cos(\Phi_i - \Psi_i) \tag{67b}$$

where i extends over *all* particles in the flow, and $a(\mathbf{x})$ and $\Phi_i = \Phi(\mathbf{x}_i(t), t)$ are given by Eqs. (29) and (25*a*), respectively. (The amplitude factor a accounts for the presence of a particle in the mv.) If we let

$$g(\mathbf{x}, t, D) = \sum_i \delta(\mathbf{x} - \mathbf{x}_i(t))\, \delta(D - D_i) \tag{68}$$

Eq. (67*b*) becomes

$$J_D(t) = \int_{\text{all } \mathbf{x}} \int_0^\infty a(\mathbf{x}) D \cos \Phi(\mathbf{x}, t)\, g(\mathbf{x}, t, D)\, d^3\mathbf{x}\, dD \tag{69}$$

where, without loss of generality, the random phase shifts Ψ_i can be ignored because the signal phases $\Phi(\mathbf{x}_i, t)$ are also random. The function $g(\mathbf{x}, t, D)$ indicates the presence of a particle at the point $\mathbf{x}$ with scattering amplitude D. When $J_D(t)$ is cast into this form, all the properties of the particles are contained in the function g.

9.3 Statistical Properties of $g(\mathbf{x}, t, D)$

Since all the random characteristics of the light flux are contained in $g(\mathbf{x}, t, D)$, this function's properties are derived in full detail. The analysis generalizes that in [65, 66] by including the effects of nonuniform concentration of particles, both in space and in time, and the distribution of signal amplitudes as determined by the particle size distribution and the optics of the LDV system.

Consider a volume V that is fixed in space (i.e., a "control volume"), and the material volume $V_m(t)$ that coincides with V at time t (Fig. 16). At any instant there may be many different particles in V_m, each with a different value of D_i. We shall refer to all particles that produce Doppler signal amplitudes such that

$$D < D_i < D + \delta D \tag{70}$$

as "D-type" particles, where D is any arbitrarily prescribed numerical value, and δD is a small range of amplitudes. At time t the number of D-type particles in V is just equal to the number of D-type particles in $V_m(t)$, and this is given by

$$n(V, t, D) = \int_V \int_D^{D+\delta D} g(\mathbf{x}, t, D)\, d^3\mathbf{x}\, dD \tag{71}$$

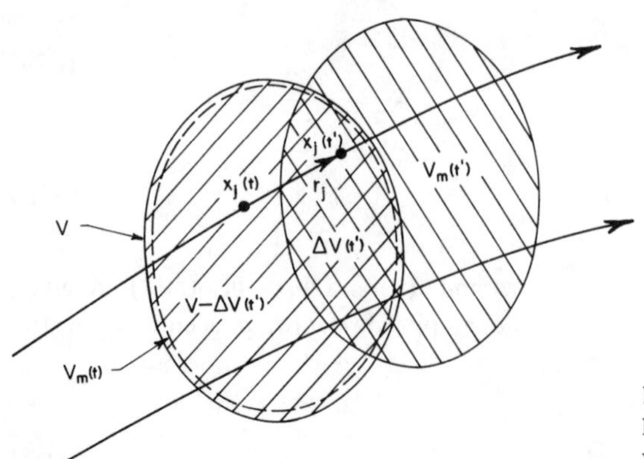

Figure 16 Motion of the particle population through the measurement volume.

since the integral of $\delta(\mathbf{x}-\mathbf{x}_i(t))\ \delta(D-D_i)$ yields a 1 if $\mathbf{x}_i(t)\in V$ and $D_j\in(D, D+\delta D)$. This is also equal to the number of D-type particles in $V_m(t)$, denoted by $n(V_m(t), D)$, and if the particles follow the fluid motion the number in V_m must be the same for all time; that is, $n(V_m(t), D)=n(V_m(t'), D)$. This assumption implies that both particle lag and particle diffusion by Brownian motion are negligible.

The statistics of g are determined by the statistics of n, and these in turn can be found under the assumption that the location of any particular particle, for example the ith particle, is entirely random. This assumption requires that the relative separation vectors between all pairs of particles are random, but it does not require that the particle concentration be constant. Let the probability that the ith particle is of type D and is located in V at time t be p_0, given by

$$p_0 = \int_V \int_D^{D+\delta D} \lambda(\mathbf{x}, t, D)\, d^3\mathbf{x}\, dD \tag{72}$$

where λ is the probability density of particles in $\mathbf{x}$-D space. Then, if the total number of particles of all types in the entire fluid volume is $\mathfrak{N}$, the probability that any n particles of type D are located in V is the binomial distribution [53]

$$\text{Prob }\{n\ D\text{-type particles in } V \text{ at time } t\} = \frac{\mathfrak{N}!}{n!(\mathfrak{N}-n)!}\, p_0^n(1-p_0)^{\mathfrak{N}-n} \tag{73}$$

The average number of D-type particles in V is

$$N(V, t, D) \equiv \langle n(V, t, D)\rangle \tag{74a}$$

$$= \sum_{n=0}^{\infty} n \frac{\mathfrak{N}!}{n!(\mathfrak{N}-n)!}\, p_0^n(1-p_0)^{\mathfrak{N}-n} \tag{74b}$$

$$= \mathfrak{N} p_0 \tag{74c}$$

The average concentration of D-type particles, i.e., the number per unit volume per unit amplitude D, is denoted by $C(\mathbf{x}, t, D)$ and is given by

$$C(\mathbf{x}, t, D) = \lim_{\substack{V \to 0 \\ \delta D \to 0}} \frac{\langle n(V, t, D) \rangle}{V\,\delta D} \tag{75a}$$

$$= \lim_{\substack{V \to 0 \\ \delta D \to 0}} \frac{\mathcal{N}}{V\,\delta D} \int_V \int_D^{D+D} \lambda(\mathbf{x}, t, D)\, d^3\mathbf{x}\, dD \tag{75b}$$

$$= \mathcal{N}\lambda(\mathbf{x}, t, D) \tag{75c}$$

where V shrinks to a point at $\mathbf{x}$.

When the total fluid volume is large compared to V, as is almost always true, $\mathcal{N}$ will be much greater than $\langle n \rangle$, and Eq. (73) can be simplified by taking its limit as $\mathcal{N} \to \infty$, keeping $N = \langle n \rangle = \mathcal{N}p_0$ fixed. The result is the Poisson distribution [53, 67]

$$\text{Prob } \{n\ D\text{-type particles in } V\} = \frac{N^n e^{-N}}{n!} \tag{76}$$

where, from Eqs. (72), (74*c*), and (75*c*),

$$N = \langle n(V, t, D) \rangle = \int_V \int_D^{D+\delta D} C\, d^3\mathbf{x}\, dD \tag{77}$$

is the mean number of D-type particles in the limit $\delta D \to 0$. The range δD may be replaced in this equation by any finite range ΔD to obtain the mean number in $(D, D + \Delta D)$; inserting this mean number into Eq. (76) gives the probability of finding n particles in V whose amplitudes are anywhere in $(D, D + \Delta D)$.

The expected value of $g(\mathbf{x}, t, D)$ is found by taking the ensemble average of Eq. (71), dividing by V and δD, and taking the limit as $V \to 0$ and $\delta D \to 0$:

$$\lim_{V, \delta D \to 0} \frac{\langle n(V, t, D) \rangle}{V\,\delta D} = \lim_{V, \delta D \to 0} \frac{1}{V\,\delta D} \int_V \int_D^{D+\delta D} \langle g(\mathbf{x}, t, D) \rangle\, d^3\mathbf{x}\, dD \tag{78a}$$

$$= \langle g(\mathbf{x}, t, D) \rangle \tag{78b}$$

Hence, from Eq. (77),

$$\langle g(\mathbf{x}, t, D) \rangle = C(\mathbf{x}, t, D) \tag{79}$$

The space-time correlation of g, defined as $\langle g(\mathbf{x}, t, D) g(\mathbf{x}', t', D') \rangle$, is also needed to describe the behavior of $J_D(t)$. It can be calculated by observing that

$$\langle n(V, t, D) n(V, t', D') \rangle = \left[\int_V \int_D^{D+\delta D} g(\mathbf{x}, t, D)\, d^3\mathbf{x}\, dD \right.$$

$$\left. \times \int_V \int_{D'}^{D'+\delta D} g(\mathbf{x}', t', D')\, d^3\mathbf{x}'\, dD' \right] \tag{80a}$$

$$= \int_V \int_V \int_D^{D+\delta D} \int_{D'}^{D'+\delta D} \langle g(\mathbf{x}, t, D) g(\mathbf{x}', t', D') \rangle \, d^3\mathbf{x} \times d^3\mathbf{x}' \, dD \, dD' \tag{80b}$$

and calculating $\langle n(V, t, D) n(V, t', D') \rangle$. A fundamental assumption implicit in Eq. (72) is that the number of D particles in V is statistically independent of the number of D' particles in V if $D \neq D'$. (This assumption would be violated if, for example, the particles were so large that the presence of one particle precluded the presence of any other particle in V, as can occur in two-phase flows. These situations are excluded from the present analysis.) Then, if $D \neq D'$,

$$\langle n(V, t, D) n(V, t', D') \rangle = \langle n(V, t, D) \rangle \langle n(V, t', D') \rangle \qquad D \neq D' \tag{81}$$

If, on the other hand, $D' = D$, it is implicit in Eq. (72) that the number of D particles in one material volume is statistically independent of the number of D particles in any other nonoverlapping fluid volume. Thus, with reference to Fig. 16, the number $n(V, t, D)$ in V at t will be independent of the number $n(V, t', D)$ in V at t' if $V_m(t')$ does not overlap any part of V, that is, if the material volume in V at t has been completely replaced by a new material volume at t'. However, for sufficiently short time differences $t' - t$, $V_m(t')$ and V must share a common volume $\Delta V_m(t')$, and $n(V, t', D)$ will be correlated with $n(V, t, D)$ because at time t', V will contain D particles that also resided in V at time t. Thus the correlation time is of the order of the time required to sweep a new volume into V. This case $(D = D')$ can be analyzed by writing $n(V, t, D)$ and $n(V, t', D)$ in terms of nonoverlapping material volumes $V_m(t') - \Delta V(t')$, $\Delta V(t')$, and $V - \Delta V(t')$:

$$n(V, t, D) = n(V_m(t), D) \tag{82a}$$

$$= n(V_m(t'), D) \tag{82b}$$

$$= n(V_m(t') - \Delta V(t'), D) + n(\Delta V(t'), D) \tag{82c}$$

$$n(V, t', D) = n(V - \Delta V(t'), D) + n(\Delta V(t'), D) \tag{83}$$

where Eq. (82*b*) follows from Eq. (82*a*) because the particles are assumed to follow the motion of the material volume. Multiplying Eqs. (82*c*) and (83) and averaging yield an expression for $\langle n(V, t, D) n(V, t', D) \rangle$ in which the average of each cross product is equal to the cross product of the averages (because the numbers in nonoverlapping volumes are statistically independent), with the exception of $\langle n^2(\Delta V(t'), D) \rangle$. For a Poisson distribution this last average is given by

$$\langle n^2(\Delta V(t'), D) \rangle = \langle n(\Delta V(t'), D) \rangle^2 + \langle n(\Delta V(t'), D) \rangle \tag{84}$$

The final expression,

$$\langle n(V, t, D) n(V, t', D) \rangle = \langle n(V, t, D) \rangle \langle n(V, t', D' \rangle + \langle n(\Delta V(t'), D) \rangle \tag{85}$$

follows after some algebraic manipulation.

The last term in Eq. (85) is the average number of particles in V at time t' that were also in V at time t, and it depends upon the statistics of the fluid motion as well as those of the particle locations. Consider a fluid volume $d\mathbf{x}^3$, located at $\mathbf{x}$ at time t, that is mapped by the motion into a volume $d^3\mathbf{x}'$ at $\mathbf{x}'$ at time t'. For incompressible flow $d^3\mathbf{x} = d^3\mathbf{x}'$ and the mean number of particles in $d^3\mathbf{x}$ or $d^3\mathbf{x}'$ is just

$$\int_D^{D+\delta D} C(\mathbf{x}, t, D)\, d^3\mathbf{x}\, dD$$

The probability that $\mathbf{x}'$ lies somewhere in V can be expressed as

$$\int_V f(\mathbf{x}', t'; \mathbf{x}, t)\, d^3\mathbf{x}'$$

where $f(\mathbf{x}', t'; \mathbf{x}, t)\, d^3\mathbf{x}'$ is defined as the probability that the fluid particle at $(\mathbf{x}, t)$ maps into $(\mathbf{x}', \mathbf{x}' + d^3\mathbf{x}')$ at t'. The average number in $d^3\mathbf{x}$ that are in V at t' is just the mean number in $d^3\mathbf{x}$ times the probability that this number is in V at t',

$$\int_D^{D+\delta D} C(\mathbf{x}, t, D)\, d^3\mathbf{x}\, dD \int_V f(\mathbf{x}', t'; \mathbf{x}, t)\, d^3\mathbf{x}'$$

and the average number in the total overlap volume is obtained by integrating over all $d^3\mathbf{x}$:

$$\langle n(\Delta V(t'), D)\rangle = \int_V \int_V \int_D^{D+\delta D} C(\mathbf{x}, t, D)\, f(\mathbf{x}', t'; \mathbf{x}, t)\, d^3\mathbf{x}\, d^3\mathbf{x}'\, dD \qquad D = D' \tag{86a}$$

$$= \int_V \int_V \int_D^{D+\delta D} \int_{D'}^{D'+\delta D} C(\mathbf{x}, t, D) \times f(\mathbf{x}', t'; \mathbf{x}, t)\, \delta(D' - D)\, d^3\mathbf{x}\, d^3\mathbf{x}'\, dD\, dD' \tag{86b}$$

Combining Eqs. (85), (77), (76*a*), (79), and (86*b*), dividing the result by $V\,\delta D$, and taking the limit as $V \to 0$ and $\delta D \to 0$ yield

$$\langle g(\mathbf{x}, t, D) g(\mathbf{x}', t', D')\rangle = C(\mathbf{x}, t, D)\, C(\mathbf{x}', t', D') + C(\mathbf{x}, t, D)\, f(\mathbf{x}', t'; \mathbf{x}, t)\, \delta(D' - D) \tag{87}$$

The time required for particles to cross the mv is clearly an upper bound on the maximum time difference $t' - t$ that is of interest in Eq. (87), and in most applications of laser velocimetry this transit time is small compared to the lagrangian time scales of the flow. Hence, the particle velocities are essentially constant, and the probability that a particle moves from $\mathbf{x}$ at time t to $\mathbf{x}'$ at time t' is just the probability that $\mathbf{u}(\mathbf{x}, t) = \mathbf{U}(\mathbf{x}, t) + \mathbf{u}'(\mathbf{x}, t) = (\mathbf{x}' - \mathbf{x})/(t' - t)$, where $\mathbf{U}$

is the mean velocity and $\mathbf{u}'$ is the fluctuating velocity. Letting Prob $\{\mathbf{c}' < \mathbf{u}'(\mathbf{x}, t) < \mathbf{c}' + d\mathbf{c}'\} = f_{\mathbf{u}'}(\mathbf{c}', \mathbf{x}, t)\ d^3\mathbf{c}'$ and Prob $\{\mathbf{c} < \mathbf{u}(\mathbf{x}, t) < \mathbf{c} + d\mathbf{c}\} = f_{\mathbf{u}}(\mathbf{c}, \mathbf{x}, t)\ d^3\mathbf{c}$ be the probability density functions for the velocity fluctuation and the total velocity, respectively, we have

$$f(\mathbf{x}', t'; \mathbf{x}, t)\ d^3\mathbf{x}' = f_{\mathbf{u}}\left(\mathbf{c} = \frac{\mathbf{x}' - \mathbf{x}}{t' - t}, \mathbf{x}, t\right) \frac{d^3\mathbf{x}'}{t' - t} \tag{88}$$

for small values of $t' - t$, wherein $d^3\mathbf{x}' = d^3\mathbf{c}(t' - t)$. Then

$$\begin{aligned}\langle g(\mathbf{x}, t, D)\, g(\mathbf{x}', t', D')\rangle &= C(\mathbf{x}, t, D)\, C(\mathbf{x}', t', D') \\ &+ C(\mathbf{x}, t, D)\, f_{\mathbf{u}}\left(\mathbf{c} = \frac{\mathbf{x}' - \mathbf{x}}{t' - t}, \mathbf{x}, t\right)(t' - t)^{-1}\ \delta(D' - D)\end{aligned} \tag{89}$$

In steady laminar flow, $f_{\mathbf{u}}(\mathbf{c}) = \delta(\mathbf{c} - \mathbf{U}(\mathbf{x}, t))$. That is, the fluctuations vanish, and the factor $f_{\mathbf{u}}/(t' - t)$ reduces to $\delta(\mathbf{x}' - \mathbf{x} - \mathbf{U}(\mathbf{x}, t)(t' - t))$. In general, if $t' = t$, Eq. (89) reduces to $\langle g(\mathbf{x}, t, D) g(\mathbf{x}', t, D')\rangle = C(\mathbf{x}, t, D')C(\mathbf{x}', t, D') + C(\mathbf{x}, t, D)\delta(\mathbf{x} - \mathbf{x}') \cdot \delta(D - D')$.

9.4 Correlation and Power Spectrum

The single time moments of the light flux $\langle (J_D - \langle J_D \rangle)^n \rangle$ contain useful statistical information, but they do not yield information on the Doppler frequency, which is the variable of primary interest. For example, the mean of J_D, found by averaging Eq. (69), taking the average inside the integral, and using Eq. (79), is

$$\langle J_D(t)\rangle = \iint a(\mathbf{x}) D \cos \Phi(\mathbf{x}, t)\ C(\mathbf{x}, t, D)\ d^3\mathbf{x}\ dD \tag{90}$$

It can be shown that this mean is virtually zero for any reasonable distributions for $a(\mathbf{x})$ and $C(\mathbf{x}, t, D)$. Specifically, if C is constant and $a(\mathbf{x})$ is gaussian, $\langle J_D \rangle \simeq \exp(-\pi^2 N_{FR}^2/8)$, which is of order 10^{-5} when $N_{FR} = 3$. Thus, we may always take J_D to be a zero-mean random variable. Likewise, higher-order moments such as $\langle J_D^2 \rangle$ are also independent of the Doppler frequency.

To extract frequency information, it is necessary to use statistical moments involving the signal's values at two or more times. The simplest quantities that accomplish this purpose are the correlation function

$$R_{J_D J_D}(\tau) = \langle J_D(t) J_D(t + \tau)\rangle \tag{91}$$

and the associated power spectrum or power spectral density, defined by

$$S_{J_D}(\omega) = \frac{1}{2\pi} \int_{-\infty}^{\infty} e^{-j\omega\tau}\ R_{J_D J_D}(\tau)\ d\tau \tag{92}$$

where $\omega = 2\pi f$. This equation assumes that $\langle J_D \rangle = 0$. The inverse Fourier transform of S_{J_D} gives $R_{J_D J_D}$ in terms of the power spectrum,

$$R_{J_D J_D}(\tau) = \int_{-\infty}^{\infty} e^{j\omega\tau}\, S_{J_D}(\omega)\ d\omega \tag{93}$$

from which it follows that the mean square is the area under the power spectral density curve

$$\langle J_D^2 \rangle = R_{J_D J_D}(0) = \int_{-\infty}^{\infty} S_{J_D}\, d\omega \tag{94}$$

As is well known [53], $S_{J_D}\, d\omega$ represents the contribution that Fourier components of $J_D(t)$ in the range $(\omega, \omega + d\omega)$ make to the mean square value of J_D; that is, $S_{J_D}\, d\omega$ is the signal "power" in a band of frequencies with width $d\omega$. All the quantities above are independent of t if J_D is a stationary random process, i.e., if the statistics of the flow and the particle population are steady in time. Stationarity is assumed throughout this section. In particular, this implies that $C(\mathbf{x}, t, D) = C(\mathbf{x}, D)$ and $f_{\mathbf{u}}(\mathbf{c}, \mathbf{x}, t) = f_{\mathbf{u}}(\mathbf{c}, \mathbf{x})$. Stationarity also implies that $R_{J_D J_D}(-\tau) = R_{J_D J_D}(\tau)$ and $S_{J_D}(-\omega) = S_{J_D}(\omega)$. The only difference between negative and positive frequencies is a 180° phase shift.

The general equation for the autocorrelation is found by forming the product $J_D(t) J_D(t + \tau)$ using the representation in Eq. (69), averaging inside the integrals in this product, and using Eq. (87):

$$R_{J_D J_D}(\tau) = \iint C(\mathbf{x}) \langle D^2(\mathbf{x}) \rangle f(\mathbf{x}', t + \tau; \mathbf{x}, t) a(\mathbf{x}) a(\mathbf{x}') \cdot \cos \Phi(\mathbf{x}, t) \cos \Phi(\mathbf{x}', t + \tau)\, d^3\mathbf{x}\, d^3\mathbf{x}' \tag{95}$$

where

$$C(\mathbf{x}) = \int_0^{\infty} C(\mathbf{x}, D)\, dD \tag{96}$$

is the total concentration of particles, and

$$\langle D^2 \rangle = \frac{\int_0^{\infty} D^2 C(\mathbf{x}, D)\, dD}{C(\mathbf{x})} \tag{97}$$

is the mean square scattering amplitude averaged over all particles. A simpler working formula, which is still accurate and general enough to encompass all effects of practical interest, is

$$R_{J_D J_D} \simeq \tfrac{1}{2} \iint C(\mathbf{x}) \langle D^2(\mathbf{x}) \rangle f_{\mathbf{u}}(\mathbf{c}, \mathbf{x}) a(\mathbf{x}) a(\mathbf{x} + \mathbf{c}\tau) \cos(\omega_S \tau + K c_1 \tau)\, d^3\mathbf{x}\, d^3\mathbf{c} \tag{98}$$

This equation uses the approximation in Eq. (88) and ignores a term of order $\exp(-N_{FR}^2)$ that arises from the cosine product in Eq. (95). It is valid whenever the particle velocities are essentially constant during the times that the particles reside in the mv.

Before investigating the case of turbulent flow, embodied in Eq. (98), it is instructive to consider the simpler situation in which the flow is steady, laminar and uniform (that is, $\mathbf{u} = \mathbf{U} = \text{const}$), the particle population is uniform [that is, $C = \text{const}$ and $\langle D^2(\mathbf{x}) \rangle = \text{const}$], and the mv is gaussian and given by Eq. (40). Then,

$$R_{J_DJ_D} = \langle J_D^2 \rangle \exp\left(-\frac{\Delta\omega_A^2}{2}\,\tau^2\right) \cos\,(\omega_S + Ku)\tau \tag{99}$$

and

$$S_{J_D} = \frac{1}{(8\pi)^{1/2}} \frac{\langle J_D^2 \rangle}{\Delta\omega_A} \left\{ \exp\left[-\frac{(\omega - \omega_s - Ku)^2}{2\,\Delta\omega_A^2}\right] + \exp\left[-\frac{(\omega + \omega_s + Ku)^2}{2\,\Delta\omega_A^2}\right] \right\} \tag{100}$$

where

$$\Delta\omega_A = 2\sqrt{2}\left(\frac{u^2}{d_m^2} + \frac{v^2}{l_m^2} + \frac{w^2}{h_m^2}\right)^{1/2} \tag{101}$$

The autocorrelation is almost a replica of the individual Doppler bursts. It oscillates at the frequency $\omega_S + Ku$, and its amplitude envelope is a gaussian function because of the assumed gaussian intensity distribution in the mv. The e^{-1} width of the gaussian envelope is $2\sqrt{2}/\Delta\omega_A$. The reader can verify from Eq. (40) that for any straight trajectory the transit time of a particle between the points where the amplitude is e^{-2} of the maximum amplitude for *that trajectory* is

$$\left(\frac{u^2}{d_m^2} + \frac{v^2}{l_m^2} + \frac{w^2}{h_m^2}\right)^{-1/2}$$

and hence that $2\sqrt{2}/\Delta\omega_A$ corresponds to the e^{-2} transit time.

The power spectrum of J_D, represented by the dashed line in Fig. 17, shows that signal power is concentrated around $\pm(\omega_s + Ku)$ in a band of frequencies that is about $\Delta\omega_A$ wide. More precisely, the first moment of S_{J_D} is

$$\mu_1 \equiv \frac{\int_0^\infty \omega S_{J_D}\, d\omega}{\int_0^\infty S_{J_D}\, d\omega} \tag{102a}$$

$$= \omega_S + Ku \tag{102b}$$

and the second central moment is

$$\mu_2 \equiv \frac{\int_0^\infty (\omega - \mu_1)^2 S_{J_D}\, d\omega}{\int_0^\infty S_{J_D}\, d\omega} \tag{103a}$$

$$= \Delta\omega_A^2 \tag{103b}$$

where $\Delta\omega_A$ is given by Eq. (101). As noted above, $\Delta\omega_A$ is inversely proportional to the transit time. Since the Doppler frequency is constant for all particles in this

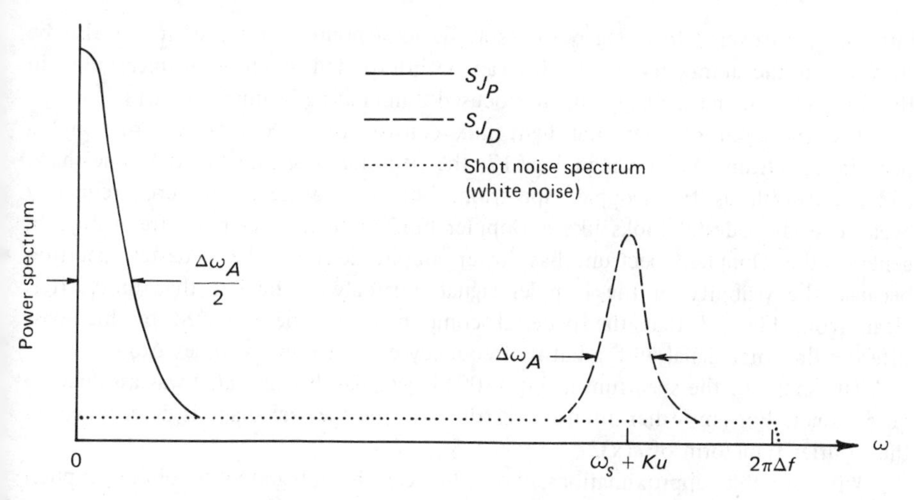

Figure 17 Power spectrum of the signal from the photodetector of an LDV.

case, it must be concluded that $\Delta\omega_A$ does not represent any variation of the Doppler frequency in time or in space. Rather, $\Delta\omega_A$ is solely a consequence of the finite duration of the individual Doppler bursts, as measured by their transit times, and for this reason the phenomenon is referred to as *transit-time broadening*. If the bursts were very long, corresponding to d_m, l_m, $h_m \to \infty$, then $\Delta\omega_A$ would approach zero, and the power spectrum would consist of Dirac delta functions at $\omega_S + Ku$ and $-\omega_S - Ku$.

The bandwidth as a fraction of the Doppler shift is

$$\frac{\Delta\omega_A}{\omega_D} = \frac{2^{1/2}}{\pi N_{FR}} \left(1 + \frac{v^2}{u^2} \tan^2 \kappa + \frac{w^2}{u^2} \sec^2 \kappa\right)^{1/2} \tag{104}$$

which is independent of the magnitude of u if the direction of u is fixed. The fractional bandwidth is minimized by aligning the mv so that $v = w = 0$. Then, typical values of $\Delta\omega_A/\omega_D$ are 0.5 to 10%, corresponding to somewhere between 5 and 100 fringes in the mv.

$\Delta\omega_A$ is also called the *ambiguity bandwidth*, a term borrowed from radar technology. This term suggests that the frequency spread in the spectrum is associated with an uncertainty in the measurement of the Doppler frequency. In practice, large transit-time bandwidths may, indeed, cause measurement errors because large bandwidths make it more difficult to determine the center frequencies of experimental spectra. In addition, it will be shown that in the case of many particles in the mv, there is an uncertainty in the frequency of the signal that is related to $\Delta\omega_A$ and is caused by the random phase superposition of signals from many particles. There is, however, no random phase superposition when there is only one particle in the mv, and therefore ambiguity broadening imposes no fundamental inaccuracies on the measurement of the frequency of an individual Doppler burst. (The reader will find confusion over this point in some early papers, but this controversy has long been resolved. As a point of interest, it can also be shown that the ambiguity bandwidth can be interpreted in terms of uncertainty in the directions of the light rays in the focused illuminating beams [28, 38]).

The spectrum of the total light flux consists of a Doppler spectrum and a pedestal spectrum. As shown in Fig. 17, the pedestal spectrum has the same shape and bandwidth as the Doppler spectrum, but it is centered at zero frequency because each pedestal looks like a Doppler burst with zero Doppler frequency. In general, the Doppler spectrum has lower amplitude than the pedestal spectrum because the visibility of the Doppler signal must always be less than unity. It is clear from Fig. 17 that the pedestal component can be removed by high-pass filtering the total signal with a cutoff frequency equal to several times $\Delta\omega_A$.

The shape of the spectrum in Eq. (100) is gaussian because $a(\mathbf{x})$ was assumed to be gaussian. For any other intensity distribution the spectral shape is determined by the Fourier transform of $a(\mathbf{x})$.

With suitable approximations, Eq. (98) can be integrated to obtain explicit expressions for $R_{J_D J_D}$ for the case of laminar flow with velocity gradients or turbulent flow with mean velocity gradients and finite turbulence intensity [8, 28, 68]. In all these analyses C and $\langle D^2 \rangle$ are approximated with constants, $a(\mathbf{x})$ is assumed to be gaussian, and it is usual to approximate the turbulent probability density function by a normal distribution. As consequences of these assumptions the envelope of the correlation function and the power spectrum are gaussian functions.

The correlation function can also be found in the most general case by performing the integrations in Eq. (98) using known functions for $C(\mathbf{x})$, $\langle D^2(\mathbf{x})\rangle$, $a(\mathbf{x})$, $U(\mathbf{x})$, and $f_{\mathbf{u}}(\mathbf{c}, \mathbf{x})$, and the power spectrum can be calculated by Fourier transformation. Explicit expressions cannot, however, be found in general, and numerical integrations are required. Fortunately, only the first and second moments of the spectrum are needed in most applications, and it appears to be possible to obtain analytical expressions for these quantities under fairly weak and nonrestrictive assumptions. The first moment of S_{J_D} can be calculated in terms of the autocorrelation by substituting Eq. (92) into the definition of μ_1 given by Eq. (102*a*) and manipulating. (The manipulations require the generalized-function Fourier transform of ω, which can be found in [69]). The result is

$$\mu_1 = \frac{\int_{-\infty}^{\infty} \tau^{-2} R_{J_D J_D}(\tau)\, d\tau}{\pi R_{J_D J_D}(0)} \tag{105a}$$

$$= \frac{\iint d^3\mathbf{x}\, d^3\mathbf{c}\, C\langle D^2\rangle f_{\mathbf{u}}(\mathbf{c}) a(\mathbf{x}) \int_{-\infty}^{\infty} d\tau\, \tau^{-2} a(\mathbf{x} + \mathbf{c}\tau) \cos(\omega_S \tau + K c_1 \tau)}{\iint C\langle D^2\rangle a^2(\mathbf{x})\, d^3\mathbf{x}} \tag{105b}$$

$$= \frac{\iint C(\mathbf{x})\langle D^2(\mathbf{x})\rangle a^2(\mathbf{x}) \langle |\omega_S + K_u|\rangle\, d^3\mathbf{x}}{\iint C\langle D^2\rangle a^2\, d^3\mathbf{x}} + 0(N_{FR}^{-2}) \tag{105c}$$

Equation (105c) is an asymptotic result that follows from Eq. (105b) by noting that $a(\mathbf{x}+\mathbf{c}\tau)$ varies slowly with τ in comparison to $\cos(\omega_S + Kc_1)\tau$ when there are many fringes in the mv, and taking the limit as $N_{FR} \to \infty$. In effect, Eq. (105c) ignores transit-time broadening of the spectra associated with individual Doppler bursts. Equation (105c) states that the mean frequency of the spectrum is a volume average of $\langle|\omega_S + Ku(\mathbf{x},t)|\rangle$ weighted by $C(\mathbf{x})D^2(\mathbf{x})a^2(\mathbf{x},t)$. It is convenient to denote this averaging operation by

$$\widetilde{(\cdot)} = \frac{\int C\langle D^2\rangle a^2(\mathbf{x})(\cdot)\, d^3\mathbf{x}}{\int C\langle D^2\rangle a^2(\mathbf{x})\, d^3\mathbf{x}} \tag{106}$$

Thus,

$$\mu_1 = \widetilde{\langle|\omega_S + Ku(\mathbf{x},t)|\rangle} \tag{105d}$$

In a properly designed experiment, ω_S is set at a value large enough to ensure that $\omega_S + Ku > 0$ for all but the most improbable velocity fluctuations. Then Eq. (105d) reduces to

$$\mu_1 = \omega_S + K\widetilde{U} \tag{105e}$$

where $\widetilde{U}$, the filtered mean velocity, is also equal to the mean of the filtered velocity; that is, $\widetilde{U} = \langle\widetilde{u}(\mathbf{x},t)\rangle$.

If the dimensions of the mv are so small that $U(\mathbf{x})$ is constant with negligible error, then $\widetilde{U} \equiv U(\mathbf{O})$, the mean velocity at the center of the mv. This is also true if $C\langle D^2\rangle$ is constant, $U(\mathbf{x})$ is at most a linear function of position, and $a(\mathbf{x})$ is symmetric about the coordinate planes [28, 68]. In all other cases the relationship between $\widetilde{U}$ and $U(\mathbf{O})$ depends on the details of the weighting factor $C(\mathbf{x})\langle D^2(\mathbf{x})\rangle a^2(\mathbf{x})$. The particle-dependent factor $C\langle D^2\rangle$ is particularly troublesome because spatial nonuniformities in the scattering particle population are usually not under the control of the experimenter, and worse, they are usually not known. Hence, the best procedure by far is to ensure that the scattering population is uniformly distributed in space. Then the volume weighting depends only upon $a^2(\mathbf{x})$, which can be estimated with fair accuracy theoretically. Mean velocity corrections for volume averaging are discussed in [28, 70]. Of course, the ideal situation is achieved when the mv is made so small that mean velocity variations are negligible.

It should be noted that the volume average in Eq. (106) is consistent with the results found in [28, 68] but differs by a factor $a(\mathbf{x})$ from the average defined in [65]. This has ramifications in the next section, where the statistics of the high-burst-density signal are considered.

The second moment of the power spectrum, defined by Eq. (103a), is a measure of the spectral bandwidth. By differentiating Eq. (93) twice with respect to τ and setting τ equal to zero, it is easy to see that μ_2 is given in terms of $R_{J_DJ_D}$ by

$$\mu_2 = -\frac{R''_{J_DJ_D}(0)}{R_{J_DJ_D}(0)} - \mu_1^2 \tag{107a}$$

Inserting Eq. (98) into Eq. (107a) yields

$$\mu_2 = -[\iint C\langle D^2\rangle a(\mathbf{x}) f_{\mathbf{u}}(\mathbf{x} + \mathbf{c}\tau)\{[\partial^2 a(\mathbf{x} + \mathbf{c}\tau)/\partial\tau^2] \cos(\omega_S + K c_1) - a(\mathbf{x} + \mathbf{c}\tau)(\omega_S + K c_1)^2\}_{\tau=0}\, d^3\mathbf{x}\, d^3\mathbf{c}] / [\int C\langle D^2\rangle a^2(\mathbf{x})\, d^3\mathbf{x}] - \mu_1^2 \quad (107b)$$

$$= \Delta\omega_A^2 + \Delta\omega_G^2 + \Delta\omega_T^2 \quad (107c)$$

where $\Delta\omega_A^2$ is the ambiguity broadening given by the first term in braces in Eq. (107*b*), and $\Delta\omega_G^2 + \Delta\omega_T^2$, the sum of the broadening due to mean velocity gradients and turbulent velocity fluctuations, is given by the last term in braces in Eq. (107*b*) minus μ_1^2. The expression for $\Delta\omega_A^2$ depends on the spatial structures of $C(\mathbf{x})$, $\langle D^2(\mathbf{x})\rangle$, $a(\mathbf{x})$, and the first two moments of $u(\mathbf{x}, t)$, all of which are accounted for in Eq. (107*b*). An approximation that is usually satisfactory is obtained by assuming that $C\langle D^2\rangle$ and $f_{\mathbf{u}}$ are independent of position (locally homogeneous flow), and $a(\mathbf{x})$ is gaussian. The result is

$$\Delta\omega_A^2 = 8\left[\frac{U^2(\mathbf{0}) + \langle u'^2(\mathbf{0})\rangle}{d_m^2} + \frac{V^2(\mathbf{0}) + \langle v'^2(\mathbf{0})\rangle}{l_m^2} + \frac{W^2(\mathbf{0}) + \langle w'^2(\mathbf{0})\rangle}{h_m^2}\right] \quad (108a)$$

which simplifies further to

$$\Delta\omega_A^2 \cong 8\,\frac{U^2(\mathbf{0})}{d_m^2} \quad (108b)$$

for low-turbulence-intensity flow with mean flow in the x direction.

The equations for $\Delta\omega_G^2$ and $\Delta\omega_T^2$ are obtained from Eq. (107*b*) by expanding $(\omega_S + Kc_1)^2$, averaging over velocity, and breaking the result into terms associated with mean flow velocity differences in space, and fluctuating velocity differences in space and in time. The equation for the mean gradient broadening is

$$\Delta\omega_G^2 = \frac{\int C\langle D^2\rangle a^2(\mathbf{x})[KU(\mathbf{x}) - (\mu_1 - \omega_S)]^2\, d^3\mathbf{x}}{\int C\langle D^2\rangle a^2(\mathbf{x})\, d^3\mathbf{x}} \quad (109a)$$

or, using Eqs. (106) and (105*e*),

$$\Delta\omega_G^2 = K^2\, \widetilde{[U(\mathbf{x}) - \tilde{U}]^2} \quad (109b)$$

Thus, $\Delta\omega_G^2$ is the volume-averaged square of the difference between the local velocity and the mean volume-averaged velocity. The turbulent broadening is given by

$$\Delta\omega_T^2 = \frac{K^2 \int C\langle D^2\rangle a^2(\mathbf{x})\langle[u'(\mathbf{x})]^2\rangle\, d^3\mathbf{x}}{\int C\langle D^2\rangle a^2(\mathbf{x})\, d^3\mathbf{x}} \quad (110a)$$

$$= K^2 \widetilde{\langle[u'(\mathbf{x}, t)]^2\rangle} \quad (110b)$$

Its interpretation is obvious.

One of the earliest applications of LDV involved the measurement of turbulence intensity by measurement of the bandwidth of S_{J_D}. According to Eqs. (107*c*) and (110*b*), this can be accomplished if $\Delta\omega_A$ and $\Delta\omega_G$ are known. Then

$$\widetilde{\langle[u'(\mathbf{x}, t)]^2\rangle} = K^{-2}(\mu_2^2 - \Delta\omega_A^2 - \Delta\omega_G^2) \quad (111)$$

can be calculated from measurements of μ_2 *for any turbulent probability density function*. If the rms is nearly constant across the mv, then $\langle[u'(\mathbf{x}, t)]^2\rangle = \langle[u'(\mathbf{O}, t)]^2\rangle$. If $\Delta\omega_A$ and $\Delta\omega_G$ can be reduced to about 20% of $\Delta\omega_T$ or less, the correction terms in Eq. (111) are quite small, and $\langle[u']^2\rangle$ can be measured with good accuracy. Note, however, that decreasing the size of the mv reduces $\Delta\omega_G$ but increases $\Delta\omega_A$, so there is an optimal size for the mv that depends on the variation of $\mathbf{U}(\mathbf{x})$ and $\langle u'^2(\mathbf{x})\rangle$ for any given flow.

For applications in which the precise shape and bandwidth of the power spectrum are not critical, it is satisfactory to model the correlation and spectrum by gaussian forms that are identical to Eqs. (99) and (100) except that the $\Delta\omega_A^2$ factors in those equations are replaced with μ_2. Further, the bandwidth μ_2 can be approximately generalized to include spectral broadening due to Brownian motion of the scattering particles

$$\Delta\omega_B = \frac{32\pi KT \sin^2 k}{3\mu_f d_p \lambda^2} \tag{112}$$

and broadening due to random frequency modulation of the laser source,

$$\Delta\omega_0^2 = \langle(\omega_{01} - \omega_{02})^2\rangle - \langle\omega_S\rangle^2 \tag{113}$$

The approximate equation for the total bandwidth becomes

$$\mu_2 = \Delta\omega_B^2 + \Delta\omega_0^2 + \Delta\omega_A^2 + \Delta\omega_G^2 + \Delta\omega_T^2 \tag{114}$$

where the terms are written in rough order of increasing importance.

9.5 Burst Density

The light flux is a superposition of Doppler bursts that occur at random times with random amplitudes; i.e., it is a shot-noise process much like the filtered-photoemission process discussed in Sec. 6. Hence, the random characteristics of the light flux are characterized by a density parameter representing the extent to which the Doppler bursts overlap with one another, on average. In the case of the photoemission process the appropriate parameter is the mean number of emissions during τ_h, a time scale characteristic of the single emission pulse. In the case of the light flux the appropriate parameter is N_e, the effective mean number of particles in the mv. For example, the Doppler bursts from two particles can overlap only if both particles reside in the mv at the same time. Likewise, if N_e particles reside in the mv on average, then N_e Doppler bursts will overlap on average. In analogy with the terminology used in the discussion of the photoemission process, we shall refer to N_e as the *burst density*.

Upon first consideration it seems reasonable to define N_e to be the mean concentration times the volume of the mv. For example, some investigators [65] use a definition of the form $\int C(\mathbf{x})[a(\mathbf{x})/a(\mathbf{o})]\, d^3\mathbf{x}$, which reduces to C times the volume defined by $\int [a(\mathbf{x})/a(\mathbf{o})]\, d^3\mathbf{x}$ when C is constant. This definition is too simple when the particles are polydisperse, for then one good scatterer may produce much more signal than ten poor scatterers, so that the effective number of particles would be one, even though eleven particles were in the mv. It is apparent that some form of amplitude weighting must be used in the definition of N_e. The appropriate choice of amplitude weighting depends somewhat on the uses one intends to make of N_e. That is to say, the definition of N_e should follow naturally from consideration of the statistical properties of J_D. In the following, we develop a particular definition by considering the way in which J_D approaches a joint normal process in the limit of high burst density.

The two-time characteristic function of J_D is defined by

$$\Phi_{J_D}(\Omega_1, \Omega_2) = \langle \exp\,[j\,\Omega_1 J_D(t) + j\,\Omega_2 J_D(t+\tau)]\,\rangle \tag{115}$$

The joint probability density function for $J_D(t)$ and $J_D(t+\tau)$ is the two-dimensional Fourier transform of $\Phi_{J_D}(\Omega_1, \Omega_2)$. Hence, the characteristic function contains all the statistical information necessary to compute all one-time and two-time moments of J_D. In particular,

$$\langle J_D^2 \rangle = -\left.\frac{\partial^2 \Phi_{J_D}}{\partial \Omega_1^2}\right|_{\Omega_1=\Omega_2=0} \tag{116}$$

and

$$\langle J_D(t) J_D(t+\tau) \rangle = -\left.\frac{\partial^2 \Phi_{J_D}}{\partial \Omega_1\, \partial \Omega_2}\right|_{\Omega_1=\Omega_2=0} \tag{117}$$

It can be shown, using the approximation in Eq. (89), that in general Φ_{J_D} is given by

$$\Phi_{J_D} = \exp\,[\textstyle\iiint f_{\mathbf{u}}(\mathbf{c}, \mathbf{x}) C(\mathbf{x}, D)(e^{j\beta} - 1)\, d^3\mathbf{x}\, dD\, d^3\mathbf{c}] \tag{118}$$

where

$$\beta = \Omega_1 D a(\mathbf{x}) \cos \Phi(\mathbf{x}, t) + \Omega_2 D a(\mathbf{x} + \mathbf{c}\tau) \cos \Phi(\mathbf{x} + \mathbf{c}\tau, t + \tau) \tag{119}$$

The interested reader can verify that applications of Eqs. (116) and (117) to this characteristic function yield the same expressions for $R_{J_D J_D}(\tau)$ as Eq. (98).

As the number of particles in the mv becomes large, J_D is a sum of a large number of independent random variables, and we expect from the central limit theorem that J_D will become a joint normal random process. This can be shown by calculating the asymptotic value of Φ_{J_D} in the limit $C \to \infty$, but it must first be recognized that $\langle J_D^2 \rangle \to \infty$ as $C \to \infty$, so that to obtain a useful asymptotic result that has finite mean square value, one must deal with the dimensionless process $J_D^* = J_D(t)/\langle J_D^2 \rangle^{1/2}$. Clearly, $\langle J_D^{*2} \rangle = 1$, independent of C. It follows after some analysis that

$$\Phi_{J_D^*} = \exp\left\{\iiint f_{\mathbf{u}}(\mathbf{c}) C(\mathbf{x}, D) \left[-\frac{\beta^2}{2\langle J_D^2 \rangle} + 0\left(\frac{\exp(-\pi^2 N_{FR}^2/24)}{\sqrt{N_e}}\right) + 0(N_e^{-1})\right] d^3\mathbf{x}\, d^3\mathbf{c}\, dD\right\} \tag{120}$$

where

$$N_e = \frac{[\iint C(\mathbf{x}, D) D^2 a^2(\mathbf{x})\, d^3\mathbf{x}\, dD]^2}{\int C(\mathbf{x}, D) D^4 a^4(\mathbf{x})\, d^3\mathbf{x}\, dD} \tag{121}$$

Ordinarily, one shows that J_D^*, and hence J_D, are asymptotically joint normal by taking the limit of Eq. (120) as $N_e \to \infty$, from which it follows that $\ln \Phi_{J_D^*}$ is quadratic in Ω_1 and Ω_2, hence that the joint probability density for $J_D^*(t)$ and $J_D^*(t+\tau)$ is asymptotically joint normal, and that the rate of convergence to a joint normal distribution is of order $N_e^{-1/2}$. (Strictly, J_D^* is a joint normal random process only if its characteristic functional is quadratic in the Ω's, but the asymptotic behavior of the characteristic functional is exactly the same as for the behavior of the two-time characteristic function, so the more complicated proof is unnecessary.) In the particular case of laser-Doppler signals the rate of convergence is faster, of order N_e^{-1}, because the second term in the brackets in Eq. (120) is extremely small for any practical value of N_{FR}.

The purpose of this derivation is to point out two facts. First, the basis for the present definition of N_e is that it is the appropriate parameter for describing the high-particle-density behavior of J_D, and it is a measure of the size of the fourth-order terms in the characteristic function, rather than the third-order term. Second, LDV signals become joint normal at much lower values of N_e than might be expected on the basis of conventional arguments (cf. [12, 65]). For example, if $N_e > 100$ assures joint normal behavior with $N_e^{-1/2}$ convergence, $N_e > 10$ assures an even closer approach to joint normal statistics with N_e^{-1} convergence. This conclusion is supported by experimental experience, which indicates that N_e greater than 5 to 10 is large enough to make the asymptotic formulas for $N_e \to \infty$ valid.

It should be noted that the equation for N_e reduces to

$$N_e = C\left(\frac{\pi^3}{512}\right)^{1/2} d_m\, l_m\, h_m \tag{122}$$

when the particle population is monodisperse and uniform in space. Equation (121) clearly weights the strong scatterers more heavily than the weak scatterers. For example, if the particle population consists of only two types, $C(\mathbf{x}, D) = C_1\,\delta(D - D_1) + C_2\,\delta(D - D_2)$, and

$$N_e = \frac{1 + (C_2 D_2^2 / C_1 D_1^2)}{1 + (C_2 D_2^4 / C_1 D_1^4)}\, C_1\left(\frac{\pi^3}{512}\right)^{1/2} d_m l_m h_m \tag{123}$$

which is proportional to C_1 for small values of $C_2 D_2^2 / C_1 D_1^2$. Thus, natural aerosols usually produce fairly low values of N_e, despite the presence of a great many fine particles. Similarly, when natural aerosols or hydrosols are artificially seeded with good scatters, N_e is approximately the number of good artificial scatters in the mv.

It is now possible to define two asymptotic cases in which $J_D(t)$ exhibits vastly different random properties: *high burst density*, $N_e \gg 1$, and *low burst density*, $N_e \ll 1$. These terms will be abbreviated to HBD and LBD, respectively. HBD signals are often referred to as "continuous" signals [66], and the term "individual realizations" is conventionally applied to LBD signals [71]. The intermediate case, $N_e = O(1)$ will be called "O(1) burst density." The reason for devoting so much time to this classification scheme and to the proper definition of N_e is that the choices of signal-processing techniques and data-processing techniques depend critically upon the type of LDV signal. Specifically, the burst density in combination with the photon density more or less determines the signal-processing technique(s) that should be used.

9.6 High-Burst-Density Signals ($N_e \gg 1$)

The most desirable characteristic of the HBD signal is that it is nearly continuous; i.e., there are no gaps in it because, according to Eq. (76), the probability of finding zero particles in the mv at any instant is $\exp(-N_e)$. Consequently, information on the fluid velocity can be obtained as a continuous function of time. In contrast the LBD signal yields velocity data only when bursts occur, and the bursts occur at random times, complicating analysis of the data.

This advantage of the HBD signal is offset by additional randomness in the signal that is created by the superposition of many random bursts. A typical HBD signal is illustrated in Fig. 18. Its amplitude, being the sum of many random amplitudes, varies randomly in time. More important, the phase of the HBD signal is random also, partly because the phases of the component bursts are random and partly because their amplitudes are random. The instantaneous frequency is defined as the time derivative of the phase, and it also fluctuates randomly about a mean value, even when the frequencies of all the bursts are identical, e.g., steady laminar flow with no velocity gradients. Fluctuations in the phase and instantaneous frequency are referred to as "phase noise" and "ambiguity noise," respectively [65]. They are inherent in the HBD signal, and they place a limitation on the accuracy of the instantaneous Doppler frequency measurement. Thus, continuity of the velocity information is accomplished only by sacrificing accuracy.

9.6.1 Rice's representation and the "measurable" velocity. The equation for the Doppler light flux is

$$J_D(t) = \iint Da(\mathbf{x}) \cos(\omega_S t + Kx) g(\mathbf{x}, t, D)\, d^3\mathbf{x}\, dD \tag{124}$$

Following the work of Rice [72] for classical shot-noise processes, it is possible to reduce this equation (which represents a generalized shot-noise process with random points in $\mathbf{x}D$ space) to the form

$$J_D(t) = A(t) \cos \Phi_D(t) \tag{125}$$

where

$$\Phi_D(t) = \omega_S t + Kx_m(t) + \phi(t) \tag{126}$$

is the aforementioned random phase of the HBD signal, and $A(t)$ is its random amplitude. This reduction is accomplished by decomposing the phase of Eq. (124) into the form $\omega_S t + Kx = \omega_S t + Kx_m(t) + K[x - x_m(t)]$, where $x_m(t)$ is *any arbitrary function of time*. At this point, we do not know how to specify $x_m(t)$, but we expect that it is related to the mean displacement of the velocity field, as determined by $u(\mathbf{x})$. It will be determined shortly by application of certain mathematical requirements that are independent of the steps needed to reduce Eq. (124) to Eq. (125). By inserting this phase decomposition into Eq. (124) and using

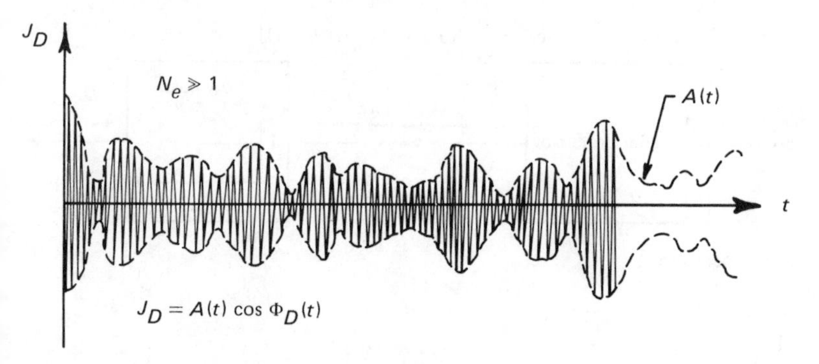

Figure 18 High-burst-density Doppler light flux.

simple trigonometric identities, it is easy to show that the amplitude and phase are given by

$$A(t) = [F^2(t) + G^2(t)]^{1/2} \tag{127}$$

$$\phi(t) = \tan^{-1} \frac{G}{F} \tag{128}$$

where

$$\begin{bmatrix} F(t) \\ G(t) \end{bmatrix} = \iint Da(\mathbf{x}) \begin{bmatrix} \cos K(x - x_m) \\ \sin K(x - x_m) \end{bmatrix} g(\mathbf{x}, t, D)\, d^3\mathbf{x}\, dD \tag{129}$$

To $O(\exp - N_{FR}^2)$, the correlation functions of F and G are given by

$$R_{FF}(\tau) = \langle F(t)F(t + \tau)\rangle = \langle G(t)G(t + \tau)\rangle \tag{130a}$$

$$= \tfrac{1}{2} \iint C(\mathbf{x})\langle D^2\rangle a(\mathbf{x})a(\mathbf{x} + \mathbf{c}\tau)f_{\mathbf{u}}(\mathbf{c}) \cos K[c_1\tau - x_m(t + \tau) + x_m(t)]\, d^3\mathbf{x}\, d^3\mathbf{c} \tag{130b}$$

and

$$\langle F(t)G(t + \tau)\rangle = \langle F(t + \tau)G(t)\rangle = 0 \tag{131}$$

These equations follow from Eq. (130*a*) by steps that are identical to those leading to Eq. (98) for $R_{J_D J_D}$. Since J_D is a joint normal process for HBD signals, $F(t)$ and $G(t)$ are also joint normal processes. From Eq. (131), they are independent as well.

The Doppler shift information is contained in $\dot{\Phi}_D$, the instantaneous frequency of the HBD signal. Given an input signal of the form $A(t) \cos \Phi_D(t)$, the ideal frequency demodulator (i.e., frequency-to-voltage convertor) would ignore the random amplitude modulation and extract from the input a signal (either a voltage or a digital number) that is directly proportional to the instantaneous frequency,

$$\dot{\Phi}_D = \omega_S + K\dot{x}_m + \dot{\phi}(t) \tag{132}$$

as shown in Fig. 19. Ideally the output would be proportional to the fluid velocity, but because of the ambiguity noise $\dot{\phi}(t)$, the real output fluctuates randomly.

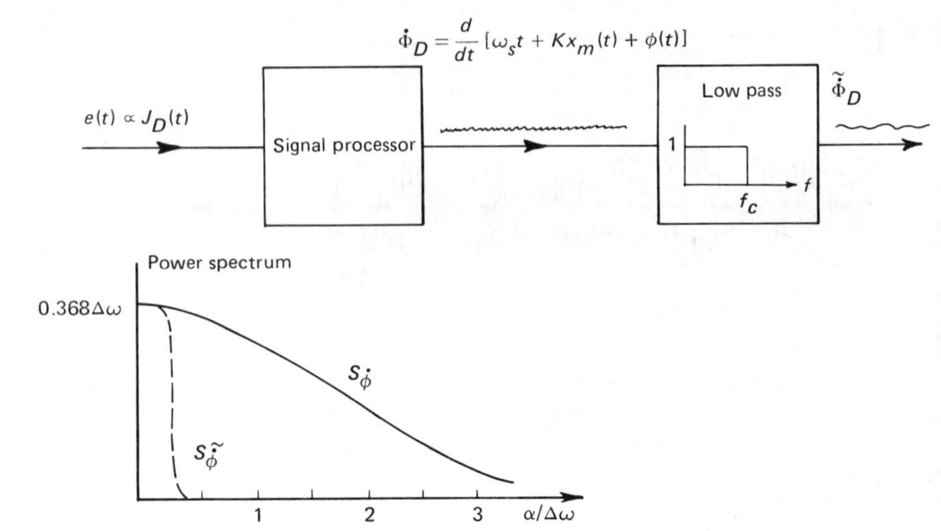

Figure 19 Frequency demodulation and the ambiguity noise spectrum.

In practice a significant part of this fluctuation is removed by passing the output of the signal processor through a low-pass filter that rejects all frequencies greater than f_c, the low-pass cutoff. The filtered signal $\tilde{\dot{\Phi}}_D$ is the quantity of primary interest. To avoid confusion, the symbols f and $\alpha = 2\pi f$ will be used to represent the frequencies at which $\tilde{\dot{\Phi}}_D$ fluctuates, i.e., the spectral frequency components of the instantaneous frequency. These should not be confused with the Doppler frequency. The reader will find it easiest, perhaps, always to think of $\dot{\Phi}_D$ and $\tilde{\dot{\Phi}}_D$ as voltages proportional to the input signal frequency. It is then obvious that these voltages may also fluctuate in time and that they will have power spectra $S_{\dot{\Phi}_D}(\alpha)$ and $S_{\tilde{\dot{\Phi}}_D}(\alpha)$.

We expect the mean value of $\tilde{\dot{\Phi}}_D$ to be simply related to the mean velocity U. This relationship can be determined by using a result derived by Rice [73], which states that the mean value of the instantaneous frequency of a joint normal signal such as J_D in Eq. [124] is equal to the centroid of the power spectrum of the signals. That is, $\langle \dot{\Phi}_D \rangle = \mu_1$, where μ_1 is given by Eq. (102a). Since low-pass filtering does not affect the mean value, $\langle \tilde{\dot{\Phi}}_D \rangle = \mu_1$ also. Taking the average of Eq. (132) and setting it equal to the value of μ_1 found in Eq. (105e) yields

$$\langle \dot{\Phi}_D \rangle = \omega_S + K\langle \dot{x}_m \rangle + \langle \dot{\phi} \rangle \tag{133a}$$

$$= \omega_S + \frac{K \iint C\langle D^2 \rangle a^2(\mathbf{x}) U(\mathbf{x})\, d^3\mathbf{x}}{\iint C\langle D^2 \rangle a^2(\mathbf{x})\, d^3\mathbf{x}} \tag{133b}$$

Comparison of Eqs. (133a) and (133b) shows that the random variable $\dot{\phi}$ will have zero mean value if and only if $x_m(t)$ is defined so that

$$\langle \dot{x}_m \rangle = \frac{\int C\langle D^2 \rangle a^2(\mathbf{x}) U(\mathbf{x})\, d^3\mathbf{x}}{\int C\langle D^2 \rangle a^2(\mathbf{x})\, d^3\mathbf{x}} \tag{134}$$

or, in other words

$$\langle \dot{x}_m \rangle = \widetilde{U} \tag{135}$$

where $\widetilde{U}$ is the mean velocity volume averaged according to Eq. (106). Clearly Eq. (134) can be satisfied by defining

$$x_m(t) = x_m(t_0) + \int_{t_0}^{t} \widetilde{u}(t')\, dt' \tag{136}$$

so that

$$\dot{x}_m(t) = \widetilde{u}(t) \tag{137}$$

and Eq. (135) follows immediately. Hence, the mean displacement is just the displacement due to the volume-averaged eulerian velocity field. The field is eulerian rather than langrangian because we are interested only in displacements caused by particles that currently reside in the mv. The instantaneous frequency is now

$$\dot{\Phi}_D = \omega_S + K\widetilde{u}(t) + \dot{\phi}(t) \tag{138}$$

If we define $\dot{x}_m = \widetilde{u}$ so that $\dot{\phi}$ has zero mean value, the signal represented by Eq. (138) is exactly the same as the signal analyzed by Rice [72, 73], even though the present signal is generated by random points in $\mathbf{x}D$ space, whereas in Rice's work the signal is initially generated by random points in time. Hence, all the results for $\dot{\phi}$ in [72, 73] can be applied to the present signal without modification. It should be noted that the present definition of $\dot{x}_m$ does not agree with the ad hoc definition of u_0, the "effective velocity seen by the velocimeter" given by George and Lumley in Eq. [2.2.5] of [65]. The difference between these two definitions is quite important, because the entire theoretical basis for interpreting HBD LDV signals ultimately rests upon the definition of the "effective" or "measurable" velocity.

Unfortunately, the reasoning needed to establish the correct definition is equally subtle because, as shown here, and as implied in Rice's [72, 73] work, the phase decomposition leading to Eq. (132) is valid for any $x_m(t)$. Hence, up to Eq. (132), $x_m(t)$ could be defined as the time integral of *any* effective velocity, which led Buchave et al. [66] to suggest that the u_0 decomposition in [65] may not be unique. However, there is only one choice, Eq. (138), which yields $\langle \dot{\phi} \rangle = 0$, and this definition must be used if one intends to make use of Rice's analyses of the statistics of $\dot{\phi}$, in which it is implicit that $\langle \dot{\phi} \rangle = 0$. Moreover, it appears that a correct analysis of the statistics of $\dot{\phi}$ for the case of $\langle \dot{\phi} \rangle \neq 0$ [that is, $\dot{x}_m$ not given by Eq. (138)] would simply lead to results that would depend on $\langle \dot{\phi} \rangle$ in such a way as to ultimately restore $\langle \dot{\phi} \rangle$ to the definition of the effective velocity. Consequently, it is the author's belief that the velocity "seen by the LDV" is uniquely given by Eq. (138).

Fortunately, it can be shown that all the results derived in [65, 66] could have been derived by starting with the definition

$$u_0 = \frac{\int a(\mathbf{x}) u \, d^3\mathbf{x}}{\int a(\mathbf{x}) \, d^3\mathbf{x}}$$

instead of Eq. [2.2.5] in [65]. This is very close in form to $\tilde{u}$ for the case $C\langle D^2 \rangle = \text{const}$, except that the weighting factor is $a(\mathbf{x})$ instead of $a^2(\mathbf{x})$. Hence, the equations in [65] can be corrected by dividing mv dimensions σ_1, σ_2, and σ_3, defined in that reference, by $\sqrt{2}$, thereby converting their $w(\mathbf{x})$ to our $a^2(\mathbf{x})$.

As mentioned above, Rice's results [72, 73] for $\dot{\phi}$ can be applied directly to the present signal. In particular, the power spectrum of the ambiguity noise $S_{\dot{\phi}}(\alpha)$ is very broad, and if S_{J_D} has a gaussian shape its value at zero frequency is

$$S_{\dot{\phi}}(0) \simeq 0.368 \, \Delta\omega, \tag{139}$$

where $\Delta\omega$ is the rms bandwidth of $F(t)$ [or $G(t)$], defined by

$$\Delta\omega^2 = \frac{\int_0^\infty \omega^2 S_F(\omega) \, d\omega}{\int_0^\infty S_F(\omega) \, d\omega} \tag{140a}$$

$$= -\frac{R''_{FF}(0)}{R_{FF}(0)} \tag{140b}$$

By inserting R_{FF} from Eq. (130*b*) into Eq. (140*b*) and using steps similar to those leading to Eq. (107*b*), it can be shown that

$$\Delta\omega^2 = \Delta\omega_A^2 + K^2 \widetilde{(U - \tilde{U})^2} + K^2 \widetilde{\langle (u' - \tilde{u}')^2 \rangle} + \Delta\omega_0^2 + \Delta\omega_B^2 \tag{141}$$

where

$$\tilde{u}' = \frac{\int C D^2 a^2(\mathbf{x}) \tilde{u}'(\mathbf{x}, t) \, d^3\mathbf{x}}{\int C \langle D^2 \rangle a^2(\mathbf{x}) \, d^3\mathbf{x}} \tag{142}$$

is the volume-averaged velocity fluctuation. The ambiguity bandwidth $\Delta\omega_A^2$ is exactly the same as in Eq. (108*a*). The second term in Eq. (141) represents the broadening due to mean velocity gradients, also the same as before, and the third term arises from turbulent velocity gradients that cause differences between the local turbulent fluctuations and the volume-averaged turbulent fluctuation, $u'(\mathbf{x}, t) - \tilde{u}'(\mathbf{x}, t)$. The bandwidths due to source broadening and Brownian motion do not follow from Eqs. (140*b*) and (130*b*), but they have been added to the result, as before.

Equation (139) pertains to $\dot{\phi}$, but in practice one is interested in the filtered output of the signal processor $\tilde{\dot{\Phi}}_D$, and hence in the properties of the filtered ambiguity noise $\tilde{\dot{\phi}}$. Rice's [73] result for the spectrum $S_{\dot{\phi}}(\alpha)$ is sketched in Fig. 19. Suppose $H(\alpha)$ is the transmission function of the filter, i.e., the ratio of the output to the input when the input is a pure sine wave at frequency α. Then the spectrum of the filtered phase noise is $S_{\tilde{\dot{\phi}}}(\alpha)|H^2(\alpha)|^2$. The cutoff frequency of the filter $\alpha_c = 2\pi f_c$ is usually selected to be greater than the highest turbulent frequencies of interest, and these are somewhat lower than $\Delta\omega$. Then, if $\alpha_c < \Delta\omega$,

$$S_{\tilde{\phi}}(\alpha) = S_{\dot{\phi}}(0)|H(\alpha)|^2 \tag{143a}$$

$$= 0.368\ \Delta\omega |H(\alpha)|^2 \tag{143b}$$

because $S_{\dot{\phi}}$ is nearly constant in the range $0 < \alpha < \Delta\omega$. The filtered ambiguity noise appears to be low-pass-filtered white noise with mean square value

$$\langle(\tilde{\dot{\phi}})^2\rangle = 0.368\ \Delta\omega\ \alpha_c \tag{144}$$

which corresponds to an equivalent velocity noise $u_{\tilde{\phi}}$ with rms

$$\langle(u_{\tilde{\phi}}^{\ 2})\rangle^{1/2} = K^{-1}\langle(\tilde{\dot{\phi}})^2\rangle^{1/2} \tag{145}$$

The ambiguity noise and u are statistically independent [65], so the correlation and power spectrum of the measured velocity $K^{-1}\dot{\Phi}_D = \widetilde{U} + \tilde{\tilde{u}}'(t) + K^{-1}\tilde{\dot{\phi}}(t)$ are the sums of the correlations and spectra of $\tilde{\tilde{u}}(t)$ and $\tilde{\dot{\phi}}(t)$, respectively. That is,

$$K^{-2}R_{\tilde{\Phi}_D\tilde{\Phi}_D}(\tau) = R_{\tilde{\tilde{u}}\tilde{\tilde{u}}}(\tau) + K^{-2}R_{\tilde{\phi}\tilde{\phi}}(\tau) \tag{146}$$

and

$$K^{-2}S_{\tilde{\Phi}_D}(\alpha) = S_{\tilde{\tilde{u}}}(\alpha) + K^{-2}S_{\tilde{\phi}}(\alpha) \tag{147}$$

where

$$S_{\tilde{\tilde{u}}}(\alpha) = |H|^2 S_{\tilde{u}}(\alpha) \tag{148}$$

is the power spectrum of the filtered values of the measurable (i.e., volume-averaged) velocity fluctuation $\tilde{\tilde{u}}'(t)$.

Turbulent motions whose scales are of the order of the dimensions of the mv are, of course, attenuated by the volume averaging that is inherent in u'. The finite mv acts as a spatial filter. Detailed calculations of this effect are presented in [65]. As a simple guideline, attentuation is less than 5% for transverse wave numbers of the velocity spectrum in the y direction (that is, the direction of l_m, the largest dimension of the mv) that are less than about $0.6/l_m$, corresponding to about 10 wavelengths in l_m.

10. SIGNAL PROCESSORS

The LDV signal processor is designed to measure the Doppler frequency plus any other necessary data from the PMT signal. With certain exceptions LDV signal processors are special-purpose instruments designed specifically to handle the peculiar characteristics of LDV signals. The primary types of signal processors are correlators, spectrum analyzers, counters, and frequency trackers, the last two being the most predominant. Less commonly used techniques included direct computer analysis and filter banks. None of these techniques provides a universally optimal solution to the LDV frequency measurement problem, so the experimentalist must select the type of device that is best suited to the particular application. The main criteria for selection are accuracy, frequency range, ability to extract signal frequency from noise, time resolution, ease of use, and ease of interpreting the output data. The various instruments differ greatly in many of these regards.

Time resolution is defined as the time required to obtain a measurement of the Doppler frequency to within some prescribed accuracy. It is essentially the inverse of the instrument's frequency response. In the following sections LDV signal processors are identified as either long-time averaging devices or time resolving devices. The measurement time of the former group of processors is long compared to the time scales of the velocity variation, so the output must be interpreted in terms of long-time averages of the velocity. The measurement time of the latter group of processors is of the order of the time for a single Doppler burst, and since this is normally short compared to the time scale of the flow, processors in this group are capable of resolving the velocity as a function of time. Broadly speaking, classical correlation analysis and spectrum analysis are long-time averaging techniques, whereas counting and frequency tracking are time resolving techniques, but exceptions are possible, especially as the speeds of correlators and spectrum analyzers improve with advances in electronic technology.

10.1 Amplitude Correlators

The conventional type of correlation analyzer computes the correlation $R_{ee}(\tau) = \langle e(t)e(t+\tau)\rangle$ between the signal *amplitudes* $e(t)$ and $e(t+\tau)$, where τ is the time delay. These devices are available commercially, but most of them are intended for general-purpose signal analysis and have not been designed specifically for LDV applications. As a consequence, they are usually too slow for most LDV experiments, although they are useful if the Doppler frequencies are less than about 2 to 20 kHz. In these cases, the correlation of the Doppler signal over long averaging times can be used to determine the mean flow velocity and the turbulence intensity, with certain corrections as discussed in Sec. 9.4. The power spectrum of the Doppler signal can be obtained also by Fourier-transforming the measured correlation function. In principle, both the correlation and the spectrum contain information on the full probability density function of the velocity $f_{\mathbf{u}}(\mathbf{c})$, but the broadening effects due to ambiguity and gradients make it difficult to obtain $f_{\mathbf{u}}(\mathbf{c})$ directly, and measurements of moments higher than $\langle (u')^2 \rangle$ require corrections that may be subject to considerable error.

To discuss the applications and limitations of correlation analyzers more fully, it is necessary to describe their operation in more detail. The conventional amplitude correlator converts the signal voltage $e(t)$ into a digital word that is B bits long at intervals of $\Delta\tau$ seconds. An estimate of the correlation function is formed by computing

$$R_{ee}(n\,\Delta\tau) = \frac{1}{(M - |n|)} \sum_{m=1}^{M} e(m\,\Delta\tau)e((m+n)\,\Delta\tau) \tag{149}$$

wherein the sum over m corresponds to an average over time, and $n = 0, \pm 1, \ldots, \pm(N-1)$ determines the time delay. $(N-1)\,\Delta\tau$ is the maximum time delay, and correlators are usually designated as "N-point" correlators. Typical values of N range from 100 to 1028. It is well known that for most signals, and certainly for LDV signals, this estimate converges to $R_{ee}(n\,\Delta\tau)$ with rms error proportional to $M^{-1/2}$ as M becomes large. Hence, the sampling times are usually longer than the time scale of the turbulent flow.

Correlator speed is determined by the rate at which the correlator can perform the multiplications in Eq. (149), an N-point correlator forming N B- $\times$ B-bit products every $\Delta\tau$ seconds. A reasonable number of correlation points is 10 per cycle of the correlation, so if the correlation frequency is f_D, $\Delta\tau$ must be $\frac{1}{10}f_D$, and the correlator product rate must be $10Nf_DB^2$ in bits per second. For example, 10-bit resolution of a 10-MHz signal with 100 lines would require a product rate of 10^{12} bits/s, far in excess of present digital technology. Clearly, it is helpful to reduce B, and, in fact, values of B of 4 and 8 bits are typical in general-purpose correlators such as those available from Nicollet Instruments, Inc. or Honeywell-Saicor, Inc. The minimum $\Delta\tau$ values for these instruments are in the range 10 to 50 μs.

Correlators for general laser velocimetry must work at the highest possible speeds because the Doppler frequencies are normally large. Correlation speed can be improved by reducing the analog-to-digital resolution to $B = 1$ bit. Let $e_c(t)$ be the 1-bit digital signal defined by

$$e_c(t) = \begin{cases} 1 & e(t) > 0 \\ -1 & e(t) < 0 \end{cases} \tag{150}$$

Clearly, $e_c(t)$ is just the signal that would be obtained by amplifying $e(t)$ with large gain and clipping it at the ± 1 levels. Hence, 1-bit correlation is referred to as *clipped correlation*. Single-clipped correlators compute $R_{ee_c} = \langle e_c(t)e(t+\tau)\rangle$, while double-clipped correlators compute $R_{e_ce_c} = \langle e_c(t)e_c(t+\tau)\rangle$. Double-clipped correlation is the simplest because it involves only the multiplication of 1s and -1s.

A well-known theorem called the *arcsine law* states that the correlation of the double-clipped signal is given by

$$R_{e_ce_c} = \frac{2}{\pi} \arcsin \frac{R_{ee}(\tau)}{R_{ee}(0)} \tag{151}$$

when $e(t)$ is a normal random process (cf. [53], for example), showing that double clipping distorts the correlation on average; that is, $R_{e_ce_c}$ is a biased estimate of R_{ee}. Fortunately, this distortion can be eliminated by adding "auxiliary signals" to $e(t)$ and $e(t+\tau)$ before they are quantized, as shown originally in [74 to 76]. *This procedure is valid for any random process* e(t) *that has bounded amplitude.* The auxiliary signals must have zero mean value, and they must be statistically independent of $e(t)$ and $e(t+\tau)$ and of each other. Finally, the probability distributions of their amplitudes should be uniform. The addition of an auxiliary signal is called *linearization*, and the function of linearization is to move $e(t)$ [or $e(t+\tau)$] up and down across the 1-bit quantization level so that each level of $e(t)$ is sampled with uniform probability. This has the same effect as quantizing the signal with a randomly varying quantization level. In this way, the average of the product of the linearized signals correctly converges to $R_{ee}(\tau)$, even though the individual samples of $e(t)$ and $e(t+\tau)$ are quantized very coarsely. In principle the auxiliary signal may be any random signal that satisfies the conditions stated above, but in practice, a simple triangular wave is used.

Reference [77] describes a 256-point correlator of this type that is capable of correlating every sample pair for $\Delta\tau > 40$ ns. Faster correlation rates ($\Delta\tau > 10$ ns) are achieved by forming time-delayed products at every kth sample, i.e., by ignoring a fraction of the samples. This does not affect the ultimate accuracy of the correlation. Also, it does not significantly lengthen the time required to achieve a certain level of sampling accuracy when the signal is correlated with itself over $k\,\Delta\tau$, for then the samples in $k\,\Delta\tau$ are not statistically independent, and therefore they contribute very little to the sampling accuracy. Correlation-time increments such as 10 ns make it possible to measure 10-MHz Doppler signals with 10 correlation points per cycle, and continuing improvements in the speed of digital electronics should lead to speeds perhaps 10 times faster. Thus, digital correlation of Doppler signals is likely to become practical for almost all frequencies of practical interest.

As electronics speeds improve, it should also become possible to obtain correlations for individual Doppler bursts, from which the Doppler frequency of the individual particle would be determined. In that case, correlation would no longer be limited to the realm of long-time averaging.

10.2 Photon Correlators

The photon correlator correlates the *rate* of photoemissions, rather than $e(t)$, which a sum of the filtered photoemission pulses. When using photon correlators, one must also use a fast PMT with no output filtering so that each photoemission appears as a narrow, discrete pulse. The ideal photon correlator output corresponds to $R_{nn}(\tau) = \langle n(t, t+\Delta\tau) n(t+\tau,\ t+\tau+\Delta\tau)\rangle$, where $n(t, t+\Delta\tau)$ is the number of photoemission pulses in $(t, t+\Delta\tau)$; that is, $n/\Delta\tau$ is a measure of the instantaneous emission rate. Recall that the conditional average of n given the total light flux J_{tot} is $\dot{\epsilon}(t)$, given by Eq. (57). It can be shown [54] that

$$R_{nn}(\tau) = \Delta\tau^2\, R_{\dot{\epsilon}\dot{\epsilon}}(\tau) + \langle\dot{\epsilon}\rangle\, \Delta\tau\, \Lambda\!\left(\frac{\tau}{\Delta\tau}\right) \tag{152}$$

where

$$\Lambda\!\left(\frac{\tau}{\Delta\tau}\right) = \begin{cases} 1 - \dfrac{|\tau|}{\Delta\tau} & |\tau| < \Delta\tau \\ 0 & |\tau| > \Delta\tau \end{cases} \tag{153}$$

and

$$R_{\dot{\epsilon}\dot{\epsilon}} = \frac{\eta_q^2}{h^2 v_0^2}\, \langle J_{tot}(t) J_{tot}(t+\tau)\rangle \tag{154a}$$

It follows from Eq. (54) that

$$R_{\dot{\epsilon}\dot{\epsilon}} = \frac{\eta_q^2}{h^2 v_0^2}\, (R_{J_B J_B} + R_{J_P J_P} + R_{J_D J_D} + R_{J_R J_R}) \tag{154b}$$

where the correlations in parenthesis are the background, pedestal, Doppler, and reference-beam light-flux correlations, respectively. R_{nn} contains the Doppler signal correlation, but it is superimposed on the pedestal correlation, which has the same shape as the envelope of $R_{J_D J_D}$, plus the background and reference-beam correlations, which will be essentially constants. In addition, there is a narrow spike at $\tau = 0$ caused by the correlation of the photoemission pulses with themselves. In contrast, these additional terms need not appear in the amplitude correlation discussed in the preceding section, because J_B, J_P, and J_R are low-frequency signals that can be removed from the analog signal by high-pass analog filters. These terms are not removed in photon correlation because digital high-pass filtering is too time-consuming. A typical correlogram obtained using a photon correlator is shown in Fig. 20.

As with amplitude correlators, photon correlators employ clipping to improve speed. In photon correlation the pulse count n is replaced by n_c, which is unity if n exceeds an adjustable clipping level q and zero otherwise. Single clipping without linearization is one common mode. In this case, if the light flux is a normal random process,

$$\frac{R_{n_c n}(\tau)}{R_{n_c n}(0)} = \frac{\langle n \rangle - q}{\langle n \rangle + 1} + \frac{1 + q}{1 + \langle n \rangle} R_{nn}(\tau) \tag{155}$$

and $R_{n_c n}$ is nearly equal to R_{nn} if q is set equal to $\langle n \rangle$. When the burst density is small, J_D is not likely to be a normal process, and the single-clipped correlator should use a linearization procedure similar to the one discussed in the preceding section. In the context of photon correlation, this procedure is referred to as *scaling*, and it is accomplished by leaving a random digital word with uniform probability distribution in the register containing n_c. Detailed theoretical analyses of photon correlation have been performed by Pike, Jakeman, Foord, and coworkers, and the reader can find references to these works plus extensive bibliographies in [12, 18, 21].

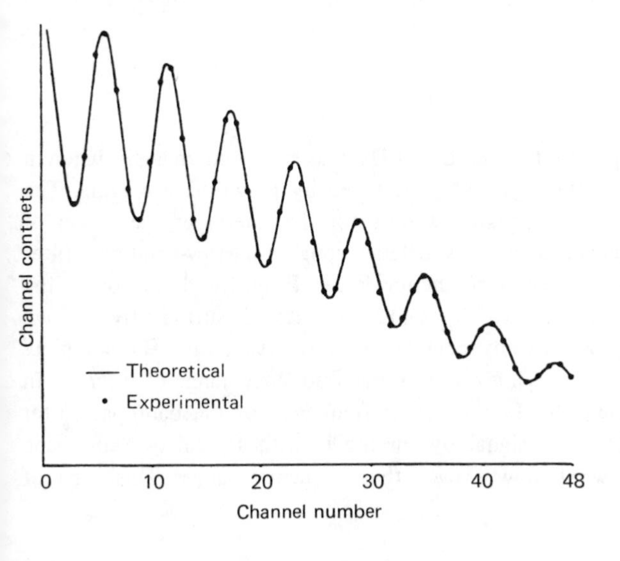

Figure 20 Correlogram obtained with a photon correlator. Solid line is a theoretical correlation function fitted through the data for the purpose of determining the Doppler frequency. *(From [47].)*

Commercial photon correlators have been developed that operate as fast as $\Delta\tau = 10$ ns with 64 correlation points (Malvern Instruments, England). These correlators are capable of reading out the correlation every 32 μs, from individual Doppler bursts. Thus, long-time averaging is not necessary if the signal is strong enough to produce accurate correlations in a single burst. As a rule, this requires more than about 50 to 100 photoemissions per burst [78].

The primary advantage of the correlation technique, either amplitude correlation or photon correlation, is its ability to process very noisy signals and extract relatively accurate frequency information. In principle, the mean photon rate can be less than one photon per Doppler cycle, or even one photon per Doppler burst, and correlations can still be formed, given a sufficiently long averaging time. (Of course, one must correlate between two photons in the same burst to obtain a measure of the frequency, but for any mean number per burst, there is always a finite, albeit small, probability that two or more photons will occur during a burst.) In this regard, there is very little difference between amplitude correlation and photon correlation. The primary differences are practical. For example, at very low photon densities, the maximum amplitude of $e(t)$ is of the order of the amplitude of a filtered photoemission pulse, which can be small and difficult to detect; in contrast, the photon correlator simply assigns a unit value to each pulse. On the other hand, as the photon density increases, for example, $\dot{\epsilon}\tau_h = O(1)$, the pulses begin to overlap and the photon correlator can no longer discriminate between single pulses and groups of pulses. This phenomenon is referred to as "photon pileup," and it seriously distorts the correlation [79]. The amplitude correlator is immune to photon pileup and, in fact, works best in this regime.

The 10-ns speeds of present amplitude correlators and photon correlators limit the maximum Doppler shift to about 10 MHz, or perhaps 20 MHz if the number of correlation points is as low as five per Doppler cycle. The maximum frequency is about one order of magnitude lower than the maximum attainable with current frequency counters. Consequently, the optical systems used in conjunction with correlator signal processors must employ relatively narrow beam angles when the flow velocities are high.

10.3 Spectrum Analysis

The earliest signal-processing technique used in LDV was spectrum analysis, wherein the power spectrum of the PMT signal was measured by long-time averaging. The sweep-type spectrum analyzer is a square-law detector that measures the power of that portion of the input signal that is transmitted through a narrow-band-pass filter whose width is $\Delta\omega_f$ and whose center frequency is ω_f. From the definition of the power spectrum, this power is equal to $S_{J_D}(\omega_f)\,\Delta\omega_f$ if $\Delta\omega_f$ is sufficiently narrow, and $\Delta\omega_f$ is defined so as to absorb any constant of proportionality. The complete power spectrum is obtained by varying ω_f over the frequency range of interest. In practice, it is more convenient to fix the center frequency of the band-pass filter and translate the spectrum of the signal by mixing it with a local oscillator sine wave whose frequency is swept slowly over the frequency range. This type of analysis is inefficient because the power detector "sees" only a $\Delta\omega_f$-wide band of the signal spectrum, and at any instant it is blind to the signal spectrum outside this band. That is, the swept spectrum analyzer processes $\Delta\omega_f$-wide bands of the spectrum serially. Consequently, the swept spectrum analysis requires about $\Delta\omega/\Delta\omega_f$ times as much time to sample a $\Delta\omega$-wide spectrum as would a device that processed the $\Delta\omega$-wide band in parallel.

Digital spectrum analyzers that operate by continuously fast-Fourier-transforming segments of the sampled input signal are parallel processors in that they use all the incoming data. These devices can obtain a spectrum in less time than comparable swept-spectrum analyzers, but current devices are limited to the audio range of frequencies, i.e., up to about 20 KHz. If this range is large enough to contain the maximum Doppler frequency, the (FFT) fast-Fourier-transform analyzer works very well and can perform the same function as a long-time averaging photon correlator or amplitude correlator followed by FFT analysis. The input can be filtered or, if the *A-D* converter is replaced by a pulse counter, unfiltered.

Two types of spectrum analyzers that are capable of time-resolving measurements are filter banks and surface acoustic-wave devices. The filter bank consists of a parallel array of band-pass filters; i.e., it is a parallel-processing spectrum analyzer. The signal is input to all of the filters simultaneously, and the instantaneous signal frequency is identified by sampling the output power of each filter to determine the location of the instantaneous spectrum. The Doppler frequency at any instant is set equal to the center frequency of the filter whose output is maximum, so the frequency resolution is determined by the bandwidths of the filters. Baker and Wigley [80] describe a filter bank consisting of 10 filters covering the range 0.63 to 9.55 MHz in a logarithmic progression. In a certain sense, the filter bank is the ideal LDV signal processor, but it has been used very little in LDV because of the expense of a large number of filters and the difficulty in setting up and maintaining their alignment.

The surface acoustic-wave (SAW) analyzer makes use of the fact that surface acoustic waves are dispersive; i.e., it takes different wave frequencies different times to travel from one end of an SAW crystal to the other. High-speed spectrum analysis is possible (51 μs/spectrum is reported in [81]), but usually at high frequencies. Development of the SAW is evolving rapidly, and the reader is referred to [81] and the references cited therein for details.

10.4 Frequency Trackers

Frequency trackers measure the instantaneous frequency $\dot{\Phi}_D$ of the LDV signal. The main types of trackers used in LDV are the phase-locked loop (PLL) and the frequency-locked loop (FLL), both of which are analog devices. These instruments work best when the signal is continuous (high burst density), but they also work with low-burst-density signals, provided that the time between bursts is not too long. Most commercial frequency trackers hold the value of the velocity measured from the last burst until a new measurement is achieved.

In its simplest form, the phase-locked loop consists of an amplitude discriminator that converts the input sinusoidal signal $e(t)$ into a square wave, a mixer that multiplies the square wave with a reference square wave, a low-pass amplifier that amplifies the square-wave product and filters the result to produce a voltage that oscillates at the frequency difference, and a voltage-controlled oscillator (VCO) that is controlled by the amplifier output and supplies the reference square wave. The PLL is "locked" onto the signal if the reference square wave from the VCO has the same frequency as the input. If the VCO and input signals are 90° out of phase, their product is equally negative and positive, and the low-pass-filtered product (corresponding to a short-time average) is zero. If the input frequency increases, the filtered product becomes positive, and since this is the feedback signal to the VCO, the VCO frequency increases until it again equals the input frequency. In this way, the VCO signal's frequency and phase are always locked onto the frequency and phase of the input signal to within feedback errors that are small if the loop gain is large.

The output of the PLL is just the VCO control voltage, which is linearly proportional to the input frequency if the VCO is linear. Successful operation of the PLL is accomplished only when the loop is locked onto the input signal's phase, so the PLL must be capable of following very rapid phase transients to stay in lock. Typical PLL slew rates are of the order of 10^{12} Hz/s. With no signal present, the VCO operates at a constant "free-running" frequency, and it must go through a complicated, nonlinear locking transient to lock onto the signal when it appears. This transient is somewhat random, and it lasts for several periods of the signal frequency. Lock-in will not occur unless the signal frequency is within the "capture bandwidth" of the PLL, which is typically ±10 to 20% of the free-running frequency. Likewise, the loop will lose lock if the signal frequency falls outside the "lock range," which is typically ±10 to 30% of the free-running frequency.

When the signal input to a PLL is a high-burst-density LDV signal, the PLL will lose lock for small time periods when the phase noise $\dot{\phi}$ changes very rapidly or when the amplitude becomes very small. (Large phase changes are highly correlated with small amplitudes in the HBD signal.) Otherwise, it will track the signal and measure $\dot{\Phi}_D$ accurately. The periods during loss of lock are called "dropouts," and they are usually only a few Doppler cycles long, so that the signal frequency changes very little and remains in the lock range of the PLL during the dropout. During loss of lock the PLL is blind to signals lying outside the capture bandwidth, so if $\dot{\Phi}_D$ changes too much during a dropout, the PLL may not recover lock until such time as $\dot{\Phi}_D$ again passes through the lock range.

The frequency-locked loop is a variation of the phase-locked loop that possesses a much larger lock range and offers some additional noise rejection. A schematic of a typical FLL is shown in Fig. 21. The input signal with frequency $f_D = \dot{\Phi}_D/2\pi$ is mixed (multiplied) with a VCO reference signal at f_M to produce a product consisting of two sine waves, one oscillating at $f_M - f_D$, and the other oscillating at $f_M + f_D$. The product signals are passed through a band-pass filter centered at f_C. Normally f_M is adjusted so that $f_M - f_D \simeq f_C$, and the signal frequency $f_M - f_D$ is transmitted, while the signal at $f_M + f_D$ is rejected. The difference between $f_M - f_D$

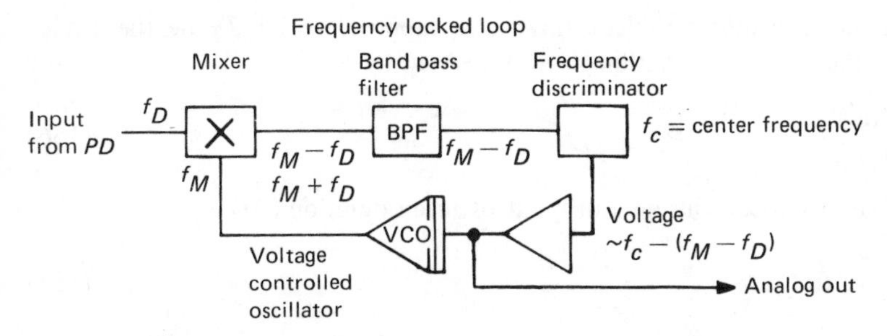

Figure 21 Schematic of a frequency-locked loop. *(Courtesy TSI, Inc.)*

and f_C is detected by a frequency discriminator (i.e., a phase-locked loop) whose voltage output is proportional to $f_C - (f_M - f_D)$. This is the output of the loop, and it is also the feedback signal that controls the VCO. The loop is locked when $f_M - f_D = f_C$. Thus, if f_D increases, $f_C - (f_M - f_D)$ also increases, causing f_M to increase until $f_M = f_D + f_C$. Hence, the feedback action of the loop is such as to always center the signal in the middle of the band-pass filter.

In effect, the filter "tracks" the signal and removes noise from the signal prior to frequency measurement by the discriminator. If the filter did not track, the noise bandwidth would necessarily be equal to the range of the Doppler frequency, but with tracking the noise bandwidth determined by the band-pass filter can be made much less. The improvement in signal-to-noise power ratio at the frequency discriminator is the ratio of the Doppler frequency range to the band-pass bandwidth. Thus, the FLL can process noisier signals than the PLL. The noise immunity of the FLL is further enhanced by the fact that the noise seen by the frequency discriminator always has a mean instantaneous frequency equal to the center frequency of the band-pass filter. In general, the mean frequency of a signal plus noise is "pulled" to a value that is intermediate between the signal frequency and the noise frequency, and this can result in a substantial error if the signal-to-noise ratio is low. In the FLL, the frequency discriminator sees a signal close to f_C and a noise with mean frequency f_C, so the difference between signal and noise frequencies is small, and the error caused by noise pulling is small. The noise immunity of the FLL and the PLL are discussed in [82].

The disadvantage of a tracking filter is that the frequency discriminator inside the feedback loop does not see the signal unless its frequency is within the pass band. Hence, its capture bandwidth at any instant is the bandwidth of the filter, and to provide good noise immunity this is small, typically ±1 to 10%. In processing low-burst-density signals, the Doppler frequency must not change (increase or decrease) by more than half the filter bandwidth during a dropout, or else the FLL will be blind to the new signal frequency when the next burst occurs, and it will not be able to resume tracking. These effects are usually minimized by holding the loop at the last known frequency when a dropout occurs. Even so, the FLL cannot track unsteady flows properly when the data rate is too low. This restriction also applies to the PLL, but it is somewhat less severe because of the wider capture bandwidth.

For example, consider turbulent flow, and suppose we let T_T be the Taylor microscale for the temporal variation of $u(t)$, defined by

$$\left\langle\left(\frac{\partial u'}{\partial t}\right)^2\right\rangle = \frac{\langle (u')^2 \rangle}{T_T^2} \tag{156}$$

The rms change in velocity during a dropout of mean duration $\langle \Delta t \rangle$ is

$$\Delta u = \frac{\langle (u')^2 \rangle^{1/2}}{T_T} \langle \Delta t \rangle \tag{157}$$

corresponding to an rms Doppler frequency change of

$$K \langle (u')^2 \rangle^{1/2} \frac{\langle \Delta t \rangle}{T_T} \tag{158}$$

in radians per second. If the capture bandwidth $\Delta\omega_C$ of the PLL or the FLL is larger than three to five times this value, the tracker will resume tracking after a dropout with high probability. That is, it is highly probable that the frequency of the new Doppler burst will fall within the capture bandwidth, and the tracker measures every burst whose amplitude is large enough to process. Otherwise, if $\Delta\omega_C < K\langle (u')^2 \rangle^{1/2} \langle \Delta t \rangle / T_T$, the tracker will miss many of the rapid, large-amplitude fluctuations in u', and statistical averages computed from the data will be biased. Since $\langle (u')^2 \rangle^{1/2}$ and T_T are determined by the flow, the only remedy is to increase $\Delta\omega_C$, which may not be possible, or to decrease $\langle \Delta t \rangle$, which is equivalent to increasing the *data rate*. The mean data rate is $\langle \Delta t \rangle^{-1}$, which can be improved by adding more particles or by improving the signal so that more particles can be processed. We shall refer to the ratio $T_T / \langle \Delta t \rangle$ as the *mean data density* and denote it by

$$\dot{N}T_T \equiv \frac{T_T}{\langle \Delta t \rangle} \tag{159}$$

This is simply the mean number of data points per Taylor microscale.

10.5 Frequency Counters

Frequency counters measure the frequency of a signal by accurately timing the duration of an integral number N of cycles of the signal. The timing is performed with respect to the zero crossings of the Doppler component of the signal, so the first step in a counter processing system is the removal of the pedestals. This is usually accomplished by passing the signal through a fixed high-pass filter, but if the high-pass frequency needed to remove the pedestals of the highest-frequency Doppler bursts exceeds the lowest Doppler frequency that is expected, it may be necessary to use a variable filter or to remove the pedestals optically. For example, suppose $\Delta\omega_A / \omega_D = 5\%$ and 1 MHz $< f_D <$ 100 MHz. Then the bandwidths of the pedestals would be 5 MHz when the Doppler frequencies are 100 MHz, and a 10-MHz high-pass filter set to remove these pedestals would also remove the Doppler signals between 1 MHz and 10 MHz. Optical pedestal removal is discussed in [83].

Once the pedestals have been removed, the processor sees only the Doppler signal plus noise. The signal plus noise is converted to a square wave by a Schmidt trigger whose output changes from a low level to a high level whenever the input increases through zero voltage, and vice versa when it decreases through zero. The leading edges of the square wave mark the zero crossings very accurately, and the square wave is compatible with digital logic circuits, which count the number of zero crossings. In the absence of any signal, the zero crossings of the noise would activate the Schmidt trigger, and the counter would measure the noise frequency, even if the noise were weak and the signal-to-noise ratio during a typical burst were high. Consequently, it is necessary to discriminate against noise by taking measurements only when a signal is present. This is no problem with high-burst-density signals because the signal is almost always present, but with low-burst-density signals, the measurements must be limited to times when a burst is present. This is accomplished by setting one or more fixed threshold levels that the signal must exceed to arm the Schmidt trigger. The sketch in Fig. 22 shows two threshold levels, one positive and one negative. The Schmidt trigger begins operation when the signal amplitude first exceeds both levels, and terminates at the first zero crossing after the signal fails to cross one of these levels. The time interval in which the signal exceeds the thresholds is called the *burst time* τ_B, and it is measured by starting a clock at the beginning of the burst detection signal and stopping it at the end. This procedure also makes it impossible to confuse the last few cycles of one burst with the beginning cycles of the next burst.

In Sec. 11, the burst detection signal is denoted by $B(t)$. The measurement volume of the LDV is defined by the burst detection thresholds according to the relations in Eqs. (45) and (46). The minimum light flux $J_{\min}$ that appears in these equations can be related to the counter threshold voltages using the photodetector sensitivity and the value of the gain associated with amplification of the signal

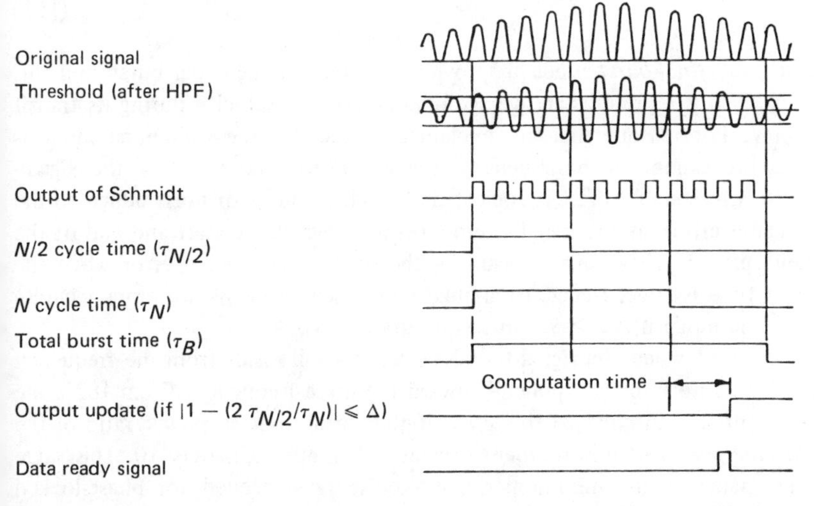

Figure 22 Timing functions performed by a typical burst counter. *(Courtesy TSI, Inc.)*

between the photodetector and the counter. Since the sensitivity and amplification are often adjustable, the effective size of the mv can be controlled by the experimenter. In addition, the data rate, i.e., the number of particles per second that produce signals exceeding the thresholds, can also be adjusted. Similar considerations apply to frequency trackers because they also contain inherent threshold voltage levels.

Having defined the burst, the counter can measure its frequency in two ways. First, the N-cycle time τ_N can be measured, and the frequency can be computed from $f_D = N/\tau_N$. The accuracy of this measurement is normally validated by comparing it with the frequency computed from measurements of a smaller number of cycles, say $N/2$. If the difference in these measurements is less than some prescribed error Δ, for example, if

$$\left| 1 - \frac{2\tau_{N/2}}{\tau_N} \right| \leqslant \Delta \tag{160}$$

the measurement is accepted; otherwise it is rejected. Comparison of $2\tau_{N/2}$ and τ_N requires a short computation time during which the counter is "dead." The error Δ is normally adjustable. The choice of N and $N/2$ is not unique. For example, many LDV counters use 5/8 comparison, i.e., comparison of the 5-cycle measurement of f_D with the 8-cycle measurement. There is, however, little discernible advantage to 5/8 comparison versus, say, 4/8 comparison.

The mode of measurement described above is called the *N-cycle* mode. This mode does not guarantee one measurement per burst. For example, if there are 50 cycles in a particular burst, an 8-cycle counter would measure the frequency six times in succession, assuming negligible dead time. Some counters are capable of limiting the measurement to one per burst.

The second method of frequency measurement uses the total burst time τ_B and the number of cycles in the burst N_B to compute the frequency from

$$f_D = \frac{N_B}{\tau_B} \tag{161}$$

This is called the *total-burst* mode. It provides one measurement per burst, and this measurement clearly represents an average of the particle's velocity during its transit across the mv. For reasons that are explained in Sec. 11, the total-burst mode is desirable in processing low-burst-density signals. Furthermore, when the signal-to-noise ratio is reasonably high, it is clear that small amounts of noise added to the signal will cause errors in the signal's zero-crossing times at the start and end of the measurement period. These errors result in the smallest frequency error when the measurement time is large. Hence, total-burst-mode measurements are more accurate than the N-cycle mode if $N_B > N$, but less accurate if $N_B < N$.

Errors incurred when the signal-to-noise ratio is small result from the frequency of the signal plus noise being "pulled" toward the noise frequency. From [82], the counter is essentially immune to this error if the signal-to-noise power ratio of the burst *during the period of measurement* is greater than approximately 10. This value is about the same as the minimum signal-to-noise ratio needed for phase-locked loops, but it is 10 to 100 times greater than the value required for frequency-locked loops. Therefore, given the same signal, a counter processor will make fewer successful measurements per second than a frequency-locked loop, assuming that it is not limited by loss-of-lock problems. The advantage of the counter-type processor is that it is ideally suited to low-burst-density signals, and it has a very large dynamic range and infinite slew rate. For example, given satisfactory pedestal removal, a counter can measure a 1-KHz signal from one burst followed by a 100-MHz signal from the next burst, which may occur a few microseconds later.

The natural output of the counter processor is a digital word corresponding to the measured frequency. In the N-cycle mode it is sufficient to output τ_N, but τ_B is also output for auxiliary use in conjunction with statistical data analysis. In the total-burst mode the outputs are τ_B and N_B. Commercial counters may also provide a voltage proportional to τ_N and/or N/τ_N for the purpose of setting up and monitoring the measurements.

While counters are primarily intended for low-burst-density signals, they work very well with high-burst-density signals, provided the signal-to-noise ratio is adequate. Lading and Edwards [84] have shown that counters and frequency trackers produce virtually identical outputs, including the ambiguity noise fluctuations, when the signal is HBD.

10.6 Selection of Signal Processors

Guidelines for the selection of a signal processor for a particular measurement problem are necessarily vague because the capabilities of various processors overlap considerably. The signal-to-noise ratio is the determining factor in most situations. If the signal is of low photon density, amplitude correlation or photon correlation is the only possibility at present. If the signal is O(1) photon density, photon correlators begin to experience distortions due to photon pileup, but amplitude correlators and spectrum analyzers will work. Frequency-locked loops and phase-locked loops that incorporate a tracking-filter capability will also work in the high end of this regime, if there are about five photons per Doppler cycle and if the signal is filtered somewhat above the Doppler frequency. High-photon-density signals may be processed with frequency-locked loops and phase-locked loops with tracking filters, and counters may also be used if there are enough bursts with signal-to-noise ratios greater than about 10. Of course, amplitude correlators and swept-spectrum analyzers work even better on high-photon-density signals than on low-photon-density signals. Photon correlators can be adapted to high-photon-density signals by the simple artifice of reducing the laser power to produce a low-photon-density signal, but this procedure is clearly wasteful of useful information.

Burst density also influences the choice of instruments. Counters are designed to work best with low-burst-density signals, but they also work very well when the burst density is high. Likewise, trackers are designed to work best with high-burst-density signals, but they will work with low-burst-density signals if the data density, defined as $T_T/\langle \Delta t \rangle$, is not much less than unity.

In terms of information obtained, the counter and the frequency tracker are superior to correlation and spectrum analysis because they provide time-resolved information. However, the current development of rapid correlation and spectral-analysis techniques may alter this situation.

11. DATA PROCESSING

Data processing refers to the procedures used to compute the desired flow properties from the data output by the signal processor. The specific procedure depends, of course, on the general type of signal processor and the type of signal, but it also depends on the type of flow and the type of quantity that is to be computed. Specifically, time-resolved measurements in unsteady flows, especially turbulent flows, require special attention, and most of this section is devoted to this general situation. Steady-flow results are special cases of the more general turbulent-flow results.

11.1 Processing Data from Time-Averaging Processors

The amplitude correlator, the photon correlator, and the spectrum analyzer are normally used in the long-time averaging mode; i.e., the correlations and spectra are averaged over many integral time scales of the turbulent motion. The relationship between these quantities and the turbulent flow field are discussed extensively in Sec. 9.4, and they may be used to calculate mean velocity and rms velocity from either correlations or spectra.

In this section we note certain techniques that permit the evaluation of higher-order statistics under special conditions. In particular, suppose that the broadening effects caused by the laser source, Brownian motion, mean velocity gradients and ambiguity broadening are all negligible compared to the broadening caused by turbulent fluctuations; i.e.,

$$\Delta\omega_0,\ \Delta\omega_B,\ \Delta\omega_G,\ \Delta\omega_A \ll \Delta\omega_T$$

Further, suppose that $f_{\mathbf{u}}(\mathbf{c}, \mathbf{x})$ is independent of $\mathbf{x}$. Then, since negligible ambiguity broadening implies that the amplitude factor $a(\mathbf{x})$ is so wide that $a(\mathbf{x} + \mathbf{c}\tau) = a(\mathbf{x})$ for all values of $\mathbf{c}\tau$ that are of interest, we have, from Eq. (98),

$$R_{J_DJ_D} \simeq \tfrac{1}{2} \int C\langle D^2\rangle a^2(\mathbf{x})\, d^3\mathbf{x} \int f_{\mathbf{u}}(\mathbf{c}, 0) \cos(\omega_s\tau + Kc_1\tau)\, d^3\mathbf{c} \tag{162a}$$

$$= \tfrac{1}{2} \int C\langle D^2\rangle a^2(\mathbf{x})\, d^3\mathbf{x} \int f_u(c_1, 0) \cos(\omega_s\tau + Kc_1\tau)\, dc_1 \tag{162b}$$

Hence, $R_{J_DJ_D}$ is proportional to the characteristic function of $f_u(c_1)$ if $\omega_S = 0$. Taking the Fourier transform of Eq. (162*b*) yields the following power spectrum:

$$S_{J_D}(\omega) = \frac{1}{2K} \int C\langle D^2\rangle a^2(\mathbf{x})\, d^3\mathbf{x}\, [f_u(\omega - \omega_S) + f_u(\omega + \omega_s)] \tag{163}$$

Thus, if either ω_s or the mean Doppler frequency is large enough compared to the spectral bandwidth to make $S_{J_D}(0) \simeq 0$, $f_u(\omega + \omega_s)$ will be negligible for $\omega > 0$, and

$$f_u(\omega - \omega_s) = 2KS_{J_D}(\omega)\ [\int C\langle D^2\rangle a^2(\mathbf{x})\, d^3\mathbf{x}]^{-1} \qquad (164)$$

The complete probability density function for u can be found, and from it higher-order moments such as $\langle (u')^3\rangle$ can be calculated. When using correlations it is convenient to take the Fourier cosine transform of the correlogram to obtain the spectrum so that Eq. (164) can be used. This procedure assumes that the Doppler component of the spectrum can be separated from the other components, such as the pedestal component and the noise component, and it ignores errors associated with the computation of the Fourier transform from a correlogram with a relatively small number of time delay points.

When the ambiguity bandwidth is not negligible, the measured spectrum is the convolution of the velocity probability density function with a spectrum whose width is determined by the ambiguity bandwidth. There have been a number of efforts, described in [85], to develop algorithms for computing f_u from broadened data, but this procedure is subject to error because, for arbitrary f_u, deconvolution amounts to solving a Fredholm integral equation of the first kind, and this problem requires special conditioning that is not necessarily valid for all forms of f_u.

It should be noted that the effects of burst density appear only in a multiplicative factor in Eq. (164), which essentially normalizes the probability density function. Burst density is irrelevant in correlation and spectral analysis techniques, except as it determines the rates at which the correlation and spectrum can be accumulated statistically.

11.2 Processing Data from Time-Resolving Signal Processors

The general-purpose LDV signal processor must be capable of processing both high- and low-burst-density signals, and the latter type of signal necessitates an ability to make measurements of the velocity within the time span of an individual Doppler burst. The outputs of processors that are designed to perform this function inevitably share certain common characteristics, and it appears to be possible to represent the outputs of all these processors (e.g., counters, trackers, burst correlators and, in the future, burst spectrum analyzers) with a fairly simple model. The model postulates an ideal signal processor whose output is equal to the instantaneous Doppler frequency while the signal amplitude exceeds a threshold level $J_{\min}$, and constant otherwise. The value of the constant may be zero, or, more commonly, it may be the value of the last known frequency. The effects of noise in the signal, locking and unlocking transients, and fringe biasing are presumed to have been reduced to negligible levels, and all other idiosyncrasies that may arise in real instruments are ignored.

The output of the ideal processor is simply $\dot{\Phi}_D$ when the burst density is large. We note in Sec. 9.6 that the spatially and temporally filtered velocity $\tilde{\tilde{u}}(t)$ defined by Eq. (142) represents the measurable velocity in this case.

When the burst density is low, the output of the ideal processor can be represented by an "observed velocity"

$$u_o(t) = \iint w(\mathbf{x}, D)u(\mathbf{x}, t)g(\mathbf{x}, t, D)\, d^3\mathbf{x}\, dD \tag{165}$$

where

$$w(\mathbf{x}, D) = H(a(\mathbf{x})D - J_{\min}) \tag{166}$$

indicates the presence of a detectable particle [$H(\cdot)$ denotes the Heaviside function]. That is, $w(\mathbf{x}, D) = 1$ if a particle is in the mv and its scattering amplitude D is large enough to make the amplitude $a(\mathbf{x})D > J_{\min}$, and it is zero otherwise. This definition follows the one used by Buchave et al. [66], but it generalizes the representation to include the effects of random D values, e.g., polydisperse scatterers. For low burst densities, the occurrence of more than one particle in the mv is negligibly improbable, so at any instant only one of the terms in $g = \Sigma_j\, \delta(\mathbf{x} - \mathbf{x}_j(t))\delta(D - D_j)$ can contribute to the integral. Then,

$$u_o(t) = u(\mathbf{x}_j(t), t) = \mathbf{v}_j(t) \tag{167}$$

during the jth Doppler burst. If no particle is present, $u_o(t) = 0$.

Figure 23 indicates the behavior of u_o. In the figure a burst-indicator signal $B(t)$ that is unity when a detectable particle is present, and zero otherwise, is also shown. $B(t)$ can be represented by

$$B(t) = \iint w(\mathbf{x}, D)g(\mathbf{x}, t, D)\, d^3\mathbf{x}\, dD$$

This is a good model of the outputs from trackers, counters, and burst correlators.

If the signal is O(1) burst density, $u_o(t)$ is not a good representation because it sums the individual particle velocities when more than one particle is present. Likewise, $u(t)$ is not a good representation either. It appears that the analysis of the O(1) burst density case is more complex than for the limiting cases, and the author is unaware of any successful attempts to fully analyze the data-processing techniques applicable to this type of signal. Consequently, only the high and low limits will be discussed here.

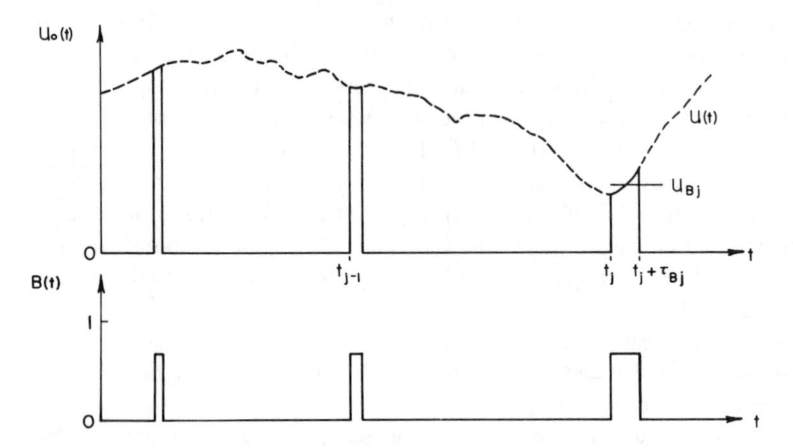

Figure 23 Velocity observed with a burst processer and a low-burst-density signal, and the burst indicator function $B(t)$.

11.2.1 Data density. The data density is defined in Eq. (159), in conjunction with the discussion of data-rate effects on frequency trackers. The data density also characterizes the output of a signal processor, and it dictates to some extent the types of data-processing methods that can, or must, be used. Data density and burst density are independent concepts because the Taylor microscale varies independently of the size of the measurement volume. Consider, for example, a turbulent flow with mean velocity in the x direction, and suppose that Taylor's time microscale T_T is related to Taylor's spatial microscale in the x direction λ_x by the frozen-field approximation: $\lambda_x = UT_T$. The burst density is $N_e \simeq CA\, d_m$, where A is the projected area of the mv in the x direction, whereas the mean data density is $\dot{N}T_T \simeq CAUT_T = CA\lambda_x$. Hence, $N/N_e = \lambda_x/d_m$. The ratio λ_x/d_m is rarely smaller than 1, and it is large in a turbulent-flow LDV experiment that has good spatial resolution. Hence, the signal can be of high data density when it is of low burst density (for example, $\lambda_x/d_m = 50$, $\dot{N} = 10$, $N_e = 0.2$), and it is always of high data density when it is of high burst density.

The important combinations of data density and burst density are depicted in Fig. 24. The low-data-density signal is always of low burst density if $\lambda_x/d_m = 0(1)$ or greater. In this case the mean time between data points $\langle \Delta t \rangle = \dot{N}^{-1}$ is greater than T_T, and the velocity cannot be resolved as a function of time. However, multiple particles hardly ever occur in the mv, so ambiguity noise is avoided, each detectable particle giving a good measurement of its velocity. The high-data-density, low-burst-density signal is even better in this regard because there is still no ambiguity noise whereas the mean time between data points is now small compared to T_T, so at least the energy containing fluctuations can be resolved in time. Finally, at large particle concentrations the data density is very high, but since the burst density is even higher, an ambiguity noise appears in the signal. The high-data-density, low-burst-density signal is optimal, but good statistical results can be obtained in the other two cases as well.

11.2.2 High-burst-density signals. In Sec. 9.6 it is shown that the instantaneous frequency of the HBD signal is

$$\tilde{\dot{\Phi}}_D = \omega_S + K\tilde{\tilde{U}} + K\tilde{\tilde{u}}' + \tilde{\dot{\phi}}$$

where $\tilde{}$ denotes the filtering after the signal processor. It is noted in Sec. 10.5 that the outputs of the counter and the frequency tracker are virtually identical for HBD signals. Both outputs will contain small dropout periods that occur when the amplitude envelope of the HBD signal drops below the threshold, coincident with very rapid phase changes. Phase changes may also cause dropouts in the FLL or PLL, owing to loss of lock, and in the counter, owing to $N/2$ or $5/8$ comparison. In either case, the dropouts are a few Doppler cycles long, and this is usually very short compared to the Taylor microscale T_T, so the signal is still essentially of high data density.

Buchave et al. [66] have analyzed the effects of the dropouts by postulating that the sequence of on (i.e., in lock) and off (i.e., dropout or out of lock) states is a Markov chain. They found that the minimum distortion of the signal statistics is

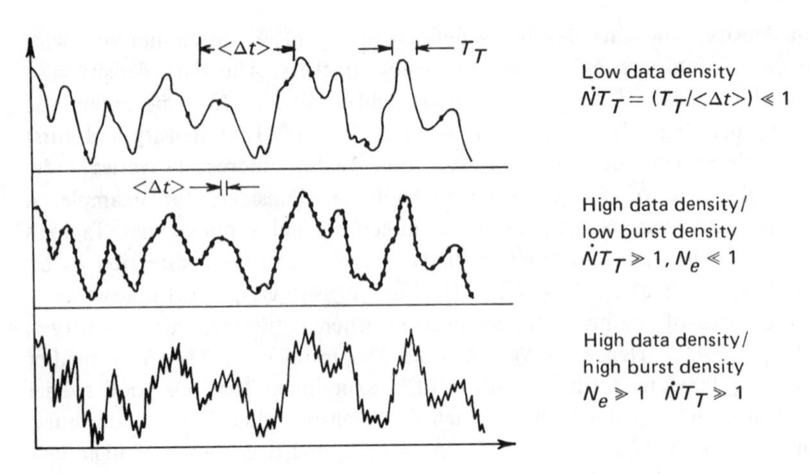

Figure 24 Data density. *(From [47].)*

achieved by holding the last known reading during a dropout. With this method all moments, including the velocity probability density function, are preserved, essentially because the dropout times are uncorrelated. Their model for short dropout times also predicts that the power spectrum of $\dot{\Phi}_D$ with dropouts is just S_{Φ_D} plus a dropout noise spectrum that is essentially constant out to the cutoff frequency of the low-pass filter. Thus, the dropout noise can be treated in the same way as the ambiguity noise.

The single time statistical moments of $\tilde{\tilde{u}}'(t)$ can be obtained from $\tilde{\dot{\Phi}}'_D = \dot{\Phi}_D(t) - \omega_s - K\tilde{\tilde{U}} = K\tilde{\tilde{u}}'(t) + \tilde{\dot{\phi}}$ by direct computation if $\tilde{\dot{\phi}}$ is small. Otherwise the statistics can be corrected for $\tilde{\dot{\phi}}$ contamination by using the fact that $\tilde{\dot{\phi}}$ and $\tilde{\tilde{u}}'$ are statistically independent. Thus,

$$\langle \tilde{\dot{\Phi}}_D \rangle = \omega_S + K\tilde{\tilde{U}} \tag{168a}$$

$$\sigma^2_{\tilde{\dot{\Phi}}_D} = \langle (\tilde{\dot{\Phi}}_D - \langle \tilde{\dot{\Phi}}_D \rangle)^2 \rangle = K^2 \langle (\tilde{\tilde{u}}')^2 \rangle + \langle \tilde{\dot{\phi}}^2 \rangle \tag{168b}$$

$$\langle (\tilde{\dot{\Phi}}_D - \langle \tilde{\dot{\Phi}}_D \rangle)^3 \rangle = K^3 \langle (\tilde{\tilde{u}}')^3 \rangle + \langle \tilde{\dot{\phi}}^3 \rangle \tag{168c}$$

$$\langle (\tilde{\dot{\Phi}}_D - \langle \tilde{\dot{\Phi}}_D \rangle)^4 \rangle = K^4 \langle (\tilde{\tilde{u}}')^4 \rangle + 2K^2 \langle (\tilde{\tilde{u}}')^2 \rangle \langle \tilde{\dot{\phi}}^2 \rangle + \langle \tilde{\dot{\phi}}^4 \rangle \tag{168d}$$

etc., can be corrected with knowledge of $\langle \tilde{\dot{\phi}}^2 \rangle$, $\langle \tilde{\dot{\phi}}^3 \rangle$, $\langle \tilde{\dot{\phi}}^4 \rangle$, and so on. These moments can be calculated using [65] the probability density

$$f_{\dot{\phi}}(x) = \frac{1}{2}\left(1 + \frac{x^2}{\Delta\omega^2}\right)^{-3/2} \tag{169}$$

and taking the filtering into account. Note that Eq. (169) is an even function of its argument, so all odd-order moments of $\dot{\phi}$ (and hence $\dot{\Phi}_D$) vanish. The even-order moments of $\dot{\phi}$ can all be calculated in terms of $\Delta\omega^2$, or, for $\tilde{\dot{\phi}}$, in terms of $\langle \tilde{\dot{\phi}}^2 \rangle$. This last quantity is not readily determined theoretically, because the turbulent gradient broadening is unknown in most experiments. Hence, $\langle \tilde{\dot{\phi}}^2 \rangle$ is best determined by direct experimental measurement.

Measurements of $\langle\tilde{\dot{\phi}}^2\rangle$ are, in principle, possible because the spectrum of $\tilde{\dot{\phi}}$ is white out to f_c. Hence, if f_c is somewhat greater than the maximum turbulent frequency, the $\tilde{\dot{\phi}}$ spectrum level is readily determined, and $\langle\tilde{\dot{\phi}}^2\rangle$ can be calculated by extrapolating this level to zero frequency and integrating under the curve. Alternatively, autocorrelation measurements of $\dot{\Phi}_D - \langle\dot{\Phi}_D\rangle$ yield the autocorrelation of the velocity plus the autocorrelation of $\dot{\phi}$. The latter is very narrow, of width f_c^{-1}, so it appears as a spike at zero time delay that extends above the zero-time-delay value of the velocity correlation by the amount $\langle\tilde{\dot{\phi}}^2\rangle$.

The foregoing procedures are good enough for finding the value of $\langle\tilde{\dot{\phi}}^2\rangle$ needed to correct single time moments, but other procedures are preferable if the velocity autocorrelation or spectrum is desired. These are based on the fact that the ambiguity noises from two different LDV systems are almost uncorrelated. Morton and Clark [86] report spatial correlation measurements of u using two LDVs with spatially separated measurement volumes. The cross correlation of the LDV signals showed virtually no ambiguity noise, even when the measurement volumes overlapped. Even more surprisingly, van Maanen et al. [87] showed that a single LDV transmitting optical system could be used, and two signals with very weakly correlated ambiguity noises could be obtained by collecting light with two photodetectors in different scattering directions.

11.2.3 Low-burst-density, low-data-density signals. Low-burst-density data consist of a series of individual particle-velocity measurements at random times of arrival. These data samples are biased toward high velocities because more particles traverse the mv when the velocity is high than when it is low. Thus, the statistical equations that one usually uses to form averages with a set of unbiased samples are invalid when applied to LBD signal data. For example, suppose the data are $u(t_j) = u_j$, $j = 1, \ldots, J$, where t_j is the arrival time of the jth particle. The customary equation for the sample mean velocity for unbiased data is

$$\frac{1}{J}\sum_{j=1}^{J} u_j$$

Now suppose that the flow is unidirectional, and its velocity is a square wave that takes the values V_0 and $2V_0$ for equal times. Then, since the arrival rate of particles is proportional to $u(t)$, there will be twice as many samples when $u = 2V_0$ as when $u = V_0$. The average calculated from the foregoing equation would be $(1 \times V_0 + 2 \times 2V_0)/(1+2) = 5V_0/3$, whereas the true average of the square wave is $3V_0/2$. This discrepancy always occurs, on average, and it is independent of the particle concentration.

The bias effect was first noted by McLaughlin and Tiederman [71], who proposed using the velocity data to statistically weight the samples, a method of correction that is correct for unidirectional flows. George [88], and later Hosel and Rodi [89], proposed using the burst time τ_{B_j} as a statistical weighting factor, on the basis that the burst time is inversely proportional to the magnitude of the three-dimensional velocity vector, and it is readily measured. An alternative method of unbiasing the samples in three-dimensional flow is to weight them with the inverse of the sampling probability [90], but this requires simultaneous measurements of all three velocity components. The present discussion concentrates on the burst time, or "residence time" weighting technique, which is the most widely accepted, albeit not totally proven, technique.

As noted earlier, the observed velocity available from an ideal signal processor with low burst density is

$$u_o(t) = \iint w(\mathbf{x}, D)u(\mathbf{x}, t)g(\mathbf{x}, t, D)\, d^3\mathbf{x}\, dD \tag{165}$$

The relationship between the statistics calculated using burst-time weighting and the statistics of $u(\mathbf{x}, t)$, can be obtained by treating $u_o(t)$ as an ordinary signal and taking time averages. The time average of the nth power of u_o is

$$\overline{u_o^n(t)} = \frac{1}{T}\int_0^T u_o^n(t)\, dt \tag{170a}$$

$$= \frac{1}{T}\sum_j u_j^n(t)\tau_{Bj} \tag{170b}$$

$$= \overline{[u^n]_o} \tag{170c}$$

where u_j, the velocity of the jth particle, is assumed to be constant during the burst time, as usual. Equation (170c) follows from the fact that

$$[u^n(t)]_o \equiv \iint w(\mathbf{x}, D)u^n(\mathbf{x}, t)g(\mathbf{x}, t, D)\, d^3\mathbf{x}\, dD \tag{171a}$$

$$= [u_o(t)]^n \tag{171b}$$

as is apparent from inspection. [Essentially the integral operators in Eqs. (165) and (171a) sample the velocity during a burst. Equation (171a) is the sample of the nth power of u, and Eq. (171b) states that this is equal to the nth power of the sample.] We now assume that the time average of u_o^n equals its ensemble average. Then,

$$\overline{u_o^n} = \overline{[u^n]_o} \tag{172a}$$

$$= \langle [u^n]_o \rangle \tag{172b}$$

$$= \langle \iint w(\mathbf{x}, D)u^n(\mathbf{x}, t)g(\mathbf{x}, t, D)\, d^3\mathbf{x}\, dD \rangle \tag{172c}$$

$$= \int C(\mathbf{x})w(\mathbf{x})\langle u^n \rangle\, d^3\mathbf{x} \tag{172d}$$

where

$$w(\mathbf{x}) = \frac{\int_0^\infty w(\mathbf{x}, D)C(\mathbf{x}, D)\, dD}{\int_0^\infty C(\mathbf{x}, D)\, dD} \tag{173}$$

showing that the burst-time-weighted sum in Eq. (170*b*) equals the volume integral given by Eq. (172*d*). The former sum is not, however, a sensible average because it averages over the dropout periods when $u_o = 0$. This defect is corrected by dividing Eqs. (170*b*) and (172*d*) by the average burst time, given by

$$\frac{1}{T}\int_0^T B(t)\,dt = \frac{1}{T}\sum_j \tau_{B_j} \tag{174a}$$

$$= \langle B(t) \rangle \tag{174b}$$

$$= \int C(\mathbf{x})w(\mathbf{x})\,d^3\mathbf{x} \tag{174c}$$

This yields

$$\frac{\sum_j u_j^n \tau_{B_j}}{\sum_j \tau_{B_j}} = \widetilde{\langle u^n \rangle}$$

where $\widetilde{\ }$ denotes the volume average:

$$\widetilde{\langle u^n \rangle} = \frac{\int C(\mathbf{x})w(\mathbf{x})\langle u^n(\mathbf{x}, t)\rangle\,d^3\mathbf{x}}{\int C(\mathbf{x})w(\mathbf{x})\,d^3\mathbf{x}} \tag{175}$$

If $\langle u^n(\mathbf{x}, t)\rangle$ is constant within the mv, then $\widetilde{\langle u^n \rangle} = \langle u^n \rangle$.

The mean velocity and mean square fluctuation with respect to the mean that are given by burst-time weighting are

$$\frac{\sum_j u_j \tau_{B_j}}{\sum_j \tau_{B_j}} = \widetilde{U} \tag{176}$$

and

$$\frac{\sum_j (u_j - U)^2 \tau_{B_j}}{\sum_j \tau_{B_j}} = (U - \widetilde{U})^2 + \widetilde{\langle (u')^2 \rangle} \tag{177}$$

respectively, where Eq. (177) follows from Eq. (175) after some manipulation.

The mv in the case of burst signal processing is determined by the volume average in Eq. (175). This average, with weighting function $C(\mathbf{x})w(\mathbf{x})$, and $w(\mathbf{x})$ given by Eq. (173), is very similar to the average that appears in the case of high-burst-density signals, where the weighting function is $C(\mathbf{x})\langle D^2\rangle a^2(\mathbf{x})$. The integral in Eq. (173) is evaluated by noting that, at fixed $\mathbf{x}$, $w(\mathbf{x}, D)$ is zero unless

$a(\mathbf{x})D \geqslant J_{\min}$, that is, unless $D \geqslant D_{\min}(\mathbf{x}) = J_{\min}/a(\mathbf{x})$. For all $D \geqslant D_{\min}$, $w(\mathbf{x}, D) = 1$. Hence, it follows that

$$w(\mathbf{x}) = \int_{J_{\min}/a(\mathbf{x})}^{\infty} \frac{C(\mathbf{x}, D)\, dD}{C(\mathbf{x})} \tag{178}$$

which states that $w(\mathbf{x})$ is just the fraction of the total number of particles for which $D > D_{\min}$, or, in other words, $w(\mathbf{x})$ is the probability that a particle at $\mathbf{x}$ has a value of $D > D_{\min}$.

If the scattering population is monodisperse, then $C(\mathbf{x}, D) = \delta(D - D_o)$, and $w(\mathbf{x}) = w(\mathbf{x}, D_o)$. This equals unity if $a(\mathbf{x}) \geqslant J_{\min}/D_o$ and zero otherwise, thereby defining an mv with a sharp boundary. If $a(\mathbf{x})$ is the usual gaussian ellipsoid, the mv will be an ellipsoid whose major axes are given by l_x, l_y, and l_z from Eq. (46).

In the case of polydisperse particles the description of the mv is simplified by assuming that $C(\mathbf{x}, D) = C(\mathbf{x}) f(D)$; that is, the normalized size distribution of particles $f(D)$ is the same everywhere, but the total number per unit volume may vary in space. Then

$$w(\mathbf{x}) = \int_{J_{\min}/a(\mathbf{x})}^{\infty} f(D)\, dD \tag{179}$$

and the $\mathbf{x}$ dependence enters only through the lower limit of integration. Clearly, $w(\mathbf{x})$ is constant when $a(\mathbf{x})$ is constant, so the mv has an ellipsoidal shape if $a(\mathbf{x})$ is ellipsoidal. The dimensions of the ellipsoid depend on $f(D)$, and this is rather troublesome since one rarely has this information. Clearly, monodisperse particles are preferable in low-burst-density flows.

It should be noted that the present equations for interpretation of burst-time-weighted statistics differ from those in [66] by the inclusion of the effects of polydispersity. It is suggested in that reference that an on-off scattering volume, that is, $w(\mathbf{x})$ equal to 0 or 1, is an excellent approximation to real LDV systems, but we see from the present results that $w(\mathbf{x})$ is a 0-1 function only if the scattering particles are monodisperse. Otherwise, $w(\mathbf{x})$ is a continuous function, much like $a^2(\mathbf{x})$.

The first term in Eq. (177) represents the apparent fluctuations caused by mean velocity gradients across the mv. These fluctuations can be significant if the gradients are large, and if they exceed about 10% of the rms turbulent fluctuations they should be eliminated by reducing the mv size.

The autocorrelation and the power spectrum can be calculated from low-data-density signals, despite the random times of the data and the gaps between data points. It is shown in [66] that the appropriate unbiased algorithm for the autocorrelation is

$$\langle u'(t) u'(t + \tau) \rangle = \frac{\sum_i \sum_i u_i' u_j' \tau_{B_{ij}}}{\sum_i \sum_j \tau_{B_{ij}}} \tag{180}$$

where $\tau = t_i - t_j$, $i < j$, and $\tau_{B_{ij}}$ is the duration of the overlap between the ith burst and the jth burst, delayed by time lag τ. This equation assumes monodisperse particles, uniform particle concentration, and vanishing mean velocity gradients across the mv. The equation should not be used at zero time lag, because at small values of τ, of the order of the mean burst time, the burst-time-weighted estimate of the autocorrelation contains a large noise spike associated with the correlation of the individual bursts with themselves. The power spectrum can be obtained by taking the Fourier transform of the autocorrelation. If the burst self-correlation spike is retained in the process, the power spectrum will contain a large spectrum that is white out to a frequency that is approximately equal to 1 divided by the mean burst time. Fortunately, the self-correlation spike is easily removed, before Fourier transformation, by replacing it with the zero-time-delay value from Eq. (177). Correlations can be obtained at very short time delays because, for any data rate, there is always a finite probability that the arrival times of two particles will be arbitrarily close. Likewise, spectra can be calculated at frequencies higher than the mean data rate. The penalty associated with low data rates is the long averaging time required to obtain many sample pairs separated by time lags that are smaller than the mean data rate.

11.2.4 High-data-density, low-burst-density signals. Data-processing algorithms for this type of signal are obtained by noting that the data points are so close together that simple interpolation schemes can be used to fill in much of the missing data between bursts. Dimotakis [90] has suggested using linear interpolation, in which case the interpolated curve of some function of the velocity $g(u)$ versus time is a series of trapezoids. Time averaging of this curve yields the following trapezoidal-rule approximation:

$$\overline{g(u(t))} = \frac{1}{t_N - t_1} \sum_j \frac{1}{2} [g(u_j) + g(u_{j-1})](t_j - t_{j-1}) \tag{181}$$

In this procedure, biasing is eliminated because the samples are weighted by $t_j - t_{j-1}$, which is inversely proportional to the data rate when the data density is high.

Equation (181) requires digital analysis of the data. Analog analysis is also possible because most signal processors hold the value of the last sample at the output until a new sample is measured. The output signal then looks like a staircase function whose steps are small if the data density is large. The steps add noise, but this can be removed in large part by low-passing filtering. In effect, the low-pass filter performs an analog interpolation. The filtered signal can be analyzed with standard analog instruments such as correlators and mean dc voltmeters, or it can be digitized for subsequent digital analysis. This case certainly offers the simplest analysis procedure and the maximum possible velocity information without contamination by ambiguity noise.

11.3 Fringe Biasing

Fringe biasing occurs when the signal processor requires a minimum number of cycles to make a measurement, and certain particle trajectories fail to provide this number. This effect is worst when the velocity is parallel to the fringes so that no particle crosses a fringe, but it is also manifested at oblique angles where particles passing through the center of the mv may cross enough fringes, but those passing near the edges do not. Thus, the data rate is greatest when the velocity vector is perpendicular to the fringes, and it decreases as the angle between the velocity and the fringes approaches zero. The resulting data are biased toward samples from perpendicular velocities. Fringe bias is analyzed in [90 to 92].

There is no satisfactory method of correcting for fringe bias analytically, but it can be reduced to negligible levels in most cases by the simple technique of frequency shifting. Frequency shifting effectively adds cycles to a Doppler burst by moving the fringes with respect to the fluid. Then, if the fringe velocity is large compared to the flow velocities, even particles traveling parallel to the fringes will produce an adequate number of cycles as the fringes move over them. Alternatively, for counters, fringe biasing is minimized by using the total-burst mode, because this mode requires the least number of cycles for a measurement. Fringe biasing is only important for low-burst-density signals, because the signal is almost always present in high-burst-density signals. The data-processing procedures discussed in Sec. 11.2 assume that fringe biasing has been eliminated.

NOMENCLATURE

A	cross-sectional area of mv
$\langle A/B \rangle$	conditional average of A given the value of B
$A(t)$	amplitude envelope of HBD signal, Eq. (127)
$a(\mathbf{x})$	amplitude distribution of Doppler burst, Eq. (29)
B	number of A-D conversion bits
$B(t)$	burst indicator
C_a	anode capacitance
$C(\mathbf{x}, t, D)$	number density of D-type particles per unit volume
$C(\mathbf{x}, t)$	number density of all particles per unit volume, Eq. (96)
$\mathbf{c}$	total velocity variable in the velocity probability density function, Eq. (88)
D	Doppler signal-amplitude factor, Eq. (26*b*)
δD	range of Doppler signal-amplitude factor
$D_{e^{-2}}$	e^{-2} diameter of unfocused illuminating beam
D_a	aperture diameter
D_c	coherence diameter
D_p	aerodynamic diameter of a particle
d_p	particle diameter
$d_{e^{-2}}$	e^{-2} diameter of focused gaussian illuminating beam, Eq. (34)
d_f	fringe spacing, Eq. (49)
d_m, l_m, h_m	e^{-2} dimensions of the LDV mv, Eqs. (41) to (43)
$\mathbf{E}$	electric vector
e	photodetector output voltage

e_c	clipped photodetector voltage
e_n	shot-noise voltage
e_N	total-noise voltage signal
$F(t)$, $G(t)$	amplitude functions, Eq. (129)
f	focal length; frequency
f_c	collecting-aperture distance from mv; low-pass cutoff frequency
f_C	center frequency
f_D	Doppler frequency
f_M	frequency of VCO
$f(\mathbf{x}', t', \mathbf{x}, t)$	probability density for particle position at t', given it was at $\mathbf{x}$ at t, Eq. (88)
$f_{3\text{-dB}}$	frequency response of particles
$f_{\dot{\phi}}$	probability density for $\dot{\phi}$
$f_{\mathbf{u}}$	probability density for $\mathbf{u}$, Eq. (88)
Δf	bandwidth of the photodetector/filter system
G	scattering gain, Eq. (65)
g	internal gain of a PMT
$g(\mathbf{x}, t, D)$	particle presence indicates function, Eq. (68)
H	transfer function of filter, Eq. (143*a*); Heaviside function
$h(t)$	impulse response function of PMT plus filters
$h_f(t)$	impulse response of the filter system after the PMT
$h_p(t)$	impulse response of the combined PMT/filter system
h	Planck's constant
I	intensity
I_{00}	fringe intensity, Eq. (48)
J	light flux
j	$\sqrt{-1}$
k	wave number, $2\pi/\lambda$; integer number, number of photoemissions, Boltzmann's constant in Eq. (112)
$\mathbf{K}$	LDV sensitivity vector, Eq. (14)
K_c	coherence factor
l_x, l_y, l_z	dimension of mv defined by threshold criterion, Eq. (46)
M	number of samples, Eq. (149)
m	refractive index ratio, integer
N	mean number of particles, Eq. (77); integer value; number of cycles
N_B	number of cycles in a burst
N_D	mean number of particles in e^{-2} mv, Eq. (52)
N_{FR}	number of fringes in mv, Eq. (51)
N_e	mean effective number of particles in mv, Eq. (121)
$N(V, t, D)$	mean number of *D*-type particles in V at t, Eq. (74)
$\dot{N}$	data rate, Eq. (159)
n	integer; refractive index of fluid
$n(V, t, D)$	number of *D*-type particles in volume V, Eq. (71)
$n(t, t + \Delta t)$	number of photon pulses in $(t, t + t)$, Eq. (152)
n_p	refractive index of particle; total number of particles in fluid

$\mathfrak{N}$	total number of particles in the entire fluid volume
P	laser beam power; pedestal amplitude factor, Eq. (26*b*)
$\hat{\mathbf{p}}$	polarization direction (unit vector)
p_0	probability, Eq. (72)
q	clipping level, Eq. (155)
q_0	quantum of electron charge
R_L	load resistance
$\mathbf{r}$	position in far field
r	$\|\mathbf{r}\|$
$R_{\mathbf{xx}}$	autocorrelation of variable $\mathbf{x}$, cf. Eq. (91)
$S_{\mathbf{x}}$	power spectrum of variable $\mathbf{x}$, cf. Eq. (92); photodetector sensitivity
SNR	signal-to-noise power ratio, Eq. (64)
$\hat{s}$	propagation direction
s_0, s_1	distance in direction of propagation
T	averaging time; absolute fluid temperature in Eq. (112)
T_T	Taylor microscale for time, Eq. (156)
t	time
t_1	particle time constant, Eq. (66)
t_j	random emission time
Δt	time between events
$\mathbf{U} = (U, V, W)$	mean velocity vector
$\mathbf{u}(\mathbf{x}, t) = (u, v, w)$	total velocity vector
$\mathbf{u}' = (u', v', w')$	fluctuating velocity vector
u_o	observed velocity
V	visibility, Eq. (30); volume
$\bar{V}$	peak visibility, Eq. (31)
V_D	volume of e^{-2} mv
V_m	material volume
V_0	constant velocity
ΔV	overlap volume
$\mathbf{v}(t)$	total lagrangian velocity, $\mathbf{u}(\mathbf{x}(t), t)$
$w(\mathbf{x}, D)$	mv indicator function for D-type particles in Eq. (166)
$w(\mathbf{x})$	mv indicator function for all particles, Eq. (173)
$w(\mathbf{x})$	mv indicator function, Eq. (173)
$\mathbf{x}$	position
$\mathbf{x}(t)$	particle trajectory
$\mathbf{x}_m(t)$	mean effective displacement in HBD signal, Eq. (126)
α	circular frequency of ambiguity noise spectrum
α_c	$2\pi f_c$
β	factor in Eq. (119)
$\delta(\ )$	Dirac delta function
Δ	comparison accuracy tolerance in a counter
$\dot{\epsilon}$	mean emission rate, Eq. (55)

$\dot{e}(t)$	conditional mean emission rate given $J_{tot}(t)$, Eq. (57)
ζ	radial distance from centerline of laser beam
η_q	quantum efficiency, Eq. (55)
θ_{12}	angle between two sources
$\theta_{e^{-2}}$	e^{-2} divergence angle of a gaussian beam
κ	half-angle between illuminating beams
Λ	triangle function, Eq. (153)
λ	wavelength of light
$\lambda(\mathbf{x}, t, D)$	particle density function in xD space, Eq. (72)
λ_x	Taylor microscale in x direction
μ_f	dynamic viscosity of fluid
μ_1	mean frequency of power spectrum, Eq. (102*a*)
μ_2	bandwidth of power spectrum, Eq. (103*a*)
ν	frequency (Hz)
ν_0	laser frequency (Hz)
ξ	dummy variable
ρ_p	particle density
σ	scattering coefficient, Eq. (3)
τ	time delay
τ_B	burst time
τ_f	time constant of filter system after the PMT
τ_h	time constant of combined PMT/filter system, Eq. (59)
τ_p	time constant of PMT anode pulses
$\Delta\tau$	time-delay increment
τ_N	time for N cycles
$\tau_{N/2}$	time for $N/2$ cycles
Φ	phase (rad)
$\dot{\Phi}$	instantaneous frequency
Φ_{J_D}	characteristic function of J_D, Eq. (115)
ϕ	phase noise, Eq. (128)
$\dot{\phi}$	ambiguity noise
Ψ	phase change
Ω	solid angle subtended by collecting aperture
Ω_1, Ω_2	argument variables in characteristic function
ω	circular frequency (rad/s)
ω_c	circular cut-off frequency, $2\pi f_c$, Eq. (144)
$\Delta\omega$	frequency bandwidth, Eq. (140*a*)
$\Delta\omega_A$	ambiguity bandwidth, Eq. (108*a*)
$\Delta\omega_C$	captive bandwidth of a PLL or an FLL
$\Delta\omega_G$	bandwidth due to mean gradient broadening, Eq. (109*a*)
$\Delta\omega_T$	bandwidth due to turbulent fluctuations, Eq. (110*a*)
$\Delta\omega_0$	bandwidth due to laser source, Eq. (113)
$\Delta\omega_B$	bandwidth due to Brownian motion, Eq. (112)
$\Delta\omega_f$	spectrum-analyzer filter bandwidth

Superscripts

B	background radiation
D	Doppler signal
E	electronics noise
e^{-2}	quantity evaluated at e^{-2} points
H	heterodyne
J_D	Doppler light flux
J_P	pedestal light flux
i, j	particles, emission times, arrival times, or other random events
$0l$	lth illuminating beam
li	wave scattered from lth illuminating beam by ith particle
m	measured quantity (high-burst-density signals); measurement volume
min	minimum detectable light flux
n	photon count
o	observed quantity
0	illuminating beam
P	pedestal quantity
peak	peak value, occurs when $\mathbf{x} = \mathbf{0}$
R	reference beam
RF	radio frequency
S	frequency shift; source
tot	total light flux
x, y, z	direction

Subscripts

$\sim$	filtered
^	unit vector
•	time derivative
—	time average
≈	volume-averaged
*	complex conjugated; normalized value

REFERENCES

1. H. Z. Cummins, N. Knable, and Y. Yeh, Observation of Diffusion Broadening of Rayleig[h] Scattered Light, *Phys. Rev. Lett.*, vol. 12, pp. 150–153, 1964.
2. Y. Yeh and H. Z. Cummins, Localized Fluid Flow Measurements with an He-Ne Lase[r] Spectrometer, *Appl. Phys. Lett.*, vol. 4, pp. 176–178, 1964.
3. D. K. Kreid, Measurements of the Developing Laminar Flow in a Square Duct: An Applica[-]tion of the Laser-Doppler Flow Meter, M.S. thesis, University of Minnesota, Minneapolis 1966.
4. R. J. Goldstein and D. K. Kreid, Measurement of Laminar Flow Development in a Squar[e] Duct Using a Laser Doppler Flowmeter, *J. Appl. Mech.*, vol. 34, pp. 813–817, 1967.

5. J. W. Foreman, Jr., R. D. Lewis, J. R. Thornton, and H. J. Watson, Laser Doppler Velocimeter for Measurement of Localized Flow Velocities in Liquids, *IEEE Proc.,* vol. 54, pp. 424–425, 1966.

6. J. W. Foreman, Jr., E. W. George, and R. D. Lewis, Measurement of Localized Flow Velocities in Gases with a Laser-Doppler Flowmeter, *Appl. Phys. Lett.,* vol. 7, pp. 77–80, 1965.

7. R. N. James, Application of a Laser-Doppler Technique to the Measurement of Particle Velocity in Gas-Particle Two-Phase Flow, Ph.D. thesis, Stanford University, Stanford, 1966.

8. R. J. Goldstein and W. F. Hagen, Turbulent Flow Measurements Utilizing the Doppler Shift of Scattered Laser Radiation, *Phys. Fluids,* vol. 10, pp. 1349–1352, 1967.

9. E. Rolfe and R. M. Huffaker, Laser Doppler Velocity Instrument for Wind Tunnel Turbulence and Velocity Measurements, NASA Rep. N68-18099, 1967.

0. F. Durst and M. Zaré, Bibliography of Laser-Doppler Anemometry Literature, Sonderforschungsberich 80, University of Karlsruhe, 1974.

1. F. Durst, A. Melling, and J. H. Whitelaw, *Principles and Practice of Laser-Doppler Anemometry,* Academic, New York, 1976.

2. C. A. Greated and T. S. Durrani, *Laser Systems in Flow Measurement,* Plenum, New York, 1977.

3. B. Chu, *Laser Light Scattering,* pp. 271–290, Academic, New York, 1974.

4. B. S. Rinkevichius, *Laser Anemometry,* Energy Publishing House, Moscow, 1979 (in Russian).

5. B. M. Watrasiewicz and M. J. Rudd, *Laser Doppler Measurements,* Butterworth, London, 1976.

6. H. D. Thompson and W. H. Stevenson (eds.), *The Use of the Laser Doppler Velocimeter for Flow Measurements, Proc. of the First Int. Workshop on Laser Velocimetry,* 1972, Project Squid Headquarters, Jet Propulsion Center, Purdue University, West Lafayette, Ind., 1972.

7. H. D. Thompson and W. H. Stevenson (eds.), *Proc. of the Second Int. Workshop on Laser Velocimetry,* 1974, Engineering Experiment Station, Purdue University, West Lafayette, Ind., 1974.

8. H. Z. Cummins and E. R. Pike (eds.), *Photon Correlation and Light Beating Spectroscopy, Proc. of NATO Advanced Study Inst.,* 1973, Plenum, New York, 1974.

9. E. R. G. Eckert (ed.), *Minnesota Symp. on Laser Anemometry Proc.,* 1975, University of Minnesota, Department of Conferences, Minneapolis, 1976.

0. P. Buchave, J. M. Delhaye, F. Durst, W. K. George, Jr., K. Refslund, and J. H. Whitelaw (eds.), *The Accuracy of Flow Measurements by Laser Doppler Methods, Proc. of the LDA-Symposium,* 1974, Hemisphere, Washington, D.C., 1977.

1. H. Z. Cummins and E. R. Pike (eds.), *Photon Correlation Spectroscopy and Velocimetry, Proc. of NATO Advanced Study Inst.,* 1976, Plenum, New York, 1977.

2. H. J. Pfeifer and J. Haertig (eds.), *Applications of Non-Intrusive Instrumentation in Flow Research,* AGARD CP-193, 1976.

3. L. S. G. Kovasznay, A. Favre, P. Buchave, L. Fulachier, and B. W. Hansen (eds.), *Proc. of the Dynamic Flow Conf. 1978 on Dynamic Measurements in Unsteady Flows,* 1978, P. O. Box 121, DK–2740, Skovlunde, Denmark, 1978.

4. H. D. Thompson and W. H. Stevenson (eds.), *Laser Velocimetry and Particle Sizing, Proc. of the Third Int. Workshop on Laser Velocimetry,* 1978, Hemisphere, Washington, D.C., 1979.

5. M. Kerker, *The Scattering of Light,* chap. 3, Academic, New York, 1969.

6. D. A. Jackson and D. M. Paul, Measurement of Hypersonic Velocities and Turbulence by Direct Spectral Analysis of Doppler Shifted Laser Light, *Phys. Lett.,* vol. 32, p. 77, 1970.

7. D. A. Jackson and D. M. Paul, Measurement of Supersonic Velocity and Turbulence by Laser Anemometry, *J. Phys. E,* vol. 4, pp. 173–176, 1970.

8. R. V. Edwards, J. C. Angus, M. J. French, and J. W. Dunning, Jr., Spectral Analysis of the Signal from the Laser Doppler Flowmeter: Time-Independent Systems, *J. Appl. Phys.,* vol. 42, pp. 837–850, 1971.

29. L. E. Drain and B. C. Moss, The Frequency Shifting of Light by Electro-Optic Techniques *Opto-electronics J.*, vol. 4, pp. 429–436, 1972.
30. E. I. Gordon, A Review of Acoustooptical Deflection and Modulation Devices, *Proc. IEEE*, vol. 54, pp. 1391-1401, 1966.
31. T. Suzuki and R. Hioki, Translation of Light Frequency by a Moving Grating, *J. Opt. Soc. Am.*, vol. 57, pp. 1551–1552, 1967.
32. M. Born and E. Wolf, *Principles of Optics*, Pergamon, New York, 1959.
33. R. J. Adrian and W. L. Earley, Evaluation of LDV Performance Using Mie Scattering Theory, pp. 426–454 in [19], 1976.
34. R. J. Adrian and K. L. Orloff, Laser Anemometer Signals: Visibility Characteristics and Application to Particle Sizing, *Appl. Opt.*, vol. 16, pp. 677–684, 1977.
35. H. Kogelnik and T. Li, Laser Beams and Resonators, *Appl. Opt.*, vol. 5, pp. 1550–1567 1966.
36. H. Weichel and L. S. Pedrotti, A Summary of Useful Laser Equation–An LIA Report *Electro-Opt. Sys. Des.*, pp. 22–36, July 1976.
37. F. Durst and W. H. Stevenson, Properties of Focused Laser Beams and the Influence on Optical Anemometer Signals, pp. 371–388 in [19], 1975.
38. R. J. Adrian and R. J. Goldstein, Analysis of a Laser Doppler Anemometer, *J. Phys. E*, vol 4, pp. 505–511, 1971.
39. M. J. Rudd, A New Theoretical Model for the Laser Doppler Meter, *J. Phys. E*, vol. 2, pp 723–726, 1969.
39a. D. B. Brayton, Small Particle Signal Characteristics of a Dual Scatter Laser Velocimeter *Appl. Opt.*, vol. 13, pp. 2346–2351, 1974.
40. W. M. Farmer, Measurement of Particle Size, Number Density and Velocity Using a Laser Interferometer, *Appl. Opt.*, vol. 11, pp. 2603–2609, 1972.
41. D. M. Robinson and W. P. Chu, Diffraction Analysis of Doppler Signal Characteristics for a Cross Beam Laser Doppler Velocimeter, *Appl. Opt.*, vol. 14, pp. 2177–2181, 1975.
42. D. W. Roberds, Particle Sizing Using Laser Interferometry, *Appl. Opt.*, vol. 16, pp 1861–1865, 1977.
43. L. E. Drain, Coherent and Non-Coherent Methods in Doppler Optical Beat Velocity Measurement, *J. Phys. D*, vol. 5, pp. 481–495, 1972.
44. V. J. Corcoran, Directional Characteristics in Optical Heterodyne Detection Processes, *J. Appl. Phys.*, vol. 36, pp. 1819–1825, 1965.
45. R. J. Adrian, Turbulent Convection in Water over Ice, *J. Fluid Mech.*, vol. 69, pp. 753–781 1975.
46. J. W. Foreman, Jr., Optical Path Length Difference Effects in Photomixing with Multimode Gas Laser Radiation, *Appl. Opt.*, vol. 6, pp. 821–829, 1967.
47. R. J. Adrian and L. M. Fingerson, *Laser Anemometry: Theory, Practice and Applications* p. SE 9/18, TSI, Inc., St. Paul, 1977.
48. G. R. Grant and K. L. Orloff, Two Color Dual-Beam Backscatter Laser Doppler Velocimeter, *Appl. Opt.*, vol. 12, pp. 2913–2916, 1973.
49. F. L. Crossway, J. O. Hornkohl, and A. E. Lennert, Signal Characteristics and Signal Conditioning Electronics for a Vector Velocity Laser Velocimeter, pp. 396–444 in [16] 1972.
50. R. J. Adrian, A Bi-Polar Two Component Laser Velocimeter, *J. Phys. E*, vol. 4, pp. 72–75 1975.
51. C. Greated, Measurement of Reynolds Stresses with an Improved Laser Flow Meter, *J. Phys E*, vol. 3, pp. 753–756, 1970.
52. W. Yanta and G. J. Crapo, Applications of the Laser Doppler Velocimeter to Measure Subsonic and Supersonic Flows, AGARD Pre-Print 193, pp. 2.1–2.8, 1976.
53. A. Papoulis, *Probability, Random Variables and Stochastic Processes*, chap. 16, McGraw Hill, New York, 1965.
54. W. T. Mayo, Modeling Laser Velocimeter Signals as Triply Stochastic Poisson Processes, pp 455–484 in [19], 1975.

55. W. H. Stevenson, R. Dos Santos, and S. C. Mettler, Fringe Model Flourescence Velocimetry, AGARD Pre-Print 193, pp. 21.1-21.9, 1976.
56. Dow Diagnostics, 1200 Madison Ave., Indianapolis, Ind. 46225.
57. Duke Standards, 445 Sherman Ave., Palo Alto, Calif. 94306.
57a. K. T. Whitby, The Physical Characteristics of Sulphur Aerosols, *Atmos. Environ.*, vol. 12, pp. 135–159, 1978.
58. W. V. Feller and J. F. Meyers, Development of a Controllable Particle Generator for LDV Seeding in Hypersonic Wind Tunnels, pp. 342-357 in [19], 1975.
59. J. K. Agarwal and L. M. Fingerson, Evaluation of Various Particles for Their Suitability as Seeds in Laser Velocimetry, pp. 50–66 in [24], 1978.
60. N. G. Jerlov and E. S. Nielsen (eds.), *Optical Aspects of Oceanography*, Academic, New York, 1974.
61. O. B. Brown and H. R. Gordon, Size-Refractive Index Distribution of Clear Coastal Water Particulates from Light Scattering, *Appl. Opt.*, vol. 13, pp. 2874–2880, 1974.
62. J. K. Agarawal and P. Keady, Theoretical Calculation and Experimental Observation of Laser Velocimeter Signal Quality, *TSI Q.*, vol. 6, pp. 3–10, 1980.
63. E. Brockmann, Computer Simulation of Laser Velocimeter Signals, pp. 328–331 in [24], 1978.
64. R. J. Adrian, Estimation of LDA Signal Strength and Signal-to-Noise Ratio, *TSI Q.*, vol. 5, pp. 3-8, 1979.
65. W. K. George and J. L. Lumley, The Laser-Doppler Velocimeter and Its Application to the Measurement of Turbulence, *J. Fluid Mech.*, vol. 60, pp. 321–362, 1973.
66. P. Buchave, W. K. George, Jr., and J. L. Lumley, The Measurement of Turbulence with the Laser-Doppler Anemometer, *Annu. Rev. Fluid Mech.*, vol. 11, pp. 443–503, 1979.
67. S. Chandrasekhar, Stochastic Problems in Physics and Astronomy, *Rev. Mod. Phys.*, vol. 15, pp. 1–89, 1943.
68. R. J. Goldstein and R. J. Adrian, Measurement of Fluid Velocity Gradients Using Laser Doppler Techniques, *Rev. Sci. Instrum.*, vol. 42, pp. 1317–1320, 1971.
69. M. J. Lighthill, *Fourier Analysis and Generalized Functions*, p. 43, Cambridge University Press, Cambridge, 1970.
70. D. K. Kreid, Laser-Doppler Velocity Measurements in Non-Uniform Flow: Error Estimates, *Appl. Opt.*, vol. 8, pp. 1872–1881, 1974.
71. D. K. McLaughlin and W. G. Tiederman, Biasing Correcting for Individual Realization of Laser Anemometer Measurements in Turbulent Flows, *Phys. Fluids*, vol. 16, pp. 2082–2088, 1973.
72. S. O. Rice, Mathematical Analysis of Random Noise, *Bell Syst. Tech. J.*, vol. 23, pp. 282–332, 1944; vol. 24, pp. 46–156, 1945.
73. S. O. Rice, Statistical Properties of a Sine Wave Plus Random Noise, *Bell Syst. Tech. J.*, vol. 27, pp. 109–156, 1948.
74. P. Jespers, P. T. Chu, and A. A. Fettweis, New Method for Computing Correlation Functions, presented at *Int. Symp. on Information Theory*, Brussels, 1962.
75. J. Ikebe and T. Sato, A New Integrator Using Random Voltage, *Electrotech. J.*, vol. 7, pp. 43–47, 1962.
76. B. P. T. Veltman and H. Kwakernaak, Theorie und Technik der Polaritats-Korrelation für die Dynamische Analyse nieder Frequentor Signale und Systeme, *Regelungstechnik*, vol. 9, pp. 357–364, 1961.
77. K. H. Norsworthy, A New High Product Rate 10 Nanosecond, 256 Point Correlator, *Phys. Scr.*, vol. 19, pp. 369–378, 1978.
78. E. R. Pike, How Many Signal Photons Determine a Velocity? pp. 285-289 in [24], 1978.
79. A. E. Smart and W. T. Mayo, Jr., Applications of Laser Anemometry to High Reynolds Number Flows, *Phys. Scr.*, vol. 19, pp. 426–440, 1978.
80. R. J. Baker and G. Wigley, Design, Evaluation and Application of a Filter Bank Signal Processor, pp. 350–363 in [20], 1975.
81. M. Alldritt, R. Jones, C. J. Oliver, and J. M. Vaughan, The Processing of Digital Signals by a Surface Acoustic Wave Spectrum Analyzer, *J. Phys. E*, vol. 11, pp. 116–119, 1978.

82. R. J. Adrian, J. A. C. Humphrey, and J. H. Whitelaw, Frequency Measurement Errors Due to Noise in LDV Signals, pp. 287–311 in [20], 1975.
83. H. H. Bossel, W. J. Hiller, and G. E. A. Meier, Noise Cancelling Signal Difference Method for Optical Velocity Measurements, *J. Phys. E,* vol. 5, pp. 893–896, 1972.
84. L. Lading and R. V. Edwards, The Effect of Measurement Volume on Laser Doppler Anemometer Measurements as Measured on Simulated Signals, pp. 64–80 in [20], 1975.
85. J. R. Abiss, The Structure of the Doppler-Difference Signal and the Analysis of Its Autocorrelation Function, *Phys. Scr.,* vol. 19, pp. 388–395, 1978.
86. J. B. Morton and W. H. Clark, Measurements of Two-Point Velocity Correlations in Pipe Flow Using Laser Anemometers, *J. Phys. E,* vol. 4, pp. 809–814, 1971.
87. H. R. E. van Maanen, K. van der Molen, and J. Blom, Reduction of Ambiguity Noise in Laser-Velocimetry by a Cross Correlation Technique, pp. 81–88 in [20], 1975.
88. W. K. George, Jr., Limitations to Measuring Accuracy Inherent in the Laser Doppler Signal, pp. 20–63 in [20], 1975.
89. W. Hösel and W. Rodi, New Biasing Elimination Method for Laser-Doppler Velocimeter Counter Processing, *Rev. Sci. Instrum.,* vol. 48, pp. 910–919, 1977.
90. P. E. Dimotakis, Single Scattering Particle Laser Doppler Measurements of Turbulence, pp. 10.1–10.14 in [22], 1976.
91. P. Buchave, Biasing Errors in Individual Particle Measurements with LDA–Counter Signal Processor, pp. 258–278 in [20], 1975.
92. M. C. Whiffen, Polar Response of an LV Measurement Volume, pp. 591–592 in [19], 1975.

CHAPTER

SIX

VOLUME FLOW MEASUREMENTS

G. E. Mattingly

1. INTRODUCTION

One of the most common types of material measurements in American industry is the determination of fluid quantity, or the measurement of fluid flow rate. Untold sums of money are spent daily on fluid custody transfers that are based upon the "read-out" information from fluid metering devices. The following examples are based upon information available in 1978–1979:

1. Amounts
   a. *Petroleum industry*. Worldwide production is 22 billion barrels of crude petroleum per year. In the United States the metering of crude is performed at 650,000 measurement sites by 50,000 employees.
   b. *Gas*. In the United States 20 trillion ft^3 of natural gas are metered each year, through 42 million fluid quantity devices.
   c. *Wastewater discharge.* The EPA has issued 90,000 discharge permits to municipalities and industries. New treatment plants cost $6 to $10 billion per year.
2. Costs
   a. *Fluids.* The petroleum-refining, chemical, and beverage industries account for $1 trillion per year. The interbasin transfer of water from Mississippi to Texas alone amounts to $1.3 billion per year.
   b. *Meters.* Fluid-flow devices cost $843 million per year.

Note, particularly, the amount of natural gas that is metered in the United States each year. At an estimated cost of \$2/ft^3, one can quickly quantify the material and dollar value that is associated with each 0.01% of systematic uncertainty that might exist in this measurement system.

The purposes of volume flow measurements are:

1. Effective regulation
2. Energy distribution
3. Custody transfer/equity in marketplace
4. International competition

A wide variety of fluid meters maintain and/or control material quality and quantity in practically every continuous industrial process. Because of the cost of the fluids and the measurements, it is essential that fluid quantity and flow-rate measurements are made as precisely and as accurately as required by the parties involved. It is also incumbent upon those involved in fluid custody transfer to establish the traceability chains that link their measurements to the appropriate standards that are involved. Only in this manner can fluid volume measurements be performed equitably, with the confidence of seller and buyer alike.

2. CLASSIFICATION OF METERING DEVICES

Every meter consists of two distinct parts, each of which performs a specific function. The first part–the primary element–is in contact with the fluid and produces an interaction with the fluid. Examples are:

1. Orifice plate
2. Turbine wheel
3. Vortex-shedding strut

The second part–the associated secondary element–converts one or more reactions received from the primary element into an observable quantity. Examples of secondary elements that would be associated with the above primary elements are:

1. A manometer to exhibit the differential pressure generated in the fluid by the orifice plate
2. An electrical system that magnetically senses the revolving turbine blade tips
3. A system that detects the vortices being shed behind the strut

When a meter is to be selected, both the primary and the secondary element must be considered, as they must operate together, as a unit, in the particular environment where the measurements are to be made. Furthermore, any other factors that may introduce variation into the measurement results should also be considered in the evaluation of the fluid metering system. Such factors might include a person who manually reads the manometer or a computer or microprocessor that calculates a final result from analog or digital data received from the secondary element. These factors and their influence are discussed further in Secs. 4 and 5.

Meters can be further classified into those that determine fluid quantity and those that indicate fluid flow rate [1]. The various types of fluid measuring devices within each of these classes are as follows:

1. Quantity meters
   a. Weighing (for liquids)
      (1) Weighers
      (2) Tilting traps
   b. Volumetric (for liquids)
      (1) Calibrated tank
      (2) Reciprocating piston
      (3) Rotating piston or ring piston
      (4) Rotating disk
      (5) Sliding and rotating vane
      (6) Gear and lobed impeller (rotary)
   c. Volumetric (for gases)
      (1) Bellows
      (2) Liquid sealed drum
      (3) Gear and lobed impeller (rotary)
2. Rate meters
   a. Differential pressure
      (1) Orifice
      (2) Venturi
      (3) Nozzle
      (4) Centrifugal
      (5) Pitot tube
      (6) Linear resistance
   b. Momentum
      (1) Turbine
      (2) Propeller
      (3) Cup anemometer
   c. Variable area
      (1) Gate
      (2) Cones, floats in tubes
      (3) Slotted cylinder and piston
   d. Force
      (1) Target
      (2) Hydrometric pendulum
   e. Thermal
      (1) Hot wire, hot film
      (2) Total heating

f. Fluid surface height or "head" type
  (1) Weirs
  (2) Flumes
g. Other
  (1) Electromagnetic
  (2) Tracers
  (3) Acoustic
  (4) Vortex shedding
  (5) Laser
  (6) Coriolis

For the most part, the quantity meters listed above are used where spatial requirements and temporal response characteristics are not as important as precision and accuracy. For example, liquid weighing devices, calibrated volumetric tanks, reciprocating-piston devices, and liquid sealed-drum systems for gases are among the devices used most as "primary" standards for the calibration of fluid flow-rate devices. The simplicity of these devices and their readily accessible and "checkable" components produce precision and accuracy and, in turn, confidence in their results. A major conclusion of Secs. 4 and 5 is that several levels of checks can and should be applied to these systems; however, here we note only that these types of systems have special status as very precise and accurate devices.

The remaining devices in the list are more commonly used today in industrial installations for measuring the rate of fluid flow; they are subdivided according to the physical principles that form the basis of their operation. In the section that follows, selected meter types are described briefly, with emphasis on principles of operation and general performance characteristics.

3. SELECTED METER PERFORMANCE CHARACTERISTICS

In this section, the characteristics of the more commonly used types of fluid metering devices, as well as several new and innovative types of metering devices that hold considerable promise for future applications, are described.

3.1 Orifice Meters

Owing to their simplicity, cost, ruggedness, and widespread acceptance, differential-pressure type meters–and notably orifice meters–deserve special attention [2]. An orifice-meter "run" is shown in Fig. 1. The primary element of an orifice meter consists of:

1. The upstream meter-tube section, including any of several types of devices intended to "condition" the flow
2. The orifice fixture and plate assembly

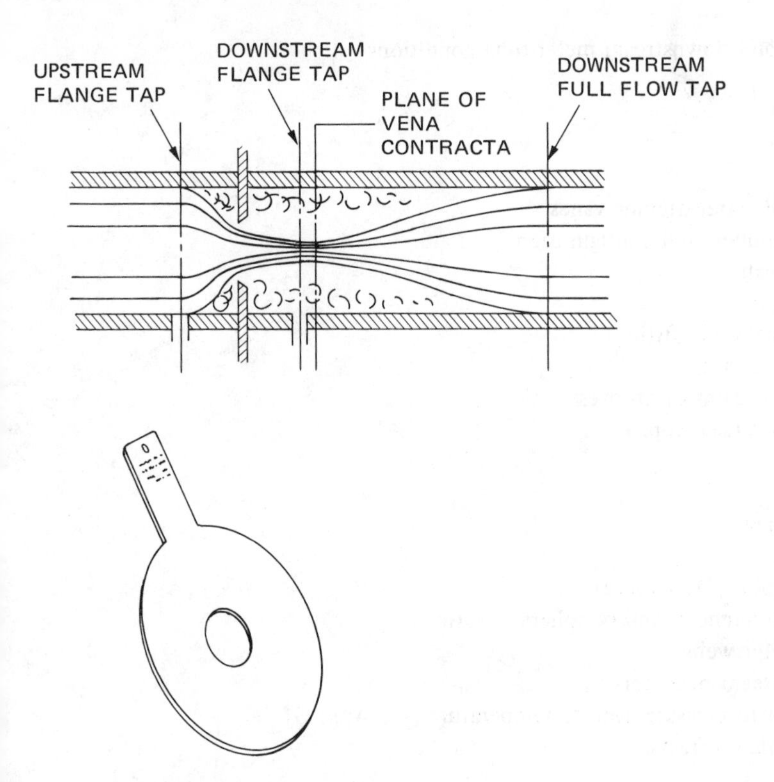

Figure 1 Orifice meter.

3. The downstream meter-tube section.
4. The secondary element, not shown in Fig. 1, might consist of a U-tube manometer in conjunction with fluid temperature measurement instrumentation.

The ideal performance characteristics for orifice meters are given in App. A for incompressible fluids. Appendix B contains ideal performance characteristics for compressible fluids, and App. C contains real-gas characteristics.

The features that affect sustained and accurate orifice-meter performance are as follows:

1. Orifice plate
   a. Dimensions
      (1) Hole diameter
      (2) Beveling
      (3) Roughness
      (4) Flatness
      (5) Edge shape and sharpness
      (6) Thickness
   b. Centeredness in pipe

2. Upstream and downstream meter-tube conditions
   a. Round
   b. Straight
   c. Smooth
   d. Length
3. Tube-bundle straightening vanes
   a. Size, number, and configuration
   b. Parallelism
   c. Ruggedness
4. Orifice flanges and fittings
   a. Pressure rating
   b. No weld seams or grooves
5. Pressure taps (see App. F)
   a. Flange
   b. Pipe
   c. D-$D/2$ taps
   d. Corner
   e. Other (pipe, Duzenburg)
   f. Burrs, tolerances, and peripheral locations
6. Thermometer wells
   a. Downstream of meters
   b. Located to measure average temperature (see App. G)
7. Flow and fluid effects
   a. Profile
   b. Swirl
   c. Pulsations
   d. Contaminants
   e. Plate bending

A paddle-type orifice plate is shown at the lower left of Fig. 1. In use, this plate is sandwiched between pipe flanges so that the orifice-plate hole is concentric with the pipe centerline. The differential pressure (upstream minus downstream) across this plate is then related to the fluid flow rate, according to

$$\dot{Q} = A_2 C_D \left(\frac{2\,\Delta p}{\rho(1-\beta^4)} \right)^{1/2} \tag{1}$$

where $\dot{Q}$ = volumetric flow rate
A_2 = throat area
C_D = discharge coefficient
Δp = upstream-to-downstream pressure differential
ρ = fluid density
β = ratio of throat diameter to internal diameter of upstream pipe

Equation (1) is based upon a series of assumptions, such as those governing Bernoulli phenomena, i.e., steady flow of an incompressible fluid in which there is no phase change.

Equivalently, Eq. (1) can be written

$$\dot{Q} = KA_2 \left(\frac{2\ \Delta p}{\rho} \right)^{1/2} \tag{2}$$

where K is termed the *flow coefficient*:

$$K = \frac{C_D}{(1 - \beta^4)^{1/2}} \tag{3}$$

Both the discharge coefficient and, correspondingly, the flow coefficient depend on the exact locations of the pressure tappings upstream and downstream of the orifice plate. For a given meter and pressure-tapping arrangement, a conventional calibration procedure, as described in Sec. 3, can determine the dependence of C_D upon the flow Reynolds number Re:

$$\text{Re} = \frac{DV\rho}{\mu} \tag{4}$$

where D = pipe internal diameter
V = average fluid velocity
μ = fluid dynamic viscosity

If conventional pressure-tapping configurations are used, it may be satisfactory to calculate the discharge or flow coefficients via such formulas as are available in the metering literature (see App. E). Figure 2 depicts the conventional American pressure-tapping convention, which uses the so-called *flange taps*. These have the advantage that (1) they are conveniently drilled radially straight through the flanges that hold the plate, and (2) they are located close to the plate, where they are capable of sensing a large differential pressure; this, in turn, is desirable because it can increase metering sensitivity. Flange taps have the disadvantage of being located

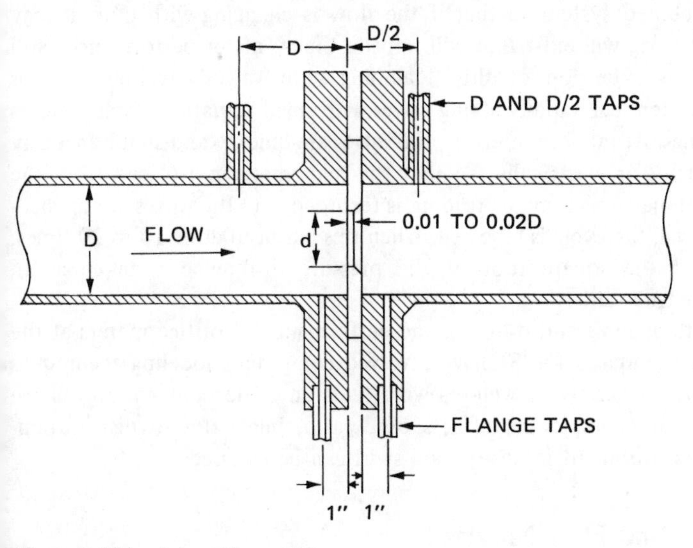

Figure 2 Thin-plate orifice meter.

1 in on either side of the orifice plate. Thus, flange-tapped meters of different sizes do not conform to geometric scaling laws.

Also shown in Fig. 2 are the so-called *D-D*/2 taps. These taps do conform to geometric scaling laws. Because of their locations, these taps sense fluid pressures that are "smoother" in the sense that turbulent pressure fluctuations at these tap locations are generally smaller than those that occur nearer the orifice plate. Nearer the plate, where the flow is stagnating against the plate (upstream and downstream surfaces) and turning so as to flow radially inward toward the orifice hole, the turbulence generally affects pressure measuring instrumentation more significantly.

U-tube or single-leg differential manometers can tend to "bounce" markedly when sensing pressures near the orifice plate. In addition, a variety of other factors influence manometer bounce. For example, the diameter of the tapping hole, the diameter, length, and configuration of the hose or tubing connecting the manometer to the meter, and the inertial characteristics of the manometer fluid all exert some influence (see App. F). This bouncing phenomenon can cause difficulty in obtaining precise metering performance from an orifice meter.

A variety of manometer reading schemes have been devised to reduce the detrimental effects of bouncing. A frequently used method is to simply "eyeball" an average value of the manometer column height after observing several cycles of the bounce. The average value of the square roots of several such values is taken to be the true value for the particular flow. A second scheme that is used on single-leg differential manometers is to read consecutive maximum and minimum values of the bouncing column of manometer fluid. After "max-min" pairs of readings have been recorded some 10 to 20 times, the average of all these pairs is taken to approximate the true average value. Another scheme is to install needle valves in the tubes connecting the manometer to the orifice meter. Partially closing these valves tends to smooth the bouncing effect. The partially closed valves also limit the time response of the combined system so that if the flow is changing with time, it may not be detected, or a lag will exist that will impede timely meter performance. Still another scheme is an extension of this "damping with valves" technique. This consists of installing identical, rapidly acting valves with small actuation displacements in the manometer lines. At random intervals, the valves are quickly and simultaneously closed, either manually or electrically, so that the manometer is isolated from the meter. The then stationary manometer reading is recorded, and the valves are opened. After a random interval, the cycle is repeated. When this has been done 10 to 20 times, the average value of the square root of the pressure difference is taken as an approximation of the true value.

Recent efforts to analyze and optimize the performance of orifice meters at the National Bureau of Standards (NBS) have involved computer modeling techniques [5]. From the results, a variety of well-known effects on orifice-meter performance can be quantified. For example the effects of β ratio, inlet-velocity distribution, orifice-plate thickness, turbulent intensity, and swirl can be assessed.

3.2 Venturi Tubes and Flow Nozzles

Figure 3 shows a Venturi tube. These differential-pressure type devices have a shape that closely approximates the streamline patterns of the flow through a reduced

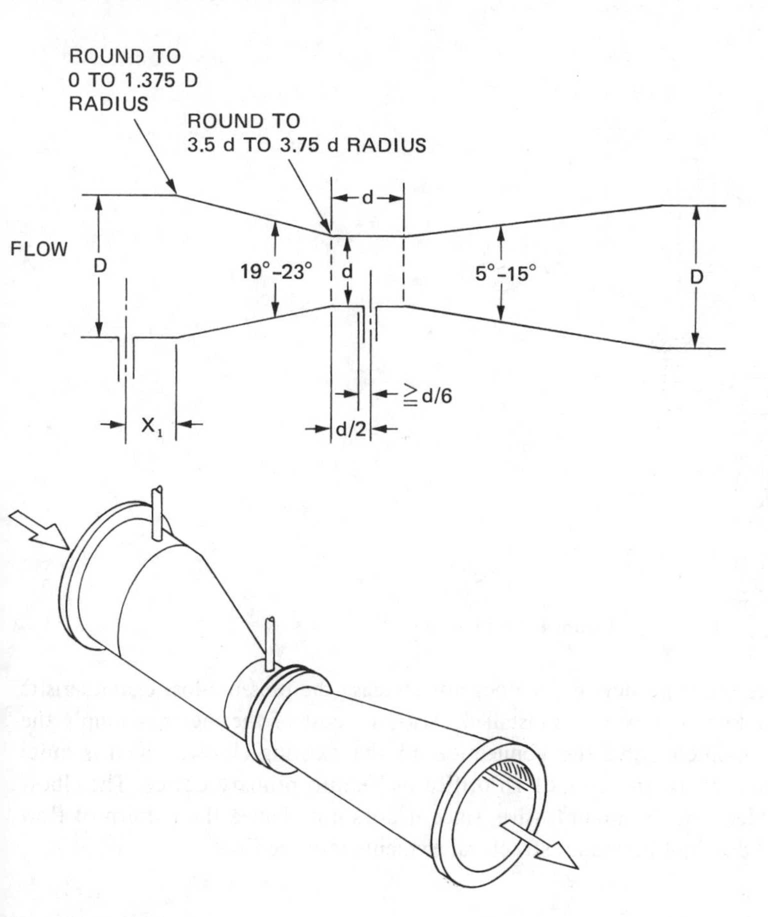

Figure 3 Venturi tube.

cross-sectional area; as a result, their pressure-loss characteristics are generally lower than those of orifice plates. Hence, the higher cost of a Venturi tube is offset in time by reduced pumping costs incurred in producing the flow through the meter. Correspondingly, the discharge coefficients of these devices tend to approximate unity more closely, and there is less variation in the performance of Venturi tubes than orifice meters.

Just as with orifice meters, Venturi meters can be of the concentric or eccentric type. Both types are shown in Fig. 3. The eccentric type provides more "self-cleaning" than the concentric type. Thus, it performs more satisfactorily on flows containing sediments or particulates that might, in horizontally installed meters, deposit in corners and thereby change the effective geometry of the meter and, in turn, its performance characteristics.

3.3 Elbow Meters

Figure 4 shows the primary and secondary components of the elbow meter. While this type of device does not generally match the precision and uncertainties of other

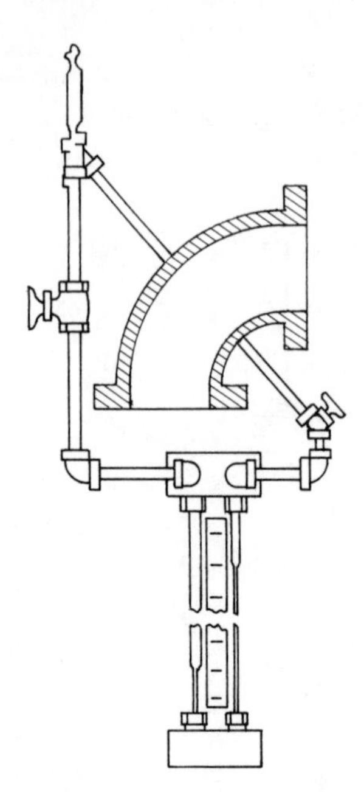

Figure 4 Elbow meter.

differential-pressure type devices, it does not increase the pressure-loss characteristic of the flow system in which it is installed. Also, the cost of the meter is simply the cost of the manometer and the connection to the existing elbow, which is quite small in comparison to the cost of an orifice or Venturi primary device. The elbow meter is considered to be nonintrusive, since it does not change the pattern of flow in the pipe and does not introduce structural elements into the flow.

3.4 Pitot Tubes

Figure 5 shows a sketch of a primary device that is based upon a differential pressure. This differential is established, separately, between the high pressure of the flow stagnating near the free tip of a cantilevered tube and the low static pressure of the flow around the side of the same tube. By displacing the sensing taps axially along the tube, any detrimental interaction between them is avoided. Because the distribution of flow in the center region of the pipe is generally quite uniform in the cross-stream directions, a good estimate is obtained of the maximum flow velocity near the centerline. By relating this value to the cross-sectional average velocity through (1) a calculation, assuming a known radial distribution of velocity in the conduit, or (2) a calibration of the device, one can generally obtain an accuracy in the range quoted in Fig. 5.

3.5 Laminar Flowmeters

Figure 6 shows a typical arrangement of the primary and secondary components of a laminar flowmeter. The basis of the device is the establishment of laminar flow

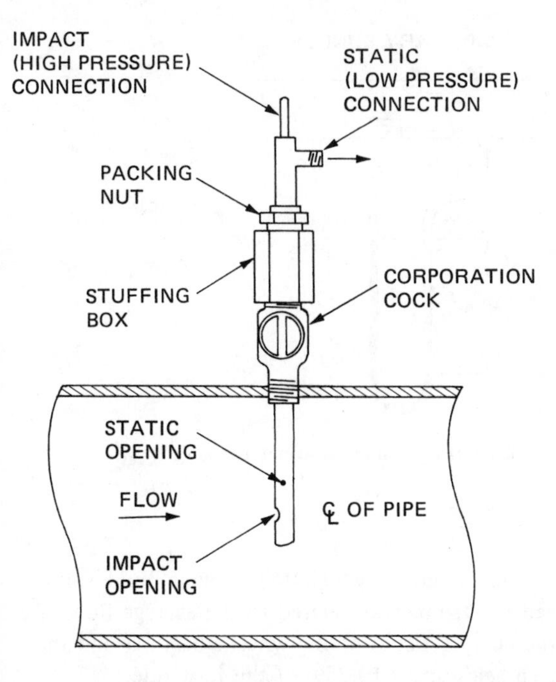

Figure 5 Pitot tube.

between the pressure taps, and use of the well-known relationship between laminar flow rate through a tube of known cross-sectional area and the pressure drop across its known length. These devices generally have the advantage that they operate bidirectionally and indicate, via the sign of the differential pressure, the direction of the flow. As shown in Fig. 6, laminar flowmeters can be nonintrusive.

Figure 7 shows other arrangements for the primary element. These can involve a single capillary tube that is a constriction in the larger pipeline, or an intrusive bundle of smaller capillary tubes in which the flow is arranged to be laminar. Other intrusive elements, such as honeycomblike structures, can also be used. Because these generally are composed of noncircular cross-sectional shapes, it is recommended that each device be calibrated to obtain assured performance. It is essential with these types of primary elements that the geometry of the conduit remain unchanged. Therefore, laminar flowmeters are not recommended for measuring flows containing particulates or materials that tend to deposit on pipe walls.

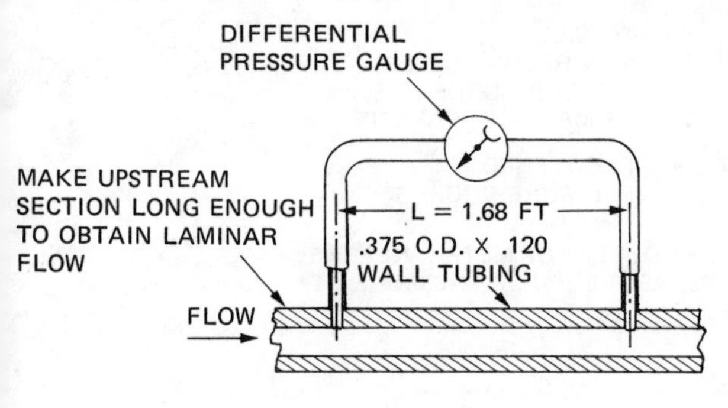

Figure 6 One possible arrangement of the components of a laminar flowmeter.

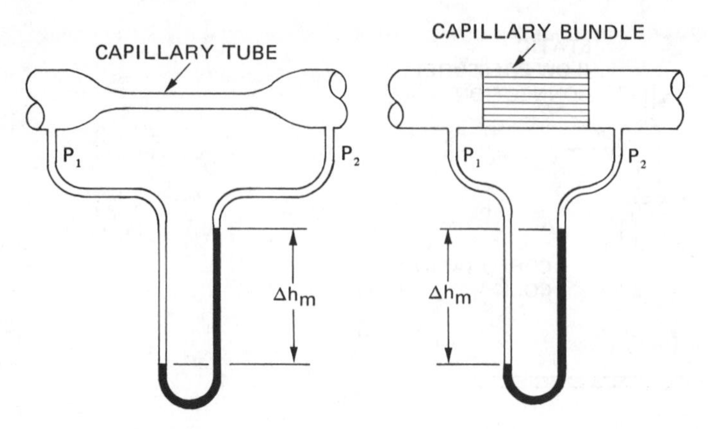

Figure 7 Two additional types of primary elements for laminar flowmeters.

3.6 Turbine Meters

Figure 8 shows a sketch of a turbine meter, in which the flowing fluid spins a propeller wheel whose angular speed is in some way related to the average flow rate in the conduit. Depending upon the characteristics of the wheel's bearings and their longevity, this device often gives accuracies up to ±0.25% of the flow rate.

The angular speed of the wheel is generally detected by the passage of the blade tips past a coil pickup on the pipe. Increased resolution can be obtained by increasing the number of pickups and summing the pulses received, by increasing the number of blades, or by shrouding the blade tips and placing magnetic buttons on the shroud so as to generate more pulses per revolution of the wheel. With

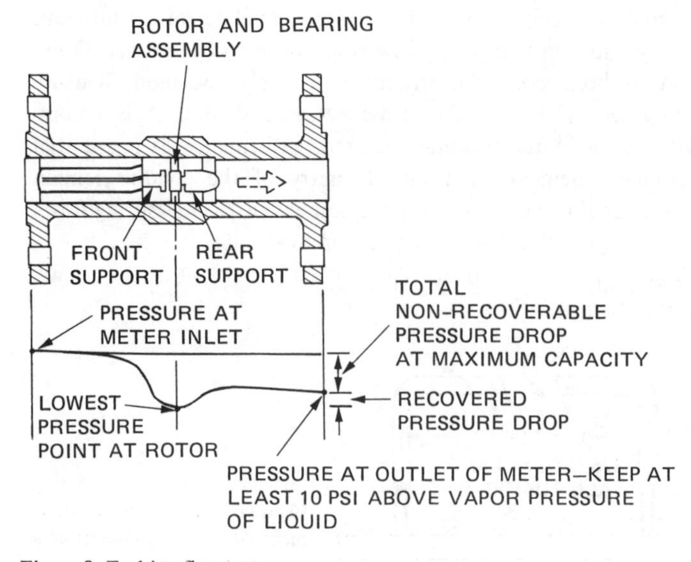

Figure 8 Turbine flowmeter.

proper arrangement of the pertinent geometry of the wheel and blades and with good bearings, turbine meters can produce a frequency output that is proportional to flow rate within a few tenths of a percent over a flow-rate range of 30 to 1. At the low end of this range, fluid viscous effects are a limiting factor. At the high end, the interaction between blade tips or magnetic buttons and the pickup are limiting.

3.7 Rotameters

A typical rotameter configuration is shown in Fig. 9. The vertically installed device operates when the upward fluid drag on the float is balanced in the upwardly diverging tube by the weight of the float. The vertical elevation of the float is then an indication of the flow rate, and this is generally read manually if the metering tube (and the fluid) is transparent. If the pressure within the fluid stream to be measured is too high to allow glass or plastic tapered tubes, metal can be used; then the float position can be sensed magnetically or electrically.

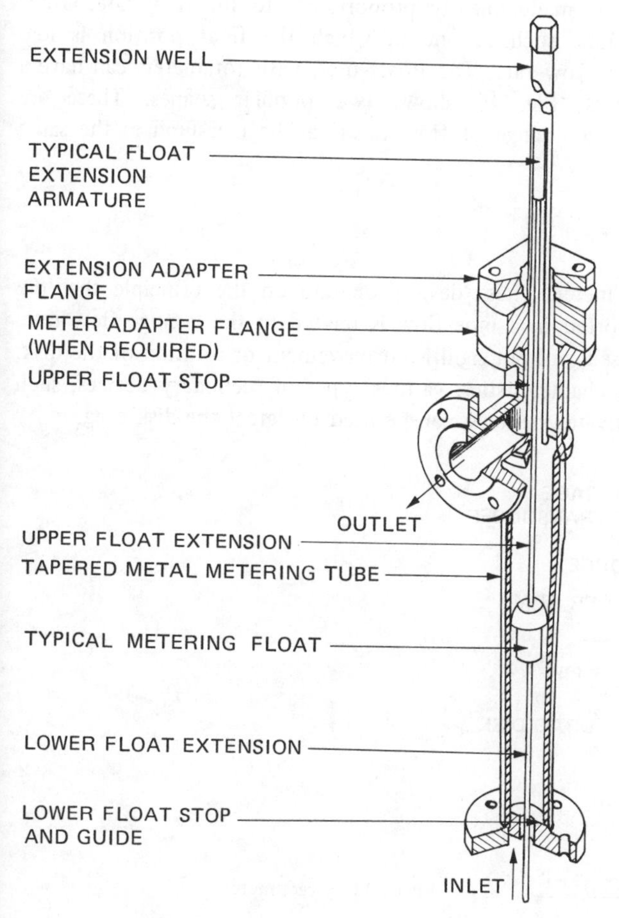

Figure 9 Typical rotameter configuration.

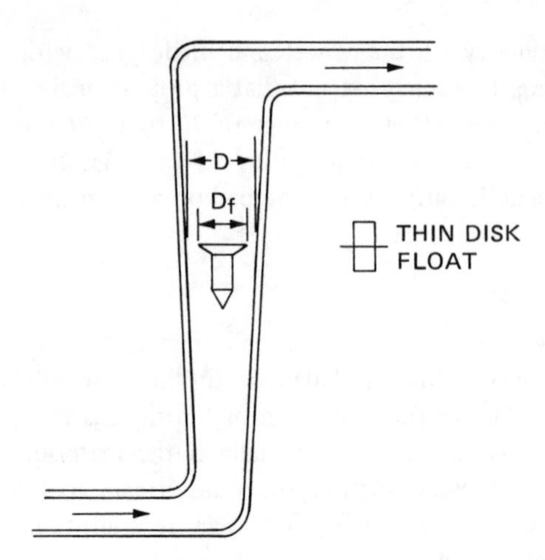

Figure 10 Rotameter tube and floats.

By appropriate arrangement of the divergence of the metering tube, the vertical position of the float may be made linearly proportional to the flow rate. Other arrangements are also feasible, such as one in which the float position is logarithmically dependent upon flow rate. The floats used with rotameters can have a variety of shapes and sizes; Fig. 10 shows two popular shapes. These are interchangeable, so that a wide range of flow rates can be measured in the same tapered tube.

3.8 Target Meters

Figure 11 shows a target meter. These devices operate on the principle that the fluid drag on a disk supported in a pipe flow is related to the average flow rate. This fluid drag can be sensed without significant movement or rotation of the disk, which would alter its drag characteristics; various types of secondary devices, such as strain gages and fluid-activated bellows, can be used to detect the disk drag.

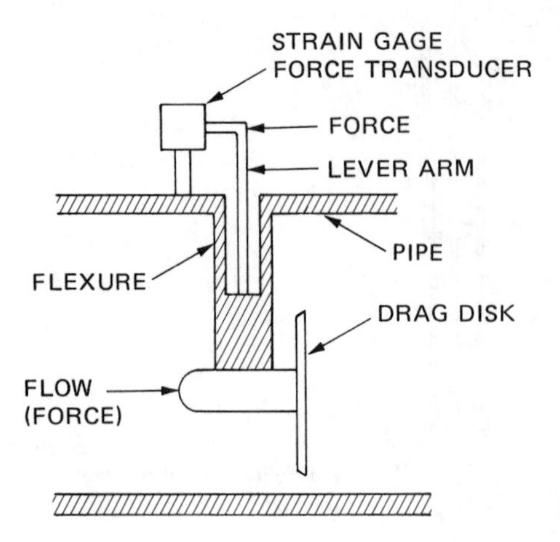

Figure 11 Target meter.

Target meters can be concentric or eccentric as shown in Fig. 11. The calibration curves vary, owing to the effects of the strut supporting the disk. These devices can also be used as bidirectional meters, with the sense of the force indicating the flow direction. Because of their simple structure, they can be used to measure the flow rate of dirty fluids, which may have entrained sediments, as long as these do not alter the critical geometrical arrangement shown in Fig. 11.

3.9 Thermal Flowmeters

Figure 12 shows a typical arrangement for a thermal flowmeter. The performance of this type of meter is based upon the increase in fluid temperature that is sensed between two thermometers when heat is added to the fluid between the thermometers. For a given fluid heating rate q, the fluid mass flow rate $\dot{M}$ is proportional to the rate of heat addition and inversely proportional to the downstream-to-upstream temperature difference ΔT. In compatible units,

$$\dot{M} = \frac{q}{c_p \, \Delta T} \tag{5}$$

where c_p is the fluid specific heat at constant pressure. If sufficiently large distances separate the temperature sensors so that complete cross-sectional mixing occurs, and if no heat is lost to or through the pipe wall, the device should not require calibration. On the other hand, if the heat addition occurs locally on the centerline

(a)

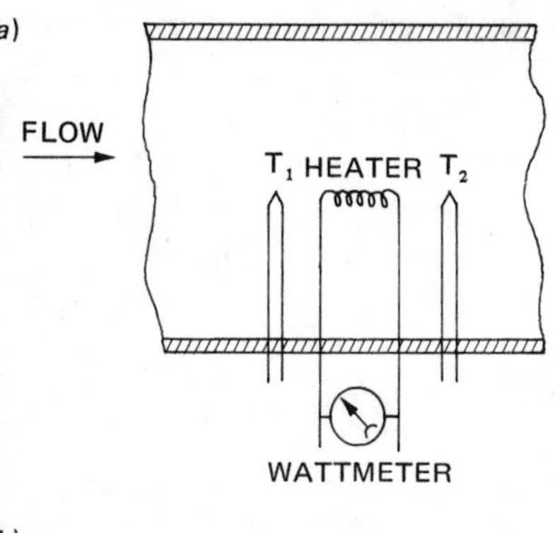

(b)

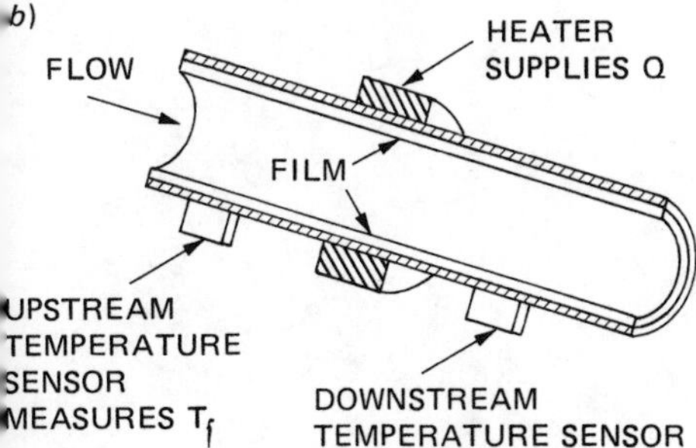

Figure 12 Thermal flowmeter. (*a*) $\Delta T = T_2 - T_1$. (*b*) $\Delta T = T_W - T_f$.

of the pipe or only in the fluid layers along the pipe wall, flow calibration is required to achieve best performance.

Thermal flowmeters can be arranged to be nonintrusive, as shown in the lower portion of Fig. 12. They can also operate through cooling, instead of heating.

3.10 Weirs and Flumes

Figure 13 shows a sharp-crested weir and the flow pattern that might exist when the weir is installed in a much wider channel. The upstream depth of water above the weir crest is related to the flow rate in the channel through the rating curve for the device. This upstream depth is generally measured with a pressure transducer sensing fluid depth via the static pressure or with a depth measurement made with a surface height gauge installed in an auxiliary chamber attached to the channel, as shown in the lower portion of Fig. 13. In this chamber surface heights can be measured on the quiescent surface more precisely than they can in the flow channel, where waves or other surface irregularities can impede good depth measurements.

For proper performance of the device, the cavity under the nappe must be aerated. Otherwise the pressure distribution through the nappe will not be atmospheric, as is assumed in the construction of the rating curve. This aeration is virtually guaranteed if the channel width is much larger than the width of the weir crest. It should be noted that a weir is essentially an open-channel version of an orifice-plate meter, wherein the differential pressure is the depth measurement. The flume (not sketched in Fig. 13) corresponds in a similar fashion to the venturi meter.

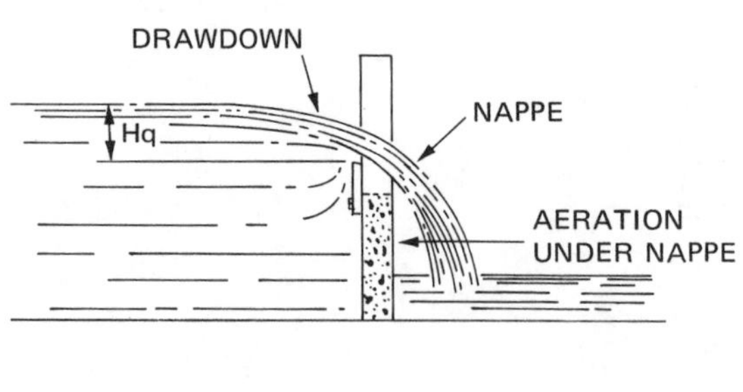

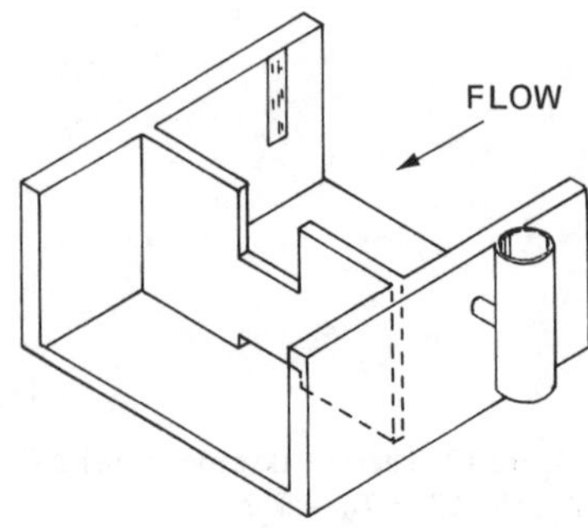

Figure 13 Sharp-crested weir.

Weirs and flumes are frequently selected to measure the flow rates in canals, streams, and rivers. It is then essential that sediment and other debris be kept from clogging the primary (or secondary) element, and to this end self-cleaning features are essential. The weir can be arranged to be self-cleaning via a small slit in the partition under the crest. This slit should be sized so as to pass any silt buildup that might deposit on the upstream side of the partition. In this manner the original geometry, on which the calculated or calibrated rating curve is based, will be preserved.

The calibration of open-channel metering devices can be very difficult—especially for the very large devices that are installed in rivers. It is necessary to determine the flow rate as accurately as is required over as wide a flow-rate range as occurs naturally, and to use these data to produce a rating curve. The process is considerably easier for prefabricated or portable units, which can be installed in laboratory facilities where precise determinations of the flow rate can be routinely made.

A calibration method that has recently been used at NBS to study the performance characteristics of Parshall flumes is computer modeling [6, 7]. With this technique the meter geometry and inlet flow distribution can be treated as boundary conditions for the problem of determining the entire flow field within the meter, together with the free-surface position. The flow field can be integrated to obtain the flow rate, and the free-surface distribution produces the depth to be measured via the secondary device, thus giving a computed rating curve. Such a technique could allow the quantitative assessment of unusual inlet flow distributions and changes in critical metering geometry, such as tilt, convergence angles, and crest heights.

.11 Magnetic Flowmeters

Figure 14 shows the arrangement of the coils and electrodes that make up the typical magnetic flowmeter. This meter operates on the principle that a conducting fluid will generate a voltage proportional to the flow rate as it passes through a magnetic field. The typical device is nonintrusive, with magnetic coils insulated from the flowing fluid and the electrodes maintaining electrical contact with the fluid. This type of device is available commercially in a wide range of pipe sizes.

.12 Acoustic Flowmeters

A variety of acoustic flowmeters are commercially available. These use several different physical principles; the most prevalent is based upon a Doppler principle. Figure 15 shows one arrangement, in which a pair of ultrasonic transmitters beam to receivers placed across a flow conduit. One of the transmitters beams downstream, while the other beams upstream. The detected differences in the times of travel are then related to the average velocity of the flow in the conduit.

These devices have been installed in closed and open conduits in a range of sizes that includes rivers. While the transmitters and receivers must maintain good acoustic contact with the fluid, this type of device is considered to be nonintrusive.

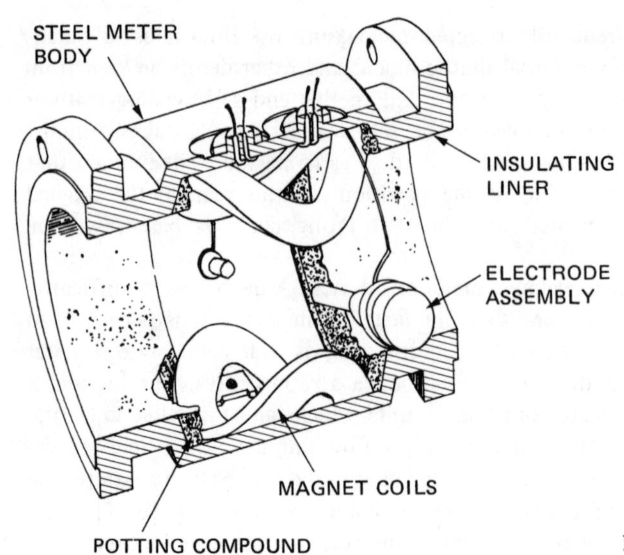

Figure 14 Magnetic flowmeter.

Recently, at NBS a new acoustic-type flowmeter was developed. This meter, which is also nonintrusive, uses a controlled source to produce sound in a closed pipe containing a flow of variable-temperature air. Via feedback circuitry, the controlled source maintains the sound wavelength at four to five times the internal pipe diameter. The sound is sensed by two microphones spaced four to five pipe diameters apart along the pipe and mounted flush with the inner pipe wall. In this configuration the difference in phase angles sensed by the two microphones allows calculation of (1) the fluid density via the sound speed–temperature relationship and (2) the average fluid velocity and thus the volumetric flow rate. The product of density and this flow rate gives the mass flow rate [8].

3.13 Vortex-Shedding Meters

A vortex-shedding meter is composed of a diametral strut or a series of chordal struts placed across the cross-sectional area of a pipe. Annular arrangements have

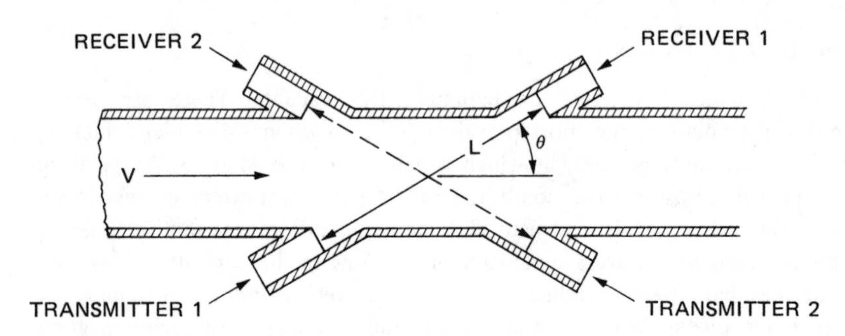

Figure 15 Ultrasonic flowmeter.

also been used. For Reynolds numbers above certain threshold values, the flow about these shapes produces vortices in the near wake regions behind the struts. These vortices are shed into the more distant wake regions at frequencies that are proportional to the flow velocity toward the strut. Any of a variety of schemes is used to sense the vortex-shedding frequency, from which the flow rate may be computed.

Sensors for vortex detection include pressure transducers mounted on one or more of the downstream surfaces of the strut, thermal sensors that detect transients produced by the vortex effect of altering the heat flux from some portion of the strut surface, and strain sensors that detect vortex-produced cross-stream oscillations of the strut itself. Since vortex shedding constitutes the basis of operation, it is important to use a strut shape that produces strong, frequent vortices. The complex nature of vortex shedding requires that the meters be calibrated to characterize their performance. Because vortex shedding takes place over a wide range of Reynolds numbers, these types of meters are specified to operate over flow-rate ranges of 100 to 1.

3.14 Laser Flowmeters

Because of the many advantages of laser-Doppler velocimetry (LDV), this velocity-measuring principle offers considerable promise as the basis of a flowmeter. By focusing two beams of the same laser light (one beam phase-shifted relative to the other) in a flowing fluid so that an interference fringe pattern is produced across the flow path in the volume of intersection of the two beams, one can determine the local fluid velocity from light scattered by particulates transported with the fluid velocity. This can be done nonintrusively through transparent sections of the flow conduit and, from measurements of the wavelength of the laser light and the relative angular displacement of the intersecting beams, the local fluid velocity can be determined without calibration. Thus, by orienting the fringe pattern in the sensing volume so as to measure the velocity component parallel to the pipe centerline, and by either traversing this so-called sensing volume over the cross-sectional area of the flow and integrating results to obtain volumetric flow rate or by determining the relationship between a particular local velocity, say the centerline value, and the average velocity, one can generate a laser-based flowmeter. Critical in all this is maintenance of the optical alignment and preservation of clean windows through which the laser light may pass into and out of the flow field.

3.15 Coriolis-Acceleration Flowmeters

Besides the gravitational and centrifugal acceleration meters described above, there are meters that use Coriolis acceleration as their basis of operation [9]. These devices are most readily (although, in principle, not solely) adapted to liquids because of their high density. To meter fluid in a closed conduit, a U-shaped tube containing the fluid flow to be metered is vibrated with the tips of the U fixed and the motion of the bend normal to the plane of the U. Within this tube the fluid flow rate produces a torquing motion about the axis of symmetry of the U. The amplitude of this torsional motion gives the fluid flow rate, and the period provides the fluid density. The meter is thus capable of mass flow-rate and density measurements in nonintrusive fashion, since no structural element enters the conduit.

3.16 Flow-Conditioning Devices

A significant factor that affects, to some degree, the performance of every type of fluid meter is the distribution of the flow into the meter. Because of the infinitude of piping arrangements that can precede and follow an installed flowmeter, a very wide range of flow anomalies can exist and can radically alter the meter's performance. Whether a meter is designed for a particular flow profile or is flow-calibrated in a laboratory, factors that cause the actual flow to differ from the expected flow will also cause a difference in meter operation. One way to ensure that the flow profile at the meter inlet is the one expected is to install a suitable flow conditioner at an appropriate distance upstream of the flowmeter.

It would seem that flow-profile anomalies could be minimized by designing the inlet piping to allow viscous and turbulent diffusion to produce the desired flow distribution. This desired distribution is produced by long (i.e., many pipe diameters), straight lengths of piping of a constant diameter that matches that of the meter. Similar piping downstream of the meter prevents any flow peculiarities from propagating upstream and affecting the meter. Several difficulties arise with this approach to establishing design flow conditions:

1. Different types of meters are affected differently by different types of flow anomalies.
2. Because of the nonlinear nature of the physical laws that govern the flow field within meters, it is difficult to predict the effects of flow anomalies on different types of meters.
3. Experiments to test all types of meters with all types of anomalous flow profiles would be extremely time-consuming and expensive.
4. Many meter users cannot spare the space needed to install long, straight lengths of upstream and downstream piping to achieve designed meter performance.

Because of such difficulties, a variety of conditioning devices have been devised to produce a desired flow condition. The device may be designed to produce the flow that is expected to occur in the particular pipe for the particular Reynolds number, after passing down a length equal to many pipe diameters. Or, it may be designed to produce the same specific flow distribution, regardless of the anomalous pattern that enters the flow conditioner. The latter is often regarded to be the easier of the two alternatives.

It is of paramount importance to satisfactorily condition the flow without producing a large static-pressure loss as the fluid flows through the conditioner. Such considerations led to the use of flow conditioners that included one or more creens, or metal plates having many holes supported across the flow cross section. ʼhe principle here was that the (generally turbulent) mixing occurring just lownstream of the screen would tend to eliminate abnormal distributions of fluid inetic energy.

Because the effects of fluid swirl are widely known to radically alter the expected erformance of many types of meters, these are of prime concern to users of fluid neters. Severe secondary motions or swirl can be introduced into pipe flow by elbows n the piping. Perhaps the elbow arrangement most likely to introduce the greatest wirl is two elbows installed close together and oriented so that they are "out of lane." This generally produces a vortex in the pipe flow, where the angular orticity vector lies along the pipe centerline. This type of swirl persists for many, nany diameters of pipe length before the effects of fluid viscosity can dissipate it. ven a single elbow can set up vortices in pipe flow. These secondary motions can be ictured as having vorticity vectors that are oppositely directed, lie parallel to but off he pipe centerline, and are symmetric with respect to the plane containing the enterlines of the pipes joined by the elbow. Again a considerable length of pipe is equired for viscous diffusion to dissipate such motions.

To remove such anomalous motions from pipe flow via flow conditioning, the rinciples of waveguiding are widely used. Thus, a wide variety of straightening ubes, baffles, and radial panels have been arranged to obstruct the swirling velocity omponents and (preferably with low pressure loss) reduce it to an acceptable level r eliminate it. In American metering practice, various arrangements for bundling umerous small tubes have been used [1, 2]. Efforts to produce an optimal flow onditioner (i.e., one that has maximum capability for removing anomalies and ninimal pressure loss) are continuing [10, 11].

. PROVING–PRIMARY AND SECONDARY TANDARDS

alibration requires flow measurement with maximum absolute accuracy and recision, usually with apparatus that collects the total flux of fluid during a neasured time interval. This flux is conventionally measured gravimetrically or olumetrically, with fluid density measurements usually required in either case. tandards of mass, length, time, and temperature are required for these flow-rate neasurements, but the uncertainty in flow-rate calibrations is greater than can be ttributed to uncertainty in the basic standards.

Uncertainties involved in flow-rate measurements result from the kinds of alibration facilities and procedures used. Uncertainties common to all facilities and rocedures result from the lack of ability to:

Set and maintain a steady flow rate.
Measure flow rate without error.
Separate imprecision in flow stability from imprecision of the calibration flow standard. A meter may be differentiated from a flow standard, as the latter is built

upon other, more basic standards directly, with minor corrections based on momentum and energy principles.
- Establish and determine the pertinent fluid properties.
- Completely remove spurious or systematic flow disturbances.

Uncertainties that are pertinent to specific facilities and procedures vary considerably; thus only generic examples are discussed here.

Depending on the fluid involved, various procedures and equipment are used to measure the bulk mass flow rate. For liquid flow, direct weighing and timing techniques are commonly used,

1. In a *static-weigh* mode, before and after the flow is collected, via a timed diverter valve in a weigh tank
2. In a *dynamic-weigh* mode, with the weighing and timing operations performed while the flow is being collected in a weigh tank

4.1 Liquid Flow: Static Weighing Procedure

Figure 16 is a schematic diagram of apparatus used for liquid flow-rate measurement. Rotation of the diverter valve allows fluid to be collected in the weigh tank and actuates a timer that measures the collection interval. This time interval, in conjunction with the collected mass as determined by weighings before and after collection of the liquid, gives the flow rate. The apparatus and procedure used should receive careful attention, to ensure accuracy of measurement.

It is desirable that the diverter cut through the liquid stream as rapidly as possible (in 30 ms or less) to help reduce the possibility of a significant diverter error. This is accomplished by rapid diverter travel through a thin liquid sheet formed by the nozzle slot. Generally, this liquid sheet has a length of 25 to 50

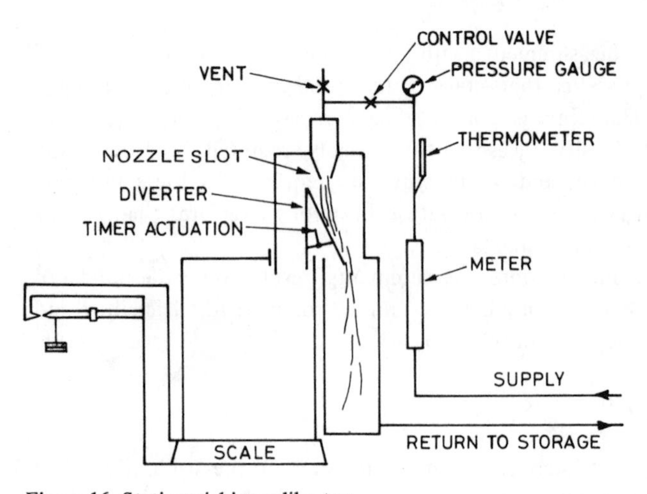

Figure 16 Static weighing calibrator.

times its thickness. Also, the pressure drop across the nozzle slot should be kept small to avoid excessive splashing and turbulence in (and on) the emitted jet and in the weigh tank.

Experience has shown that, for a well-designed system, the switching error for one start-stop cycle of the diverter may correspond to an error of 0 to 10 ms. This error is dependent upon the flow rate, the velocities of traverse (in each direction) of the diverter tip through the liquid sheet, and the exact location of timer actuation with respect to the liquid sheet emerging from the nozzle slot. Appendix D describes a convenient procedure for determining the switching error introduced by the diverter start-stop cycle.

Atmospheric buoyancy should be considered in a precision weighing procedure. Its magnitude may be in the range 0.1 to 0.2% of the actual weight, depending upon the densities of the ambient air and the liquid, and the weight of the tank.

In those calibration applications requiring a conversion between volumetric and gravimetric units, the measurement of liquid density is of extreme importance. Density error may range from 0.002 to 0.1%, depending on the particular technique employed and the accuracy of temperature measurements at the flowmeter during the density determination and in the collection tank.

Because loss by evaporation is a source of error, the volatility of the liquid should be considered whenever a liquid-flow-calibration system is vented to the atmosphere. However, for liquids of high vapor pressure such as gasoline, refrigerants, and liquefied gases, considerable refinement of the techniques described here is necessary to eliminate both loss by evaporation and vapor formation in the flowmeter and the meter discharge lines.

Typical systematic errors for current state-of-the-art liquid-flowmeter, static-weigh calibration systems are as follows:

Source	Error, %
Diverter switching	0.025
Time	0.005
Mass from weighing scale	0.025
Mass-to-volume conversion	0.03

These errors combine to form a possible overall systematic error in the range 0.04 to 0.08%, depending upon the procedure used to combine the individual errors, and on whether or not conversion from mass to volume units is required. The precision of such a system is best evaluated by performing repeated observations to obtain a measure of the closeness of repeated observations, or a value for the standard deviation. Generally, the value of the standard deviation for one observation is in the range 0.03 to 0.15%. This includes the imprecision (sensitivity and repeatability) of the flowmeter under test, which, of course, varies considerably among different types of meters.

4.2 Liquid Flow: Dynamic Weighing Procedure

The static-weighing method of flowmeter calibration is time-consuming and thus not well suited for those applications in which convenience and speed of operation are important. Therefore, dynamic weighing is utilized frequently. In this procedure, the time interval required to collect a preselected mass of liquid is measured; the weighing is performed while liquid is flowing into the weigh tank.

A diverter is not used in dynamic weighing procedures. Rather, under a condition of steady flow, the weigh-tank dump valve is closed. As the weight of liquid in the tank increases, it overcomes the resistance of a tare-value counterpoise mass on the end of the weigh beam, which then rises, starting the timer (Fig. 17). An additional preselected mass is added to the pan, depressing the weigh beam. When it rises again, the timer is stopped. This procedure requires acceleration of the weighing scale just prior to both the start and stop actuations of the timer.

Four important dynamic phenomena take place during the dynamic-weighing cycle. They are:

- A change in the impact force of the falling liquid between the initial and final weigh points
- Collection of an extra amount of liquid from the falling column by the rising level in the tank
- A change in the inertia of the scale and liquid in the weigh tank, with a resultant change in the time required to accelerate the weigh beam to the timer actuation point
- Forces due to waves in the tank.

Generally, the decrease in impact forces is equal and opposite to the additional weight of liquid collected from the vertically falling column. Thus, these two effects

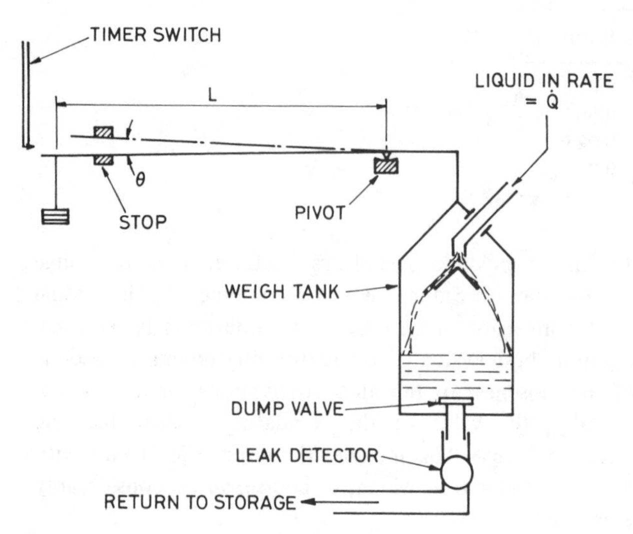

Figure 17 Dynamic-weighing calibrator.

cancel each other. The change in inertia between the initial and final weight points can affect the indicated flow rate by as much as 0.5% if this inertial error is not accounted for [12]. On smaller dynamic weighing systems, the inertia effect can be eliminated by using a substitution weighing technique. Surging or oscillations of liquid within the weigh tank may have a strong influence on the precision of the weighing procedure. Baffles and other arrangements within the tank can reduce, but not eliminate completely, this undesirable phenomenon, which is always most pronounced at higher rates of flow.

From this discussion, it may be seen that dynamic weighing procedures may introduce unknown systematic errors and weighing imprecision into the calibration procedure. Thus, a static check and calibration of the instrumentation is not sufficient to prove absolute accuracy, because the response time of the instrumentation under transient conditions is very important.

Another type of dynamic procedure for determining liquid flow rate is the so-called standpipe system. The principle of operation involves determining the liquid quantity that dynamically fills the left leg of the manometerlike apparatus sketched in Fig. 18. The dynamic elevations of the mercury column in the right leg of the device can then be related to the volume of liquid in the left leg. This type of device offers the obvious advantage of speed in determining liquid flow rate. It is important that the relationship between the liquid quantity in the left leg of the device and the mercury-column height in the right leg be very well characterized. A disadvantage of this system is that the collection of fluid in the standpipe increases the pressure in the test pipeline. This is transmitted to the output of the pump (or head tank) producing the desired flow rate and tends to decrease it. Thus the calibration of meters in such systems can be affected by the decreasing flow rate. Of course, the magnitude of this effect depends upon the geometric features of the standpipe, the amount of fluid being collected, and the characteristics of the pump and reservoir system.

At times, load cells are used on flowmeter calibration systems instead of mechanical weigh scales. These have the advantages of simplicity of installation and digital readout. Like scale systems, load-cell systems have to be properly calibrated at frequent intervals to assure accuracy.

4.3 Gas Flow: Static Procedure

Figure 19 shows the layout of an NBS facility that can be used to measure gas flow via a static procedure similar to that described above for liquids. Figure 20 shows the vacuum collection tank depicted in Fig. 19. The measurements are called "static," as both the collected gas volume and the density change over the collection interval and can be measured at presumably stationary conditions. Dynamics are involved only in the timed opening and closing of a diverter valve. The collection volume is found from the weight and density of water and/or gaseous nitrogen and argon fillings of the tank. (The density is determined from temperature and pressure measurements.) Results of multiple filings of the tank indicate that a volume uncertainty as small as 0.02% can be achieved. The

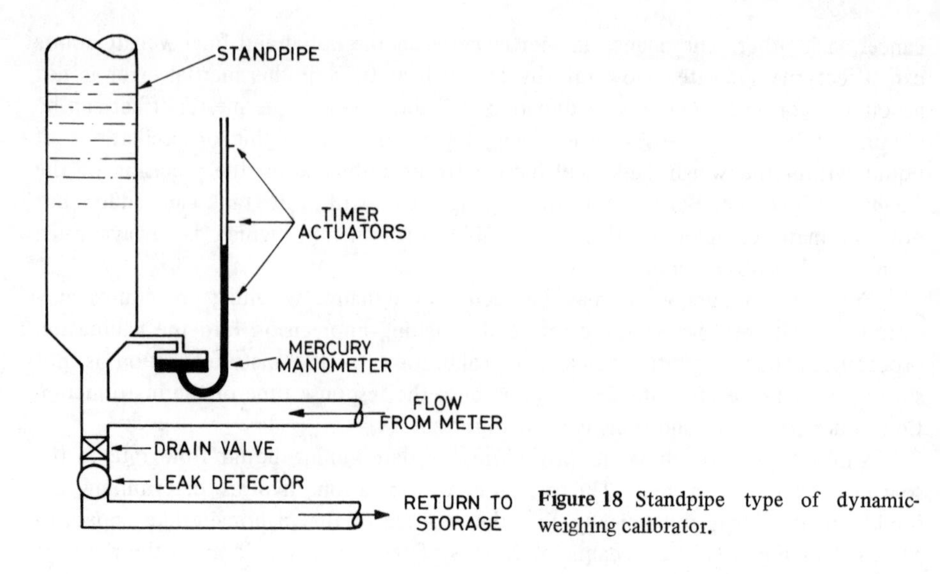

Figure 18 Standpipe type of dynamic-weighing calibrator.

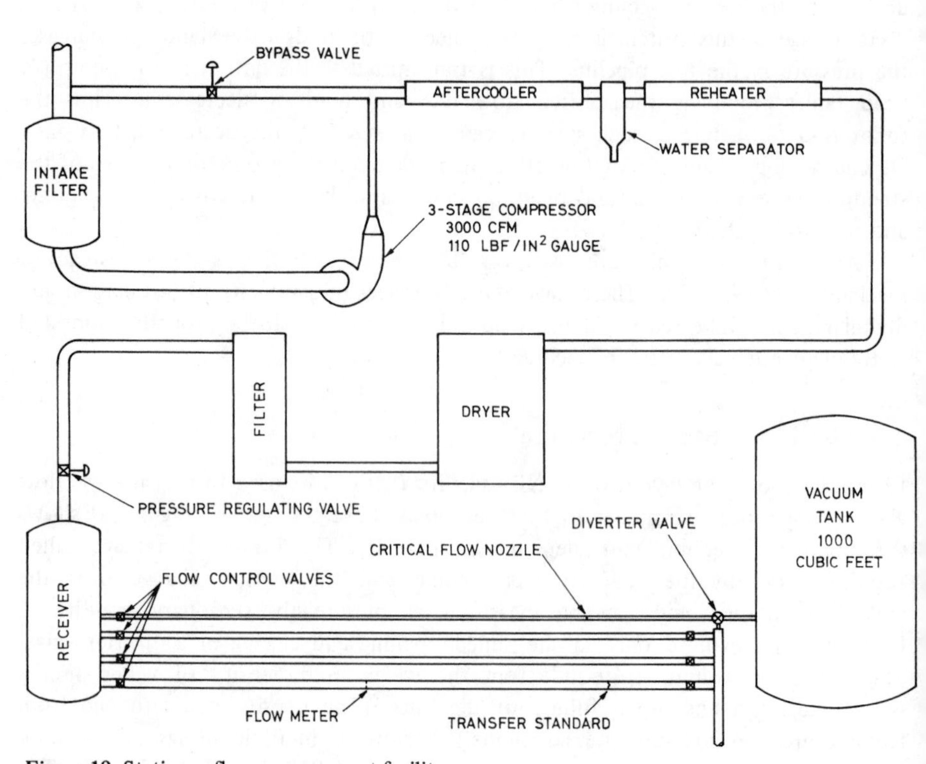

Figure 19 Static gas-flow measurement facility.

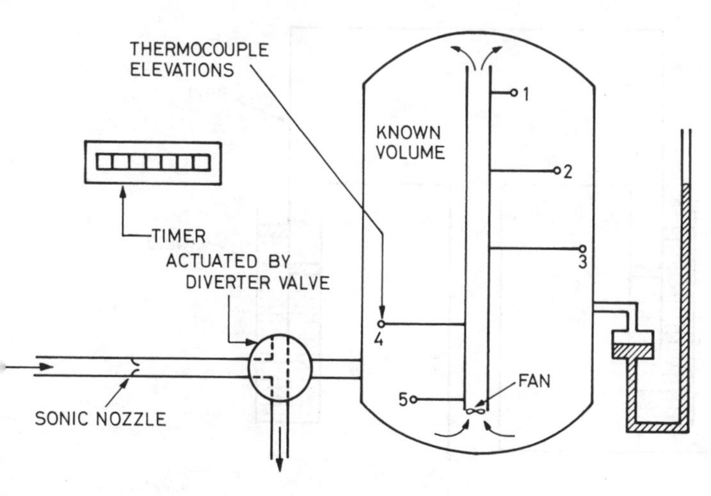

Figure 20 The collection tank of Fig. 19.

emperature of gas in the larger tank is measured with 25 copper-constantan hermocouples, all made from a single batch of wire, located over the interior of the ank. A sonic nozzle prevents the tank pressure (which is practically zero at the nitiation of collection) from affecting the back pressure on the meter under test. It s estimated that facilities of this type are capable of determining gas flow rate with n uncertainty of ±0.26% [12].

.4 Gas Flow: Dynamic Procedure

Gas flow can also be measured with bell-type provers, as illustrated in Fig. 21. imilarly, gas can be collected via the motion of mercury-sealed pistons in vertical lass tubes; the collection process is timed, after an initial acceleration period, by a imer actuated by two photosensors when their light beams are interrupted by the iston. The light beams that traverse the glass tube are placed a known vertical istance apart. Figure 21 shows that the vertical motion of the bell in its annular ath of sealing liquid is similarly timed by a switch that is actuated successively by wo arms mounted on the side of the bell. Internal-diameter measurements are used o derive volume per unit of distance traveled for both types of provers; additional ieasurements are required to account for displacement motion of the sealing liquid n the bell-type prover. The transfer of air between a standard volumetric measure nd a bell prover is sometimes used for its calibration, but dimensional calibration is ecommended for large provers.

Careful attention to numerous details is necessary to avoid measurement ifficulties. These arise from small rates of flow and small collected volumes, from he dynamics of the measurement process, and from the difficulty of making ieaningful gas temperature measurements in small gas flow systems. A nearly onstant laboratory temperature, both spatially and temporally, is used in conunction with sufficient piping upstream from the meter and calibrator to bring the

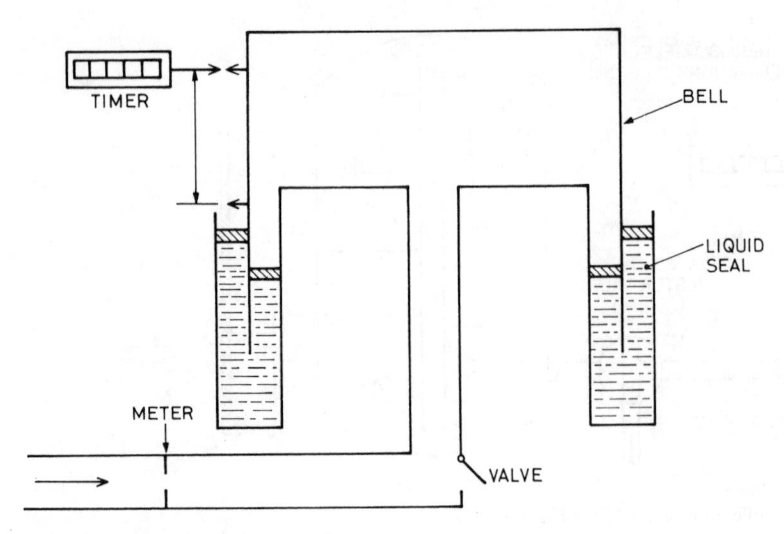

Figure 21 Bell-type dynamic gas-flow prover.

gas temperature to that of the calibration system, meter, and laboratory. This not only reduces temperature measurement problems but also prevents heat transfer in the meter and prover, a very important requirement. Thermal insulation and/or heat exchangers are also used and are recommended to ensure equal meter and gas temperature in difficult environments.

It is estimated that, with facilities such as these, gaseous flow rates may be determined to within 0.26% [12].

4.5 Ballistic Calibrators

A type of piston prover that is quite similar in operation to that described above for gas flow is the ballistic calibrator, which can be used on either gases or liquids. This system has several advantages, including simplicity, bidirectionality, compactness, and ruggedness, which permit it to be installed in the field on pipelines that transport large quantities of costly fluids such as petroleum. A typical arrangement is shown in Fig. 22.

The ballistic calibrator, sometimes referred to as the "ball prover" determines fluid flow rate via the following sequence:

1. The desired flow rate is produced by the flow control valve, downstream of the meter. The fluid flows from the reservoir, through the pump to the meter through valve *A*, and back to the reservoir. All other valves designated by letters in Fig. 22 are closed. The entire piping system is completely filled with the test fluid. All valving is electrically operated.
2. At the instant at which fluid collection is to begin, valves *B, D, E,* and *C* are opened, and valve *A* is closed. This initiates the motion of the spheroid (ball) from behind the detector switch, which, when actuated, starts a timer. The flow

from the meter now passes through valves *B* and *D* and into the region behind the ball. The fluid in front of the moving ball passes through valves *E* and *C* and back into the reservoir.

3. When the ball reaches the second detector switch, the timer is stopped, valves *D* and *E* close, and valve *A* opens so that the flow from the meter is returned to the reservoir.
4. Quickly, before the desired flow rate changes, valve *A* is closed as valves *G* and *F* are opened. This initiates ball motion in the opposite direction. When the ball passes the first detector switch, the timer is again started. Fluid from the meter passes through valves *B* and *G* and into the region behind the ball. The fluid in front of the ball passes through valves *F* and *C* and into the reservoir.
5. When the ball reaches the other detector switch, the timer is stopped, valves *G* and *E* close, and valve *A* opens to return the flow to the reservoir.

Each of the two "ball travels" displaces a volume of fluid between the detector switches that start and stop the timer. Thus, each pass enables one to calculate the volumetric flow rate. Any "diverterlike" error in the system (see App. D) that would cause a systematic error to be associated with the displaced volume in one direction and its corresponding time interval tends to be canceled when the two displaced volumes and the two respective time intervals are summed and used to determine the flow rate.

It is, of course, necessary to assume that there is no leakage past the ball during its travel. Also, one must accurately calculate or measure the volume displaced by the ball as it moves between the switches. By approximately sizing the ball-pipeline

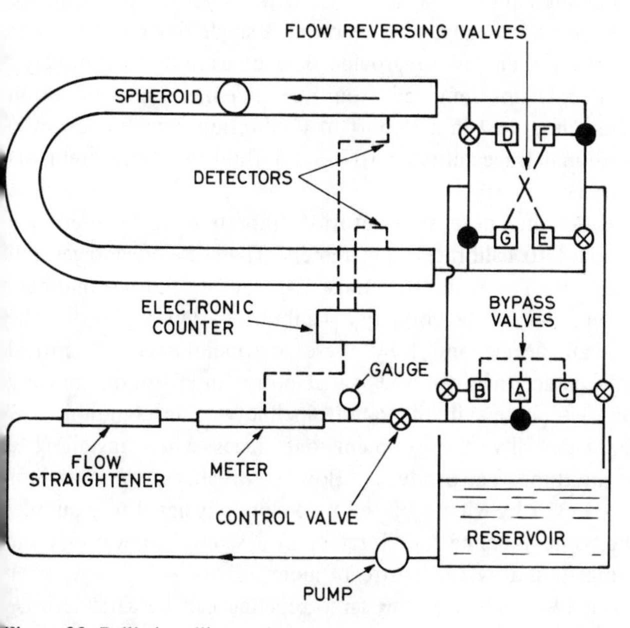

Figure 22 Ballistic calibrator.

diameter in relation to the size of the meter under test, the speed of the ball can be kept low. This adds plausibility to the assumption that the displaced volume can be measured by slowly filling or emptying the ball pipeline through the motion of the ball and measuring the fluid displaced. It allows the critical volume of the ball prover to be determined or checked in the same dynamic manner in which the device is actually used.

The integrity of the ball must also be maintained. If it becomes scratched so that leakage occurs, it should be replaced or repaired immediately. In addition, the switching must be carefully arranged and maintained so that any switching error is either canceled out or kept negligible.

4.6 NBS Facilities and Secondary Standards

The conventional calibration capabilities at NBS include facilities for calibrating flowmeters and wind-velocity sensors. Flowmeters of many types and line sizes can be calibrated in water, air, and hydrocarbon fluids. A summary of these facilities is given in Table 1. The usual calibration procedure at NBS consists in testing the device repeatedly at each of five flow rates spanning the normal range of usage of the meter, in the fluid actually metered by the instrument. If NBS is unable to calibrate a device in the fluid actually metered, sometimes a surrogate fluid is used. How well this works is dependent upon specific details, and each case is treated individually.

In the repeated testing of a meter, the fluid flow rate is determined five consecutive times by gravimetric or volumetric means, together with the meter response. The 25 points generated provide a data base with which the performance of the device can be predicted statistically, as a result of a single day's testing. The entire process is repeated on another day to provide data concerning "day-to-day" repeatability when the device is not removed from the pipeline. The calibration procedure and the results are described in a Report of Calibration, which allows the requestor to claim "conventional traceability" to NBS for fluid flow measurements made with the calibrated device.

A flow-metering device that has been satisfactorily calibrated can be used as a secondary, or transfer, standard to calibrate other meters. There are advantages and disadvantages to such a scheme. The main advantage is speed and correspondingly lower cost. The quality of such a "secondary" calibration depends upon the characteristics of the original device and how these are maintained. A critical disadvantage of such a scheme involves the satisfactory interruption of any "communication" between two meters in the same pipeline via the common flow rate. This is, of course, essentially the problem one faces when installing a flowmeter downstream of anything that produces "flow disturbances," or variations of the inlet flow profile expected by virtue of the pipe flow physics. For example, two turbine meters in the same pipeline can interact in a significant way via the swirl imparted to the flowing stream by the upstream meter.

Communication between two meters in the same pipeline can be satisfactorily interrupted in several ways. For example, if long lengths of straight, constant-diameter piping separate the two meters, then viscous diffusion of anomalous effects

Table 1 NBS flow-calibration facilities and performance characteristics

Facility	Capabilities	Flow determination system, features
	Gas and Liquid Metering in Closed Conduits	
Low air flow	1.5 m^3/min (50 scfm) max.; 3440 kPg (500 psig) max.; ambient temperature (20°C) ± 0.25% uncertainty	Bell-type provers, mercury-sealed piston devices, critical nozzles
High air flow	83 m^3/min (3000 scfm) max.; 861 kPg (125 psig) max.; ambient temperature (20°C) ± 0.25% uncertainty	Constant-volume collection tank, critical nozzles
Low water flow*	2.5 kg/s (40 gpm) max.; 345 kPg (50 psig) max.; ambient temperature (20°C) ± 0.13% uncertainty	Dynamic weighing
High water flow	6.25 kg/s (10,000 gpm) max.; 517 kPg (75 psig) max.; ambient temperature (20°C) ± 0.13% uncertainty	Static weighing
Low-liquid-hydrocarbon flow	10 kg/s (200 gpm) max.; 207 kPg (30 psig) max.; ambient temperature (20°C) ± 0.13% uncertainty	Dynamic weighing
High liquid-hydrocarbon flow	100 kg/s (2000 gpm) max.; 348 kPg (50 psig) max.; ambient temperature (20°C) ± 0.13% uncertainty	Static weighing
	Wind Sensors	
Wind tunnel	0.9- × 0.9-m test section; 4.5 cm/s to 9 m/s ± 1.0% uncertainty	Velocities measured using laser velocimetry, hot-wire anemometry
Wind tunnel (dual test section)	1.5- × 2.1-m test section; 46 m/s max. ± 0.3% uncertainty	Low stream turbulence, adjustable pressure gradient; velocities measured using hot-wire anemometry and pitot tube
	1.2- × 1.5-m test section; 82 m/s max. ± 0.3% uncertainty	Low stream turbulence; velocities measured using hot-wire anemometry and pitot tube
Wind tunnel	1.4- × 1.4-m test section; 27 m/s max.	Can be operated in steady or fluctuating mode; gusts range from 0.1 to 25 Hz; amplitudes range to 50% of mean flow

*Antifreeze additives may be used.

can occur, resulting in the "proper" inlet flow entering the downstream meter. The specific length of piping required depends critically upon

1. How susceptible the downstream meter is to particular disturbances in the inlet flow

2. The nature of the disturbances imparted to the flow by the upstream meter
3. The nature of the flow in the pipe connecting the two meters

It is generally found, however, that testing laboratories and laboratories operated by meter manufacturers do not contain enough space to accommodate the "long, straight lengths" required by the installation specifications for the meter in question. Thus, one confronts the problem of artificially "conditioning" the flow so that either (1) the effect of long, straight lengths of piping on the pipeflow is somehow duplicated, or (2) the anomalous flow effects are reduced (or eliminated) without incurring excessive fluid pressure loss, so that the flow entering the downstream meter does not contain disturbances or anomalies that would impair meter performance. The latter is the purpose of flow conditioners. Then, the downstream unit, operating in conjunction with its upstream conditioner, becomes a reliable product for sale, or a satisfactory secondary, or transfer, flow standard.

Flow conditioners have in the past and will in the future play a critical role in fluid metering. They are available in a wide variety of shapes and sizes, as the result of the very wide range of flow disturbances that are known to impair meter performance and the very diverse range of intuition and imagination that have produced them [10, 11].

5. TRACEABILITY TO NATIONAL FLOW STANDARDS–MEASUREMENT ASSURANCE PROGRAMS FOR FLOW

Conventional practice for establishing the traceability of a flow measurement includes many activities. For example, the owner of a newly purchased and properly installed flowmeter might define traceability through the flowmeter manufacturer's quoted performance for the device. The manufacturer, in turn, might define traceability as the spot-checking of finished products against a secondary standard meter that is installed in the pipeline of the calibration facility. Calibration against an in-line standard meter has the significant benefits mentioned earlier. Of course, from time to time the manufacturer should check the performance of the secondary standard against a "master" standard. The master standard probably is on a shelf most of the time and thus is not degraded through normal use, as the in-line standard may be. Again, from time to time the manufacturer should check the performance of the master meter.

The master meter can be calibrated at NBS, or it might be checked in house if facilities for gravimetric or volumetric calibration as described above are available. Such a calibration can, however, be a rather slow process that may impair the normal production routine. The traceability that the manufacturer might cite for its gravimetric or volumetric test facility would probably be a calibration of its system by the state office of weights and measures. For a gravimetric system this generally means that state weights were brought to the producer's laboratory and used to calibrate its scale system. To check its timing system, the manufacturer would either calibrate its timers against a timing standard or use the WWV time signal from NBS.

Further, the manufacturer would check its diverter system (or inertial corrections if the system were dynamic) and pertinent fluid properties in some appropriate manner. Since the weights and timing standards would all be solidly traceable to NBS, the new meter owner would be assured of the traceability of flow measurements to NBS.

However, this traceability chain can have some very weak links, as indicated by the following questions:

1. What statistics justify the manufacturer's reliance on spot checks of its products?
2. For what period of time should one rely on manufacturer's specifications for the performance of a new meter?
3. Are the flow and fluid parameters properly bracketed by the spot-checking calibration procedure?
4. Is the meter purchaser using the secondary element in the manner in which it was used in the spot-check calibration?
5. Is the fluid flow profile at the manufacturer's lab the same as that at the user's installation?

The link between the user and the manufacturer may not be the only weak one. For example, the manufacturer might ask:

1. How do I justify my spot-checking arrangement?
2. When I check two flowmeters in the same pipeline, is there any interaction that affects either meter?
3. When I check my flow determination system with a static (or nearly static) process, how well do I account for the dynamics involved in its actual use?
4. What is the flow profile of the fluid entering my meters in the calibration facility?
5. Is the calibration I receive from NBS for my master meter indicative of the performance of this device when I use it as I do in my lab?

Some questions can be asked at NBS as well:

1. When a manufacturer sends a master meter for calibration, how much of the manufacturer's auxiliary pipework and secondary element should be included to get the calibration that is wanted?
2. When NBS personnel operate and read the secondary element, is the technique that used by the operators in the manufacturer's laboratory?
3. Is the flow profile entering the meter at NBS indicative of that prevailing in the manufacturer's lab?
4. How good is the NBS Flow Determination System?
5. How often should the NBS Flow Determination System be checked?

Thus it seems apparent that the traceability chain can have some weak links, if not some broken sections. Several of these questions could perhaps be avoided if users sent their units directly to NBS for calibration. However, this might not solve all the problems.

So the question remains: What should be done, and by whom, to establish traceability? To answer this question, it is necessary to decide on what is meant by traceability and what constitutes traceability for flow measurements. Following this, it should be decided how flow traceability should be established and maintained.

The following definitions of traceability have been proposed [13]:

1. Traceability is the ability to demonstrate conclusively that a particular instrument or artifact standard has either been calibrated by NBS at accepted intervals, or has been calibrated against another standard in a chain or echelon of calibrations, ultimately leading to a calibration performed by NBS.
2. Traceability to designated standards (national, international, or well characterized reference standards based upon fundamental concepts of nature) is an attribute of some measurements. Measurements have traceability to the designated standards if and only if scientifically rigorous evidence is produced on a continuing basis to show that the measurement process is producing measurement results (data) for which the total measurement uncertainty relative to national or other designated standards is quantified.
3. Traceability means the ability to relate individual measurement results to national standards or nationally accepted measurement systems through an unbroken chain of comparisons.
4. Traceability implies a capability to quantitatively express the results of a measurement in terms of units that are realized on the basis of accepted reference standards, usually national standards.

Although there are similarities among these four definitions, the differences among them are more interesting. The salient difference between the first two is the involvement with calibrations of instruments. The second definition doesn't include the word "instrument" or "calibration," whereas the first definition is intrinsically based upon instruments and their calibrations. The second definition stresses the *results of measurement processes.*

This contrast–between instruments and calibrations on the one hand and the results of measurement processes on the other–is considered by this author to be the basis for distinguishing between "static" and "dynamic" traceability.

5.1 Static Traceability

By static traceability is meant the collection of activities that metrologists must perform to quantify the performance of the various components of their measurement facilities. For example, in the United States, in the case of flow measurement this might mean checking a weighing facility (or volumetric prover) with weights or volumetric test measures from a state office of weights and measures (OWM) that is itself traceable to NBS. Also, it would mean checking the laboratory's timing mechanisms against NBS traceable timing standards. Furthermore, if liquid density were required to convert mass measurements to volume, then static traceability would necessitate establishing the relationship between density-measuring instruments and NBS standards.

However, if the diverter mechanism in the metrologist's flow facility were faulty, if a peculiar flow profile entered a flowmeter under test, if a significant human error recurred during flowmeter tests, or if the processing of the raw data to final results were wrong, then evidently static traceability would not be sufficient to establish the resulting flow measurements in the laboratory. Something more would be required–namely, quantitative proof that the entire flow measurement process in the laboratory continuously produces results for which the uncertainty relative to specified standards is quantified. It is this extra sufficiency condition that constitutes dynamic traceability.

5.2 Dynamic Traceability

Properly established dynamic traceability is intended to assess, in a realistic manner, all the pertinent factors influencing the flow measurement results produced in a facility. These factors include:

1. The particular scheme or instruments arranged for determining mass or volumetric flow rate
2. The environmental conditions prevailing during the test period
3. Operator idiosyncracies, if present
4. The calculation procedures used for determining final test results from raw data

To establish the dynamic traceability of flow measurements to national standards (and between the standards of nations on an international basis), measurement assurance programs (MAPs) for fluid flow are required [14 to 16].

5.3 Measurement Assurance Programs

The purpose of a flow MAP is to:

1. Quantitatively establish the total uncertainty of flow measurement processes
2. Provide proof that flow measurements are as good as specified
3. Evaluate the entire flow measurement system in question (operators, environment, methods, instruments, etc.)

The advantages of a flow MAP are that it:

1. Provides, clearly and quantitatively, the limitations of a particular method of measurement
2. Gives a clear description of the factors affecting the uncertainty of the measurement
3. Provides guidance for obtaining satisfactory results where small uncertainties are required

4. Provides the rationale for simplifying existing procedures
5. Provides a means for cyclically monitoring the performance of measurement processes

A flow MAP can be arranged within individual countries in which the national laboratory is a participant and, usually, the organizer of the activity. A MAP is composed of two essential elements. The first is an artifact (defined as "any object made by man, especially with a view to subsequent use"); the second is an algorithm (defined as "a particular rule or procedure devised for solving a specific problem"). Artifacts consist of a flow meter or meters, adjacent pipework, and the auxiliary instrumentation that is required to establish the dynamic traceability of flow measurements. Algorithms consist of the testing procedure devised to produce the desired data, the analyses of these data, and the scheduling through which the measurements traced are continually checked. The process is sketched in Fig. 23 alongside a conventional calibration procedure.

To initiate the flow MAP, the originating laboratory (possibly the national standards organization) generally selects a simple, widely used, reliable type of transfer standard, perhaps a flowmeter that is capable of performance adequate to the intended task, i.e., making good, repeatable measurements for a particular fluid, flow-rate range, and pipeline size. Confidence can be added to the test results by

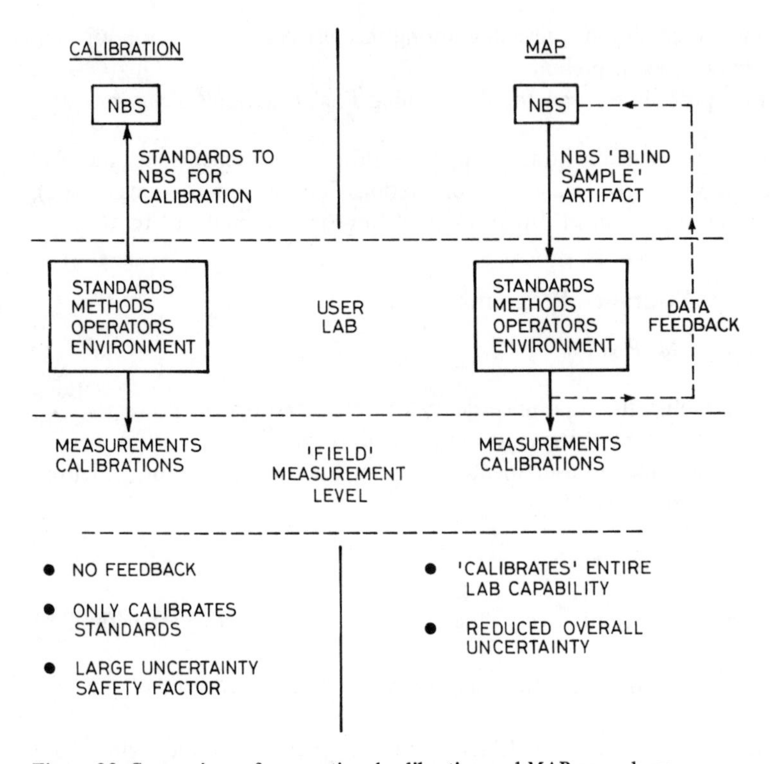

Figure 23 Comparison of conventional calibration and MAP procedures.

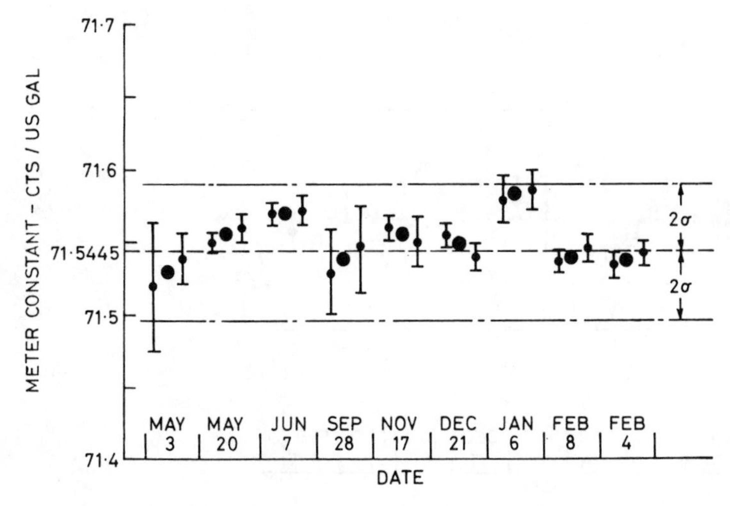

Figure 24 Typical turbine-meter control chart for meter factor.

using two flowmeters connected in series in the flow facility. To ensure that the particular flowmeters are performing as expected, it is recommended that the performance of these meters be cyclically checked at a particular laboratory, such as the originating laboratory. Control-chart records should be maintained and analyzed to quantify any changes observed in performance. A typical chart for an NBS turbine meter is shown in Fig. 24.

Tandem meter arrangements, composed of different types of meters or meters of the same type but of different sizes connected with appropriate fittings, should also be capable of rendering realistic data to assure flow measurements. The use of two meters in the same pipeline, so that the same flow passes through both, produces data in which considerable credibility can be placed [17]. This tandem meter-testing configuration has the obvious advantage of providing redundancy for the data generated. The tandem arrangement also provides the opportunity to quickly and efficiently monitor the relative responses of the two meters to the flow through both of them. For example, if the two meters were identical turbine meters, their relative responses to the flow could be the ratio of their respective frequencies. In such a case, when the ratio meter monitoring the two signals continually displays a value that lies within an acceptably small tolerance of an expected value, increased confidence can be placed in the data being taken. This expected value (which should approximate unity for identical meters) should be obtained from previous testing conducted in the originating laboratory.

Experience at NBS has shown that two flowmeters that are identical in make, manufacturer, and model can perform quite differently [17, 18]. A typical control chart for the ratio performance of a particular pair of NBS turbine meters is shown in Fig. 25. Deviation of the monitored ratio of frequencies from the expected value by more than the preset tolerance would be interpreted as indicating that one of the two meters had been changed somehow—an event that would obviously impair

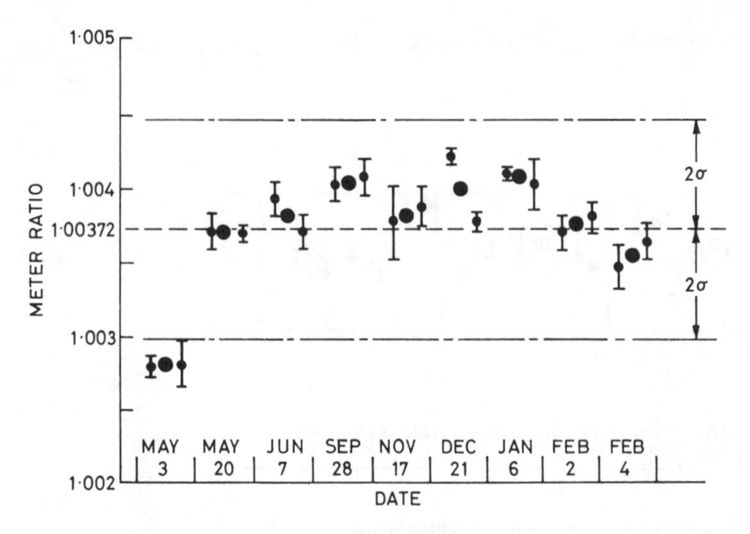

Figure 25 Typical turbine-meter control chart for ratio.

the credibility of the test results. The testing should then not proceed until the ratio could be made to conform to the expected value within the specified tolerance.

It is possible that identical malfunctions could occur simultaneously to both meters in such a way as to shift their performance by the same amount and in the same direction; however, the probability of this event is assumed to be negligibly small.

Figure 26 shows a chart of flowmeter performance. For a flow rate $\dot{Q}$ the respective flowmeter factors are K_1 and K_2. Figure 27 shows the ratio performance R_{12} for these two meters at the flow rate $\dot{Q}$ and the tolerance $\pm\delta$ for the expected ratio value $R_{12,\dot{Q}}$. When, as shown in Fig. 28, the ratio display indicates a value R such that

$$R_{12,\dot{Q}} - \delta \leqslant R \leqslant R_{12,\dot{Q}} + \delta \tag{6}$$

The testing of the two meters continues and the meter factors are calculated via the

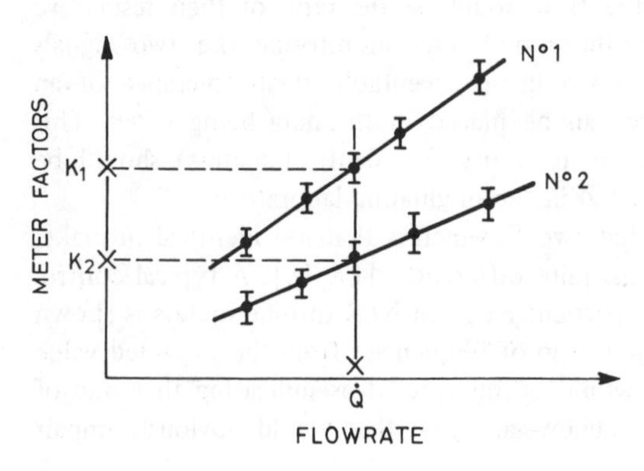

Figure 26 Normal meter performance.

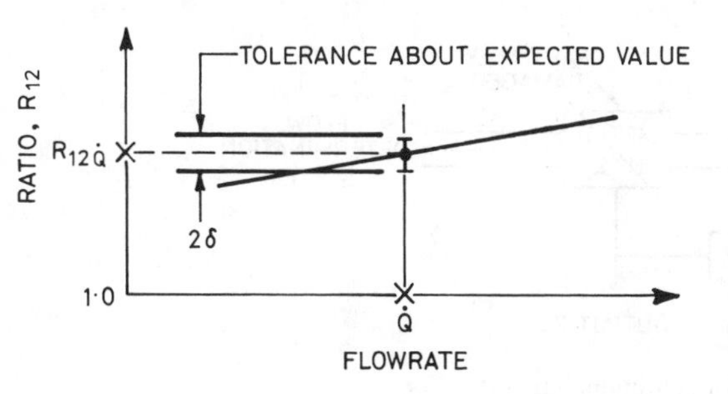

Figure 27 Expected ratio for flow rate $\dot{Q}$.

flow determination system of the laboratory under test. These meter factors then contribute to the data base for this laboratory.

Figure 29 illustrates circumstances in which testing should not be continued. Here the ratio display indicates a value R such that

$$R \gg R_{12,\dot{Q}} \tag{7}$$

and exceeding the specified tolerance. The reason for the unacceptably high value of R is illustrated as a "partially damaged" meter 2. A particle of dirt may have gotten into the bearing of this turbine meter, so that the rotational frequency of its impeller was decreased. This would decrease the denominator of the ratio R, producing the observed high value. In this case, a decision to continue the test on the basis that the upstream meter 1 is probably functioning properly is detrimental to (1) the performance of the laboratory under test, and (2) confidence in the data base that will be used to establish dynamic traceability among the participating laboratories.

There is an additional advantage to testing flowmeters in tandem. At each flow rate, one obtains repetitive and simultaneous determinations of the performance of the two meters–i.e., the respective meter constants or flow factors. Correlation of these results enables one to decompose the total variation observed for each meter into components [19]. These component variations are, for each meter, those attributable (1) to the meters themselves, and (2) to the flow facility in which the meters are tested. Figures 30 through 33 illustrate how the total variation observed in the meter-factor results for each meter can be decomposed into these two components.

In Fig. 30 are shown the actual meter performance and expected value of meter factor for the two meters. Figure 31 shows the piping configuration under test; the

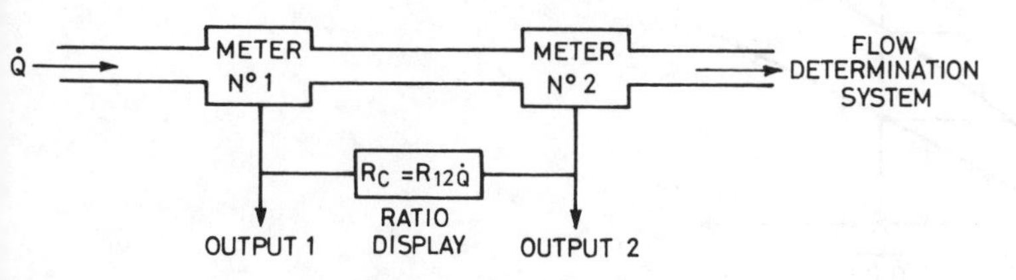

Figure 28 Satisfactory condition for conducting the test.

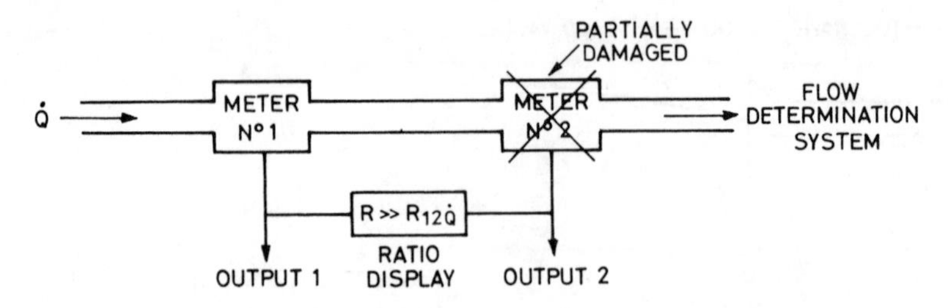

Figure 29 Unsatisfactory conditions for continuing the test.

flow rate $\dot{Q}(t)$ is taken to be a function of time. The time-averaged value of this unsteady flow rate is $\overline{\dot{Q}(t)}$, as shown in Fig. 30. Figure 32 shows a possible flow-rate $\dot{Q}(t)$ as varying sinusoidally with mean value $\overline{\dot{Q}(t)}$. The figure also shows the time interval over which the laboratory under test determines flow rate. For these conditions, it is apparent that this laboratory's meter-factor results will be markedly scattered, in spite of the fact that the meters are very repeatable, high-resolution devices, as shown in Fig. 30. This variation is displayed in Fig. 33, which shows the scattering of the meter-factor data about mean values of K_1 and K_2. The standard deviations $\sigma_{1,T}$ and $\sigma_{2,T}$ represent the quantified scattering about the mean values.

The correlation coefficient r_{12} is computed for these simultaneously determined meter factors at a particular flow rate such as $\overline{\dot{Q}(t)}$. Under the assumption that the sources of variation are statistically independent, one can interpret the square of this correlation coefficient as that fraction of the total observed meter-factor variation that can be associated with both meters. The two meters are assumed statistically independent from each other except for their physical connection via piping and the (unsteady) flow passing through them. These two connections are taken together and referred to as the *facility component* of the total meter-factor variation observed for each meter; that is,

$$\begin{aligned} \sigma_{1,T}^2 &= \sigma_{1,M}^2 + \sigma_F^2 \\ \sigma_{2,T}^2 &= \sigma_{2,M}^2 + \sigma_F^2 \end{aligned} \tag{8}$$

where $\sigma_{1,T}$ and $\sigma_{2,T}$ are, as before, the standard deviations for the total

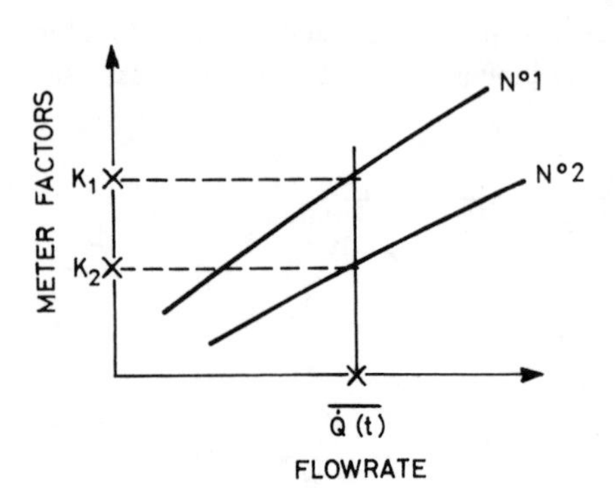

Figure 30 Normal meter performance.

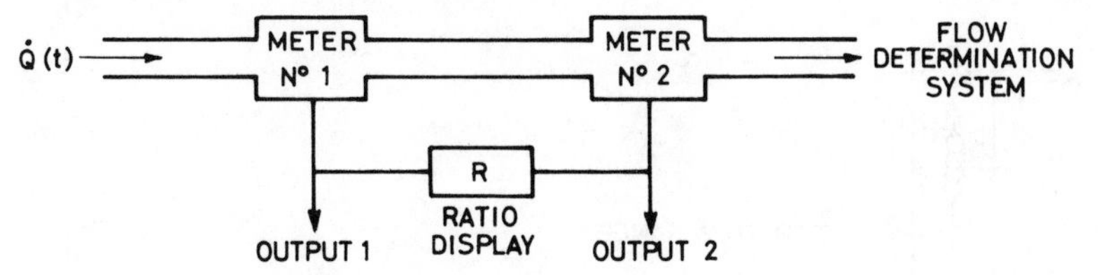

Figure 31 Flow-test configuration.

meter-factor variation. The quantities $\sigma_{1,M}^2$ and $\sigma_{2,M}^2$ are, respectively, the portions of the total variation in meter factor that are due to the meters themselves. The quantity σ_F^2 is that portion of the total observed meter-factor variation that is attributable to the facility. According to the assumptions described above,

$$r_{12} = 1 \tag{9}$$

and

$$r_{12}^2 = \frac{\sigma_F^2}{\sigma_{1,T}^2} = \frac{\sigma_F^2}{\sigma_{2,T}^2} = 1 \tag{10}$$

Therefore, it is concluded that

$$\sigma_{1,M} = \sigma_{2,M} = 0 \tag{11}$$

that the variation observed in the meter-factor results for each meter is due entirely to the facility, and that none of the total variation can be attributed to the meters (see also App. H).

In real situations, the circumstances are not so simple. However, this analysis is one more way in which increased confidence can be placed in the resulting data base. When, through the course of testing the same set of meters in the prescribed manner in a series of laboratories, the same (or similar) results are continually obtained for meter variation, all participants are more confident of the implications of the tests—that is, the traceability sought.

It should be noted that the temporal variation of $\dot{Q}(t)$ assumed via Fig. 32 should not necessarily violate the ratio criteria of Fig. 27; this is because of the nature of the meter characteristics in Fig. 30. In fact, for the situation indicated by Figs. 30 through 33, the laboratory under test appears to obtain the "correct" values for K_1 and K_2 in spite of the undesirable fact that the flow cannot be

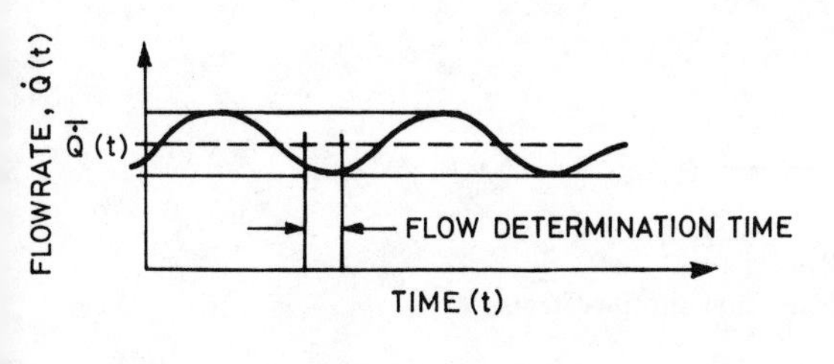

Figure 32 Assumed flow-rate variation.

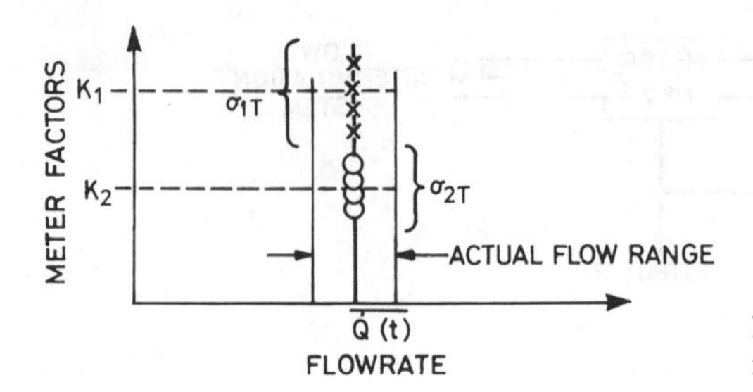

Figure 33 Meter-factor results for single flow rate.

steadied. However, for different meter characteristics, or for different flow-rate variations, a wide range of alternative results are feasible. An important example is the situation wherein the inability to produce a steady flow results in both excessive variations in, and systematically wrong values for the meter factor.

5.4 The Role of Flow Conditioning in the Artifact Package

The flow velocity profile is widely known to have a significant effect on the performance of flowmeters. For this reason, the incorporation of flow conditioners into the MAP algorithm has been found to be very effective [17, 18]. Furthermore, when present development efforts have finally evolved optimal flow conditioners, it is expected that entirely new arrangements of devices and meters will become available for inclusion in flow MAPs.*

The role of flow-conditioning devices in flow MAPs will be to aid in determining whether participating laboratories have anomalous flow profiles in their facilities. Figures 34 through 36 indicate how flow conditioners could quantitatively

*At NBS, computer modeling efforts have produced preliminary flow-conditioner configurations that appear to have very good conditioning capabilities. Laser-based experiments are planned to validate these results. Dr. F. Kinghorn at the National Engineering Laboratory (NEL) in the United Kingdom, and Dr. M. Sens at Société Nationale des Gaz du Sud-Ouest in France are actively developing other flow-conditioning ideas.

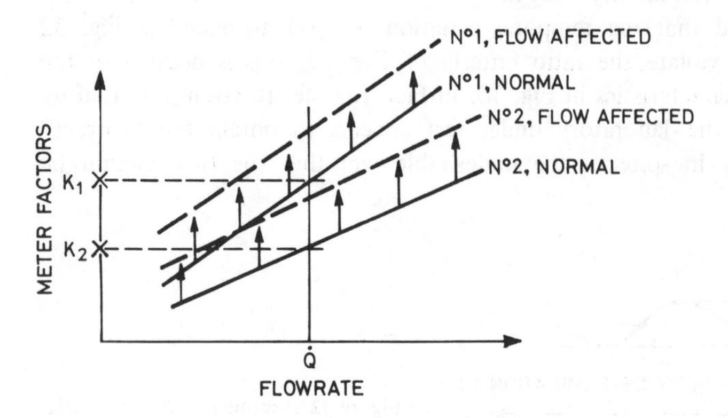

Figure 34 Assumed meter performance and flow-profile effects.

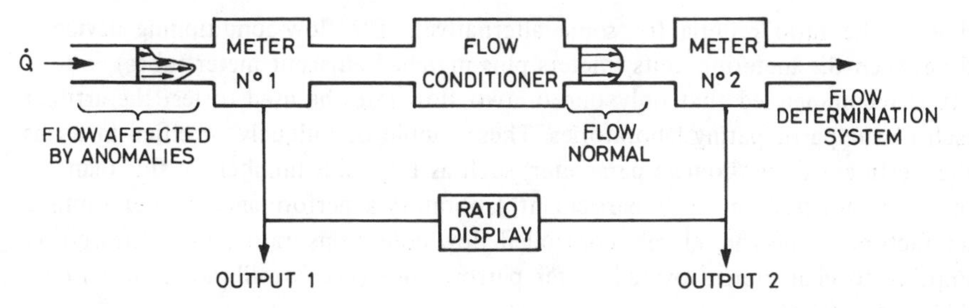

Figure 35 Flow-test configuration.

assess flow-profile problems in laboratories. For the assumed meter performance characteristics shown in Fig. 34, the meter factor for a flow affected with "anomalies" is increased from the value measured in "normal" profiles. When the proper flow-conditioning element is bolted between* the tandemly piped meters, only the upstream meter (meter 1) is affected by flow anomalies (Fig. 35). The downstream meter, being screened from these anomalies (and any introduced into the pipe flow by the upstream meter), exhibits normal performance. The results of the tests of Fig. 35 are shown in Fig. 36. The unexpected deviation represents, quantitatively, the effect of the flow anomaly on meter 1.

5.5 Test Program

The test program must be devised so that (1) high confidence can be placed upon the artifact package, and (2) the data base produced is adequate to the task of clearly evaluating the significant components of the system in question. To establish and maintain high confidence in the artifact package, an adequate testing program is carried out in the originating laboratory, with close monitoring of meter performance. The results, in control-chart form, are updated in timely fashion and disseminated to the participants† in the MAP. The tandem-meter configuration is

*Suggestion attributed to Dr. E. A. Spencer of the Flow Measurement Division, National Engineering Laboratory, United Kingdom.

†The participants in the program have to agree on the level of anonymity they collectively wish to maintain through the operation of the MAP.

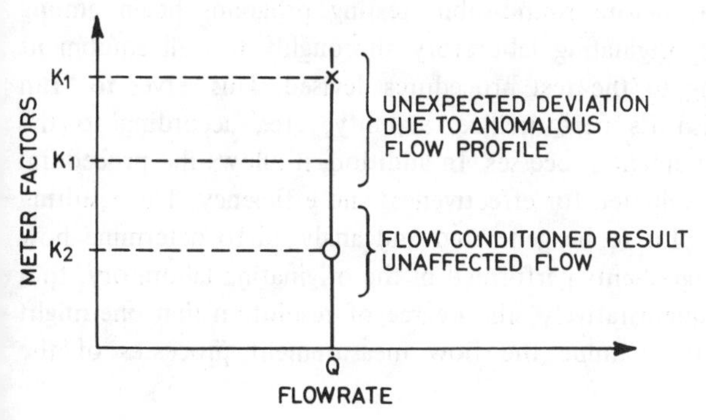

Figure 36 Meter-factor results for single flow rate.

used with the ratio criteria (or some alternative). The flow-conditioning device is used between the metering units (meters plus matched adjacent meter tubes).

It is recommended that only one or two flow rates be used to test the artifact in each of the participating laboratories. These should be uniquely specified in terms of the pertinent dimensionless parameter, such as Reynolds number. In this manner, attention is focused on each participant laboratory's performance in determining meter factors at specific, closely controlled flow conditions, rather than diffused by attempting to characterize overall meter performance (i.e., a calibration curve) over a wide range of flows.

Each of the two flow meters is tested in the upstream and downstream positions. This generates two statistically independent sets of data for each participant in the MAP, for each flow rate specified for testing. More than one flow rate is recommended to allow a laboratory facility with, say, a low-flow-rate capability, to participate at least partially. A sketch of the piping configurations is shown in Fig. 37. In the upper portion of the figure, meter 2 occupies the upstream position, and meter 1 is downstream. With this arrangement, all flows are tested, and sufficient repeat testing is performed to evaluate the "switch-off, switch-on" repeatability and/or the "day-to-day" or "running" repeatability. When testing in this configuration is completed, the tests are duplicated for the arrangement shown in the lower portion of Fig. 37.

5.6 Data Analysis

The method of analysis used to assess the variance in flow measurement results among participating laboratories was formulated by W. J. Youden [20] and has been used at NBS for years in interlaboratory testing programs. In this procedure, the statistically independent results for each flow condition for the respective meters in the same position are plotted, one meter factor on the ordinate and the other on the abscissa (see Fig. 38). Thus, each laboratory is represented by a single point that is labeled with a coded number, letter, or symbol to maintain the anonymity of the laboratories. Usually the originating laboratory is custodian of such information and of the data used to perform this analysis. Plots such as that shown in Fig. 38 have been referred to as *Youden plots*, in honor of Youden's significant contributions to NBS efforts to develop round-robin testing programs [20, 21].

It is recommended that, before round-robin testing programs begin among participating laboratories, the originating laboratory thoroughly test all equipment and instrumentation according to the test procedures devised. This serves to "run in" the artifact and establish its performance, stability, etc., according to the originating laboratory's measurement processes. In addition, it allows the procedural steps in the algorithm to be evaluated for effectiveness and efficiency. The resulting data base enables a Youden plot to be generated and analyzed to determine how well the whole set of MAP ingredients performed in the originating laboratory. In a sense, these results indicate, quantitatively, the degree of resolution that one might expect in using this MAP to examine the flow measurement processes of the participating laboratories.

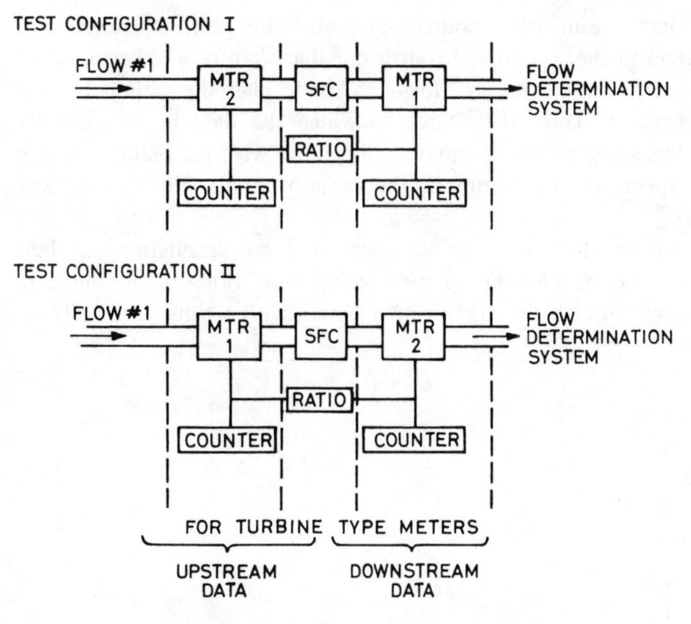

Figure 37 Schematic test procedure for a single flow rate.

The Youden analysis is performed by drawing vertical and horizontal lines, respectively, through the medians of the abscissa and ordinate data (see Fig. 38). The intersecting median lines then divide the data into four quadrants. In cartesian notation, data lying in the first of these quadrants is, to some extent, systematically inaccurate, since each point is "high" relative to the best available estimate of the true values of the turbine-meter constants, namely, the coordinates of the inter-

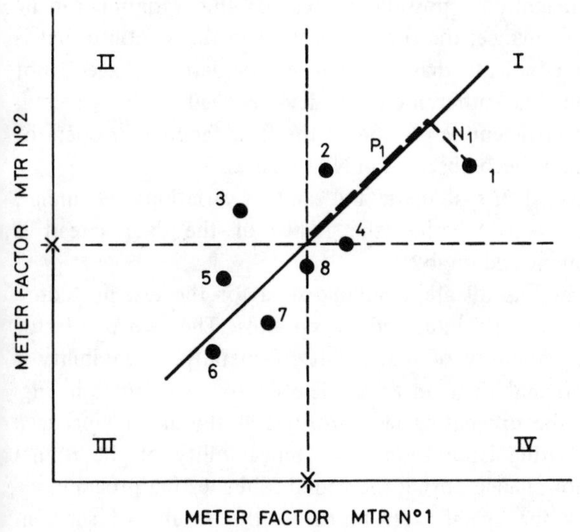

Figure 38 Graphic representation of the Youden analysis.

section of the medians. Similarly, points lying in the third quadrant are systematically low. Data in the second and fourth quadrants are termed *inconsistent* or *random*, since points in these areas are "low-high" and "high-low" relative to the intersection of the medians. Thus, the degree to which the data lie in a circular pattern about this intersection or are distributed elliptically with the major axis at a slope of +1 or −1 quantifies the nature of the variation in the flow-facility and turbine-meter systems.

The total variation in the data can be categorized by calculating standard deviations based upon the parallel and perpendicular projections of all the data points onto a line drawn through the median intersection with a slope of +1. These are

Systematic:

$$\sigma_s = \left[\frac{1}{N-1}\sum_{i=1}^{N} P_i^2\right]^{1/2} \tag{12}$$

Random:

$$\sigma_r = \left[\frac{1}{N-1}\sum_{i=1}^{N} N_i^2\right]^{1/2} \tag{13}$$

The ratio of these parallel and perpendicular standard deviations gives the degree and orientation of the ellipticity of the data:

$$e = \frac{\sigma_s}{\sigma_r} \tag{14}$$

If this ratio is larger than unity, the variation can be interpreted as predominantly systematic. Depending on its magnitude, possible causes for this variation can be sought in either the meter's performance, the flow facility, or both. A ratio e that is very much less than unity could indicate that the transfer standard package is not capable of sufficient resolution, and other metering devices need to be selected. Alternatively, there could be insufficient resolution in the flow facility, or operator inconsistencies. This pattern has not been observed in NBS studies.

If the ratio is close to unity, then systematic and random variations are similar. Decisions can then be made as to whether the radius of the data spread is acceptable or whether improvement is desirable.

The typical results shown in Fig. 39 are simulated data for the test procedure of Fig. 37. The various participants are indicated by symbols. The two points for each laboratory indicate the repeatability of that facility. Generally, repeatability is tested by switching the flow off and then on again. The darkened symbols in Fig. 39 denote results obtained in the originating laboratory after the artifact package was returned from the participating laboratories. The repeatability of the results indicates that the artifact did not change during the round-robin testing program.

It is not possible to prescribe specific responses to the results of such an

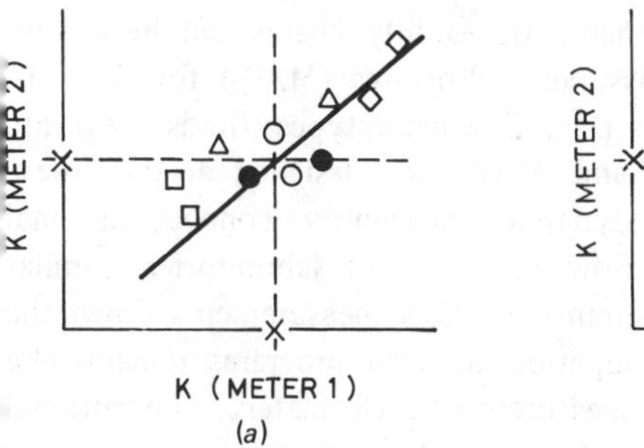

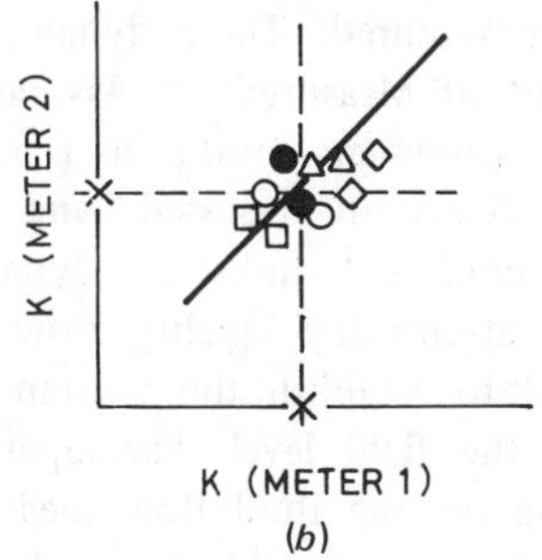

Figure 39 Typical Youden results for a single flow rate. (*a*) Upstream meters. (*b*) Downstream meters.

analysis. At present, the individual participants are entirely on their own. Their positions on the Youden plots can provide the impetus for reexamining some (or all) of the ingredients of the flow measurement processes.

When other participants are the competitors of a particular laboratory whose results lie outside the main body of results, the impetus for improvement can be immense. On the other hand, when a facility's performance is comfortably close to that of the national standards laboratory, a very satisfied feeling develops in that facility. In fact, this situation can, in some senses, be more detrimental than being an outlier–for example, if excellent agreement is used as the excuse to postpone or eliminate the periodic checking of a laboratory's measurement processes. It is highly erroneous to believe that, because good (or excellent) agreement was obtained once via a flow MAP, the system should "forever after" be left alone.*

To avoid such faulty thinking, the flow MAP should be designed so that a repeat test is performed at a later date. The scheduling should be based upon meter performance as shown by the control charts (Figs. 24 and 25). That is, the artifact package should exhibit stable performance over the time interval required to test a whole round of laboratories.

If such testing intervals are excessively long, then the set of participating laboratories should be subdivided such that the testing interval is commensurate with the stability shown by the control charts. When a group of flow laboratories is subdivided into smaller groups, the testing efforts of the originating laboratory are increased, owing to the fact that the "before-and-after" testing must be done for each subround. To reduce the amount of extra testing, MAPs can be arranged regionally. In such a scheme, local originating laboratories are chosen by geographical location. To distribute the work load, this duty can rotate among the participants. Cross checks among regional originating laboratories can be used to unite the results of all participants.

To satisfactorily and realistically demonstrate that one's future fluid flow measurements will be good as specified, "high-confidence," realistic, dynamic-

*This is similar to the fluid-metering adage that, "if one plans to make a very accurate, conventional calibration of a very accurate flowmeter, one should plan to do this only once!"

traceability chains are required. These dynamic-traceability chains can be set up with the establishment of Measurement Assurance Programs (MAPs) for flow. In these, generic flow measurement conditions (i.e., flowmeter types, fluids, flow-rate ranges, and pipeline sizes) are assessed, and MAPs are designed around these conditions, so that confident links of dynamic traceability connect national laboratories to the nation's high-quality flow measurement laboratories. Similar programs are designed to establish the assurance of flow measurements down the traceability chain to the field level. Participation in such programs remains the responsibility of those making fluid flow measurements. Ultimately, conventional flowmeter calibration practice should give way to the enhanced assurance attainable via flow MAPs.

The national laboratories of countries that depend upon fluid resources would be well advised to establish flow MAPs within their borders and set up dynamic-traceability links with other national laboratories. In this way, the increasingly critical and costly transfer of fluid resources can occur in the world's marketplaces with satisfactory and demonstrable equity for all concerned.

APPENDIX A. IDEAL PERFORMANCE CHARACTERISTICS FOR DIFFERENTIAL-PRESSURE TYPE METERS: INCOMPRESSIBLE FLUIDS

For the steady, incompressible (i.e., isentropic) flow of an ideal fluid through two cross-sectional areas $A_1 > A_2$ in a closed conduit, the continuity equation is

$$\dot{m}_1 = \rho_1 A_1 V_1 = \rho_2 A_2 V_2 = \dot{m}_2 \tag{A1}$$

where $\dot{m}_1, \dot{m}_2$ = constant mass flow rates
ρ_1, ρ_2 = fluid densities
V_1, V_2 = average fluid velocities through A_1, A_2

Conservation of energy in the absence of mechanical work and elevation change gives

$$0 = \frac{dp}{\xi} + \frac{V\,dV}{g} \tag{A2}$$

where p = absolute pressure
ξ = specific weight
V = average fluid velocity normal to cross-sectional area
g = acceleration of gravity

Integrating Eq. (A2) between A_1 and A_2 gives

$$\frac{V_2^2 - V_1^2}{2g} = \frac{p_1 - p_2}{\xi} \tag{A3}$$

where, if $\beta = D_2/D_1$, and D_1 and D_2 are the respective diameters at A_1 and A_2,

$$V_2 = \left[\frac{2g(p_1 - p_2)}{\xi(1 - \beta^4)}\right]^{1/2} \tag{A4}$$

Thus, the ideal mass flow rate $\dot{M}_I$ is

$$\dot{M}_I = A_2 \left[\frac{2\rho(p_1 - p_2)}{1 - \beta^4}\right]^{1/2} \tag{A5}$$

APPENDIX B. IDEAL PERFORMANCE CHARACTERISTICS FOR DIFFERENTIAL-PRESSURE TYPE METERS: COMPRESSIBLE FLUIDS

For an ideal gas, the equation of state is

$$p = \rho RT \tag{B1}$$

where p = absolute pressure
ρ = fluid density
R = gas constant
T = absolute temperature

The isentropic relationship between pressures and densities at the respective stream locations is

$$\rho_2 = \rho_1 \left(\frac{p_2}{p_1}\right)^{1/\gamma} \tag{B2}$$

where $\gamma = c_p/c_v$
c_p = specific heat at constant pressure
c_v = specific heat at constant volume

Thus, continuity and energy considerations give

$$V_2 = \left[\frac{2\gamma p_1(1 - r^{1-1/\gamma})}{(\gamma - 1)\rho_1(1 - \beta^4 r^{2/\gamma})}\right]^{1/2} \tag{B3}$$

where $r = p_2/p_1$. Thus, the ideal mass flow rate $\dot{M}_I$ for compressible fluids is

$$\dot{M}_I = \frac{A_2 p_1}{T_1^{1/2}} \left[\frac{r^{2/\gamma}(r^{2/\gamma} - r^{1+1/\gamma})}{1 - \beta^4 r^{2/\gamma}}\right]^{1/2} \left(\frac{g}{R}\frac{2\gamma}{\gamma - 1}\right)^{1/2} \tag{B4}$$

where the bracketed term is called the *fluid meter function* X, and the term in parentheses is called the *fluid meter constant* K_X. Then Eq. (B4) can be written

$$\dot{M}_I = \frac{A_2 p_1}{T_1^{1/2}} X K_X \tag{B5}$$

APPENDIX C. REAL, COMPRESSIBLE ORIFICE–FLOW CALCULATION

The equation characterizing the orifice metering of gases is [2]

$$\dot{Q}_h = C'\sqrt{h_w p_f} \tag{C1}$$

where $\dot{Q}_h$ = volumetric rate of flow at reference conditions, L^3/T
C' = orifice flow constant*
h_w = differential pressure referred to reference temperature, L
p_f = absolute static pressure, F/L^2

The orifice flow constant* is computed via

$$C' = F_b F_r Y F_{pg} F_{tb} F_{tf} F_g F_{pv} F_m F_a F_l \tag{C2}$$

where F_b = basic orifice factor
F_r = Reynolds-number factor
Y = expansion factor
F_{pb} = pressure-base factor
F_{tb} = temperature-base factor
F_{tf} = flowing-temperature factor
F_g = specific-gravity factor
F_{pv} = supercompressiblity factor
F_m = mercury-manometer factor
F_a = orifice thermal-expansion factor
F_l = gauge-location factor

APPENDIX D. DIVERTER EVALUATION AND CORRECTION

Experience has shown that diverter systems can contribute significantly to errors in flow measurement processes where static weighing schemes are used to determine liquid flow rates. In a well-designed diverter system, the timing error (the error in starting and stopping the liquid collection interval) may be reduced to less than 10 ms. On the other hand, if such systems are not properly evaluated and periodically checked, significant errors can result.

This error is dependent upon

1. The liquid flow rate and its distribution within the diverter system
2. The motion of the diverter through the liquid jet
3. The location of the timing switches activated by the moving diverter

To evaluate the diverter system and quantify its performance characteristics, a flow sensor such as a turbine meter is first installed in the facility to monitor flow

*Not to be confused with the discharge coefficient.

rate. This is used to vary the flow rate in some reasonable manner over the range of operation of the diverter. If W_1 is the liquid weight collected during time interval t, the collection process may be plotted in terms of the collected flow rate $\dot{W}_c$, as shown in Fig. D1*a*.

The time $t = 0$ denotes the instant at which the diverter mechanism is activated. The time T_1 is the starting point for the timer recording the diversion interval. Time T_2 is the instant at which the collected flow rate $\dot{W}_c$ is $\dot{W}_1$, which is being monitored by the turbine meter, and T_3 is the instant at which the diverter system is activated to stop the collection. Time T_4 is the instant at which the timer stops, and T_5 denotes the end of the collection, including dripping into the collection tank.

The area under the collection curve is then

$$W_1 = \int_0^{T_5} \dot{W}_c(t)\, dt$$

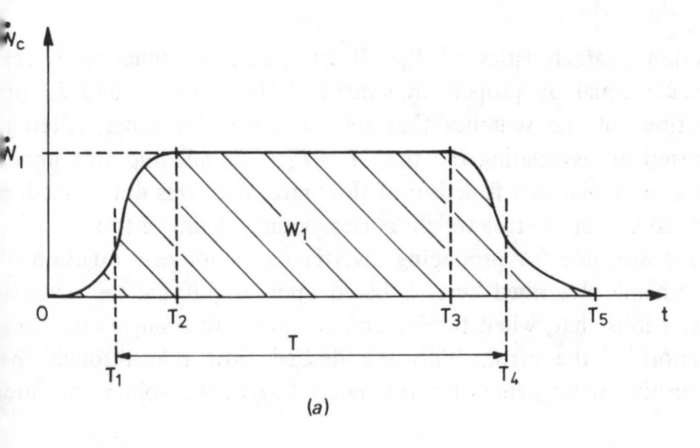

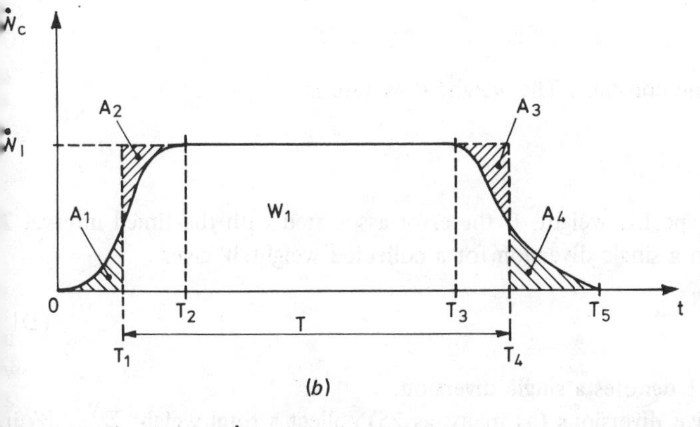

Figure D1 (*a*) Graph of $\dot{W}_c$ versus time. (*b*) The areas of interest are shaded.

This weight will be the correct one for the diversion interval $T = T_4 - T_2$, where $W_1 = \dot{W}_1 T$, when

$$\int_0^{T_1} \dot{W}_c(t)\, dt + \int_{T_4}^{T_5} \dot{W}_c(t)\, dt = \int_{T_1}^{T_2} [\dot{W}_1 - \dot{W}_c(t)]\, dt + \int_{T_3}^{T_4} [\dot{W}_1 - \dot{W}_c(t)]\, dt$$

This equality is shown graphically in Fig. D1b as

$$A_1 + A_4 = A_2 + A_3$$

which is satisfied if

$$A_1 = A_2 \quad \text{and} \quad A_3 = A_4$$

or if

$$A_1 = A_3 \quad \text{and} \quad A_2 = A_4$$

Given the collection characteristics of the diverter, i.e., the function $\dot{W}_c(t)$, these areas can be made equal by proper adjustment of the times T_1 and T_4 or equivalently, the positions of the switches that start and stop the timer. Alternatively, the error incurred by associating the time $T = T_4 - T_1$ and the area under the curve W_1 can be determined as a function of flow rate. With this determined, a correction can be made to obtain the true weight corresponding to time interval T.

Several schemes are available for producing diverter correction as a function of flow rate. One such scheme, described here, is based upon amplifying the diverter error by repetitive diversions that, when totaled and compared to a single diversion permit simple evaluation of the error. With the desired flow rate through the diverter system, the turbine-meter generating frequency f gives the volumetric flow rate $\dot{Q}$ via

$$\dot{Q} = \frac{f}{K}$$

where K is the turbine constant. The weight flow rate is

$$\dot{W} = \frac{f\xi}{K}$$

where ξ is the fluid specific weight. If the error associated with the timed interval T is denoted by ϵ, then a single diversion for a collected weight W gives

$$f_1 = \frac{K_1}{\xi_1} \frac{W_1}{T + \epsilon} \qquad \text{(D1)}$$

Where the subscript 1 denotes a single diversion.

When N repetitive diversions (as many as 25) collect a total weight $\Sigma_{i=1}^{N} W_i$ in the total time interval $\Sigma_{i=1}^{N} (T_i + \epsilon)$, then

$$f_2 = \frac{K_2}{\xi_2} \frac{\sum_{i=1}^{N} W_i}{\sum_{i=1}^{N} (T_i + \epsilon)}$$

where the subscript 2 denotes values pertaining to repetitive diversions. If the N repetitive diversions are equal in duration, then

$$T = \sum_{i=1}^{N} T_i = NT_i$$

$$f_2 = \frac{K_2}{\xi_2} \frac{\sum_{i=1}^{N} W_i}{\sum_{i=1}^{N} T_i} \frac{1}{1 + N\epsilon/T} \tag{D2}$$

The ratio of Eqs. (D1) and (D2) produces

$$\frac{f_1 \xi_1 / K_1}{f_2 \xi_2 / K_2} = \frac{W_1/(T + \epsilon)}{\Sigma_{i=1}^{N} W_i / \Sigma_{i=1}^{N} T_i [1/(1 + N\epsilon/T)]}$$

$$= \frac{(W_1/T)[1/(1 + \epsilon/T)]}{\Sigma_{i=1}^{N} W_i / \Sigma_{i=1}^{N} T_i [1/(1 + N\epsilon/T)]}$$

For convenience, let

$$\dot{W}_1 = \frac{W_1}{T} \qquad \dot{Q}_1 = \frac{f_1 \xi_1}{K_1}$$

$$\dot{W}_2 = \frac{\sum_{i=1}^{N} W_i}{\sum_{i=1}^{N} T_i} \qquad \dot{Q}_2 = \frac{f_2 \xi_2}{K_2}$$

where $\dot{Q}_1$ and $\dot{Q}_2$ are the respective volumetric flow rates. Then

$$\frac{\dot{Q}_1}{\dot{Q}_2} = \frac{\dot{W}_1 [1/(1 + \epsilon/T)]}{\dot{W}_2 [1/(1 + N\epsilon/T)]}$$

Solving for ϵ/T produces the fractional correction for the time T, or

$$\frac{\epsilon}{T} = \frac{\dot{Q}_1 \dot{W}_2 / \dot{Q}_2 \dot{W}_1 - 1}{N - \dot{Q}_1 \dot{W}_2 / \dot{Q}_2 \dot{W}_1}$$

When the dependency of ϵ/T upon flow rate is determined, corrections can be made if the error is significant. Alternatively, the switch position can be changed until subsequent evaluations of this diverter error indicate it is negligibly small for all flow rates, in which case it can be justifiably excluded from the calculations performed on calibration data.

APPENDIX E. EMPIRICAL FORMULAS FOR ORIFICE DISCHARGE COEFFICIENTS

American conventional practice for locating pressure taps in an orifice meter includes formulas for calculating discharge coefficients [1]. Because these formulas are quite involved, we present instead a simpler relationship produced by J. Stolz [3]. This empirical result can be used to calculate discharge coefficients for any tapping arrangements. It is

$$C_d = 0.5959 + 0.0312\beta^{2.1} - 0.184\beta^8 + 0.0029\beta^{2.5}\left(\frac{10^6}{\mathrm{Re}_{D_1}}\right)^{0.75} + 0.09\,\frac{L_1\beta^4}{1-\beta^4} - 0.0337\,L_2'\beta^3$$

where L_1 and L_2', respectively, are the distances of the centerlines of the pressure-tap holes from the upstream surface of the orifice plate, nondimensionalized by the internal diameter of the upstream pipe. When

$$L_1 \geqslant 0.4333$$

it is recommended that the value 0.039 be used for the coefficient of the $\beta^4/(1-\beta^4)$ term.

This relationship produced a fit to a data base of orifice discharge coefficients to within 0.2%.

APPENDIX F. PRESSURE MEASUREMENTS

The measurement of fluid pressure consists in determining one or more of the three quantities P_0, P_s, and P_v, where, from energy considerations,

$$P_0 = P_s + P_v \tag{F1}$$

where P_0 = fluid stagnation pressure when stagnation is ideal
P_s = fluid static pressure
P_v = fluid dynamic pressure

The interrelationships among these three pressures depend [4] upon

1. Fluid characteristics and properties
2. The nature of the real or ideal stagnation process
3. Pertinent laminar or turbulent flow characteristics

F.1 Sensing Static Pressure

Since static-pressure sensing schemes often constitute the critical link between the primary and secondary elements of a flowmeter, it is essential to good metering that they allow optimal overall performance. Conventionally it is assumed that very small, square-edged holes drilled through flow conduits normal to parallel flow give the correct static pressure. Should this assumption not be completely true, it is further assumed that meter calibrations will account for discrepancies. Figures F1 and F2 illustrate qualitatively the kinds of dynamics that can occur and the corresponding effect on static-pressure measurements.

F.2 Sensing Total Pressure

Ideally, total pressures can be sensed only by stagnating the flow isentropically. Thus, one may define the Pitot coefficient C_p as

$$C_p = \frac{P_{tI} - P_s}{P_v}$$

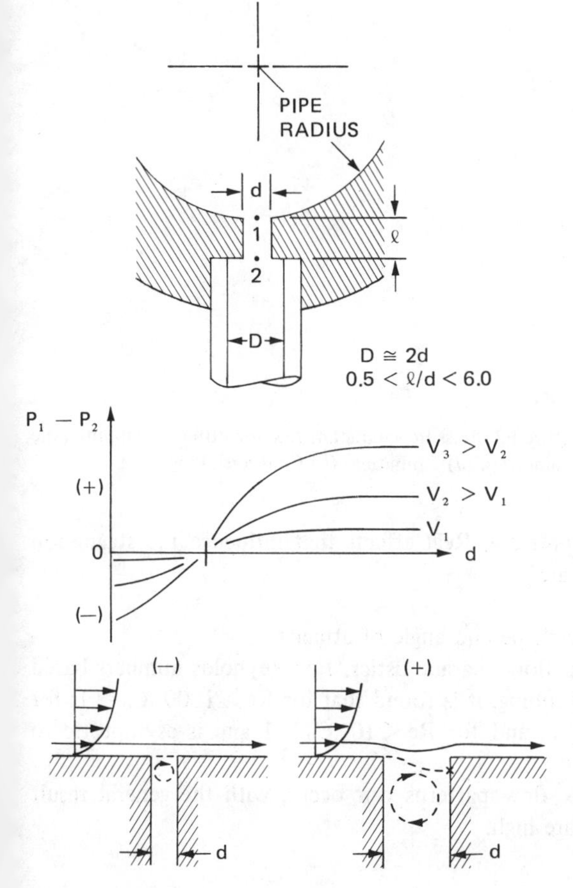

Figure F1 Pressure characteristics for different tapping hole diameters.

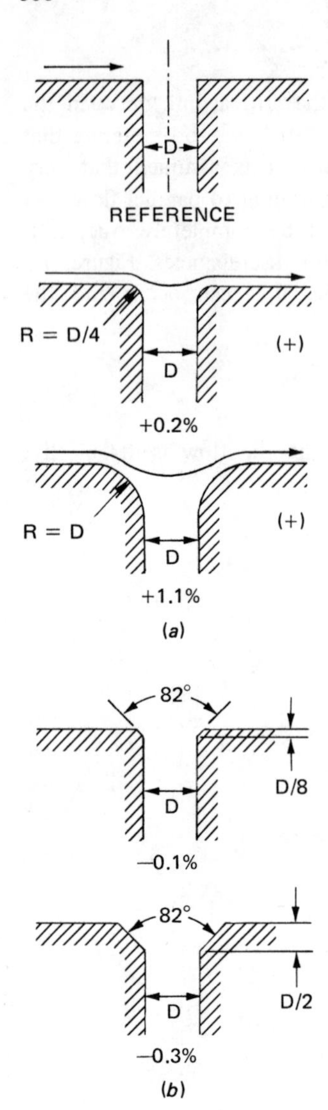

Figure F2 Pressure characteristics for different tapping hole geometries. (*a*) Rounding. (*b*) Countersinking.

where P_{tI} is the indicated total pressure. Real effects that influence the stagnation process for total-pressure sensors are:

1. Geometry–for example, probe shape and angle of attack.
2. Viscosity–fluid properties and flow characteristics. For Reynolds numbers based on the outside radius of Pitot tubing, it is found that for $\mathrm{Re} > 1000$, $C_p \simeq 1$; for $50 < \mathrm{Re} < 1000$, $0.99 < C_p < 1$; and for $\mathrm{Re} < 10$, $C_p > 1$ and is asymptotic to $5.6/\mathrm{Re}$.
3. Transverse gradients. Complex flow patterns can occur, with the general result that indicated total pressures are high.

4. Turbulence. Both total-pressure and static-pressure sensors are affected by turbulence via

$$P_{tI} = P_s + \rho \frac{V^2}{2} + \frac{\rho(v')^2}{2} \tag{F3}$$

where ρ = fluid density
V = mean axial fluid velocity
v' = rms magnitude of the turbulence velocity

Thus, the indicated static pressure P_{Is} becomes

$$P_{Is} = P_s + \frac{\rho(v')^2}{2} \tag{F4}$$

It should be noted that v' must be 20% of the mean velocity V before the turbulence terms amount to 4% of the mean dynamic pressure P_v.

APPENDIX G. TEMPERATURE MEASUREMENT AND RECOVERY FACTOR

The measurement of the temperature of a real fluid generally consists in reading the output signal of a thermometer as the fluid stagnates against the sensor surface and equilibriates thermally with this sensor surface. Because of the viscous and thermal diffusion properties of real fluids, the thermometer generally does not indicate the thermodynamically ideal stagnation temperature

$$T_0 = T_s + T_v \tag{G1}$$

where T_0 = thermodynamically ideal stagnation temperature
T_s = fluid static temperature
$T_v = V^2/2Jgc_p$ = ideal dynamic temperature
V = initial local fluid velocity of stagnating fluid
J = mechanical equivalent of heat

The Prandtl number Pr expresses the ratio of fluid viscous to thermal diffusion effects; that is,

$$\mathrm{Pr} = \frac{\text{viscous effects}}{\text{thermal diffusion}} = \frac{\mu c_p}{k} \tag{G2}$$

where μ = dynamic viscosity of fluid
k = thermal conductivity

Thus when, as in air, with

$$0.65 < \mathrm{Pr} < 0.7 \tag{G3}$$

the thermal diffusion effects are greater than the viscous effects, the real dynamic temperature is less than the ideal. Similarly, when, as in water, with

$$1 < \mathrm{Pr} < 13 \tag{G4}$$

the viscous effects are greater than the thermal diffusion effects, the real dynamic temperature is greater than the ideal. Of course, for

$$\mathrm{Pr} = 1 \tag{G5}$$

the viscous heating effects generated by the stagnating real fluid are exactly diffused thermally so that the real dynamic temperature is the ideal.

Thus [4],

$$T_{pi} = T_s + rT_v \tag{G6}$$

where T_{pi} = ideal probe temperature
r = recovery factor

The recovery factor can then be written

$$r = \frac{T_{pi} - T_s}{T_0 - T_s} \tag{G7}$$

It has been found that

$$r = \mathrm{Pr}^{1/2} \tag{G8}$$

Therefore the temperature sensed by a sensor in a fluid stream depends upon

1. The viscous and thermal diffusion properties
2. The characteristics of the sensor, i.e., shape and orientation to the flow
3. The nature of the fluid stagnation process, including any pertinent fluid motion effects, such as turbulence, that might alter the molecular viscous and thermal diffusion phenomena described above
4. Any other pertinent thermal losses from the temperature sensor

APPENDIX H. ANALYSIS OF VARIANCE WITH TWO FLOWMETERS IN SERIES*

When two flowmeters are calibrated in the same line, analysis of variance techniques may be applied for the purpose of assessing the portion of the variation that is due to the flow facility and each meter.

Suppose two meters that respond similarly to flow phenomena are installed in series on a pipeline. They need not be identical meters or even meters of the same type, but for present purposes let us consider that they are identical turbine meters. For N determinations of the flow rate for a single flow setting via, say, primary standards, we can compute the respective average meter constants as

$$\bar{K}_x = \frac{1}{N}\sum_{i=1}^{N} K_{xi} \qquad \bar{K}_y = \frac{1}{N}\sum_{i=1}^{N} K_{yi} \tag{H1}$$

*See [19].

where K_{xi} and K_{yi} are the turbine-meter constants determined N times via calibration techniques at a flow rate, and overbars denote averages. The standard deviations are

$$\sigma_x = \left[\frac{1}{N-1}\sum_{i=1}^{N}(K_{xi}-\bar{K}_x)^2\right]^{1/2} \tag{H2}$$

$$\sigma_y = \left[\frac{1}{N-1}\sum_{i=1}^{N}(K_{yi}-\bar{K}_y)^2\right]^{1/2} \tag{H3}$$

The correlation coefficient is

$$r = \frac{\sum_{i=1}^{N}(K_{xi}-\bar{K}_x)(K_{yi}-\bar{K}_y)}{(N-1)\sigma_x\sigma_y} \tag{H4}$$

Now assume that the variances σ_x^2 and σ_y^2 are composed of portions σ_f^2 due to the facility and portions σ_{my}^2 and σ_{mx}^2 due to the meters:

mfmfmfmfm

mfmfmfmfm

That fraction of the total variance for each meter that correlates with the other is assumed to be due to the variation attributable to the facility, or

$$r^2 = \frac{\sigma_f^2}{\sigma_y^2} = \frac{\sigma_f^2}{\sigma_x^2}$$

so

$$1 = r^2 + \frac{\sigma_{my}^2}{\sigma_y^2}$$

or

$$\frac{\sigma_{my}^2}{\sigma_y^2} = 1 - r^2$$

is that fraction of the total variance that remains after the correlated portion is removed. Thus, the standard deviations of the meters themselves are

$$\sigma_{my} = \sigma_y\sqrt{1-r^2}$$

$$\sigma_{mx} = \sigma_x\sqrt{1-r^2}$$

In this manner, the meter performance can be quantified in any such test, and the performance of the facility can be assessed as viewed through the response of the two meters. For example, if the variations in the meter constants correlate perfectly with each other (that is, $r = 1$), then the interpretation would be that

$$\sigma_{my} = 0 \qquad \sigma_{mx} = 0$$

The total variation observed in the meter constants would be due to the facility.

On the other hand, if no correlation were observed between the respective meter constants, then $r = 0$ and

$$\sigma_{my} = \sigma_y \qquad \sigma_{mx} = \sigma_x$$

The interpretation here would be that all the observed variation in the meter constants is due to the meters themselves or to factors that were completely uncorrelated.

NOMENCLATURE

A	area
C'	flow constant
C_D, C_d	dimensionless discharge coefficients
c_p, c_v	specific heat at constant pressure, constant volume
D	inside pipe diameter
e	ellipticity
F	meter factor
f	frequency (Hz)
g	gravitational acceleration
h	height, differential pressure
J	Joule's constant
K	flow coefficient, flowmeter constant
k	thermal conductivity
L	characteristic length
M	molecular weight
$\dot{m}$, $\dot{M}$	mass rate of flow
N	number of samples
Pr	Prandtl number
p	pressure
$\dot{Q}$	volume flow rate, volumetric flow rate
q	heating rate
Re	Reynolds number
R	ratio of meter constants, gas constant
r	recovery factor, correlation coefficient, pressure ratio
St	Strouhal number
T	temperature, absolute temperature, time
t	time
V	average fluid velocity in conduit, point fluid velocity
$\dot{W}$	weight flow rate
β	ratio of orifice hole to pipe diameter
γ	specific-heat ratio

δ	allowable tolerance
ϵ	expected error
μ	viscosity
ν	kinematic viscosity
ρ	density
σ	standard deviation
ξ	specific weight

Subscripts

1,2	positions along conduit, specific meters in same pipe
a	orifice thermal expansion
b	basic
c	collected
F	facility in which meter is tested
f	fluid, facility in which meter is tested
g	gravity
I	ideal
i	sample number
l	gauge location
M	specific flowmeter
m	manometer, specific flowmeter
p	probe
pb	pressure base
pv	supercompressibility
$\dot{Q}$	specific value of flow rate
r	random, specific Reynolds number
s	systematic, static
T	total
t	total
tb	temperature base
tf	flowing temperature
v	dynamic
x,y	collected particular meters
0	stagnation

Superscripts

$\bar{}$	average of a set of values
$\overline{}$	time averaged
$'$	deviation from time-smoothed value
$\cdot$	time derivative

REFERENCES

1. H. S. Bean (ed.), *Fluid Meters–Their Theory and Applications,* 6th ed., ASME, New York, 1971.

2. American Gas Association, *Orifice Metering of Natural Gas,* Rep. 3, Arlington, Va., 1972.
3. R. W. Miller, National and International Orifice Coefficient Equations Compared to Laboratory Data, *Proc. of the ASME Winter Annu. Meeting,* New York, 1979.
4. R. P. Benedict, *Fundamentals of Temperature, Pressure and Flow Measurement,* Wiley, New York, 1969.
5. R. W. Davis and G. E. Mattingly, Numerical Modeling of Turbulent Flow through Thin Orifice Plates, U.S. National Bureau of Standards Spec. Publ. 484, *Proc. of Symp. on Flow on Open Channels and Closed Conduits,* 1977.
6. R. W. Davis, Numerical Modeling of Two-Dimensional Flumes, U.S. National Bureau of Standards Spec. Publ. 484, *Proc. of Symp. on Flow in Open Channels and Closed Conduits,* 1977.
7. R. W. Davis and S. Deutsch, A Numerical-Experimental Study of Parshall Flumes, to appear in *J. Hydraul. Res.*
8. B. Robertson and J. E. Potzick, A Long Wavelength Acoustic Flow Meter (in preparation).
9. K. Plache, Coriolis/Gyroscopic Flow Meter, *Proc. of the ASME Winter Annu. Meeting,* Atlanta, 1977.
10. F. C. Kinghorn and K. A. Blake, The Design of Flow Straightener-Nozzle Packages for Discharge Side Testing of Compressors, *Proc. of the Conf. on Design and Oper. of Ind. Compressors,* Glasgow, 1978 (published by the Institute of Mechanical Engineering).
11. F. C. Kinghorn, A. McHugh, and W. D. Dyet, The Use of Etoile Flow Straighteners with Orifice Plates in Swirling Flow, *Proc. of the ASME Winter Annu. Meeting,* New York, 1979.
12. M. R. Shafer and F. W. Ruegg, Liquid-Flowmeter Calibration Techniques, *Proc. of the ASME Winter Annu. Meeting,* New York, 1957.
13. B. C. Belanger, Traceability–An Evolving Concept, *Am. Soc. Test. Mater. Standardization News,* February 1979.
14. J. M. Cameron, Measurement Assurance, U.S. National Bureau of Standards Intern. Rep. 77-1240, 1977.
15. P. E. Pontius and J. M. Cameron, Realistic Uncertainties and the Mass Measurement Process–An Illustrated Review, U.S. National Bureau of Standards Monograph 103, 1967.
16. P. E. Pontius, Measurement Assurance Programs–A Case Study: Length Measurements, Pt. I, Long Gage Blocks, U.S. National Bureau of Standards Monograph 149, 1975.
17. G. E. Mattingly, P. E. Pontius, H. H. Allion, and E. F. Moore, A Laboratory Study of Turbine Meter Uncertainty, U.S. National Bureau of Standards Spec. Publ. 484, *Proc. of Symp. on Flow In Open Channels and Closed Conduits,* 1977.
18. G. E. Mattingly, W. C. Pursley, R. Paton, and E. A. Spencer, Steps toward an Ideal Transfer Standard for Flow Measurement, *FLOMEKO Symp. on Flow,* Groningen, The Netherlands, 1978.
19. W. Strohmeier, *Notes on Turbine Meter Performance,* TN 17, Fischer and Porter Co., Warminster, P., 1971.
20. W. J. Youden, Graphical Diagnosis of Interlaboratory Test Results, *J. Ind. Qual. Control,* vol. 15, no. 11, pp. 133–137, 1959.
21. C. Eisenhart, "W. J. Youden," *Dictionary of Scientific Biography,* vol. XIV, Scribners, New York, 1976.

CHAPTER

SEVEN

FLOW VISUALIZATION BY DIRECT INJECTION

Thomas J. Mueller

A man is not a dog to smell out each individual track; he is a man to see, and seeing, to analyze.

F. N. M. Brown

1. INTRODUCTION

Throughout the history of aerodynamics and hydrodynamics there has been great interest in making flow patterns visible. The visualization of complex flows has played a uniquely important role in the improvement of our understanding of fluid-dynamic phenomena. Flow visualization has been used to verify existing physical principles and, in the process, has led to the discovery of numerous flow phenomena. In addition to obtaining qualitative global pictures of the flow, the possibility of acquiring quantitative measurements without introducing probes that invariably disturb the flow has provided the necessary incentive for the development of a large number of visualization techniques. Although clean air and water are transparent, smoke or other particles in air, and dye or other particles in water, provide the necessary contamination for flow visualization. For very practical reasons, the study of fluid mechanics was concerned with the flow of water and other liquids until relatively recently. Man's interest in flight, however, pointed out the necessity of visualizing

It is a pleasure to thank H. Werlé (ONERA), R. W. Hale (Sage Action, Inc.), W. C. Wells and J. M. Hample (USAF/WAL), C. R. Smith and T. Wei (Lehigh University), and W. J. McCroskey, K. W. McAlister, and L. W. Carr (U.S. Army Aeromechanics Laboratory, AVRADCOM) for generously providing photographs and written descriptions of their research. My sincere thanks go to my colleagues at the University of Notre Dame–R. C. Nelson, S. M. Batill, T. L. Doligalski, and A. A. Szewczyk–and to our graduate students–J. T. Kegelman, B. J. Jansen, Jr., and S. J. Elsner–for sharing their research in flow visualization as well as for their helpful comments during the writing of this chapter.

airflows to understand the mechanics of objects moving through the air. Many substances have been used to visualize the flow of air and water. Smoke, helium bubbles, dust particles, and even glowing iron particles have been used in air; a variety of dyes, particles, neutrally buoyant spheres, and both air and hydrogen bubbles have been employed in water.

An important consideration for any method of flow visualization is: What does the picture show about the fluid motion and how can the flow patterns in the picture be interpreted? Does the injected substance (e.g., smoke, dye, or bubbles) trace out a streamline, streakline, or pathline? In boundary layers, separated regions, and other regions of relatively high vorticity gradients, why does the injected substance accumulate in some places and not in others? To begin to answer these questions, one must review several fundamental notions of fluid mechanics, namely, the definitions of streamline, streakline, and pathline.

Streamlines, streaklines, and pathlines are three curves that have been defined to help describe the flow of a fluid. A streamline is a curve everywhere tangent to the instantaneous velocity vectors or, in other words, everywhere parallel to the instantaneous direction of the flow. A streakline is the locus of all fluid particles that have passed through a prescribed fixed point during a specified interval of time. A pathline is the curve traversed by a particular fluid particle during a specified interval of time. When the flow is steady (i.e., not dependent upon time), the streamline, streakline, and pathline that pass through the same point are identical. Conversely, when the flow is unsteady (i.e., dependent upon time), these three lines are, in general, different.

If individual particles or bubbles can be followed for a given length of time, then the trajectory of these particles or bubbles may be recorded, and a pathline obtained. Smokelines or dyelines emanating continuously from each opening of the smoke or dye injector are streaklines. If the smoke or dye line can be interrupted for a specific time interval, then a time streakline can be created. The type of fluid curve that may be obtained, and/or how the visual flow pattern may be used to gain insight into the problem under study, is discussed for each method presented here.

The discussion of flow visualization by direct injection in this chapter considers only the use of smoke and helium bubbles in air, and dye and hydrogen bubbles in water. Because of the author's experience with the use of smoke, a somewhat more extensive treatment of this subject is presented. The purpose of the presentation is to acquaint the reader with the types of equipment and procedures necessary to use these flow-visualization techniques, as well as the advantages and disadvantages of each technique.

2. AERODYNAMIC FLOW VISUALIZATION

It was recognized as early as 1759 by John Smeaton that the natural wind was variable in direction and speed and, therefore, unreliable for aerodynamic research. An artificial and controllable wind was required. Although many early investigations used the whirling machine, invented in 1746 by Ellicott and Robins, or some variation of this basic idea, it was not until the wind tunnel was invented by F. H. Wenham in Great Britain about 1871 that systematic aerodynamic experiments were possible. This first wind tunnel had a 260-mm square cross section with air blown into the inlet by a steam-driven fan at speeds of up to 18 m/s [1]. Flow visualization in wind tunnels closely followed the development of these facilities. In this section, the use of smoke and helium bubbles as visualizing agents is discussed. Two methods of introducing smoke are presented, namely, the smoke-tube and smoke-wire methods.

2.1 Smoke-Tube Method

The use of smoke to visualize the flow in wind tunnels began around the turn of the century and was due to L. Mach (Vienna, 1893) and E. J. Marey (Paris, 1899). Some years later, Prandtl and his associates (Gottingen, 1923), L. F. G. Simons and N. S. Dewey (Teddington, 1930) and W. S. Farren (Cambridge, 1932) experimented with smoke visualization. The important advances toward the eventual use of smoke visualization as a research tool began in the 1930s with the work of A. M. Lippisch (Darmstadt, 1937) and F. N. M. Brown (Notre Dame, 1937). A more complete history of the use of smoke in wind tunnels is given in [2, 3].

Although Brown and Lippisch established subsonic smoke-tunnel techniques that still represent the state of the art, Brown's equipment had important advantages not possible with that built by Lippisch. For example Brown's three-dimensional tunnels had much larger inlet contraction ratios (24:1, compared to 12:1) and more antiturbulence screens, which produced lower turbulence levels at the higher velocities. In fact, by making smaller test sections while increasing the contraction to 48:1 and 96:1, Brown was able to produce low turbulence levels at speeds up to about 60 m/s. Furthermore, with the smoke introduced upstream of the screens instead of inside the tunnel, as in the apparatus of Lippisch and most of the others, the smoke-injection rake did not add to the disturbances in the test section. Expanding the techniques of Brown, V. P. Goddard (Notre Dame, 1959) was able to produce the world's first supersonic smoke tunnel. With this equipment, smoke photographs were taken at speeds up to 404 m/s. It is not surprising that these wind tunnels, smoke generators, and photographic techniques have been copied extensively.

The most important requirement for a smoke flow-visualization tunnel is that it should have a low turbulence level in the test section. This is readily accomplished by using several screens followed by a large contraction in area ahead of the test section. In the next section, a description of the equipment and techniques developed by Brown and Goddard is presented. These facilities have been used continuously by faculty and students, and for sponsored research projects, over the past three decades. The equipment is in the Aerospace Laboratory at the University of Notre Dame.

2.1.1 Low-turbulence subsonic wind tunnels. There are four nonreturn or *indraft* subsonic wind tunnels in the Notre Dame Aerospace Laboratory. All were designed in accordance with the basic concepts developed by Brown. A summary of the specifications of all these tunnels is given in [2]. The two largest tunnels are identical indraft tunnels, each powered by a 15-hp ac motor, with a double-cone transmission for speed control (see Fig. 1). The motor drives an eight-bladed fan that is 1220 mm in diameter. The motor-and-fan assembly is mounted on a concrete base and is isolated from the diffuser. To eliminate mechanical vibrations, the diffuser is isolated from the test section by a 101-mm section of sponge rubber. The motor is enclosed in a ventilated shelter. The open area of the shelter walls may be varied by raising or lowering canvas curtains. This shelter protects the motor from the elements and provides a windbreak for the tunnel exhaust.

Several interchangeable working sections are available for these tunnels. Both constant-area and tapered working sections are on hand. These tunnels can be adapted quickly to use square test sections of dimensions 610 × 610, 431 × 431, and 305 × 305 mm. The corresponding contraction ratios are 24:1, 48:1, and 96:1. All these test sections are 1828 mm in length and have 6.35-mm plate glass on at least one side, with a flat black or black velvet back. However, the 431- × 431- and 305- × 305-mm test sections have additional contractions and diffuser sections attached to them. An externally mounted two-component strain-gage balance may be used in conjunction with any desired working section.

Antiturbulence screening is used upstream of the contraction cone. There are five 5.51 × 7.09 meshes/cm bronze screens, followed by seven 7.87 × 7.87 meshes/cm nylon screens. The bronze screens are made of wire 0.305 mm in diameter, and the nylon screens have a thread diameter of 0.076 mm. The main contraction cone (see Fig. 1) has a square cross section and a shape given by Smith and Wang [4].

A study of the longitudinal component of turbulence intensity and spectrums of turbulence intensity was made for the 610- × 610- and 431- × 431-mm configurations using a hot-wire anemometer. During this study, a condenser microphone was used to measure tunnel noise spectrums. It was concluded from this investigation that (1) the major portion of the turbulence present in the two test sections

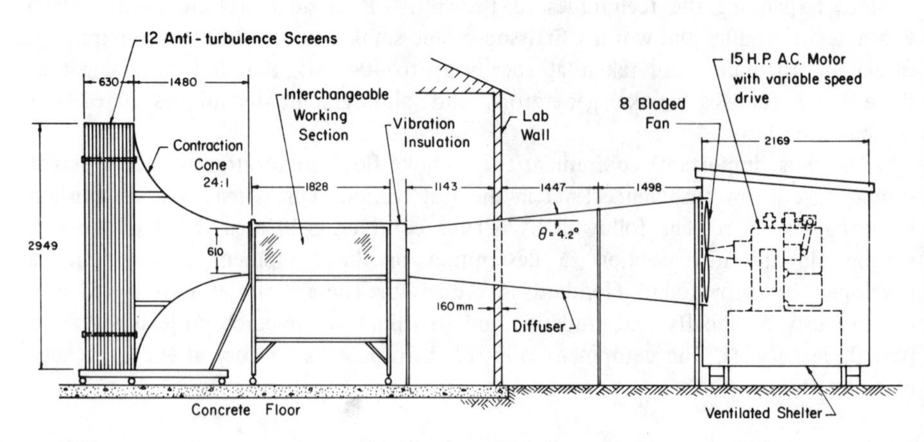

Figure 1 Low-turbulence subsonic smoke tunnel (all dimensions in millimeters). *(Courtesy of T. J. Mueller, University of Notre Dame.)*

is due to tunnel noise, (2) the turbulence intensity was found to vary with test-section velocity,* (3) the average turbulence intensity over a 7-kHz frequency is 0.1% for the 610- × 610-mm test section and 0.08% for the 431- × 431-mm test section, and (4) the frequency of blade passage of the wind-tunnel fan was the cause of the major peaks in the turbulence intensity and in the tunnel noise spectrums.

2.1.2 Smoke. The word "smoke" is used in a very broad sense in flow-visualization techniques and includes a variety of smokelike materials such as vapors, fumes, and mists. The smoke used must be generated in a safe manner and must possess the necessary light-scattering qualities so that it can be readily photographed. It is also important that the smoke not adversely affect the wind tunnel into which it is introduced, nor the model being studied. Another desirable but not absolutely necessary qualification is that the smoke be nontoxic, in the unlikely event the experimenters are exposed to it. Finding a smokelike substance that meets all these criteria is not an easy task.

A large number of materials have been used to generate smoke, e.g., the combustion of tobacco, rotten wood, and wheat straw, the products of reaction of various chemical substances such as titanium tetrachloride and water vapor, and the vaporization of hydrocarbon oils. The smokelike materials used may be referred to as aerosols, since aerosols are composed of colloided particles suspended in a gas. A great deal of interest has been focused on aerosol generation and its properties because of its close relationship to meteorology, air pollution, cloud chambers, smokes, combustion of fuels, colloid chemistry, etc.

Two very practical items must be carefully examined before the choice of smokelike material is made for flow visualization. The smoke or aerosol particles must be as small as possible so they will closely follow the flow pattern being studied. The smoke particles must also be large enough to scatter a sufficient amount of light so that photographs of the smoke pattern can be obtained. Although many materials and substances have particle sizes below 1 μm, the most practical ones for flow visualization are tobacco smoke, rosin smoke, carbon black, and oil smoke [5]. Tobacco-smoke and carbon-black particles range from about 0.01 to about 0.20 μm while rosin-smoke particles range from about 0.01 to about 1.0 μm and oil-smoke particles from 0.30 to about 1.0 μm. There is no doubt that all these particles are small enough to follow the flow. It should be noted that the flow particles in water vapor (fog) are generally much larger than 1 μm (i.e., 1 to about 50 μm). If one now considers the light-scattering ability of the particles used, another constraint becomes apparent. Particles should be larger than about 0.15 μm to scatter a sufficient amount of light to be readily seen. This light-scattering criterion indicates that tobacco-smoke and carbon-black particles, which are mostly lower than 0.15 μm, would be more difficult to photograph.

*A recent study of the turbulence intensity in the 610- × 610-mm test section using the latest solid-state hot-wire anemometer shows only a slight variation of turbulence intensity with tunnel velocity.

Resin is a semisolid, organic substance exuded from various plants and trees or prepared synthetically, whereas rosin is the hard, brittle resin remaining after oil of turpentine has been distilled. Maltby and Keating [6] describe an electrically fused, pyrotechnic resin smoke generator developed for wind-tunnel use up to a speed of 15 m/s. This smoke generator was manufactured by Brocks Fireworks Company, Ltd., in England and was based on the vaporization of resin. Maltby and Keating also mention that some ammonium chlorate is present in the resin canisters, which must be stored away from heat since this substance is unstable. The device is simply a smoke bomb adapted for wind-tunnel use!

There are, of course, many hydrocarbon mixtures, i.e., oils, that undoubtedly could be used to produce smoke by combustion or vaporization. From the point of view of laboratory safety, it would be desirable to use vaporization rather than the combustion technique. Furthermore, it would also be safer to use an oil that would vaporize at the lowest possible temperature and be the least flammable. Of the oils most commonly used to produce smoke, mineral oil requires the highest temperature for vaporization, while charcoal lighter fluid requires the lowest. The second lowest temperature for vaporization is that for kerosene [5]. Since kerosene is less flammable than charcoal lighter fluid, it is the obvious choice. Kerosene seems to offer the best compromise when particle size, light-scattering ability, low vaporization temperature, and low flammability are considered.

2.1.3 Kerosene smoke generation and rake system. Although the first oil smoke generator was developed by Preston and Sweeting [7], one of the most successful oil smoke generators was designed by Brown in 1961. Large quantities of dense kerosene smoke were produced quickly and safely with this four-tube kerosene generator, shown schematically in Fig. 2.

A flat electric strip heater is located inside a 51-mm square thin-wall conduit tube. The entire unit is set at a convenient angle (about 60°), and a sight-feed oiler is mounted on the unit at the upper end of each tube so the oil drips onto the upper end of the strip heater. It has been determined that a drip rate of approximately two drops per second is more than sufficient to produce the desired amount of smoke. Faster rates result in inefficient and wasteful operation. Furthermore, an extremely fast drip rate can result in backfiring of the unit. A squirrel-cage blower mounted at the low end of the unit is used to force the smoke through the system. The squirrel-cage blower is more or less mandatory–in the event of backfiring the sudden increase in pressure is easily transmitted through the rotor.

Before entering the smoke rake, the smoke is allowed to pass through a heat exchanger made of 42-mm-diameter pipe, as shown in Fig. 3. The prime function of this heat exchanger is to cool the smoke down to room temperature. (The system has conveniently located drain cocks, one at the bottom of each tube of the generator itself to remove excess oil not converted into smoke, and others at the bottom of the heat exchanger to remove whatever oil might have been condensed.) After passing through the heat-exchanger condenser system, the smoke flows into a 117-mm manifold and is passed through an absorbent cloth filter. This filter serves a

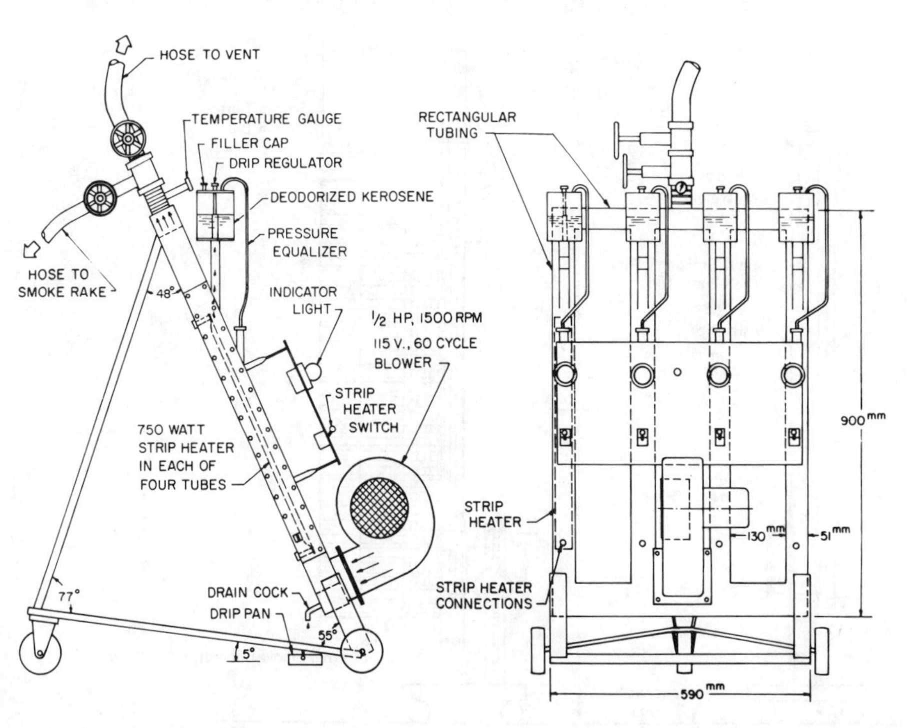

Figure 2 Schematic of Brown's four-tube oil smoke generator. *(Courtesy of T. J. Mueller, University of Notre Dame.)*

dual purpose; it removes most of the remaining lighter tars and aids in distributing the smoke uniformly into the evenly spaced 19-mm tubes that extend from the manifold. These evenly spaced tubes determine the initial smokeline spacing. Such an array of tubes with the manifold has a rakelike appearance, which is why the assembly is referred to as a *smoke rake*. Appropriate measures should be taken to guard against leaks in the system and to provide an additional outside exhaust for the smoke generator, since the excessive and prolonged inhalation of oil smoke could be a health hazard.

Oil smoke generators of this type have been constructed in single, double, and quadruple units. The grouping of units is basically a method of increasing the volumetric output of smoke. The smoke rake shown in Fig. 3 can be moved up or down, as desired, from the subsonic smoke-tunnel control panel. A fixed-position smoke rake of similar design is used in conjunction with the supersonic tunnels.

2.1.4 Steam. The only reason for trying to replace the kerosene smoke-generation technique of Brown would be to produce a nontoxic and nonflammable smokelike substance. *All* products of combustion, reactions of chemical substances, oil and paraffin, vapors, and aerosols are toxic to some degree. Many of these substances are flammable, and some are corrosive or chemically active. The only available technique that does not have one or more of these undesirable properties appears to

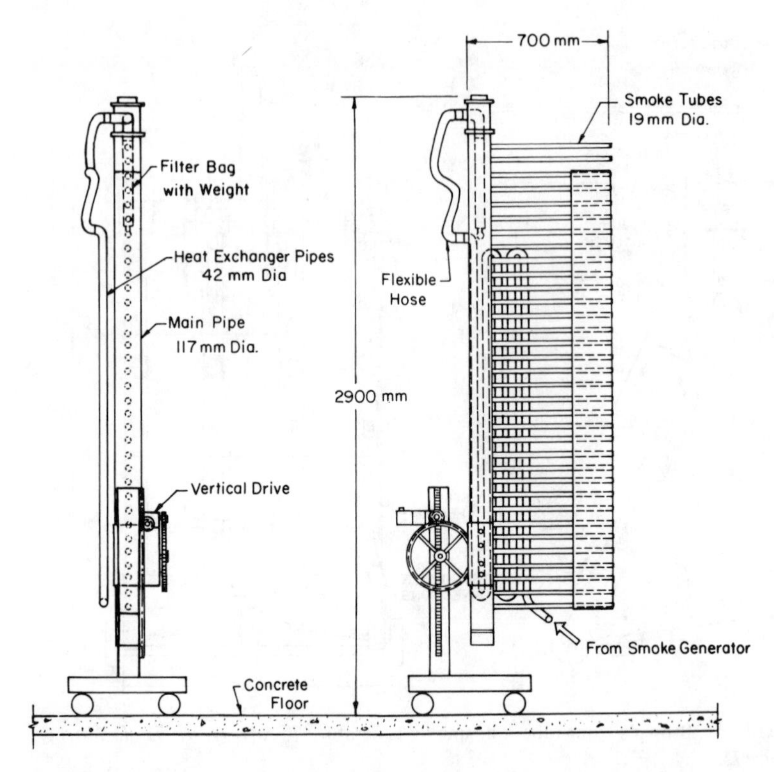

Figure 3 Schematic of smoke rake with vertical movement used for subsonic tunnels. *(Courtesy of T. J. Mueller, University of Notre Dame.)*

be the steam–liquid nitrogen method; however, generating large quantities of steam in some type of boiler presents a different safety hazard. After generation, the steam is mixed with liquid nitrogen and introduced into the wind tunnel. The apparatus used to generate the steam is somewhat different in [8, 9], but the end product is the same. Both the MIT and Iowa State University laboratories have used steam with reasonable success. A stream of steam–liquid nitrogen mixture has been used in both indraft and closed-circuit subsonic wind tunnels.

Although steam used for flow visualization has the advantages of being clean and nontoxic, it does have disadvantages. For example, the system temperature must be controlled carefully if a neutrally buoyant fog is to be obtained. As pointed out earlier, water vapor is composed of much larger particles on the average than oil smoke. It also has the disadvantage that the steam condenses to water on cold model surfaces, walls, and other protrusions in the test section. The steam photographs reasonably well and can produce usable visualization records.

2.1.5 Photographic techniques. Photographs can be taken with a variety of cameras: still, stereo, cinematic, and stereo-cinematic. Photographs may be made with steady light but are most commonly made with short-interval (20-μs) strobolumes, so placed as to illuminate the flow field under investigation. To obtain sufficient intensity and to provide uniform illumination, three to five lamps are used. The strobolumes are triggered by the camera shutter or other means, as dictated by the problem.

Still photographs are made with a 101.6- × 127-mm (4- × 5-in) view camera using an f4.5 or f6.3 coated lens. Polaroid Type 57 film and Kodak Royal-X Pan film, thrice overdeveloped in D-11, have been used to obtain maximum contrast. The Royal-X Pan negatives are projection-printed on contrast 5 bromide paper. If linear measurements are to be made directly from prints, a dimensionally stable paper is used.

Three-dimensional pictures may be taken with a stereographic camera designed and constructed by Brown and his colleagues. This apparatus consists of two 101.6- × 127-mm (4- × 5-in) cameras mounted on a base casting supported by a heavy but mobile base. The lens separation can be varied from 203 to 610 mm. A stereo comparator is used to extract data from the photographs obtained. This apparatus has also been used with two 16-mm movie cameras for stereo-cinematic investigation.

High-speed movies are taken with a Wollensak WF-3 Fastex 16-mm camera, capable of speeds from 1000 to 8000 frames/s, or a DMB-5 Milliken 16-mm camera, capable of speeds up to 500 frames/s. With Eastman 4-X negative film 7224, camera speeds of up to 4000 frames/s can be used with good results. Various combinations of high-intensity lights (that is, 1000 and 2000 W) are used for taking high-speed movies. It is important for the experimenter to have access to a darkroom so that film may be developed quickly. If unsatisfactory results are obtained, a new series of photographs may be taken with little time lost. Since photography is a fine art, a great deal of experimentation is necessary in both taking and processing photographs and movie film.

A computer-compatible video camera system offers a great deal of potential for extracting quantitative data from flow-visualization experiments. However, limitations based upon the speed suggest that such a system be used in conjunction with the proven methods of obtaining still and high-speed cinematic photographs.

2.1.6 Low-turbulence supersonic and transonic wind tunnels. The Aerospace Laboratory at Notre Dame also houses three supersonic wind tunnels and one transonic nonreturn tunnel, developed by extending the successful low-speed smoke-tunnel concepts of Brown to much higher speeds.

Notre Dame's supersonic installation consists of three separate diffusers, connected to a common manifold and three Allis-Chalmers rotary vacuum pumps. Each of these positive-displacement vane vacuum pumps is driven by a 125-hp induction motor at a speed of 435 rpm, delivering 94 m^3/min at 457 mm of mercury vacuum. By suitable arrangement of the tunnel exhaust valves, any one of the three tunnel stations may be operated in a continuous mode. The pressure difference essential for operating a given tunnel is produced by turning on one, two, or all three of the rotary vacuum pumps. Operation of any of these tunnels is limited by the moisture content of the ambient air. Dryers are not feasible because they would interfere with the introduction of smoke into the tunnel. It has been determined that good data can be obtained with ambient dew points as high as about −4°C.

The world's first supersonic smoke tunnel [10], called the "pilot tunnel," was designed by the method of Foelsch for a Mach number of 1.38 and is shown in Fig. 4. The aluminum block test section is about 101 mm long and has a square cross-sectional area of 4032 mm^2. The upper nozzle block has a piece of 12.7-mm lucite sandwiched in the center so that smokelines may be lighted from above when necessary. The side walls consist of 19-mm-thick Parallel-O-Plate glass. The glass extends over almost the entire nozzle length (i.e., from upstream of the throat to the end of the test section). The converging portion of the supersonic nozzle and the inlet contraction cone were designed by the method of Smith and Wang [4]. The overall contraction ratio from the beginning of the inlet to the nozzle throat is about 93:1. Seven screens are located at the upstream side of the contraction cone: a single brass screen with 5.51 × 7.09 meshes/cm and six nylon screens with 7.87 × 7.87 meshes/cm. The pilot tunnel can be operated continuously, using a single rotary vacuum pump. Several other supersonic nozzle configurations and sizes are also available. A transonic smoke tunnel with slotted upper and lower walls and clear plastic side walls has recently been fabricated and calibrated [5].

2.1.7 Visualization techniques for supersonic flow. The same oil smoke generator, basic smoke rake, and photographic techniques have been used with the supersonic tunnels as with the subsonic tunnels for direct smokeline visualization. The most critical item is, of course, lighting.

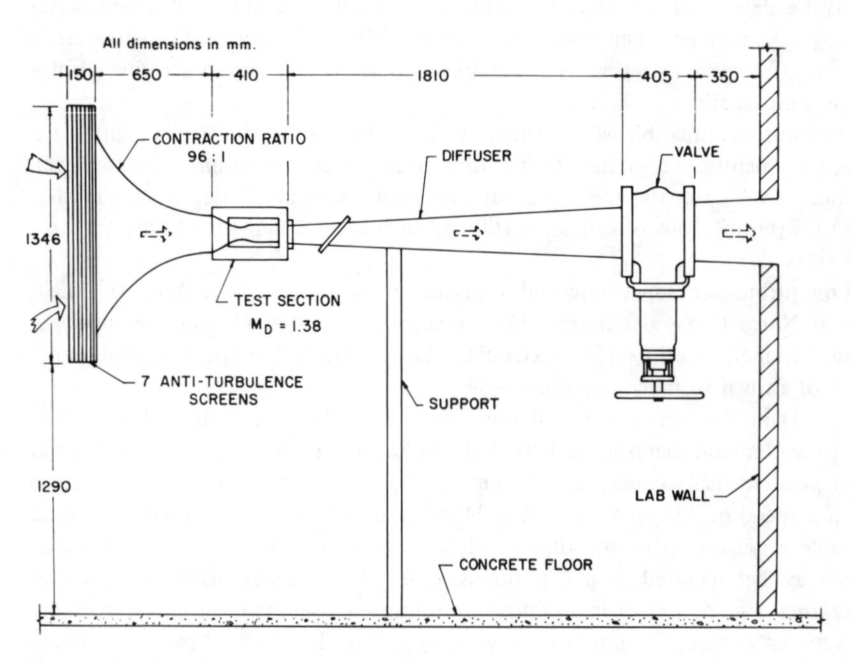

Figure 4 Pilot supersonic smoke tunnel. *(Courtesy of T. J. Mueller, University of Notre Dame.)*

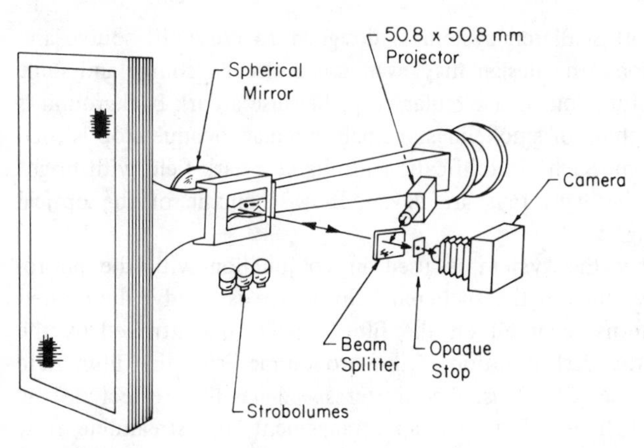

Figure 5 Modified schlieren system and lighting arrangement for simultaneous schlieren-smokeline photographs. *(Courtesy of T. J. Mueller, University of Notre Dame.)*

For taking pictures in the pilot tunnel, three lamps are arranged as shown in Fig. 5. For the other larger nozzles, four (and sometimes five) lamps are used. The basic lighting technique can be varied for taking an ordinary direct photograph of the smokelines; e.g., front or back lighting may be used. On some occasions, the back-lighting technique is most useful, as maximum scattering occurs when the light is brought in from behind at about 130° from the direct line of sight. However, this results in a loss of contrast, because the black background is sacrificed with this technique.

As the oil smoke is not visible through the schlieren system (i.e., its index of refraction is not much different from that of air), the use of other materials was studied by Goddard [11]. Although other gases might be used, nitrous oxide was found to be convenient because of its availability and relatively low cost. The nitrous oxide can be admitted directly into the smoke rake, although it would be better to pass the gas through some sort of piping system to return the gas to room temperature. The cloth bag inside the rake is allowed to remain as an aid to the uniform dispersion of the nitrous oxide. The gas flow is regulated during tunnel operation until the minimum flow to produce satisfactory streaklines (as observed through the schlieren) is obtained. This method of supersonic streakline visualization is quite attractive because of its simplicity, low cost, and adaptability to any existing installation where the proper inlet reduction ratio and antiturbulence screening can be provided.

Goddard [11] developed a modified nonparallel, double-pass schlieren system for use with the supersonic smoke tunnel. It is fabricated from a single spherical mirror where the light source is located at the center of curvature. This type of schlieren system can also be further classified as to whether it is truly coincident or noncoincident (i.e., whether the source and knife-edge are located on or slightly off the optical axis). To accomplish the desired results (i.e., to take a combined smoke and schlieren photograph), the true coincident type is a requisite, from the standpoint of light-gathering capabilities.

The ordinary coincident schlieren system is designed to use a slit source and knife-edge stop. On occasions, the design may even use a circular source and some sort of iris mechanism to function as a circular stop. Because a dark background is necessary for the photographing of smokelines, a small, circular, opaque stop is used in place of an iris mechanism. Such a modification results in a dark field with bright lines appearing wherever the light rays are deviated. A schematic of the optical arrangement is shown in Fig. 5.

Figure 5 indicates how the system is used in conjunction with the photographing of the streamlines. In use, the schlieren light source is steady; the desired contrast of the wave patterns recorded on the film negative is controlled by the camera shutter speed. As the flash duration of the strobolume lights that illuminate the streamlines is on the order of 20 μs, the shutter speed has little effect on the photographing of the smokelines. With such an arrangement, the streamline flow and wave patterns are recorded simultaneously on the same photographic film.

Originally, a red light source and a greeen stop were used, with the idea that the green stop would cause little interference with the photographing of the smokelines. It was found that a small, opaque stop, 2.381 mm in diameter, did not interfere with the light-gathering capabilities of the camera. Moreover, it is far easier to produce various sizes of opaque stops than it is to cut small circular green stops from a gelatin filter. Opaque stops are easily produced by photographing a white circle on a black background at the desired magnification. The negative so produced is then mounted in a 50.8- $\times$ 50.8-mm glass slide and, as a unit, is ready for use as a stop. The red filter was not removed, even after it was determined that an opaque stop was satisfactory; the red light gives some semblance of monochromatic light. Even more important is the fact that, as an absolutely dark field cannot be fully realized, the red light provides a better background for the smokelines; the photographs produced then have a good contrast between the smokelines and the background.

2.1.8 Application of smoke-tube method. The smoke tubes or smoke filaments or, as many prefer, smokelines, that emanate from each opening of the smoke rake are streaklines. Since a continuous tube or streak of smoke is usually used, the patterns are the result of all previous motion and, thus, produce an integrated effect that must be considered as such. The smoke can hinder as well as serve the experimenter in defining flow phenomena. For example, as is pointed out later in this section, very shallow waves appear in the transition region. To make these waves visible, a very thin layer of smoke must be used. If the smoke is too thick, the waves cannot be seen, and it is very easy to conclude that no disturbances exist. Yet, if the very limited quantity of smoke necessary to define the shallow waves is used, then phenomena that take place at a larger distance above the surface of the model are not defined, and erroneous conclusions can likewise be reached. It is imperative, therefore, to use various quantities of smoke and to take enough photographic data to completely define the flow phenomena. It is also important to remember that the patterns represent a material deformation, or lagrangian representation of the motion; therefore, care must be taken in interpreting smoke streakline photographs.

Furthermore, it is always desirable to compare the data from smoke and dye techniques with pressure, velocity, and force measurements, as well as with theoretical or numerical analysis.

Transition in attached shear layers. The transition process in the boundary layer of an axisymmetric body is an important factor that determines the magnitude and direction of the aerodynamic forces acting on the body. These forces are closely related to how rapidly the boundary layer grows and whether or not it separates from the body surface. Both these factors have a significant effect on the aerodynamic forces. Although the transition process in attached shear layers has received a great deal of attention, there is, at the present time, no theory of transition to turbulence. There are some excellent flow-visualization experiments, however, that show the important physical features of transition.

It is generally agreed that the transition from laminar to turbulent flow may be described as a series of events that take place more or less continuously, depending on the problem being studied. Since turbulence is a diffusive and dissipative phemomenon with large-scale, three-dimensional vorticity fluctuations, the breakdown of a two-dimensional or three-dimensional laminar flow may be viewed as the process whereby finite-amplitude velocity fluctuations, or traveling wave disturbances, acquire significant fluctuations and three-dimensionality. The velocity fluctuations or traveling wavefront, which is initially straight (for the flow over a flat plate), develops spanwise undulations that are enhanced by second-order effects. For the flow over a nonspinning axisymmetric body, the traveling wavefront is axisymmetric.

Studies of the boundary layers on spinning and nonspinning axisymmetric bodies were performed under the direction of the author and R. C. Nelson at the University of Notre Dame. The baseline model for these flow-visualization, pressure, and force studies was an axisymmetric model consisting of a 3-caliber secant ogive nose, a 2-caliber cylindrical midsection, and a 1-caliber 7° conical boattail. Two smooth baseline models were designed, one to be used in the flow-visualization and force tests, and the other for measuring the pressure distribution along the body. The flow-visualization model was designed to be constructed in three parts. The cylindrical midsection contained the bearings, mounting supports, and drive system. The flow-visualization models were anodized black to improve photographic contrast between the model and smoke filaments. For zero spin and zero angle of attack, a single smoke tube was positioned to impinge on the sharp nose of the baseline model in a symmetrical fashion. The experiments covered the range of Reynolds numbers based upon body length between 315,000 and 1,030,000.

The high-speed smoke photographs and pressure data indicated several trends that occurred as Reynolds number was varied during the study of spontaneous transition at zero free-stream pressure gradient [12]. The smoke photographs showed that the point where axisymmetric vortex rings were formed (downstream of where separation occurred) moved up the boattail as the Reynolds number was increased from 315,000. These axisymmetric vortex rings formed in the free shear layer downstream of the separation point and accelerated rapidly with the flow. These

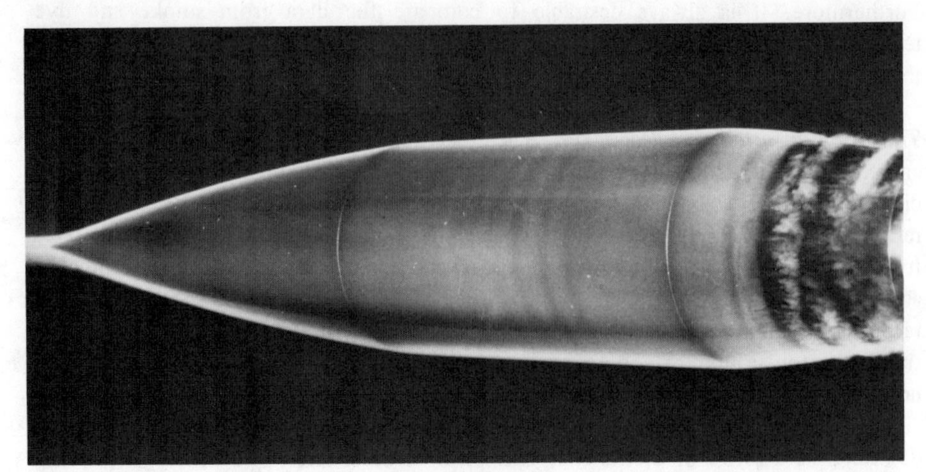

Figure 6 Smoke photograph of nonspinning axisymmetric body for $\alpha = 0°$ and $Re_L = 815{,}000$. *(Courtesy of T. J. Mueller, R. C. Nelson, and J. T. Kegelman, University of Notre Dame.)*

became turbulent more quickly as Reynolds number was increased. Two-dimensional (i.e., axisymmetric) Tollmien-Schlichting waves appeared sporadically along the body at $Re = 0.631 \times 10^6$ and appeared continuously at all higher Reynolds numbers, as shown in Fig. 6. Vortex trusses were formed on the body for Reynolds number 0.928×10^6 and 1.030×10^6 (see Fig. 7). At the highest Reynolds number there was continuous formation of the two-dimensional waves.

The transition phenomenon observed at the highest Reynolds number showed many of the characteristics of transition described in the earlier works conducted at Notre Dame by Brown [13]. The formation of the two-dimensional Tollmien-Schlichting waves and the subsequent breakdown of these waves into vortex trusses

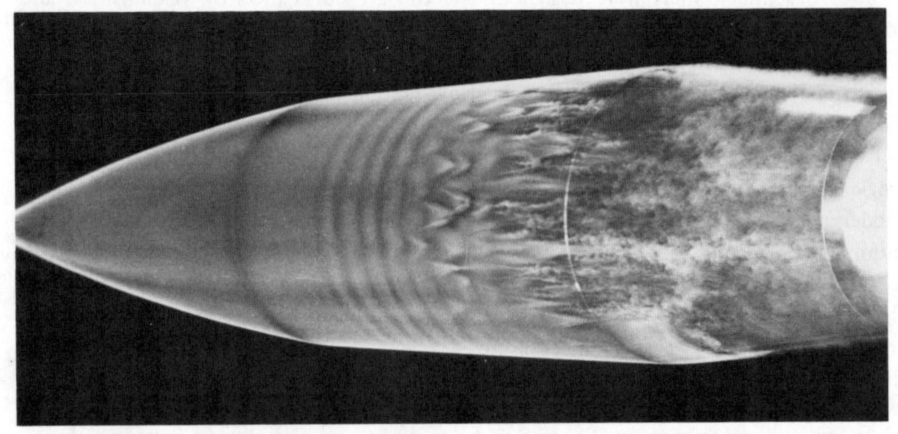

Figure 7 Smoke photograph of nonspinning axisymmetric body for $\alpha = 0°$ and $Re_L = 1{,}030{,}000$. *(Courtesy of T. J. Mueller, R. C. Nelson, and J. T. Kegelman, University of Notre Dame.)*

was strikingly similar to the previous works. The intermittent nature of the phenomenon observed in this work, however, was not entirely identical to that observed by Brown and his associates on tangent ogive models. Although it was observed that the two-dimensional waves were formed continuously at the same point on the model, the formation of the vortex trusses was somewhat intermittent. The trusses were formed in groups, periodically appearing anywhere from 0.9-caliber to 1.75-caliber along the body. It should be noted that the tunnel speed at which the transition process occurred entirely on the body was significantly higher than that used in the previous works, where the model was longer [13]. These higher velocities yield a much thinner boundary layer at the transition point, resulting in higher frequencies and smaller amplitudes of the two-dimensional waves, as well as smaller trusses.

For a spinning axisymmetric body, vortices originate in the crossflow and spiral around the body [14]. These crossflow vortices eventually break down into turbulence, but do so in a distinctly different manner than axisymmetric waves. Depending on the length Reynolds number and spin ratio, the transition process is initiated by either the breakdown of axisymmetric waves or the breakdown of the vortices generated in the crossflow. Furthermore, for certain combinations of these parameters, both the axisymmetric waves and the crossflow vortices occur. The simultaneous occurrence of these two phenomena was first discovered at the University of Notre Dame using smoke visualization. Because of the complex nature of the transitional process and the sensitivity of the individual events in this process, experiments are very difficult. The most important recent contributions to understanding the physics of the transition process have come from flow-visualization experiments. The nonintrusive nature of flow visualization and its global view have been important factors in its success.

The baseline model was studied at spin rates from 0 to 4500 rpm at angles of attack from 0° to 10° for Reynolds numbers based upon total body length of between 315,000 and 1,030,000. For the spinning model at 0° angle of attack, the phenomenon was primarily related to the ratio of the peripheral velocity to the free-stream velocity V/U_∞, and relatively independent of Reynolds numbers (i.e., it was not significantly affected by changes in Reynolds number for a given V/U_∞) [14]. Tests were conducted for a range of V/U_∞ between 0 and 1.67. There were no notable changes in the boundary-layer characteristics for V/U_∞ less than 0.4, with the exception of a slight skewness in the tips of the vortex trusses. When vortex trusses were present, this skewness could be seen for V/U_∞ values as slow as 0.1. As V/U_∞ increased, striations in the smoke (manifestations of the crossflow vortices) appeared at an angle approximately equal to $\tan^{-1} V/U_\infty$, as shown in Fig. 8. The wavelength of the striations λ/D was approximately 3.8×10^{-2} and remained constant regardless of spin ratio or Reynolds number. As the Reynolds number was increased, the transition process took place over a shorter distance. The striations broke down into a helicoidal shape just before becoming turbulent (see Fig. 8). The transition zone moved forward with increasing spin rate. Furthermore, when the striations appeared toward the end of the midsection, they were superimposed (Fig. 9) on the two-dimensional (axisymmetric) Tollmien-Schlichting

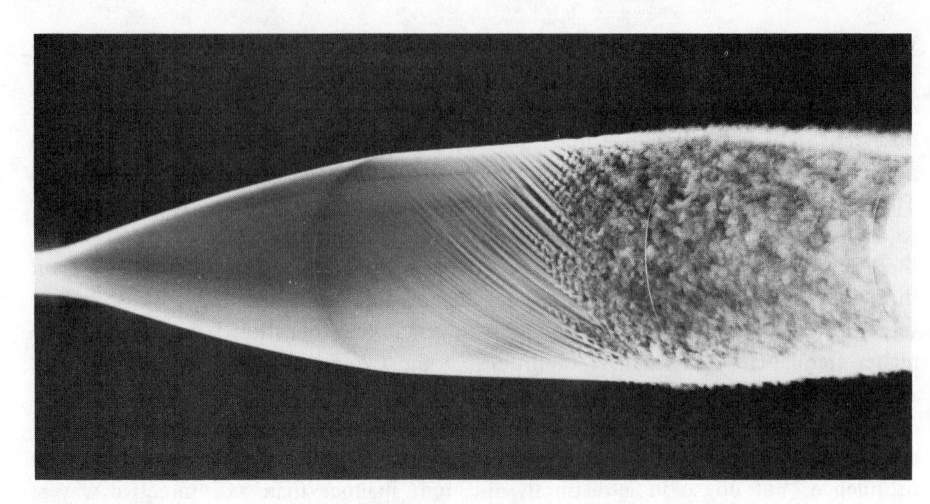

Figure 8 Typical striations in the smoke resulting from crossflow vortices for $\alpha = 0°$, $V/U_\infty = 0.848$ (1250 rpm), and $Re_L = 315{,}000$. *(Courtesy of T. J. Mueller, R. C. Nelson, and J. T. Kegelman, University of Notre Dame.)*

waves, which are similar to those that appear on the nonspinning body. At high values of V/U_∞ (greater than 1.0), the boundary layer was fully turbulent along the entire midsection, regardless of Reynolds number.

Magnus-force and flow-visualization data were also obtained for the baseline model for Reynolds numbers of 315,000 and 1,030,000 based upon the body length. The angle of attack of the model was varied from 0° to 10° in 2° increments, and the nondimensional spin rate V/U_∞ was varied from 0 to 1.67. The

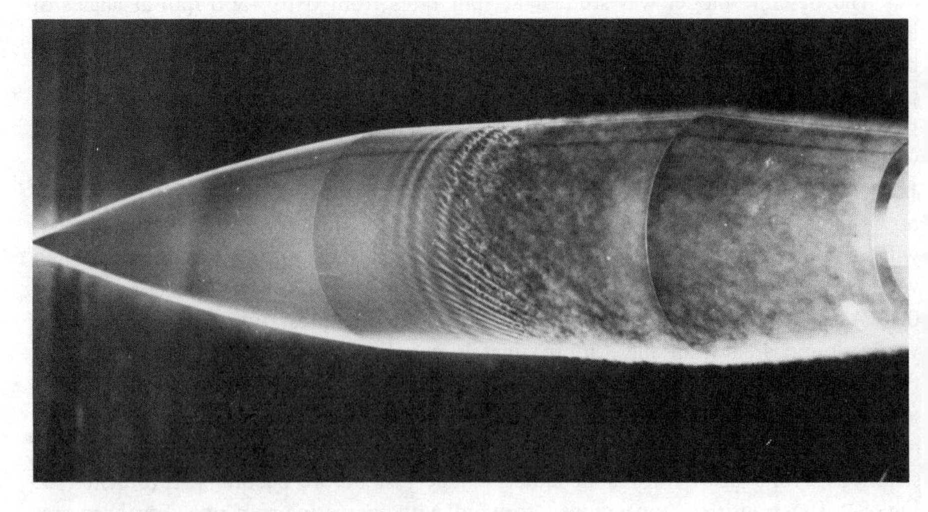

Figure 9 Simultaneous appearance of Tollmien-Schlichting waves and crossflow vortices for $\alpha = 0°$, $V/U_\infty = 0.61$ (2900 rpm), and $Re_L = 1{,}030{,}000$. *(Courtesy of T. J. Mueller, R. C. Nelson, and J. T. Kegelman, University of Notre Dame.)*

data exhibited both positive and negative side forces over the spin rates tested. Several trends were observed, which agreed with both the visual data and force measurements of the spinning model at angle of attack [14]. At 2° and 4° angles of attack, transition took place via the formation and breakdown of the striations, closely resembling the 0° angle-of-attack transition process. These striations occurred symmetrically about the axis of rotation; however, they were antisymmetric about the plane of angle of attack. The striations moved forward and became shorter as spin was increased. They occurred at an angle $\theta = \tan^{-1} V/U_\infty$. At a moderate angle of attack ($\alpha = 6°$), the transition region was confined to a small patch that passed across the starboard side of the model as spin was increased, corresponding exactly with a rise in the side-force-versus-V/U_∞ curve for $\alpha = 6°$. The first mechanism of transition in the patch region appeared as the formation and breakdown of the waves skewed in the same direction as the striations, but it did not occur at an angle equal to $\tan^{-1} V/U_\infty$. For high spin at an angle of attack of 6°, the striations appeared superimposed over the large-amplitude waves. At the high angle of attack of $\alpha = 10°$, a patch region only formed at 0 spin rate (Fig. 10) and passed over the starboard side of the model as the spin rate increased (Figs. 11 to 13), coincident with the magnitude of the negative Magnus force. No striations were formed. At high spin rates and high angles of attack, the side force became negative, corresponding to a positive Magnus force.

Several experiments were also performed to study the influence of acoustical disturbances on the transition process. Sound from a sine-wave generator was introduced upstream of the wind-tunnel inlet. Still and high-speed movie photography of the smoke clearly indicates the influence of the acoustical disturbances on the boundary layer of the axisymmetric body. The case studied is shown in Fig. 6 with no sound. Figures 14 and 15 show what happens for sound frequencies of 551 and 785 Hz, respectively. For example, it has been found that the Tollmien-

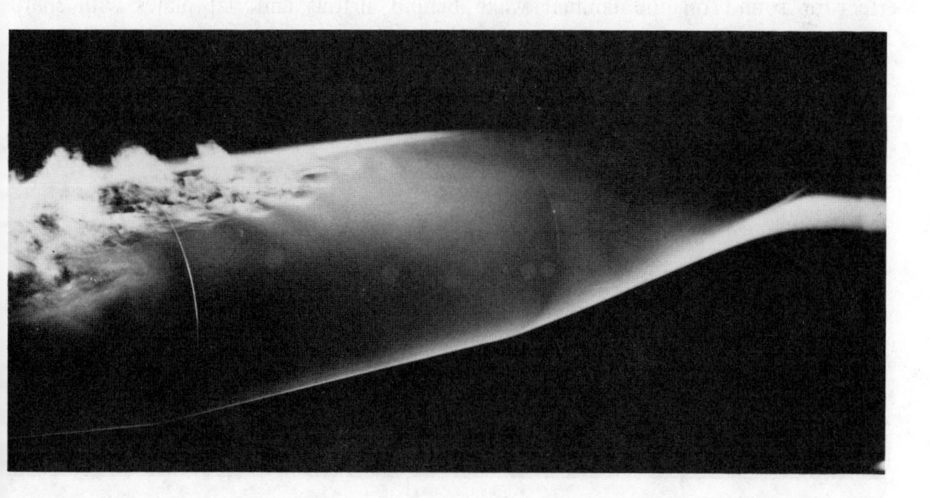

Figure 10 Starboard view at $\alpha = 10°$, $V/U_\infty = 0$ (0 rpm), and $Re_L = 315{,}000$. *(Courtesy of T. J. Mueller, R. C. Nelson, and J. T. Kegelman, University of Notre Dame.)*

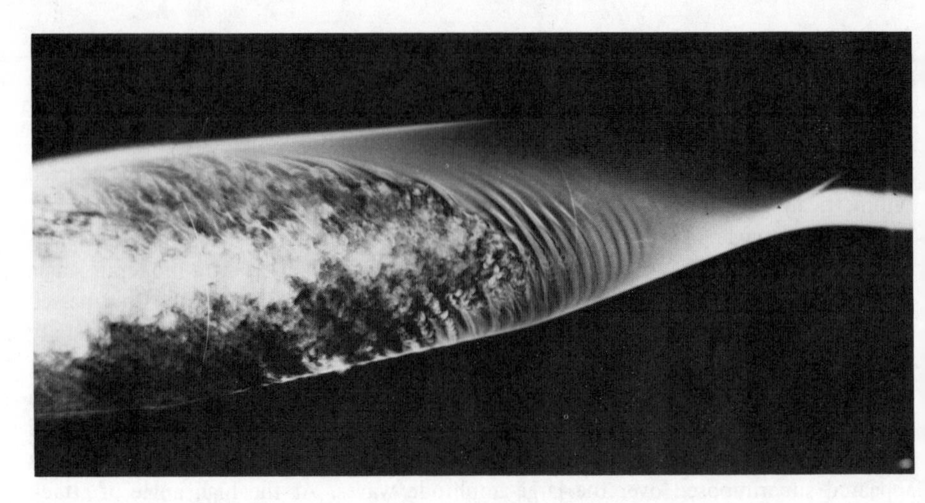

Figure 11 Starboard view at $\alpha = 10°$, $V/U_\infty = 0.65$ (1000 rpm), and $\mathrm{Re}_L = 315{,}000$. *(Courtesy of T. J. Mueller, R. C. Nelson, and J. T. Kegelman, University of Notre Dame.)*

Schlichting wave formation can be altered substantially and controlled to some degree with the use of an external disturbance such as sound. Sound also produces a significant change in the truss formation and wavelength.

It is clear that with this flow-visualization technique, it is now possible to conduct a detailed investigation of the three-dimensional deformations that initiate the breakdown of both Tollmien-Schlichting waves and the vortex tubes resulting from the crossflow instability.

Influence of sound on a laminar wake. Brown and Goddard [15] investigated the effect of sound on the laminar wake behind airfoils and flat plates with sharp

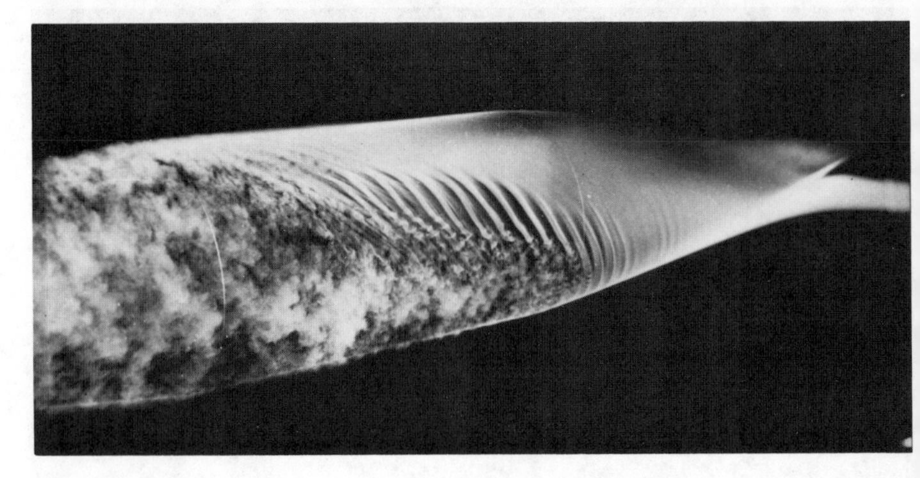

Figure 12 Starboard view at $\alpha = 10°$, $V/U_\infty = 1.0$ (1500 rpm), and $\mathrm{Re}_L = 315{,}000$. *(Courtesy of T. J. Mueller. R. C. Nelson. and J. T. Kegelman, University of Notre Dame.)*

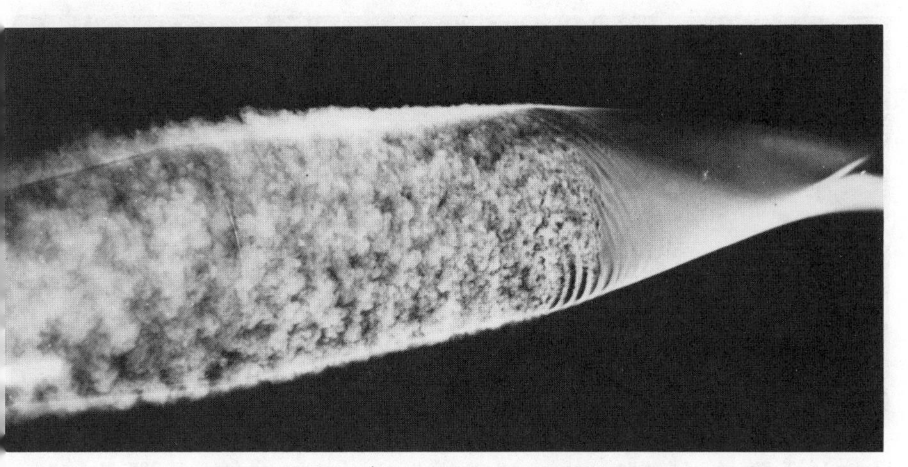

Figure 13 Starboard view at $\alpha = 10°$, $V/U_\infty = 1.65$ (2500 rpm), and $Re_L = 315{,}000$. *(Courtesy of T. J. Mueller, R. C. Nelson, and J. T. Kegelman, University of Notre Dame.)*

leading and trailing edges. In the visual study of the flow past an airfoil or flat plate, the three basic characteristics of the wake flow with or without sound are: the frequency of vortex-pair formation, the geometry of the vortices (i.e., the spacing ratio h/λ, where h is the vertical distance between rows, and λ is the horizontal distance between vortices in the same row), and the wake speed V_w as compared to the free-stream velocity. Since the wake becomes visible with the introduction of smoke, the frequency of vortex formation can be determined using standard stroboscopic techniques. The geometry of the wake can be determined by linear measurements made from photographic prints on dimensionally stable photo-

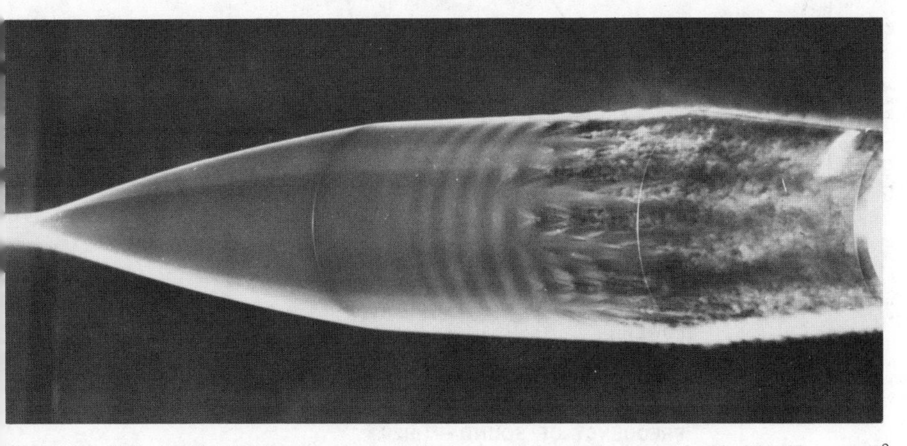

Figure 14 Tollmien-Schlichting waves and their breakdown for sound of 551 Hz, $\alpha = 0°$, $V/U_\infty = 0$ (0 rpm), and $Re_L = 815{,}000$. *(Courtesy of T. J. Mueller, R. C. Nelson, and J. T. Kegelman, University of Notre Dame.)*

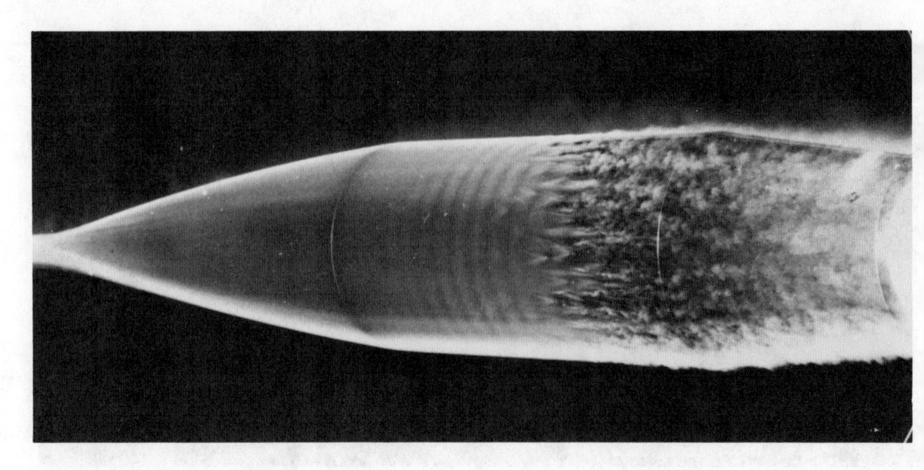

Figure 15 Tollmien-Schlichting waves and their breakdown for sound of 785 Hz, $\alpha = 0°$, $V/U_\infty = 0$ (0 rpm), and $Re_L = 815{,}000$. *(Courtesy of T. J. Mueller, R. C. Nelson, and J. T. Kegelman, University of Notre Dame.)*

graphic paper. If both the free-stream velocity and the frequency of vortex formation are known, the wake speed can be determined.

The introduction of sound has a direct influence upon the frequency of vortex formation. The wake frequency follows the sound frequency for a limited range both above and below the natural frequency of formation. The vortex spacing ratio varied with the sound frequency. However, the wake speed was invariant with change in sound frequency. The effects of sound on the wake flow behind a typical flat plate are summarized in Fig. 16, where results for two different free-stream speeds are shown. The sound control limits, upper and lower, obtained from

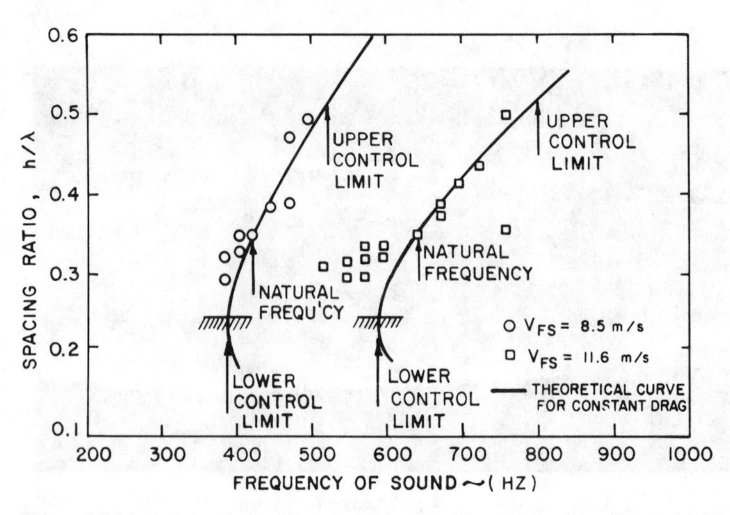

Figure 16 Variation of spacing ratio with sound frequency for a flat plate. *(Courtesy of University of Notre Dame.)*

experiments are indicated in the figure. Theoretical curves of the spacing-ratio variation with change in sound frequency are also shown. These curves were generated by making use of von Karman's work, in which he showed that the drag of an object can be determined from wake characteristics. Since the drag is unchanged when sound is introduced, if the wake characteristics without sound are known, then the theoretical curve can be generated from the change in wake characteristics with sound. Figures 17 and 18 are smoke photographs of the flat-plate wakes used in Fig. 16; they show the difference in spacing ratio when sound is introduced.

Dynamic stall on an oscillating airfoil. In unsteady-flow experiments, useful information concerning the gross behavior of a complicated flow field can often be obtained, even when individual streaklines cannot be identified. An example of such a situation is dynamic stall on an oscillating airfoil.

The main obstacle to the continuing development and improvement of the conventional helicopter configuration has been the dynamic retreating blade stall that accompanies high loading and advance ratios [16]. The most important characteristics of dynamic stall are the unusually high lift coefficients, angles of attack α, and pitching moments C_m. A large amount of research has been directed at understanding and documenting this complex unsteady-flow problem in the past decade [16 to 18]. In research under the direction of W. J. McCroskey, smoke visualization was very helpful in studying this problem. Thin layers of smoke were emitted through slits in the airfoil's leading edge, and downstream through slits in the upper surface of the airfoil. High-speed movies and still photographs were taken from directly above the airfoil, using high-intensity quartz lamps and high-intensity strobe lights, respectively [16, 17].

It was found that the predominant feature of dynamic still is the shedding of a strong vortexlike disturbance from the leading-edge region. For an airfoil whose incidence or angle of attack is increasing rapidly, the onset of stall can be delayed to incidences much higher than the static stall angle, as shown in Fig. 19. The vortex shed from the leading-edge region moves downstream over the upper surface

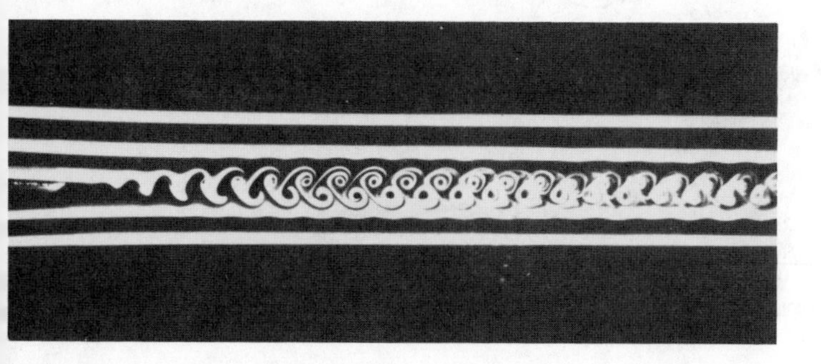

Figure 17 Flat plate of Fig. 16; $U_\infty = 8.5$ m/s and sound frequency of 0 Hz. *(Courtesy of University of Notre Dame.)*

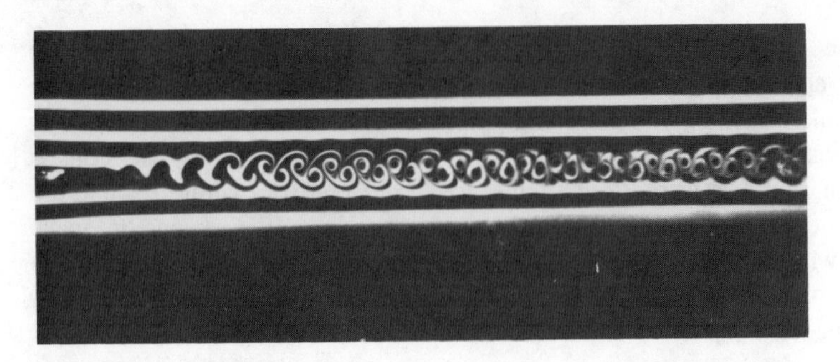

Figure 18 Flat plate of Fig. 16; $U_\infty = 8.5$ m/s and sound frequency of 470 Hz. *(Courtesy of University of Notre Dame.)*

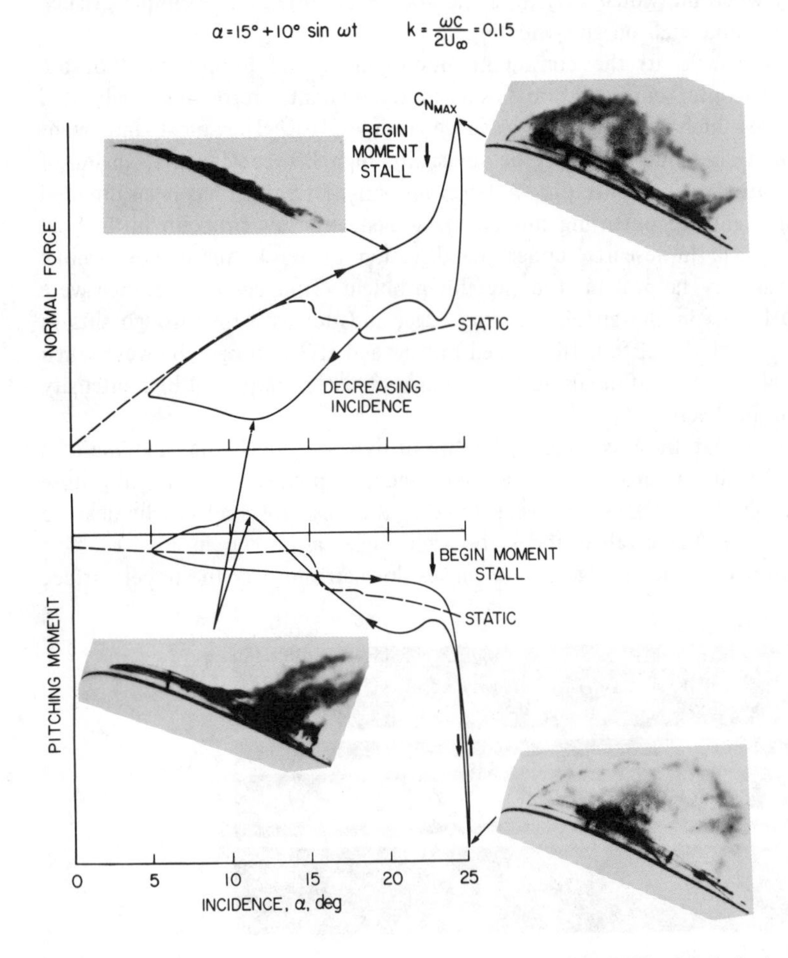

Figure 19 Normal force and pitching moment on NACA 0012 airfoil during dynamic stall ($\alpha = 15° + 10° \sin \omega t$, $k = \omega c/2U_\infty = 0.15$, $Re = 2.5 \times 10^6$) [17]. *[Courtesy of W. J. McCroskey, U.S. Army Aeromechanics Laboratory (AVRADCOM).]*

of the profile, as shown in Fig. 20, for a Reynolds number based upon chord c and free-stream velocity U_∞ of 10^6. The passage of this vortex distorts the chordwise pressure distribution and produces transient forces and moments that are basically different from their static counterparts. The pressure coefficient C_p is graphed versus position along the airfoil chord X/c in Fig. 20, where ω is the angular velocity, t is the time, and k is the reduced frequency, defined as angular velocity times the chord divided by two times the free-stream velocity.

Supersonic flow visualization. Figure 21 shows the supersonic flow past a wedge. This photograph was made with a modified schlieren system to obtain simultaneous schlieren lines and smokelines [10]. The photograph shows the two fundamental ways in which supersonic flow tends to follow parallel surfaces: flow through a shock wave and expansion around a corner. In the flow through the shock wave, the abrupt deflection of the streamline to flow parallel to the front surface of the wedge can be observed. The expansion at the shoulder of the wedge to follow parallel to the main body of the wedge is clearly seen. The Mach number of the flow can be determined from the photograph by measuring the shock angle and the streamline deflection. In a similar way, streamlines can be followed throughout the flow field to map the entire flow field.

To study wake flows, a lucite wedge-shaped plug with a rounded leading edge was inserted slightly upstream of the nozzle throat [19]. This configuration resembles the plug nozzle that is referred to as the "expansion-deflection nozzle," or it may be thought of as representing a strut, a flame holder, a scramjet fuel injector, etc. A simultaneous smokeline and opaque-stop schlieren photograph of the wake flow using a laser light source is presented in Fig. 22. Smoke is not visible in a

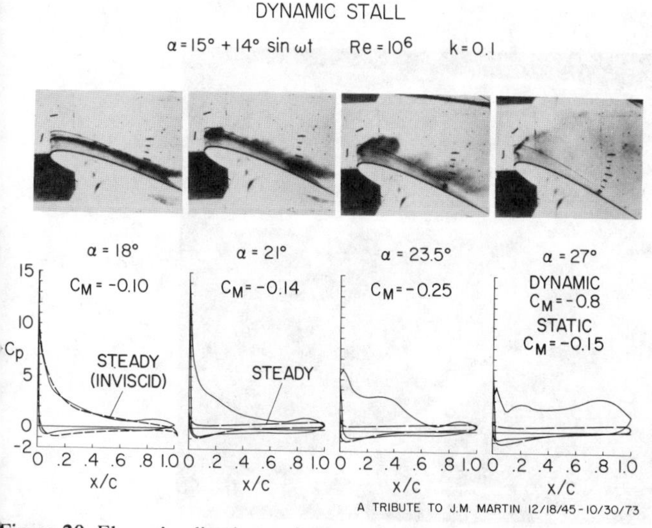

Figure 20 Flow visualization and pressure measurements of the vortex-shedding phase of dynamic stall on an oscillating airfoil [18]; pitch axis at $X/c = 0.25$. *[Courtesy of W. J. McCroskey, U.S. Army Aeromechanics Laboratory (AVRADCOM).]*

Figure 21 Supersonic flow past a 5° half-angle wedge; Ma = 1.38. *(Courtesy of University of Notre Dame.)*

schlieren system, since it has approximately the same index of refraction as air. By measuring the local deflection of the smokelines passing through the recompression shock wave and the wave angles, the Mach number immediately ahead of and behind the recompression shock can be determined. Another series of experiments was run with an actual expansion-deflection nozzle. A comparison of the Mach numbers obtained from smokeline-shock wave patterns and from total- and static-pressure measurements is shown in Fig. 23. This correlation is for the Mach

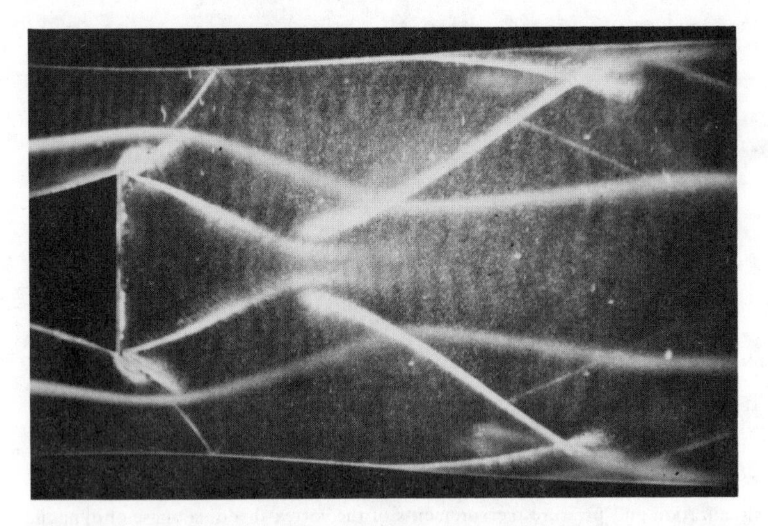

Figure 22 Simultaneous smokeline and opaque-stop schlieren photograph of wake flow with laser light source. *(Courtesy of T. J. Mueller, University of Notre Dame.)*

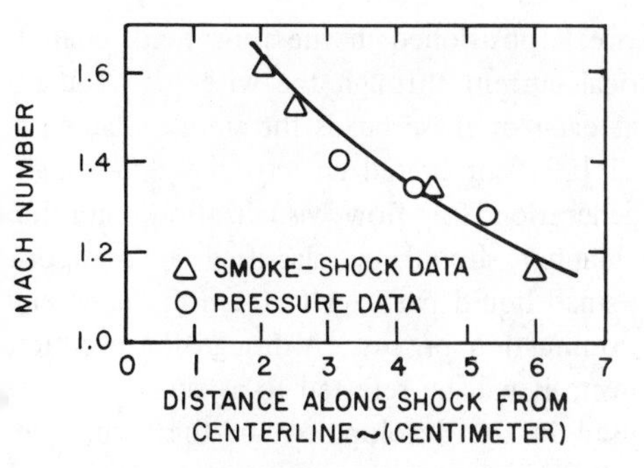

Figure 23 Correlation of smokeline-shock data and pressure data for Mach number immediately downstream of recompression shock for E-D nozzle. *(Courtesy of T. J. Mueller, University of Notre Dame.)*

number immediately downstream of the recompression shock versus the distance measured along the recompression shock from the nozzle centerline.

These high-speed flow-visualization techniques have also been used to study a transonic cascade problem [20]. More recently, the techniques have been used to develop design criteria. for improved high-speed smoke wind tunnels [5].

2.2 Smoke-Wire Method

Although the smoke-tube method has been used successfully to study many complex flow problems, there are fundamental flow phemomena that require the ability to produce small but discrete smoke filaments (streaklines) and to locate these filaments accurately within the flow field so that small-scale details may be studied.

The smoke-wire technique, developed by Raspet and Moore in the early 1950s and subsequently improved and extended [21 to 25], is capable of producing very fine smoke filaments and can be used to study the detailed structure of complex flow phenomena. The "smoke" is produced by vaporizing oil from a fine (~0.1-mm) wire by the use of resistive heating. The technique was initially applied to the measurement of velocity profiles in a boundary layer. Recent applications have included the investigation of the large eddy structure in turbulent shear flows [25 to 27]. Corke et al. [24] indicate a number of benefits associated with the smoke-wire technique, but, for the applications documented here, the method's primary benefit is the fine structure of the smoke streaklines.

A study of the separation bubble near the leading edge of an airfoil at low Reynolds numbers including transition is discussed here. This particular phenomenon presents some unique challenges that indicate the need for flow visualization, and the smoke wire is an ideal candidate. It can be used to study the physics of this complex flow phenomenon in regions of slow, recirculating flow, where other measurement techniques such as hot-wire and pressure transducers are very difficult to use. The wake behind a bluff body in a linear shear flow is also discussed.

The smoke-wire technique is limited to (and, fortunately, ideally suited to) applications where the Reynolds number based on wire diameter is small (~20.0). For practical applications, this requires wind-tunnel speeds on the order of 4 to 6 m/s. In the smoke-wire method, a fine wire is positioned in the flow field, coated with oil, and heated by passing an electrical current through the wire [28]. Small beads of the oil form on the wire, and at each of these beads the smoke filaments originate when the wire is heated. Cornell [29] conducted an extremely thorough study of materials used in "smoke" generation for flow visualization, and he indicated that particles formed in this manner should be classified as a vapor-condensation aerosol, so they are actually small liquid particles (~1 μm in diameter) and not solid particles or products of a combustion process. Although the product is not smoke in a strict definition of the term, it will be referred to as such.

A number of different wires were used in the development of the technique [28]. These included 0.025-, 0.076-, and 0.152-mm-diameter stainless steel and tungsten wires. The strength, resistive heating characteristics, and size are all important factors in choosing the "best" wire for a given application. To minimize the disturbance produced by the wire, which must be close to the model in the test section, the Reynolds number based upon the wire diameter was maintained at less than 20.0.

As the wire is heated, it expands and sags, which is not desirable if accurate placement of the smoke streaklines is required. Therefore, the wire was prestressed so that, when heated, there was no noticeable sagging. This required a prestress of approximately 1.03×10^9 Pa for the stainless wire. Since this is quite near the yield stress for the wire, it was important that once it was stressed it was handled carefully.

2.2.1 Applying oil to the wire. A number of different liquids have been used to coat the smoke wire to produce the smoke filaments. These included several types of lubricating and mineral oils and a commercially available product, Life-Like Model Train Smoke, produced by Life-Like Products of Baltimore, Md. The results were similar to those achieved in [29], which indicated that the model-train smoke produced the best smoke filaments. This product is composed of a commercial-grade mineral oil to which a small amount of oil of anise and blue dye have been added. The oil is very easy to work with, and such small quantities are used that the commercial product is ideal and inexpensive.

There would be obvious safety problems associated with large quantities of the smoke, but since such small amounts are produced and, in this application, the open-circuit tunnel exits to the outside of the laboratory, there were no safety problems. Reference [24] even indicates that since the amount of smoke is so small the method is quite suitable for limited use in closed-circuit tunnels.

There appear to be a number of methods that can be used to coat the wire, each with its own advantages and disadvantages. Reference [24] documents a pressurized gravity-feed method that might be suitable for a vertical wire. A method similar to this was tried, but there were problems with the large droplets running down the wire and coating the surface. These droplets would periodically be blown off the wire and would wet the surface of the airfoil model. Nagib [25] has developed a "windshield-wiper" device that automatically coats the wire by wiping oil along the wire, but this apparatus would have to be located within the test section, which was unsuitable for the types of tests to be conducted. The technique used for these applications was manual coating. The rear wall of the test section was fitted with an easily removable section. Before each use of the wire, the section was removed and the wire carefully wiped with a cotton applicator soaked with oil. This provided a uniform coating with no model fouling. The method is somewhat cumbersome, but since each coating could be accomplished within a few seconds, it was adequate.

2.2.2 Timing circuit. As the coated wire is heated, fine smoke streaklines are formed at each droplet on the wire (~8 lines/cm for the 0.076-mm wire). Depending upon the current through the wire, the beads can be vaporized very rapidly or, if a lower current is used, continuous filaments of adequate density will emanate from the wire for as long as 2 s. With even lower current, the streaklines become fainter and cannot be photographed. For example, the 0.076-mm-diameter, 302 stainless wire of 0.40 m length was heated using a power-supply setting of approximately 0.7 amp. Both ac and dc power supplies were tried. However, the dc power supply was selected for ease of control and because the steady current results in smoother and better-defined smokelines. Because of the relatively short duration of smoke generation for a single wire coating, it was important that the event being photographed, the lighting, the camera, and the smoke be properly controlled and synchronized. To accomplish this, a timing circuit was designed. The design is a modification of those shown in [24, 28] and is included here in Fig. 24.

The power-set potentiometer controls the amount of current passing through the wire, and the pulse-length potentiometer controls the length of time this current

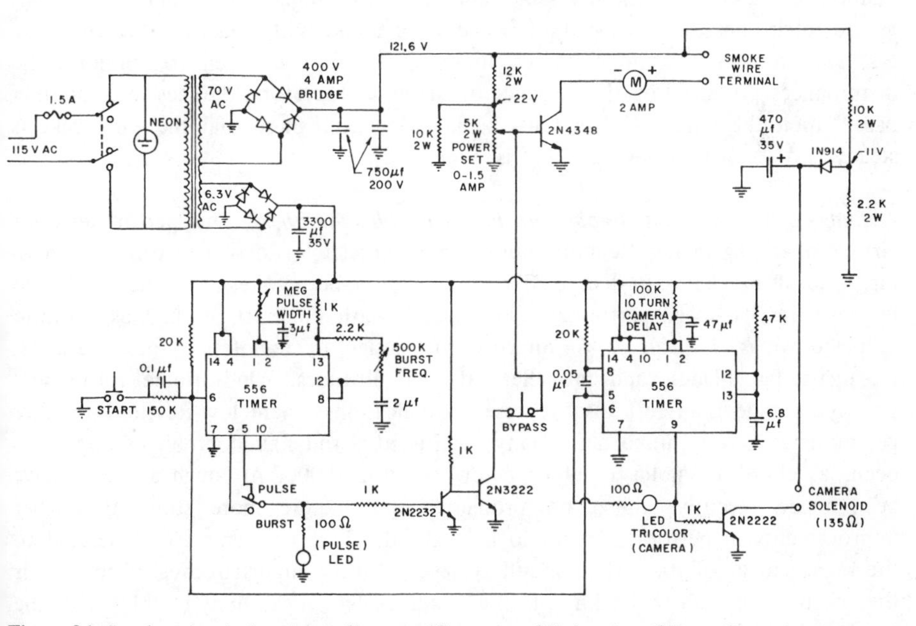

Figure 24 Smoke-wire timing-circuit diagram. *(Courtesy of University of Notre Dame.)*

is applied. The burst potentiometer can be switched into the circuit to enable the operator to control the frequency of current pulses during the burn, resulting in an intermittent pattern of smoke. The wire is heated and cooled because of the pulsed voltage; the smoke density varies in a similar manner, and streak-time lines can be formed. An additional potentiometer in this "switched" circuit controls the on-off period of these bursts. The camera-delay potentiometer controls the time delay before the camera is triggered.

All these controls can be preset, reducing the photographing of a given event to a single-step operation. The circuit applies power to the smokewire, and, with the appropriate user-set delay, it activates a camera and lights using a solenoid attached to a camera trigger. The controls are set, the wire is oiled, and the start button is momentarily depressed. This causes the smoke to be generated and the camera to be triggered when the smoke has reached the desired intensity and location. A similar circuit could be designed so the camera is triggered by some event within the tunnel, such as an oscillating flap or airfoil. This would allow for a conditional photographic sampling. The timing circuit is invaluable in the practical application of the smoke-wire technique.

2.2.3 Application of the smoke-wire method. The smoke tube makes use of rather large quantities of smoke that are produced in a smoke generator outside the wind tunnel and introduced into the tunnel just upstream of the inlet. Visualization data acquired using this method are characterized by a few (one to about six) rather thick smoke streaklines (i.e., about 16 mm in diameter). The smoke-wire method, however, is capable of producing very thin smoke streaklines (i.e., about 1 mm in diameter). These very thin streaklines are, of course, subject to the same limitations as the thick ones discussed earlier. Because the smoke wire is usually located in the test section near the model under study, care must be taken to minimize the disturbances produced by the presence of the wire. Therefore, the Reynolds number based upon the wire diameter must be kept low enough so that the wire wake is steady and as small as possible (that is, $\mathrm{Re}_d < 40$).

Leading-edge separation bubble at low Reynolds numbers. The performance of airfoils operating in low-Reynolds-number incompressible flows has been of increasing interest in the past decade. This interest has been a result of the desire to improve the low-speed performance of general-aviation aircraft and high-aspect-ratio sailplane wings, as well as to improve the design of remotely piloted vehicles, jet-engine fan blades, and propellers at high altitudes. Wind-turbine rotors and free-flying model aircraft also represent applications where low-Reynolds-number performance is very important. Many significant aerodynamic problems appear to occur at chord Reynolds numbers below about 200,000. Although advances have been made recently, there are problems that require more study if further improvements in performance are to be realized. These problems are all related to the management of the airfoil boundary layer. A very important area of concern is the occurrence and behavior of the leading-edge separation bubble and the associated transition phenomena in the free-shear layer. It is well known that the development and characteristics of this separation bubble are highly sensitive to Reynolds number, pressure gradient, and the disturbance environment. The separation bubble plays a critical role in determining the development of the boundary layer, which, in turn, affects the overall performance of the airfoil [30].

Prior to this study, the lowest speed attainable in the tunnel shown in Fig. 1 was approximately 7.6 m/s; the present study required a minimum speed of about half that value to achieve the required airfoil and wire Reynolds numbers. This was accomplished, with no adverse influence on the flow upstream of the test section or costly modification to the impeller power transmission, by introducing a device, similar to a flow straightener, into the diffuser section of the tunnel. An additional section, fitted with ordinary plastic drinking straws (~5 mm inside diameter and 200 mm long) was attached to the tunnel, aft of the test section and upstream of the diffuser. Two such inserts were fabricated and used singly or as a pair. This arrangement allowed for a minimum tunnel speed of 3.0 m/s and an actual improvement of the flow uniformity within the test section.

An airfoil model with a 0.25-m chord and a 0.40-m span with a NACA 66_3-018 airfoil section was fitted with end plates and mounted in the 0.6- × 0.6-m square-cross-section test section of the wind tunnel shown in Fig. 1. A schematic representation of the model and end plates is shown in Fig. 25. Also shown in Fig. 25 are the two smoke-wire locations that were used. The horizontal wire *AA* was used to introduce a sheet of fine smoke streaklines in a plane along the span of the airfoil model. The wire was located 65 mm forward of the leading edge of the airfoil and was parallel to the leading edge. The vertical location of the wire was adjustable through a screw-track device attached to the outside of each end plate. This allowed for accurate positioning of the sheet of smoke relative to the airfoil. The vertical wire *BB* was used to produce a sheet of streaklines in a plane normal to the leading edge of the airfoil. This wire was located 430 mm forward of the leading edge of the airfoil.

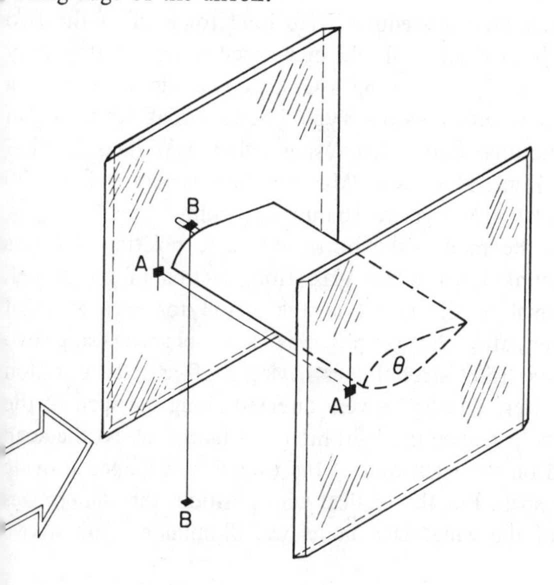

Figure 25 Schematic of wind-tunnel model end plates and wire locations. *(Courtesy of T. J. Mueller and S. M. Batill, University of Notre Dame.)*

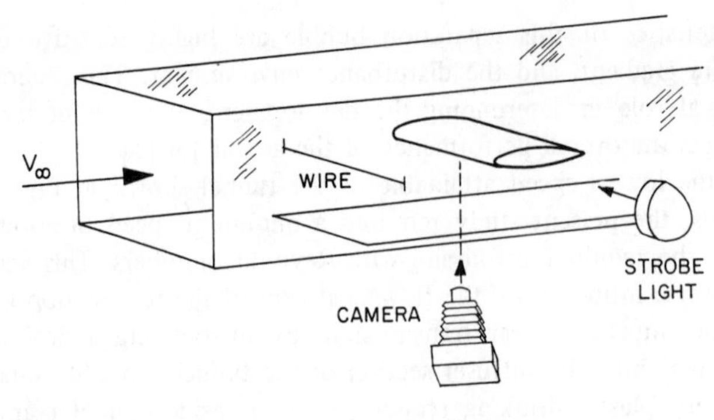

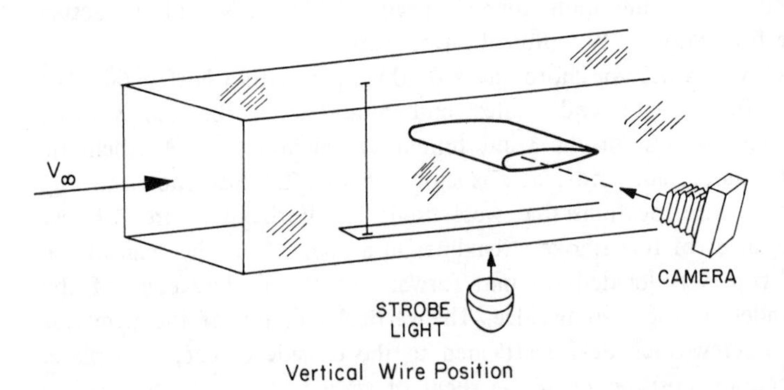

Figure 26 Wire, camera, and light locations for planform and profile views. *(Courtesy of T. J. Mueller and S. M. Batill, University of Notre Dame.)*

Different lighting and photographic procedures were used for each of the two wire orientations (Fig. 26). In both cases, still and high-speed movie photography were used. The still photographs were taken using a Graflex Graphic View camera with an ACU-Tessar 210 (f6.3) lens with Polaroid Type 57 and Kodak Royal-X Pan films. The high-speed motion pictures were taken using either a Wollensak WF-3 Fastex camera (1000 to 3000 frames/s) or a DBM-5 Milliken camera (64 to 500 frames/s). Both cameras used Kodak 4-X negative 16-mm 7224 film.

For the planform views of the model, the horizontal wire position *AA* was used, and the cameras were mounted below the glass floor section of the model. The camera was positioned normal to the chord of the airfoil for each angle of attack studied. Lighting for the still photographs was accomplished using two high-intensity General Radio Type 1532 strobolumes having a 20-μs flash duration and triggered by the camera shutter. The lights were directed along the span of the airfoil, as shown in Fig. 26. To help reduce the light-intensity falloff across the span of the wing, a mirror was placed on the back wall of the tunnel; this helped provide uniform illumination across the span. For the vertical wire position, the camera was aimed along the spanwise axis of the wing. The model was illuminated from above

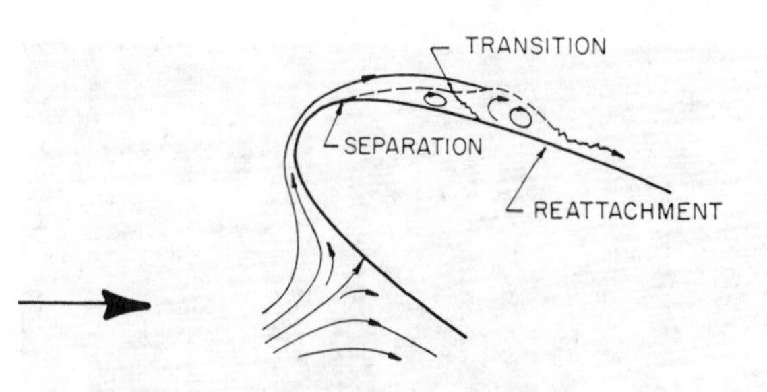

Figure 27 Schematic of leading-edge separation bubble. *(Courtesy of University of Notre Dame.)*

and below the wing, normal to the wing, through 25.0-mm slits in the top and bottom of the test section. For the Fastex camera, continuous lighting was supplied by a single 1000-W quartz lamp. The Milliken camera was synchronized with two GENRAD Type 1540 strobolumes triggered by the camera. Each of these methods provides adequate illumination and contrast for photographing the smoke streaklines.

The leading-edge separation bubble, shown in Fig. 27, is formed when the laminar boundary layer separates from the surface as a result of the strong adverse pressure gradient downstream of the point of minimum pressure [30]. This separated shear layer is very unstable, and transition usually begins a short distance downstream of separation as a result of the amplification of velocity disturbances present immediately after separation. Reattachment can occur while the shear layer is undergoing transition or after the transitional process is complete and the flow is turbulent. The region between separation and reattachment is referred to as the *separation bubble*. Figure 28 shows the leading-edge separation bubble on the NACA 66_3-018 airfoil at a chord Reynolds number of 40,000 and an angle of attack of 12°. The vertical smoke-wire configuration *BB* was used to obtain this photograph [28].

It was noted earlier that the transition from laminar to turbulent flow may be described as a series of events that take place more or less continuously, depending on the flow problem being studied. Since turbulence is essentially a three-dimensional phenomenon, the breakdown of a two-dimensional laminar flow may be viewed as the process whereby finite-amplitude velocity fluctuations, or traveling wave disturbances, acquire significant three-dimensionality [31]. Transition has been very graphically described as the process by which the straight and parallel vortex lines of a two-dimensional flow deform into a constantly changing and twisting three-dimensional mess called "turbulence" [32].

The flow-visualization study of this phenomenon demonstrates the usefulness of the smoke-wire technique. Using both still and high-speed motion-picture photography, detailed visual records of the physics of the separation bubble were acquired. The actual photographic records yielded detailed streakline dynamics that provide significant insight into the dynamics of the flow field.

A series of experiments was conducted in which the horizontal *AA* wire was used for both profile and planform views of the airfoil. The airfoil and relative wire positions used are shown in Fig. 29. For the profile views, only a small center section of the smoke wire was coated. The angle of attack was 12°, and Re_c was

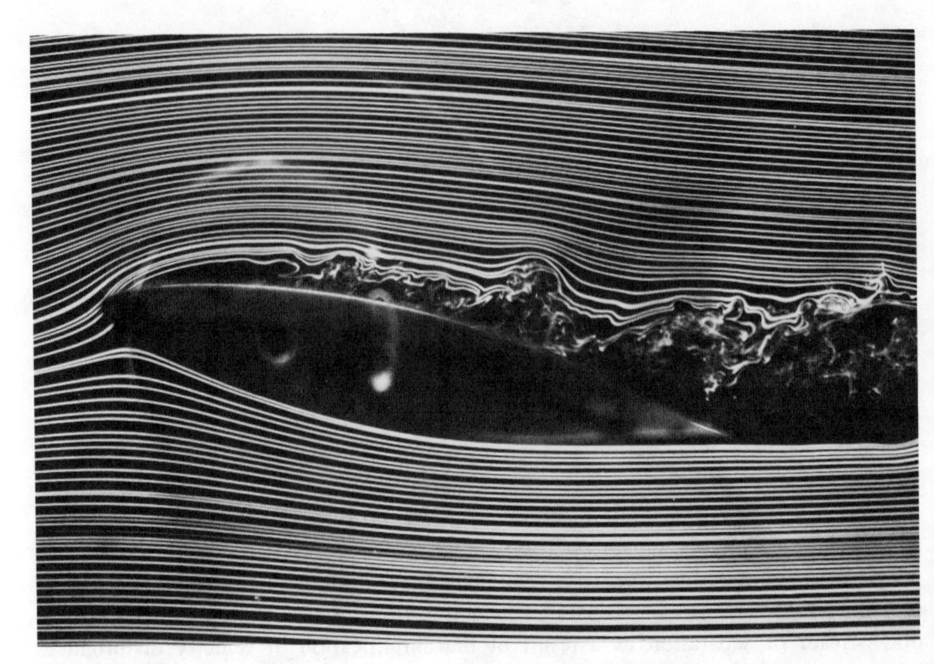

Figure 28 Smoke-wire visualization for a smooth NACA 66_3-018 airfoil at an angle of attack of 12° and chord Reynolds number of 40,000. *(Courtesy of T. J. Mueller and S. M. Batill, University of Notre Dame.)*

55,000. In Fig. 30, the wire is located so that the smoke sheet is close to the stagnation streamline (wire position 1). In Fig. 31, the smoke wire has been raised approximately 1 cm to wire position *A*2, and the smoke sheet lies above the top surface of the airfoil. The contamination of the laminar flow above the airfoil is evident as the effect of transitioning and reattaching as the turbulent boundary layer propagates away from the airfoil surface. These photographs represent only two of a series that graphically illustrates the streakline dynamics in this complex flow field and that is made possible by accuracy in positioning the smoke filaments.

An examination of the smoke photographs substantiates the notion of a highly unstable two-dimensional flow that breaks down in a very definite manner to a three-dimensional chaotic turbulent flow. The smoke photographs represent the

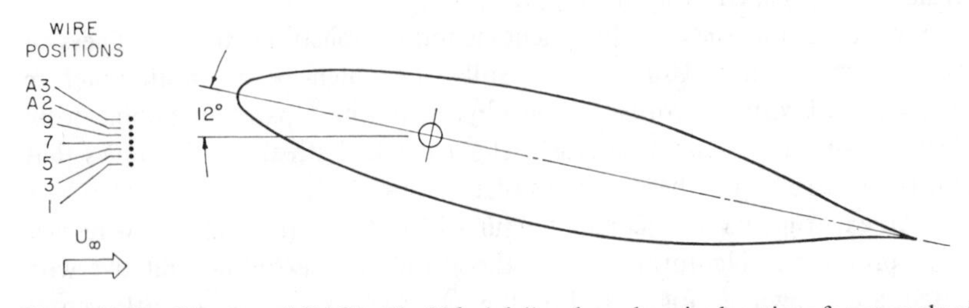

Figure 29 Profile view of NACA 66_3-018 airfoil and smoke-wire locations for an angle of attack of 12°. *(Courtesy of T. J. Mueller and S. M. Batill, University of Notre Dame.)*

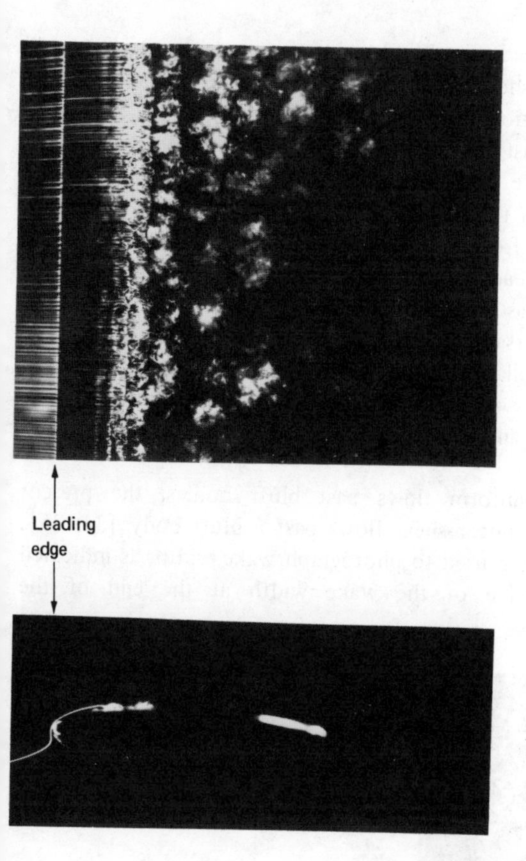

Figure 30 Smoke-wire position 1 for $\alpha = 12°$ and $Re_C = 55{,}000$. *(Courtesy of T. J. Mueller, S. M. Batill, and B. J. Jansen, Jr., University of Notre Dame.)*

most definitive visual description of separated shear-layer transition available. Some structure is also visible in the developing turbulent flow. Although this visual technique has been applied only for chord Reynolds numbers below 120,000, the basic transition process should follow the same series of events at higher Reynolds numbers. For example, for the same airfoil, the beginning of the transition process moves toward the separation location as the free-stream velocity is increased. The length of the transition region also decreases with higher free-stream velocities. Thus, the understanding gained at low Reynolds numbers can definitely help develop a physical model of transition that will be useful at high Reynolds numbers.

Bluff body in wind shear. Tall buildings and other high structures cause wakes in the natural wind that present a number of problems to architects and structural engineers. A problem of considerable recent interest focuses around the fluctuating pressures on building surfaces and fluctuating wind loadings caused by vortices being shed on the leeward side of buildings. These fluctuating loads can cause windows to be pulled out and create such severe downdrafts that pedestrian traffic near the base of the buildings is jeopardized. A continuing study of this problem, under the direction of A. A. Szewczyk at the University of Notre Dame, makes use of the smoke-wire method to study the wake of a bluff body of rectangular cross section in a linear shear flow. The linear shear flow is produced by an S-shaped screen upstream of the wind-tunnel test section.

It is often the case that flows past bluff bodies or other unstreamlined structures produce oscillating instabilities. These instabilities lead to an organized and periodic shedding of vortices in the wake as the flow separates alternately from the body. The flow field exhibits a dominant frequency; hence, the drag and pressure forces acting on the body are also unsteady. For uniform flows past stationary and vibrating structures, Griffin [33] has found that universal similarity exists in the wakes of such structures. A universal Strouhal number proposed for such wake flows was shown to successfully collapse the characteristic parameters for various bluff bodies over a Reynolds-number range of 10^2 to 10^7. In addition, the product of the universal Strouhal number and the drag coefficient was found to be a constant.

While these results are for uniform flows past bluff bodies, the present investigation seeks an extension to linear shear flows past a bluff body [34, 35]. Smoke-wire flow visualization is being used to photograph wake width, as indicated in Fig. 32. Measurements are made of the wake width at the end of the

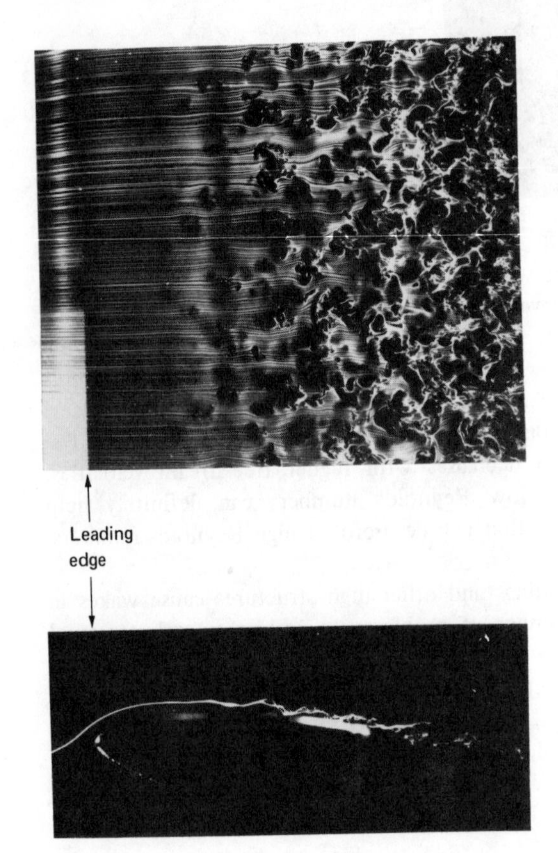

Figure 31 Smoke-wire position $A2$ for $\alpha = 12°$ and $\mathrm{Re}_c = 55{,}000$. *(Courtesy of T. J. Mueller, S. M. Batill, and B. J. Jansen, Jr., University of Notre Dame.)*

vortex-formation region, where a vortex is fully formed, is shed, and then moves away from the base, as shown in Figs. 33 and 34. The wake width is the characteristic length needed for a universal Strouhal number, if such a universal Strouhal number can actually be found. Further research into this complex problem is in progress.

2.3 Helium-Bubble Method

The most popular agent for airflow visualization in wind tunnels has been smoke. As discussed earlier, the use of smoke requires a low turbulence level in the wind-tunnel test section, to minimize diffusion. Furthermore, the smoke particles are so small that they cannot be observed or photographed individually. A larger, neutrally buoyant tracer agent is needed to follow the paths of individual particles. Fortunately, particles can be used in existing wind tunnels with moderate turbulence.

The soap bubble is an *ideal* particle because its size and buoyancy can be controlled. The scientific interest in soap bubbles dates back to Robert Hooke and Isaac Newton [36]. The first use of soap bubbles for flow visualization in wind tunnels appears to be by Redon and Vinsonneau [37] in 1936 at Marseille, France. In 1938, Kampé de Fériet [38] reported the use of bubbles to make statistical measurements of turbulence in a wind tunnel. Interest in this method of flow visualization seems to have disappeared until B. V. Johnson [39] generated small bubbles to study the flow within a cylindrical vortex tube in 1961. The most significant and practical bubble-generation system for wind-tunnel use developed from this latter work without knowledge of the earlier work of Redon.

The development of this modern system was begun in 1967 by Hale, Tan,

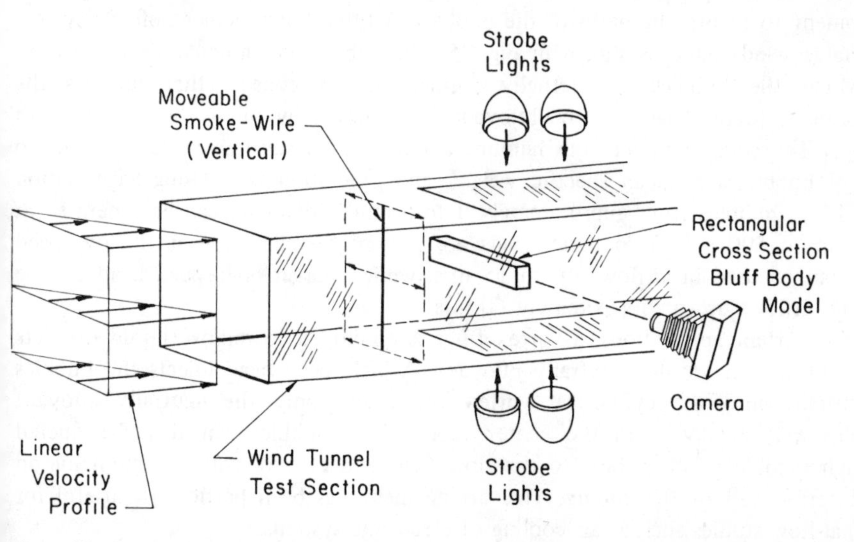

Figure 32 Arrangement for smoke-wire study of bluff body in wind shear showing model, wire, camera, and light locations. *(Courtesy of S. J. Elsner and A. A. Szewczyk, University of Notre Dame.)*

Figure 33 Vortex-formation region behind a rectangular bluff body (smoke-wire technique). *(Courtesy of S. J. Elsner and A. A. Szewczyk, University of Notre Dame.)*

Stowell, and Ordway. The complete system consists of a bubble generator,* lighting and optical components* for illuminating the bubbles, and the photographic equipment to record the paths of the bubbles. A typical arrangement of this system in a large wind tunnel is shown in Fig. 35. The bubble generator consists of a head in which the bubbles are actually formed and a console that supplies the constituents to the head. Neutral buoyancy is achieved by filling the bubbles with helium. The console meters the helium, a bubble film solution (BFS), and air to control bubble size, mean specific weight, and generation rate. Using Sage Action BFS 1035, bubbles from approximately 1 to 5 mm in diameter can be generated at rates up to 500 s^{-1}. Two basic heads are presently available. A small, low-speed head can be used in airflows up to 15 m/s, while a larger high-speed head can be used in airflows up to approximately 60 m/s.

For certain applications, a vortex filter is employed to remove liquid droplets and bubbles that are not neutrally buoyant. A high-speed head injects the bubbles tangentially into the cylindrical vortex filter, and only the neutrally buoyant bubbles exit axially from the center tube. These bubbles can then be ducted through flexible plastic tubing to the flow field under study without impinging on the interior wall of the tubing. This arrangement has been particularly useful for internal-flow studies such as air cooling of electronic systems.

*Available from Sage Action, Inc., P. O. Box 416, Ithaca, N.Y. 14850.

Figure 34 The separated flow from the sharp edges of the bluff body and the vortex-formation region (smoke-wire technique). *(Courtesy of S. J. Elsner and A. A. Szewczyk, University of Notre Dame.)*

2.3.1 Lighting and photography. Because the bubbles reflect about only 5% of the incident light, careful selection and placement of the light sources is necessary. The principal aim is to shine as much light as possible on the bubbles while keeping the background dark. This may require painting the models and background areas with low-reflectivity paint (i.e., flat black), even when they are not in the direct beam of the source. Usually, a well-defined light beam is directed along the mean airflow direction, and the line of sight of the observer or camera is essentially perpendicular to the light-beam axis. Since each application is somewhat different, an all-purpose lighting system cannot be chosen. For small illumination, however, Sage Action, Inc. [40] suggests the Eimac 150-W xenon arc lamp; an optical shroud and arc-lamp modulator were specifically designed for this lamp. The advantages of this lighting system are the lamp's compactness and its adaptability to electronic modulation or chopping for quantitative measurements. This modulation makes the system very useful for a wide range of air velocities, since continuous variation of the chopping frequency is possible. Illumination of a larger field is possible with a variety of incandescent light sources that are available at reasonably low cost. Mechanical modulation can be used with these light sources if necessary. A detailed description of the various illumination arrangements and modulation capabilities is presented in [40].

Bubble-trace photography, in common with smoke photography, is quite different from conventional photography. For streak photographs it is desirable to obtain a high trace intensity and a low background exposure. In addition to using increased illumination, it is necessary to use highly sensitive film. Two types of film have been used [40]: Kodak 2475 with ASA 1000 and Kodak Royal-X Pan with ASA 1250. As discussed in the section on smoke photography, the sensitivity of such film may be further increased through special processing. The image intensity of a bubble streak is independent of camera shutter speed but dependent, instead, on bubble velocity. Furthermore, the shutter speed determines the streak length and number of bubbles in a given photograph. If the bubble-generation rate is R and the shutter speed is T, the number of bubbles in the photograph is RT. The number of bubbles that pass all the way through the camera's field of view is $RT-RL/V$, where L is the field dimension and V is the air speed. Hence, a large value of R allows high shutter speeds, which, in turn, reduce exposure of the model and background [40].

In a number of tests, bubble traces have been recorded by using closed-circuit television. There are several important advantages to this method. Certain television cameras provide higher sensitivity for bubble-trace recording than the fastest available photographic emulsions, and such cameras are readily available. It is possible to record sequentially with closed-circuit television or motion-picture photography, and this is especially vital in unsteady flows where the events are

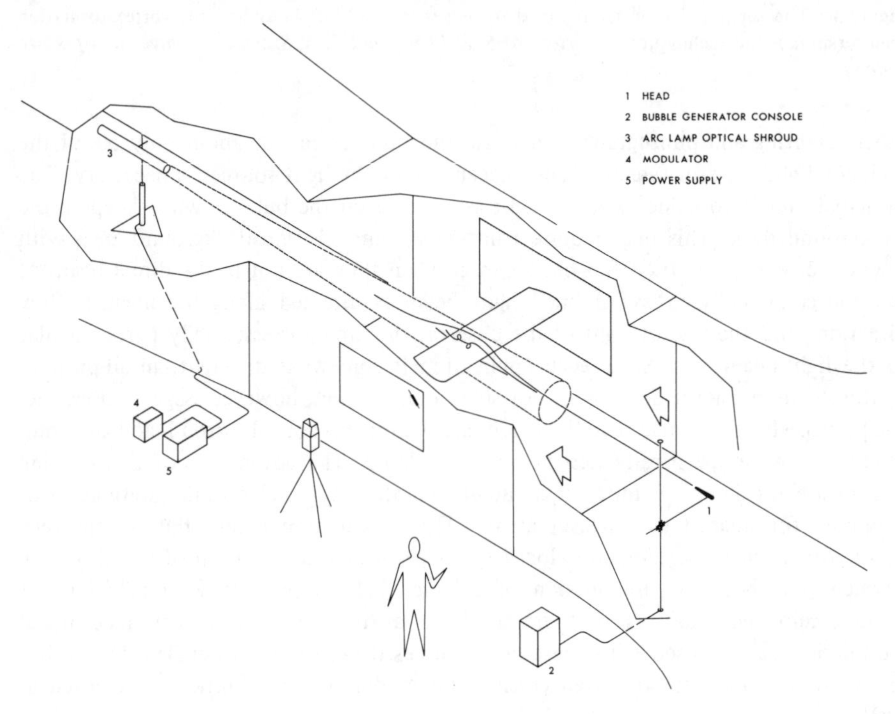

Figure 35 General arrangement of the helium-bubble visualization system in a wind tunnel [36, 40]. *(Courtesy of Sage Action, Inc.)*

often not repeated. Another advantage of using television is that the videoscanning process records the complete path of every bubble that passes through the camera's field of vision; a disadvantage of using a movie camera is that about one-half of the bubble path information is lost because of the mechanical shutter. As television also permits immediate replay, on-the-spot adjustments can be made to the bubble generator, the camera, and/or the lighting. This and the ability to monitor the live camera view result in effective use of the researcher's time.

2.3.2 Applications of helium-bubble techniques. According to the manufacturer, approximately 320 bubble-generator systems have been sold to universities, government facilities, and industry. Almost 100 of these systems have been delivered outside the United States, primarily to Europe and Japan. This equipment and the associated techniques have been used for aerodynamic wind-tunnel testing and the design and evaluation of ventilation systems, wind effects on structures, airflow through blowers and fans, natural-convection systems, and many other practical flow situations.

Early in the development and application of the bubble visualization techniques, a comparison was made between the potential flow solution and streak photography for the two-dimensional Karman-Trefftz airfoil [40]. The potential flow solution and the visualization experiment were obtained for an angle of attack of -5°. The tunnel speed used was 30.48 m/s to minimize boundary-layer effects including separation. The results of this study are shown in Fig. 36. Except for the slight separation near the trailing edge, the agreement between the two flow fields is very good. The crossing of the bubble paths indicates that the flow was not completely steady. The effect of using different shutter speeds for the flow over the two-dimensional Karman-Trefftz airfoil is shown in Fig. 37. The angle of attack was 0°, and the free-stream velocity was 15.24 m/s for both photographs. The shutter speed was 0.50 s in Fig. 37*a*, and 0.10 s in Fig. 37*b*. While the overall flow pattern is visible in Fig. 37*a*, it is difficult to follow individual streaks. Although individual bubble streaks are evident in Fig. 37*b*, some of the streaks are a little out of focus as a result of the small depth of field obtained by using a low f-stop. This depth-of-field problem is common to most flow-visualization techniques using direct injection.

The helium-bubble technique was used together with force and pressure measurements to study a close-coupled-canard (CCC) configuration [41, 42]. The CCC configuration has the principal advantage of high total lift at large angles of attack, combined with reduced trim drag. This is an important capability for advanced air-superiority fighter aircraft. The objective of this study was to obtain a good physical understanding of the canard-wing flow field at high angles of attack so that a simplified theory could be formulated.

Figure 38 is a drawing of the CCC half-span model. The wing section was a NACA 66A008 airfoil laid out in a direction inclined 25% outward relative to the chordal direction, or nearly normal to the trailing edge. The canard was a flat plate, 0.48% thick at the root, with a semicircular leading edge and a blunt trailing edge. For the flow-visualization studies, videotape recordings were made to supplement

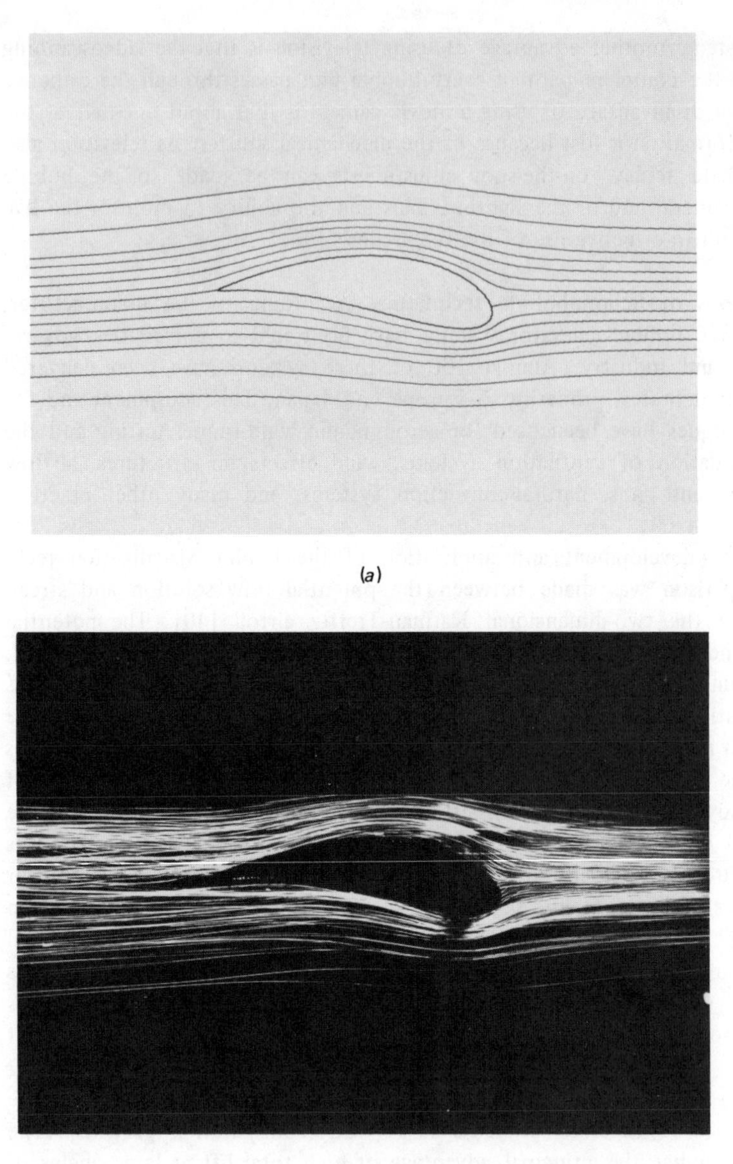

(a)

(b)

Figure 36 Comparison of the potential flow streamline pattern with a helium-bubble streak photograph for a two-dimensional Karman-Trefftz airfoil [40]. (*a*) Streamlines of potential flow field. (*b*) Streak photograph from wind-tunnel test (f2). *(Courtesy of Sage Action, Inc.)*

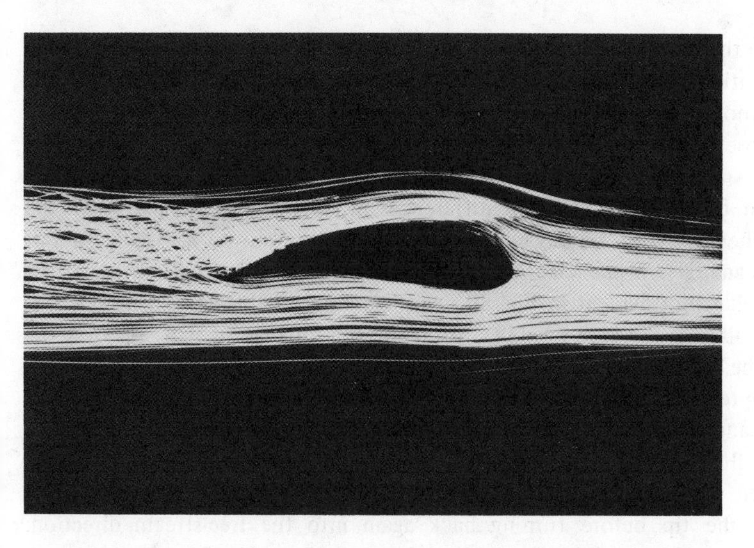

(a)

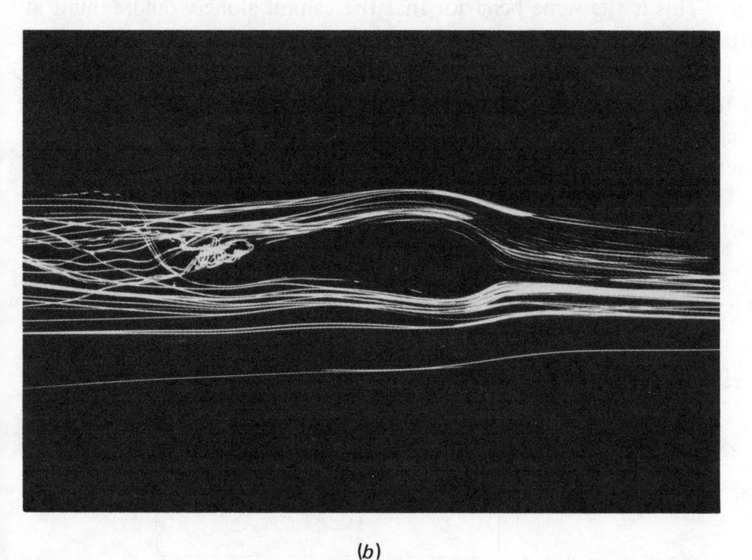

(b)

Figure 37 Helium-bubble photographs of the flow over a two-dimensional airfoil for different exposures [40]. (*a*) Photograph taken at 1/2 s (f4). (*b*) Photograph taken at 1/10 s (f4). *(Courtesy of Sage Action, Inc.)*

the still photography, to identify unsteady effects. Surface oil-flow visualization using titanium dioxide was also employed.

Figure 39 shows two representative streak photographs for an angle of attack of 25°. This angle was chosen because it was found from force data to be in the regime of favorable interaction. Figure 39*a* shows the wing without the canard, and

Fig. 39*b* with the canard. The wing is badly stalled without the canard, and a turbulent separation region covers the whole upper surface. The remainder of the flow follows smoothly around the separation region, which is fed by only a small stream tube close to the wing root. No evidence of an organized leading-edge vortex or a tip vortex is apparent for this case.

There seem to be two important differences in the wing flow field when the canard is attached. A tight leading-edge vortex, extending out to a distance about equal to the canard semispan, is formed inboard. As this vortex is in a narrow shadow region, it is not evident in Fig. 39*b*. The bursting of this vortex, however, is shown dramatically, as the core becomes turbulent and enlarges as it passes over the wing and off the trailing edge. The flow is laminar outside this turbulent "funnel," but it has appreciable rotation. The other notable occurrence is the unusual *spanwise flow* sandwiched between the burst vortex system and the upper surface of the wing. This flow originates inboard as flow that has gone over the leading-edge vortex and then reattached behind. After traveling underneath the burst vortex, it almost reaches the tip before turning back again into the free-stream direction. Again, there is no evidence of tip vortex.

The leading-edge vortex on the canard bursts near the apex, although this is not shown in Fig. 39*b*. This is the same behavior that the canard alone would exhibit at this angle of attack. The core of this burst vortex is very turbulent and increases considerably as it moves rearward to the trailing edge and into the wake above the wing. It appears that the presence of the wing holds this burst trailing-vortex system down near the canard and, thus, closer to the wing plane. However, the flow is quite smooth between the canard and the wing, which suggests a "channeling"

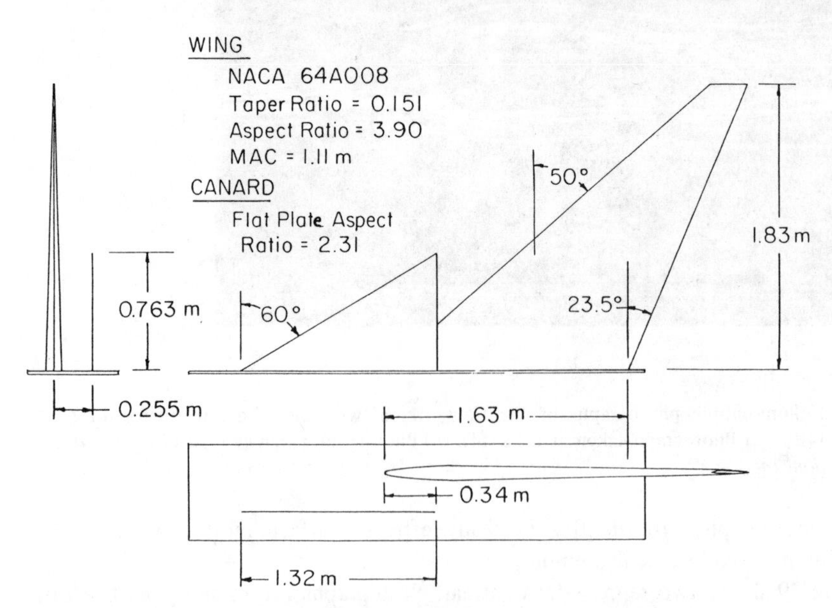

Figure 38 Three-view drawing of semispan close-coupled-canard model for wind-tunnel tests; scale 1:24 [41, 42]. *(Courtesy of Sage Action, Inc.)*

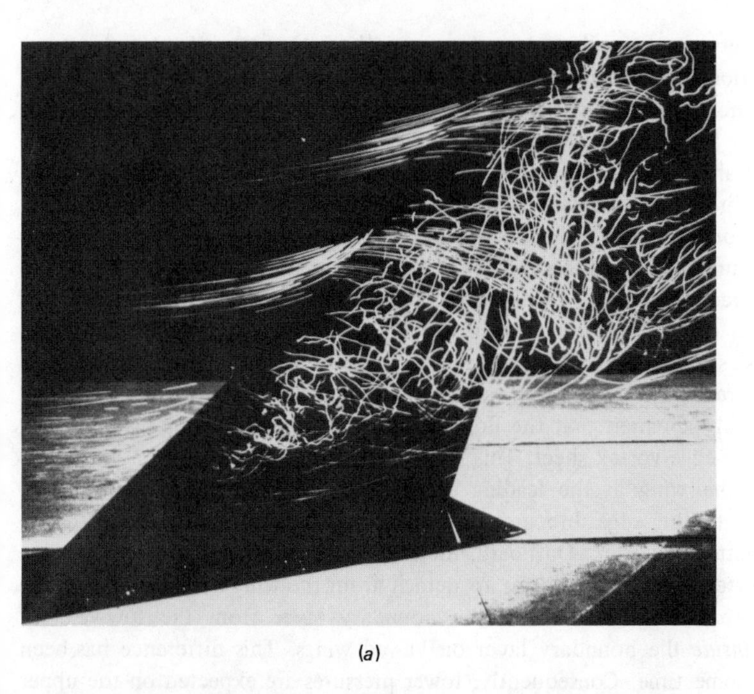

(a)

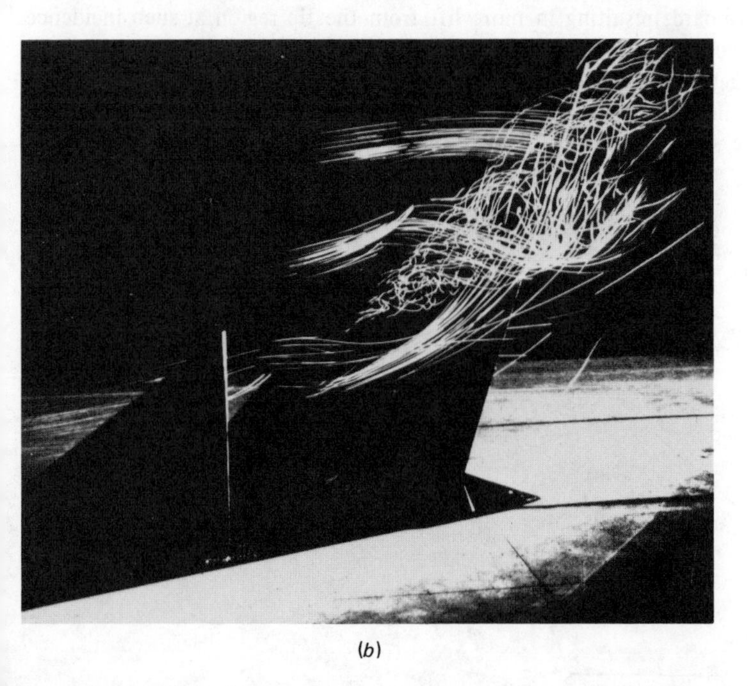

(b)

Figure 39 Plan view of flow pattern over upper surface of wing. (*a*) Without canard. (*b*) With canard. Reynolds number $= 0.73 \times 10^6$ based upon wing mean aerodynamic chord [41, 42]. *(Courtesy of Sage Action, Inc.)*

effect that preserves the leading-edge vortex on the inboard portion of the wing.

The separation line ahead of the leading-edge vortex on the inboard portion of the wing was shown quite distinctly by surface flow visualization. However, in a number of other locations, a comparison of the surface patterns with those of the outer flow revealed significant differences in flow direction, indicating a drastic change in flow direction through the boundary layer.

An average or quasisteady flow field was constructed from the superposition of streaks from a number of photographs [41, 42]. Considerable effort was needed to establish the three-dimensional relationship between the streaks. The result of this superposition for $\alpha = 25^\circ$ is shown in Fig. 40. The broad arrows describe the overall flow, which is steady and well ordered; the narrow arrows are actual bubble trajectories that represent the random motion in the burst vortex core.

It is reasonably apparent that the flow separates all along the wing leading edge, creating a leading-edge vortex sheet. This implies that there is no pressure difference or load across the wing at the leading edge, equivalent to the Kutta-Joukowski condition at the trailing edge. Inboard, the leading-edge vortex sheet rolls up tightly to form a concentrated vortex. Outboard, the sheet wraps around the turbulent core of the burst vortex as the core seems to detach from the wing surface, causing the strong, visible spanwise flow *outside* the boundary layer from the low-velocity spanwise flow *inside* the boundary layer on yawed wings. This difference has been recognized for some time. Consequently, lower pressures are expected on the upper wing surface outboard, resulting in more lift from the tip region at such incidence.

By comparison, the inside of the burst vortex is nearly a dead-air region. The velocities are much lower and almost completely random, implying that the pressure is practically constant. The vortex sheet around this stagnant region is roughly conical in shape from the point of bursting to the wing trailing edge. These two

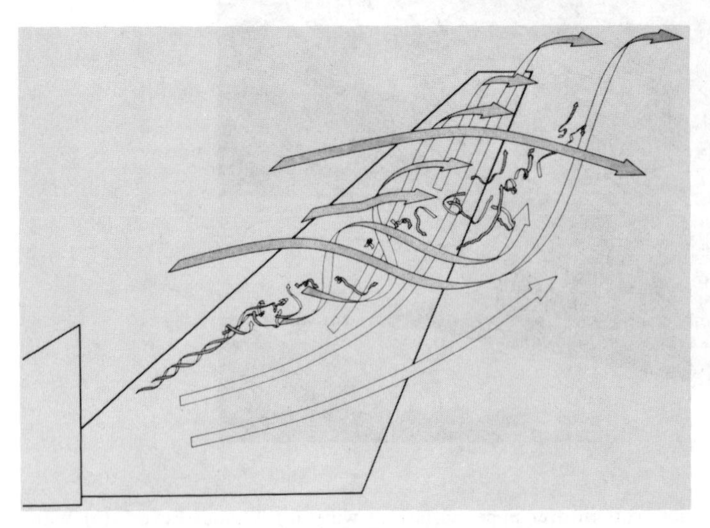

Figure 40 Composite drawing of average wing flow field for close-coupled canard [42]. *(Courtesy of Sage Action, Inc.)*

properties of the burst vortex system make theoretical formulation much easier. Furthermore, aerodynamic measurements on a full-span model, including pressure contours on the upper surface of the wing, correlate with the observed burst vortex.

The helium-bubble method has been used to study a variety of complex flow problems. For example, the flow field surrounding a parachute during the opening process has been investigated using this method [43]. Velocity profiles were obtained from bubble-streak photographs for a case late in the inflation process and for the fully inflated case. These data suggested that a potential-flow mathematical model would suffice to provide a reasonable description of the opening process.

An extensive study was performed to determine the feasibility of using the neutrally buoyant bubbles at transonic speeds [44]. The bubbles were successfully injected and photographed in a transonic flow with $Ma = 0.9$. New lighting techniques and methods of strengthening the bubble film were examined.

The complex flow fields related to rotary-wing aerodynamics have also been studied using the helium-bubble method [45, 46]. The flow-visualization portion of the research, documented in [45], revealed a small, well-defined ground vortex at moderate and high wind velocities. The force data obtained showed that this ground vortex and the trailing-vortex systems of the main-rotor flow field produced tail-rotor thrust perturbations. In [46], the flow field about an isolated rotor blade in rectilinear flow was examined. The results of these experiments revealed details of the formation of the tip vortex and the overall rollup of the vortex sheet.

2.4 Concluding Remarks–Aerodynamic Flow Visualization

Smoke visualization in wind tunnels has been developed to the point where it can produce a global view of very complex subsonic and supersonic flow fields. This global picture is helpful in understanding complicated fluid-dynamic phenomena, as well as indicating specific regions where quantitative measurements should be taken. When the flow is steady, the smoke streaklines are identical to streamlines and can be readily compared with analytical or numerical results for the same flow. When the flow under investigation is periodic, as the vortex shedding from a sharp, flat plate, the vortex spacing obtained from smoke photographs agrees well with theoretical predictions. Shedding frequencies for this type of flow may also be determined by using a mechanical or electronic strobe during the wind-tunnel experiments. Even when the flow is unsteady, the streakline pattern obtained is of considerable value in helping to understand complicated flow phenomena. Continuing attempts to pulse the smokeline and produce streak-time lines indicate promise for the future.

The use of helium bubbles in wind-tunnel experiments is increasing. The ability to follow helium bubbles photographically and obtain particle paths and velocities has provided the motivation for the development of this visualization method. Preliminary studies indicate that this method can also be used at transonic speeds. It should be clear that many fluid-dynamic and heat-transfer problems can be studied using the smoke-tube, smoke-wire, and helium-bubble techniques described above.

3. HYDRODYNAMIC FLOW VISUALIZATION

Flow visualization in water has been important to the understanding of fluid motions. It is clear that visual observations of flow phenomena were the first and, for a long time, the only experimental techniques available. It is difficult to imagine that da Vinci, Galileo, Newton, Bernoulli, Euler, and many others did not take advantage of flow visualization in their studies of fluid mechanics.

One of the most important discoveries in the history of fluid mechanics was a result of using aniline dye to produce colored water. This was the Osborne Reynolds experiment (1883) with a small filament of colored water in the center of a tube filled with clear water. As the velocity was increased through the tube, the transition from laminar to turbulent flow was observed by watching the formation of eddies and the subsequent diffusion of the colored water [47]. Many other experiments, with a variety of visualizing agents, have been developed and performed since 1883. Excellent reviews of the many available techniques have been presented by Clayton and Massey [48], Merykirch [49], and Werlé [50]. A brief description of many of the water-tunnel facilities existing worldwide has been given by Erickson [51]. In this section, the use of dye and hydrogen bubbles as visualizing agents is discussed.

3.1 Dye Method

A large variety of experimental equipment and procedures has been used in connection with the dye-injection method of flow visualization. From the famous experiment of Reynolds to the present, the use of dye injection as a visualizing agent has grown steadily. The most significant developments in this field, however, began at the Office National d'Études et de Recherches Aerospatiales (ONERA) in France with the work of Roy and Werlé [50] in the early 1950s. The equipment and techniques developed at ONERA established the state of the art and have been copied by many laboratories throughout the world. For this reason, the ONERA vertical water tunnel and visualization techniques, and a somewhat similar, though smaller, facility at the U.S. Air Force Wright Aeronautical Laboratories (USAF/WAL) are discussed here.

3.1.1 Vertical water tunnel at ONERA. As with the wind tunnels used for smoke visualization, the most desirable characteristics of a water tunnel to be used for visualization purposes are uniform flow and a low turbulence level in the test section. The ONERA water tunnel [52] is vertical, of the open-circuit type, and operates under the action of gravity, as shown in Figs. 41 and 42. It consists of a constant-head reservoir, flow straighteners, a contraction section, a test section (220 × 220 × 700 mm), a test-section extension (200 mm long), and a discharge section. The capacity of the reservoir provides for a useful testing time of 2 min at a speed of 10 cm/s. Speeds of 50 cm/s for 12 min are obtained in a test section of 10 × 160 mm, and for 2 min in a test section of 100 × 100 mm. A variety of model-mounting and dye-injection methods are available.

Figure 41 Photograph of the ONERA vertical water tunnel. *(Courtesy of H. Werlé, ONERA.)*

3.1.2 Vertical water tunnel at USAF/WAL. This facility was built as a pilot tunnel to develop components for a much larger tunnel. The USAF/WAL pilot water tunnel (Fig. 43) is vertical and of the open-circuit type; as in the ONERA setup, the tunnel operates intermittently under the action of gravity. It is made up of a reservoir, turbulence damper, contraction section, test section (146 × 146 × 457 mm), and a discharge section. The reservoir capacity allows for a run time of 2 min at 15 cm/s and an extended run time with the addition of water from the spray bar. The open-pore foam used as the flow straightener appears to adequately damp any turbulence from the spray bar. Currently, models are mounted from the side walls, and up to eight dye tubes (four colors) are available. The four-dye reservoirs are pressurized from a nitrogen bottle through a pressure regulator. Each dye reservoir holds 0.12 liter and has two valved tubes available.

3.1.3 Dye. A large number of dyes and dye solutions have been used for the marking of filament lines in water tunnels [49]. The most popular are milk, food coloring, and ink. Many other substances, including fluorescent materials, have also been used [53]. It has been found that the stability of the filaments may be improved by mixing the dye with milk [48, 49]; the fat content of milk is presumed to retard its diffusion.

The injection of dye without significantly altering the flow under study is a primary concern. It is important that the velocity of the injected dye be equal to the velocity of the surrounding fluid. This helps maintain a stable dye filament and reduces the disturbance to the surrounding flow. The problems of injecting dye

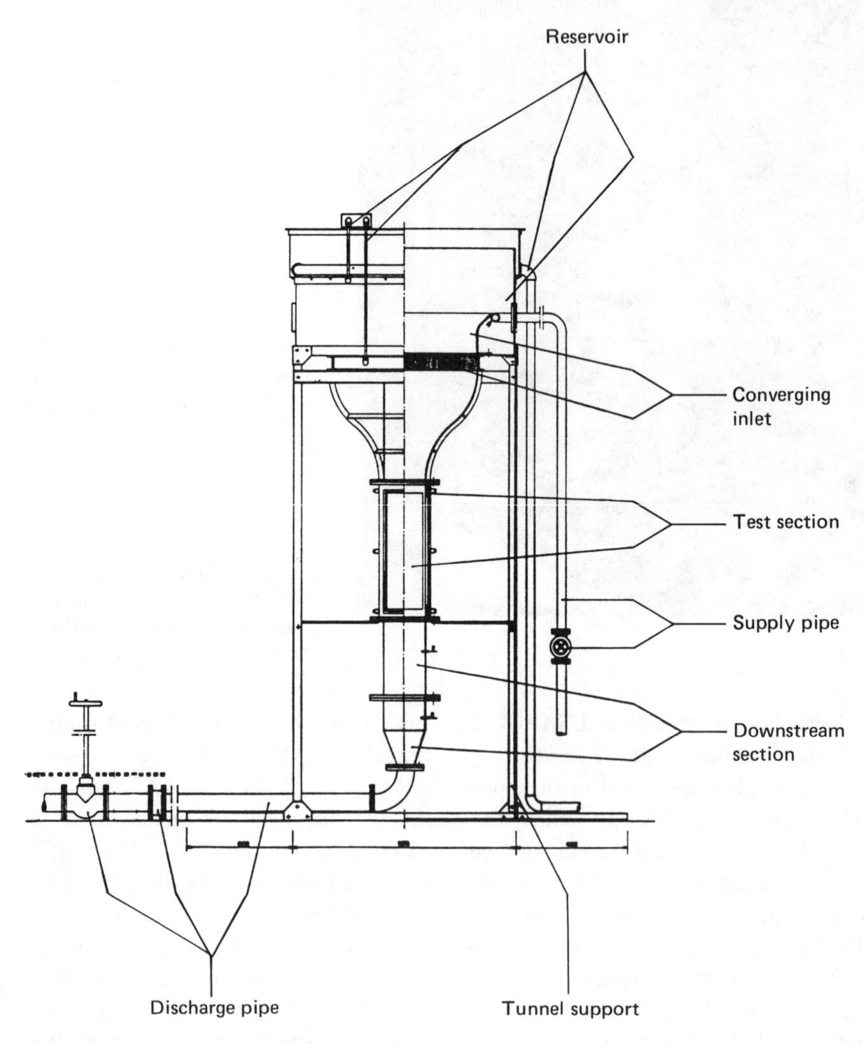

Figure 42 Schematic of the ONERA vertical water tunnel. *(Courtesy of H. Werlé, ONERA.)*

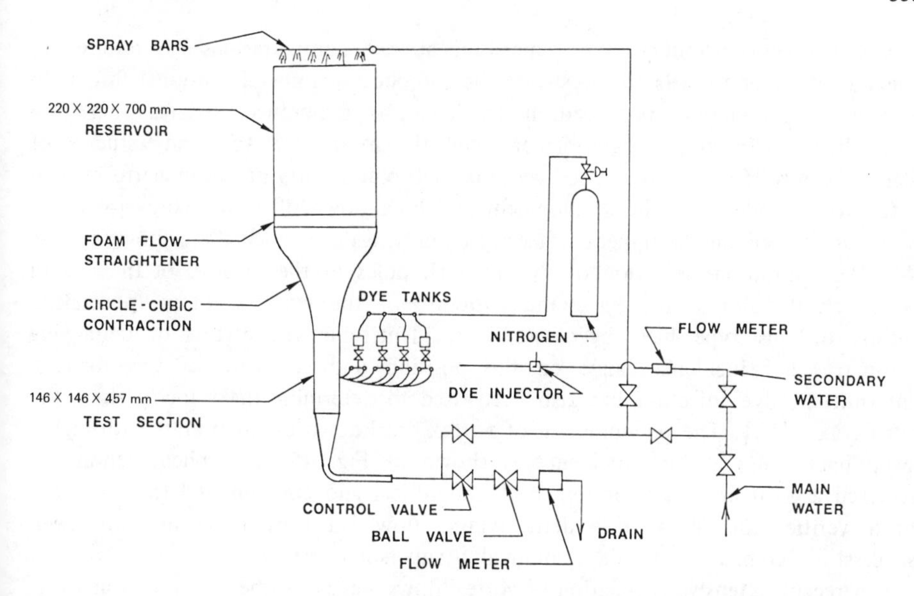

Figure 43 Schematic of the USAF/WAL pilot vertical water tunnel. *(Courtesy of USAF/WAL.)*

from a rake upstream of the model are the same as with the early smoke tunnels. The presence of the injection rake or tube in the flow disturbs the tunnel flow, which, in turn, disturbs the dye filaments. To circumvent this problem, the dye is usually injected from holes in the surface of the model. In using this method, care must be taken that the dye has as small a velocity component perpendicular to the surface as possible, or the effect will be equivalent to blowing into the boundary layer. It is difficult to accurately pulse dye filaments for direct velocity measurements, and thus continuous dye filaments are generally used.

3.1.4 Photographic techniques. Because the velocities involved in most water-tunnel experiments are quite low and there is great contrast between the colored dyes and the surrounding water flow, no exotic photographic techniques are necessary. Photo floodlights and still or movie cameras with color film are usually used. Readily available and relatively inexpensive color or black-and-white video systems may also be used.

3.1.5 Application of dye method. The use of dye filaments in a water tunnel is analogous to the use of smoke filaments in a wind tunnel and, thus, is subject to the same limitations. The dye method has been used to study a very large number of two- and three-dimensional steady and unsteady flow [50]. These flows may be further classified according to type of phenomenon studied: jets, wakes, boundary-layer structure, vortices, etc. Only a few representative flows are discussed here.

Vortices from a swept-back wing (ONERA). Because of their importance in a large number of fluid-dynamic situations, vortices have been studied for many years. The relatively recent introduction of supersonic aircraft configurations has resulted in the use of water tunnels to investigate the complicated three-dimensional flow field and vortex interactions produced. In particular, vortex-induced lift resulting from controlled leading-edge separation provided the motivation for many studies of slender wings [51]. A variety of water-tunnel experiments has been performed in attempts to understand the development and breakdown of vortex cores [50, 51]. Vortices formed on the upper surfaces of delta wings at incidence are shown in Fig. 44. The continuous injection of dye through holes in the surfaces of these delta wings clearly shows the shape and structure of the vortices. Flow-visualization results of this type have been useful in identifying the effects of delta-wing geometry and incidence angle on the position and structure of the vortices. Intermittent dye injection has also been used to determine the velocity along the vortex axis [50]. The phenomenon of vortex breakdown on an aircraft with highly swept-back wing at high incidence is shown in Fig. 45. This phenomenon has received a great deal of attention, both theoretical and experimental [54], and has been verified in flight. For delta wings, flow-visualization results have been successfully compared with data obtained at supersonic speeds.

A recent extensive correlation of vortex flows was performed by Erickson [51]. In this study several parameters were identified, which permitted the correlation of vortex flow simulations in water tunnel and wind tunnel with flight data. One of the conclusions of this work was: "Vortex generation, vortex sheet and core location, and vortex strength on thin slender wings are accurately represented in a water tunnel due to the insensitivity of separation point location to changes in Reynolds number." This is not a surprising conclusion, since vortices of this type have been treated theoretically as inviscid, with reasonable success. Erickson [51] has also been able to correlate the location of vortex bursting, obtained from dye studies in water tunnels, with results obtained from flight tests.

Forward-swept wing with canard (USAF/WAL). Interest in aircraft with forward-swept wings is growing rapidly. This type of configuration, made possibly by lightweight, nonmetallic composite materials, would have greater range and fuel economy, lower stall speeds, spin resistance, and improved low-speed control for easier landings and takeoffs than the present swept-back wing design. Some of the present aircraft designs use canards to improve maneuverability. The forward-swept wing design concept is relatively new, and, therefore, a large number of analytical and experimental studies are necessary to understand the complex flow over such a geometry.

Water-tunnel experiments, using dye injection, were performed at the Wright Aeronautical Laboratories to study the influence of the canard on the main-wing flow. For these experiments, a half-span model of a Grumman Aerospace Corp. design was placed in the USAF/WAL vertical water tunnel, as shown in Fig. 46. The model was studied at pitch angles α from -2° to $+17^\circ$, and at each model pitch angle the canard was pitched at angles β from -5° to $+25^\circ$. The influence of the canard on the flow over the main wing is shown in the series of photographs in Fig. 47, where $\alpha = 5^\circ$ and $\beta = 2.5^\circ$, 5°, 10°, and 20°. The water-tunnel velocity was 6

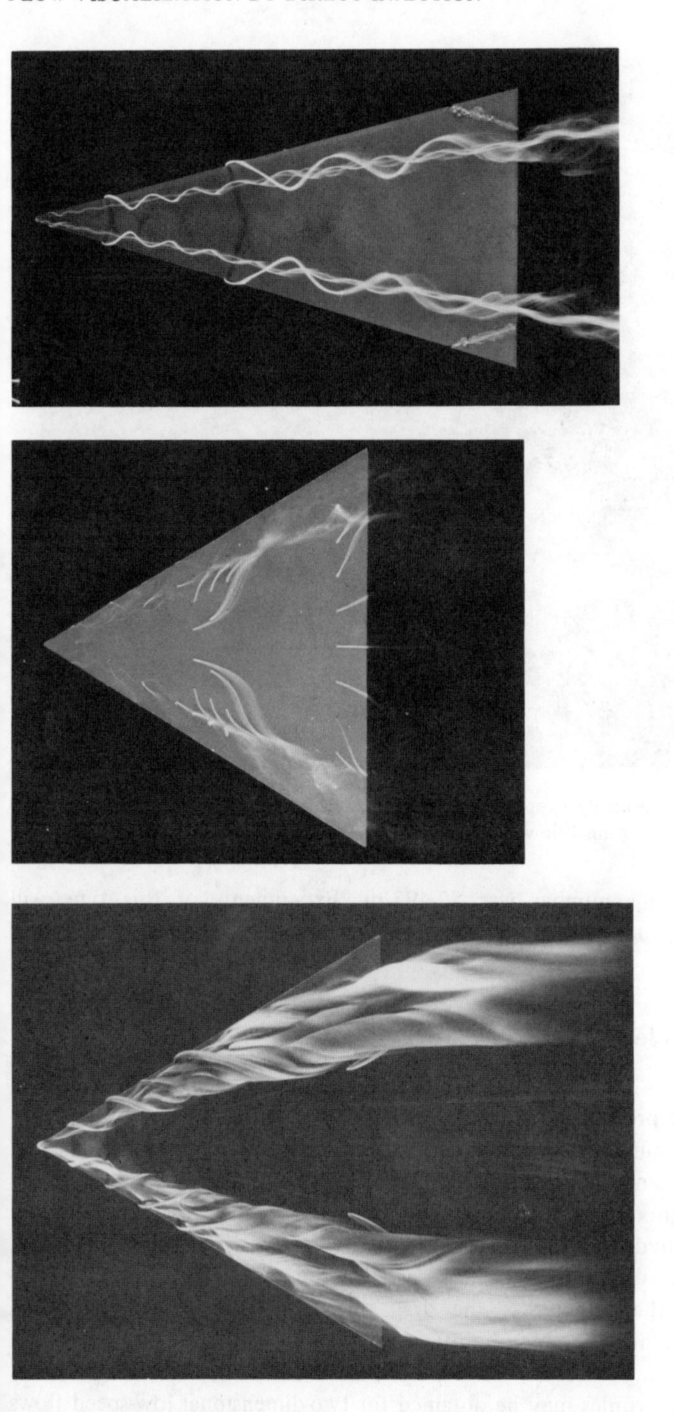

Figure 44 Slender delta wing configurations at incidence. *(Courtesy of H. Werlé, ONERA.)*

Figure 45 Vortex flow on a highly swept-back-wing aircraft from flight tests (upper) and from a 1/72-scale model in the water tunnel (lower). *(Courtesy of H. Werlé, ONERA.)*

cm/s, and the Reynolds number was 52,493/m. Experiments of this type will undoubtedly continue, together with analytical and numerical studies, as this new configuration evolves.

3.2 Hydrogen-Bubble Method

Visualizations of flow fields using dye-injection techniques that yield streakline patterns must be interpreted very carefully [55]. This is especially true in unsteady flows. Although streamlines cannot be visualized directly by any technique [56], quantitative measurements of unsteady velocity profiles can be obtained using the hydrogen-bubble method first introduced by Geller [57] in 1954. This method generates very small hydrogen bubbles from a fine wire that is part of a dc circuit, similar to the circuit used for the smoke-wire method (see Fig. 24). The predominant force on the bubbles is the drag due to the local water motion. Decreasing the bubble diameter has the effect of lowering the buoyancy force at a greater rate than the drag force. In this situation, the bubbles follow the local water motions, and velocity profiles may be obtained for two-dimensional low-speed flows [58, 59]. An excellent review of the use of the hydrogen-bubble method to determine velocity profiles is given by Schraub et al. [58] and Mattingly [60]. The

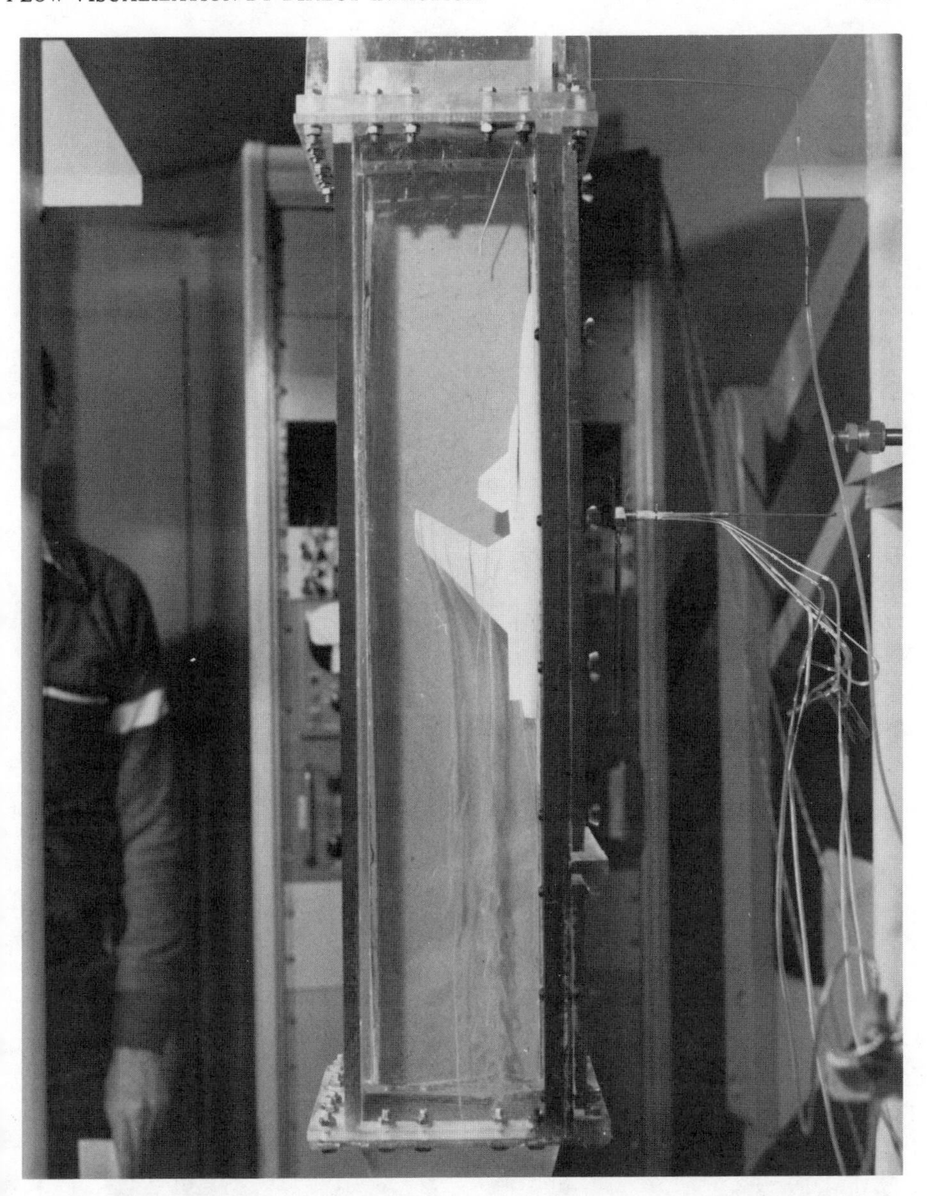

Figure 46 Forward-swept-wing model with canard in the USAF/WAL vertical water tunnel. *(Courtesy of USAF/WAL.)*

hydrogen-bubble technique is also useful in the qualitative study of global flow patterns around bodies. For flow problems of this type, separation, transition, and unsteady wakes have been investigated using this technique. Because it provides both quantitative and qualitative flow-field information, this technique has been used increasingly steadily since its discovery, for both internal and external flows.

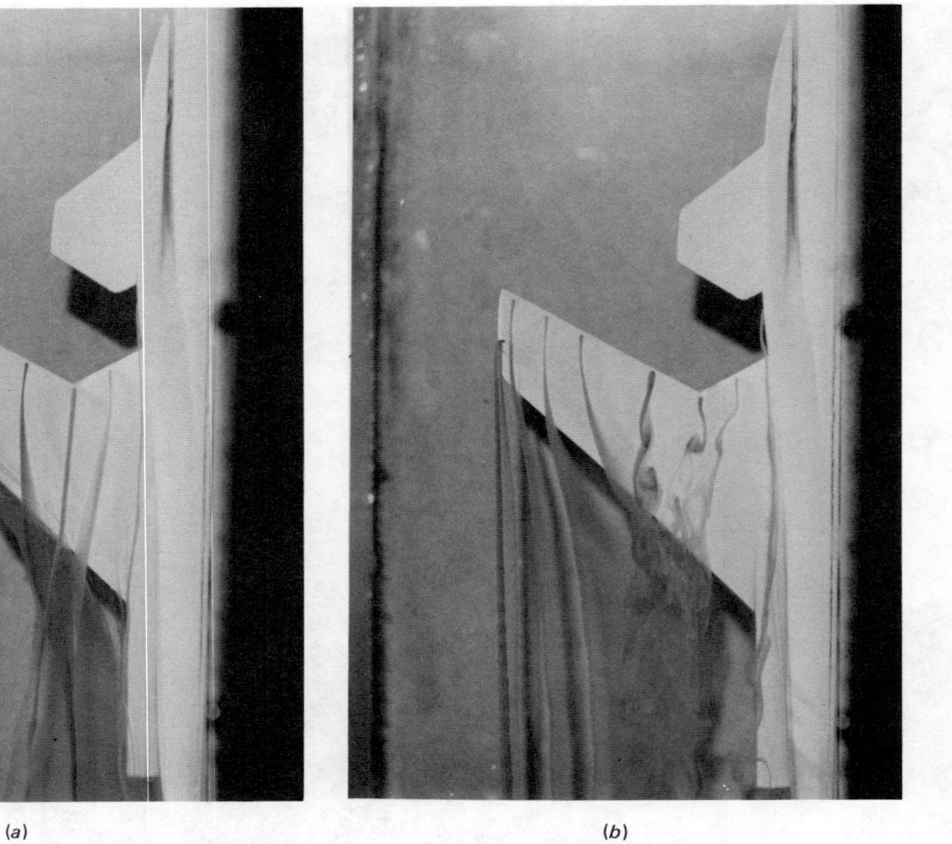

Figure 47 Forward-swept-wing model with canard at (*a*) $\alpha = 5°$ and $\beta = 2.5°$ and (*b*) $\alpha = 5°$ and $\beta = 5°$. *(Courtesy of USAF/ WAL.)*

(c)

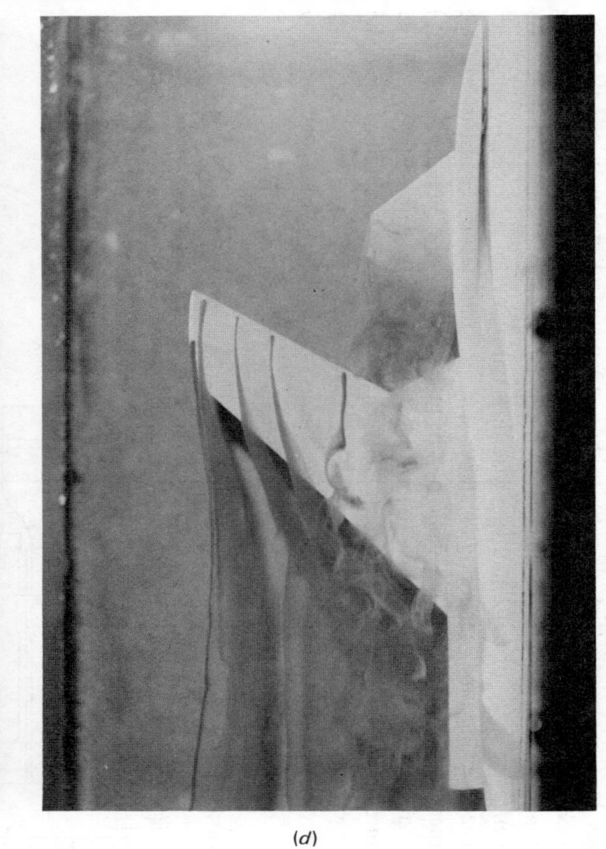

(d)

Figure 47 Forward-swept-wing model with canard at (*continued*) (*c*) $\alpha = 5°$ and $\beta = 10°$ and (*d*) $\alpha = 5°$ and $\beta = 20°$. *(Courtesy of USAF/WAL.)*

3.2.1 Free-surface water channel at Lehigh University. The Lehigh University facility has a free-surface plexiglass water channel with a 5-m working section, 0.9 m wide by 0.3 m deep (Fig. 48). By using a speed-control unit with a shaft-speed feedback circuit, stable speeds from 0.1 to 0.60 m/s may be attained. By combining a settling sponge, flow straightener–screen combination, and 5:1 inlet contraction, a turbulence intensity of 0.4% and spanwise flow uniformity of ±2% can be achieved.

The viewing and recording system used in this facility consists of two high-speed, closed-circuit video cameras (manufactured by the Video Logic Corp.) and synchronized strobe lights to provide 120 frames/s with effective frame exposure times of 10^{-5} s. An exceptionally clear picture can be obtained from the high-resolution screen using 250 horizontal, direct-overlay rasters with a sweeping frequency of 25.2 kHz. With conventional lenses, fields of view as small as 6 × 6

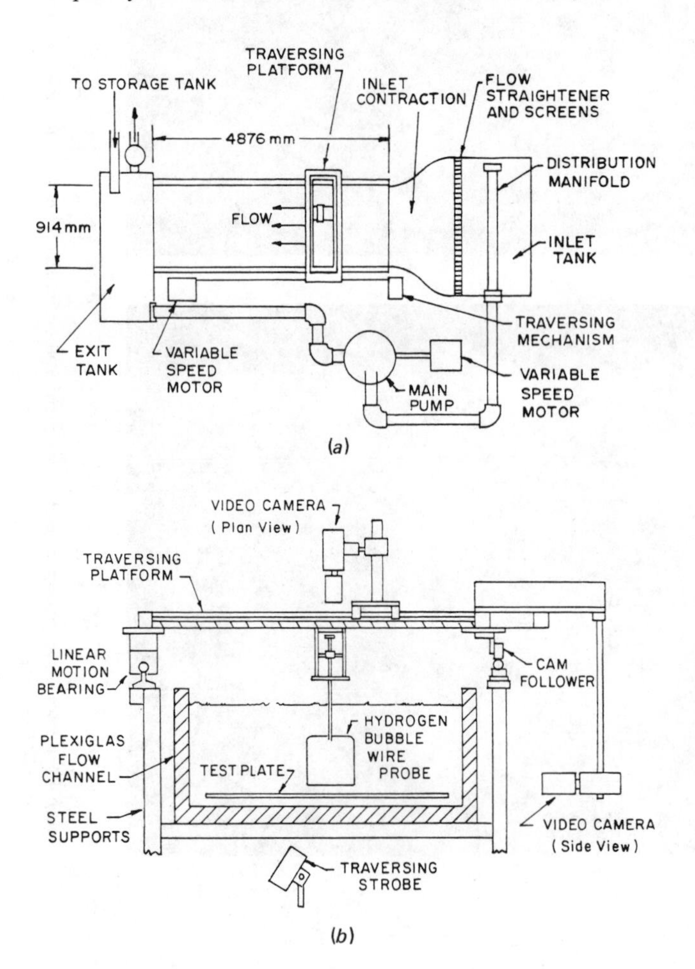

Figure 48 Lehigh University free-surface water channel. (*a*) Schematic of flow facility. (*b*) End-on view of channel (looking downstream) and traversing platform. *(Courtesy of C. R. Smith and T. Wei, Lehigh University.)*

mm can be obtained at distances of 0.5 m. A split-screen capability allows two different fields of view to be displayed simultaneously and recorded. All recorded data can be played in flicker-free slow motion (both forward and reverse), as well as in single-frame mode for detailed data analysis. Once a video sequence has been recorded, still photographs of individual stop-action frames can be taken directly from the video screen, using conventional photography or via a videographic copier that interfaces directly with the video recorder.

3.2.2 Horizontal water tunnel at the U.S. Army Aeromechanics Laboratory. The tunnel at the U.S. Army Aeromechanics Laboratory (AVRADCOM) is a closed-circuit design [61]. It contains approximately 4000 liters of water and, with the exception of the fiberglass contraction section and the plexiglass windows, is constructed from type 304 stainless steel. The tunnel passages are predominantly circular, except for the rectangular test section, which measures 0.2 m wide by 0.3 m high by 1.0 m long. The transition from circular to rectangular cross section takes place in the contraction and diffuser sections adjoining the test section.

The velocity that can be attained in the test section is continuously variable from 0 to 6 m/s. Prior to entering the 10:1 contraction, the flow is straightened by two sets of honeycombs, and the turbulence is reduced by four sections of screening. Two large storage tanks are used for dissolving fresh chemicals before filling the tunnel and for holding that portion of the water withdrawn from the tunnel when model changes are made. Whenever water is pumped back into the tunnel, it must first pass through a filter system designed to remove contaminates down to 5 μm. Continuous filtration is also possible while the tunnel is in operation.

The presence of cavitation-induced air bubbles represents a serious limitation on the maximum usable speed in a test program, since air bubbles can severely interfere with the viewing of the smaller bubbles intentionally generated during flow-visualization studies. Under atmospheric conditions, the amount of air dissolved in water is generally found to be about 2% by volume. This air remains in solution until drawn out by cavitation to form a lingering air bubble of detectable size. Typical low-pressure sources responsible for cavitation include the impeller, turning vanes, points of tunnel-wall separation, and the leading edge of a lifting airfoil. The problem can be alleviated, however, by subjecting the water in the tunnel to a vacuum and extracting a majority of the dissolved air. Degassing is normally necessary only with each initial filling of the tunnel. Once the air has been removed, the water can be transferred to the holding tanks (while model changes are made) and returned to the tunnel with negligible reingestion of air.

3.2.3 Bubble generation. Flow visualization by the hydrogen-bubble method involves the placement of a fine metallic wire within the flow to serve as the cathode of a dc circuit. An anode (also submerged), consisting of any suitable conductive object, is placed nearby. Supplying voltage to the circuit causes the liberation of hydrogen at the cathode through electrolysis, with oxygen being released at the anode. Although the composition of the ions produced depends on the particular electrolyte used, the significant reaction is the decomposition of the water according to $2H_2O \rightarrow 2H_2 + O_2$. Hydrogen bubbles forming on the wire are swept off by the flow to form a continuous sheet; if the voltage is supplied in the form of square-wave pulses, discrete lines of bubbles are formed. Since the hydrogen bubbles formed are smaller than the oxygen bubbles, better flow visualization is obtained using the cathode. The size of the bubbles released from the wire is on the order of one-half the wire diameter; the use of an extremely fine wire (0.025 to 0.05 mm) renders buoyancy effects negligible [58]. Because the hydrogen bubbles are generated in a region where there is a velocity defect due to the wire, some distance is required before the bubbles reach the velocity of the fluid motion under investigation. Although common water supplies generally contain enough impurities to provide a current path, the addition of a small amount of electrolyte such as sodium sulfate (0.15 g of Na_2SO_4 per liter of water works well) greatly enhances bubble generation, producing denser bubble lines with a consequent increase in clarity and contrast of the visual data obtained.

3.2.4 Lighting and photographic techniques. The technique used to obtain still and high-speed movies in the water tunnel at the U.S. Army Aeromechanics Laboratory (AVRADCOM) are typical of those used in many facilities of this type and similar to those described for the smoke-tube method. The bubbles are illuminated by a narrow sheet of light directed through the upper test-section window, as shown in Fig. 49. Two baffles are used to control the width of the light beam passing into the test section. The path of the light is canted 10° from the plane of the bubbles to provide a component backlighting without compromising the required vertical spread of illumination above the model [61].

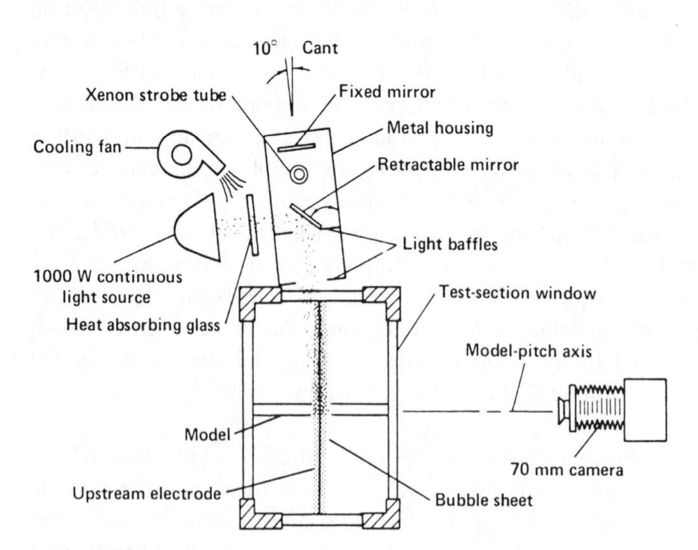

Figure 49 Apparatus (as viewed from upstream) required to illuminate hydrogen bubbles in horizontal water tunnel. *(Courtesy of K. W. McAlister and L. W. Carr, U.S. Army Aeromechanics Laboratory, AVRADCOM.)*

Both continuous and flash sources of light are produced over the needed length. Continuous lighting is provided by two lamps totaling 1000 W. These lamps are outside the optical housing so that the infrared content of the light spectrum can be removed conveniently by a row of 6-mm-thick heat-absorbing glass plates. This heat, in turn, is convected away by a high-velocity stream of air directed at the absorbent glass plates. Once inside the optical housing, the light is turned 90° by a mirror that can be retracted when this mode of lighting is no longer required. Continuous lighting is used for general viewing, for high-speed movies, and for long exposures (normally 1/60 s) with a single-frame camera.

When an instantaneous visualization of the flow field is needed at a given phase angle of model oscillation, the mirror is retracted and a 30-cm-long xenon strobe tube is activated. Connected across the tube is a 2000-V (maximum) supply, capable of delivering up to 8 kW. A flash is produced each time a 20,000-V pulse is applied to the wire coiled around the tube. The pulse can either be executed manually or be automatically synchronized with the motion of the model [61]. More recently, a circuit has been developed that uses a pulse-forming network with variable inductance to protract the photoemission of the strobe. This permits particle-path visualization to be obtained over a Reynolds-number range requiring high levels of bubble illumination that are normally beyond the practical limits of continuous light sources.

The still-photograph system consists of an automatic 70-mm film magazine, a bellows type of focusing body, and a 240-mm lens (f4.5 minimum), coupled with an electronically controlled aperture and shutter (1/60 s minimum). This composite camera is mounted on a rigid pipe so that the film plane is located a nominal distance of 152 cm from the center of the test section. This combination of lens and film plane-to-subject distance was found to offer the best compromise of image size, maximum depth of field, and minimum perspective distortion. The black-and-white film used in this camera is a high-speed, medium-grain roll film that yields normal-density exposures at f8 when processed in a high-contrast developer.

The cinematic system consists of a Milliken high-speed precision motion-picture camera, capable of indexing 16-mm film up to 500 frames/s with ±1.5% speed stability. This camera is mounted on a rotatable arm that allows the lens axis to be coaxially positioned in front of the lens of the first camera system. A minimum setting of 64 frames/s is required for motion analysis of various flow phenomena. A 160° shutter, equivalent to 1/140 s at 64 frames/s and with an aperture opening no longer than f2.8, is used to maintain a reasonable depth of field. Because of its high sensitivity to light, 400 ASA film is used. A less sensitive color film can be used with an aperture setting of f1.4 [61]. More recently, a specially tailored discharge circuit has been developed for the strobe that produces superior bubble illumination at camera speeds of up to 250 frames/s.

3.2.5 Application of hydrogen-bubble method. Since its introduction in 1954, the hydrogen-bubble method has been used to study a very large number of low-speed fluid-dynamic phenomena. Although, with care, the hydrogen-bubble method can be used to obtain measurements of velocity, it has also been used in many instances to obtain global flow patterns. This method has been especially useful in the area of transition and turbulent boundary-layer structure, and unsteady flow phenomena.

Wake of a circular cylinder. The objective of this study, under the direction of C. R. Smith at Lehigh University, was to visually examine the wake of a circular cylinder and to examine spanwise variations in the flow. Evidence of a spanwise variation in the wake region had been previously noted by several investigators using dye injection [62, 63], but the characteristics of the structure and its Reynolds-number dependence had not been established since dye could not be injected in a spanwise sheet away from the surface of the cylinder. The use of a hydrogen-bubble wire alleviates the shortcomings of dye injection and reveals an ordered spanwise structure of the near wake of a cylinder.

A 1.2-cm-diameter stainless steel circular cylinder, 0.9 m in length, was mounted spanwise across the channel at the half-depth of the channel. The cylinder was polished to eliminate roughness and any irregularities that might give rise to spanwise flow irregularities. To examine the effect of end conditions, the cylinder was mounted alternately using rigid end supports and a guy-wire support arrangement. Identical results were obtained for both support techniques.

Platinum wire, 0.025 mm in diameter, was used. Bubble lines were made visible by illuminating the flow with a light source of high intensity directed at an oblique angle to the line of sight. Placement and orientation of the bubble wire within a flow was accomplished with a bubble-wire probe consisting of two insulated, conductive metal prongs, between which the platinum wire was stretched, with the ends secured by soldering. The probe was constructed of brass and insulated with heat-shrink tubing and red glyptol. The probe was mounted to a calibrated traversing mechanism that could be adjusted in 0.05-mm increments.

The bubble wire was energized by square-wave voltage pulses supplied by a voltage generator with a range of 0 to 90 V and a maximum output of 2.5 A. The pulse duration was continuously adjustable, and the pulsing frequency could be varied from 0 to 340 Hz. The pulsing frequency was monitored by a multifunction counter that could measure frequencies from 5 Hz to 100 MHz.

After about a minute of operation, the quality of the bubble lines shed from the wire began to degenerate. This was due to a buildup on the wire of positively charged ions, which may be removed by reversing the polarity of the voltage supply for a few seconds and then switching it back. However, abrupt changes in the voltage output can damage the pulse generator and/or cause a breakage of the bubble wire. Also, the residual capacitance present in the generator causes the square-wave signal to persist for a few seconds after the generator has been shut down. Therefore, to avoid potential damage to the equipment, it was necessary to wait for the residual pulses to fade away before switching the polarity in either direction.

Using the flow-visualization system, both vertical and horizontal visualizations were performed over a Reynolds-number range of $50 < \mathrm{Re}_d < 2500$. Figure 50 shows a typical side view of the vortex-shedding pattern for $\mathrm{Re}_d = 300$ with a pulse rate of 30 Hz. The picture in Fig. 51 is a single video frame obtained directly from

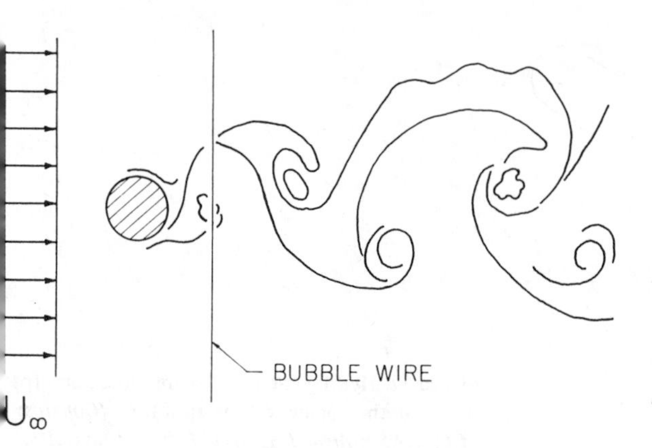

Figure 50 Schematic of flow over a circular cylinder (side view). *(Courtesy of C. R. Smith and T. Wei, Lehigh University.)*

the recorded video signal using a videographic copier. This figure illustrates the very organized, periodic vortex-shedding pattern that has become a well-known fluid-dynamic phenomenon and gives just a hint of spanwise behavior (as indicated by the folded edges of the vortex wake region).

To visualize the spanwise structure, the hydrogen bubble wire was located parallel to the cylinder, as shown in Fig. 52. As indicated, the wire could be moved relative to the cylinder to facilitate visualization of the flow structure. With the wire

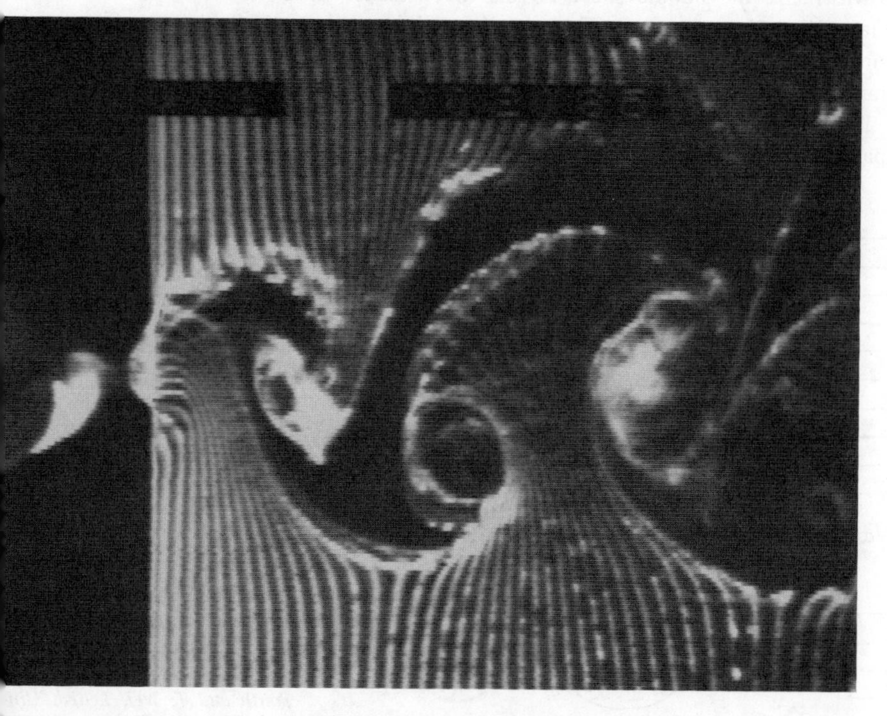

Figure 51 Photograph using a videographic copier of the periodic vortex-shedding pattern behind a circular cylinder. *(Courtesy of C. R. Smith and T. Wei, Lehigh University.)*

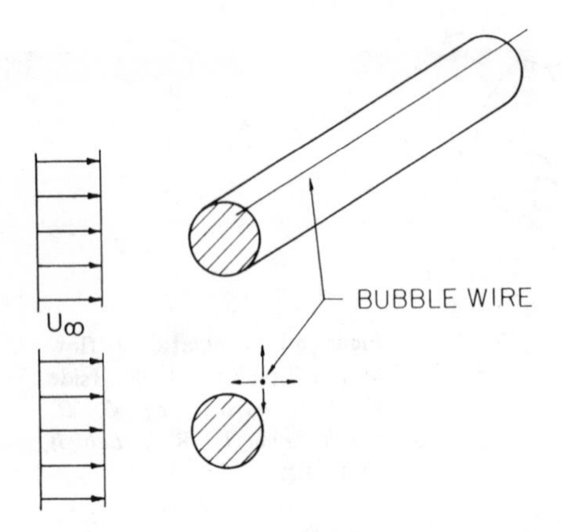

Figure 52 Hydrogen-bubble wire location for study of the spanwise flow structure. *(Courtesy of C. R. Smith and T. Wei, Lehigh University.)*

positioned essentially in line with both the trailing edge and the top of the cylinder, Fig. 53 shows schematically the top-view pattern that could be observed. The corresponding side-view vortex pattern is also shown for reference. What came as a total surprise was the appearance of spanwise "pockets" of inflow normal to the plane of the bubbles, which occur between the transverse vortices. In the top view of Fig. 53, these pockets appear just behind the cylinder. As the pockets convect downstream, the bubble sheet reveals the presence of axially oriented vortices of relatively uniform spacing. These vortices (shown schematically between transverse vortices of like rotation) were of small core size and strong vorticity relative to the transverse vortices and were observed to occur in counterrotating pairs.

Figure 54 is a four-picture top-view sequence illustrating the formation of the spanwise structure. These pictures are single video frames obtained by taking a

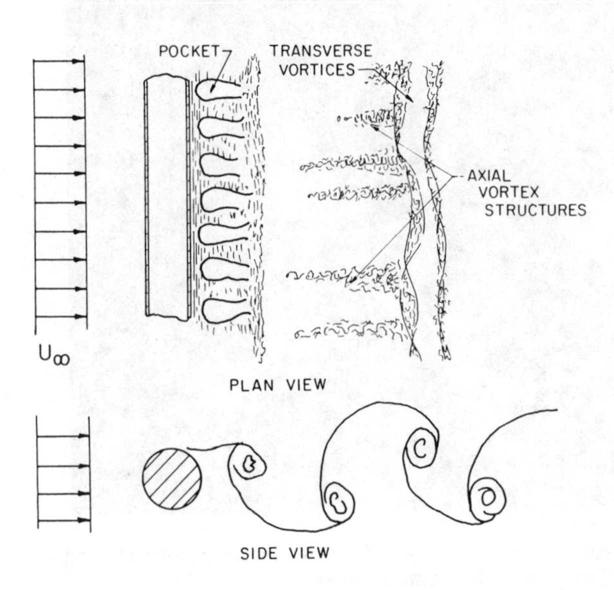

Figure 53 Sketch of the hydrogen-bubble patterns from the top and side of the circular cylinder. *(Courtesy of C. R. Smith and T. Wei, Lehigh University.)*

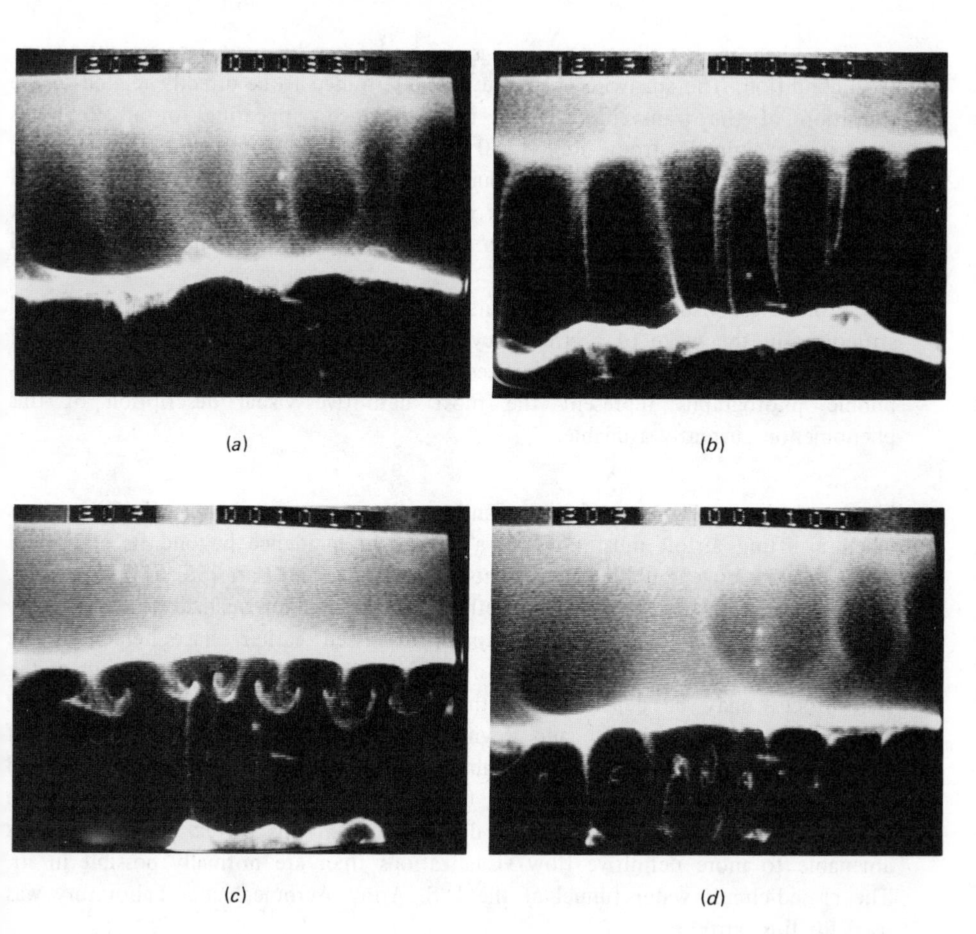

Figure 54 Hydrogen-bubble photographs from video monitor of the spanwise structure. (*a*) and (*b*) Formation of spanwise pockets. (*b*) and (*c*) Evolution to axial vortices. (*d*) Formation of a new set of spanwise pockets has just started. *(Courtesy of C. R. Smith and T. Wei, Lehigh University.)*

Polaroid photograph directly from the video monitor. The orientation of each picture is the same as the top view in Fig. 53, with the cylinder just out of the field of view to the left. For these pictures, the bubble wire was running in a dc mode such that a continuous sheet of bubbles was generated. The formation of the spanwise pockets is illustrated in Fig. 54*a* and *b*, with the evolution to axial vortices shown in Fig. 54*b* and *c*. Note the "mushrooms" that appear in Fig. 54*c* as a result of the penetration of the bubble sheet by the axial vortices. In Fig. 54*d*, the remnants of the visualization bubbles depicting the spanwise structure have dissipated at the downstream side of the transverse vortex (although their presence is still in evidence), and the formation of a new set of spanwise pockets has just started.

The phenomenon illustrated in Figs. 53 and 54 is by no means unique to one flow condition. The spanwise structure was determined to be directly related to the transition of the transverse vortices to turbulence, appearing for all Reynolds numbers examined, from $Re_d = 150$ (the accepted transition value) up to $Re_d = 2500$. The phenomenon was found not to be a function of the cylinder end condition, nor peculiar to free-stream conditions in the channel. This latter point was confirmed by *towing* the cylinder through a quiescent channel, with identical structures appearing in the wake.

From these visualizations, it appears that the formation of counterrotating axial structures in the wake of a circular cylinder is a necessary part of the process of transition to turbulence of the vortices shed from the cylinder. These hydrogen-bubble photographs represent the most definitive visual description of this phenomenon currently available.

Flow over an oscillating airfoil. Dynamic stall occurs in many practical situations when a lifting airfoil must undergo a change in incidence beyond its static-stall angle. The hydrogen-bubble experiments of McAlister and Carr [56, 61] at the U.S. Army Aeromechanics Laboratory, together with the smoke-visualization experiments of McCroskey (of the same laboratory), discussed earlier, have helped in the understanding of the phenomenon of dynamic stall. Although certain details of the stall process may be dependent on the airfoil geometry and the particular flow environment (e.g., pitch rate and Reynolds number), the general characteristics of the stall vortex are believed to be qualitatively invariant. An important step toward better understanding of dynamic stall, therefore, would be to examine the collapse of the boundary layer and the initial development of this vortex under conditions amenable to more definitive flow visualizations than are normally possible in air. The closed-circuit water tunnel at the U.S. Army Aeromechanics Laboratory was used for this purpose.

To study the flow pattern around the airfoil, a wire cathode was stretched across the test section and oriented normal to the direction of flow. Although both platinum and stainless steel are good noncorrosive materials, stainless steel was selected because of its superior tensile strength. A nonconductive model of a NACA 0012 airfoil with a modified leading edge, a chord of 10 cm, and a span of 21 cm was used for these experiments. Electrodes were also placed at nine chordwise locations along the upper surface of the model during its construction. Sinusoidal pitching motion of the airfoil was accomplished by a flywheel, connecting rod, and rack-and-gear mechanism that functioned to transform a circular motion first to reciprocating motion and then to airfoil pitch oscillations.

Typical results, using both the fine wire electrode upstream of the airfoil and the electrodes built into the airfoil's upper surface, are shown in Fig. 55. This figure shows the location of flow separation, the wake vortex patterns, shear-layer vortices, and the dynamic-stall vortex. For this airfoil at $Re = 21{,}000$, a reduced frequency of $k = 0.25$, and $\alpha = 10^\circ + 10^\circ \sin \omega$, the hydrogen-bubble experiments were important factors in arriving at the following conclusions:

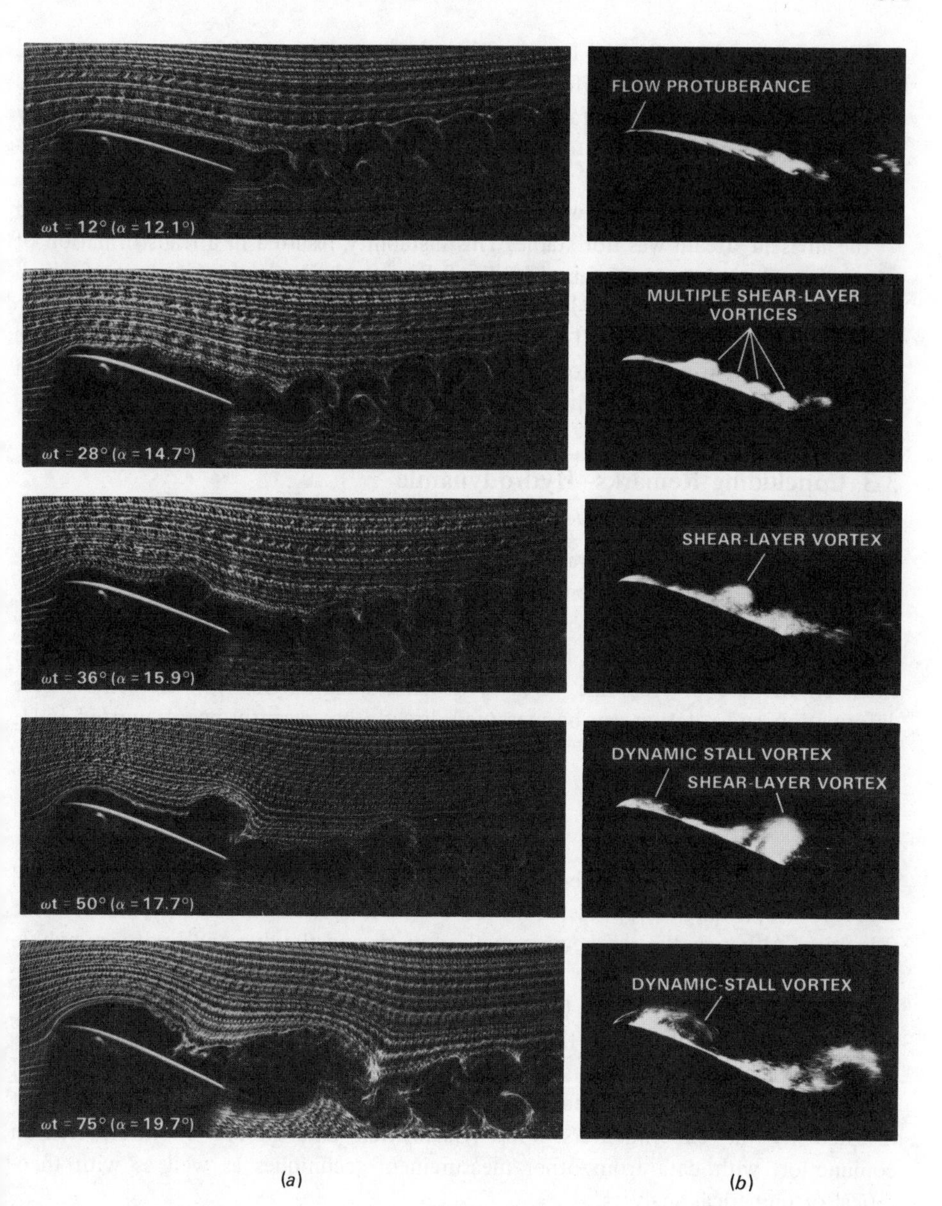

Figure 55 Hydrogen-bubble visualization of the critical stages of dynamic stall. (*a*) Free-stream electrode. (*b*) Model electrodes. *(Courtesy of K. W. McAlister and L. W. Carr, U.S. Army Aeromechanics Laboratory, AVRADCOM.)*

1. The onset of dynamic stall began with a rapid movement of flow reversal toward the leading edge of the airfoil. At one point, a thin layer of reversed flow was observed to exist momentarily throughout the entire upper-surface boundary layer without causing any appreciable disturbance to the viscous-inviscid interface.
2. The free shear layer that was created between the region of reversed flow and the inviscid stream was not stable. This instability resulted in a transformation of the free shear layer into a multitude of discrete clockwise vortices, out of which emerged a dominant "shear-layer vortex."
3. Once the flow had reversed up to the leading edge of the airfoil, a protuberance grew and eventually developed into the "dynamic-stall vortex" that has been observed in high-Reynolds-number experiments.

3.3 Concluding Remarks–Hydrodynamic Flow Visualization

Although restricted to comparatively low-speed flows, the dye and hydrogen-bubble visualization methods have been responsible for much of our understanding of many basic steady- and unsteady-flow phenomena. These techniques have been widely used for almost three decades, but new ways of using them continue to emerge. For example, the use of cross-line wires to visualize quadrilateral flow elements whose deformation and rotation can be measured directly has recently been reported [64].

4. CONCLUSIONS

The smoke-tube and helium bubble methods in air have been found to be useful over the widest range of speeds, from low subsonic to transonic and even supersonic velocities. Although the smoke-wire method in air and the dye and hydrogen-bubble methods in water are usually used at low and moderate velocities, they provide greater detail of the flow than the other methods.

While the usefulness of the direct-injection methods of flow visualization is apparent, each flow problem should be approached with care. The importance of the results obtained for a given problem are directly related to the time and effort invested by the experimenters. Flow-visualization data should be interpreted in conjunction with data from other measurement techniques as well as with theoretical or numerical analyses.

REFERENCES

1. N. H. Randers-Pehrson, Pioneer Wind Tunnels, *Smithson. Misc. Collect.*, vol. 93, no. 4, 1935.
2. T. J. Mueller, Smoke Visualization of Subsonic and Supersonic Flows (The Legacy of F. N. M. Brown), University of Notre Dame Rep. UNDAS TN-3412-1 (AFOSR TR-78-1262), 1978.
3. T. J. Mueller, On the Historical Development of Apparatus and Techniques for Smoke

Visualization of Subsonic and Supersonic Flows, AIAA Paper 80-0420-CP, presented at the AIAA 11th Aerodynamic Testing Conference, 1980.

4. R. H. Smith and C. T. Wang, Contracting Cones Giving Uniform Throat Speeds, *J. Aeronaut. Sci.*, vol. 11, no. 4, pp. 356–360, 1944.
5. S. M. Batill, R. C. Nelson, and T. J. Mueller, High Speed Smoke Flow Visualization, Air Force Wright Aeronautical Laboratories Rep. AFWAL/TR-3002, 1981.
6. R. L. Maltby and R. F. A. Keating, Smoke Techniques for Use in Low Speed Wind Tunnels, AGARD-ograph 70, pp. 87–109, 1962.
7. J. H. Preston and N. E. Sweeting, An Improved Smoke Generator for Use in the Visualization of Airflow, Particularly Boundary Layer Flow at High Reynolds Numbers, Aeronautical Research Council Reports & Memoranda 2023 (ARC 7111), 1943.
8. R. L. Bisplinghoff, J. B. Coffin, and C. W. Haldeman, Water Fog Generation System for Subsonic Flow Visualization, *AIAA J.*, vol. 14, no. 8, pp. 1133–1135, 1976.
9. S. A. Brandt and J. D. Iversen, Merging of Aircraft Trailing Vortices, *AIAA J. Aircr.*, vol. 14, no. 12, pp. 1212–1220, 1977.
10. V. P. Goddard, J. A. McLaughlin, and F. N. M. Brown, Visual Supersonic Flow Patterns by Means of Smoke Lines, *J. Aerosp. Sci.*, vol. 26, no. 11, pp. 761–762, 1959.
11. V. P. Goddard, Development of Supersonic Streamline Visualization, Report to the National Science Foundation on Grant 12488, 1962.
12. J. T. Kegelman, R. C. Nelson, and T. J. Mueller, Smoke Visualization of the Boundary Layer on an Axisymmetric Body, AIAA Paper 79-1635, presented at the AIAA Atmospheric Flight Mechanics Conference for Future Space Systems, Boulder, Colo., 1979.
13. F. N. M. Brown, The Physical Model of Boundary Layer Transition, reprinted from *Proc. of the Ninth Midwestern Mechanics Conf.*, University of Wisconsin, pp. 421–429, 1965.
14. J. T. Kegelman, R. C. Nelson, and T. J. Mueller, Boundary Layer and Side Force Characteristics of a Spinning Axisymmetric Body, AIAA Paper 80-1584-CP, presented at the AIAA Atmospheric Flight Mechanics Conference, 1980.
15. F. N. M. Brown and V. P. Goddard, The Effect of Sound on the Separated Laminar Boundary Layer, Final Rep., NSF Grant G11712, University of Notre Dame, pp. 1–67, 1963.
16. J. M. Martin, R. W. Empey, W. J. McCroskey, and F. X. Caradonna, An Experimental Analysis of Dynamic Stall on an Oscillating Airfoil, *J. Am. Helicopter Soc.*, vol. 19, pp. 26–32, 1974.
17. W. J. McCroskey, L. W. Carr, and K. W. McAlister, Dynamic Stall Experiments on Oscillating Airfoils, *AIAA J.*, vol. 14, no. 1, pp. 57–63, 1976.
18. W. J. McCroskey, Notes on Unsteady Aerodynamics, AGARD Lecture Series 94, "Three-Dimensional and Unsteady Separation at High Reynolds Numbers," Von Karman Institute for Fluid Dynamics, Brussels, 1978.
19. T. J. Mueller, C. R. Hall, Jr., and W. P. Sule, Supersonic Wake Flow Visualization, *AIAA J.*, vol. 7, no. 11, pp. 2151–2153, 1969.
20. W. B. Roberts and J. A. Slovisky, Location and Magnitude of Cascade Shock Loss by High-Speed Smoke Visualization, *AIAA J.*, vol. 17, no. 11, pp. 1270–1272, 1979.
21. J. J. Cornish, A Device for the Direct Measurements of Unsteady Air Flows and Some Characteristics of Boundary Layer Transition, Mississippi State University, Aerophysics Res. Note 24, p. 1, 1964.
22. C. J. Sanders and J. F. Thompson, An Evaluation of the Smoke-Wire Technique of Measuring Velocities in Air, Mississippi State University, Aerophysics Res. Rep. 70, 1966.
23. H. Yamada, Instantaneous Measurements of Air Flows by Smoke-Wire Technique, *Trans. Jpn. Mech. Eng.*, vol. 39, p. 726, 1973.
24. T. Corke, D. Koga, R. Drubka, and H. Nagib, A New Technique for Introducing Controlled Sheets of Smoke Streaklines in Wind Tunnels, *Proc. of the Int. Congr. on Instrum. in Aerosp. Simulation Facilities,* IEEE Publ. 77 CH 1251-8 AES, p. 74, 1974.
25. H. M. Nagib, Visualization of Turbulent and Complex Flows Using Controlled Sheets of Smoke Streaklines, *Proc. of the Int. Symp. on Flow Visualization,* Tokyo, pp. 181–186, 1977 (also Suppl. 29-1 to 29-7).

26. N. Kasagi, M. Hirata, and S. Yokobori, Visual Studies of Large Eddy Structures in Turbulent Shear Flows by Means of Smoke-Wire Method, *Proc. of the Int. Symp. on Flow Visualization,* Tokyo, pp. 169–174, 1977.
27. T. Torii, Flow Visualization by Smoke-Wire Technique, *Proc. of the Int. Symp. on Flow Visualization,* Tokyo, pp. 175–180, 1977.
28. S. M. Batill and T. J. Mueller, Visualization of the Laminar-Turbulent Transition in the Flow over an Airfoil Using the Smoke-Wire Technique, AIAA Paper 80-0421, presented at the AIAA 11th Aerodynamic Testing Conference, Colorado Springs, Colo., 1980.
29. D. Cornell, Smoke Generation for Flow Visualization, Mississippi State University, Aerophysics Res. Rep. 54, 1964.
30. T. J. Mueller and S. M. Batill, Experimental Studies of the Laminar Separation Bubble on a Two-Dimensional Airfoil at Low Reynolds Numbers, AIAA Paper 80-1440, presented at the AIAA 13th Fluid and Plasma Dynamics Conference, Snowmass, Colo., 1980.
31. T. Cebeti and P. Bradshaw, *Momentum Transfer in Boundary Layer,* Hemisphere, Washington, D.C., 1977.
32. R. Betchov, Transition, in W. Frost and T. H. Moulden (eds.), *Handbook of Turbulence,* vol. 1, pp. 147–164, Plenum, New York, 1977.
33. O. Griffin, Universal Similarity in the Wakes of Stationary and Vibrating Bluff Structures, ASME Paper 80-WA/FE-4, 1980.
34. C. Fiscina, An Investigation into the Effects of Shear on the Flow Past Bluff Bodies, M.S. thesis, University of Notre Dame, 1977.
35. C. Fiscina, An Experimental Investigation of the Flow Field around a Bluff Body in a Highly Turbulent Shear Flow, Ph.D. thesis, University of Notre Dame, 1979.
36. R. W. Hale, P. Tan, and D. E. Ordway, Experimental Investigation of Several Neutrally Buoyant Bubble Generations for Aerodynamic Flow Visualization, *Nav. Res. Rev.,* vol. 24, no. 6, pp. 19–24, June 1971.
37. M. H. Redon and M. F. Vinsonneau, Étude de l'écoulement de l'air autour d'une maquette, *Aeronautique,* vol. 18, no. 204, pp. 60–66, 1936.
38. J. Kampé de Fériet, Some Recent Researches on Turbulence, *Proc. of the Fifth Int. Congr. for Appl. Mech.,* Cambridge, Mass., 1938, pp. 352–355, Wiley, New York, 1939.
39. F. S. Owen, R. W. Hale, B. V. Johnson, and A. Travers, Experimental Investigation of Characteristics of Confined Jet-Driven Vortex Flows, United Aircraft Res. Lab. Rep. R-2494-2, AD-328 502, 1961.
40. R. W. Hale, P. Tan, R. C. Stowell, and D. E. Ordway, Development of an Integrated System for Flow Visualization in Air Using Neutrally Buoyant Bubbles, Sage Action, Inc. Rep. SAI-RR 7107, 1971.
41. R. W. Hale and D. E. Ordway, High-Lift Capabilities from Favorable Flow Interaction with Close-Coupled Canards, Sage Action, Inc. Rep. SAI-RR 7501, 1975.
42. R. W. Hale, P. Tan, and D. E. Ordway, Prediction of Aerodynamic Loads on Close-Coupled Canard Configurations–Theory and Experiment, Sage Action Inc. Rep. SAI-RR 7702 (ONR-CR215-194-3F), 1977.
43. P. C. Klimas, Helium Bubble Survey of an Opening Parachute Flow Field, *AIAA J. Aircr.,* vol. 10, no. 9, pp. 567–569, 1973.
44. L. S. Iwan, R. W. Hale, P. Tan, and R. C. Stowell, Transonic Flow Visualization with Neutrally Buoyant Bubbles, Sage Action, Inc. Rep. SAI-RR 7304, 1973.
45. R. W. Empey and R. A. Ormiston, Tail-Rotor Thrust on a 5.5-foot Helicopter Model in Ground Effect, Preprint 802, presented at the 30th Annual National Forum of the American Helicopter Society, 1974.
46. R. W. Hale, P. Tan, R. C. Stowell, L. S. Iwan, and D. E. Ordway, Preliminary Investigation of the Role of the Tip Vortex in Rotary Wing Aerodynamics through Flow Visualization, Sage Action, Inc. Rep. SAI-RR 7402, 1974.
47. G. A. Tokaty, *A History and Philosophy of Fluid Mechanics*, Foulis, Oxfordshire, U.K., 1971.
48. B. R. Clayton and B. S. Massey, Flow Visualization in Water: A Review of Techniques, *J. Sci. Instrum.,* vol. 44, pp. 2–11, 1967.

49. W. Merykirch, *Flow Visualization,* Academic, New York, 1974.
50. H. Werlé, Hydrodynamic Flow Visualization, *Annu. Rev. Fluid Mech.,* vol. 5, pp. 361–382, 1973.
51. G. E. Erickson, Vortex Flow Correlation, Air Force Wright Aeronautical Lab. Rep. AFWAL TR 80-3143, 1981.
52. H. Werlé, Methodes d'etude par analogie hydraulique des ecoulements subsonique, supersonique et hypersonique, (NATO) AGARD Rapport 399, 1960.
53. D. R. Campbell, Flow Visualization Using a Selectively Sensitive Fluorescent Dye, Aerospace Res. Lab. Rep. ARL 73-005, 1973.
54. P. Poisson-Quinton and H. Werlé, Water Tunnel Visualization of Vortex Flows, *Astronaut. Aeronaut.,* vol. 5, pp. 64–66, June 1967.
55. F. R. Hama, Streaklines in a Perturbed Shear Flow, *Phys. Fluids,* vol. 5, no. 6, 1962.
56. K. W. McAlister and L. W. Carr, Water Tunnel Visualization of Dynamic Stall, *ASME J. Fluids Eng.,* vol. 101, pp. 376–380, 1979.
57. E. W. Geller, An Electrochemical Method of Visualizing the Boundary Layer, *J. Aeronaut. Sci.,* vol. 22, no. 12, pp. 869–870, 1955.
58. F. A. Schraub, S. J. Kline, J. Henry, P. W. Runstadler, Jr., and A. Littel, Use of Hydrogen Bubbles for Quantitative Determination of Time-Dependent Velocity Fields in Low-Speed Flows, *ASME J. Basic Eng.,* vol. 87, pp. 429–444, 1965.
59. E. Kato, M. Suita, and M. Kawamata, Visualization of Unsteady Pipe Flows Using Hydrogen Bubble Technique, *Proc. of the Int. Symp. on Flow Visualization,* Bochum, West Germany, 1980, pp. 342–346.
60. G. E. Mattingly, The Hydrogen-Bubble Flow-Visualization Technique, David Taylor Model Basin Rep. 2146 (AD 630 468), 1966.
61. K. W. McAllister and L. W. Carr, Water Tunnel Experiments on an Oscillating Airfoil at R = 21,000, NASA TM 78446, 1978.
62. F. R. Hama, Three-Dimensional Vortex Pattern behind a Circular Cylinder, *J. Aeronaut. Sci.,* vol. 24, p. 156, 1957.
63. J. H. Gerrard, The Wakes of Cylindrical Bluff Bodies at Low Reynolds Numbers, *Phil. Trans. R. Soc. London Ser. A,* vol. 288, p. 351, 1978.
64. T. Matsui, H. Nagata, and H. Yasuda, Some Remarks on Hydrogen Bubble Technique for Low Speed Water Flows, in T. Asanuma (ed.), *Flow Visualization,* pp. 215–220, Hemisphere, Washington, D.C., 1979.

CHAPTER

EIGHT

OPTICAL SYSTEMS FOR FLOW MEASUREMENT: SHADOWGRAPH, SCHLIEREN, AND INTERFEROMETRIC TECHNIQUES

R. J. Goldstein

1. INTRODUCTION

Optical techniques that have been used in the measurement of flow include (1) direct visualization, where some type of marker (e.g., dye, bubbles, solid particles) is followed along with the fluid motion; (2) laser-Doppler systems, in which the frequency shift of scattered illumination from such a marker–usually a particle–is measured; and (3) what might be called index-of-refraction methods, in which the index of refraction or a spatial derivative of the index of refraction of a medium is measured, and from this some property or properties of the flow are determined.

Only the methods in the last category are examined in this paper.* These include schlieren, shadowgraph, and interferometric techniques† that are used to study density fields in transparent media, usually gases or liquids. References [1 to 8] have considerable information and extensive bibliographies on these methods.‡ Although all three methods depend on variation of the index of refraction in a transparent medium and the resulting effects on a light beam passing through the test region, quite different quantities are measured with each one. Shadowgraph systems are used to indicate the variation of the second derivative (normal to the

Mr. H. D. Chiang was of great assistance during the preparation of this material–in particular, with the flow-visualization photographs and the bibliography. E. R. G. Eckert, A. G. Hayener, A. Roshko, and A. B. Witte kindly supplied photographs from their research.

*Techniques falling into categories (1) and (2) are described in Chaps. 7 and 5, respectively.

†Birefringent fluids, which are doubly refractive when subject to shear stresses, are discussed in Chap. 9.

‡Much of the material in this chapter is taken from [8].

light beam) of the index of refraction. With a schlieren system, the first derivative of the index of refraction (in a direction normal to the light beam) determines the light pattern. Interferometers respond directly to differences in optical path length, essentially giving the index-of-refraction field within the flow.

Optical measurements have many advantages over other techniques. Perhaps the major one is the absence of an instrument probe that could influence the flow field. The light beam can also be considered as essentially inertialess, so that very rapid transients can be studied. The sensitivities of the three optical methods are quite different, so that they can be used to study a variety of systems. Thus, interferometers are often used to study flows in which density gradients are small, while schlieren and shadowgraph systems are often employed in studying shock and flame phenomena, in which very large density gradients are present.

All three techniques are valuable when visualizing flows in which density differences occur naturally or are artificially induced. When used quantitatively, these techniques can often be used to determine density, pressure, and/or temperature variations in the flow. From these, other properties of the flow field (e.g., laminar versus turbulent nature of the motion, boundary-layer thickness, shock angles, points of separation, and reattachment) can often be inferred.

Shadowgraph, schlieren, and interferometric measurements are essentially integral; they integrate the quantity measured over the length of the light beam. For this reason they are well suited to measurements in two-dimensional fields, where there is no index of refraction or density variation in the field along the light beam, except at the beam's entrance to and exit from the test (disturbed) region. These latter variations can be considered as sharp discontinuities, or appropriate end corrections can be made. Axisymmetric fields can also be studied [8]. If the field is three-dimensional, an average (along the light beam) of the measured quantity can still be determined. Since both schlieren and shadowgraph systems are primarily used for qualitative studies, this is often acceptable; and even in interferometric studies, the averaging done by the light beam can sometimes be advantageous. Observations of local three-dimensional variations in the flow have been made of such phenomena as the rise of thermal plumes and turbulent bursts [9].

In the three methods to be studied, the index of refraction (or one of its spatial derivatives) determines the resulting illumination or light pattern. The index of refraction of a homogeneous medium is a function of the thermodynamic state, often only the density. According to the Lorenz-Lorentz relation, the index of refraction of a homogeneous transparent medium can be obtained from

$$\frac{1}{\rho}\frac{n^2 - 1}{n^2 + 2} = \text{const} \tag{1}$$

When $n \simeq 1$, this reduces to the Gladstone-Dale equation

$$\frac{n - 1}{\rho} = C \tag{2}$$

or

$$\rho = \frac{n - 1}{C} \tag{3}$$

which holds quite well for gases. The constant C, called the Gladstone-Dale Constant, is a function of the particular gas and varies slightly with wavelength. Usually, instead of using C directly, the index of refraction at a standard condition n_0 is given:

$$n - 1 = \frac{\rho}{\rho_0}(n_0 - 1) \tag{4}$$

or

$$\rho = \rho_0 \frac{n - 1}{n_0 - 1} \tag{5}$$

When the first or second derivative (say, with respect to y) is determined as in a schlieren or shadowgraph apparatus, then, using from Eqs. (3) and (4),

$$\frac{\partial \rho}{\partial y} = \frac{1}{C}\frac{\partial n}{\partial y} = \frac{\rho_0}{n_0 - 1}\frac{\partial n}{\partial y} \tag{6}$$

$$\frac{\partial^2 \rho}{\partial y^2} = \frac{1}{C}\frac{\partial^2 n}{\partial y^2} = \frac{\rho_0}{n_0 - 1}\frac{\partial^2 n}{\partial y^2} \tag{7}$$

The question arises as to what phenomena in a flowing field cause the density variations that are observed. Naturally occurring variations are present in compressible flows–for example, high-speed flows involving shock waves. Natural-convection flow fields are nonuniform in density, as are the flows in flames and combustion systems. The mixing of fluids of different density can be studied, as can forced-convection flow fields, when heat or mass transfer produces nonuniform densities. An artificial density distribution can be introduced into a flow through local heating–often of a transient nature–from a heating wire or a spark. This "tracer" can then be followed.

Many of the measurements using index-of-refraction methods involve temperature variation in a two-dimensional flow field. It is of interest to see how the temperature affects the index of refraction and its derivatives if the flowing medium is a gas.

If the pressure can be assumed constant and the ideal gas equation of state ($\rho = P/RT$) holds, then

$$\frac{\partial n}{\partial y} = -\frac{CP}{RT^2}\frac{\partial T}{\partial y} = -\frac{n_0 - 1}{T}\frac{\rho}{\rho_0}\frac{\partial T}{\partial y} \tag{8}$$

or

$$\frac{\partial T}{\partial y} = -\frac{T}{n_0 - 1}\frac{\rho_0}{\rho}\frac{\partial n}{\partial y} \tag{9}$$

and

$$\frac{\partial^2 n}{\partial y^2} = C\left[-\frac{\rho}{T}\frac{\partial^2 T}{\partial y^2} + \frac{2\rho}{T^2}\left(\frac{\partial T}{\partial y}\right)^2\right] \tag{10}$$

Note that Eq. (9), which applies to a schlieren study, shows a relatively simple relationship between the gradient of the temperature and the gradient of the index of refraction. For a shadowgraph, the equivalent relation, Eq. (10), is more complicated, although under many conditions the second term may be small.

The index of refraction of a gas as measured in an interferometer can indicate the temperature directly. From Eqs. (3) and (4), assuming constant pressure and the perfect-gas equation of state,

$$T = \frac{C}{n-1}\frac{P}{R} = \frac{n_0 - 1}{n-1}\frac{P}{P_0}T_0 \tag{11}$$

Low-speed flows in which heat transfer occurs can often be approximated as constant-pressure systems in which Eqs. (8) to (11) apply.

For a reversible, adiabatic (isentropic) process in an ideal gas, the pressure, temperature, and density all vary with

$$P\left(\frac{1}{\rho}\right)^k = \text{const}$$

or

$$\frac{P}{P_0} = \left(\frac{\rho}{\rho_0}\right)^k \tag{12}$$

where k is the ratio of specific heats (constant pressure to constant volume), and ρ_0 and P_0 refer to some reference condition. If the pressure variation is derived from the optical measurements, then from Eqs. (5) and (12),

$$\frac{P}{P_0} = \left(\frac{n-1}{n_0 - 1}\right)^k \tag{13}$$

and

$$\frac{\partial P}{\partial y} = P\frac{k}{n-1}\frac{\partial n}{\partial y} \tag{14}$$

or

$$\frac{\partial n}{\partial y} = \frac{1}{P}\frac{\partial P}{\partial y}\frac{n-1}{k} \tag{15}$$

Similarly, the expression relating the second derivatives of the pressure and index of refraction could be obtained for an isentropic flow.

The index of refraction of a liquid is primarily a function of temperature and, for accurate results, should be obtained from direct measurement rather than from Eq. (1). Many tabulated results and empirical expressions for the index of refraction of liquids are available. For comparison, Table 1, derived from [10, 11], cites values at 20°C and 1 atm for air and water. The two wavelengths chosen are a commonly

Table 1 Index of refraction for air and water at 20°C and 1 atm

λ, nm	$n_{air} - 1$	n_{H_2O}	$\frac{dn}{dT}_{air}$, $(°C)^{-1}$	$\frac{dn}{dT}_{H_2O}$, $(°C)^{-1}$
546.1	2.733×10^{-4}	1.3345	-0.932×10^{-6}	-0.895×10^{-4}
632.8	2.719×10^{-4}	1.3317	-0.927×10^{-6}	-0.880×10^{-4}

used mercury line (546.1 nm) and the visible line from a CW He-Ne laser (632.8 nm).

Of the three systems to be discussed, two of them–schlieren and shadowgraph–can be described by geometric or ray optics, although under certain conditions diffraction effects can be significant. Interferometers, as the name implies, depend on the interference of coherent light beams, and some discussion of physical (wave) optics is required.

2. SCHLIEREN SYSTEM

2.1 Analysis by Geometric or Ray Optics

To study both schlieren and shadowgraph systems, the path of a light beam in a medium whose index of refraction is a function of position must be analyzed. Consider Fig. 1, which shows a light beam, traveling initially in the z direction, passing through a medium whose index of refraction varies (for simplicity) only in the y direction. At time τ the beam is at position z, and the wavefront (surface normal to

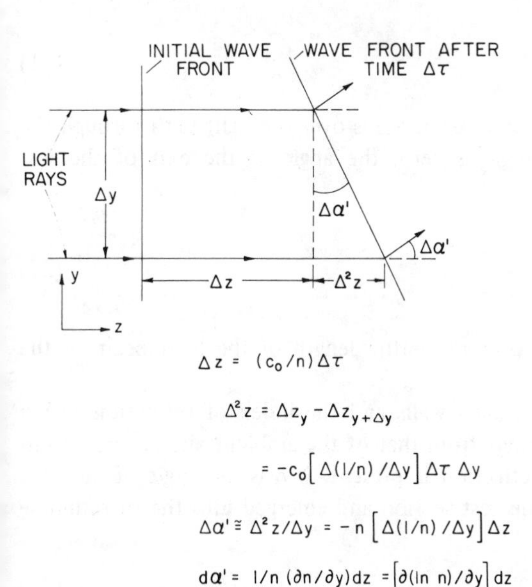

Figure 1 Bending of light rays in an inhomogeneous medium.

the path of the light) is as shown. After a time interval $\Delta\tau$, the light has moved a distance of $\Delta\tau$ times the velocity of light, which, in general, is a function of y so the wavefront or light beam may have turned an angle $\Delta\alpha'$. The local value of the speed of light is c_0/n. With reference to Fig. 1, and assuming that only small deviations occur, the distance Δz that the light beam travels during time interval $\Delta\tau$ is

$$\Delta z = \frac{c_0}{n}\,\Delta\tau \tag{16}$$

Now

$$\Delta^2 z = \Delta z_y - \Delta z_{y+\Delta y} = -\frac{\Delta(\Delta z)}{\Delta y}\,\Delta y \tag{17}$$

or

$$\Delta^2 z = -c_0\,\frac{\Delta(1/n)}{\Delta y}\,\Delta\tau\,\Delta y \tag{18}$$

The angular deflection of the ray is

$$\Delta\alpha' \approx \frac{\Delta^2 z}{\Delta y} = -n\,\frac{\Delta(1/n)}{\Delta y}\,\Delta z \tag{19}$$

In the limit, if Δy and Δz are considered to be very small,

$$d\alpha' = \frac{1}{n}\frac{\partial n}{\partial y}\,dz = \frac{\partial(\ln n)}{\partial y}\,dz \tag{20}$$

For small deflections, the angle of the light beam is the slope $\partial y/\partial z$ of the light beam, and thus

$$\frac{\partial^2 y}{\partial z^2} = \frac{1}{n}\frac{\partial n}{\partial y} \tag{21}$$

If the angle remains small, this expression holds over the light path through the disturbed region. If the entering angle is zero, the angle at the exit of the test region is

$$\alpha' = \int \frac{1}{n}\frac{\partial n}{\partial y}\,dz = \int \frac{\partial(\ln n)}{\partial y}\,dz \tag{22}$$

where the integration is performed over the entire length of the light beam in the test region.

If the test region is enclosed by glass walls and the index of refraction within the test section is considerably different from that of the ambient air, n_a, then, from Snell's law, an additional angular deflection is present. If α is the angle of the light beam after it has passed through the test section and emerged into the surrounding air, then

$$n_a \sin\alpha = n \sin\alpha' \tag{23}$$

If the test-section windows are plane and of uniform thickness, then for small values of α and α',

$$\alpha = \frac{n}{n_a} \alpha' \tag{24}$$

Equation (22) gives

$$\alpha = \frac{n}{n_a} \int \frac{1}{n} \frac{\partial n}{\partial y} \, dz \tag{25}$$

If the $1/n$ within the integrand does not change greatly through the test section, then

$$\alpha = \frac{1}{n_a} \int \frac{\partial n}{\partial y} \, dz \tag{26}$$

or, since $n_a \simeq 1$,

$$\alpha \simeq \int \frac{\partial n}{\partial y} \, dz \tag{27}$$

Note that if a gas (not at extremely high density) is the test fluid, $\alpha' \simeq \alpha$.

The angle α in Eq. (27) is in the yz plane. If there is a variation of index of refraction in the x direction, then, again assuming small angular deviation, a similar expression would give the deflection in the xz plane proportional to $\partial n/\partial x$.

If variations of n in the x and y directions as well as the effect of significant angular deflection are included, the resulting equations for the path of the light beam, equivalent to Eq. (21), are [2]

$$y'' = \frac{1}{n} [1 + (x')^2 + (y')^2] \left(\frac{\partial n}{\partial y} - y' \frac{\partial n}{\partial z} \right) \tag{28a}$$

$$x'' = \frac{1}{n} [1 + (x')^2 + (y')^2] \left(\frac{\partial n}{\partial x} - x' \frac{\partial n}{\partial z} \right) \tag{28b}$$

where primes refer to differentiation with respect to z.

Note that the light beam is turned in the direction of increasing index of refraction. In most media this means that the light is bent toward the region of higher density.

A schlieren system is basically a device to measure or indicate the angle α, typically of the order of 10^{-6} to 10^{-3} rad, as a function of position in the xy plane normal to the light beam. Consider the system shown in Fig. 2. A light source, which we assume to be rectangular (of dimensions a_s by b_s) and at the focus of lens L_1 provides a parallel light beam entering the field of disturbance in the test section. The deflected rays, when the disturbance is present, are indicated

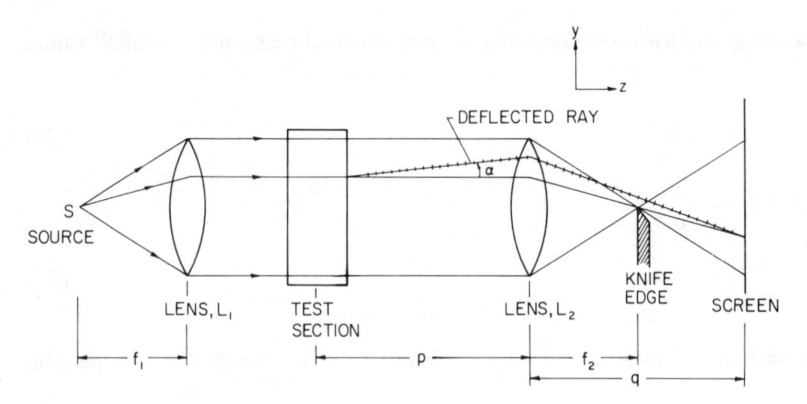

Figure 2 Typical schlieren system using lenses.

by cross-hatched lines. The light is collected by a second lens L_2, at whose focus a knife-edge is placed, and then passes onto a screen located at the conjugate focus of the test section. As is shown below, if the screen is not at the focus of the disturbance, shadowgraph effects will be superimposed on the schlieren pattern.

If no disturbance is present, then ideally the light beam at the focus of L_2 would be as shown in Fig. 3, with dimensions a_0 by b_0 which are related to the initial dimensions by

$$\frac{a_0}{a_s} = \frac{b_0}{b_s} = \frac{f_2}{f_1} \tag{29}$$

where f_1 and f_2 are the focal lengths of L_1 and L_2, respectively.

As shown in the figure, the source is usually adjusted so that the shorter dimension a_0 is at right angles to the knife-edge to maximize sensitivity.* The knife-edge (typically a razor blade) is adjusted, when no disturbance is present, to cut off all but an amount a_K (typically, $a_K = a_0/2$) of the height a_0. When the knife-edge is moved across the beam exactly at the focus, the illumination at the screen should decrease uniformly; if the knife-edge is not in the focal plane, the image at the screen will not darken uniformly. The illumination at the screen when no knife-edge is present is I_0, and with the knife-edge inserted in the focal plane the illumination is

$$I_K = \frac{a_K}{a_0} I_0 \tag{30}$$

The light passing through each section of the test region comes from all parts of the source. Thus, at the focus not only is the image of the source composed of light coming from the whole field of view, but light passing through every point in

*If a light source with sharply defined edges as in Fig. 3 is not available, a condensing lens is sometimes used with the source. At the conjugate focus of this lens, an auxiliary knife-edge (or knife-edges) is aligned so that the image of the source adjacent to the main knife-edge (in the beam following the test section) is straight and sharply defined.

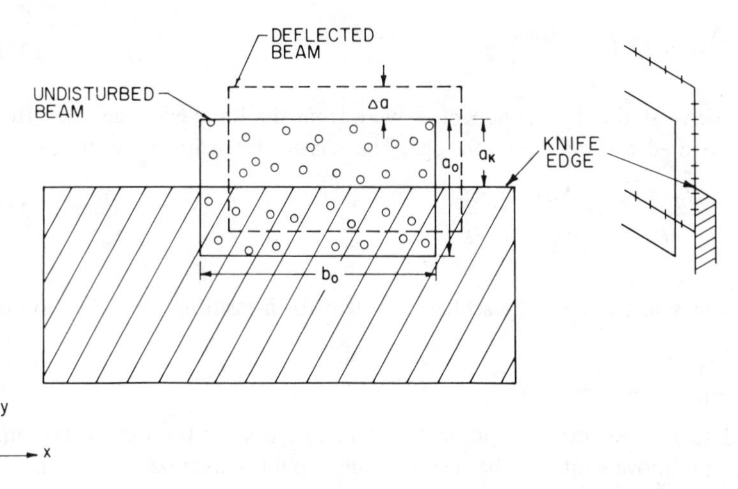

Figure 3 View of deflected and undisturbed beams at the knife-edge of a schlieren system.

the field of view gives an image of the source at the knife-edge. If the light beam at a position x, y in the test region is deflected by an angle α, then, from Fig. 4, the image of the source coming from that position will be shifted at the knife-edge by an amount

$$\Delta a = \pm f_2 \alpha \tag{31}$$

where the sign is determined by the orientation of the knife-edge; it is positive when (as in Fig. 4) $\alpha > 0$ gives $\Delta a > 0$, and negative if the knife-edge is reversed so that $\alpha > 0$ leads to $\Delta a < 0$. The illumination at the image of position x, y on the screen will be (see Fig. 3)

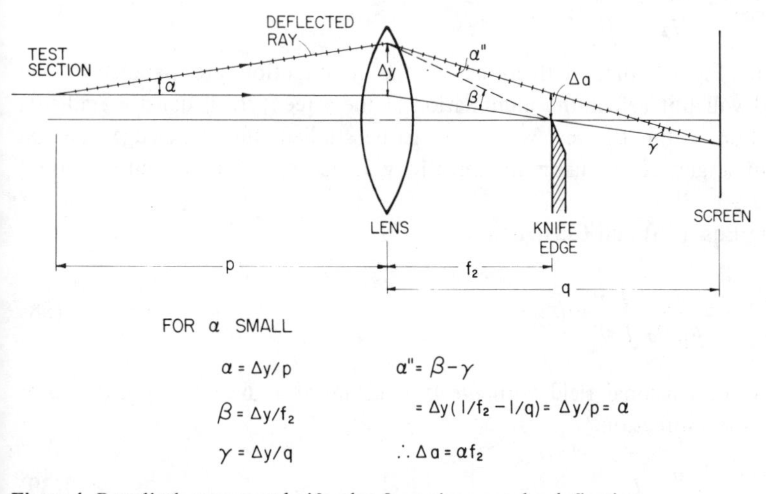

Figure 4 Ray displacement at knife-edge for a given angular deflection.

$$I_d = I_K \frac{a_K + \Delta a}{a_K} = I_K \left(1 + \frac{\Delta a}{a_K}\right) \tag{32}$$

where Δa is positive if the light is deflected away from the knife-edge, and negative if the light is deflected toward the knife-edge. The relative intensity or contrast is

$$\text{Contrast} = \frac{\Delta I}{I_K} = \frac{I_d - I_K}{I_K} = \frac{\Delta a}{a_K} = \pm \frac{\alpha f_2}{a_K} \tag{33}$$

using Eq. (31).

Note that the sensitivity of the schlieren system for measuring the deflection is

$$\frac{d\,(\text{contrast})}{d\alpha} = \frac{f_2}{a_K} \tag{34}$$

or proportional to f_2 and inversely proportional to a_K. For a given optical system, minimizing a_K by movement of the razor blade would maximize the contrast. However, this would limit the range for deflection of the beam toward the knife-edge to

$$\alpha_{\max,\text{neg}} = \frac{a_K}{f_2} \tag{35}$$

as all deflections this large or larger would give (neglecting diffraction) no illumination. The maximum angle of deflection away from the knife-edge that could be measured is

$$\alpha_{\max} = \frac{a_0 - a_K}{f_2} \tag{36}$$

as a deflection of this magnitude would permit all the source illumination to pass to the screen. For equal range in both directions, $a_K = a_0/2$ and

$$\alpha_{\max,\text{neg}} = \alpha_{\max} = \frac{a_0}{2f_2} = \frac{a_s}{2f_1} \tag{37}$$

Note from Fig. 3 that deflections in the x direction are parallel to the knife-edge and will not affect the illumination at the screen; so, if density gradients in the x direction within the test region are to be studied, the knife-edge must be turned at right angles. For maximum sensitivity (since $a_s < b_s$) the source should also be rotated 90°.

Combining Eqs. (26) and (33) gives

$$\text{Contrast} = \frac{\Delta I}{I_K} = \pm \frac{f_2}{a_K n_a} \int \frac{\partial n}{\partial y}\, dz \tag{38}$$

Assuming a two-dimensional field with $\partial n/\partial y$ constant at a given x, y position over the length L in the z direction,

$$\text{Contrast} = \pm \frac{f_2}{a_K} \frac{1}{n_a} \frac{\partial n}{\partial y} L \tag{39}$$

This equation holds for every x, y position in the test section and gives the contrast at the equivalent position in the image on the screen.

If the deflection is toward the knife-edge, the field will darken and the contrast will be negative. Using the coordinate system of Fig. 3, if the knife-edge covers up the region $y<0$ (that is, knife-edge pointing upward) at the focus, then

$$\frac{\Delta I}{I_K} = +\frac{f_2}{a_K} \frac{1}{n_a} \frac{\partial n}{\partial y} L \tag{40}$$

If the knife-edge is reversed and covers the region $y>0$, then

$$\frac{\Delta I}{I_K} = -\frac{f_2}{a_K} \frac{1}{n_a} \frac{\partial n}{\partial y} L \tag{41}$$

Changing the knife-edge reverses the dark and light images on the screen. The brighter areas of the image represent regions in the test section where the index of refraction (and thus usually the density) increases in the direction away from the knife-edge (Fig. 5). For a gas, Eq. (38) can be rewritten, using Eq. (6),

$$\frac{\Delta I}{I_K} = \pm \frac{f_2}{a_K n_a} \frac{n_0 - 1}{\rho_0} \int \frac{\partial \rho}{\partial y}\, dz \tag{42}$$

and equivalent to Eq. (39),

$$\frac{\Delta I}{I_K} \simeq \pm \frac{f_2}{a_K} \frac{n_0 - 1}{\rho_0} \frac{\partial \rho}{\partial y} L \tag{43}$$

taking $n_a \simeq 1$.

For a gas at constant pressure,

$$\frac{\Delta I}{I_K} = \mp \frac{f_2}{a_K n_a} \frac{n_0 - 1}{\rho_0} \int \frac{\rho}{T} \frac{\partial T}{\partial y}\, dz \tag{44}$$

and, as in Eq. (43),

$$\frac{\Delta I}{I_K} \simeq \mp \frac{f_2}{a_K} \frac{n_0 - 1}{\rho_0} \frac{P}{RT^2} \frac{\partial T}{\partial y} L \tag{45}$$

For a liquid, where n is a function only of T,

$$\frac{\Delta I}{I_K} = \pm \frac{f_2}{a_K n_a} \int \frac{\partial T}{\partial y} \frac{dn}{dT}\, dz \tag{46}$$

If the field is two-dimensional and n does not change greatly, then

$$\frac{\Delta I}{I_K} = \pm \frac{f_2}{a_K n_a} \frac{\partial T}{\partial y} \frac{dn}{dT} L \tag{47}$$

$$\frac{\Delta I}{I_K} \simeq \pm \frac{f_2}{a_K} \frac{\partial T}{\partial y} \frac{dn}{dT} L \tag{48}$$

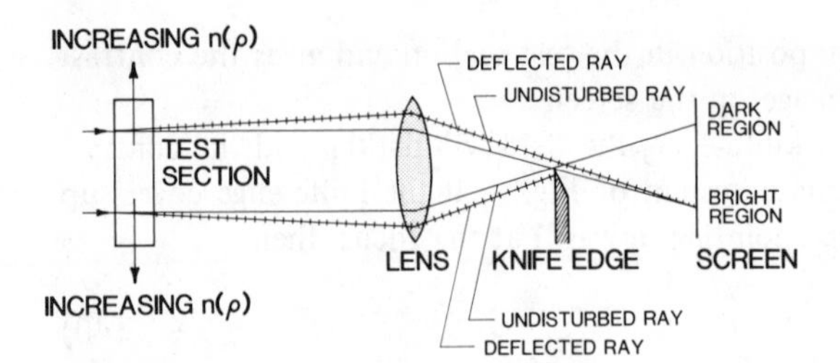

Figure 5 Effect of index-of-refraction gradient on illumination at screen.

In a quantitative study, measurements of the illumination or contrast, usually of the image on a photographic negative, must be made. These are quite time-consuming, and the resulting accuracy has not usually warranted the effort. Standard schlieren systems are employed for qualitative studies of density or temperature fields, although quantitative measurements of such things as shock-wave angles or positions can be made. The minimum value of the contrast that can easily be observed is of the order of 0.05, a value that can be used in determining the sensitivity of the system. Since n and its derivatives vary somewhat with wavelength, it is preferable that the light source be relatively monochromatic, although for both schlieren and shadowgraph systems this is usually not of major importance.

2.2 Applications and Special Systems

The high cost of large aberration-free lenses usually precludes construction of the system shown in Fig. 2, but the optically similar system using concave mirrors shown in Fig. 6 is widely used. The source and knife-edge should be in the same plane and on opposite sides of the axes of the two mirrors in the "Z" arrangement shown. This eliminates the aberration coma, although astigmatism is still present in the off-axis system. The legs of the Z should each be at the same angle to the line between the two mirrors; this angle should be as small as possible to minimize astigmatism [12]. Schlieren photographs taken with a system similar to that of Fig. 6 are shown in Figs. 7, 8, and 9.

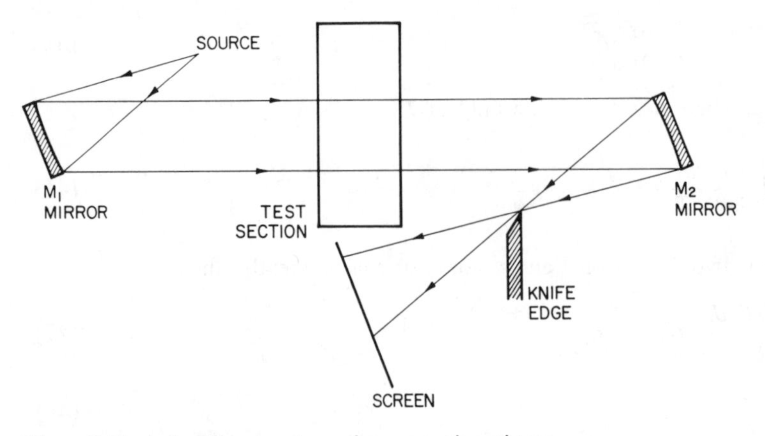

Figure 6 Typical schlieren system using converging mirrors.

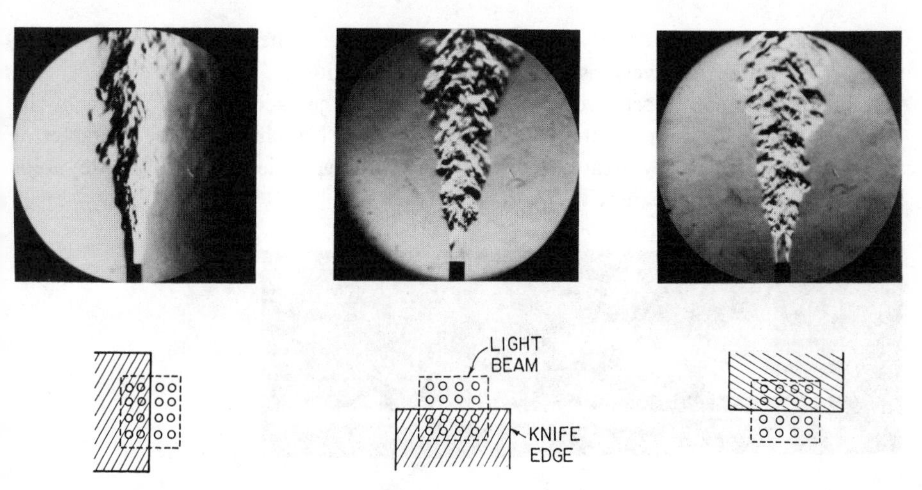

Figure 7 Schlieren images of a helium jet entering an atmosphere of air: The effect of knife-edge orientation (Re = 630).

Figure 7 is the image of a helium jet entering a still atmosphere of air. Different images of the flow field in the jet are shown. Although the Reynolds number is essentially constant for the different photographs, the difference in knife-edge orientation, shown for each view, dramatically changes the image. As mentioned above, the schlieren image lightens when the index of refraction increases in the direction away from the knife-edge. Thus the lighter regions on the photo

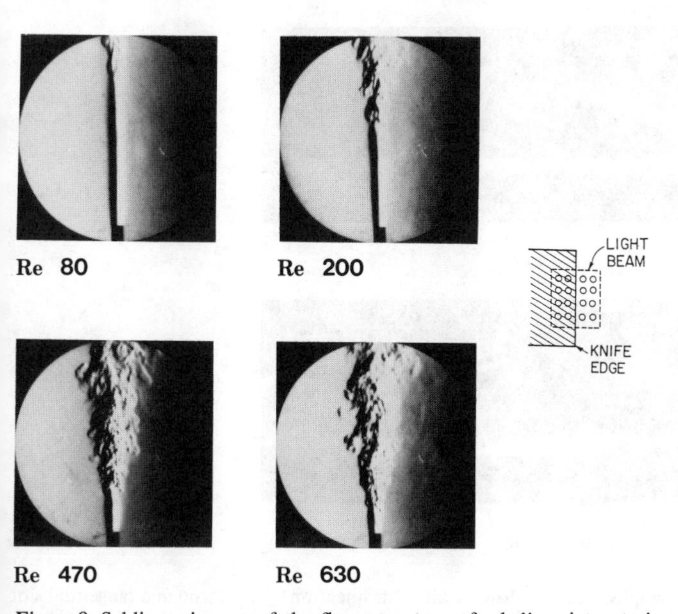

Figure 8 Schlieren images of the flow structure of a helium jet entering air at different Reynolds numbers.

show that the local density gradient is positive in the direction toward the knife-edge, while the reverse is true for the darker regions.

Figure 8 presents schlieren images of the same jet geometry as shown in the previous figure for four different Reynolds numbers. The change in the character of the jet from a relatively steady flow to flow with large-scale eddies as the Reynolds number increases is apparent.

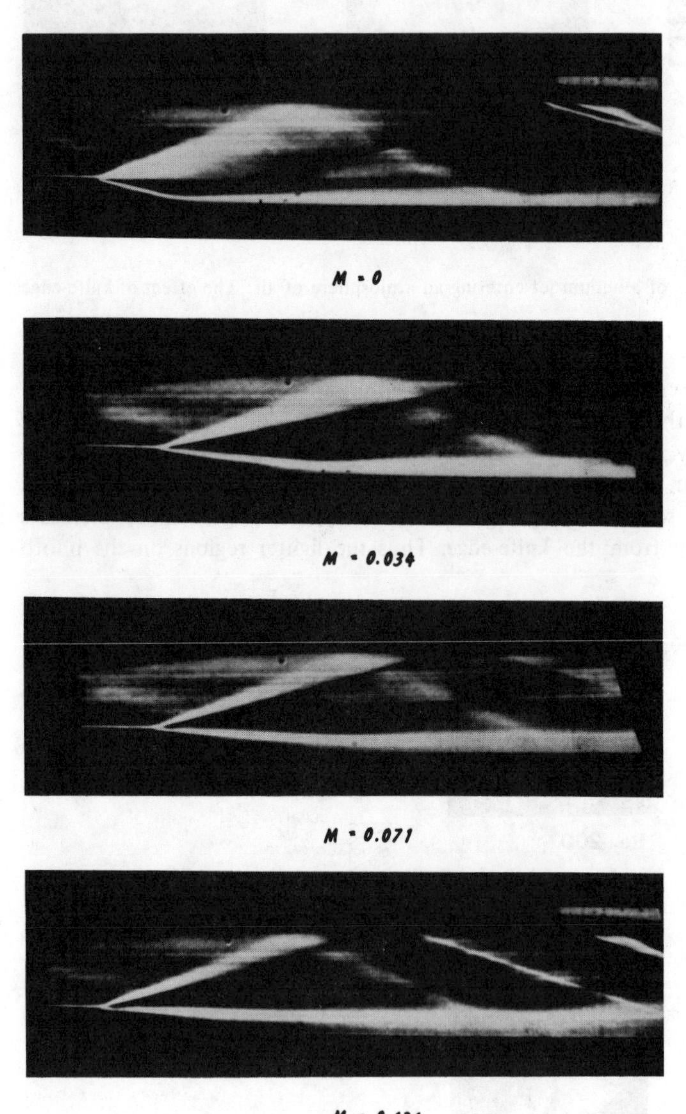

Figure 9 Schlieren photographs: Mach-3 flow of air with injection of air through a tangential slot (M = ratio of mass velocity of injected flow to mainstream flow). $(Ma)_1 = 3.01$; $h = 0.239$ in; $s = 0.182$ in.

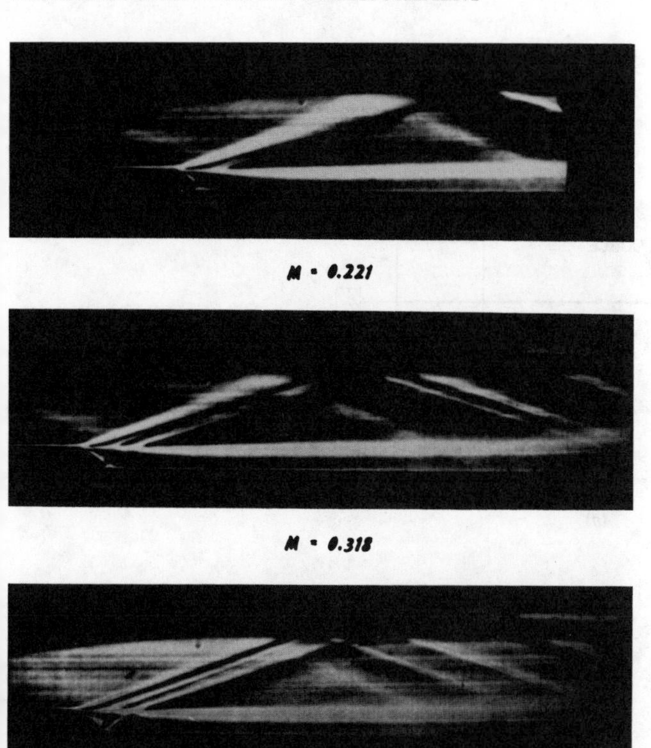

Figure 9 (*Continued*) Schlieren photographs: Mach-3 flow of air with injection of air through a tangential slot (M = ratio of mass velocity of injected flow to mainstream flow). $(Ma)_1 = 3.01$; $h = 0.239$ in; $s = 0.182$ in.

Figure 9 contains a series of schlieren photographs taken of a Mach-3 flow of air into which air is injected through a tangential slot [13]. The relative amount of air injected is given by M, the ratio of the mass velocity of the injected flow to that of the mainstream. Various flow phenomena and their dependence on M can be observed in the different photos. These phenomena include the expansion fan at the top edge of the splitter plate (between the main and injected flows), a separation region, lip shock, recompression shock, and alternative expansions and compressions.

A number of variations on the schlieren systems shown in Figs. 2 and 6 have been used, and two of these are shown in Fig. 10. In Fig. 10*a*, one plane and one converging mirror are used. Since plane mirrors are easier to make than converging mirrors, this apparatus would be somewhat less expensive than the one in Fig. 6. The main advantage of this system, however, is the amplification of the angle representative of the disturbance. As the beam passes through the test section twice, the deflection angle is doubled and, all other parameters being the same, so is the sensitivity. However, this double passage causes, in general, a slight blurring of the image, as the beam does not go through the exact same part of the test section on

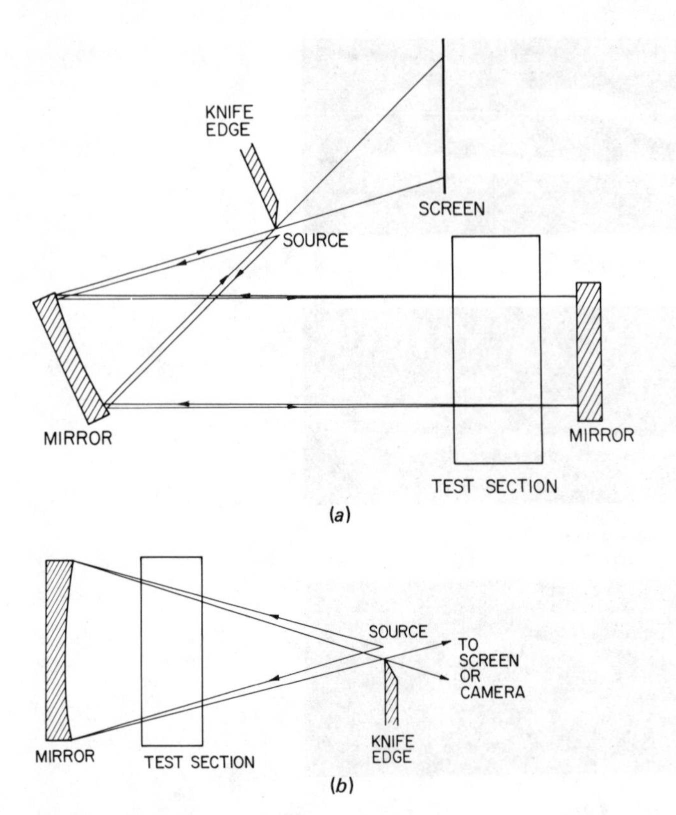

Figure 10 Alternative schlieren systems. (*a*) One converging and one plane mirror. (*b*) One converging mirror.

each passage. In addition, since the screen cannot be at the focus of both views of the test section, some shadowgraph effects are present.

A single-mirror system, shown in Fig. 10*b*, is still simpler, although the blurring of the image is still more serious than in Fig. 10*a* because the light is not parallel. The source and knife-edge are at conjugate foci, and for convenience they are often kept close together and thus at a distance twice the focal length f from the mirror. If α is the deflection of the beam after a single transit of the test section, the sensitivity [cf. Eq. (34)] is $4f/a_K$. Since the source and knife-edge in the two systems shown in Fig. 10 are in such close proximity, a splitter plate is sometimes used to give them a larger physical separation. It should be apparent that, in all these systems, a camera placed in the beam after the knife-edge and focused on the test section can be used in place of the screen.

Other schlieren systems have used one of the optical arrangements shown above, usually that of Fig. 6, but without the knife-edge. In a color schlieren, the knife-edge is replaced by colored filters held at the focus, and the deflection of the light beam (necessarily nonmonochromatic) gives rise to different colors. In another system, an aperture is placed at the focus, giving a darker image for any deflection or density gradient irrespective of the direction of the gradient. An opaque disk at the focus gives a brightening of the image for a light deflection in any direction.

Quantitative studies with standard schlieren systems require measurement of the local light intensity. This has been done with a photomultiplier to study density fluctuations in a supersonic turbulent jet [14]. Experiments similar to those of [14] have been performed in a convecting liquid with a grating rather than a single knife-edge [15]. These systems for studying the fluctuations of the index of refraction are closely related to crossed-beam measurements to study turbulence in shear flow.

Crossed-beam systems [16, 17] depend on fluctuations in the absorption or scattering coefficients in the flow. The local statistical properties of the flow are determined from cross correlation of the fluctuating outputs of two photodetectors measuring the intensity of two orthogonal light beams that cross the flow. In an early study, the light-beam fluctuation was produced by a water mist in a subsonic airflow [16]. Other studies used what are essentially crossed schlieren beams to study flow in subsonic jets [18] and supersonic jets [19]. These systems relate closely to the single-beam device described in [14, 15].

Standard schlieren systems can also be used to derive quantitative information on the flow field. For example, a scanning microdensitometer has been used to determine the local light intensity across the image of a flow field [20]. In general, measurements of this type require a calibration relating the light intensity to the light-ray displacement at the knife-edge.

A special apparatus for quantitative studies is the Ronchi or grid-schlieren system [21 to 23]. The grid, shown as part of a focusing-lens schlieren system in Fig. 11, could be used in any of the optical arrangements previously described. In general, however, quantitative studies are best performed using parallel light that passes through the test section only once. The grid has equally spaced opaque lines (whose widths are usually set equal to the spacing) on a transparent sheet. It is placed before the focus, as shown in Fig. 11. The resulting schlieren image is a series of lines that are parallel if no disturbance is present. When a disturbance is present, the displacement of these fringes is directly proportional to the local angular deflection α.

The grid could also be placed right at the focus where the knife-edge is located in a standard schlieren system; then, the beam just passes through one of the gaps between two opaque lines when no disturbance is present. Thus, the spacing

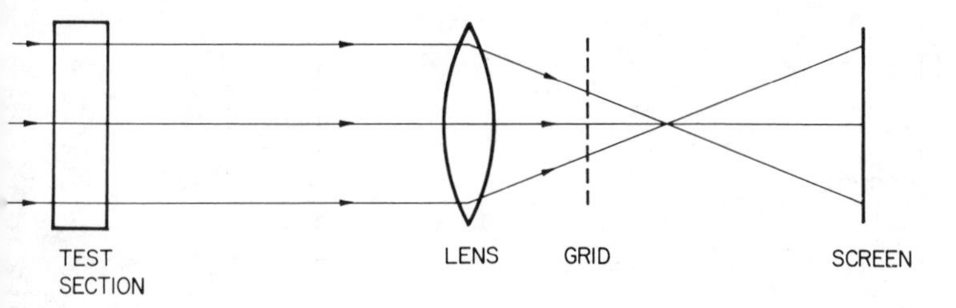

Figure 11 Grid-schlieren system.

between the lines equals a_0 (Fig. 3), and the screen is uniformly illuminated. In the presence of a disturbance, the deflection of the light beam at the focus Δa causes the beam to traverse the grid, producing a light or dark image depending on the magnitude of Δa. The resulting image is a series of fringes, called *isophotes*, representing regions of constant angular deflection (often approximating constant density gradient). Placement of the grid at the focus of the second lens is often impractical. The required line spacing on the grid may be so small as to cause significant diffraction effects.

Other systems that may be of interest in flow studies include a self-illuminated schlieren system for the study of plasma-system jets [24], a sharp-focusing schlieren system [25], and a stereoscopic schlieren [26]. The latter two, though cumbersome to use, have application to the study of three-dimensional fields. A modification of the sharp-focusing system using holography permits local measurements all along the light beam by using reconstruction from a single hologram [27].

Most of these special schlieren systems, including the sharp-focusing systems, the Ronchi system, and other quantitative schlieren systems described, have not been widely used. Most common are systems similar to the one shown in Fig. 6, which provide qualitative flow visualization, although measurement of shock angles and locations, flow-separation positions, and even the region of transition to turbulence can be made. Schlieren interferometers are described in Sec. 4.

3. SHADOWGRAPH SYSTEM

In a shadowgraph system the linear displacement of the perturbed light is measured, rather than the angular deflection as in a schlieren system. The shadowgraph image can be understood by referring to Fig. 12, which shows a parallel light beam entering a nonuniform test section. To simplify the derivation, index of refraction variations in the test section are assumed to exist only in the y direction.

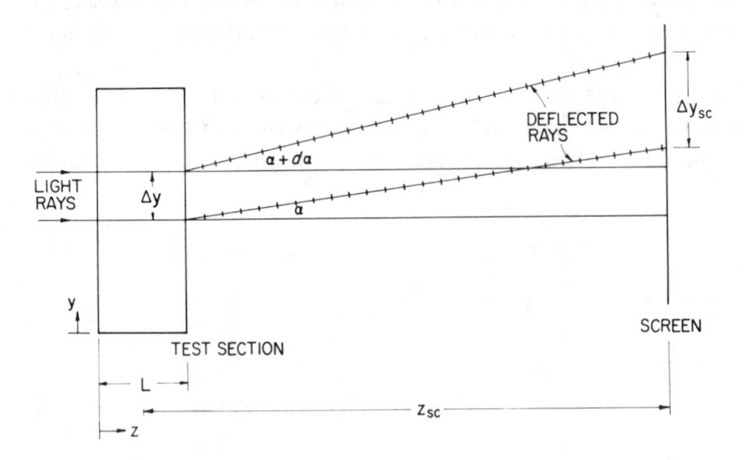

Figure 12 Displacement of light beam for shadowgraph evaluation.

Consider the illumination at the exit of the test section. The linear displacement of the light beam is probably not large there, because of the relatively short distance the light has traveled. If the illumination is uniform entering the test section, it should still be closely uniform there. The beam, however, is no longer parallel, having been deflected by an angle α, which is a function of y. The illumination within the region defined by Δy at this position is within the region defined by Δy_{sc} at the screen. If the initial intensity is I_T, then at the screen,

$$I_0 = \frac{\Delta y}{\Delta y_{sc}} I_T \tag{49}$$

If Z_{sc} is the distance to the screen, then

$$\Delta y_{sc} = \Delta y + z_{sc}\, d\alpha \tag{50}$$

The contrast is

$$\frac{\Delta I}{I_T} = \frac{I_0 - I_T}{I_T} = \frac{\Delta y}{\Delta y_{sc}} - 1 \simeq -z_{sc} \frac{\partial \alpha}{\partial y} \tag{51}$$

Combining this with Eq. (26) gives

$$\frac{\Delta I}{I_T} = -\frac{z_{sc}}{n_a} \int \frac{\partial^2 n}{\partial y^2}\, dz \tag{52}$$

If the index of refraction is only a function of density, then

$$\frac{\Delta I}{I_T} = -\frac{z_{sc}}{n_a} \int \frac{\partial^2 \rho}{\partial x^2} \cdot \frac{\partial n}{\partial \rho}\, dz \tag{53}$$

assuming $dn/d\rho$ is constant. For a gas, Eq. (10) could be substituted into Eq. (52). If there is also a variation of n in the x direction, then, equivalent to Eq. (52),

$$\frac{\Delta I}{I_T} = -\frac{z_{sc}}{n_a} \int \left(\frac{\partial^2 n}{\partial x^2} + \frac{\partial^2 n}{\partial y^2} \right) dz \tag{54}$$

Shadowgraphs, like schlieren and interferometer photographs, are often taken of phenomena that can be approximated as two-dimensional. As with all integrating optical systems, good qualitative flow visualization can be obtained, even of three-dimensional phenomena. Note that variations of the index of refraction in both the x and y directions are obtained from a single shadowgraph image, while schlieren systems usually only indicate variations normal to the knife-edge.

Different optical geometries are possible for shadowgraph systems, as shown in Fig. 13. Parallel light systems, as in Fig. 13*a*, are easiest to understand, although the lensless and mirrorless system of Fig. 13*b* is also usable if the distance from the (small) source to the test region is large. Other combinations of mirrors and lenses analogous to the schlieren systems of Figs. 2, 6, and 10 have been used.

It should be noted that if a mirror or lens is used after the test region in a

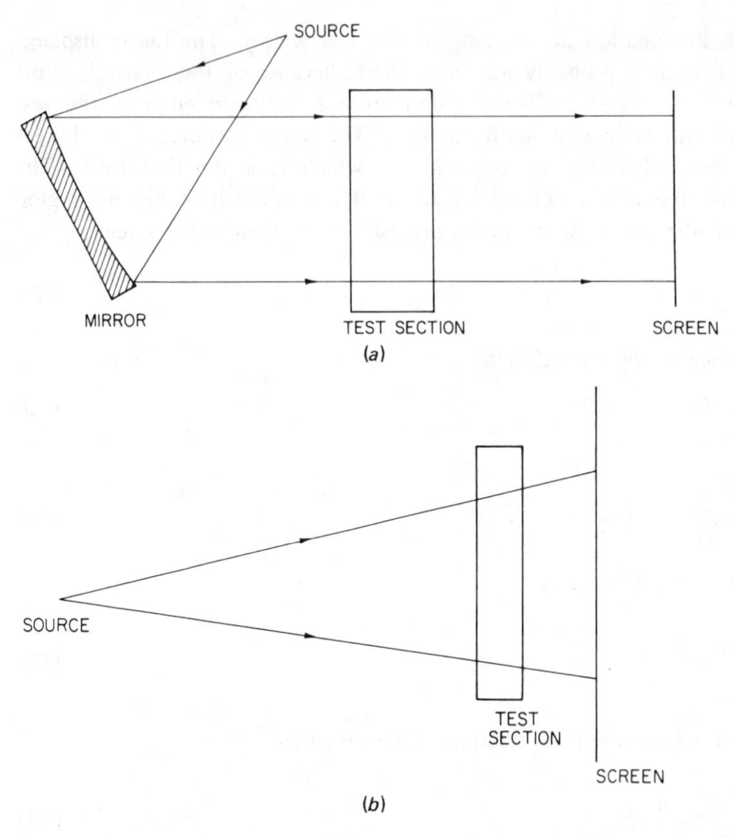

Figure 13 Alternative shadowgraph systems. (*a*) One converging mirror. (*b*) No lens or mirror.

shadowgraph, it should not be placed so that the conjugate focus of the test section is in the plane of the screen. At the conjugate focus the parts of the beam deflected at different angles in the test region are all brought back together, so there is no linear displacement there and thus no shadowgraph effect. In fact, schlieren systems are usually focused on the test region to eliminate shadowgraph effects.

Standard shadowgraph systems are rarely used for quantitative density measurements. The contrast would have to be measured accurately, and Eq. (53), for example, integrated twice to determine the density distribution. Even the density or temperature gradient, which is of interest in heat-transfer studies, would require one integration. If, however, large gradients of density are present, as in a shock wave or a flame, shadowgraph pictures can be very useful. Quantitative measurements of such things as shock angles and the location of boundary-layer transition can be made (see page 28 of [3]).

Figure 14 shows shadowgraphs of the helium jets whose schlieren photos are shown in Fig. 8. The same optical system was used as is described in reference to Figs. 7 and 8 (i.e., a system similar in layout to that in Fig. 6), with the knife-edge removed and the camera set out of focus to obtain a shadowgraph image.

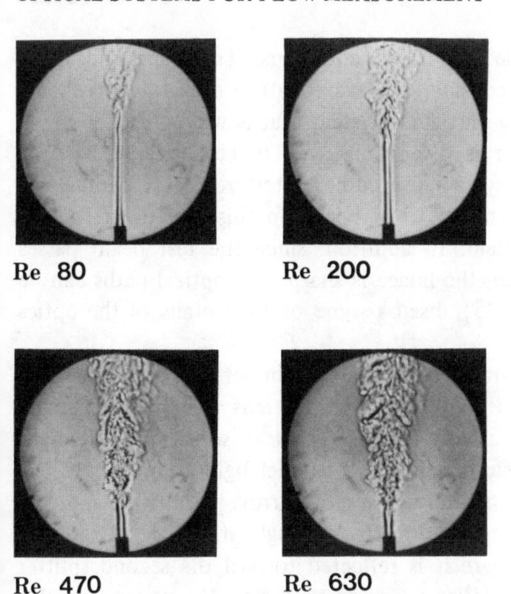

Figure 14 Shadowgraphs of a helium jet entering an atmosphere of air.

Figure 15 is a shadowgraph showing the large-scale structure in the mixing layer between parallel streams of helium (upper region) and nitrogen (lower region) [28]. The helium velocity is about three times the velocity in the nitrogen layer.

4. INTERFEROMETERS

4.1 Basic Principles

Interferometers are often used in quantitative studies. Unlike the schlieren and shadowgraph systems, an interferometer does not depend upon the deflection of a light beam to determine density or index of refraction variation. In fact, refraction effects are usually of second order and undesirable in interferometry, as they

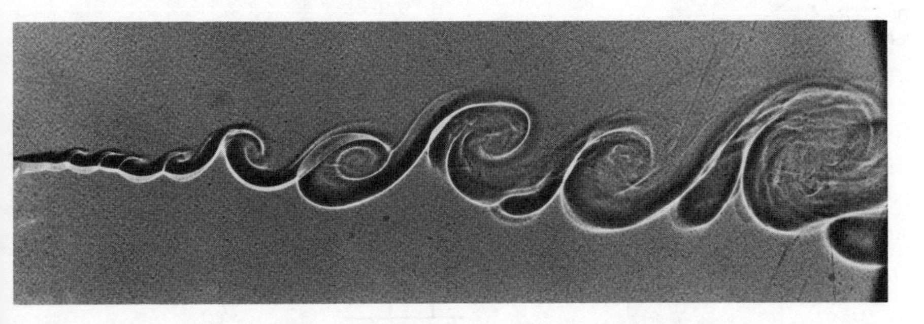

Figure 15 Shadowgraph of mixing of parallel-flowing streams of helium (above) and nitrogen (below). *(From [28].)*

introduce deviations or errors in the evaluating equations. To understand interferometric measurements, one must consider the wave nature of light, and this is perhaps best done by first examining a particular system that is widely used.

The Mach-Zehnder interferometer is often employed in aerodynamic studies. One of the main advantages of this system over other interferometers is the large displacement of the reference beam from the test beam. In this way the reference beam can pass through a uniform field. In addition, since the test beam passes through the disturbed region only once, the image is sharp, and optical paths can be clearly defined. References [7, 29 to 35] discuss some of the details of the optics of Mach-Zehnder interferometers.

Figure 16 is a schematic diagram of a Mach-Zehnder interferometer. A monochromatic light source is used in conjunction with a lens to obtain a parallel beam of light. The requirement of a very narrow spectral width for the light source is critical when using an interferometer. The parallel light beam strikes the first splitter plate SP_1, which is a partially silvered mirror permitting approximately half the impinging light to pass directly through it. This transmitted light follows path 1 to mirror M_1, where it is reflected toward the second splitter plate SP_2. The light reflected by SP_1 follows path 2 to mirror M_2, where it is also reflected toward SP_2. The second splitter plate also transmits about half the impinging light and reflects most of the rest. The recombined beams from paths 1 and 2 pass on to the screen. Note that there would also be another recombined beam leaving SP_2, but in general only one beam from the final splitter is used. The mirrors and splitter plates are usually set at corners of a rectangle [33]. They should then all be closely parallel and at an angle of $\pi/4$ to the initial parallel beam.

Let us assume that in Fig. 16 the mirrors are perfectly parallel and that in both paths 1 and 2 there are no variations in optical properties normal to either beam.

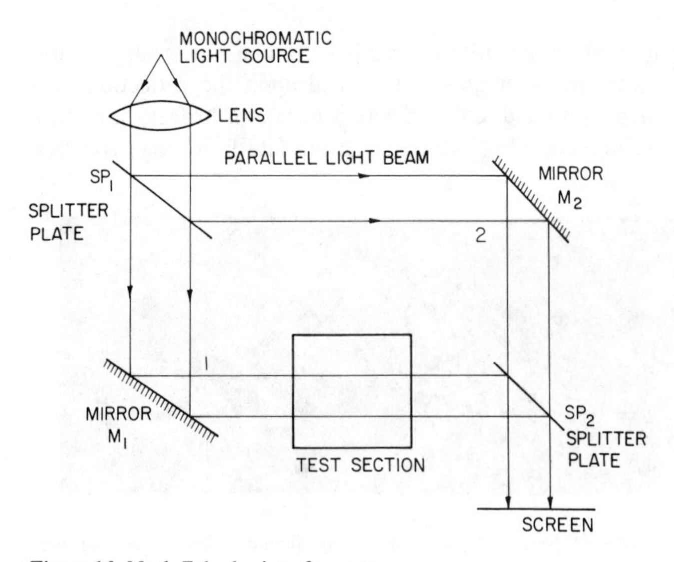

Figure 16 Mach-Zehnder interferometer.

This requires uniform properties, not only between the mirrors but also in the splitter plates. (Note that the effect of a variation in the thickness of a splitter plate–i.e., a slight wedge–can be corrected by a rotation of one of the mirrors if both surfaces of the plate are flat.) The two beams, from paths 1 and 2, emerging from SP_2 are then parallel.

The amplitude of a plane light wave in a homogeneous medium can be represented by

$$A = A_0 \sin \frac{2\pi}{\lambda}(c\tau - z) \tag{55}$$

where A_0 = peak amplitude
c = speed of light
τ = time
z = distance
λ = wavelength

Consider the amplitude of beam 1 at a fixed position past SP_2 to be

$$A_1 = A_{01} \sin \frac{2\pi c\tau}{\lambda} \tag{56}$$

The other beam could be represented at the same position by

$$A_2 = A_{02} \sin \left(\frac{2\pi c\tau}{\lambda} - \Delta\right) \tag{57}$$

where the phase difference Δ appears because the two beams will probably not be exactly in phase because of a difference in their path lengths. Before the first splitter plate, the two beams (1 and 2) were, of course, one beam and in phase.

Since beams 1 and 2 come from the same source, they are coherent and can interfere with each other. This is implicit in Eqs. (56) and (57) if Δ is not a function of time. Adding Eqs. (56) and (57), and assuming that $A_{01} = A_{02} = A_0$,

$$A_T = A_1 + A_2 = A_0 \left[\sin\left(\frac{2\pi c\tau}{\lambda} - \Delta\right) + \sin \frac{2\pi c\tau}{\lambda}\right] \tag{58}$$

which can be rewritten

$$A_T = 2A_0 \cos \frac{\Delta}{2} \sin \left(\frac{2\pi c\tau}{\lambda} - \theta\right) \tag{59}$$

where θ is a new phase difference. Thus the sum of the two waves is a new wave of the same frequency and wavelength.

The intensity of the combined beam is the quantity observed visually or measured on a photographic plate. The intensity I is proportional to the square of the peak amplitude, or

$$I \sim 4A_0^2 \cos^2 \frac{\Delta}{2} \tag{60}$$

Note that when $\Delta/2\pi$ is an integer (say j), the peak intensity is four times that of either of the two beams, but when $\Delta/2\pi$ is a half-integer $(j+\frac{1}{2})$, the intensity is zero. The interesting yet not too difficult paradox in this latter case is: Where did the energy go?

The optical path along a light beam is defined by

$$PL = \int n\, dz \tag{61}$$

or

$$PL = \int \frac{c_0}{c}\, dz = \lambda_0 \int \frac{dz}{\lambda} \tag{62}$$

Thus the optical path length is the vacuum wavelength times the real light path in wavelengths (which can vary along the path). In Fig. 16, the difference between paths 1 and 2 is

$$\overline{\Delta PL} = \int_1 n\, dz - \int_2 n\, dz \tag{63}$$

$$\overline{\Delta PL} = PL_1 - PL_2 = \lambda_0 \left(\int_1 \frac{dz}{\lambda} - \int_2 \frac{dz}{\lambda} \right) \tag{64}$$

The phase difference between two points a distance of z apart [cf. Eq. (55)] is $2\pi\, dz/\lambda$. The difference in phase of the two beams upon recombination is

$$\Delta = 2\pi \left(\int_1 \frac{dz}{\lambda} - \int_2 \frac{dz}{\lambda} \right)$$

or

$$\frac{\Delta}{2\pi} = \frac{\overline{\Delta PL}}{\lambda_0} \tag{65}$$

If $\overline{\Delta PL}/\lambda_0$ is zero or an integer, then from Eqs. (60) and (65), there is constructive interference, and the field on the screen in Fig. 16 is uniformly bright.

4.2 Fringe Pattern with Mach-Zehnder Interferometer

Consider a Mach-Zehnder interferometer with beams 1 and 2 passing through homogeneous media so that, initially, the recombined beam is uniformly bright ($\overline{\Delta PL}$ is assumed to be zero). If a disturbance (inhomogeneity) were put in part of the field of light beam 1, the path difference $\overline{\Delta PL}$ would no longer be zero, nor would the field be uniform. At any position on the cross section of the beam (neglecting refraction), Eq. (63) could be used to obtain ϵ, the path-length difference in terms of vacuum wavelengths,

$$\epsilon = \frac{\overline{\Delta PL}}{\lambda_0} = \frac{1}{\lambda_0} \int (n - n_{\text{ref}})\, dz \tag{66}$$

where n_{ref} is the reference value of the index of refraction in reference beam 2. If $\overline{\Delta PL}/\lambda_0$ is an integer, the field will be bright, while if $\overline{\Delta PL}/\lambda_0$ is a half-integer, the field will be dark. Thus the initially uniformly bright field will have a series of bright and dark regions (fringes), each one representative of a specific value of $\overline{\Delta PL}/\lambda_0$ and differing in magnitude from the adjacent fringe of the same intensity by a value $\epsilon = \overline{\Delta PL}/\lambda_0 = 1$. If there is a gas in light beam 1, the Gladstone-Dale relation, Eq. (2), can be used in Eq. (66):

$$\epsilon = \frac{C}{\lambda_0} \int (\rho - \rho_{\text{ref}})\, dz \tag{67}$$

If the field is two-dimensional in that the only variations in the index of refraction along the light beam (i.e., in the z direction) are the sharp discontinuities at the entrance and exit of the test section (see [8] for a treatment of axisymmetric density fields), and ρ only varies over a length L, the fringe shift is given by

$$\epsilon = \frac{n - n_{\text{ref}}}{\lambda_0} L \tag{68}$$

For a gas,

$$\epsilon = \frac{C}{\lambda_0} (\rho - \rho_{\text{ref}}) L \tag{69}$$

or

$$\rho - \rho_{\text{ref}} = \frac{\lambda_0 \epsilon}{CL} = \frac{\lambda_0 \epsilon}{n_0 - 1} \rho_0 \tag{70}$$

If the pressure is constant and the ideal gas law is used, then

$$\frac{1}{T} = \frac{\lambda_0 R}{PCL} \epsilon + \frac{1}{T_{\text{ref}}} \tag{71}$$

$$T = \frac{PCLT_{\text{ref}}}{PCL + \lambda_0 R \epsilon T_{\text{ref}}} \tag{72}$$

or

$$T - T_{\text{ref}} = \left[\frac{-\epsilon}{PCL/(\lambda_0 R T_{\text{ref}}) + \epsilon} \right] T_{\text{ref}} \tag{73}$$

For a two-dimensional field in a liquid, Eq. (68) could be written

$$n = \frac{\lambda_0 \epsilon}{L} + n_{\text{ref}} \tag{74}$$

where n and n_{ref} would have to be known as functions of temperature. For small temperature differences,

$$\epsilon = \frac{L}{\lambda_0} \frac{dn}{dT} (T - T_{\text{ref}}) \tag{75}$$

and

$$T - T_{\text{ref}} = \frac{\epsilon \lambda_0}{L} \frac{1}{dn/dT} \tag{76}$$

If $\lambda_0 = 546.1$ nm and L is 30 cm, each fringe (that is, $\epsilon = 1$) represents a temperature difference of about 2°C in air at 20°C and 1 atm. In water under the same conditions, each fringe represents a temperature difference of about 0.02°C.

If the initial optical path lengths of beams 1 and 2 are not exactly equal (but their difference is less than the coherence length of the light source), the above equations can still be used to determine density and temperature differences between different parts of the cross section of the test beam 1, assuming uniform optical path length over the cross section of the reference beam. Then n_{ref}, ρ_{ref}, and T_{ref} refer to a uniform portion of the test region, and all properties at other locations in the test section are measured relative to the properties there. This is how interferograms–the fringe patterns obtained with an interferometer–are normally evaluated.

When the interferometer beams are recombined parallel to each other (called *infinite fringe setting*), each fringe is the locus of points in a two-dimensional field where the optical path length or density is constant. These fringes are useful for qualitative flow-field visualization as well as for quantitative studies; they can delineate boundary layers, mixing regions, etc.

A Mach-Zehnder interferometer is not always used with the beams parallel upon recombination, as in the discussion above. Consider two beams, each of which is uniform (in phase) normal to the direction of its propagation, although diverging slightly, at a small angle θ, from the other beam–as represented by the two wave trains in Fig. 17. Lines are shown drawn through the crests (maxima of amplitude) for each wave train to represent the planes (wavefronts) normal to the direction of propagation. Constructive interference occurs where the maxima of the two beams coincide; dashed lines representing the loci of these positions are shown. If a screen is placed approximately normal to the two beams, the intensity distribution on the screen follows a cosine-squared law [Eq. (60)], as shown in Fig. 17. Thus, parallel, equally spaced, alternately dark and bright fringes (called *wedge* fringes) appear on the screen when there are no disturbances in either field. The difference in optical path length between the two beams varies linearly across the field of view with wedge fringes, so that only one fringe in the field (the *zero-order* fringe) can represent equal path lengths for the two beams. From Fig. 17, the spacing between the fringes is

$$d = \frac{\lambda/2}{\sin \theta/2} \tag{77}$$

or

$$d \sim \frac{\lambda}{\theta}$$

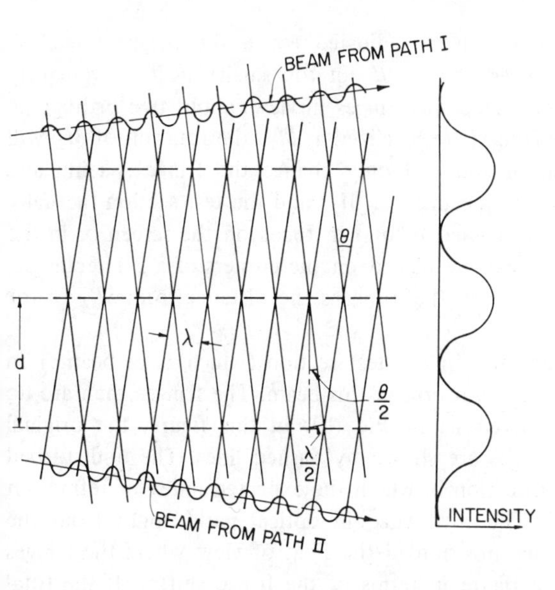

Figure 17 Intensity pattern from two intersecting plane light beams.

for small θ. For fringes to be observable, θ must be very small. For example, if d is about 5 mm, θ (using the green mercury line) is about 10^{-4} rad. As θ is decreased to zero, the fringes get further and further apart, approaching the "infinite fringe pattern" found when the two beams are parallel.

When a Mach-Zehnder interferometer is adjusted to give wedge fringes, the fringes are localized as shown in Fig. 18. Only a pair of rays is shown; after diverging, they are brought to a focus on the screen (or film in a camera) by a focusing lens or mirror. The angular separation of the beams is greatly exaggerated on the figure. The dashed lines represent the paths from the virtual object of the beams in paths 1 and 2 as they would appear along the other path. The fringes are localized where, tracing backward along the real and imaginary paths, the rays

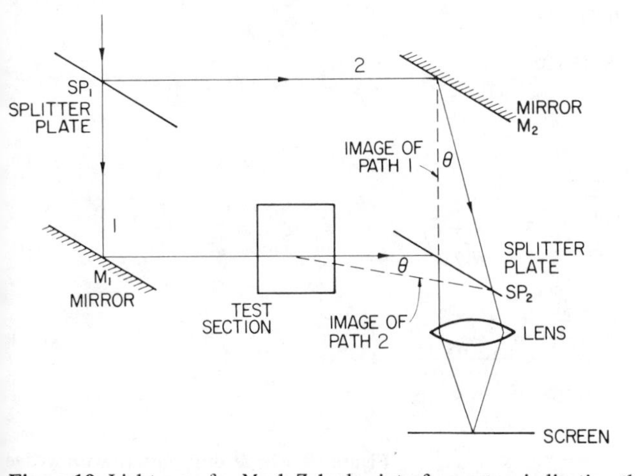

Figure 18 Light rays for Mach-Zehnder interferometer, indicating the preferred position of focus.

intersect. The plane of localization can be adjusted (once the beams are close enough to parallel to produce fringes—that is, d not too small) most conveniently by rotation of SP_2 (and of M_2, to keep the fringes in view) about two orthogonal axes in the plane of SP_2. If the fringes are localized at M_2, the rotation of M_2 will not affect the plane of localization and will only change the fringe spacing and orientation. To get both the fringes (localized at M_2) and the test section (actually the center of the test section, as discussed below) in focus on the screen or in the camera, the interferometer mirrors can be placed on the corners of a 2-1 rectangle, the distance from SP_2 to M_2 then being the same as the distance from SP_2 to the middle of the test section.

When a disturbance is present within the test section (which is in beam 1 in Fig. 18), the optical path is no longer uniform in this beam. The fringes then are no longer straight, but rather are curved as in Fig. 19. In this figure, the original (undisturbed) positions of the fringes are shown by dashed lines. The undisturbed fringes are normally aligned in a direction in which the expected index of refraction (density) gradient will be large. The difference in optical path length from the original value or from the reference position in the field of view where the fringes have not changed is shown in the figure in terms of the fringe shift ϵ. If the total fringe shift is large, only integral values of ϵ are usually measured; for small differences in optical path length, fractional values of ϵ can be measured as shown.

The possibility of measuring fractional values of ϵ is one advantage of wedge fringes. In addition, it is difficult to be certain with infinite fringes that the

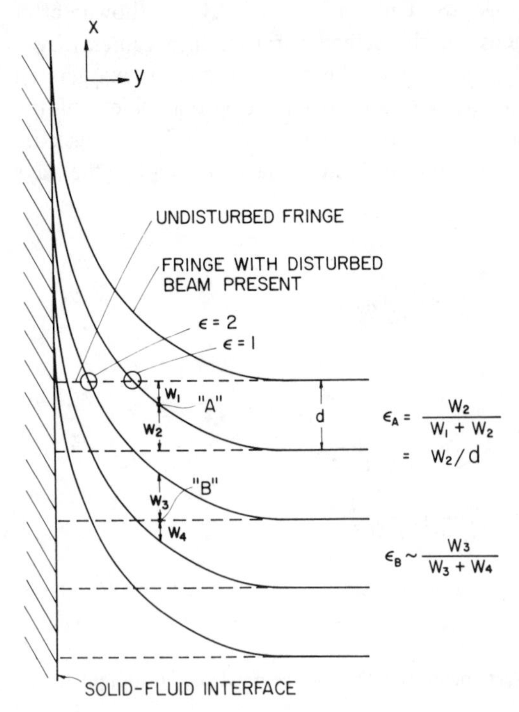

Figure 19 Fringe shift pattern with wedge fringes.

undisturbed (or reference) field is at the maximum brightness, so there is an uncertainty in the reference position. This problem does not occur with wedge fringes as long as there is a region of known uniform properties in the field of view. Contours of constant optical path length can also be obtained with wedge fringes by superposing the disturbed interference pattern over the undisturbed pattern. The resulting moiré fringe pattern gives grey lines that can represent constant-density lines in a two-dimensional test region. If there are irregularities in the undisturbed image due to faulty optical parts, this superposition method can still be used to obtain quantitative results.

Interferograms are often evaluated by using a microscope with a moving bed to determine the locations of individual fringes. The fringe positions can be put in digital form by using a digitizing tablet or its equivalent [36, 37] on which the image of the interferogram is projected and a cursor is used to record the fringe locations. No matter how the fringe locations are evaluated, some reference marks are normally required in the field to clearly indicate the position of test surfaces and to evaluate the scale of the image.

4.3 Examples of Interferograms

Interferograms are often used for quantitative studies of heat transfer [8]. In direct flow measurements they have been used to study the mixing of fluids of different densities (often of different compositions), of compressible (often supersonic) flows, and of flows in which some effective tracer, perhaps a hot spot from a local heat source, can be used to change the index of refraction or density of the flowing field.

The flow of a jet of helium at low Reynolds number into a still atmosphere of air is shown in Fig. 20. The upper set of figures represents the infinite fringe setting of the interferometer, so that the lines represent contours of constant optical path length. The lower set of figures are wedge fringes, which are often more convenient for quantitative evaluation, particularly when the total fringe shift in the field is small. The jet used in these photos is axisymmetric, and a transformation [8] would have to be made to convert the fringe position into the axisymmetric density field.

Figure 21 shows the impingement of a helium jet on a flat plate. Figure 22 also shows impingement of a jet, but near the edge of the plate, where the flow field of the helium around to the opposite side of the plate can be observed.

Figure 23 contains interferograms of a high-speed projectile (sphere) and a cone with a cylindrical projection, taken with a holographic interferometer (see below) [38, 39]. Figures similar to these have been used to determine the turbulent fluctuations in the wake of objects in a ballistic range [38].

Figure 24 compares a shadowgraph photo and two holographic interferograms of a sharp-tipped spike with conical flare in a Mach-3 flow [40]. Figure 25 shows the transition from laminar to turbulent flow in the boundary layer on a slender sharp-tipped cone [41]. The interferograms of these figures were taken with holographic interferometers in a fashion that yields results similar to those obtained with a Mach-Zehnder interferometer.

Figure 26 contains two interferograms of the flow through a cascade of

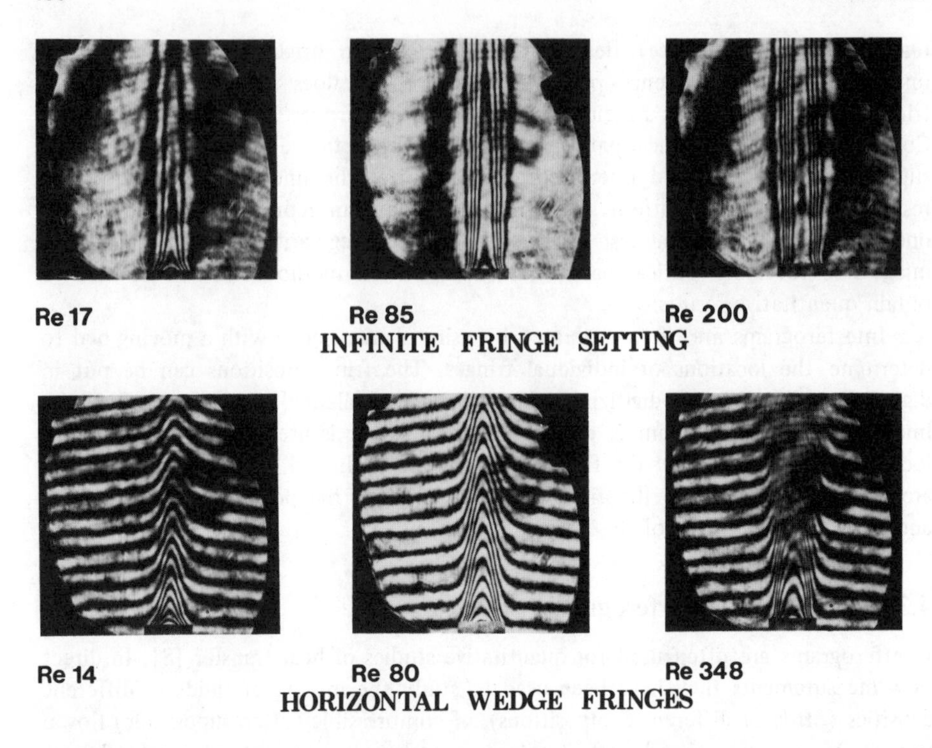

Figure 20 Interferograms of a low-Reynolds-number helium jet entering a still atmosphere of air.

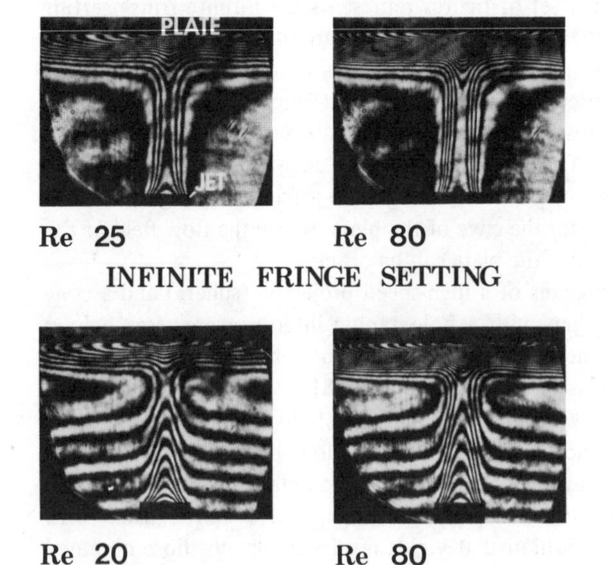

Figure 21 Interferograms of a helium jet impinging on a flat plate.

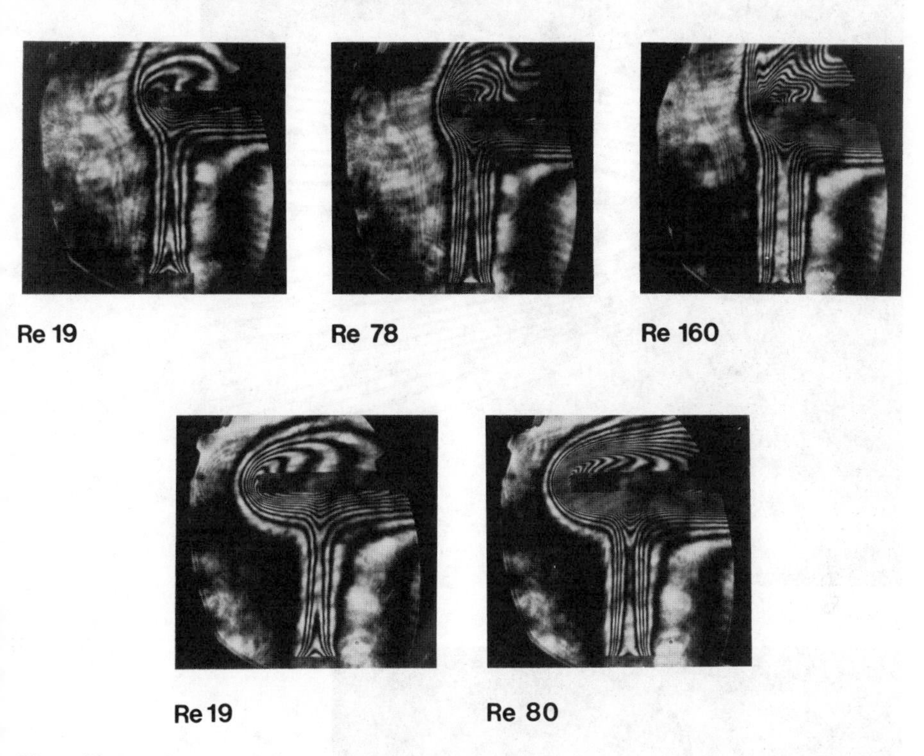

Figure 22 Interferograms of an impinging helium jet with flow around edge of a plate (infinite-fringe setting).

gas-turbine blades [42]. In each photo, one of the blades is heated (note the heating-wire connection) for better visualization of the flow around the blades. The incoming flow is parallel to the heating-wire connection. Note the difference in the boundary-layer growth on the concave and convex sides of the blades. Also, there is a marked difference in the boundary layer, depending on the direction of the incoming flow relative to the blade orientation. Close examination of the flow shows boundary-layer thickening, and even separation.

4.4 Design and Adjustment

Figures 16 and 18 are schematics of a typical Mach-Zehnder interferometer. Instead of lenses to form the parallel light beam and to bring the wide beam down into a smaller region for focusing on a screen or through the aperture of a camera, mirrors are often used to decrease the cost of the system.

For a sufficiently long coherence length (see below), a relatively monochromatic light source must be used. In the past, low-pressure lamps with filters were often used, and these are still quite common. Today, lasers are also used as the light source for interferometers. In fact, they are necessary for some special types. Lasers offer a quite large coherence length, which can be of great value. With other light sources, if thick windows must be used on the test section, or if the fluid in

Figure 23 Holographic interferograms of high-speed flow over an object. (*a*) One-half-inch-diameter sphere; 1 atm pressure; Ma $\simeq$ 6. (*b*) Shock interaction on 60° cone cylinder projected at Ma $\simeq$ 3.5; spark-produced blast wave. *(From [38, 39].)*

the test section has a refractive index very different from that of air, it may be necessary to have a compensating tank to ensure that the path lengths of the beams are not too different. With a laser light source, such tanks are normally not needed. With a laser source, however, spurious fringes may be present, owing to multiple reflections in some of the optical components.

Note that the test section should be placed in path 1, as shown in Figs. 16 and 18, rather than in the other light beam. In this way, the beam representing the shadow image of the test section does not pass through the last splitter plate, which could cause considerable astigmatism [43].

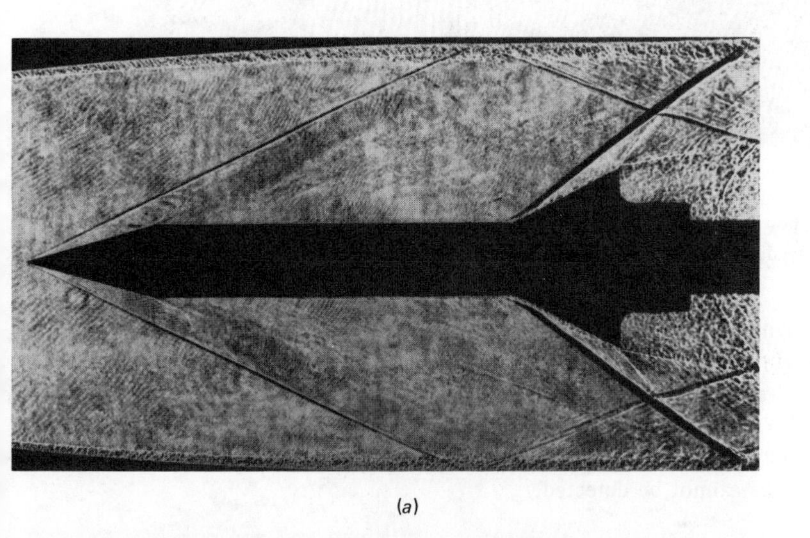

(*a*)

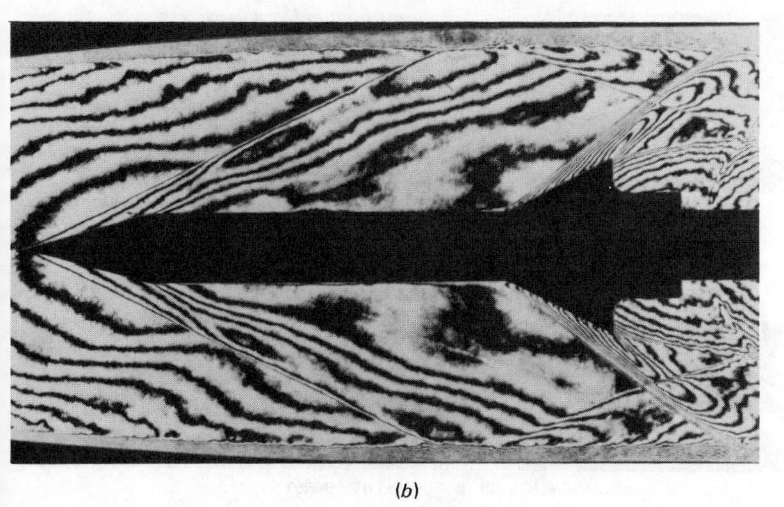

(*b*)

Figure 24 Flow over sharp-tipped spike with conical flare; pressure 100 psia; Ma = 2.98. (*a*) Shadowgraph. (*b*) Infinite-fringe interferogram. *(From [40].)*

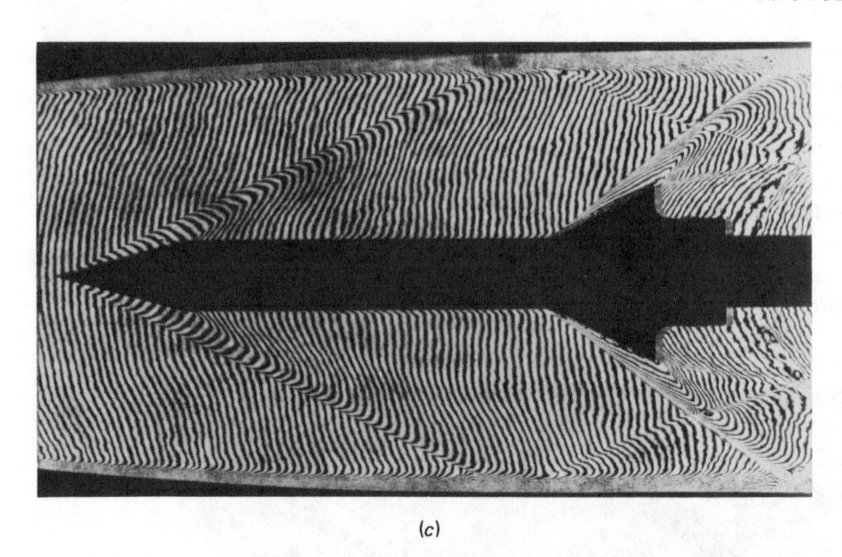

(c)

Figure 24 Flow over sharp-tipped spike with conical flare; pressure 100 psia; Ma = 2.98. (*Continued*) (*c*) Wedge-fringe interferogram. *(From [40].)*

The alignment of an interferometer, such as is shown in Figs. 16 and 18, is somewhat complicated and time-consuming but is not so horrendous a task as is often perceived. A number of methods that greatly simplify the task have been described [29, 33, 44, 45]. The chief concern is to align the reference and test beams so that they are closely parallel. If they are not, the fringe spacing is so small that the fringes cannot be detected.

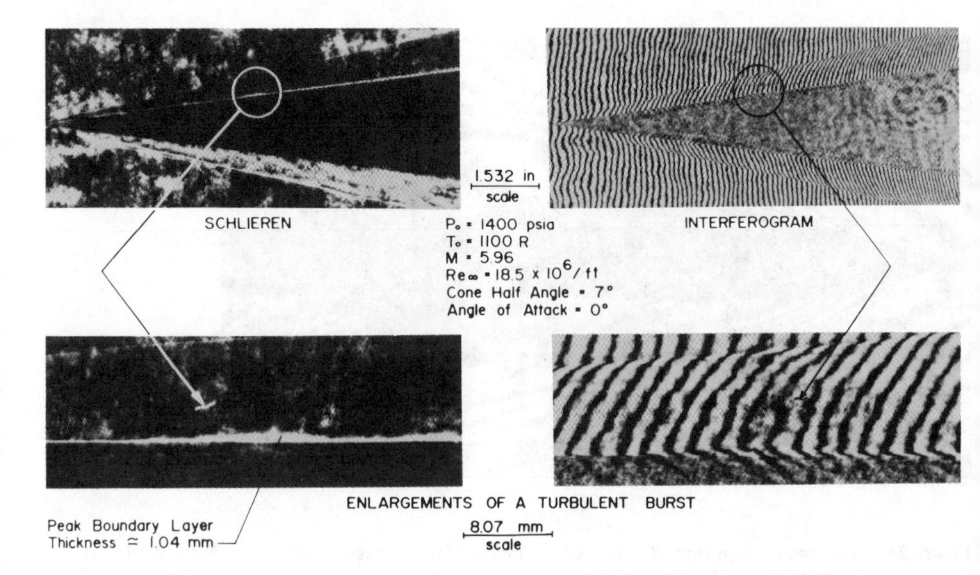

Figure 25 Holographic schlieren and interferometric depiction of boundary-layer transition on a slender, sharp-tipped cone. *(From [41].)*

The parallelism of the beam leaving the paraboloidal mirror F can be determined by measuring its dimensions at various positions along its path or, more accurately, by reflecting it back with a plane mirror and observing the focus of this returned beam. The two path lengths are then usually set approximately equal. The difference in length must be smaller than the coherence length of the source-filter combination, which increases as the light becomes more monochromatic. If the light intensity variation with wavelength is gaussian with a bandwidth of $\Delta\lambda$, the coherence length or optical path difference over which fringes can still be observed is approximately $\lambda^2/\Delta\lambda$ [46].

To obtain fringes, beams 1 and 2 (Fig. 18) must be nearly parallel to each other following SP_2. This can be accomplished by aligning the images of two

Figure 26 Wedge-fringe interferograms used for visualizing flow over a heated gas-turbine blade held in a cascade; oncoming flow direction is parallel to the visible wire carrying the heating current. *(From [42].)*

Figure 26 (*Continued*) Wedge-fringe interferograms used for visualizing flow over a heated gas-turbine blade held in a cascade; oncoming flow direction is parallel to the visible wire carrying the heating current. *(From [42].)*

objects observed by looking back through SP_2, preferably with a small telescope. The two objects examined should be far apart and must both be located before the first splitter SP_1, so that images are obtained for paths 1 and 2. Rotation of M_2 and SP_2 about two orthogonal axes moves each pair of virtual images of the two objects. When both images of each object are superimposed, the two beams leaving SP_2 are closely parallel, and fringes should appear in the field of view. While focusing on the center of the test section and M_2 with the telescope, camera, or screen, fringes are made as sharp as possible by further rotation of SP_2 while rotating M_2 to keep the fringe spacing from getting too small.

When the plane of focus of the fringes is in M_2 (and the center of the test section), the final adjustment of optical path length can be made. It is advantageous to keep the interferometer set close to the zero-order fringe (zero optical path length between the two light beams), since that is the position for sharpest fringes, even for filtered light. In addition, white light and the zero-order fringe are useful for measuring the index of refraction for fluids whose optical properties are unknown or in tracing fringes through regions of large density gradients. In many interferometers the path length of one of the beams can be altered by translation of a mirror (M_1). If white light (an incandescent bulb suffices) replaces the monochromatic light on half the field of view, the mirror can be translated until the zero-order white-light fringe is observed. In practice, it is often helpful, particularly if the initial setting is far from the zero-order fringe, to use light of varying coherence lengths, from the most monochromatic to white, to make each set of fringes as sharp as possible while translating the mirror.

Once adjusted, the interferometer, if properly constructed, usually needs only minor adjustment. Placing the unit in a vibration-free constant-temperature area helps to maintain alignment.

4.5 Errors in a Two-Dimensional Field

Since the Mach-Zehnder interferometer is of great value for quantitative studies in two-dimensional fields, considerable attention [8, 47 to 55] has been directed toward the corrections that must be applied when the idealizations assumed in the derivation of Eqs. (70), (73), and (76) are not strictly met. The two most significant errors usually encountered are due to refraction and end effects.

Refraction occurs when there is a density (really, index of refraction) gradient normal to the light beam that causes the beam to "bend." The resulting error increases with increasing density gradient and with increasing path length in the disturbed region L. It is refraction that usually hinders accurate interferometric measurements in thin forced-convection boundary layers.

End effects are caused by deviation from two-dimensionality in the density field, particularly where the light beam enters and leaves the disturbed region. The end effects are usually large when the disturbed region is large normal to the light-beam direction, and the test-section length along the light beam is relatively short. End effects are often significant with thick thermal boundary layers. Thus, if an experimental apparatus is designed to minimize refraction error, the end effect may be large, and vice versa. Further details and a simplified analysis of the errors are contained in [8].

4.6 Other Interferometers

Interferometric systems other than the Mach-Zehnder have been used in flow studies. These, in general, produce interferograms that can be evaluated in a manner similar to that for the Mach-Zehnder patterns. In some systems, a grating divides the initial beam into two coherent beams, one of which traverses the test region. The beams are recombined on another grating, yielding an interference pattern. One system [56, 57] uses two gratings, with the reference and test beams close to each other. In a four-grating apparatus [58], the beams are further apart, but residual fringes are often superimposed on the pattern. A laser light source has also been used with a grating interferometer [59].

Several special laser-light-source interferometers have been used or suggested for flow measurement [60 to 62]. Systems that are used with laser sources include schlieren or shearing interferometers. In one such system [63 to 65], a very small wire or stop is placed at the focus of a standard schlieren system to block the central maximum of the Fraunhofer diffraction pattern. Then the phase distribution produced when the beam passes through a disturbance in the test region can be observed in reference to an undisturbed part of the test beam. The result is a fringe pattern similar to that observed with a Mach-Zehnder interferometer set for infinite fringe spacing. Another system [66] uses a small glass shearing plate in place of the knife-edge. When the angle between the incoming light beam and the normal to the plate is approximately 50°, the first two reflected beams interfere, resulting in fairly straight parallel fringes. A Wollaston prism [67, 68] and a grating [69] have also been used to produce finite fringe interference patterns.

Shearing interferometers in which the reference beams are sheared slightly [68, 70] in a lateral direction have been used, with conventional and laser light sources. A polarization interferometer [7, 71] is a wave-shearing interferometer in which two coherent beams, polarized at right angles to one another, are produced from a single incoming beam. The two beams often diverge at a finite angle, and when they are recombined after passage through the test section, a fringe pattern results. Either a finite-fringe or infinite-fringe field can be obtained using three Wollaston prisms [72, 73].

In a shearing interferometer, the test and reference beams are often separated by only a small amount δ. Let us assume this separation is in the y direction and

$$\Delta y' = \delta \tag{78}$$

When the beams are recombined, the interference field is representative of the local difference in optical path length $\overline{\Delta PL}$ between positions y and $y + \delta$ (or $y + \Delta y'$). Equivalent to Eq. (68), the corresponding fringe shift is

$$\epsilon = \frac{\overline{\Delta PL}}{\lambda_0} = \frac{n_{y+\Delta y'} - n_y}{\lambda_0} L \tag{79}$$

$$\epsilon = \frac{\Delta n}{\Delta y} \Delta y' \frac{L}{\lambda_0} = \frac{\Delta n}{\Delta y} \delta \frac{L}{\lambda_0} \tag{80}$$

If the beam separation is small,

$$\epsilon = \frac{\partial n}{\partial y} \frac{1}{\lambda_0} L\delta \tag{81}$$

or

$$\frac{\partial n}{\partial y} = \frac{\lambda_0 \epsilon}{L\delta} \tag{82}$$

For a gas,

$$\frac{\partial \rho}{\partial y} = \frac{\lambda_0 \epsilon}{CL\delta} = \frac{\lambda_0 \rho_0 \epsilon}{(n_0 - 1)L\delta} \tag{83}$$

Assuming constant pressure, the perfect gas law gives

$$\frac{\partial T}{\partial y} = -\frac{T\lambda_0}{n_0 - 1}\frac{\rho_0}{\rho}\frac{\epsilon}{L\delta} \tag{84}$$

Note that δ is a constant (often adjustable) of the shearing interferometer. As long as it is small, the local derivative of the index of refraction (or density or temperature) in a two-dimensional field is proportional to the fringe shift.

An advantage of a shearing interferometer over a Mach-Zehnder interferometer is that the density or temperature gradient near the surface of an object can be measured directly. The sensitivity is proportional to the shear spacing as well as to the length of the test object in the light-beam direction. With a small shear spacing the errors due to refraction and thick-window effects tend to vanish. Among the disadvantages is the need to integrate the fringe displacement to obtain density distributions. Also, only the density gradient in the direction of the beam shear can be measured, although this direction may be changed. The closest to a surface that an accurate measurement of the density gradient can be made is of the order of one-half the shear displacement.

4.7 Holography

Holography can also be used in flow measurement [74 to 79]. In a holographic interferometer, the light beam (in this case, almost necessarily a highly uniform beam from a laser source) is split into two coherent beams, one of which passes through the test section while the other (reference beam) bypasses the test section. The two beams are recombined on a photographic plate. The resulting hologram is a diffraction grating formed by the emulsion on the plate. If the hologram on the developed plate is viewed via the reference beam, the interference pattern can be interpreted in a manner analogous to a Mach-Zehnder interferogram.

An example of a holographic system is shown in Fig. 27 [80]. Although holographic interferometry is normally done using a double exposure, real-time holographic interferograms can be obtained for flow studies using a precise and adjustable hologram mount. With this system, either infinite fringes or, for more quantitative measurement, wedge fringes can be obtained.

Although somewhat more inconvenient to use than a Mach-Zehnder interferometer, a relatively low-cost system can be set up for holographic interferometry. This is because most of the optical components of the holographic apparatus need not be extremely uniform or of very high quality to produce good interferograms.

What would be highly desirable for interferometric measurement is a means of determining a three-dimensional density or temperature field. Such systems have been discussed over the years, one concept being to use a number of different beams from an interferometer passing through a test section at different angles.

With the arrival of holographic systems, measurement of three-dimensional fields is brought closer to practicality. To obtain three-dimensional information, diffuse illumination rather than parallel light can be used to traverse the test section [81, 82]. A phase grating can also be used before the test section to obtain images of light beams traversing the test section at a number of different angles [83].

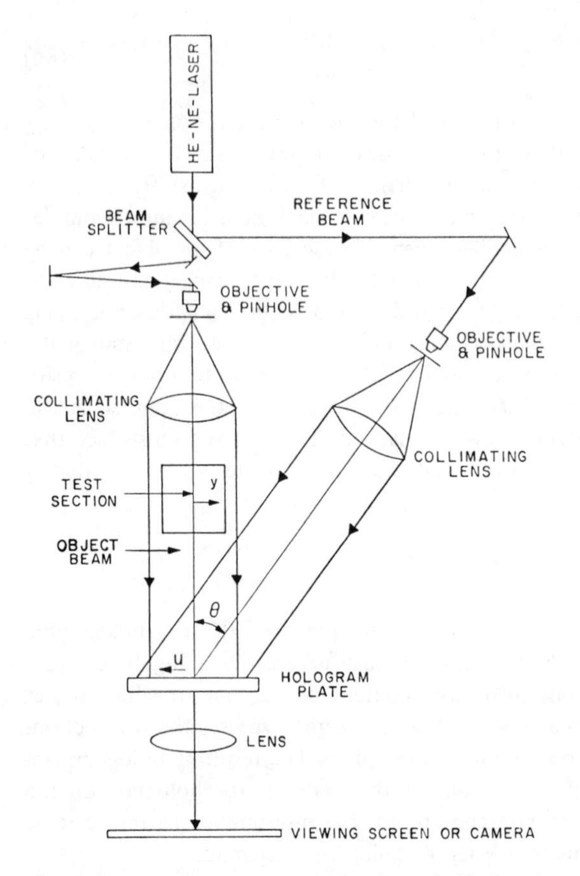

Figure 27 Holographic interferometer.

Evaluation of three-dimensional flow fields has been obtained for transonic flow out of a nozzle [36], supersonic flow over a yawed cone [84], and natural convection flow in interacting plumes [83].

The advantage of using a holographic system is that a single photograph (hologram) can give information on the interference patterns that would otherwise have to be observed by looking in different directions through the test section. This follows from the large amount of information that is available on the hologram. The hologram contains the record of very complex wavefronts produced in passing through the test section. Viewing the hologram from a specific direction provides essentially the equivalent of the interference pattern that would be obtained by a beam passing through the test section in that direction. The inversion of the data obtained from the interference patterns from a number of directions through the test section into the three-dimensional distribution is complex. This has limited the application of three-dimensional holograms, to date, to a few simple systems. Improvements in simplifying the inversion procedure should make the technique more valuable.

5. CONCLUSION

The properties of shadowgraph, schlieren, and interferometric systems have been described. The basic operating equations for all three systems have been derived with emphasis on their use in determining density and temperature distributions in two-dimensional fields.

Shadowgraph and schlieren systems are used principally for qualitative descriptions of a density field. Because schlieren and shadowgraph photographs yield information on the first and second derivatives of density, their widest application can be found in systems with steep gradients of density and temperature, such as flame fronts and shock waves. Interferometers can be used in quantitative studies of two-dimensional (including axisymmetric) density and temperature fields.

The advent of lasers permits novel and useful interferometric designs. Lasers can be used in schlieren interferometers and in holography for temperature and density measurements. Holographic interferometers using laser light sources can be used in the evaluation of three-dimensional density fields.

NOMENCLATURE

A	amplitude of light beam
A_1	amplitude of light beam from path 1
A_2	amplitude of light beam from path 2
A_0	maximum of amplitude
A_{01}	maximum of amplitude of beam from path 1
A_{02}	maximum of amplitude of beam from path 2
A_T	amplitude of recombined light beams from paths 1 and 2
a_s, a_0, a_K	dimension of schlieren beam normal to knife-edge at source, at knife-edge, and above knife-edge when no disturbance is present, respectively
b_0, b_s	dimension of schlieren beam parallel to knife-edge at source and knife-edge, respectively
C	Gladstone-Dale constant
c	speed of light
c_0	speed of light in a vacuum
d	fringe spacing
f	focal length of lens or mirror
I	light intensity
I_d	illumination at screen of disturbed field when knife-edge is present
I_K	illumination at screen of undisturbed field when knife-edge is present
I_0	illumination at screen with no knife-edge
I_T	illumination at exit of test section (and at screen if no deflection) in shadowgraph system
j	integer
K	knife-edge
k	ratio of constant-pressure to constant-volume specific heats
L	length of test section in light-beam direction z
M	mirror, ratio of mass velocity of injected flow to mainstream flow
n	index of refraction
n_a	index of refraction of air outside test section
n_0	index of refraction at standard conditions
n_{ref}	index of refraction in reference region

n_y	index of refraction at x, y
$n_{y+\Delta y'}$	index of refraction at x, $y + \Delta y'$
P	pressure
P_0	pressure at standard conditions
p	object distance from lens or mirror
PL	optical path length
$\overline{\Delta PL}$	difference in optical path length
q	image distance from lens or mirror
R	gas constant in terms of mass
S	source
SP	splitter plate
T	temperature
T_{ref}	temperature in reference region
W	distances defined in Fig. 19
x	direction normal to y and z
y	direction perpendicular to z, usually the direction in which the gradient of density or temperature lies
y_{sc}	height of light ray at screen
z	direction along light beam
z_{sc}	distance from test section to screen
α	angular deflection of light ray as measured in air outside test section; same as α' if $n \simeq n_a$
$\alpha_{max,neg}$	maximum deflection angle toward knife-edge that can be measured with schlieren system
α_{max}	maximum deflection angle away from knife-edge that can be measured with schlieren system
α'	angular deflections of light ray within test fluid
α''	angle defined in Fig. 4
β	angle defined in Fig. 4
γ	angle defined in Fig. 4
Δ	phase difference, change of
Δa	deflection of light beam away from schlieren knife-edge
Δy	width of light beam
$\Delta y'$	displacement of beams in shearing interferometer, δ
δ	characteristic length in shearing interferometers, displacement of beams, $\Delta y'$
ϵ	interferometer fringe shift, optical path-length difference in vacuum wavelengths
θ	angle between interferometer beams when recombined
λ	wavelength of light
λ_0	vacuum wavelength
τ	time
ρ	density
ρ_0	density at standard conditions
ρ_{ref}	density in reference region

REFERENCES

1. F. J. Weinberg, *Optics of Flames,* Butterworth, London, 1963.
2. R. W. Ladenburg, B. Lewis, R. N. Pease, and H. S. Taylor, *Physical Measurements in Gas Dynamics and Combustion,* chaps. A1, A2, A3, Princeton University Press, Princeton, N.J., 1954.
3. D. W. Holder, R. J. North, and G. P. Wood, Optical Methods for Examining the Flow in High-Speed Wind Tunnels, pts. I and II, AGARD, 1965; also Schlieren Methods, *Notes Appl. Sci.,* no. 31, National Physical Laboratory, London, 1963.
4. N. F. Barnes, Optical Techniques for Fluid Flows, *J. Soc. Motion Pict. Tel. Eng.*, vol. 61, pp. 487–511, 1953.
5. R. C. Dean, Jr., Aerodynamic Measurements, MIT Gas Turbine Laboratory, Cambridge, Mass., 1953.
6. H. Shardin, Toepeler's Schlieren Method: Basic Principles for Its Use and Quantitative Evaluation, Navy Translation 156, 1947.
7. W. Hauf and U. Grigull, Optical Methods in Heat Transfer, in J. P. Hartnett and T. F. Irvine, Jr. (eds.), *Advances in Heat Transfer,* vol. 6, p. 133, Academic, New York, 1970.
8. R. J. Goldstein, Optical Techniques for Temperature Measurement, in E. R. G. Eckert and R. J. Goldstein (eds.), *Measurements in Heat Transfer,* 2d ed., p. 241, Hemisphere, Washington, D.C., 1976.
9. T. Y. Chu and R. J. Goldstein, Turbulent Convection in a Horizontal Layer of Water, *J. Fluid Mech.,* vol. 60, pp. 141–159, 1973.
10. Landolt-Bornstein, *Physikalisch-Chemische Tabellen*, Suppl. 3, p. 1677, 1935.
11. L. Tilton and J. Taylor, Refractive Index and Dispersion of Distilled Water for Visible Radiation at Temperatures 0 to 60°C, *J. Res. Nat. Bur. Stand.,* vol. 20, pp. 419–477, 1938.
12. G. S. Speak and D. J. Walters, Optical Considerations and Limitations of the Schlieren Method, ARC Tech. Rep. 2859, London, 1954.
13. R. J. Goldstein, E. R. G. Eckert, F. K. Tsou, and A. Haji-Sheikh, Film Cooling with Air and Helium Injection through a Rearward-Facing Slot into a Supersonic Flow, *AIAA J.,* vol. 4, pp. 981–985, 1966.
14. M. R. Davis, Quantitative Schlieren Measurements in a Supersonic Turbulent Jet, *J. Fluid Mech.,* vol. 51, pp. 435–447, 1972.
15. G. E. Roe, An Optical Study of Turbulence, *J. Fluid Mech.,* vol. 43, pp. 607–635, 1970.
16. M. J. Fisher and F. R. Krause, The Crossed-Beam Correlation Technique, *J. Fluid Mech.,* vol. 28, pp. 705–717, 1967.
17. M. Y. Su and F. R. Krause, Optical Crossed-Beam Measurements of Turbulence Intensities in a Subsonic Jet Shear Layer, *AIAA J.,* vol. 9, pp. 2113–2114, 1971.
18. L. N. Wilson and R. J. Damkevala, Statistical Properties of Turbulent Density Fluctuations, *J. Fluid Mech.,* vol. 43, pp. 291–303, 1970.
19. B. H. Funk and K. D. Johnston, Laser Schlieren Crossed-Beam Measurements in a Supersonic Jet Shear Layer, *AIAA J.,* vol. 8, pp. 2074–2075, 1970.
20. B. Hannah, Quantitative Schlieren Measurements of Boundary Layer Phenomena, *Proc. 11th Int. Cong. of High Speed Photography,* pp. 539–545, 1974.
21. V. Ronchi, Due Nuovi Metodi per lo Studio delle Superficie e dei Sistemi Ottici, *Ann. Regia Sc. Normale Super Pisa,* vol. 15, 1923 (bound 1927).
22. P. F. Darby, The Ronchi Method of Evaluating Schlieren Photographs, *Tech. Conf. Opt. Pheno. Supersonic Flow,* NAVORD Rep., pp. 74–76, 1946.
23. D. A. Didion and Y. H. Oh, A Quantitative Schlieren-Grid Method for Temperature Measurement in a Free Convection Field, Mechanical Engineering Department Tech. Rep. 1, Catholic University of America, Washington, D.C., 1966.
24. L. A. Watermeier, Self-Illuminated Schlieren System, *Rev. Sci. Instrum.,* vol. 37, pp. 1139–1141, 1966.
25. A. Kantrowitz and R. L. Trimpi, A Sharp Focusing Schlieren System, *J. Aero Sci.,* vol. 17, pp. 311–314, 1950.

26. J. H. Hett, A High Speed Stereoscopic Schlieren System, *J. Soc. Motion Pict. Tel. Eng.*, vol. 56, pp. 214–218, 1951.
27. R. D. Buzzard, Description of Three-Dimensional Schlieren System, *Proc. 8th Int. Congress on High Speed Photography,* pp. 335–340, 1968.
28. Anatol Roshko, Structure of Turbulent Shear Flows: A New Look, *AIAA J.*, vol. 14, pp. 1349–1357, 1976.
29. E. R. G. Eckert, R. M. Drake, Jr., and E. Soehngen, Manufacture of a Zehnder-Mach Interferometer, Wright-Patterson AFB, Tech. Rep. 5721, ATI-34235, 1948.
30. F. D. Bennett and G. D. Kahl, A Generalized Vector Theory of the Mach-Zehnder Interferometer, *J. Opt. Soc. Am.*, vol. 43, pp. 71–78, 1953.
31. T. Zobel, The Development and Construction of an Interferometer for Optical Measurement of Density Fields, NACA Tech. Note 1184, 1947.
32. H. Shardin, Theorie und Anwendung des Mach-Zehnderschen Interferenz-Refraktometers, *Z. Instrumentenkd.*, vol. 53, pp. 396, 424, 1933 (DRL Trans. 3, University of Texas).
33. L. H. Tanner, The Optics of the Mach-Zehnder Interferometer, ARC Tech. Rep. 3069, London, 1959.
34. L. H. Tanner, The Design and Use of Interferometers in Aerodynamics, ARC Tech. Rep. 3131, London, 1957.
35. D. Wilkie and S. A. Fisher, Measurement of Temperature by Mach-Zehnder Interferometry, *Proc. Inst. Mech. Eng.*, vol. 178, pp. 461–470, 1963.
36. L. T. Clark, D. C. Koepp, and J. J. Thykkuttathil, A Three-Dimensional Density Field Measurement of Transonic Flow from a Square Nozzle Using Holographic Interferometry, *J. Fluid Eng.*, vol. 99, pp. 737–744, 1977.
37. G. Ben-Dor, B. T. Whitten, and I. I. Glass, Evaluation of Perfect and Imperfect Gas Interferograms by Computer, *Int. J. Heat Fluid Flow*, vol. 1, pp. 77–91, 1979.
38. A. B. Witte, J. Fox, and H. Rungaldier, Localized Measurements of Wake Density Fluctuations Using Pulsed Laser Holographic Interferometry, *AIAA J.*, vol. 10, pp. 481–487, 1972.
39. A. B. Witte and R. F. Wuerker, Laser Holographic Interferometry Study of High-Speed Flowfields, *AIAA J.*, vol. 8, pp. 581–583, 1970.
40. A. G. Havener and R. J. Radley, Jr., Supersonic Wind Tunnel Investigations Using Pulsed Laser Holography, ARL 73-0148, 1973.
41. A. George Havener, Detection of Boundary-Layer Transition Using Holography, *AIAA J.*, vol. 15, pp. 592–593, 1977.
42. E. R. G. Eckert, personal communication.
43. D. B. Prowse, Astigmatism in the Mach-Zehnder Interferometer, *Appl. Opt.*, vol. 6, p. 773, 1967.
44. E. W. Price, Initial Adjustment of the Mach-Zehnder Interferometer, *Rev. Sci. Instrum.*, vol. 23, p. 162, 1952.
45. D. B. Prowse, A Rapid Method of Aligning the Mach-Zehnder Interferometer, *Aust. Def. Sci. Serv.*, Tech. Note 100, 1967.
46. M. Born and E. Wolf, *Principles of Optics,* 3d ed., p. 319, Pergamon, New York, 1965.
47. E. R. G. Eckert and E. E. Soehngen, Studies on Heat Transfer in Laminar Free Convection with the Zehnder-Mach Interferometer, Air Force Tech. Rep. 5747, ATI-44580, 1948.
48. G. P. Wachtell, Refraction Effect in Interferometry of Boundary Layer of Supersonic Flow along Flat Plate, *Phys. Rev.*, vol. 78, p. 333, 1950.
49. R. E. Blue, Interferometer Corrections and Measurements of Laminar Boundary Layers in Supersonic Stream, NACA Tech. Note 2110, 1950.
50. E. R. G. Eckert and E. E. Soehngen, Distribution of Heat-Transfer Coefficients around Circular Cylinders in Crossflow at Reynolds Numbers from 20 to 500, *Trans. ASME,* vol. 74, pp. 343–347, 1952.
51. M. R. Kinsler, Influence of Refraction on the Applicability of the Zehnder-Mach Interferometer to Studies of Cooled Boundary Layers, NACA Tech. Note 2462, 1951.

52. W. L. Howes and D. R. Buchele, A Theory and Method for Applying Interferometry to the Measurement of Certain Two-Dimensional Gaseous Density Fields, NACA Tech. Note 2693, 1952.
53. W. L. Howes and D. R. Buchele, Generalization of Gas-Flow Interferometry Theory and Interferogram Evaluation Equations for One-Dimensional Density Fields, NACA Tech. Note 3340, 1955.
54. W. L. Howes and D. R. Buchele, Practical Considerations in Specific Applications of Gas-Flow Interferometry, NACA Tech. No. 3507, 1955.
55. W. L. Howes and D. R. Buchele, Optical Interferometry of Inhomogeneous Gases, *J. Opt. Soc. Am.*, vol. 56, pp. 1517–1528, 1966.
56. R. Kraushaar, A Diffraction Grating Interferometer, *J. Opt. Soc. Am.*, vol. 40, pp. 480–481, 1950.
57. J. R. Sterrett and J. R. Erwin, Investigation of a Diffraction Grating Interferometer for Use in Aerodynamic Research, NACA Tech. Note 2827, 1952.
58. F. J. Weinberg and N. B. Wood, Interferometer Based on Four Diffraction Gratings, *J. Sci. Instrum.*, vol. 36, pp. 227–230, 1959.
59. J. R. Sterrett, J. C. Emery, and J. B. Barber, A Laser Grating Interferometer, *AIAA J.*, vol. 3, pp. 963–964, 1965.
60. R. J. Goldstein, Interferometer for Aerodynamic and Heat Transfer Measurements, *Rev. Sci. Instrum.*, vol. 36, pp. 1408–1410, 1965.
61. A. K. Oppenheim, P. A. Urtiew, and F. J. Weinberg, On the Use of Laser Light Sources in Schlieren-Interferometer Systems, *Proc. R. Soc. London Ser. A.*, vol. 291, pp. 279–290, 1966.
62. L. H. Tanner, The Design of Laser Interferometers for Use in Fluid Mechanics, *J. Sci. Instrum.*, vol. 43, pp. 878–886, 1966.
63. E. L. Gayhart and R. Prescott, Interference Phenomenon in the Schlieren System, *J. Opt. Soc. Am.*, vol. 39, pp. 546–550, 1949.
64. E. B. Temple, Quantitative Measurement of Gas Density by Means of Light Interference in a Schlieren System, *J. Opt. Soc. Am.*, vol. 47, pp. 91–100, 1957.
65. J. B. Brackenridge and W. P. Gilbert, Schlieren Interferometry. An Optical Method for Determining Temperature and Velocity Distributions in Liquids, *Appl. Opt.*, vol. 4, pp. 819–821, 1965.
66. C. J. Wick and S. Winnikow, Reflection Plate Interferometer, *Appl. Opt.*, vol. 12, pp. 841–844, 1973.
67. R. D. Small, V. A. Sernas, and R. H. Page, Single Beam Schlieren Interferometer Using a Wollaston Prism, *Appl. Opt.*, vol. 11, pp. 858–862, 1972.
68. W. Merzkirch, Generalized Analysis of Shearing Interferometers as Applied to Gas Dynamic Studies, *Appl. Opt.*, vol. 13, pp. 409–413, 1974.
69. S. Yokozeki and T. Suzuki, Shearing Interferometer Using the Grating as the Beam Splitter, *Appl. Opt.*, vol. 10, pp. 1575–1580, 1971.
70. O. Bryngdahl, Applications of Shearing Interferometry, in E. Wolf (ed.), *Progress in Optics,* vol. 4, Wiley, New York, 1965.
71. R. Chevalerias, Y. Latron, and C. Veret, Methods of Interferometry Applied to the Visualization of Flows in Wind Tunnels, *J. Opt. Soc. Am.*, vol. 47, pp. 703–706, 1957.
72. W. Z. Black and W. W. Carr, Application of a Differential Interferometer to the Measurement of Heat Transfer Coefficients, *Rev. Sci. Instrum.*, vol. 42, pp. 337–340, 1971.
73. W. Z. Black and J. K. Norris, Interferometric Measurement of Fully Turbulent Free Convective Heat Transfer Coefficients, *Rev. Sci. Instrum.*, vol. 45, pp. 216–218, 1974.
74. M. H. Horman, An Application of Wavefront Reconstruction to Interferometry, *Appl. Opt.*, vol. 4, pp. 333–336, 1965.
75. L. O. Heflinger, R. F. Wuerker, and R. E. Brooks, Holographic Interferometry, *J. Appl. Phys.*, vol. 37, pp. 642–649, 1966.
76. L. H. Tanner, Some Applications of Holography in Fluid Mechanics, *J. Sci. Instrum.*, vol. 43, pp. 81–83, 1966.

77. O. Bryngdahl, Shearing Interferometry by Wavefront Reconstruction, *J. Opt. Soc. Am.*, vol. 58, pp. 865–871, 1968.
78. F. P. Kupper and C. A. Dijk, A Method for Measuring the Spatial Dependence of the Index of Refraction with Double Exposure Holograms, *Rev. Sci. Instrum.*, vol. 43, pp. 1492–1497, 1972.
79. Franz Mayinger and Walter Panknin, Holography in Heat and Mass Transfer, *Heat Transfer 1974 (Proc. 5th Int. Heat Trans. Conf.)*, vol. VI, pp. 28–43, Japan Society of Mechanical Engineers/Society of Chemical Engineers, Tokyo, 1974.
80. W. Aung and R. O'Regan, Precise Measurement of Heat Transfer Using Holographic Interferometry, *Rev. Sci. Instrum.*, vol. 42, pp. 1755–1759, 1971.
81. R. D. Matulka and D. J. Collins, Determination of Three-Dimensional Density Fields from Holographic Interferograms, *J. Appl. Phys.*, vol. 42, pp. 1109–1119, 1971.
82. D. W. Sweeney and C. M. Vest, Measurement of Three-Dimensional Temperature Fields above Heated Surfaces by Holographic Interferometry, *Int. J. Heat Mass Transfer*, vol. 17, pp. 1443–1454, 1974.
83. C. M. Vest and P. T. Radulovic, Measurement of Three-Dimensional Temperature Fields by Holographic Interferometry, in E. Marom, A. A. Friesem, and E. Wiener-Avnear (eds.), *Applications of Holography and Optical Data Processing*, pp. 241–249, Pergamon, London, 1977.
84. Tse-Fou Zien, William C. Ragsdale, and W. Charles Spring III, Quantitative Determination of Three-Dimensional Density Field by Holographic Interferometry, *AIAA J.*, vol. 13, pp. 841–842, 1975.

CHAPTER

NINE

FLUID MECHANICS MEASUREMENTS IN NONNEWTONIAN FLUIDS

Christopher W. Macosko

1. INTRODUCTION

Many liquid flow problems of interest today deal with nonnewtonian fluids. The main types of nonnewtonian liquids are polymer solutions, molten polymers, and concentrated dispersions such as slurries, food products, pastes, sealants, inks, and paints.

The departures of nonnewtonian fluids from typical newtonian behavior can be severe. The viscosity of a typical nonnewtonian fluid is strongly shear-thinning. If low-shear-rate viscosity data are used for pump selection and pressure calculations, considerable error can result. Nonnewtonian fluids can even behave *opposite* to newtonian fluids in certain flows. Typically, newtonian jets contract after emerging from a tube, while polymer solutions can expand. Both droplet breakup and the onset of turbulence can be suppressed by the addition of very small amounts of polymer. Lubrication, flow instabilities, mixing, and tank draining become considerably more complicated with nonnewtonian fluids. Some of these and other examples are reviewed by Bird [1], Bird et al. [2], and Middleman [3].

Why do nonnewtonian fluids present such a problem for fluid mechanical analysis? For incompressible newtonian liquids there is just one constant, the viscosity μ, which relates stress to rate of deformation:

$$\boldsymbol{T} = -p\boldsymbol{I} + 2\mu\boldsymbol{D} \tag{1}$$

where $\boldsymbol{T}$ is the total-stress tensor p is the isotropic pressure, and $\boldsymbol{D}$ is the rate-of-deformation tensor:

$$D = \frac{1}{2}(\nabla\mathbf{v} + \nabla\mathbf{v}^T) \quad \text{or} \quad D_{ij} = \frac{1}{2}\left(\frac{\partial v_i}{\partial x_j} + \frac{\partial v_j}{\partial x_i}\right) \tag{2}$$

Since the newtonian viscosity is a function only of temperature (and a weak function of pressure), generally if we measure the velocity distribution we can determine the stresses, and vice versa. However, for nonnewtonian fluids stress can have a complex dependence on rate of deformation and time and even the total deformation or strain:

$$B = \int D\,dt: \qquad T = -pI + \mathfrak{F}(D, t) \tag{3}$$

where $\mathfrak{F}$ may be an integral or a differential relation. This relation, the rheological constitutive equation for the fluid, is not easy to obtain, and even if it is known it can be extremely difficult to apply to anything but the simplest flows. This means that usually both stress and velocity measurements are needed to understand a nonnewtonian flow problem. The experimental problem is compounded further by the fact that many nonnewtonian materials of interest are hard to work with. They are often very viscous, opaque, require high temperature, or contain volatile solvents.

The goal of this chapter is to show how some of these experimental difficulties can be overcome. We point out the special problems encountered with nonnewtonian fluids in trying to utilize the methods described in the previous chapters. We also described some flow measurement methods that are unique to nonnewtonian materials, particularly flow birefringence.

Before describing the measurement methods, however, we summarize some results from the field of rheology. We first look at some of the basic types of flow behavior we can expect from nonnewtonian fluids. Then we examine some of the more useful constitutive relations and discuss how these relations can be verified.

2. MATERIAL FUNCTIONS

A newtonian fluid requires only a single material *constant* to relate stress to deformation. For a nonnewtonian fluid a material *function* is required. To determine this material function can be an impossible experimental problem. Fortunately a limited number of material functions are required to completely characterize a given simple flow. To help determine which rheological measurements should be made, it is valuable to see which functions can possibly arise and what they look like for typical materials. This is an important aspect of rheology. Here we are only able to summarize the main points without proof. Most advanced rheology texts [2, 4, 5] discuss this problem in more detail.

2.1 Steady Shear Flows

For homogeneous simple shear the only velocity is in the flow direction, and it is only a function of the cross direction: $\mathrm{v}_1 = f(\mathrm{x}_2)$. For steady simple shear at any x_2 position, velocity is a constant:

$$v_1 = \dot{\gamma} x_2 \tag{4}$$

For such a simple flow, the rate-of-deformation tensor reduces to two equal components,

$$2D = \begin{bmatrix} 0 & \dot{\gamma} & 0 \\ \dot{\gamma} & 0 & 0 \\ 0 & 0 & 0 \end{bmatrix} \tag{5}$$

Because of the symmetry of a shear deformation, the stress tensor can only have four different components,

$$T = \begin{bmatrix} T_{11} & T_{12} & 0 \\ T_{12} & T_{22} & 0 \\ 0 & 0 & T_{33} \end{bmatrix} \tag{6}$$

For an incompressible material we can determine the normal stresses only to within the arbitrary pressure p. Thus the deformation can affect only the normal stress differences. It follows then that there are only three independent stress quantities in shear:

$$T_{12} \qquad T_{11} - T_{12} \qquad T_{22} - T_{33}$$

These stresses will be, in general, a function of the history of the shear deformation. The simplest history is that of *steady* shearing. In this case the three stress quantities can only be a function of the shear rate, $2D_{12}$ or $\dot{\gamma}$, and are independent of time:

$$T_{12} = \tau(\dot{\gamma}) \tag{7}$$

$$T_{11} - T_{22} = N_1(\dot{\gamma}) \tag{8}$$

$$T_{22} - T_{33} = N_2(\dot{\gamma}) \tag{9}$$

Since in the limit of low shear rates these stresses go with $\dot{\gamma}$ and $\dot{\gamma}^2$, respectively, it is most common to define viscosity and normal-stress coefficients:

Viscosity:

$$\eta(\dot{\gamma}) = \frac{T_{12}}{\dot{\gamma}} \tag{10}$$

First normal-stress coefficient:

$$\psi_1(\dot{\gamma}) = \frac{T_{11} - T_{22}}{\dot{\gamma}^2} \tag{11}$$

Second normal-stress coefficient:

$$\psi_2(\dot{\gamma}) = \frac{T_{22} - T_{33}}{\dot{\gamma}^2} \tag{12}$$

These three functions are called the *viscometric material functions*. They have been fairly well studied for polymeric fluids. Figure 1 illustrates all the three functions for (1) a polyisobutylene solution in oil and (2) polyethylene at typical extrusion or molding temperature. Note that for both materials the second normal-stress coefficient is negative and considerably smaller than the first. Also note that at low shear rates, particularly for the polymer solution, all three material functions approach constant limiting values:

$$\lim_{\dot{\gamma}\to 0} \eta(\dot{\gamma}) = \eta_0 \tag{10a}$$

$$\lim_{\dot{\gamma}\to 0} \psi_1(\dot{\gamma}) = \psi_{10} \tag{11a}$$

$$\lim_{\dot{\gamma}\to 0} \psi_2(\dot{\gamma}) = \psi_{20} \tag{12a}$$

These limiting values are often used in phenomenological and molecular constitutive equations. The trends in Fig. 1 seem to be generally true for polymeric liquids. For dispersions very little normal-stress data are available, although they are predicted theoretically [7]. For dilute suspensions the viscosity function does look like Fig. 1, but concentrated dispersions often do not show a constant limiting viscosity. Instead, viscosity seems to increase continuously at low rates, indicating a yield stress. This "plastic" behavior is discussed further in Sec. 3.2. Some concentrated suspensions (corn starch is one example) have a shear-thickening viscosity.

2.2 Transient Shear

It is often in transient response that nonnewtonian fluids behave so differently from newtonian fluids. There are an infinite number of possible time histories of deformation; however a few types that occur frequently in process flow problems can be generated fairly easily in various rheometers.

One of the most important is start-up for steady shear flow. Shear rate is set essentially instantaneously at a fixed value, and the shear and normal stresses are recorded as they come to equilibrium. This is illustrated in Fig. 2. In Fig. 3 the data have been normalized by their equilibrium viscosity values. Note the large overshoot that occurs on start-up and the faster relaxation at high shear rates. This is characteristic of most polymeric materials. We can define start-up and relaxation shear-stress material functions η^+ and η^- as

$$\eta^+(t, \dot{\gamma}_0) = \frac{\tau_{12}^+(t, \dot{\gamma}_0)}{\dot{\gamma}_0} \tag{13}$$

$$\eta^-(t, \dot{\gamma}_0) = \frac{\tau_{12}^-(t, \dot{\gamma}_0)}{\dot{\gamma}_0} \tag{14}$$

Similar functions can be defined for the normal stresses.

Another important transient test is a step application of strain, illustrated in Fig. 4. This shows clearly the viscoelastic character typical of many polymeric

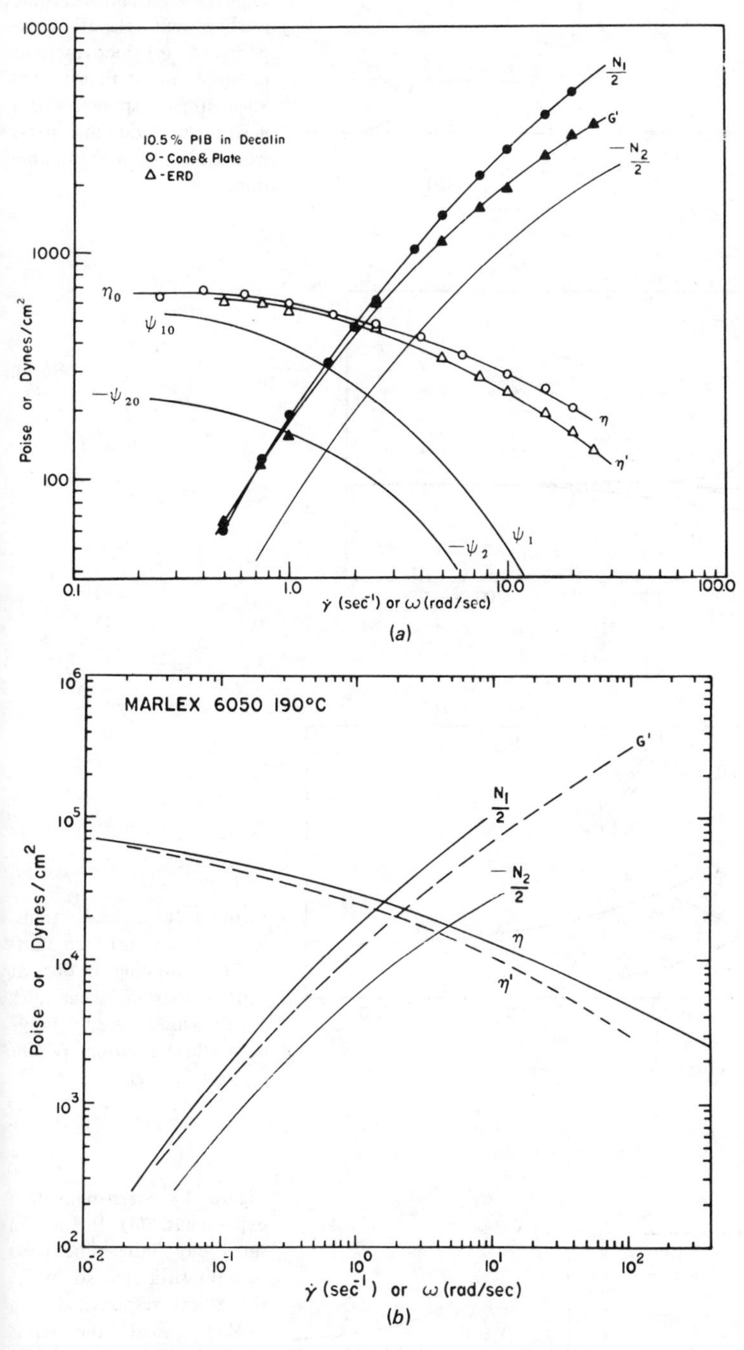

Figure 1 Examples of the three steady-shear (viscometric) material functions: (*a*) 10.5% polyisobutylene in decalin. (*b*) A high-density polyethylene melt, Marlex 6050, made by the Phillips Petroleum Co. *(From [6].)* Also shown are the two linear viscoelastic material functions G' and η', Eqs. (20) and (22).

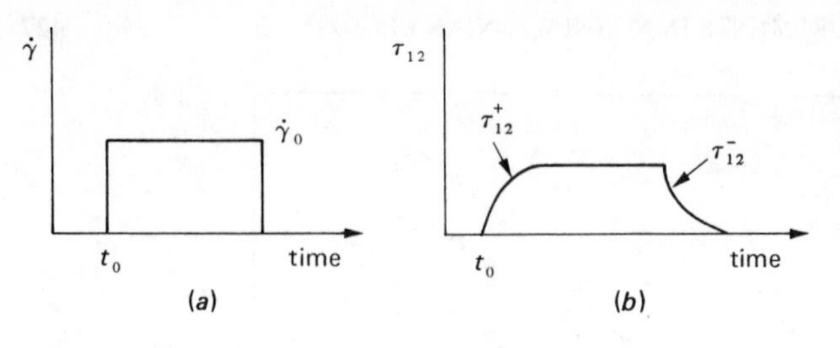

Figure 2 Constant-shear-rate start-up and relaxation experiment. (*a*) Step increase in shear rate at time t_0. (*b*) Shear-stress response. For a newtonian fluid the stress would also be a step function.

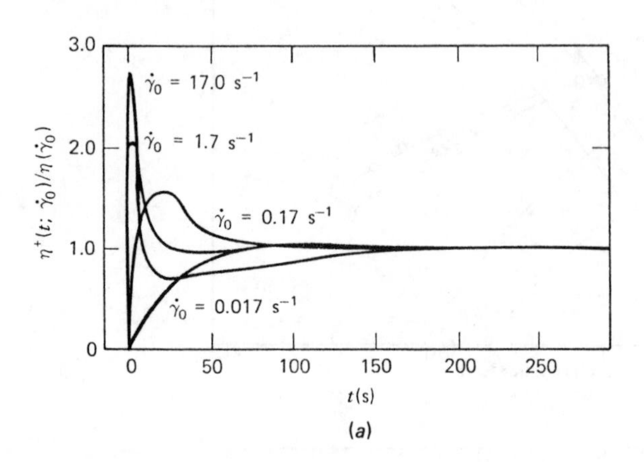

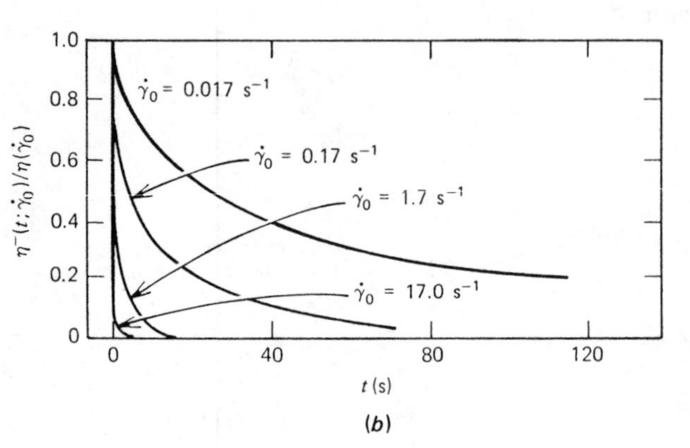

Figure 3 Normalized transient viscosity. (*a*) Step start-up. (*b*) Stopping of flow at various constant shear rates $\dot{\gamma}_0$. Data are for a 2.0% polyisobutylene solution. *(From [2], pp. 158, 161.)*

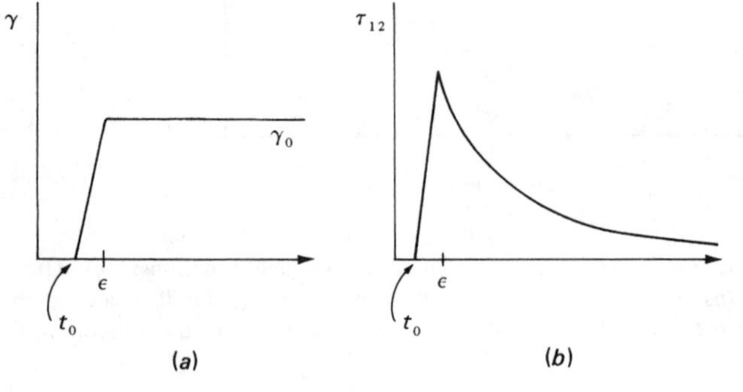

Figure 4 Stress-relaxation experiment. (*a*) It takes a small but finite time ϵ to accomplish a step strain γ_0. (*b*) Stress response. For a hookean solid, the stress would remain constant at its maximum value; for a newtonian liquid, relaxation would be instantaneous.

solutions. At short times the stress is large, typical of a rubbery solid. At long times the stress relaxes back to zero as expected for a purely viscous fluid. To describe the data, we define a stress relaxation modulus

$$G(t - t_0, \gamma_0) = \frac{\tau_{12}(t, \gamma_0)}{\gamma_0} \tag{15}$$

This material function is illustrated in Fig. 5 for a 20% polystyrene solution. We note that for small strains $G(t - t_0)$ is independent of strain, and at large strains it has roughly the same shape but is just shifted vertically. This behavior is typical of other polymeric liquids. We find these observations particularly important in development viscoelastic constitutive relations in the next section.

It is also possible to subject a sample to a step increase in stress. This is known as *creep* and is frequently used for solids testing. Figure 6 shows typical data. Initially the creep is rapid, but then it slows to a steady viscous flow characterized by a constant shear rate $\dot{\gamma}_\infty$. The data are usually described in terms of the creep compliance $J(t, \tau_0)$:

$$\gamma(t) = \int_0^t \dot{\gamma}(t')\, dt' = J(t, \tau_0)\tau_0 \tag{16}$$

The γ_0 intercept at t_0 in Fig. 6*b* is used to define the equilibrium creep compliance

$$J_e = \frac{\gamma_0}{\tau_0} \tag{17}$$

For many polymeric liquids the limit of J_e at very small stresses approaches a constant that is often used as a basic measure of the elasticity of a material, just as η_0 [Eq. (10*a*)] is a measure of viscosity [8a]. The same limit is obtained from the recoverable strain after removing the stress, Fig. 6*b*.

$$J_e^0 = \lim_{\tau_0 \to 0} \frac{\gamma_0}{\tau_0} = \lim_{\tau_0 \to 0} \frac{\gamma_r}{\tau_0} \tag{18}$$

Another transient testing method adapted from solids testing is the use of sinusoidal oscillations. Because it can be used for a wide range of liquid and solid materials and can be highly automated, sinusoidal testing, often called dynamic mechanical analysis, is very popular. As illustrated in Fig. 7, the sample is subjected to a continuous small-amplitude sinusoidal strain oscillation, and the stress is monitored. For a hookean solid the stress wave will be in phase with the strain, while for a newtonian liquid it will be 90° out of phase. The viscoelastic liquid (or solid) will lie between the two. To characterize this "betweenness," we decompose the stress into two waves; one wave τ' is in phase with the strain, and the other wave τ'' is exactly 90° out of phase with the strain. The prime notation arises from the fact that the decomposition of the τ wave can be conveniently expressed as a complex number:

$$\tau_{12}(t) = \tau^* = \tau' + i\tau'' \tag{19}$$

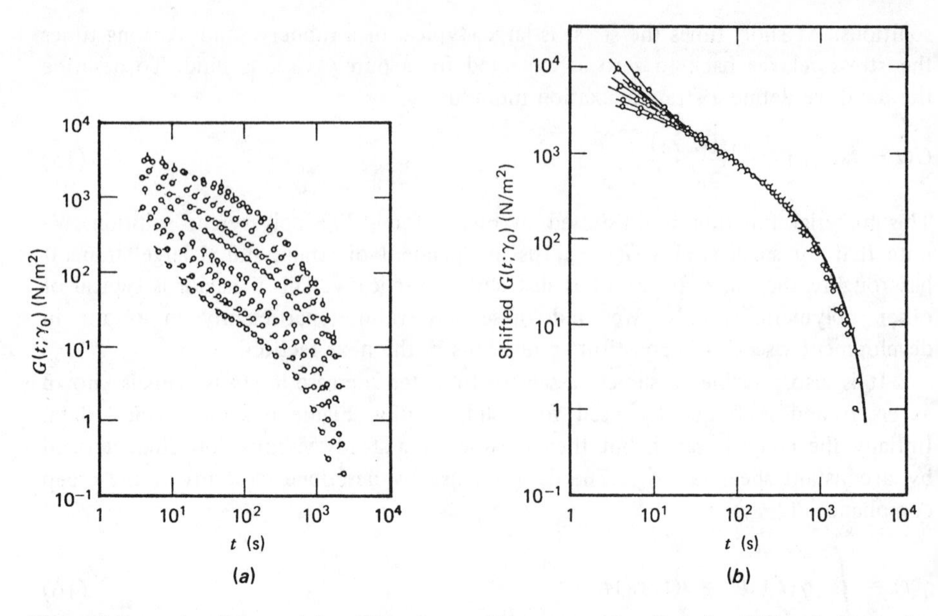

Figure 5 Log relaxation modulus $G(t, \gamma_0) = \tau_{12}(t)/\gamma_0$ versus log time for 20% polystyrene in Aroclor. Shear magnitudes are (o) 0.41, (o̍) 1.87, (σ)3.34, (o–) 5.22, (q) 6.68, (ϙ) 10.0, (ρ) 13.4, (–o) 18.7, and (ъ) 25.4. In (*b*) the data of (*a*) are shifted vertically. *(From [2], p. 164.)*

The magnitudes of the τ' and τ'' waves are used to define two dynamic moduli,

Elastic or in-phase modulus:

$$G'(\omega) = \frac{\tau_0'}{\gamma_0} \tag{20}$$

Viscous or out-of-phase modulus:

$$G''(\omega) = \frac{\tau_0''}{\gamma_0} \tag{21}$$

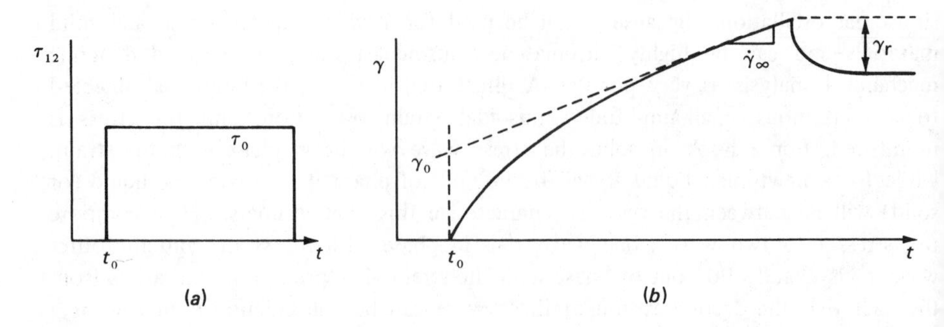

Figure 6 Creep experiment. (*a*) Step increase in shear stress. (*b*) Resultant shear strain. The equation for the asymptote is $\gamma_0 + \dot{\gamma}_\infty t$.

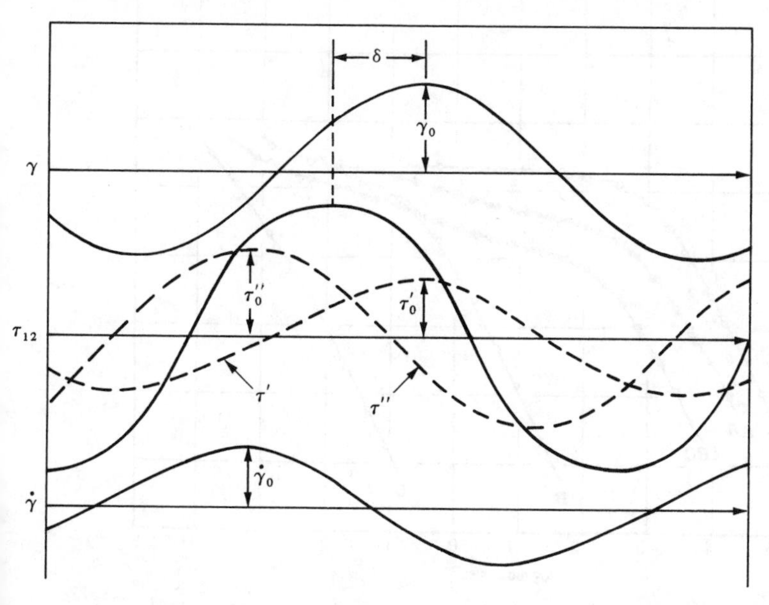

Figure 7 Sinusoidal oscillation experiment. The stress wave is decomposed into two waves: τ', which is in phase with the strain, and τ'', which is 90° out of phase with the strain. Note that τ'' is in phase with the rate of strain $\dot{\gamma} = d\gamma/dt$.

The phase shift δ between the strain and stress waves is

$$\tan \delta = \frac{G''}{G'}$$

Note that we can also define a dynamic viscosity $\eta'(\omega)$ by looking at the part of the stress that is in phase with the *rate* of strain $\dot{\gamma}$. As can be seen in Fig. 7, τ'' is in phase with the $\dot{\gamma}$ wave; thus

$$\eta'(\omega) = \frac{\tau_0''}{\dot{\gamma}_0} \tag{22}$$

The G' and G'' or G' and η' material functions are then sufficient to characterize a material's response to small strain oscillation. At larger strains, however, higher-order terms and harmonics can enter. The linear moduli are usually measured as a function of oscillation frequency and temperature. Typical data for several polymeric liquids are shown in Figs. 1 and 8. The shapes are quite characteristic for most polymers of high molecular weight: G' has a limiting low-frequency slope of 2, which decreases to nearly a plateau and then rises again. G'' has a limiting slope of 1, goes through a local maximum, and then rises again. The blends in Fig. 8 show that the high-molecular-weight component has the greatest influence on the rheology.

2.3 Material Functions in Extension

Besides shear, the other major classification of flows is extension. Many processes involve extensional flow—for example, fiber spinning, foaming, and flow in any

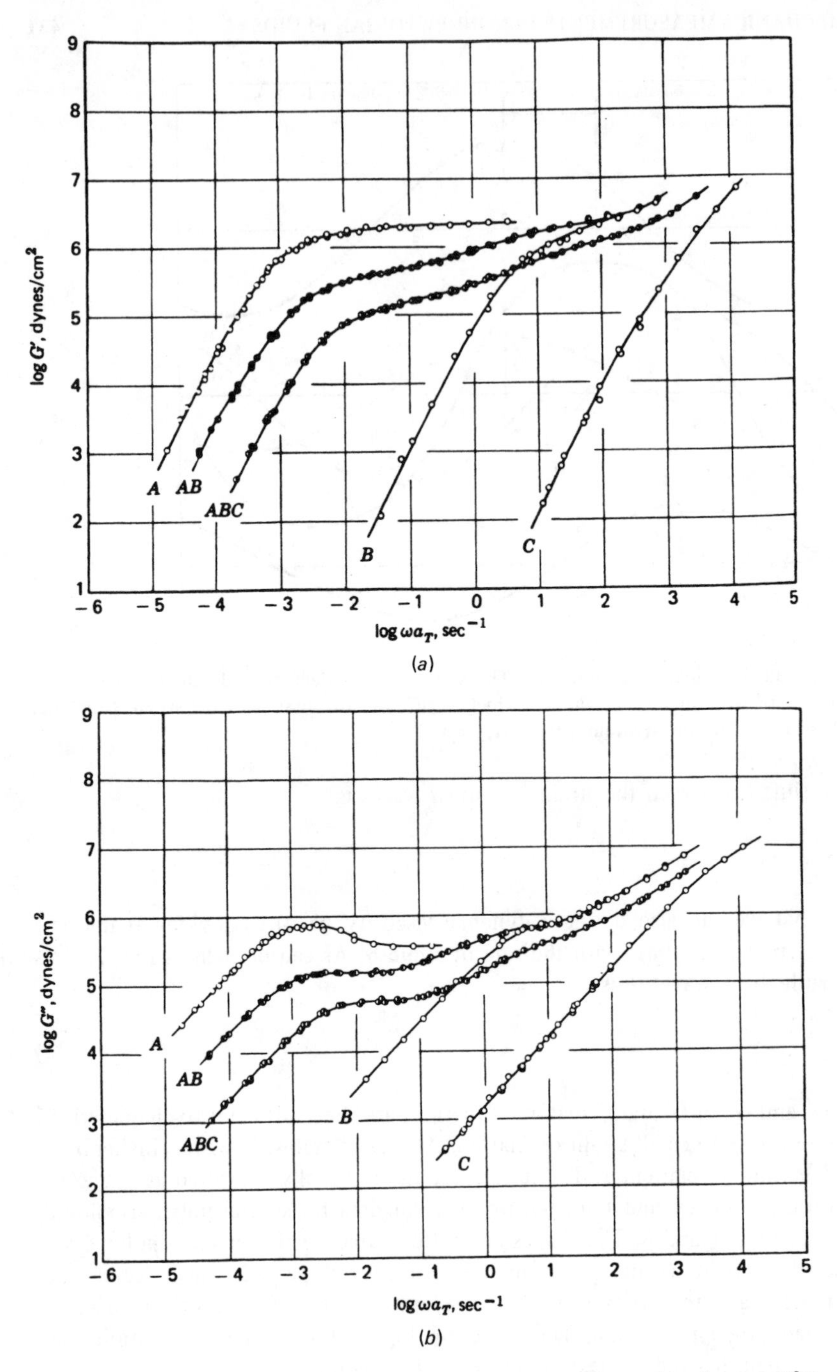

Figure 8 (*a*) Storage shear modulus and (*b*) loss modulus reduced to 160°C, for three narrow-distribution polystyrenes and two blends. Viscosity-average molecular weights $\times 10^{-4}$ are *A*, 58; *B*, 5.9; *C*, 0.89. Sample *AB* is a blend of equal parts of *A* and *B*; sample *ABC* is an equal blend of all three. a_T is an empirical factor used to shift data from other temperatures to 160°. *(From [8], p. 394.)*

converging or diverging channel. All real flows can be made up of some combination of shear and extension. To a nonnewtonian fluid, extension is fundamentally different from shear, and in general it is not possible to predict one from the other.

In pure extension (irrotational flow), velocities can vary only in the flow direction:

$$v_i = a_i x_i \tag{23}$$

For incompressible fluids, by the continuity relation only two velocities can be independent; i.e.,

$$\Sigma a_i = 0 \tag{24}$$

The rate-of-deformation tensor has only diagonal components,

$$D = \begin{bmatrix} a_1 & 0 & 0 \\ 0 & a_2 & 0 \\ 0 & 0 & -(a_1 + a_2) \end{bmatrix} \tag{25}$$

as does the stress tensor,

$$T = \begin{bmatrix} T_{11} & 0 & 0 \\ 0 & T_{22} & 0 \\ 0 & 0 & T_{33} \end{bmatrix} \tag{26}$$

As discussed with regard to shear flow, for incompressible fluids we can measure normal stress only to within an arbitrary constant. Thus there are only two normal-stress differences, and at most two material functions, required for general extensional flow:

$$T_{11} - T_{22} = f_1(D_{ii}, t) \tag{27}$$

$$T_{22} - T_{33} = f_2(D_{ii}, t) \tag{28}$$

Usually, extensional experiments are broken down into three types: uniaxial, biaxial, and planar. However, as is discussed further in Sec. 4, it is much more difficult to carry out purely extensional tests on fluids than shear tests. Only for uniaxial deformation is much information available, and those data appear reliable only, so far, for very viscous materials where solidlike test methods can be used.

For steady uniaxial extension,

$$\begin{aligned} v_1 &= \dot{\epsilon} x_1 \\ v_2 &= -\tfrac{1}{2}\dot{\epsilon} x_2 \end{aligned} \tag{29}$$

where $\dot{\epsilon}$ is the constant extension rate. Thus the rate-of-deformation tensor becomes

$$D = \begin{bmatrix} \dot{\epsilon} & 0 & 0 \\ 0 & -\frac{1}{2}\dot{\epsilon} & 0 \\ 0 & 0 & -\frac{1}{2}\dot{\epsilon} \end{bmatrix} \tag{30}$$

This additional symmetry of uniaxial extension leads to $T_{22} = T_{33}$. Thus only one material function is necessary to describe uniaxial data. This is usually done in terms of the extensional viscosity

$$\eta_e(\dot{\epsilon}) = \frac{T_{11} - T_{22}}{\dot{\epsilon}} \tag{31}$$

For a newtonian fluid η_e is just three times the shear viscosity.

Extensional viscosity data versus stress for two molten polystyrene samples are shown in Fig. 9. We see that at low stress levels the newtonian relation $\eta_e = 3\eta_0$ does hold. We also note that there is much less "extensional thinning" of the polystyrene than there is shear thinning. This appears to be generally true. In fact, a number of workers have reported extensional "thickening," i.e., an increase in η_e with stress or extension rate. These results are confounded by the difficulty in determining whether steady extension has, in fact, been achieved during the experiment. Figure 10 shows recent results at very high extensions for a low-density

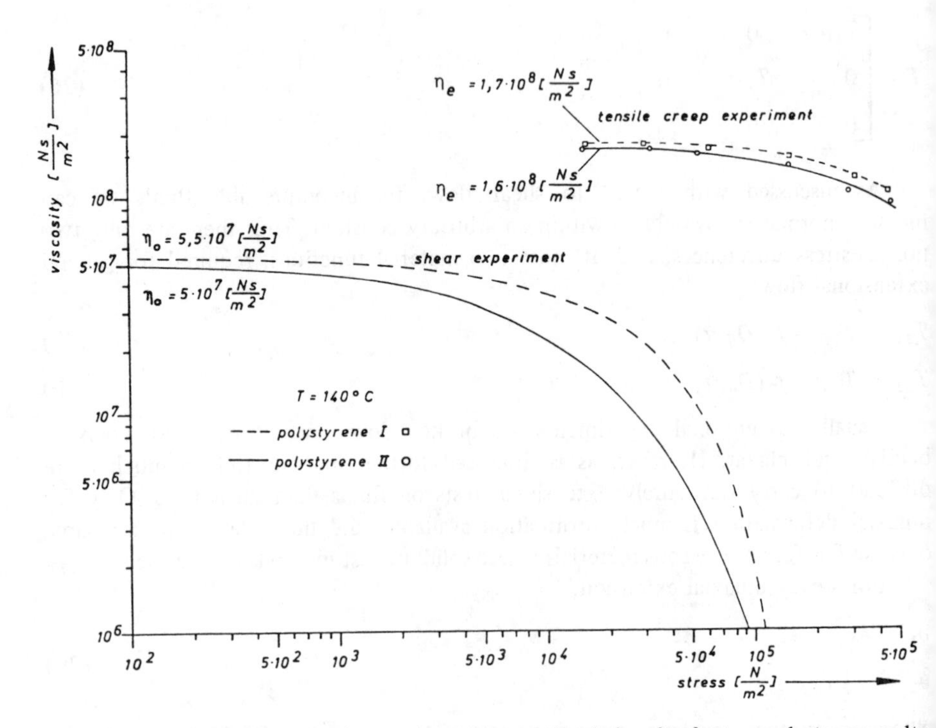

Figure 9 Comparison of steady shear and steady extensional viscosity for two polystyrene melts. *(From [9].)*

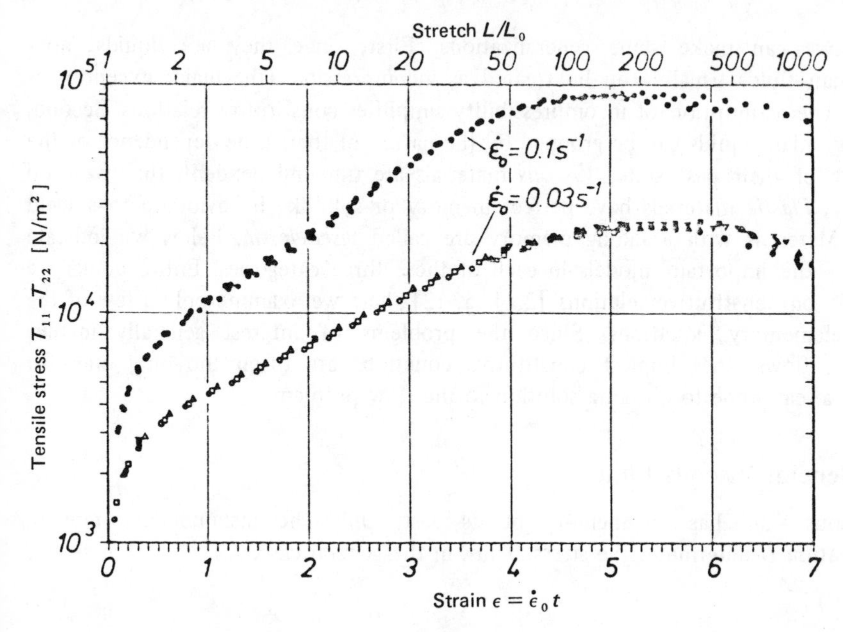

Figure 10 Tensile stress versus strain at constant rate of extension for a low-density polyethylene melt. *(From [10].)*

polyethylene melt. We see that even at total strain of over 7 (extension ratios of 10^3) the steady state has not been attained.

In light of the data in Fig. 10 and the fact that in most real flows the extensional strain seldom exceeds 2, it may be more appropriate to focus on transient characterization in extension. The deformation histories defined for shear in the previous section can be applied to extension. For small strains the same interrelationships hold as given for shear, and the time-dependent extensional modulus is just $3G(t)$. Thus small strain extension can be predicted from shear. For example, $E' = 3G'$ and $E'' = 3G''$, and extensional stress growth can be predicted from shear data for strains up to 2. At larger strains these linear viscoelastic results no longer hold, and actual extensional measurements are needed to predict behavior [11]. Some of the nonlinear constitutive equations given in the next section are fairly successful in describing this nonlinear behavior.

3. CONSTITUTIVE RELATIONS

The goal of fluid mechanics is to understand particular flow problems. To analyze a problem and to compare results against experimental measurements, the fluid mechanist uses the balance equations, boundary conditions, and a constitutive equation. The rheologist's goal is to determine that constitutive relation.

What type of constitutive equations can we expect? Although this area continues to be the subject of much research and there is much that we do not know, we can make some generalizations. First, since they are liquids, nonnewtonian fluids usually can be treated as incompressible (the major exception is foam). The assumption of incompressibility simplifies constitutive relations. Second, nonnewtonian liquids can be grouped by the nature of their time dependence or the memory of their past state. *Viscous* materials are time-independent; they have no memory. *Plastic* materials have perfect memory or solidlike behavior up to a yield stress. Materials with a fading memory are called *viscoelastic*. Below we indicate some of the important models in each of these three categories. Entire books are available on constitutive relations [2, 4, 5, 12]; here we examine only a few of the more elementary equations. Since the problems of interest generally involve complex flows, the simplest constitutive equations are often the only ones for which we can hope to obtain a solution to the flow problem.

3.1 General Viscous Fluid

A viscous fluid has no memory of its past. Only the instantaneous rate of deformation $\boldsymbol{D}$ determines the stress. Thus, in the general case,

$$\boldsymbol{\tau} = f(\boldsymbol{D}) \tag{32}$$

Most fluids are *isotropic* (liquid crystals are an exception). This means [2, 13, 14] that the function f can be expressed with three scalar functions α_i of the invariants of $\boldsymbol{D}$.

$$\boldsymbol{\tau} = \alpha_0 \boldsymbol{I} + \alpha_1 \boldsymbol{D} + \alpha_2 \boldsymbol{D}^2 \tag{33}$$

The principal invariants of a tensor are defined [14] as

$$\mathrm{I}_D = \mathrm{tr}\, \boldsymbol{D} \qquad \mathrm{II}_D = \tfrac{1}{2}(\mathrm{I}_D^2 - \mathrm{tr}\, \boldsymbol{D}^2) \qquad \mathrm{III}_D = \det \boldsymbol{D} \tag{34}$$

For an incompressible fluid, α_0 can be lumped in with the arbitrary pressure p:

$$\boldsymbol{T} = -p\boldsymbol{I} + \alpha_1 \boldsymbol{D} + \alpha_2 \boldsymbol{D}^2$$

The newtonian fluid is just a special case of the general viscous fluid, with $\alpha_1 = 2\mu$ and $\alpha_2 = 0$. In general $\alpha_1 = \alpha_1(\mathrm{I}_D, \mathrm{II}_D, \mathrm{III}_D)$. It is this feature that is most useful about the general viscous model. Often the most important rheological effect is the shear-rate dependence of viscosity. In polymeric liquids and concentrated suspensions, the viscosity can change by 10^3 over the accessible shear-rate range. In polymer processing operations, particularly flow through channels and dies, simple equations that accurately describe the shear-rate-dependent viscosity are essential for process modeling.

Normally α_2 is neglected. In steady shear flow it predicts only a second normal-stress coefficient and zero for the first coefficient [Eq. (59) with $\beta_2 = 0$], which is essentially opposite to all experimental results. There are several useful forms for α_1. As indicated above, α_1 should be a function of all the invariants; however, for an incompressible fluid $\mathrm{I}_D = 0$, and for simple shear $\mathrm{III}_D = 0$. Thus, α_1 is usually assumed to be a function $\eta(\mathrm{II}_D)$ of II_D only, but in general it can be a function of II_D and III_D. Therefore, we are interested in constitutive equations of the form

$$T = -pI + \eta(\mathrm{II}_D)2D \qquad \text{or} \qquad \tau = \eta(\mathrm{II}_D)2D \tag{35}$$

Below several common expressions for $\eta(\mathrm{II}_D)$ are given.

3.1.1 Power-law model. The most widely used form of the general viscous constitutive relation is the power-law model

$$\tau_{ij} = m(4\mathrm{II}_D)^{(n-1)/2}(2D_{ij}) \tag{36}$$

This equation is most often applied to steady, simple, shear flows, where the second invariant becomes

$$4\mathrm{II}_D = \dot{\gamma}^2$$

Thus, for steady shear, the power law becomes

$$\tau_{12} = \tau_{21} = m\dot{\gamma}^n \qquad \text{or} \qquad \eta = m\dot{\gamma}^{n-1} \tag{37}$$

with no other stress components. Equation (37) is often how the power law appears in the literature, but it is important to remember that this is only for simple shear. For other flows, for example, radial flow between plates [15], the full three-dimensional equations (36) must be used.

In the processing range of many polymeric liquids and dispersions, the power law is a good approximation to viscosity shear-rate data. Figure 11 shows viscosity versus shear rate for an acrylonitrile-butadiene-styrene (ABS) polymer melt. At high shear rate, $\dot{\gamma} > 1$, the power law fits the data well, with m a function of temperature. The power law has been used extensively in polymer process models [2, 3, 17].

One of the obvious disadvantages of the power law is its failure to describe the low-shear-rate region. Since n is usually less than 1, at low shear rate η goes to infinity rather than to a constant η_0, which is usually observed experimentally (Fig. 11).

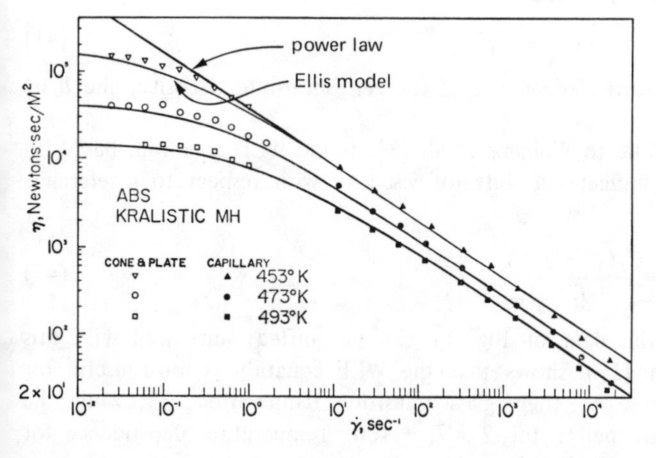

Figure 11 Viscosity versus shear rate for an ABS polymer melt. *(From [16].)*

3.1.2 Ellis model. Three parameter models, like the Ellis model below, have been proposed to provide a newtonian region at low shear rate and power-law dependence at high rates:

$$\frac{\eta}{\eta_0} = \frac{1}{1 + m(4\mathrm{II}_D)^{(1-n)/2}} \tag{38}$$

Though it is somewhat more complex than the power law, the Ellis model has been shown to fit a much wider range of viscosity versus shear rate data. This can be seen in Fig. 11. The Ellis model has been used in a number of complex flow problems [1, 2].

3.1.3 Other models. A number of other empirical expressions for $\eta(\mathrm{II}_D)$ are available. Some nonnewtonian liquids show a limiting viscosity η_∞ at high shear rates. This can be treated with the modified Ellis model of Carreau [2]:

$$\frac{\eta - \eta_\infty}{\eta_0 - \eta_\infty} = \frac{1}{(1 + 4\lambda^2 \mathrm{II}_D)^{(1-n)/2}} \tag{39}$$

where λ has units of time and can be considered a time constant of the fluid. Figure 12 shows viscosity data for a soap and three polymer solutions compared with the Carreau model.

To fit experimental data more accurately, a power series is frequently used [18]:

$$\eta = a_0 + a_1(\mathrm{II}_D) + a_2(\mathrm{II}_D)^2 + \cdots \tag{40}$$

This equation seems to have its greatest use in numerical modeling. Bird et al. [2] give a number of other empirical models for $\eta(\mathrm{II}_D)$.

The temperature and pressure dependence of viscosity can also be critical to understanding some processing problems. Van Krevelan [19] reviews both these areas, and Goldblatt and Porter [20] have reviewed the problem of pressure dependence. Exponential relations have been found useful to correlate both temperature and pressure dependence:

$$\eta_0 = K e^{E_\eta / RT} e^{bp} \tag{41}$$

where E_η is the temperature dependence of the zero-shear-rate viscosity, and b its pressure dependence.

Another approach, due to Williams et al. [8], is the WLF equation, based on free-volume changes. It indicates a shift for viscosity with respect to a reference state

$$\log a_T = \log \frac{\eta(T)}{\eta(T_{\mathrm{ref}})} = \frac{C_1(T - T_{\mathrm{ref}})}{C_2 + T - T_{\mathrm{ref}}} \tag{42}$$

Figure 13 shows that the data of Fig. 11 can be shifted quite well with this approach. Van Krevelan [19] shows that the WLF equation is most useful for amorphous polymers close to their glass transition temperature, T_g, while the exponential form appears better for $T \geqslant T_g + 100$. Temperature dependence for suspensions has not been studied extensively.

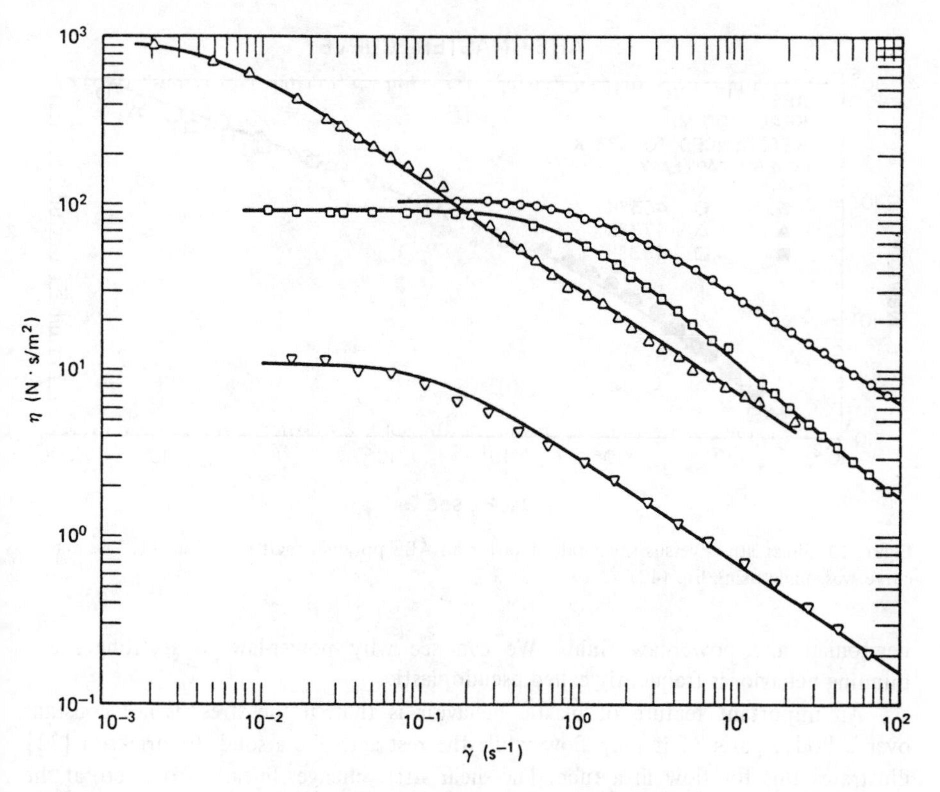

Figure 12 Fit of Carreau model to viscosity versus shear rate data: (△) 2.0% polyisobutylene in Primol; (□) 7% aluminum laurate in a mixture of decalin and *m*-cresol; (▽) 0.75% polyacrylamide in 95/5 mixture of water and glycerin; (○) 5% polystyrene in Aroclor. *(From [2].)*

3.2 Plastic Behavior

A plastic material is one that shows little or no deformation up to a certain level of stress. Above this *yield stress*, the material flows like a liquid. Plasticity is common to widely different materials. Many metals yield at strains of less than 1%. Concentrated suspensions of solid particles in newtonian liquids also show a yield stress followed by nearly newtonian flow. Such materials are often called *Bingham plastics* after E. C. Bingham, who first described paint in this way in 1919. House paint and food substances like margarine, mayonnaise, and ketchup are good examples of Bingham plastics.

A simple model for a plastic material is hookean behavior at stresses below yield, and newtonian behavior above. For one-dimensional deformations,

$$\begin{aligned} \tau &= G\gamma \qquad \text{for } \tau \leqslant \tau_y \\ &\mu\dot{\gamma} + \tau_y \qquad \text{for } \tau > \tau_y \end{aligned} \tag{43}$$

Figure 14 illustrates this behavior and compares the Bingham plastic flow to

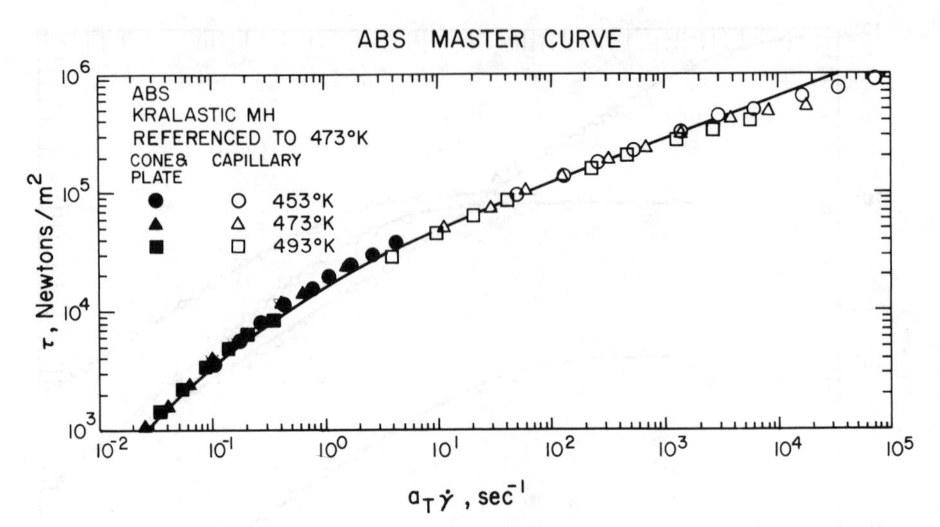

Figure 13 Shear stress versus shear rate data for an ABS polymer melt from Fig. 11. The master curve was made using Eq. (42). *(From [21].)*

newtonian and power-law fluids. We can see why power-law or a strong shear-thinning behavior is frequently called pseudoplastic.

An important feature of plastic behavior is that, if the stress is not constant over a body, parts of it may flow while the rest acts like a solid. Fredrickson [22] illustrates this for flow in a tube. The shear stress changes linearly from zero at the center of the tube to a maximum at the wall. Thus the central portion of the material flows like a solid plug. Neck formation during uniaxial extension of a solid at constant strain rate is another example of this type of behavior. At the smallest sample cross section or at an inhomogeneity, the stress during the test just exceeds the yield stress, and large deformation can occur.

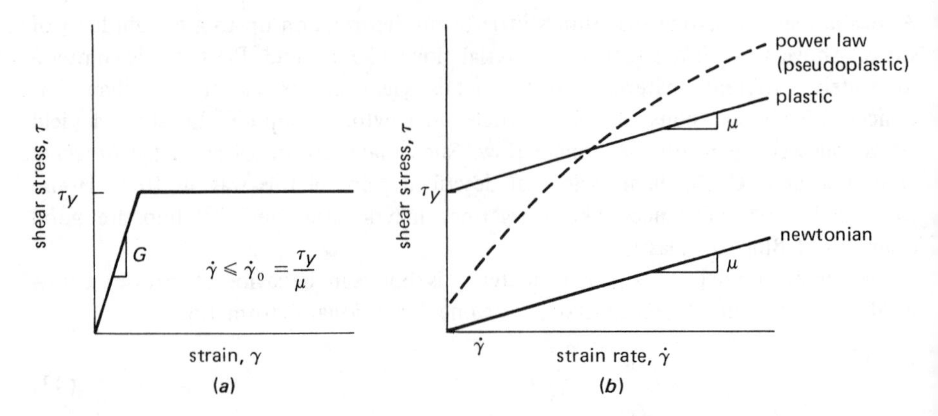

Figure 14 (*a*) Shear stress versus strain and (*b*) shear stress versus strain rate for a Bingham plastic material.

Example: The Ketchup Bottle

We have all been frustrated by that malevolent Bingham plastic, ketchup. To exceed its yield stress in the neck of the bottle, one must frequently tap the bottle, and then, when the shear stress at the wall exceeds τ_y, flow is rapid. Figure 15 shows shear stress versus shear rate data for ketchup and several other food products. $\tau_y \simeq 200$ dyn/cm² for this ketchup sample. (Note that, in contrast to Fig. 14, this is a log-log plot.) Will ketchup empty under gravity from a typical bottle?

In the neck of the bottle the wall shear stress τ_W will be balanced by the pressure head of ketchup in the bottle. If we can approximate the bottle as a tube of length L and diameter D and a cylindrical reservoir of height H, then [note Eq. (87)]

$$\tau_W(\pi DL) = p\,\frac{\pi D^2}{4} \qquad \text{or} \qquad \tau_W = \frac{\rho g H D}{4L} \tag{44}$$

The density is $\rho \simeq 1$ g/cm³, the gravitational acceleration is $g = 980$ cm/s², and a bottle with a "standard" neck has $D \simeq 1.5$ cm and $L \simeq 6$ cm. If the bottle is partially full, $H \simeq 4$ cm. Then $\tau_W \simeq 200$ dyn/cm², and the ketchup should not flow without some thumping. Note that the situation is probably worse, since we have assumed atmospheric pressure above the ketchup in the bottle. It is typically less, owing to the partial vacuum created as the bottle is inverted. For a "wide mouth" bottle, $D \simeq 3$ cm and $\tau_W \simeq 400$, which may make mealtimes flow more smoothly.

To handle deformations occurring in more than one direction, Eq. (43) should be put into three-dimensional form. The only significant change is to replace the one-dimensional τ in the yield criterion with some scalar function of the invariants of $\boldsymbol{\tau}$. There are a number of yield criteria in the literature [14]. The von Mises criterion, which uses the second invariant of $\boldsymbol{\tau}$ is the most common:

$$\boldsymbol{\tau} = G\boldsymbol{B} \qquad \text{for } \mathrm{II}_\tau \leqslant \tau_y^2$$

where $\boldsymbol{B}$ is the Finger large-strain-deformation tensor, and

$$\boldsymbol{\tau} = 2\left(\mu + \frac{\tau_y}{2\mathrm{II}_D^{1/2}}\right)\boldsymbol{D} \qquad \text{for } \mathrm{II}_\tau > \tau_y^2 \tag{45}$$

Note that postyield behavior other than newtonian flow can readily be substituted. The most common is a power-law viscosity.

Other constitutive equations for plastic materials are described by Fredrickson [22] and Argon [12]. However, these relations are less well developed than the viscoelastic constitutive equations described in the next sections. This state of affairs is partially due to the difficulty in obtaining accurate data, independent of rheometer geometry, on plastic materials.

3.3 Linear Viscoelasticity

Most polymeric liquids show the phemomenon called "fading memory." At short times they behave like a rubbery solid, and at long times like a viscous liquid. This can perhaps be best seen in the simple stress-relaxation experiment that was discussed in Sec. 2.2. Recall that when a polymeric liquid is subjected to a sudden, small step strain, the stress rises quickly and then relaxes with time. This is illustrated in Fig. 16.

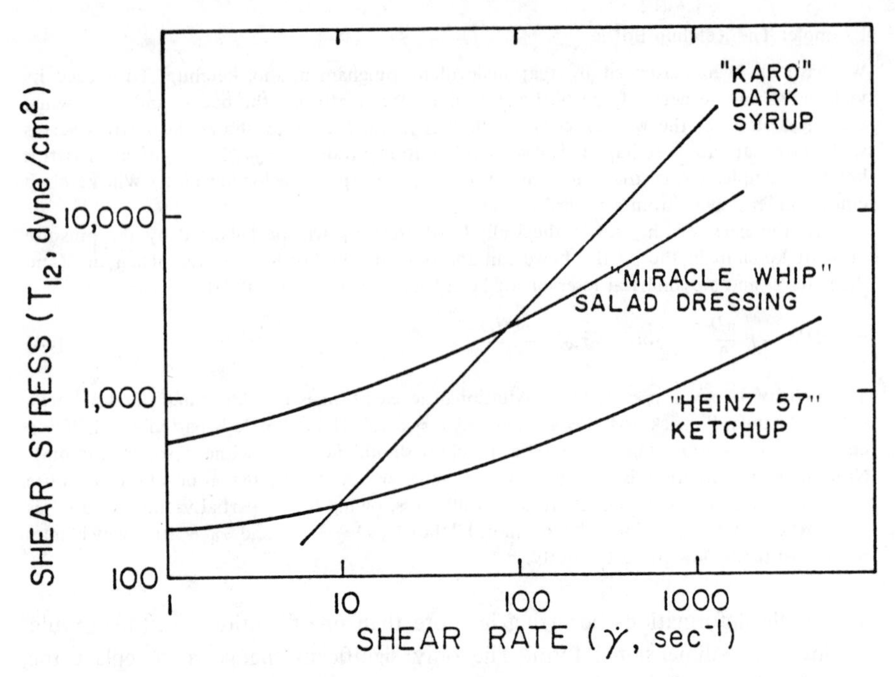

Figure 15 Flow data for several food products. *(Courtesy Graco Co., Minneapolis, Minn.)*

Maxwell, over a century ago, suggested that this type of behavior could be modeled by a linear combination of the ideal elastic or hookean solid and the newtonian liquid:

$$\tau + \frac{\eta_0}{G_0}\frac{\partial \tau}{\partial t} = \eta_0 \dot{\gamma} \tag{46}$$

The Maxwell model is often represented as a series combination of springs (elastic elements) and dashpots (viscous) elements as in Fig. 17. From Eq. (46) and the spring-and-dashpot representation we see that for slow motions the dashpot or

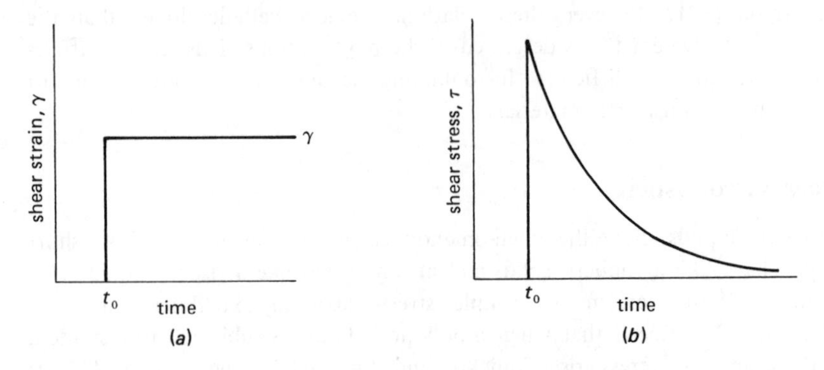

Figure 16 Sudden application of shear strain to a sample, and resultant stress response.

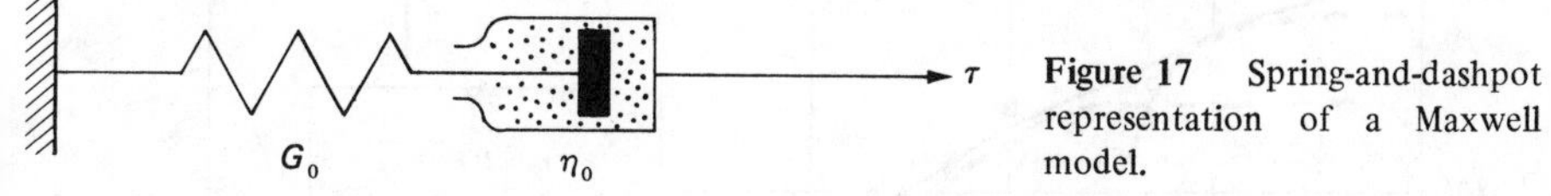

Figure 17 Spring-and-dashpot representation of a Maxwell model.

newtonian behavior dominates. For rapidly changing stresses the derivative term dominates, and thus at short times the model approaches elastic behavior. The Maxwell model is a simple linear combination of viscous and elastic effects and is called the *linear viscoelastic model.*

The term η_0/G_0 in Eq. (46) has units of time and is the characteristic relaxation time λ_0 of the system. This can be seen more clearly if we put Eq. (46) into an integral form:

$$\tau = \int_{-\infty}^{t} [G_0 e^{-(t-t')/\lambda_0}]\, \dot{\gamma}(t')\, dt' \tag{47}$$

The function inside the brackets is called a *relaxation modulus* $G(t)$ [note Eq. (15)]. It is multiplied by the strain rate, which can be a function of time $\dot{\gamma}(t)$.

An equivalent integral representation is in terms of a memory function $M(t)$ times the strain γ:

$$\tau = -\int_{-\infty}^{t} \frac{G_0}{\lambda_0}\, e^{-(t-t')/\lambda_0}\, \gamma(t')\, dt' = \int_{-\infty}^{t} M(t-t')\gamma(t')\, dt' + G_0\gamma(t) \tag{48}$$

Note that the memory function is just the derivative of the modulus:

$$\frac{-dG(t)}{dt} = M(t) \tag{49}$$

To better fit data, several relaxation times are generally used:

$$G(t) = \sum_{k=1}^{N} G_k e^{-t/\lambda_k} \tag{50}$$

Usually five to ten relaxation times are adequate to fit the typical experimental data range. This is illustrated in Fig. 18, in which Laun [23] uses eight relaxation times to fit the shear relaxation modulus for a low-density polyethylene. We clearly see the contribution of each relaxation time. The values of λ_k and G_k are given in Table 1.

Polymer molecular theories like the Rouse model [2, 8] suggest an infinite-series form for the relaxation spectra:

$$G_k = G_0 \frac{\lambda_k}{\Sigma_k \lambda_k} \qquad \lambda_k = \frac{\lambda_0}{k^2} \tag{51}$$

Clearly one could also construct a continuous relaxation-modulus function.

These integral models can be made three-dimensional by simply substituting the

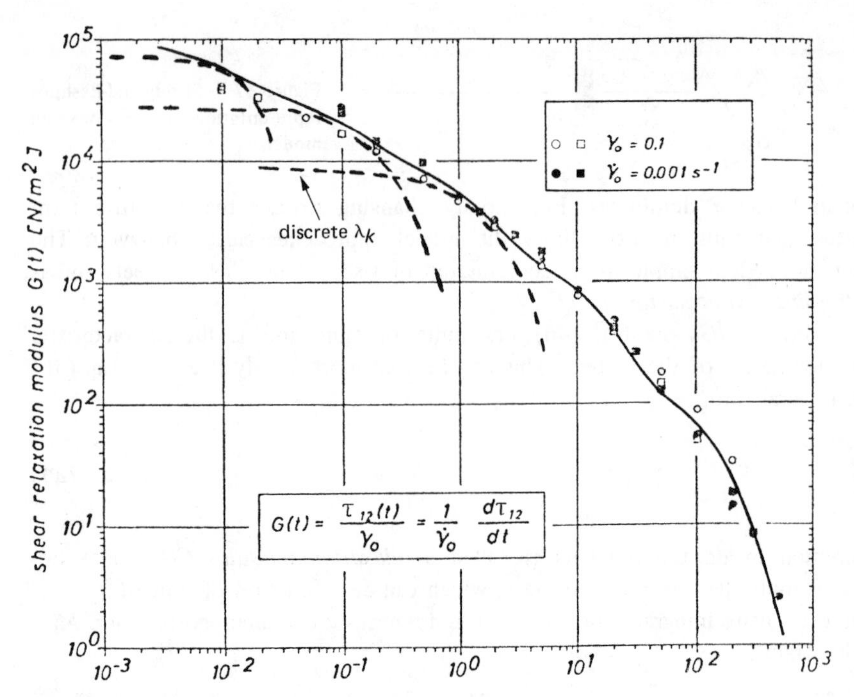

Figure 18 Shear relaxation modulus versus time for a low-density polyethylene at 150°C. The heavy line is the sum of eight exponential relaxation times given in Table 1. *(From [23].)*

stress tensor τ for the shear stress, and the rate-of-deformation tensor D for the shear-rate. Thus, in general, we can write

$$\tau = \int_{-\infty}^{t} G(t - t')2D(t')\, dt' \tag{52}$$

or the small-strain tensor E for the shear strain γ

$$\tau = -\int_{-\infty}^{t} M(t - t')E(t')\, dt' \tag{53}$$

These equations should be applicable to any type of flow, shear or extension, at small strains or strain rates as long as the material has a fading memory.

The general linear viscoelastic model (generalized Maxwell model) has been quite successful in fitting small-strain data for polymeric liquids. Figures 18 and 19 illustrate this. The distribution of relaxation times given in Table 1 was used to calculate all three curves. For sinusoidal oscillations, Eq. (52) gives

$$G'(\omega) = \Sigma_k G_k \frac{\omega^2 \lambda_k^2}{1 + \omega^2 \lambda_k^2} \tag{54}$$

$$G''(\omega) = \Sigma_k G_k \frac{\omega \lambda_k}{1 + \omega^2 \lambda_k^2} \tag{55}$$

Table 1 Relaxation times for low-density polyethylene at 150°C

k	λ_k (s)	G_k, Pa
1	10^3	1.00
2	10^2	1.80×10^2
3	10^1	1.89×10^3
4	10^0	9.8×10^3
5	10^{-1}	2.67×10^4
6	10^{-2}	5.86×10^4
7	10^{-3}	9.48×10^4
8	10^{-4}	1.29×10^5

Source: From [23].

For small strains or strain rates, all the various transient shear tests can be interrelated using linear viscoelasticity. The interrelations are summarized in terms of J_e^0 and λ_0 in Table 2.

3.4 Nonlinear Viscoelasticity

A major drawback of the Maxwell model is that although it is reasonable for small strain transients, it predicts a newtonian viscosity in shear and does not predict

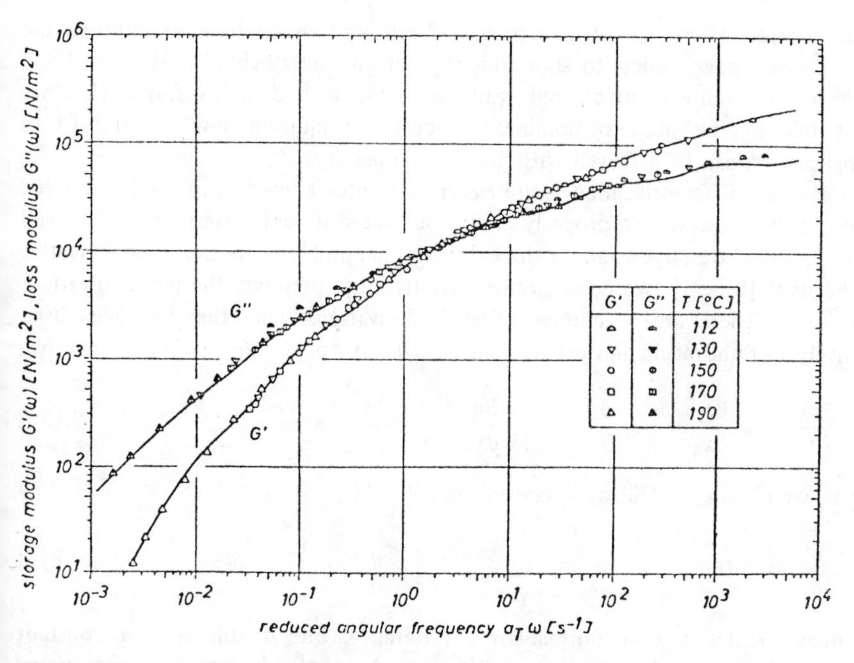

Figure 19 Dynamic shear moduli for the same low-density polyethylene as in Fig. 18. The lines were calculated from the same relaxation spectra. *(From [23].)*

Table 2 Limiting relations for linear viscoelasticity

Transient shear test		Equilibrium creep compliance J_e^0		Longest relaxation time λ_0
Steady shear	$\lim_{\dot\gamma \to 0}$	$\dfrac{\psi_1}{2\eta^2}$	$\lim_{\dot\gamma \to 0}$	$\dfrac{\psi_1}{2\eta}$
Sinusoidal oscillations	$\lim_{\omega \to 0}$	$\dfrac{G'}{(G'')^2}$	$\lim_{\omega \to 0}$	$\dfrac{G'}{G''\omega}$
Creep	$\lim_{\tau_0 \to 0}$	$\dfrac{\gamma_0}{\tau_0}$	$\lim_{\tau_0 \to 0}$	$\dfrac{\gamma_0}{\dot\gamma_\infty}$
Constrained recoil	$\lim_{\tau_0 \to 0}$	$\dfrac{\gamma_r}{\tau_0}$	$\lim_{\tau_0 \to 0}$	$\dfrac{\gamma_r}{\dot\gamma_\infty}$
Stress relaxation	$\lim_{\dot\gamma_0 \to 0}$	$\int_0^\infty \dfrac{\eta^-}{\eta_0^2}\,dt$	$\lim_{\dot\gamma_0 \to 0}$	$\int_0^\infty \dfrac{\eta^-}{\eta_0}\,dt$
Stress growth	$\lim_{\dot\gamma \to 0}$	$\int_0^\infty \dfrac{\eta_0 - \eta^+}{\eta_0^2}\,dt$	$\lim_{\dot\gamma_0 \to 0}$	$\int_0^\infty \left(1 - \dfrac{\eta^-}{\eta_0}\right) dt$

normal stresses. Hundreds of papers have been written on how to improve the model or write new models to show this type of nonlinear behavior. Many of these models become quite complex and really unusable with complex flows. Here we present two of the simplest nonlinear models and indicate how a better fit to rheological data can be achieved with greater complexity.

In general, differential models are easier than integral models to use in complex flow problems. However, to properly apply even the differential form of the Maxwell model, Eq. (46), to a large strain or three-dimensional problem, another time derivative must be used [2 to 5]. We need a time derivative that preserves the tensor invariant nature of τ. There are a number of such derivations. One that has been used extensively in fluid mechanics problems is the *Oldroyd* or *codeformational* derivative,

$$\frac{\delta \tau_{ij}}{\delta t} = \frac{\partial \tau_{ij}}{\partial t} + v_k \frac{\partial \tau_{ij}}{\partial x_k} - \tau_{kj} \frac{\partial v_i}{\partial x_k} - \tau_{ik} \frac{\partial v_j}{\partial x_k} \tag{56}$$

Using it, we obtain the Oldroyd-Maxwell model:

$$\tau + \lambda_0 \frac{\delta \tau}{\delta t} = 2\mu D \tag{57}$$

This does predict a first normal-stress difference, but it still gives a constant viscosity and no second normal-stress difference. Using the *Jaumann* or *corotational* derivative [2, 4] gives shear thinning that is much too strong to be realistic. To achieve more realistic shear thinning, one can replace μ in Eq. (57) with a power-law viscosity [4], Eq. (37).

Many other differential constitutive models have been developed. One approach is to formally expand the stress in derivatives of the rate of strain. The result is similar to that for the general viscous fluid except that, owing to the time dependence of the fluid, the series is infinite. To solve actual problems, the series is truncated, usually at second order [2, 4]:

$$T = -pI + \beta_1(2D) + \beta_2 \frac{\delta(2D)}{\delta t} + \beta_3(2D)^2 \tag{58}$$

where the β_i are material constants. Clearly, β_1 gives the shear viscosity, and β_2 and β_3 are elasticity parameters. For steady shear, the limiting values of the material functions are simply related to the β_i:

$$\eta_0 = \beta_1 \qquad \psi_{10} = -2\beta_2 \qquad \psi_{20} = 2\beta_2 + \beta_3 \tag{59}$$

The steady extensional viscosity is

$$\eta_e = 3\beta_1 + 3(\beta_2 + \beta_3)\dot{\epsilon} \tag{60}$$

The *second-order fluid* model, as this model is called, is strictly valid for small departures from newtonian behavior, flows that have both low shear rates and slow transients. Because it is simple and valid in this limit, it has frequently been used to get a qualitative idea of how elasticity affects a steady flow, such as the direction of a secondary flow [1] and the magnitude of the rod climbing effect [24].

The other approach to improving upon the Maxwell model is to work with the integral equations. Although these are more difficult to apply to flow problems, they have been more successful in fitting rheological data.

If the infinitesimal strain tensor in Eq. (53) is replaced with the finite strain tensor B (the Finger tensor), which arises in the theory of rubber elasticity [25, 26], time-dependent response can be modeled to larger strains [27]:

$$\tau = \int_{-\infty}^{t} M(t - t')B(t')\,dt \tag{61}$$

However, this *rubberlike liquid* model still does not have a shear-thinning viscosity. This requires some strain dependence in the memory function. Many forms have been suggested for this strain dependence. One of the simplest is the factorized memory function proposed by Wagner [11]. He noticed that for many polymeric materials, as illustrated in Fig. 5, the shape of the relaxation modulus (or memory function) does not change significantly with strain. This suggests that for large strains it can be factored into the time-dependent small-strain modulus and a strain-dependent damping function h. When this idea is combined with the rubberlike liquid, we obtain

$$\tau = \int_{-\infty}^{t} M(t - t')h(\mathrm{I}_B, \mathrm{II}_B)B(t)\,dt \tag{62}$$

where I_B and II_B are the first and second invariants of the strain tensor $\boldsymbol{B}$.

This model has been found to fit a wide variety of shear and extensional data for polymer melts. Laun [23] used the strain dependence of the stress relaxation modulus for low-density polyethylene (shown in Fig. 20) to determine the damping function shown in Fig. 21. With these data and the linear viscoelastic relaxation times on the same material given in Table 1, he was able to use Eq. (62) to predict the start-up and steady-state viscosity and first normal-stress coefficient data given in Figs. 22 and 23. For the start-up viscosity, Eq. (62) becomes

$$\eta^+(t, \dot{\gamma}_0) = \sum_k \frac{G_k \lambda_k}{(1 + n\dot{\gamma}_0\lambda_k)^2} \left[1 - e^{-t_{r,k}}\left(1 - n\dot{\gamma}_0\lambda_k t_{r,k}\right)\right] \tag{63}$$

where $t_{r,k} = t/\lambda_k + n\dot{\gamma}_0 t$
$n = 0.18$

This success is encouraging. The factorized memory-function model is of the same form as that of Doi and Edwards [28], developed recently from molecular theory.

A problem with Wagner's model is that it is not very accurate in small strains, in the linear viscoelastic regime, or in very large strains. An improvement can be achieved by substituting a sigmoidal damping function for the exponential one [28a]. Extensive

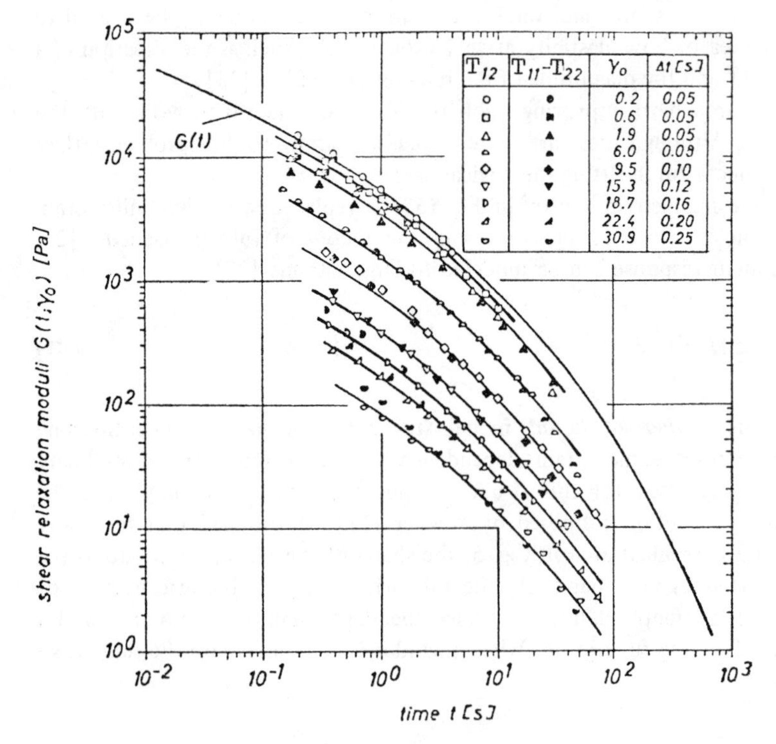

Figure 20 Time and strain dependence of the stress relaxation modulus for a low-density polyethylene melt at 150°C. *(From [23].)*

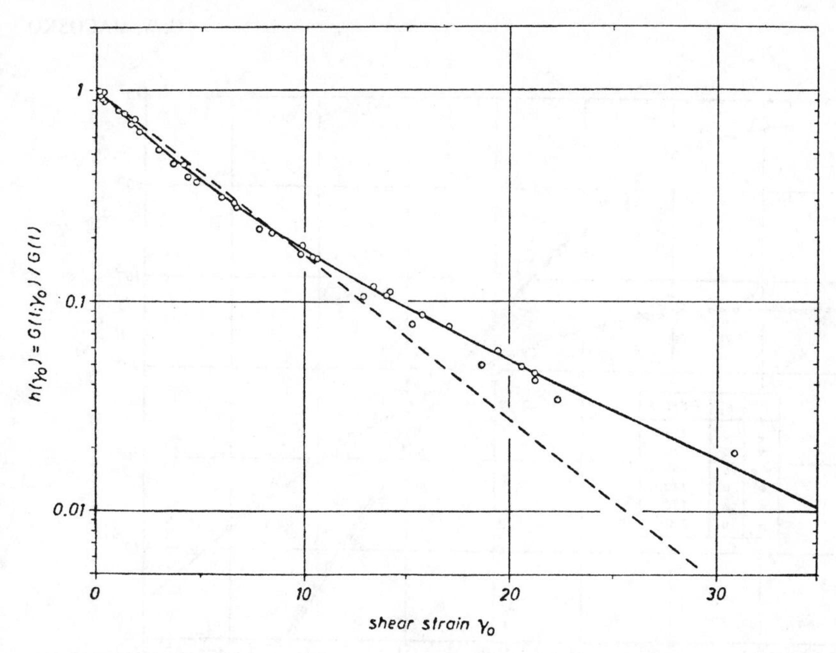

Figure 21 Damping function versus strain calculated from the data of Fig. 20. Broken line: $h = e^{-0.18\gamma_0}$. Solid line: $h = 0.57\ e^{-0.310\gamma_0} + 0.43\ e^{-0.106\gamma_0}$. *(From [23].)*

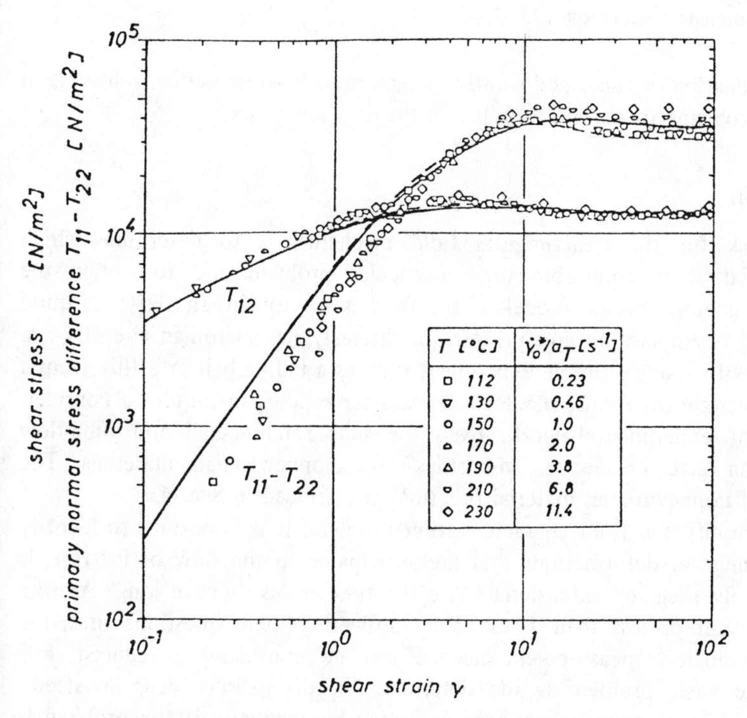

Figure 22 Start-up of shear stress and first normal-stress difference versus strain for a low-density polyethylene melt. *(From [23].)*

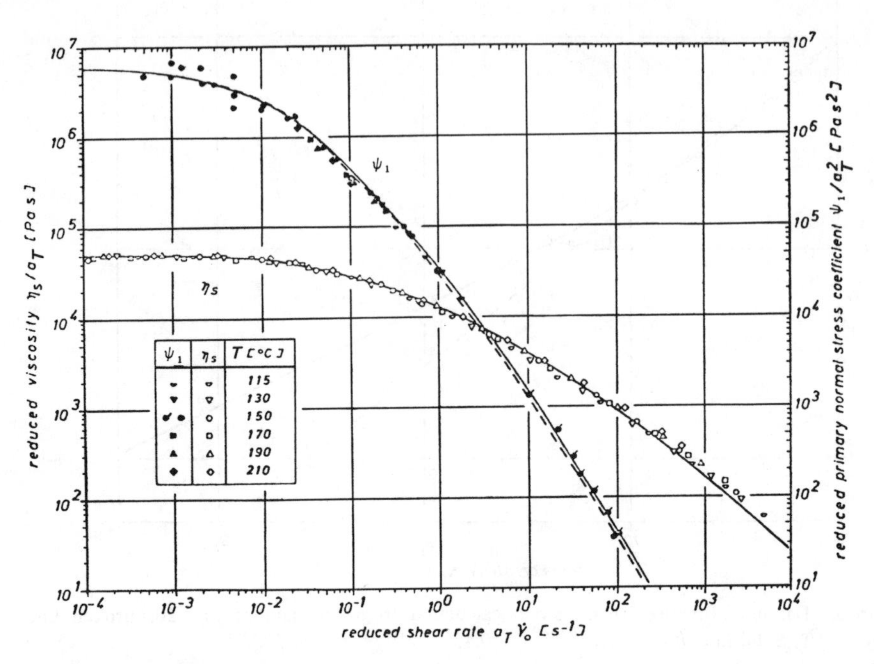

Figure 23 Steady-state shear viscosity and first normal-stress coefficient versus shear rate for a low-density polyethylene melt. *(From [23].)*

testing and application of these and similar integral models to numerical solutions of complex flow problems are certain to follow in the next few years.

3.5 Discussion

A difficult task for the experimental fluid mechanist is to determine which constitutive relation is applicable to a particular problem and to obtain the appropriate parameters for the model. If the fluid is newtonian, the latter is quite simple. Given the temperature and pressure of interest, the newtonian viscosity can be determined with a single simple experiment such as a falling ball or efflux from a tube. For nonnewtonian fluids, rheological characterization can require a considerable amount of experimental work. Even the simple falling-ball and tube-flow experiments can give erroneous "viscosities" for nonnewtonian materials. The measurement of nonnewtonian material functions is discussed in Sec. 4.

To help simplify the fluid characterization problem, it is important to identify the type and range of deformations that are anticipated in the flow of interest. Is the flow primarily shear or extensional? Are the time scales short or long? Are the total strains greater or less than 1 to 2? By answering these questions first, the amount of rheological measurement needed can be considerably reduced. For example, if the basic problem is the determination of pressure drop in steady conduit flows, viscosity versus shear rate data may be adequate. If the problem is surface instability in a coating flow, the time dependence of the stresses to changing deformations may be more important than fitting the shear-rate dependence.

A useful dimensionless group in this regard is the *Deborah number* [3], which is the ratio of the fluid relaxation time λ_0 to some appropriate flow time t_r:

$$D_e = \frac{\lambda_0}{t_r} \tag{64}$$

If $t_r < \lambda_0$, transient, often solidlike response may dominate the flow behavior. Steady-state data may have little relevance, and a purely viscous model will not be likely to explain the phenomenon. If $t_r > \lambda_0$, a test to determine whether normal stresses may be relevant is the *Weissenberg number*, which is the product of a relaxation time and a characteristic deformation rate:

$$W_s = \lambda_0 \dot{\gamma}_c$$

$$= \frac{\psi_1}{2\eta} \tag{65}$$

where $\dot{\gamma}_c$ is evaluated at a characteristic shear rate of the process. When W_s is large, normal stresses can be a significant factor in the flow behavior.

4. RHEOMETRY

In Sec. 2 we discussed the stresses that need to be measured to completely characterize a fluid in shear or extensional flow. The real job of the experimental rheologist is to actually achieve the kinematics of these simple flows and then measure the stresses. In some sense this task is the reverse of that of the experimental fluid mechanicist. In principle, the latter studies a known fluid in a complex flow. The rheologist begins with a very simple flow, one in which the kinematics are completely determined regardless of the type of fluid, and then uses it to characterize a complex fluid. Clearly in any study of nonnewtonian fluids these roles must merge. The rheologist must use fluid mechanics methods to verify that a new rheometer does indeed produce the kinematics assumed. The fluid mechanicist must get involved with rheometry to do serious work with non-newtonian flows.

A rheometer, then, is a flow device in which the stresses can be measured and the kinematics are known or can be determined from a few simple measurements, regardless of the fluid's constitutive equation. Below we briefly survey the most common and useful rheometers. We give but not derive the working equations for data analysis, and point out the advantages and disadvantages of each. However, there are several potential systematic errors in making measurements with these flow devices. For these and further information on rheometers, the reader is referred to the several texts on rheometry [29 to 32] and to the specific references given below.

As with our discussion of the material functions, the treatment of rheometers can be broken into the two basic flows: shear and extension. Shear measurements have constituted most of rheometry; however, one of the most active areas of research today is in developing new extensional devices.

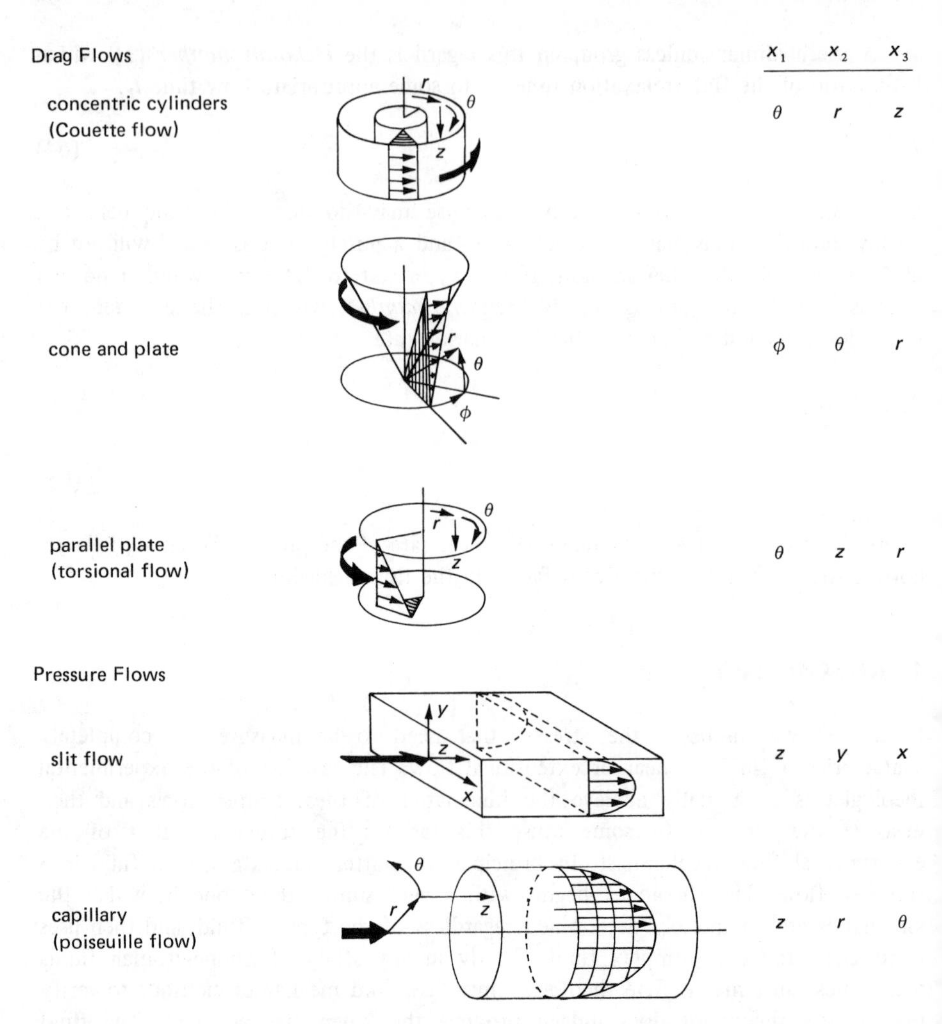

Figure 24 Common shear-flow geometries. In each sketch x_1 is the flow direction, x_2 is the direction of the shear gradient, and x_3 the neutral direction. *(Adapted from [28b].)*

4.1 Shear Rheometers

There are five basic shear geometries in common use as rheometers. These are shown with their coordinate systems in Fig. 24. The working equations for these geometries are given in the following subsections.

4.1.1 Concentric-cylinders rheometer

Working equations Refer to Fig. 25 for definitions of the symbols:

Shear stress:

$$T_{21} = \tau_{r\theta}(R_i) = \frac{M_i}{2\pi R_i^2 L} \tag{66}$$

Shear strain:

$$\gamma = \frac{\theta \bar{R}}{R_0 - R_i}$$

or (67)

$$\gamma = \frac{\Omega t \bar{R}}{R_0 - R_i} \qquad \text{(narrow gap)}$$

where $\theta = \Omega t$ is the angular displacement for steady rotation, and $\bar{R} = (R_0 + R_i)/2$ is the mean radius.

Shear rate:

$$\dot{\gamma}(R_i) = \dot{\gamma}(R_0) = \frac{\Omega_i \bar{R}}{R_0 - R_i} = \frac{2\Omega_i}{1 - \kappa^2} \qquad \text{for } \kappa = \frac{R_i}{R_0} > 0.99 \tag{68}$$

(narrow gap, homogeneous) or

$$\dot{\gamma}(R_i) = \frac{2\Omega_i}{n(1 - \kappa^{2/n})} \qquad n = \frac{d(\ln M_i)}{d(\ln \Omega_i} \qquad \text{for } \kappa < 0.99 \tag{69}$$

(wide gap, nonhomogeneous) or

$$\dot{\gamma}(R_0) = \frac{2\Omega_0}{n(1 - \kappa^{-2/n})} \qquad n = \frac{d(\ln M_0)}{d(\ln \Omega_i)} \qquad \text{for } \kappa < 0.99 \tag{70}$$

Normal stress:

$$T_{11} - T_{22} = T_{\theta\theta} - T_{rr} = \frac{[T_{rr}(R_0) - T_{rr}(R_i)]\bar{R}}{R_0 - R_i} \qquad \text{(narrow gap)} \tag{71}$$

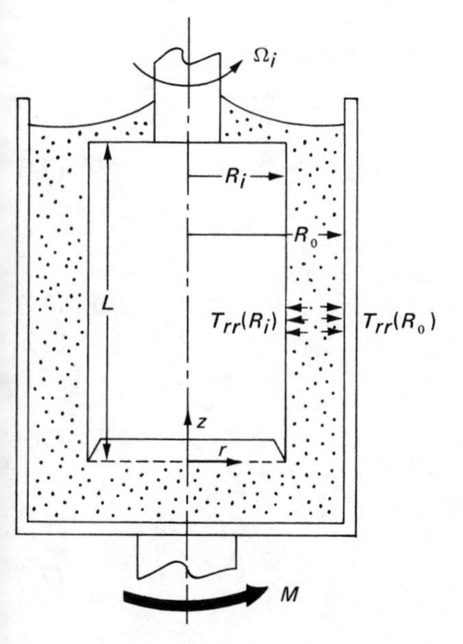

Figure 25 Schematic of concentric-cylinder rheometer.

Corrections The most common corrections are due to end effects and fluid inertia problems. The contribution to torque from the bottom can be greatly reduced by using a large gap or by trapping a gas pocket, as indicated in Fig. 25. Conical bottoms with the cone angle to match the shear rate between the cylinders are also used [31]. Fluid inertia can be a problem in transient studies of low-viscosity fluids [32, 33]. It also leads to secondary flows (Taylor vortices) if the inner cylinder rotates [33, 34].

Criteria for secondary flow:

$$\frac{\rho^2\Omega^2(R_0 - R_i)^3 R_i}{\eta_0} < 1700 \tag{72}$$

Shear heating:

$$\frac{M}{M_0} = 1 - b\,\frac{Br}{12} \tag{73}$$

where

$$\eta(T) = \eta_0(T_0)e^{bT} \qquad Br = \frac{\eta_0 R^2 \Omega^2}{k_T T_0}$$

Utility This is the best geometry for lower-viscosity systems ($\eta_0 < 100$ Pa•s), but it is hard to load and clean out high-viscosity materials. The device is good for high shear rates. The gravity settling of suspensions has less effect than in cone-and-plate rheometers. Normal stresses are hard to measure, owing to the curvature and the need to transmit the signal through a rotating shaft. Rod climbing with a large gap can also be used to measure normal stresses [24, 34].

4.1.2 Cone-and-plate rheometer

Working equations Refer to Fig. 26 for definitions of the symbols:

Shear stress:

$$T_{12} = T_{\phi\theta} = \frac{3M}{2\pi R^3} \tag{74}$$

Shear strain:

$$\gamma = \frac{\phi}{\beta} \qquad \text{(homogeneous)} \tag{75}$$

Shear rate:

$$\dot{\gamma} = \frac{\Omega}{\beta} \tag{76}$$

Normal stress:

$$N_1 = T_{\phi\phi} - T_{\theta\theta} = \frac{2F_z}{\pi R^2} \tag{77}$$

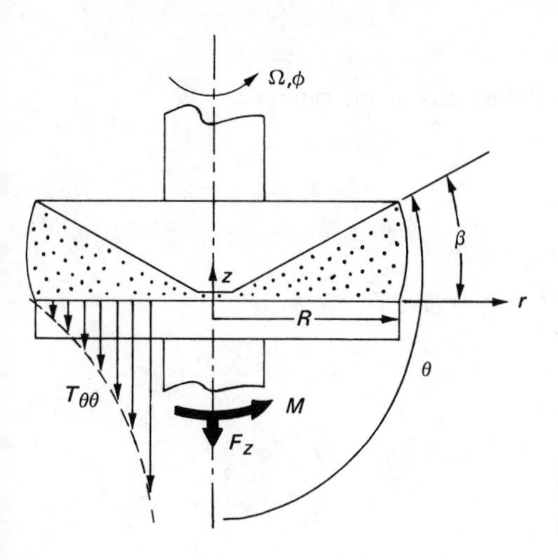

Figure 26 Schematic of cone-and-plate rheometer.

$$N_1 + 2N_2 = -\frac{\partial T_{\theta\theta}}{\partial(\ln r)} \tag{78}$$

Corrections

Inertia and secondary flow:

$$N_1 = \frac{2F_z}{\pi R^2} - 0.15\rho\Omega^2 R^2 \tag{79}$$

$$\frac{M}{M_0} = 1 + 6 \times 10^{-4}\ \mathrm{Re}^2 \tag{80}$$

$$\mathrm{Re} = \frac{\rho\Omega^2\beta^2 R^2}{\eta_0}$$

Edge failure for high-viscosity samples occurs around $\dot{\gamma} = 1\ \mathrm{s}^{-1}$.

Gap opening, oscillations:

$$\frac{6\pi R\eta_0}{K\beta^3} < \text{material relaxation time} \tag{81}$$

Shear heating:

$$\frac{M}{M_0} = 1 - b\frac{Br}{20} \tag{82}$$

where *Br* is given in Eq. (73).

Utility This is the most common instrument for normal-stress measurements. It has the simplest working equations for homogeneous deformation. It requires a stiff, well-aligned instrument and is useful for low- and high-viscosity materials. High-viscosity use is limited by elastic edge failure; low-viscosity use, by inertia corrections, secondary flow, and loss of sample at edges.

4.1.3 Parallel-plates rheometer

Working equations Refer to Fig. 27 for definitions of the symbols:

Shear strain:

$$\gamma = \frac{\theta r}{h} \tag{83}$$

(nonhomogeneous, depends on position)

Shear rate (at perimeter):

$$\dot{\gamma}_R = \frac{\Omega R}{h} \tag{84}$$

Shear stress:

$$T_{12} = T_{\theta z} = \frac{M}{2\pi R^3}\left[3 + \frac{d(\ln M)}{d(\ln \dot{\gamma}_R)}\right] \tag{85}$$

Normal stress:

$$N_1 - N_2 = \frac{F_z}{2\pi R^2}\left[2 + \frac{d(\ln F_z)}{d(\ln \dot{\gamma}_R)}\right] \tag{86}$$

Corrections For inertia and secondary flow, use Eqs. (79) and (80) with $h/R = \beta$. For edge failure, the corrections are the same as for the cone-and-plate rheometer. For shear heating, they are similar to those for cone-and-plate rheometers [2, 31].

Utility The key advantage over the cone-and-plate rheometer is the ability to independently vary shear rate (and shear strain) by the rotation rate Ω or by changing the gap h. This permits an increased range with a given experimental setup. For very viscous materials and soft solids, sample preparation and loading are simpler. Edge failure can be delayed to higher shear rates by decreasing the gap during an experiment. This same effect requires a change of cone angle in cone-and-plate rheometers.

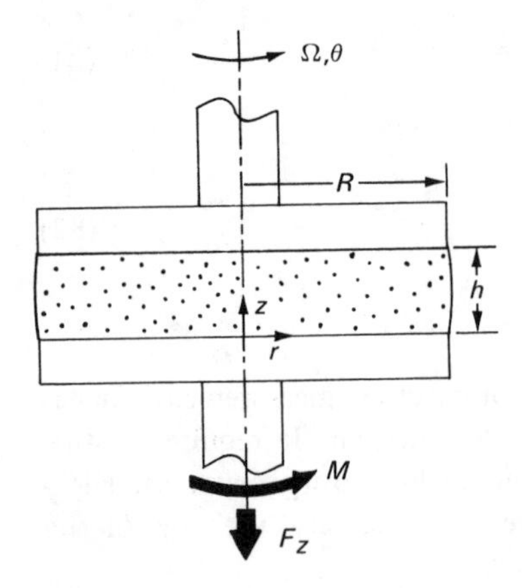

Figure 27 Schematic of parallel-plate rheometer.

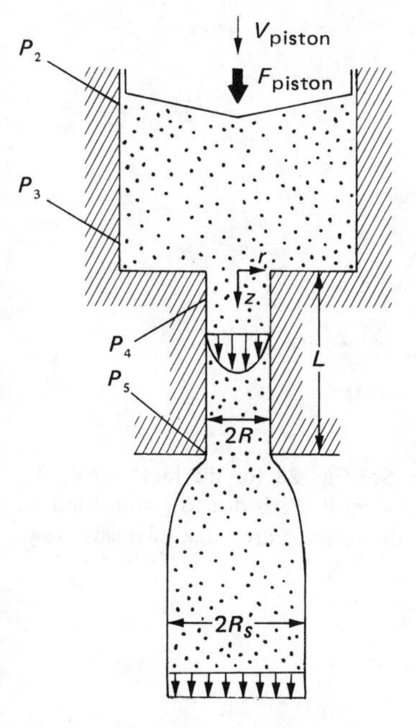

Figure 28 Cross section of a typical capillary rheometer. The extrudate swell is shown. The pressure-transducer readings at P_2, P_3, etc. are indicated in Fig. 29.

The main disadvantage is the nonhomogeneous strain field. But if only small strain or a steady rate of strain material functions is required, this is not a problem.

4.1.4 Capillary rheometer

Working equations Refer to Figs. 28 and 29 for definitions of the symbols. A key assumption in the analysis is that there is fully developed steady flow.

Wall shear stress:

$$\tau_W = \frac{R\,\Delta P}{2L} \qquad \Delta P = P_4 - P_5 \tag{87}$$

Wall shear rate:

$$\dot{\gamma}_W = \frac{4Q}{\pi R^3}\left[\frac{3}{4} + \frac{1}{4}\,\frac{d(\ln Q)}{d(\ln \Delta P)}\right] \qquad \text{(nonhomogeneous deformation)} \tag{88}$$

$$\dot{\gamma}_W = \left[\frac{3n+1}{n}\right]\frac{Q}{\pi R^3} \qquad \text{(power law)} \tag{89}$$

Viscosity:

$$\eta = \frac{\tau_W}{\dot{\gamma}_W} \tag{90}$$

First normal-stress difference:

$$(T_{11} - T_{22})^2 = 8\tau_{12}^2\,(B^6 - 1) \tag{91}$$

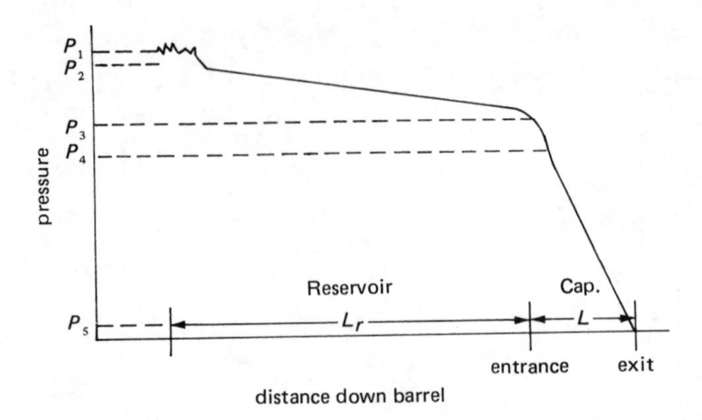

Figure 29 Pressure profile in a capillary or slit rheometer. See Fig. 28 for the locations of the readings. P_1 is the total force per unit area acting on the piston; P_1-P_2 is due to piston friction; P_2-P_3 to reservoir losses; P_3-P_4 to entrance loss; P_4-P_5 is the pressure drop due to steady flow; P_5 is the exit pressure drop.

(which requires constitutive assumptions [31])

$$B = \frac{R_S}{R} - 1.10$$

(the extrudate swell ratio, corrected for swell observed in slow newtonian flow).

Corrections The measurement of ΔP from piston force requires correction for friction, reservoir loss, and entrance loss (note Fig. 29). The measurement of Q from piston velocity requires correction for compressibility [31]. Leakage around the piston leads to errors. The value of Q from extrudate weight requires melt-density data. Shear heating may be a problem [2, 16, 21, 36]. Finally, $2R_S$ is very difficult to measure accurately, owing to density changes and the long time it takes to reach equilibrium [31].

Utility The capillary rheometer is frequently chosen over drag-flow rheometers because it can achieve higher shear rates with high-viscosity systems. The shape is similar to many dies and pipes in process flows, and it is often used as a process simulator. It is inexpensive, and high accuracy can often be achieved with long capillaries.

The main disadvantages include variation of shear rate and residence time across the flow, melt fracture and shear heating with high-viscosity samples, and the other corrections given above.

4.1.5 Slit rheometer

Working equations Refer to Fig. 30 for definitions of the symbols. Key assumptions in the analysis are fully developed steady flow and $A/B \leqslant 0.1$.

Wall shear stress:

$$\tau_w = \frac{A\,\Delta P}{2L} \tag{92}$$

Wall shear rate:

$$\dot{\gamma}_w = \frac{6Q}{A^2 B}\left[\frac{2}{3} + \frac{1}{3}\frac{d(\ln Q)}{d(\ln \Delta P)}\right] \tag{93}$$

Viscosity:

$$\eta = \frac{\tau_w}{\dot{\gamma}_w} \tag{94}$$

$$\eta = \frac{A^3 B\,\Delta P}{4QL}\,\frac{n}{2n+1} \qquad \text{(power law)} \tag{95}$$

Corrections These are the same as for the capillary rheometer.

Utility The main advantage of this device over the capillary rheometer is the flat side wall. This permits the use of flush-mounted pressure transducers and provides for flow visualization. Disadvantages include more difficult construction and possible errors due to side walls.

4.2 Extensional Rheometry

As indicated in Sec. 2.3 for nonnewtonian fluids, extension is fundamentally different from shear; extensional viscosity is a different material function from shear viscosity. Constitutive equations that are similar in shear can predict quite different results in extension. Furthermore, many important flows are highly extensional—fiber spinning, film blowing, and bubble growth, for example. Thus, there has been great interest recently in making extensional rheological measurements.

Another reason for this research activity is that it is so difficult to generate homogeneous extensional flow, especially for low-viscosity liquids. The basic

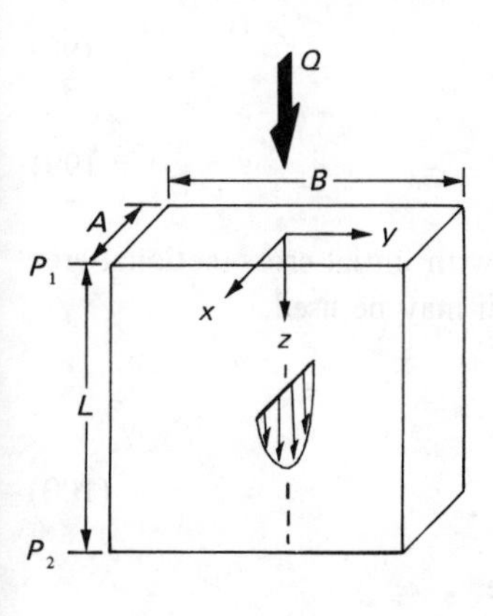

Figure 30 Coordinates and dimensions for the slit rheometer.

problem is that flow over stationary boundaries results in shear stresses, but without such boundaries it is difficult to control the deformation of a low-viscosity fluid. Surface tension, gravity, and flow instabilities all conspire to change the streamlines. A further problem arises from the large strains that are often required for stresses in memory fluids to reach their steady straining limit. In shear flows streamlines are parallel so that large strains can be achieved with long residence times. The streamlines in extensional flow diverge (or converge), meaning that to achieve infinitely large strain a sample must become infinitely thin in one direction. It may not be possible to attain a steady rate of extension in some materials, since they may rupture or deform unstably at high strains.

Many different methods have been tried to circumvent these problems and generate purely extensional flows. These are described in [29, 31, 37, 38]. Here we summarize those extensional rheometers that appear most promising. They are shown schematically in Fig. 31. At present, only the first geometry, simple tension, is accepted as an extensional rheometer, and then only for high-viscosity liquids. Below we give the working equations for two methods of generating uniaxial extension using the tensile geometry. Following this is a short discussion of the other geometries, with pertinent references.

4.2.1 Tension-translating clamp

Working equations See Fig. 32. Key assumptions are uniform drawdown, no end effects, and no surface-tension effects.

Strain:

$$\epsilon = \ln \frac{L}{L_0} \tag{96}$$

Strain rate:

$$\dot{\epsilon} = \frac{d(\ln L)}{dt} \tag{97}$$

Stress:

$$T_{11} - T_{22} = \frac{F}{A} = \frac{FL}{A_0 L_0} \tag{98}$$

$$T_{11} - T_{22} = \frac{FL}{\pi R_0^2 L_0} \qquad \text{(cylinder)} \tag{99}$$

The last equations assume an incompressible material with initial cross-sectional area A_0 and length L_0. Cross sections other than cylindrical may be used.

Corrections For surface tension [39]:

$$T_{11} - T_{22} = \frac{F}{A} - \frac{\gamma}{R}\left(1 - \frac{2R}{L}\right) \tag{100}$$

For $L/R \geqslant 10$, end effects are believed to be negligible.

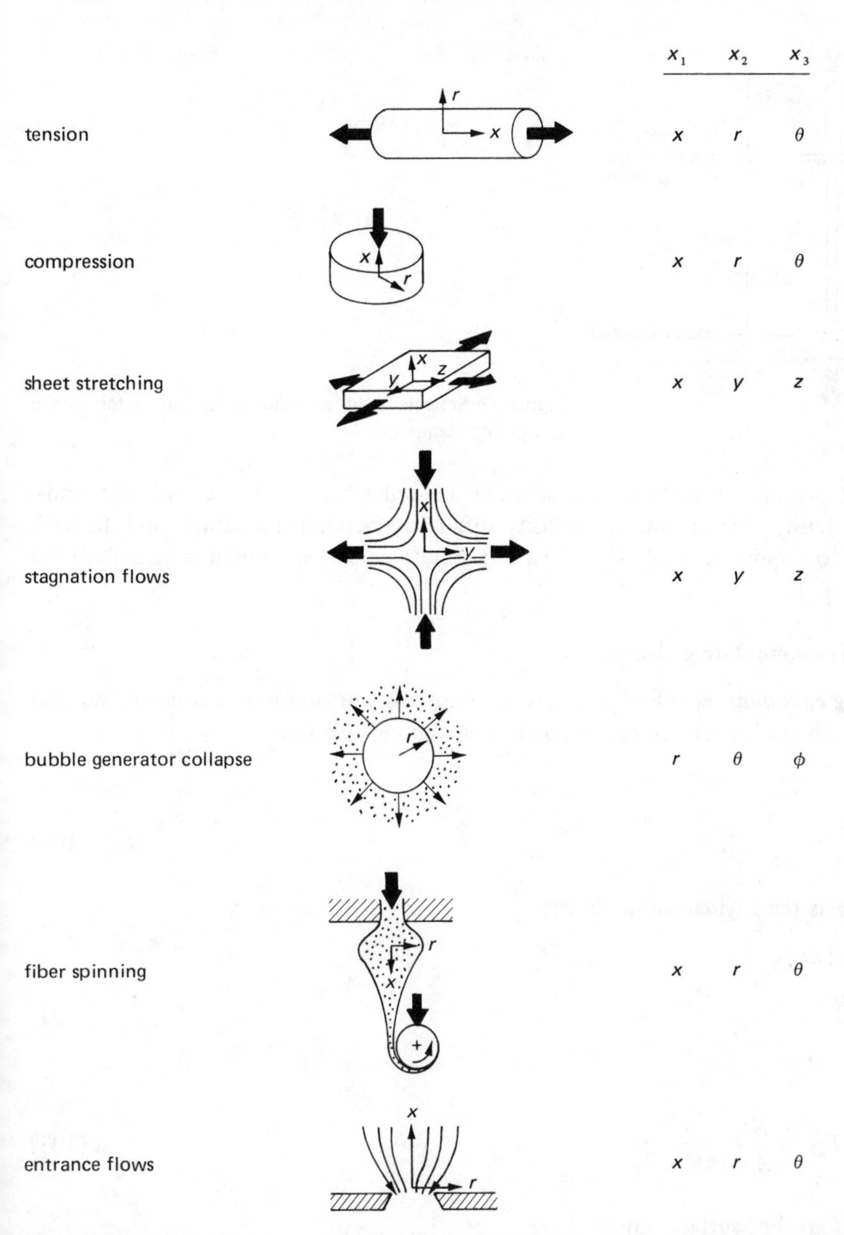

Figure 31 Extensional flow geometries. Only the first three geometries give homogeneous deformations.

Utility This is the most widely used geometry, and instruments are available commercially [40]. Homogeneous deformation and servo control of the clamp permit a wide range of steady and transient tests: start-up, recovery creep, and stress relaxation [41, 42]. The constant strain rate requires an exponentially increasing velocity. Sample preparation requires very high viscosity ($\geqslant 10^6$ Pa·s) or

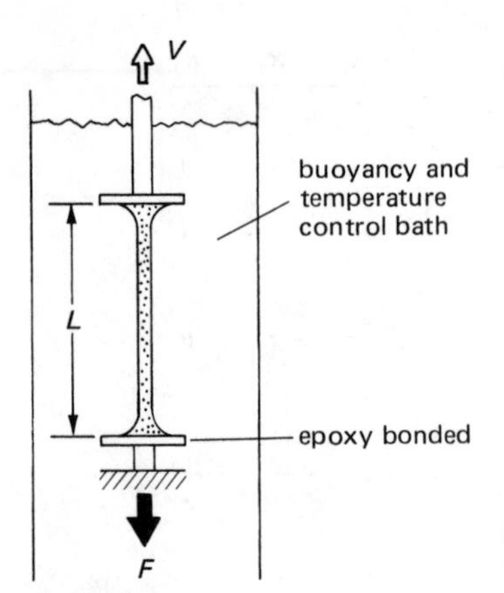

Figure 32 Schematic of an extensional rheometer with a translating clamp.

a solid sample at ambient temperature in order to clamp or glue the ends. Surface-tension effects and the density difference between the sample and the bath appear to require viscosity $\geqslant 10^4$ Pa·s. Limitations on bath length restrict the total strain $\epsilon \leqslant 4$.

4.2.2 Tension-rotating clamp

Working equations See Fig. 33. Key assumptions are uniform drawdown, no end effects, no surface-tension effects, and an incompressible fluid.

Strain:

$$\epsilon = \frac{R\theta}{L} \tag{101}$$

where θ is the angular displacement.

Strain rate:

$$\dot{\epsilon} = \frac{\Omega R}{L} \tag{102}$$

Stress:

$$T_{11} - T_{22} = \frac{F}{A} = \frac{F}{\pi R_0^2 e^{-\epsilon}} \tag{103}$$

Corrections For surface tension [39]:

$$T_{11} - T_{22} = \frac{F}{A} - \frac{\gamma}{R} \tag{104}$$

Utility The advantages of this device over the translating clamp are: (1) a simpler apparatus is required to generate constant strain rate [43 to 45], (2) a much shorter bath is required to achieve high strains [10, 39, 46 to 48], (3) higher strain rates may be possible, and (4) density match may not be so critical, since the sample floats. However, tests other than constant rate are generally more difficult. For recoverable strain the sample must be cut and measured. The sample size is generally larger, and sample preparation is more critical. End and thermal effects may reduce the stability of the flow [44] unless pairs of rotating clamps are used at each end and special temperature control is provided [10, 48].

4.2.3 Other extensional rheometers. The other geometries shown in Fig. 31 are either less well established or are not able to generate steady extension [Eq. (23)]. However, because the simple tension rheometers are apparently limited to high-viscosity samples, there is continued interest in these other test geometries.

Compression is the opposite of tension, but in practice the test requires a short sample to prevent buckling. If the ends are clamped as in the tension tests, considerable shear deformation will develop. Thus, the ends must be maintained parallel but allowed to slip. This can be achieved by lubricating the end plates with a low-viscosity liquid, as Chatraei et al. [49] have recently demonstrated. This method looks quite promising but will probably be limited to relatively high-viscosity liquids.

Equal biaxial stretching of a sheet can also be used to generate steady compression. Stephenson and Meissner [50] have recently built a sophisticated device for sheet stretching, with eight pairs of rotating clamps arranged in a circle and eight automated scissors to cut the sheet between the rollers. Controlling the rollers at the same speed generates equal biaxial stretching, while other programs can give any combination of extension rates. Again, this method appears limited to high-viscosity samples with $\eta \geqslant 10^5$ Pa•s.

Flow into a converging or diverging channel has a strong extensional component and is not limited to high-viscosity materials. Cogswell [51] has recently reviewed attempts to use converging flows to determine extensional viscosity. However, for nonnewtonian fluids, the entrance flow pattern can change dramatically with different materials, as shown in Fig. 34 [52]. Steady extension can be achieved by a properly shaped die and the use of a thin lubricant layer at the wall [53, 54]. Macosko et al. [54] measured steady planar extensional viscosities on a polystyrene melt with such a lubricated die.

Bubble growth, shown in Fig. 31, gives uniaxial compression, while bubble collapse gives extension. This geometry is attractive for low viscosity liquids. The deformation is not homogeneous throughout the sample, and experiments so far have not been able to measure the steady, extensional viscosity for elastic liquids [55 to 57].

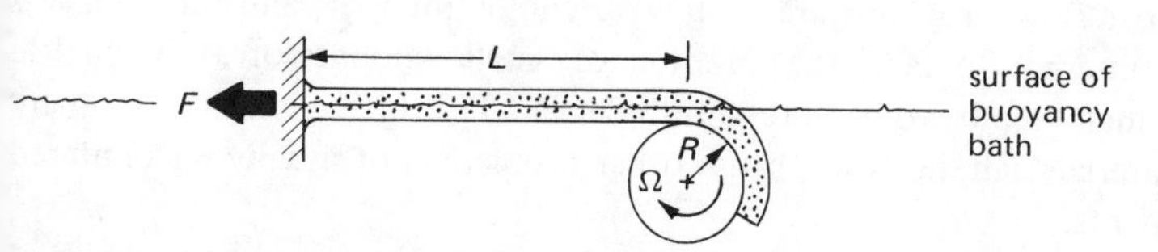

Figure 33 Schematic of a rotating-clamp extensional rheometer. Other signs employ pairs of rotating clamps at each end [46 to 48].

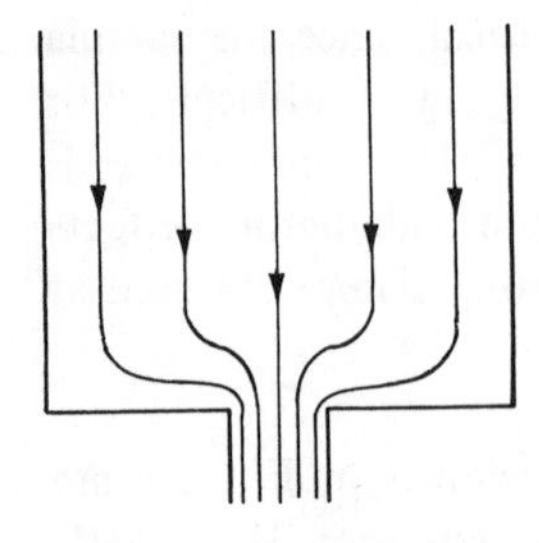

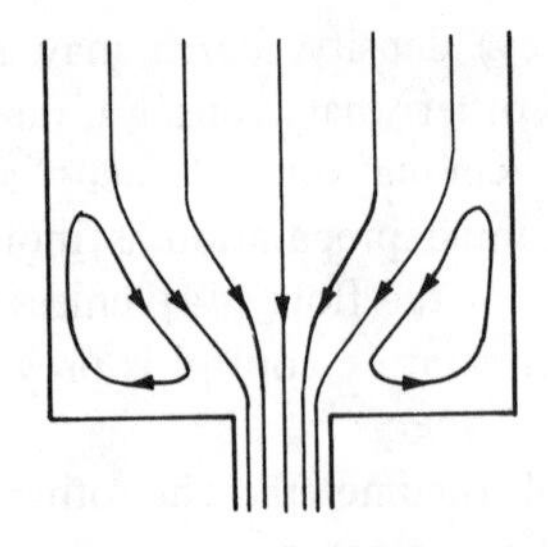

Figure 34 Streamlines for entrance flow of two different fluids. (*a*) No recirculation, typical of high-density polyethylene. (*b*) With recirculation, typical of low-density polyethylene and polystyrene melts. *(From [32].)*

The fiber-spinning experiment is also not in general a steady extension; however, owing to its applicability to industrial processes and its ability to test lower-viscosity liquids, there has been considerable interest in fiber-spinning methods [37]. Figure 35 shows a typical experimental setup [58]. Ideally, the fiber radius should decrease with the square root of the distance from the exit, but photographic results show that it usually decreases with a different power [37, 58, 59]. Surface tension, gravity, inertia, and the elastic memory of the fluid can all cause deviation from ideality.

5. MEASUREMENTS IN COMPLEX FLOWS

As discussed above, in a rheometer the kinematics of the flow are determined from some simple external measurement such as flow rate or angular velocity. Now we turn our attention to complex flows in which the kinematics are unknown. As pointed out in the introduction, experimental analysis of a complex nonnewtonian flow usually requires measurement of both the velocity field and the stresses. In this section we discuss special problems encountered making these measurements in nonnewtonian fluids. The basic techniques, such as pressure transducers and hot-wire or laser-Doppler anemometry, are well discussed in the other chapters of this text. Here we concentrate on the differences in using these methods with nonnewtonian fluids. Before looking at each method we make some general comments on working with nonnewtonian materials.

Essentially all nonnewtonian liquids are either dispersions of solid or liquid particles or polymer solutions and melts. Since nonnewtonian fluids tend to be of high viscosity, most flow problems are of low Reynolds number, and turbulence is usually not encountered. In fact, the addition of small amounts of axisymmetric particles or polymers tends to suppress turbulence. The high viscosity of many nonnewtonian materials can, however, cause other problems not usually encountered in flow measurements.

Two common assumptions in newtonian studies, isothermal flow and no slip at the wall, need to be checked carefully. Because of their high viscosity and relatively low thermal conductivity, nonnewtonian fluids, and in particular polymer melts, can generate significant heat through viscous dissipation. Temperature increases of over 50°C have been measured for polymer melts flowing through a slit or tube [16, 21]. Temperature profiles can be highly nonuniform, since generation is proportional to the square of the velocity gradient. Winter [36] has done an extensive review of this problem area.

With high-viscosity polymer melts and concentrated suspensions, particularly those that show a yield stress, it appears that the no-slip wall boundary condition can fail to hold. Chauffoureaux et al. [60] reported direct evidence for wall slip with polyvinyl chloride melts. Snelling and Lontz [61] and Uhland [62] indicated that slip occurs for other polymer melts in die flows. Kraynik [63] has shown evidence for wall slip with aqueous solutions of polyvinyl alcohol and sodium borate using a hot-film anemometer at the wall. Polymeric materials can undergo an elastic flow instability at high flow rates, which may involve a wall slip mechanism [64]. Although the no-slip assumption is valid in most cases, these counterexamples suggest that it must be checked in each case where high-viscosity, nonnewtonian materials are being studied.

Another problem with nonnewtonian fluids is composition instability. Since they are generally prepared from two or more components, it is possible for relative concentrations to change during experiments, owing to solvent evaporation, degradation of polymer, or settling out of suspended particles. Water-soluble polymer molecules are particularly subject to bacteriological attack. Fungicides should be added to samples that must be used for more than several days. The high temperatures necessary to process molten polymers can lead to thermal and

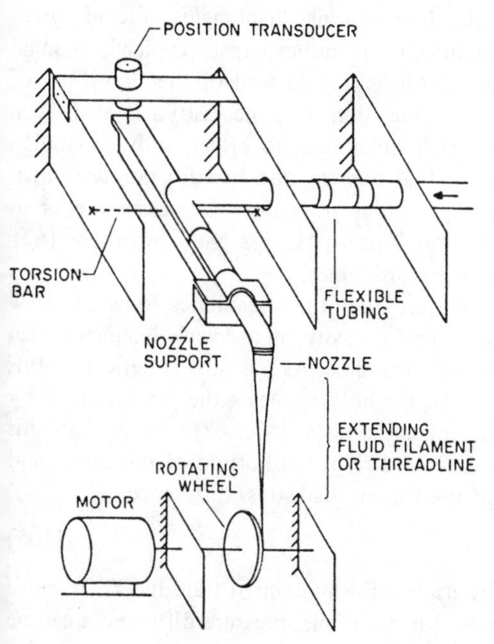

Figure 35 Typical fiber-spinning apparatus. *(From [58].)*

oxidative degradation. High-molecular-weight polymer chains are also susceptible to mechanical degradation in strong flow fields [65].

With concentrated suspensions, settling and sample uniformity are major problems. In fact, the effect of changing sample homogeneity during a complex flow may be as important as the nonnewtonian properties. As is discussed below, particles migrate during flow. Suspended particles can settle out completely in bends and expansions. In flows of rodlike particles, such as glass-fiber suspensions, it can be particularly difficult to obtain reproducible results.

5.1 Pressure Measurements

Special problems arise in pressure measurements on nonnewtonian fluids, due to high viscosity, high temperature, and elastic effects.

High viscosity means high pressures. Over 1000 atm is not uncommon in polymer processing equipment. Sensitivity is typically adequate down to one-hundredth or one-thousandth of full scale; thus, several transducers with different ranges are often needed for a given study. With molten polymers, temperatures can be as high as 350°C. Zero shift due to temperature changes can be particularly troublesome and should be checked carefully in setting up an experiment. The calibration constant is less sensitive to temperature, but this should also be verified and checked periodically during service. A number of pressure transducers that have been designed for the plastics industry can operate accurately under high temperature and wide pressure-range conditions [66].

One of the surprising discoveries a few years ago in nonnewtonian fluid mechanics concerns the influence of mounting holes on pressure readings. Typically, pressure transducers are mounted with their sensing diaphragms not in direct contact with the flow. Often, these diaphragms are rather larger. Typically a small fluid-filled hole is used to connect the transducer to the wall of the flow field of interest. In rising manometers, such a connection is a necessity. It had been established that the size and shape of such holes had no effect with newtonian fluids, and it was assumed for many years that this was also true for nonnewtonian fluids. However, in 1968, by comparing various normal-stress measurements in rheometers with pressure holes and by total thrust, Lodge and coworkers [67] discovered a significant influence of these pressure holes.

Figure 36 illustrates the problem. We see that the streamlines for shear flow near a wall are bent as the flow passes over a cavity in the wall. Nonnewtonian fluids can have a normal tension along the streamline. As the flow is deflected, this normal tension tends to "pull" fluid out of the hole, reducing the pressure read by the transducer. Experimental and theoretical studies [67, 68] show that this pressure error, the difference between the pressure at the bottom of the cavity and at a flush transducers, is about −20% of the first normal-stress difference:

$$p_H \simeq -0.2(T_{11} - T_{22}) \tag{105}$$

The hole error can be eliminated by using a flush-mounted transducer. Or, since the error will be approximately the same for each hole, pressure differences can be

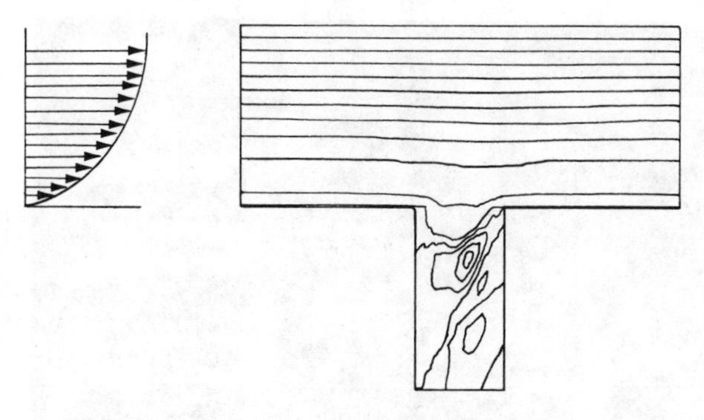

Figure 36 Streamlines for Couette flow near a wall cavity from numerical calculations of Crochet and Bezy.

correctly measured. The pressure-hole error seems to be less significant for high-viscosity liquids, probably because the shear stresses are so much larger than the normal stresses; i.e., the flow has small Weissenberg number [70, 71].

One advantage of the pressure-hole error is that it can be used to measure normal stresses. This has been applied by Lodge [68, 72] to the design of an on-line rheometer.

5.2 Velocity Measurements

Most velocity-measurement methods rely on optics. This means that the fluid must be transparent to the wavelength of interest. This can be a problem, since many nonnewtonian fluids are opaque or contain too many small particles. If gas bubbles become entrained in a high-viscosity, nonnewtonian fluid, they can be very difficult to remove. A few can be tolerated or even used for tracers, but a large number will scatter too much light.

Hydrogen-bubble generation by water hydrolysis from a wire stretched across the flow can be used with water-based fluids. The same techniques as with nonnewtonian fluids can be applied [73]. However, with polymer solutions, polymer degrades on the wire, fouling it and reducing the uniformity of the generation pattern. This is shown by the streaks in Fig. 37. If the wire is carefully cleaned periodically, the method can be utilized. Another problem is wire bending and breakage, owing to the high stresses generated by viscous liquids. The wire curvature is apparent in Fig. 37.

Similar problems are encountered in using hot-film or hot-wire anemometry—i.e., fouling due to thermal degradation of the sample and mechanical failure due to high viscosity. As discussed above, high-viscosity fluids can lead to viscous dissipation around the probe, altering the heat-transfer character and thus the interpretation of the results. The nature of laminar heat transfer from nonnewtonian fluids has not been well studied. Kraynik [63] discusses some of these problems with hot-film anemometry.

Particle tracer methods for flow visualization have been used extensively to measure velocity profiles in nonnewtonian fluids. Most materials already have many (often too many) impurity particles that can serve as tracers. An important problem

Figure 37 A 3.5% polyacrylamide solution in water flowing over a 25-μm-diameter tungsten wire. 100 V was used to generate the hydrogen bubbles shown. *(From [53].)*

in these studies is that of particle migration. Leal [74] has recently reviewed this subject. Due to inertia or deformability, particles in newtonian fluids can migrate. If the particles are small enough, Brownian motion and viscosity counterbalance the migration, and the tracers follow fluid path lines. However in nonnewtonian fluids additional stresses can act on particles. Karnis and Mason [75] report that particles in nonnewtonian Couette flow actually migrated opposite to the direction expected for newtonian fluids.

Laser-Doppler anemometry is subject to the same contamination and tracer problems as discussed above. Nonisothermal conditions can cause distortion, owing to the temperature dependence of the index of refraction. Since most nonnewtonian fluids are strongly shear-thinning, velocity profiles are steeper near walls than for newtonian fluids. This can present special measurement problems, both in focusing the crossed beams near the wall and in dealing with the finite scattering volume. The problems can be compounded by the relatively low velocities that are often of interest in these more viscous materials. Special low-frequency techniques need to be employed. Kramer and Meissner [76] discuss these problems in a recent laser-Doppler study on polymer melts.

5.3 Flow Birefringence

To fully characterize a complex flow of a nonnewtonian fluid, the stress as well as velocity distribution is needed. Flow birefringence can be used to obtain stress distributions in transparent polymeric liquids [77].

Nearly all asymmetric molecules also show optical anisotropy; i.e., their index of refraction is different in different directions along the molecule. For example, in polystyrene, light travels faster across the chains, through the large benzene rings, than along them (Fig. 38). Thus, if polarized light is transmitted through an oriented polystyrene sample, it will be separated into two mutually perpendicular components that are out of phase with each other and rotated with respect to the incident beam. This effect is called *double refraction* or *birefringence*.

Birefringence has been used extensively to study stress patterns in solids. This field is often referred to as photoelasticity [78]. To see how the effect is used, consider a solid rectangle subjected to simple shear (Fig. 39). The stress distribution can be represented by an ellipsoid with principal axes p_{I} and p_{II}. The stress ellipsoid can be related to the stresses measured on the boundaries of the solid by the difference between the principal stresses $\Delta p = p_{\mathrm{I}} - p_{\mathrm{II}}$, and the angle χ_p that p_{I} makes with respect to the shear direction x_1.

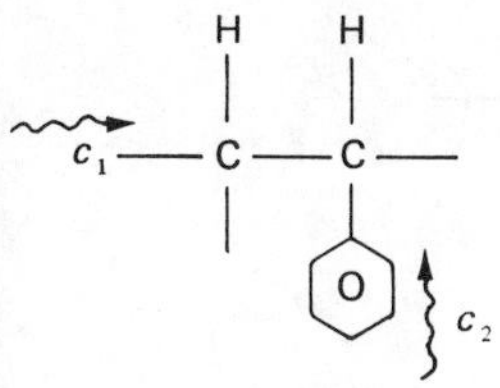

Figure 38 Light travels faster across a polystyrene chain than along it: $c_2 \gg c_1$.

Shear stress:

$$T_{12} = \tfrac{1}{2}\, \Delta p \sin 2\chi_p \tag{106}$$

Normal-stress difference:

$$T_{11} - T_{22} = \Delta p \cos 2\chi_p \tag{107}$$

Since this stress field will orient the molecules of the solid, we will see birefringence (if it is transparent). In principle, since stress and the index of refraction are both second-rank tensors, we need a fourth-rank tensor (81 components) to relate them. However, for many homogeneous, amorphous materials such as glass and many polymers, the two tensors are related by a simple constant called the *stress optical coefficient C*. Thus the two principal axes of the polarized light in the plane of the sheared sample in Fig. 39 will be aligned with the principal stresses, and their difference will be proportional to Δp:

$$\chi_n = \chi_p \tag{108}$$

$$\Delta n = C\, \Delta p \tag{109}$$

These same relations should hold for homogeneous liquids. They can be tested by measuring both the principal stresses and birefringence in the same flow. This has been done for a number of newtonian and some polymeric liquids [77].

Figure 40 shows χ_n and Δn versus shear stress for decalin, a newtonian oil, and polyisobutylene in decalin. For the decalin we see that χ_n is 45° and is independent of shear stress, which by Eq. (103) means there are no normal shear stresses, as expected for a newtonian fluid. For the polymer solution we see that χ_n decreases, indicating normal stresses. χ_p from cone-and-plate thrust measurements [see Eq. (78)] is in good agreement with χ_n. As expected from combining Eqs. (109) and (106), $\Delta n \sin 2\chi$ gives a straight line versus T_{12}. The slope is C, the stress optical coefficient. The only exceptions to $C =$ const appear to be suspensions or dilute polymer solutions, in which the solvent and polymer indexes of refraction are not

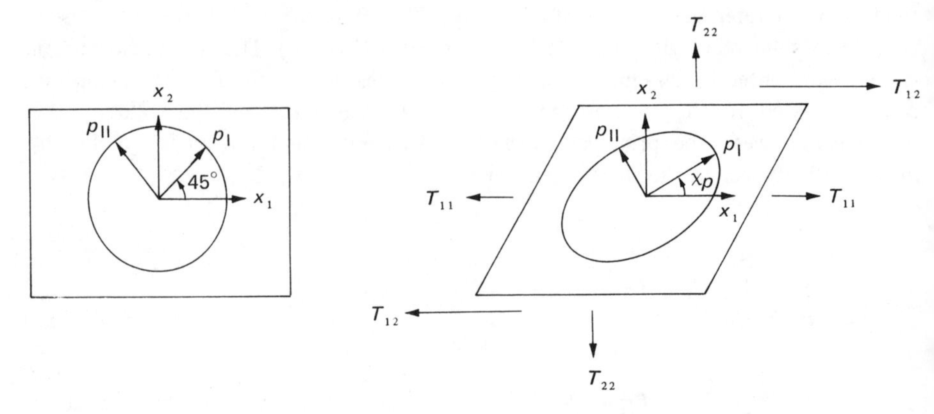

Figure 39 Stress ellipsoid in simple shear.

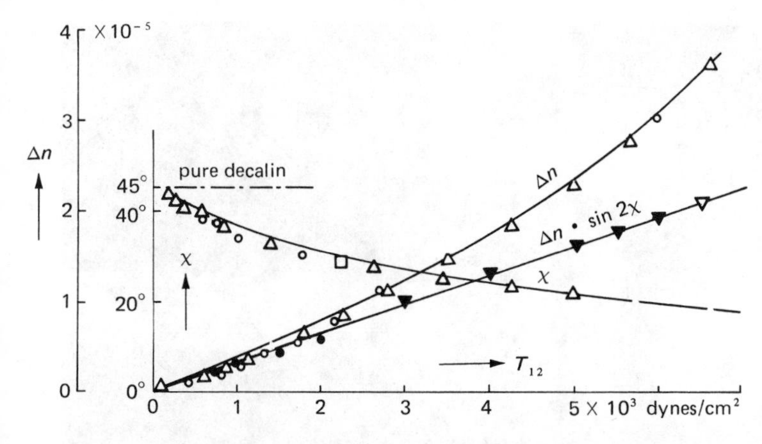

Figure 40 Test of stress-optical law for a 15% polyisobutylene solution in decalin. (□) χ_p from total thrust in cone and plate and χ_n, from birefringence data, (○) at 30°C, and (Δ) at 50°C. *(From [77].)*

closely matched. In these cases the shape of the particle or of the polymer will result in an additional "form" birefringence that cannot readily be removed.

The great advantage of flow birefringence is that it can provide the needed stress distribution in a nonnewtonian flow. The technique can also be sensitive to low stress levels. For example, in Fig. 40 we see that 10^3 dyn/cm^2 (~0.02 psi) gives $\Delta n \simeq 3 \times 10^{-7}$. Such changes in Δn are readily measurable. This can be seen from the basic relation between Δn and the measured phase shift δ:

$$\Delta n = \frac{\delta\lambda}{2\pi L} \tag{110}$$

where λ is the wavelength of the light used, and L is its path length through the sample. Thus, for a mercury lamp (560 nm) and a 1-cm sample thickness, $\Delta n = 3 \times 10^{-7}$ means $\delta \simeq 2°$, which can readily be detected with a suitable compensator. Another advantage of birefringence is that it can be combined with laser-Doppler anemometry. Some of the same optical systems can be used for both techniques [79].

5.3.1 Birefringence methods. The birefringence effect can be demonstrated by simply placing crossed polars over a section of the flow with transparent boundaries. Figure 41 shows the birefringence around the entrance to a slit die. If white light is used, the fringes are colored; with monochromatic light they are black and white. From Fig. 41 we can estimate Δn by counting the fringes. If there are no more fringes outside the field of the photo, then for the small circular fringe at the entrance to the slit, $\delta = 9(2\pi)$. Thus, by Eq. (110), assuming 560-nm light and $L \simeq 8$ mm gives $\Delta n \simeq 4 \times 10^{-4}$. Using $C \simeq 2 \times 10^{-9}$ Pa^{-1} for polyethylene, the difference in principal stresses at the entrance is $\Delta p \simeq 2 \times 10^5$ Pa $\simeq 2$ atm.

To measure δ more precisely, a photomultiplier can be used. If the flow rate is

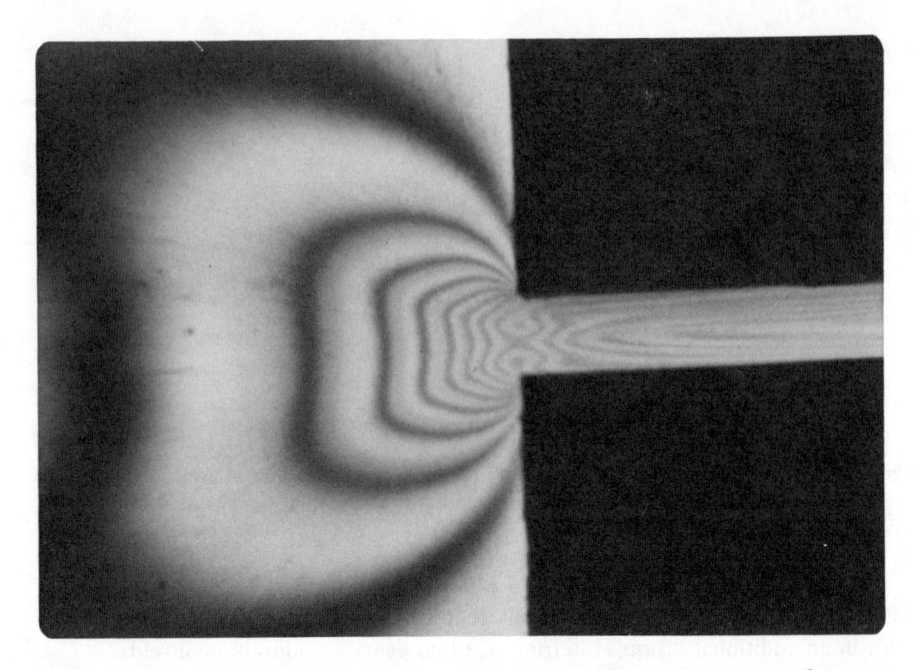

Figure 41 Photograph through crossed polarizers of high-density polyethylene at 200°C entering a slit die at 17 cm^3/min. *(From [80].)*

slowly increased from zero, the trace of intensity from the photomultiplier will look like Fig. 42, an oscillation with each 2π change in δ. Calibration of the oscillation pattern can give δ fairly accurately; however, at the extrema the resolution is not as good. Janeshitz-Kriegl [77] and Gortemacher [81] discuss a modulation method for improving resolution with the photomultiplier technique.

The classical method for precise birefringence measurements is to use a slit of light and a compensator that can accurately measure χ_n and δ. Such an optical system is shown in Fig. 43. With this arrangement the light is in the x_3 direction, and the birefringence is proportional to the principal stresses in the x_1x_2 plane, the plane of Fig. 39. Other planes can be studied if the light can be introduced in another direction. However, great care must be taken against distortion of the beam. The two major problems are temperature gradients and "parasitic" birefringence

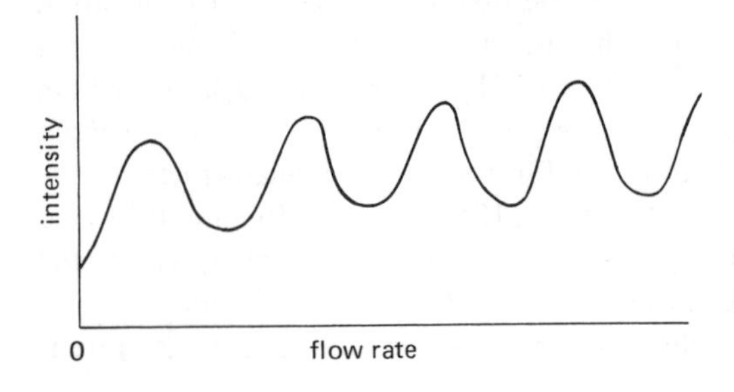

Figure 42 Representative output from a photocell during a steady increase in flow rate for a birefringence experiment.

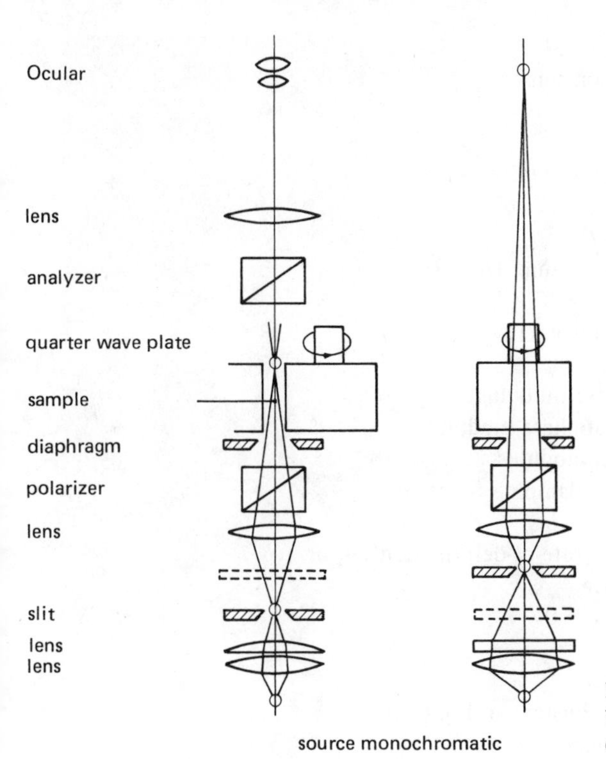

Figure 43 Optical alignment of birefringence apparatus for couette flow. *(From [54].)*

from the windows and from end effects in the flow near them. The index of refraction is a function of temperature; typically $dn/dT \simeq 10^{-3}$ $(°\mathrm{C})^{-1}$. Thus, a relatively small temperature gradient, $dT/dx \simeq 0.1$ to $1°\mathrm{C/mm}$, can cause the incident light to bend and reflect off one of the walls. Janeschitz-Kriegl [54] analyzes this problem for the Couette rheometer. He shows that reflections from metallic walls are particularly bad, since they have high adsorption coefficients. The use of glass walls provides nearly perfect reflection and allows one to obtain data despite some temperature gradients. Of course, better temperature control and shorter path lengths reduce the problem. Janeschitz-Kriegl used a slow crossflow in his Couette apparatus to improve temperature control.

Parasitic birefringence can come from stresses inherent in the glass windows and from thermal stresses and pressure on the glass that can arise during operation. The shear flow over the windows will also generate birefringence. The main approach is to match each window and the flow field over it as closely as possible in an attempt to cancel out the birefringence from both. Stress-free glass can be obtained, and the uses of rubber gaskets can reduce the thermal-stress problem.

NOMENCLATURE

a_0, a_1, a_2	constants
a_T	temperature-shift factor in Eq. (42)

b	constant
$\boldsymbol{B}$	finite-deformation tensor
C_1	constant
C_2	constant
$\boldsymbol{D}$	rate-of-deformation tensor
D_e	Deborah number
D_{ij}	components of $\boldsymbol{D}$
$\boldsymbol{E}$	small-strain tensor, $\lim_{\gamma \to 0} (\boldsymbol{B} - \boldsymbol{I})$
E_n	viscosity activation energy
G	elastic modulus
G'	elastic or in-phase modulus
G''	viscous or out-of-phase modulus
$G(t - t_0, \gamma_0)$	stress relaxation modulus
h	strain-dependent damping function
$\boldsymbol{I}$	unit tensor
$\mathrm{I}_D, \mathrm{II}_D, \mathrm{III}_D$	invariants of the rate-of-deformation tensor
J_e	creep compliance
J_e^0	limit of J_e as $\tau_0 \to 0$
K	constant
M	memory function
m	preexponential constant in Eq. (36)
n	power-law exponent
$N_1(\dot{\gamma})$	first normal-stress difference
$N_2(\dot{\gamma})$	second normal-stress difference
p	pressure
$\boldsymbol{T}$	stress tensor
T_{ij}	stress-tensor components
$\mathbf{v}$	velocity vector
$\alpha_0, \alpha_1, \alpha_2$	coefficients
$\dot{\gamma}$	shear rate
$\dot{\epsilon}$	extension rate
η	nonnewtonian shear viscosity
$\eta^+(t, \dot{\gamma}_0)$	start-up shear-stress material function
$\eta^-(t, \dot{\gamma}_0)$	relaxation shear-stress material function
$\eta_e(\dot{\epsilon})$	elongational viscosity
η^*	complex viscosity
η'	real part of complex viscosity
η''	imaginary part of complex viscosity
η_0, η_∞	low-shear-rate and high-shear-rate limiting viscosities
λ_k	relaxation time
μ	newtonian viscosity
τ	viscous or extra-stress tensor
τ_{ij}	components of $\boldsymbol{\tau}$
τ^*	complex stress

τ' real part of complex stress
τ'' imaginary part of complex stress
ψ_1 first normal-stress coefficient
ψ_2 second normal-stress coefficient
χ angle of principal stress with respect to shear direction x_1

REFERENCES

1. R. B. Bird, Useful Non-Newtonian Models, *Annu. Rev. Fluid Mech.*, vol. 8, pp. 13–34, 1976.
2. R. B. Bird, R. C. Armstrong, and O. Hassager, *Dynamics of Polymeric Liquids*, vol. I, chap. 3, Wiley, New York, 1977.
3. S. Middleman, *Fundamentals of Polymer Processing*, McGraw-Hill, New York, 1977.
4. G. Astarita and G. Marrucci, *Principles of Non-Newtonian Fluid Mechanics*, McGraw-Hill, New York, 1974.
5. W. R. Schowalter, *Fundamentals of Non-Newtonian Fluid Mechanics*, Wiley, New York, 1979.
6. C. W. Macosko and W. M. Davis, Flow between Eccentric Rotating Discs. *Rheol. Acta*, vol. 13, pp. 814–829, 1974.
7. H. Brenner, Rheology of a Dilute Suspension of Axisymmetric Brownian Particles, *Int. J. Multiphase Flow*, vol. 1, pp. 195–341, 1974.
8. J. D. Ferry, *Viscoelastic Properties of Polymers*, 3d ed., Wiley, New York, 1980.

8a. W. W. Graessley, The Entanglement Concept in Polymeric Liquids, *Adv. Polym. Sci.*, vol. 16, pp. 1–179, 1974.

9. H. Münstedt, Viscoelasticity of Polystyrene Melts in Tensile Creep Experiments, *Rheol. Acta*, vol. 14, pp. 1077–1088, 1975.
10. T. Raible, A. Demarmels, and J. Meissner, Stress and Recovery Maxima in LDPE Melt Elongation, *Polym. Bull.*, vol. 1, p. 397, 1979.
11. M. Wagner, Analysis of Stress-Growth Data for Simple Extension of a Low-Density Branched Polyethylene Melt, *Rheol. Acta*, vol. 15, p. 136, 1976.
12. A. S. Argon (ed.), *Constitutive Equations in Plasticity*, MIT Press, Cambridge, Mass., 1975.
13. P. Chadwick, *Continuum Mechanics*, chap. 4, Halsted-Wiley, New York, 1976.
14. L. E. Malvern, *An Introduction to the Mechanics of a Continuous Medium*, Prentice-Hall, Englewood Cliffs, N.J., 1969.
15. P. A. Good, A Schwartz, and C. W. Macosko, Analysis of the Normal Stress Extruder, *AIChE Jr.*, vol. 20, pp. 67–73, 1974.
16. H. W. Cox and C. W. Macosko, Viscous Dissipation in Die Flows, *AIChE J.*, vol. 20, pp. 785–795, 1974.
17. Z. Tadmor and C. G. Gogos, *Principles of Polymer Processing*, Wiley, New York, 1979.
18. Z. Tadmor and I. Klein, *Engineering Principles of Plasticating Extrusion*, Reinhold, New York, 1970.
19. K. van Krevelan, *Properties of Polymers*, 2d ed., North Holland, Amsterdam, 1977.
20. P. H. Goldblatt and R. S. Porter, A Comparison of Equations for the Effect of Pressure on the Viscosity of Amorphous Polymers, *J. Appl. Polym. Sci.*, vol. 20, pp. 1199–1207, 1976.
21. H. W. Cox and C. W. Macosko, Effect of Shear Heating in Capillary Flow, *Soc. Plast. Eng. Tech. Pap.*, vol. 20, pp. 28–32, 1974; also H. W. Cox, Ph.D. thesis, University of Minnesota, 1974.
22. A. G. Fredrickson, *Principles and Applications of Rheology*, p. 178, Prentice-Hall, Englewood Cliffs, N.J., 1964.
23. H. M. Laun, Description of the Non-Linear Shear Behavior of a Low Density Polyethylene

Melt by Means of an Experimentally Determined Strain Dependent Memory Function, *Rheol. Acta,* vol. 17, pp. 1–15, 1978.
24. D. D. Joseph and R. Fosdick, The Free Surface on a Liquid between Cylinders Rotating at Different Speeds. Part I, *Arch. Rat. Mech. Anal.,* vol. 49, pp. 321–380, 1972.
25. R. S. Rivlin, in F. R. Eirich (ed.), *Rheology,* vol. I, p. 351, Academic, New York, 1956.
26. L. R. G. Treloar, *The Physics of Rubber Elasticity,* 3d ed., Oxford, New York, 1974.
27. A. S. Lodge, *Elastic Liquids,* Academic, London, 1964.
28. M. Doi and S. F. Edwards, Dynamics of Concentrated Polymer Systems. Parts 1–3, *Chem. Soc., London. J.: Faraday Transactions 2,* vol. 74, pp. 1789–1832, 1978.
28a. A. C. Papanastasiou, L. E. Scriven, and C. W. Macosko, An Integral Constitutive Equation for Mixed Flows, *J. Rheol.,* in press.
28b. S. Middleman, *The Flow of High Polymers,* Wiley, New York, 1968.
29. K. Walters, *Rheometry*, Halsted-Wiley, London, 1975.
30. R. W. Whorlow, *Rheology Techniques,* Ellis Horwood, Chichester, U.K., 1980.
31. C. W. Macosko, Rheological Measurements, Department of Chemical Engineering and Materials Science, University of Minnesota, short course notes, 1982.
32. J. M. Deally, *Rheometers for Molten Plastics,* Van Nostrand Reinhold, New York, 1982.
33. J. L. Schrag, Deviation of Velocity Grandient Profiles from the "Gap Loading" and "Surface Loading" Limits in Dynamic Simple Shear Experiments, *Trans. Soc. Rheol.,* vol. 21, pp. 399–413, 1977.
34. M. M. Denn and J. Roisman, Rotational Stability and Measurement of Normal Stress Functions in Dilute Polymer Solutions, *AIChE J.,* vol. 15, pp. 545–459, 1969.
35. G. S. Beavers and D. D. Joseph, The Rotating Rod Viscometer, *J. Fluid Mech.,* vol. 69, pp. 475–511, 1975.
36. H. H. Winter, Viscous Dissipation in Shear Flows of Molten Polymers, *Adv. Heat Transfer,* vol. 13, pp. 205–267, 1977.
37. C. J. S. Petrie, *Extensional Flows,* Pittman, London, 1979.
38. J. M. Dealy, Extensional Rheometers for Molten Polymers: A Review, *J. Non-Newt. Fluid Mech.,* vol. 4, pp. 9–21, 1978.
39. H. M. Laun and H. Münstedt, Elongational Behavior of a Low Density Polyethylene Melt. I, *Rheol. Acta,* vol. 17, pp. 415–425, 1978.
40. Rheometrics, Inc., Union, NJ; Göttfert, Stuttgart, W. Germany.
41. H. Münstedt, New Universal Extensional Rheometer for Polymer Melts. Measurements on a Polystyrene Sample, *J. Rheol.,* vol. 23, p. 421, 1979.
42. V. S. Au-Yeung and C. W. Macosko, Extensional Rheometry of Several Blow Molding Polyethylenes, in G. Astarita, G. Marrucci, and L. Nicolias (eds.), *Proc. of the VIII Int. Congr. on Rheology,* Naples, vol. 3, pp. 717–722, Plenum, New York, 1980.
43. C. W. Macosko and J. M. Lorntson, Rheological Comparison of Two-Blow Molding Polyethylenes, *Soc. Plast. Eng. Tech. Pap.,* vol. 19, pp. 461–467, 1973.
44. R. W. Connelly, L. J. Garfield, and G. H. Pearson, Local Stretch History of a Fixed-End-Constant-Length-Polymer-Melt Stretching Experiment, *J. Rheol.,* vol. 23, pp. 651–652, 1979.
45. Y. Ide and J. L. White, The Spinnability of Polymer Fluid Filaments, *J. Appl. Polym. Sci.,* vol. 20, p. 2511, 1976; vol. 22, pp. 1061–1079, 1978.
46. J. Meissner, Rheometer Zur Untersuchung der Deformationsmechanischen Eigenschaften von Kunststoff-Schmelzen unter Definierter Zugbeanspruchung, *Rheol. Acta,* vol. 8, pp. 78–88, 1969.
47. J. Meissner, Development of a Universal Extensional Rheometer for the Uniaxial Extension of Polymer Melts, *Trans. Soc. Rheol.,* vol. 16, pp. 405–420, 1972.
48. T. Raible and J. Meissner, Uniaxial Extensional Experiments with Large Strains Performed with Low Density Polyethylene (LDPE), in G. Tstarita, G. Marrucci, and L. Nicolias (eds.), *Proc. of the VIII Int. Congr. on Rheology,* Naples, vol. 2, pp. 425–430, Plenum, New York, 1980.
49. S. Chatraei, C. W. Macosko, and H. H. Winter, Lubricated Squeezing Flow: A New Biaxial Extensional Rheometry, *J. Rheol.,* vol. 25, pp. 433–443, 1981.

50. S. E. Stephenson and J. Meissner, Large Homogeneous Biaxial Extension of Polyisobutylene and Comparison with Uniaxial Behavior, in G. Astarita, G. Marrucci, and L. Nicolias (eds.), *Proc. of the VIII on Rheology,* Naples, vol. 2, pp. 431–436, Plenum, New York, 1980.
51. F. N. Cogswell, Converging Flow and Stretching Flow: A Compilation, *J. Non-Newt. Fluid Mech.,* vol, 4, pp. 23–28, 1978.
52. A. E. Everage and R. L. Ballman, The Extensional Flow Capillary As a New Method for Extensional Viscosity Measurement, *Nature,* vol. 273, pp. 213–215, 1978.
53. H. H. Winter, C. W. Macosko, and K. E. Bennett, Orthogonal Stagnation Flow; A Framework for Steady Extensional Flow Experiments, *Rheol. Acta,* vol. 18, pp. 323–334, 1979.
54. C. W. Macosko, M. A. Ocansey, and H. H. Winter, Steady Planar Extension with Lubricated Dies, in G. Astarita, G. Marrucci, and L. Nicolias (eds.), *Proc. of the VIII Int. Congr. on Rheology,* Naples, vol. 3, pp. 723–728, Plenum, New York, 1980.
55. G. H. Pearson and S. Middleman, Elongational Flow Behavior of Viscoelastic Liquids: Parts 1-2, *AIChE J.,* vol. 23, pp. 714–725, 1977.
56. E. D. Johnson and S. Middleman, Elongational Flow of Polymer Melts, *Polym. Eng. Sci.,* vol. 18, p. 963, 1978; H. Münstedt and S. Middleman, Comparison of Elongational Rheology as Measured in the Universal Rheometer and by the Bubble Collapse Method, *J. Rheol.,* vol. 25, pp. 24–40, 1981.
57. R. Y. Ting and D. L. Hunston, Some Limitations on the Detection of High Elongational Stress Effects in Dilute Polymer Solutions, *J. Appl. Polym. Sci.,* vol. 21, pp. 1825–1833, 1977.
58. K. Baid and A. B. Metzner, Rheological Properties of Dilute Polymer Solutions Determined in Extensional and in Shearing Experiments, *Trans. Soc. Rheol.,* vol. 21, pp. 237–260, 1977.
59. C. B. Weinberger and J. D. Goddard, Extensional Flow Behavior of Polymer Solutions and Particle Suspensions in a Spinning Motion, *Int. J. Multiphase Flow,* vol. 1, pp. 465–486, 1974.
60. J. C. Chauffoureaux, C. Dehennau, and J. Vanrijckevorsel, Flow and Thermal Stability of Rigid PVC, *J. Rheol.,* vol. 23, pp. 1–24, 1979.
61. G. R. Snelling and J. F. Lontz, Mechanism of Lubricant-Extrusion of Teflon TFE-Tetrafluoroethylene Resins, *J. Appl. Polym. Sci.,* vol. 3, pp. 257–265, 1960.
62. E. Uhland, The Anomalous Flow Behavior of High Density Polyethylene, *Rheol. Acta,* vol. 18, pp. 1–24, 1979.
63. A. M. Kraynik and W. R. Schowalter, Slip at the Wall and Extrudate Roughness with Aqueous Solutions of Polyvinyl Alcohol and Sodium Borate, *J. Rheol.,* vol. 25, pp. 95–114, 1981.
64. J. R. A. Pearson, *Principles of Polymer Melt Processing,* Academic, London, 1966.
65. A. Casale and R. S. Porter, *Polymer Stress Reactions,* Academic, New York, 1978.
66. Dynisco, Waltam, Mass.; Kulite, Richfield, N.J.; Sensotec, Columbus, Ohio.
67. J. M. Broadbent, A. Kaye, A. S. Lodge, and D. G. Vale, Possible Systematic Error in the Measurement of Normal Stress Differences in Polymer Solutions in Steady Shear Flow, *Nature,* vol. 217, pp. 55–56, 1968.
68. K. Higashitani and A. S. Lodge, Hole Pressure Error Measurements in Pressure-Generated Shear Flow, *Trans. Soc. Rheol.,* vol. 19, pp. 307–335, 1975; also A. S. Lodge, University of Wisconsin-Madison, Rheology Research Center Rep. 60, 1980.
69. J. M. Crochet and M. J. Bezy, Numerical Solution for the Flow of Viscoelastic Fluids, *J. Non-Newt. Fluid Mech.,* vol. 5, pp. 201–218, 1979.
70. C. D. Han, *Rheology in Polymer Processing,* Academic, New York, 1976.
71. G. Ehrmann and H. H. Winter, Druckmessung mit Hilfe von "pressure holes," *Kunststofftechnik,* vol. 12, no. 6, pp. 156–159, 1973.
72. A. S. Lodge, U.S. Patent 3,777,549, 1973; Seiscor Corp., Tulsa, Okla.
73. F. A. Schraub, S. J. Kline, J. Henry, P. W. Runstadler, and A. Littel, Use of Hydrogen Bubbles for Quantitative Determination of Time-Dependent Velocity Fields, *Trans. ASME, J. Basic Eng.,* vol. 87, pp. 429–444, 1965.

74. L. G. Leal, Particle Motions in a Viscous Fluid, *Annu. Rev. Fluid Mech.*, vol. 12, pp. 435–476, 1979.
75. S. Karnis and S. G. Mason, Particle Motions in Sheared Suspensions. XIX. Viscoelastic Media, *Trans. Soc. Rheol.*, vol. 10, pp. 571–592, 1966.
76. H. Kraemer and J. Meissner, Application of the Laser Doppler Velocimetry to Polymer Melt Flow Studies, in G. Astarita, G. Marrucci, and L. Nicolias (eds.), *Proc. of the VIII Int. Congr. of Rheology*, Naples, vol. 2, pp. 463–468, Plenum, New York, 1980.
77. H. Janeschitz-Kreigl, Flow Birefringence of Viscoelastic Polymer Systems, *Adv. Polym. Sci.*, vol. 6, pp. 170–318, 1969.
78. A. W. Hendry, *Photoelastic Analysis*, Pergamon, New York, 1966.
79. G. G. Fuller and L. G. Leal, Effect of Molecular Weight and Flow Type on Flow Birefringence of Dilute Polymer Solutions, in G. Astarita, G. Marrucci, and L. Nicolias (eds.), *Proc. of the VIII Int. Congr. of Rheology*, Naples, vol. 2, pp. 393–398, Plenum, New York, 1980; *Rheol. Acta*, vol. 19, p. 580, 1980; *J. Polym. Sci.: Phys. Ed.*, vol. 19, p. 557, 1981.
80. C. D. Han, On Slit- and Capillary-Die Rheometry, *Trans. Soc. Rheol.*, vol. 18, pp. 163–190, 1974.
81. F. H. Gortemacher, Ph.D. thesis, T. H., Delft, 1976.

CHAPTER

TEN

TWO-PHASE FLOW MEASUREMENT TECHNIQUES IN GAS-LIQUID SYSTEMS

Owen C. Jones, Jr.

1. INTRODUCTION

In the past 30 years, new chemical processing systems, modern power-generation methods, and space propulsion devices have increasingly involved multiphase flows. This is especially true of nuclear-reactor systems, where off-normal and accident situations have been intensively studied to provide assurance of public safety under extreme conditions. Multiphase flows will become still more important in the future development of advanced energy systems, owing to the attractiveness of utilizing the latent heat of phase change to enhance the energy intensiveness of these systems. Currently, for instance, two-phase, gas-liquid flows are found in such diverse systems as ocean thermal-energy conversion equipment, liquid-metal magnetohydrodynamic generators, geothermal wells and turbine generating equipment, oil-gas pipelines, boilers, nuclear-reactor systems, liquid-metal blankets of fusion power systems, droplet combustors, distillation towers, turbomachinery, refrigeration systems, and coal liquefaction systems.

The overriding conclusion one draws from an examination of two-phase flow literature is that multiphase flow is such an exceedingly complex physical situation that the general analytical effort has had practically no impact on this field.

Prepared for the U.S. Nuclear Regulatory Commission, Office of Nuclear Regulatory Research, under contract DE-AC02-76H00016, NRC FIN no. A-3045.

This manuscript was originally prepared for the Minnesota Short Course on Fluid Mechanics Measurements, given by the University of Minnesota Department of Mechanical Engineering. Certain portions of this manuscript were extracted from the author's prior publications, especially [9]. This work was performed under the auspices of the U.S. Nuclear Regulatory Commission.

Empiricism abounds. Experimental methods have been both borrowed from other fields and newly developed to allow examination of just one or two of the phenomena. Workers in general are in the position of the blind men attempting to describe an elephant by feel. One has hold of its trunk, another its leg, and yet a third the tail. Each arrives at a different conclusion due to his own vantage point, and none perceives the elephant as a whole.

As was true at first of single-phase fluid mechanics, the overall picture remains elusive due to lack of insight into the intricacies of the phenomena involved. When the experiments of J. Osborne Reynolds demonstrated the two major categories of single-phase flow fields, significant progress was made in piecing together various conflicting results. Similarly, in two-phase flow, while people are cognizant of the vagaries of laminar versus turbulent conditions, few have paid more than lip service to other areas of demarcation. Flow regimes are of overriding importance. One would not expect the same equations to accurately describe pressure drop in both laminar and turbulent flow. No one would then expect a single equation to do the same for completely different two-phase regimes. And, in fact, virtually no distinction between two-phase flows of laminar or turbulent character has been seriously considered.

Many workers are still trying to treat this phenomenon as a single-phase fluid. It is not. Two-phase immiscible fluids are decidedly different. They behave differently. They are not generally microscopic intermixtures but are macroscopic conglomerations. Formulations and methodology based on single-phase technology can be of only limited utility. In the final analysis, new methods, new viewpoints, and new technology must be developed. We must view the entire field from as many points of view as possible in hope of determining the true characteristics of two-phase flow.

Realizing that two fluids flowing together are different, we must ask how they differ from a single fluid flowing in a pipe. First, one phase is usually lighter than the other, the result being diffusion of one fluid with respect to the other. This is due in part to Archimedes' principle, a principle almost as old in concept as the study of fluid mechanics itself. This is basic. The movement of one fluid with respect to the other depends on the individual phase flow rates as related through the void fraction. No one would think of running a basic single-phase experiment without measuring the flow rates. But for two phases, the void fraction is just as basic a parameter as the flow rates. No basic experiment in two-phase flow should be run without measuring the void fraction and phase flow rates where possible as an absolute minimum.

The need for determining the void fraction, that is, the area or volume fraction of the flow occupied by the vapor phase, has been widely recognized and has led to the development of numerous methods for obtaining these measurements. Briefly, these methods may be divided into four general categories, depending on the spatial scale in relation to the duct size over which measurements are taken. The categories are:

1. Point-average probe-insertion techniques such as the conductivity probes used by Neal [153], hot-film anemometers used by Hsu et al. [68], and impedance probes used by Bencze and Orbeck [63]. These are discussed in Sec. 3.

2. Chordal-average void measurements commonly obtained through the use of particle or photon-beam attenuation techniques. These methods were used by Cravarolo and Hassid [154], Petrick [155], and Pike et al. [156] and are discussed in Sec. 5.
3. Area or small-volume average void fraction measurements by such means as impedance-sensing devices similar to those employed by Orbeck [64] and Cimorelli and Premoli [66] (Sec. 3) or flow-sampling devices (Sec. 4).
4. Large-volume average measurements such as accomplished by Lockhart and Martinelli [151] and by Hewitt et al. [152], through the use of quick-closing valves at various points in a tube (not discussed here).

The methods listed above and the techniques employed are by no means exhaustive. Many will be discussed in what follows. The interested reader is directed to an excellent survey of the field of void-fraction measurement compiled by Gouse [157].

A second way in which two-phase flow differs from single-phase flow is due to the immiscibility of the phases, each separated from the other by interfaces. It can be shown that the existence of these interfaces and their number and location are of fundamental importance in the analytical description of the movement of two-phase flows. Hydrodynamic and thermal transfer and relaxation phenomena in two-phase gas-liquid flows, in addition to being dependent on the departure from equilibrium for phase interactions, are strongly dependent on the interfacial area density available for transfer [1 to 3]. Such interactions include mass, momentum, and energy exchanges for which both the transfer coefficients and the areas available for transfer must be known accurately to specify transfers based on driving potential [4]. For instance, closure of a problem involving mass transfer between the phases requires specification of the volumetric vapor generation rate Γ_v to predict the vapor volume growth rates. Vapor generation rates are usually calculated in general terms by means of a constitutive relation [1] given by

$$\Gamma_v = \frac{1}{A\,\Delta i_{fg}} \int_{\xi_i} \sum_{k=l,v} \mathbf{q}_k'' \cdot \mathbf{n}_k \, \frac{dA_i}{dz} \tag{1}$$

where surface tension, shear, and relative kinetic-energy effects are considered negligible. In this case the net heat flux to the interface is given by the summation of the normal phase fluxes $\mathbf{q}_k'' \cdot \mathbf{n}_k$ at the point on the interface A_i having a density of $dA_i/A\,dz$. The interaction then sums the effects over all interfaces in A along the interfacial perimeter ξ_i. Similarly, specification of the momentum transfer between phases is important in the prediction of the relative velocity between phases, which is needed to couple void fraction to the quality in two-phase flows. The latter is especially important in low-velocity situations such as with natural convective flows, or under conditions of countercurrent flows. The degree of momentum transfer (and hence the relative velocity) is very sensitive to the flow regime. In highly coupled flows such as bubbly and slug flows, the transfer rates are high and relative velocities are low. Conversely, in weakly coupled flows such as in the annular or droplet regimes, momentum transfer is small and the relative velocities can be quite large [2]. In both cases, a knowledge of the interfacial area density is required. Thus, basic experiments should also include some method of monitoring the passage and density of interfaces.

Is this sufficient? Certainly, measurements of the gross flows in basic single-phase work were insufficient. Detailed velocity profiles were required as confirmation of theoretical models. It can likewise be assumed that both void and velocity profiles as well as interfacial area-density profiles would eventually be required for analysis of two-phase flows. Likewise, differential pressures are considered as essential measurements coupled with measurements relating to the frequency of void appearance. Lastly, as a minimum, certain characteristics of the flow must be identified as related to the structural behavior of the mixture, and levels of turbulence should be determined.

As in most other areas of scientific endeavor, purely theoretical analysis and prediction of the behavior of gas-liquid systems such as suggested above is clearly inadequate. Rather, experimental and analytical efforts must go hand in hand. While the development of single-phase thermofluid measurement techniques has been intensively pursued during the last century, special problems are encountered in attempting to observe mixtures of fluids, and it has been only over the last two to three decades that these problems have been considered. Measurement in two-phase gas-liquid systems is very much still in the empirical stage. Researchers generally try to find something that "wiggles when tickled" and then try to calibrate and interpret the response. In short, it can be said that measurements in two-phase gas-liquid systems carry all the complexities of similar measurements in single-phase systems plus additional difficulties, some of which are as follows:

1. Interface deformation due to sensor presence causes temporal distortion of the true pointwise behavior. Unequal distortions are generally observed in going from gas to liquid rather than from liquid to gas.
2. Inadequate transference from liquid-phase response to gas-phase response due to film retention causes bias in resulting interpretations unless precautions are taken.
3. Destruction of metastability due to sensor presence, which, for instance, causes cavitation on a thermal probe, which may then record saturation rather than superheat conditions.
4. Inaccurate response of pressure tap lines may result, owing to liquid-vapor interfaces caused by gas entrapment or flashing. The former can occur in two-component systems even in steady flow conditions, owing to the normal fluctuations, while the latter can occur owing to rapid decompression in heated lines during transients.
5. Improper averaging of nonlinear signals occurring due to fluctuations, with or without consideration of damping factors, causes interpretation errors.
6. Errors due to distortion of signals by the transmitting medium, such as shunting effects in sheathed high-temperature thermocouples.
7. Simple lack of understanding of the physical phenomena affecting a sensor response mass lead to misinterpretation of the results.

Several excellent summaries of two-phase measurement techniques have been written for steady-state measurements [5 to 8] and for transient or statistical techniques [9, 10]. The purpose of this review is to summarize the fundamental techniques that can now be considered classical in nature.

This chapter is divided into three sections:

1. In-stream electrical sampling techniques
2. In-stream mechanical sampling techniques
3. Global sampling techniques

2. TWO-PHASE GAS-LIQUID FLOW PATTERNS

Since this chapter deals exclusively with measurements in two-phase flows, it is appropriate to briefly describe the basic configurations or patterns that these flows can attain.

Just as there can be two different regimes in single-phase flows, within which different descriptions of momentum and heat transfer apply, so too in two-phase flows do we find laminar and turbulent situations. Unfortunately, at this stage, any distinctions made on the basis of these two classifications are purely qualitative. Rather, the major distinctions that have been used, and which are probably the first-order contributors to system behavior, are space-time phase distributions. These can probably best be described as a sequence of developing situations within a vertical-tube evaporator.

As described by Collier [11] and shown in Fig. 1, the changes with vertical upflow of an initially all liquid system are mainly due to "departure from thermodynamic equilibrium coupled with the presence of radial temperature profiles and departure from hydrodynamic equilibrium throughout the channel."

In the initial *single-phase* region the liquid is being heated to the saturation temperature. A thermal boundary layer forms at the wall, and a radial temperature profile is set up. At some position up the tube the wall temperature exceeds the saturation temperature and the conditions for the formation of vapor (nucleation) at the wall are satisfied. Vapor is formed at preferred positions or sites on the surface of the tube. Vapor bubbles grow from these sites, finally detaching to form a *bubbly flow.* With the production of more vapor due to continued heat addition, the bubble population increases with length, and coalescence takes place to form *slug flow*, which in turn gives way to *annular flow* as the elongated bubbles increase in length and merge further along the channel. Close to this point the formation of vapor at sites on the wall may cease, and further vapor formation takes place as a result of evaporation at the liquid film–vapor core interface. Increasing velocities in the vapor core cause entrainment of liquid from the surface waves on the film, which results in *droplets* in the central vapor stream. The depletion of the liquid from the film by this entrainment and by evaporation finally causes the film to dry out completely. Droplets continue to exist (some of which may even be remnants of the upstream destruction of liquid slugs) and are slowly evaporated until only *single-phase vapor* is present.

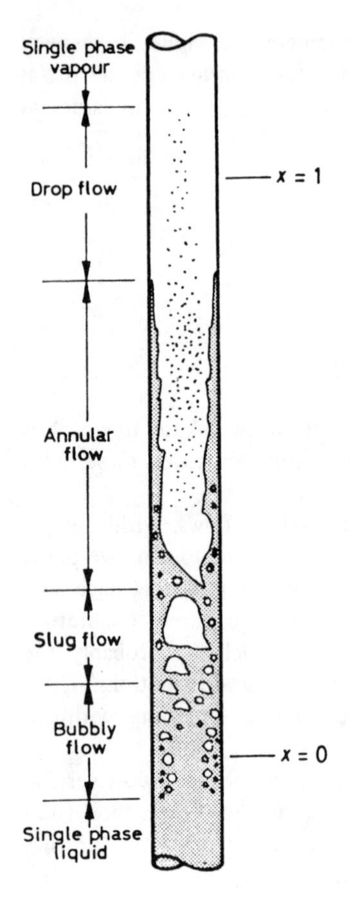

Figure 1 Flow patterns in a vertical evaporator tube. *(From [11].)*

While this description presents a commonly perceived and encountered evolution of flow patterns in a heated duct, it should be kept in mind that these patterns may, in fact, be encountered in other situations and evolved differently.

From a researcher's viewpoint, then, for both analytical and experimental purposes, the phenomena of interest are generally associated with those detailed characteristics that distinguish one pattern from another. Of course, as in single-phase flow, temperature, pressure, velocity, and mass flow rates are of interest. In addition, specific items that deal with particular aspects include:

1. Bubbly flows: boiling inception, bubble size, bubble trajectory, bubble boundary-layer thickness, liquid superheat, bubble agglomeration
2. Slug flows: slug lengths, bubble sizes, bubble spacings, film thicknesses, vapor entrainment, bubble velocity, slug velocity
3. Annular flows: wave height, wavelength, wave celerity, film thickness, film velocity, entrainment rate, deentrainment rate
4. Drop flows: dryout inception, drop size, drop trajectories, drop impingement

In all cases, departures from mechanical or thermal equilibrium as evidenced by relative velocities of one phase with respect to the other and by differences between phase temperatures are important variables for both design and analysis, as they can affect operating conditions of engineering equipment. Phase residence time fractions at a point in space and volume fractions at an instant in time, as well as the distributions and alternative averages of each, are of overriding importance and interest. The balance of this chapter is devoted to methods for obtaining the measurements identified above.

3. IN-STREAM SENSORS WITH ELECTRICAL OUTPUT

3.1 Conductivity Devices

The general principle of operation for conductivity-sensing devices is for two electrodes to be immersed in a two-phase mixture. A potential difference is created between the two electrodes, so that the current flow is a direct measurement of the conductivity of the fluid between the two electrodes. This current flow may be measured by means of a voltage drop across a calibrated resistor connected to some steady-state or transient measuring device, such as an oscilloscope, oscillograph, or recording voltmeter. Within certain limits, the relative amounts of conducting and nonconducting fluid–in other words, the amounts of liquid and gas–will determine the amount of current flow between the two electrodes. In this manner, current may be calibrated as a function of the vapor fraction and/or phase distribution between the two electrodes. (These devices may also be used in the impedance mode, where the capacitive reactance is the primary variable, but they are still called conductivity devices.) Such devices may be used for indications or measurements of void fraction, liquid level, film thickness, and flow patterns; with some modifications, they can be arranged to give quantitative information on wave frequencies, heights, and velocities.

3.1.1 Level probe. Figure 2 shows an early liquid-level transducer, designed and built by Kordyban and Ranov [12] for use in their air-water experiments. This probe consisted of a pair of insulated copper wires inserted through a hypodermic tubing coated with insulating paint. The wire ends were stripped of insulation and attached to the outside of the hypodermic tubing, thus being electrically insulated from themselves and from the tubing. When a potential was placed between the two wires, the amount of current flow was a measure of the percentage of the sensing element covered by the conducting fluid. These transducers were used in series along the test section to measure the time interval of slug travel between sensing stations, to obtain the velocity of slugs and variations along the length of the tube. The authors were also able to use the transducer to make qualitative measurements of flow regime within their horizontal tube.

Figure 3 shows a typical output trace from a pair of transducers located 20 in apart in the horizontal tube during conditions of slug flow. Not only is the passage

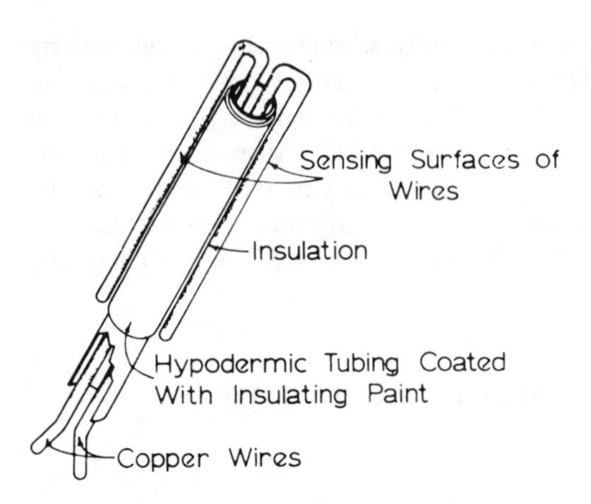

Figure 2 Liquid-level transducer of Kordyban and Ranov [12].

of slugs immediately evident from the trace, but also when the two traces are compared they show that the flow regimes retained their identity along the 20-in length so that slug velocities could immediately be measured. Kordyban and Ranov mention that "the main difficulty with the transducers ... was the fact that the exposed copper electrodes deteriorated gradually, affecting the range and linearity of output and require frequent cleaning and recalibration." In addition, 60-cycle noise on the transducer output made measurements difficult.

3.1.2 Needle probe. The needle-type conductance probe as designed by Solomon [13] was further developed and used by Griffith [14], Neal and Bankoff [15], and Nassos [16]. This probe, shown schematically in Fig. 4, consists of a single 0.8-mm-diameter steel wire, completely insulated and bent at an angle of 90°, giving a point of ~30 mm length. This needle forms the ungrounded conductor in a conductivity measurement system. The probe is inserted into a tube such that the point is aligned parallel to and opposing the flow streams. The noninsulated tip of the probe, when immersed in a conducting medium, allows current to travel between the tip and a grounded electrode imbedded in the tube wall. Thus, the distribution of fluid between the probe tip and the grounded conductor on the wall determines the amount of electric current flow. In this manner, information on the distribution, number, and velocity of bridges of liquid traveling in the tube, as well as qualitative information on flow regime, can be determined.

This type of instrument also seems readily adaptable to the qualitative determination of the transition between slug flow and annular flow. Further, Neal and Bankoff describe a method of using this type of probe for the determination of local void fractions in mercury-nitrogen flow [15]. The nonwetting characteristics of mercury made nearly instantaneous make-and-break contacts possible, such that there was no surface-tension effect on void-fraction measurements. Nassos, on the other hand, when attempting to adapt this to an air-water system, experienced difficulty due to the surface-tension effects at low temperature [16], which today are well accepted. In other words, when entering a small void, the gas phase would deform, keeping the probe tip within the conducting medium. Thus, the measure-

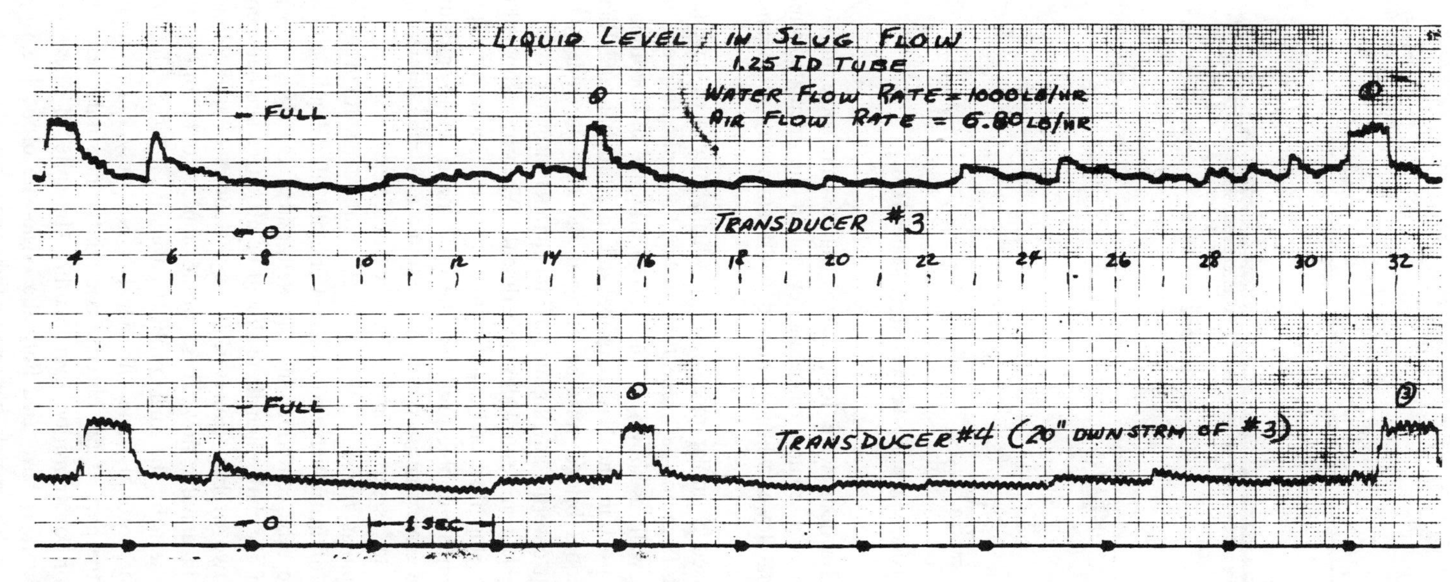

Figure 3 Liquid-level readings from two transducers at different axial locations. *(From [12].)*

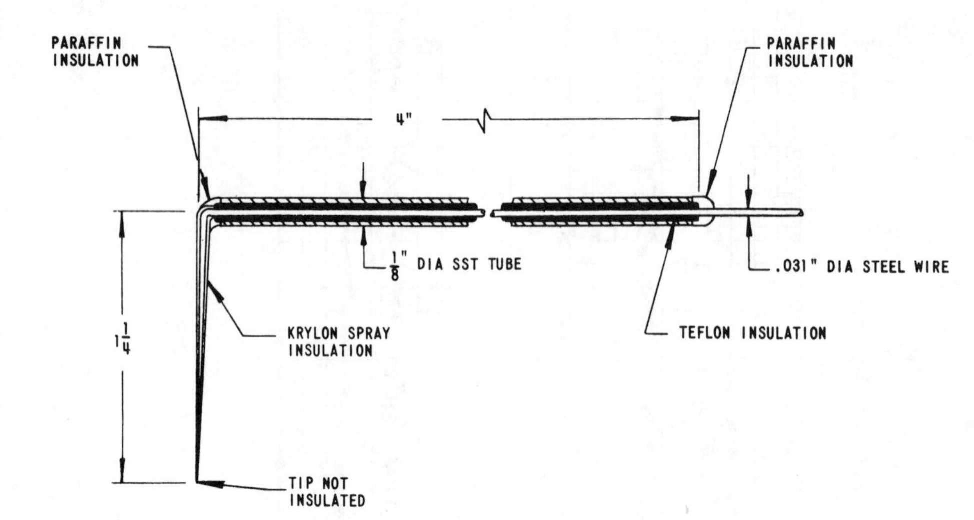

Figure 4 Electrical conductivity probe developed by Solomon [13].

ments for vapor fractions were low. Nassos stated that some improvement was obtained by means of a separate triggering device, but the calculated vapor fractions were still somewhat lower when compared with gamma-ray attenuation measurements.

Wallis [17] used a similar mechanism in which, instead of having the ground electrode attached flush to the inside tube surface, a duplicate probe was inserted from the opposite wall, as shown in Fig. 5. The principle of operation was exactly the same as that used in the Nassos probe and in Kordyban and Ranov's probe, except that the liquid bridge must be between the two electrodes, which can be placed at any point covering any amount of the channel diameter.

An interesting fact arose in Wallis' experiments. When he compared data on the transition between slug flow and annular flow from the Nassos-type probe on the one hand, and from his double probe on the other hand, in all cases under similar conditions and the same transition criteria, he found a definite discrepancy between the two sets of data, as shown in Fig. 6. The discrepancy was due to the way in which the slug-annular transition was defined, both conceptually and implicitly through the type of instrumentation used. Certainly, if a liquid conduction path exists between the tube centerline and the wall, and if the signal is a pure binary signal regardless of the path, then there should be no difference between transitions indicated by the two probes. However, if large globules of liquid float down the core of the pipe in annular flow, these would appear to be bridges to the dual sensor but not be seen by the single probe. Conversely, if the conductance path for the single probe is tortuous, and the signal is not truly binary in nature, then subjectiveness on the part of the observer and arbitrariness must result in variations in one's estimation of transition over another. In spite of these drawbacks, a number of other workers have used both the single-wire probe [18 to 23] and the double probe [24 to 26], mainly because of their simplicity.

Considerable other information may be obtained from conductivity devices. If local velocity is obtained, bubble diameter distributions can also be measured [48 to 50], as well as gas and liquid slug lengths [48]. Uga [51] obtained histograms of bubble sizes in the riser and downcomer sections of a Boiling Water Reactor (BWR). However, he used average values of bubble velocities rather than instantaneous values corresponding to the dewetting signals, so that his size distributions were really normalized inverse dry-time distributions, where a constant value of velocity was the normalized factor.

Similarly, Ibragimov et al. [52] used miniature traversing probes to obtain bubble frequency profiles in water-nitrogen flow similar to those obtained by Jones [53] with the anemometer. Sekoguchi et al. [49] reported histograms of bubble sizes using a double-tipped probe with tips 4 mm apart and 30 μm in diameter. Using the transport time averaged over 10 observations, the bubble rise velocity was obtained, which was then used to obtain the bubble size from the dewetting time. Similar to the method of Uga [51], this method requires the assumption that there is no correlation between bubble size and rise velocity. While this may be true for the larger than 1 mm bubble size observed by Sekoguchi [49], it certainly is not true in general. It is interesting to note, however, that these workers obtained excellent agreement between void residence-time profile and the volume-fraction profile calculated from the bubble diameter and frequency measurements, in agreement with the theoretical analysis by Serizawa [54] for long sampling times.

In their studies of thin-film characteristics of annular two-phase flow, Telles and Dukler [55], Dukler [56], and Chu and Dukler [57] used the straight electrical probe to determine information on wave structure. Such information included probability density functions for wave height, where the contributions due to the

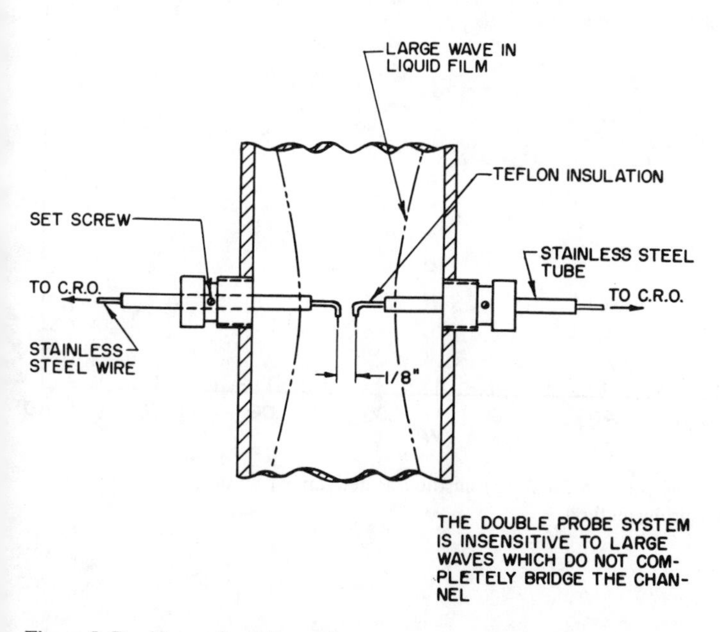

Figure 5 Double-conductivity-probe arrangement tried by Wallis [17].

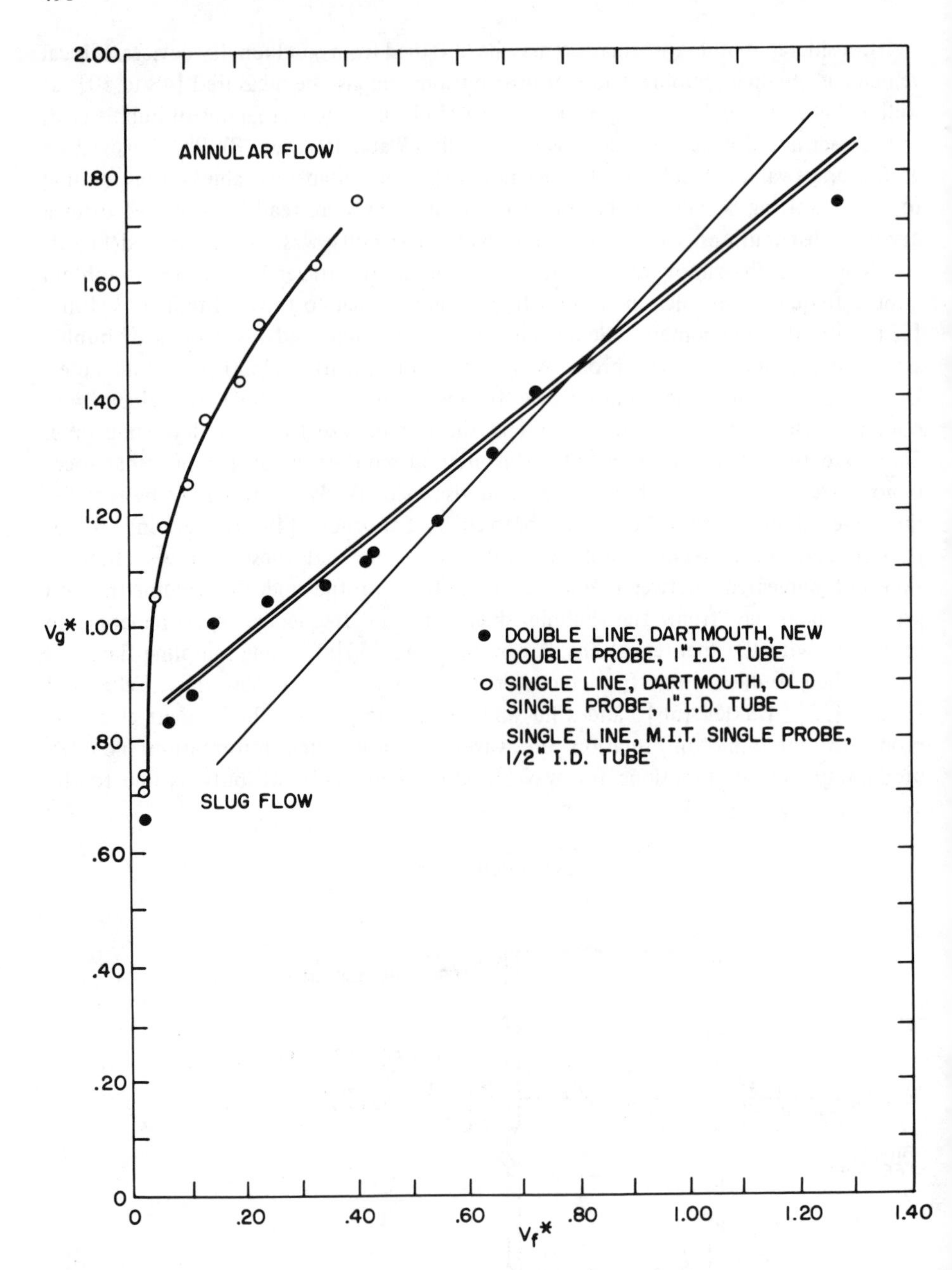

Figure 6 Comparison of the slug-annular flow transition as determined by the single conductivity probe and by the double-probe method.

film substrate and due to large waves were identified along with wave frequency. Also, the spectral and cross-spectral densities of film-thickness fluctuations were determined.

Other statistical data can be obtained by using a double probe. Several authors tentatively described the granulometry of bubble flows with sophisticated models [51, 54, 58]. A bubble displacement velocity has been sought by the same authors and by Lecroart and Porte [59], Kobayasi [40], and Galaup [50].

Transport methods have now begun to be utilized to obtain velocity information, although no one to date has specifically identified the correlated quantity as interfacial passage, so that the velocities thus obtained are *interfacial velocities*. Serizawa [54], using a double-tip probe with tips 5 mm apart, utilized both correlation and pulse-height methods to determine bubble velocity spectrums. Correlation of the outputs of the two probes after passing through Schmitt triggers provides a function that exhibits a well-defined maximum at a time delay corresponding to the transport time between probes. Dispersion of the amplitude of the correlation function is representative of the probability distribution of this velocity, and hence of the fluctuations. Also, by using one probe signal as a starter and the other as a stopper, ramp functions are generated during the transport time; when stopped and differentiated, they yield pulses whose heights are proportional to the transport time delay. Height analysis of this pulse train is assumed to yield the bubble velocity spectra.

3.1.3 Wall probes. A second type of conductivity instrument is a flush-mounted type developed at the Atomic Energy Research Establishment (AERE) at Harwell, England, and used in a number of studies in that laboratory [27 to 32]. This device, which is shown in Fig. 7, can be mounted for use in both tubes and rods. The operational principle is similar to that previously discussed, except that a high-frequency alternating current (ac) carrier is used. Thus, when the probes are covered by a thin liquid film, the signal is amplitude-modulated by the film itself, such that the output is related uniquely to the film thickness. This type of probe was used to determine the transition between slug and annular flow and to measure film thicknesses in annular flow when the probes are calibrated.

A similar type of sensing device was also used at AECL in Canada [33, 34], where a kicksorter circuit, shown in Fig. 8, was used to ignore all waves of thicknesses less than a predetermined value. In this manner, a wave-thickness spectrum may be determined. The signal can also be half-wave rectified as shown in Fig. 9, and combined as shown in Fig. 10, such that the recordings from four separate measuring stations can be obtained simultaneously on a dual-beam oscilloscope [35]. Then, by comparing the outputs from probes at different stations, and keeping in mind that the flow patterns tend to retain their identity along the length of the tube, the velocities and frequencies of the waves may be determined.

Another type of surface conductance probe was developed at Centro Informazione Studi Esperienze (CISE) [36]. As shown in Fig. 11, the tube is entirely surrounded at various axial locations by metal rings flush with the inner tube surface. Whenever a potential is generated between any two rings, the conductance

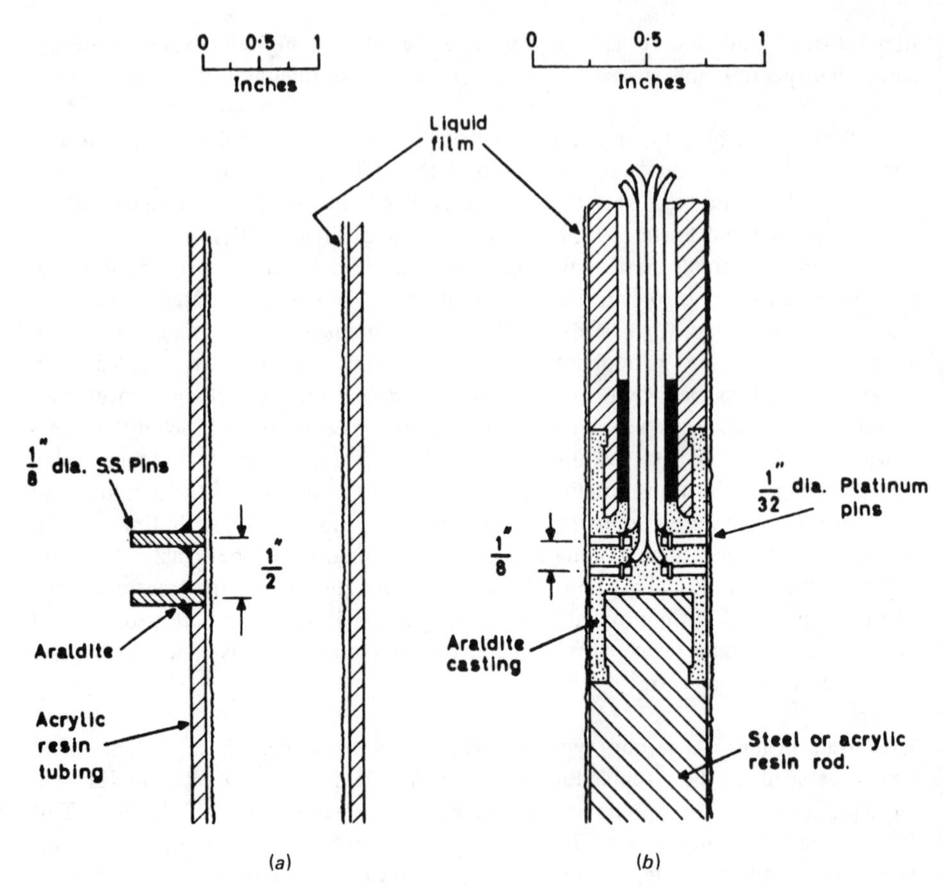

Figure 7 Installation characteristics of the Harwell flush-mounted conductance probe. (*a*) Measurements inside tubes. (*b*) Measurements outside rods. *(From [27].)*

is a function of the amount of fluid between the rings. This type of instrument is insensitive in annular flow to rapid changes in film thickness and high, sharply peaked waves.

Reference [37] describes a test assembly (shown in Fig. 12) that was used to compare both the AERE and CISE type conductance probes. In addition to the CISE and AERE conductance devices, the test assembly contained quick-acting, simultaneously operated valves on both ends, which enabled the air-water mixture to be completely isolated so that a measurement of the total amount of each phase could be made. The results of these tests are shown in Fig. 13. In the figure, several things are immediately evident. First, the CISE probe gave measurements definitely lower than those produced by the AERE instrument. This is probably due to the fact that the CISE probe is not sensitive to sharp or high wave peaks and thereby underestimates the average film thickness. This is evidenced by the approach of the CISE data to the AERE data with increasing film thickness, i.e., where the wave peaks start to make less and less difference in the overall film thickness. Second, the holdup method appears to show film thicknesses that are larger on the average than the AERE measurements. This is because the holdup method measures not only the liquid in the film itself, but also the liquid that has been entrained in the gaseous core. In some instances, a large amount of the total liquid in the system is in the form of entrained droplets. Thus, any film-thickness measurement that does not

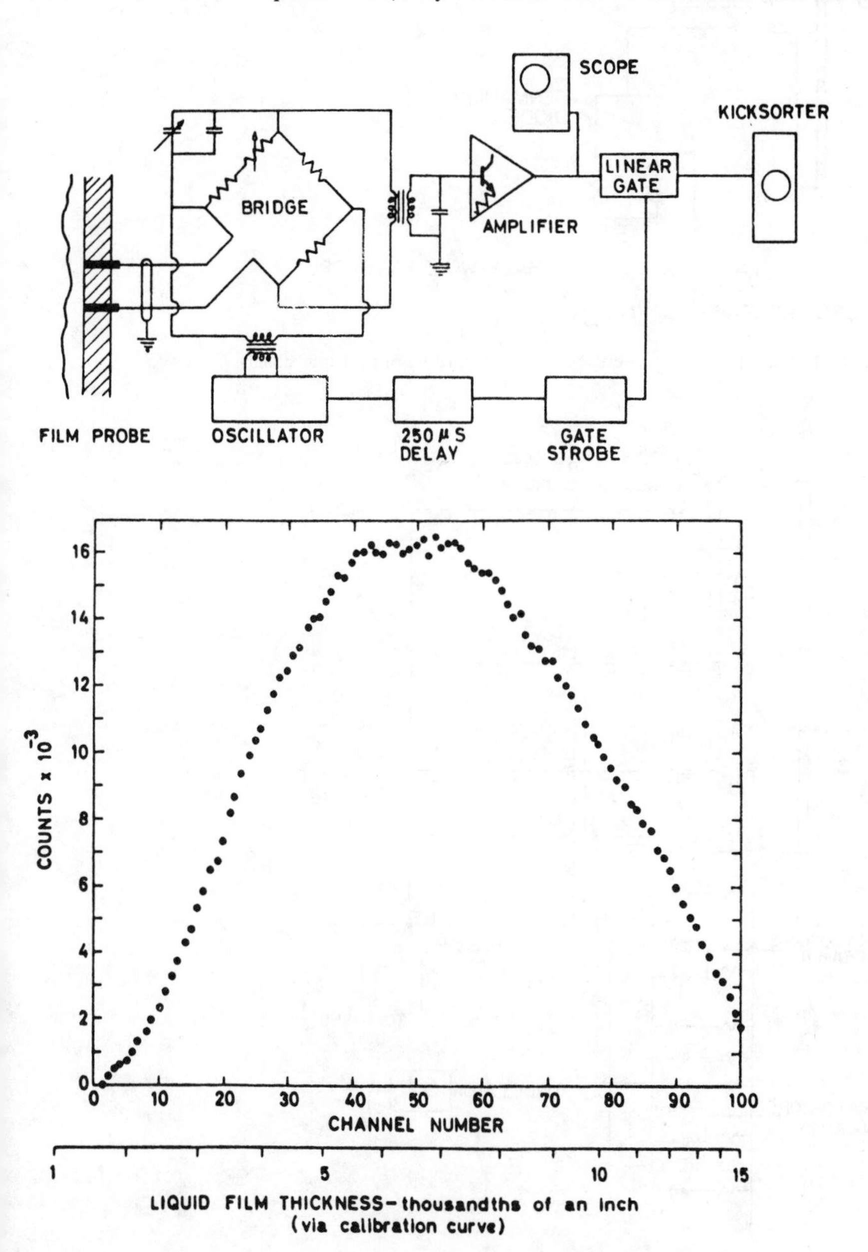

Figure 8 Kicksorter circuit and representative liquid-film amplitude distribution. *(From [33].)*

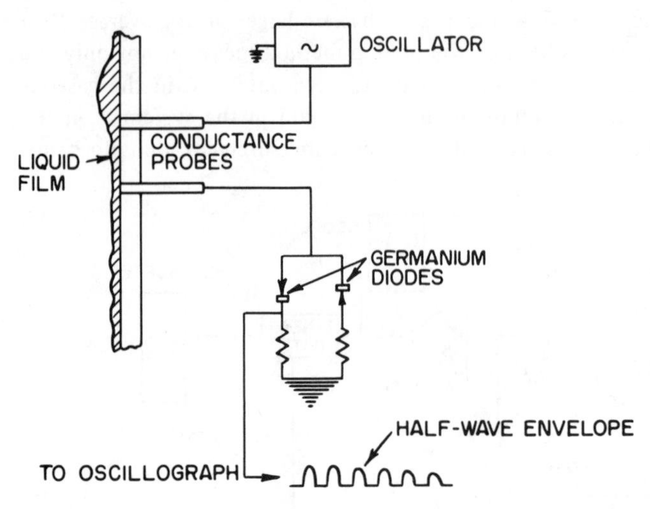

Figure 9 Method of half-wave rectification for recording of multiple signals. *(From [35].)*

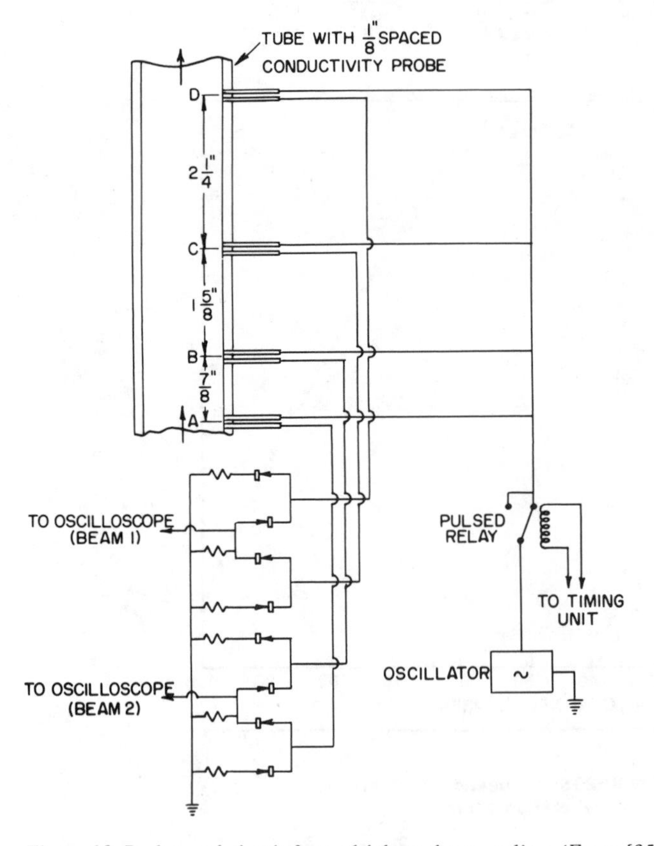

Figure 10 Probes and circuit for multiple-probe recording. *(From [35].)*

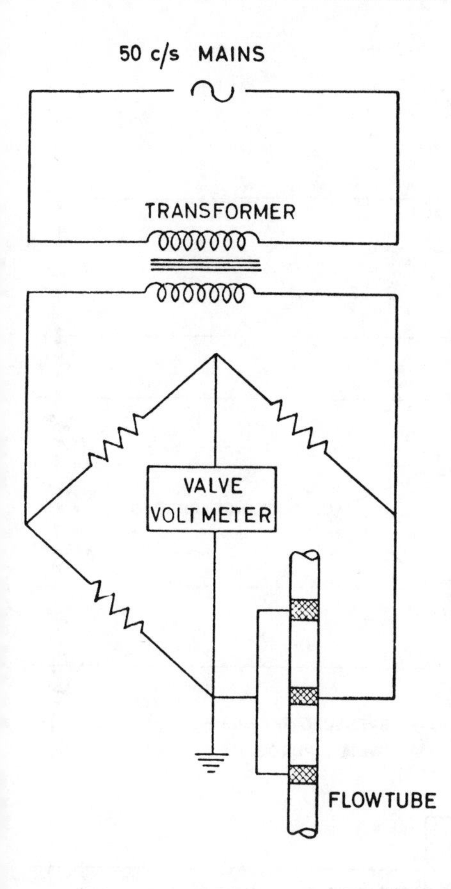

Figure 11 Schematic of film-conductance method developed at CISE [36].

separate the entrained liquid from the film liquid must necessarily be overestimating the average film thickness. It is also probable that the AERE probes miss large waves, thereby underestimating the average and instantaneous film thickness.

Problems encountered in the use of these instruments show that, as mentioned previously, wave peaks are not seen by the CISE instrument and, in some cases, not by the Harwell instrument. This may make the CISE instrument underestimate the average film thickness by as much as 40% in some cases. In addition, these instruments both require conducting mediums and thus may be difficult to use in systems having high-resistivity water-chemistry requirements. The AERE method requires an extremely well-designed geometry mockup for calibration of the instrument. Any misalignment or nonconcentricity in the calibration unit itself can have a significant effect on the calibration.

An excellent review of conductance measuring devices is given in [37]. In general, the characteristics of such devices have restricted their use to studies where the continuous phase is a conducting medium, their major application being in film-thickness studies. Bergles and his coworkers at Dynatech had some success in the use of needle-type probes as flow-regime detectors [60], but the discussion here seems to lend some uncertainty to this application.

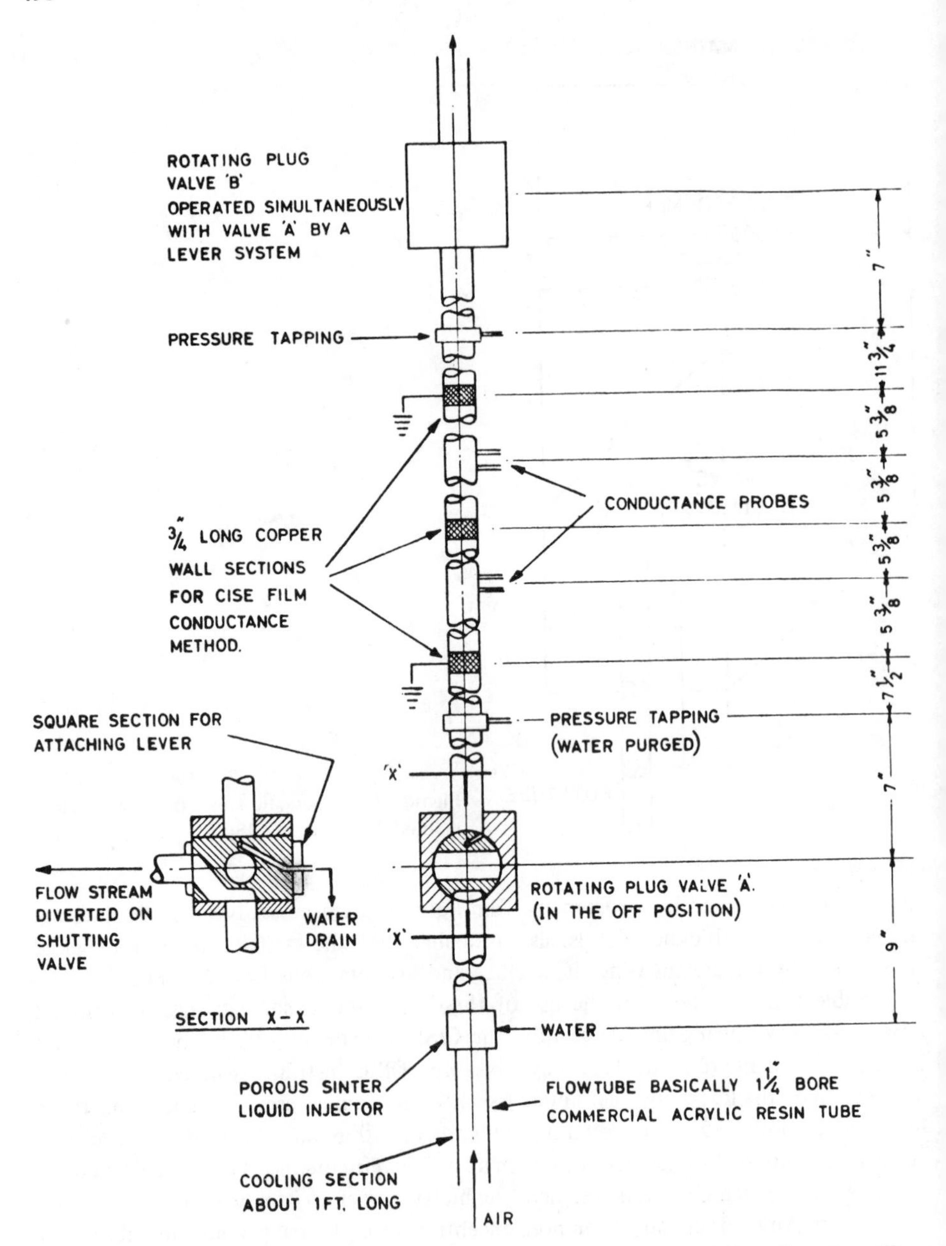

Figure 12 Layout of test assembly used at Harwell to compare CISE and Harwell probes. *(From [37].)*

3.1.4 Summary–conductivity probes. Conductivity devices of both the probe type and the wall type have been used for many years for various purposes, including local void fraction (residence-time fraction) measurement, flow-regime indication, and film-thickness wave celerity, and when coupled with a velocity indication, bubble-size and slug-length distributions. Two used in tandem can also be inter-

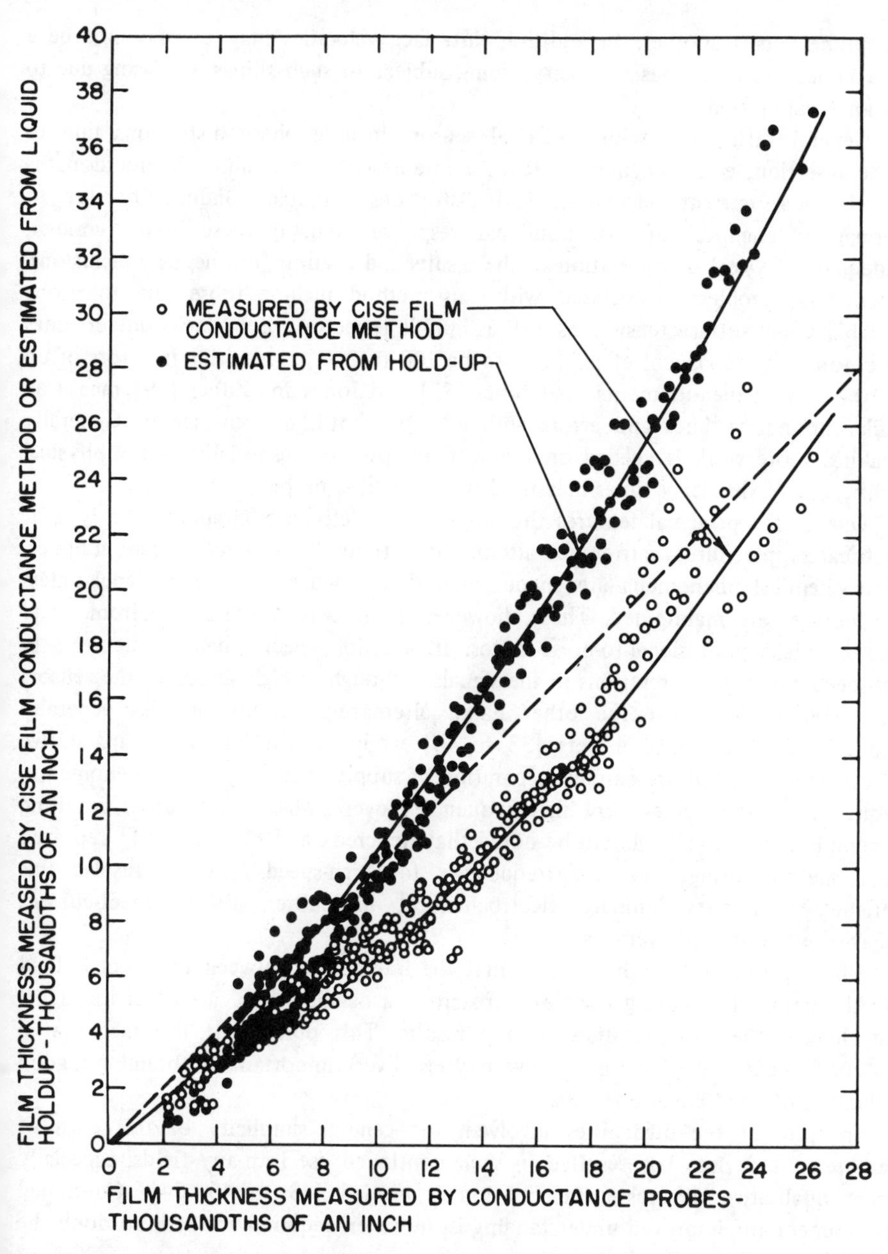

Figure 13 Comparison of data obtained by CISE and Harwell conductance probes. *(From [37].)*

preted to give a velocity indication. It should be understood that since the probe-type device is basically a phase indicator, any transport information interpreted as a velocity between two points is really propagation of a phase delimiter or interface, i.e., interfacial velocity. As such, it is subject to all the effects that tend to produce errors at a single probe and may as well include selective distortion at one probe versus another. In addition, interface velocities may or may not be a good indication of a phasic velocity, being subject to such things as biasing due to slip and evaporation.

General difficulties with electrical sensors include phase distortions due to probe insertion, electrochemical effects, nonrepresentative conductive liquid paths for electrical current, electrical field distortions, variable conductivity due to thermal or chemical effects, nonlinear response, and, if these aren't enough, inadequate physical interpretation of the results and resulting erroneous conclusions. Specifically, problems associated with this method include those due to probe wettability and surface tension, as well as bubble trajectory. Boundary contact times were noted by Sekoguchi et al. [45] to be 100 to 200 μs, resulting in errors of up to 10% in void measurements, and Jones [53] and Jones and Zuber [69] measured similar boundary times and errors with a 50-μm hot-film anemometer. Generally speaking, more work is needed on this difficult problem, especially on the physical significance of the delay times measured with a double probe.

One of the principal features that differentiates electrical circuits is the type of electrical supply, direct current or alternating current. With direct current supplies, electrochemical phenomena are encountered that obscure the desired signal unless low voltages are maintained. These, however, lead to complicated electronics and tend to yield poor signal-to-noise ratios. In addition, electrochemical deposits in low-speed flows give alterations in the signals, although at high velocities the sensors may remain clean. On the other hand, alternating current supplies generally eliminate electrochemical effects [38 to 40] while substituting stray capacitative effects. One must ensure suitable separation of supply frequency from phenomenon frequency. In some cases, very high frequencies (over 1 MHz) are required, resulting in complex circuitry. It should be noted that Reocreux and Flamand [41] reported on a method using very low frequencies for high-speed flows to resolve this difficulty and yet eliminate electrochemical difficulties. Pseudo-direct-current behavior was obtained every half-wave.

Finally, Tawfik [42] has shown that the interaction between the electric field and the liquid field during bubble approach to a needle probe can affect the signal and, hence, the interpretation of the results. This points out the fact that a thorough knowledge of the interactive physics is very important in obtaining a good evaluation of experimental results.

In spite of the difficulties involved, the general simplicity of this class of measurement devices has resulted in their continued use in many fields, especially where qualitative diagnostic information is needed [43 to 46, 54]. Continued development for improved understanding is, however, required and will no doubt be undertaken as the need arises.

3.2 Impedance Void Meters

Another in-stream electrical measuring instrument is the impedance-type void sensor [60]. While these devices may be considered a subset of the previous types, significant differences in geometry exist, and they are usually used in the impedance mode, relying on phase variations of the capacitative reactance. Several types have been designed and used. The concentric-cylinder meter, shown in Fig. 14, consists of concentric, thin-walled, short cylinders that are alternately connected to one of two electrodes. When connected to an air supply, the relative impedance may be measured as a function of the vapor fraction.

The void meter may be calibrated by means of a system such as is shown in Fig. 15. Measured amounts of water and air are injected into the base of a vertical section. The water velocity is measured initially by means of a turbine meter. In the upper portion of the channel, another turbine meter is placed immediately downstream of the impedance void gauge. Thus, the mixture velocity, coupled with an accurate knowledge of the air and water flow rates, may be used to calculate the void fraction. This calculation is then compared with the impedance of the void meter and used as a calibration point. In this manner a curve of relative impedance versus vapor fraction can be easily obtained. There does, however, seem to be a sensitivity to the void distribution between the plates, as shown in Fig. 16. This is in accordance with established theory [61, 62]. Here there may be an effect amounting to 15% or more, depending on whether the voids are distributed

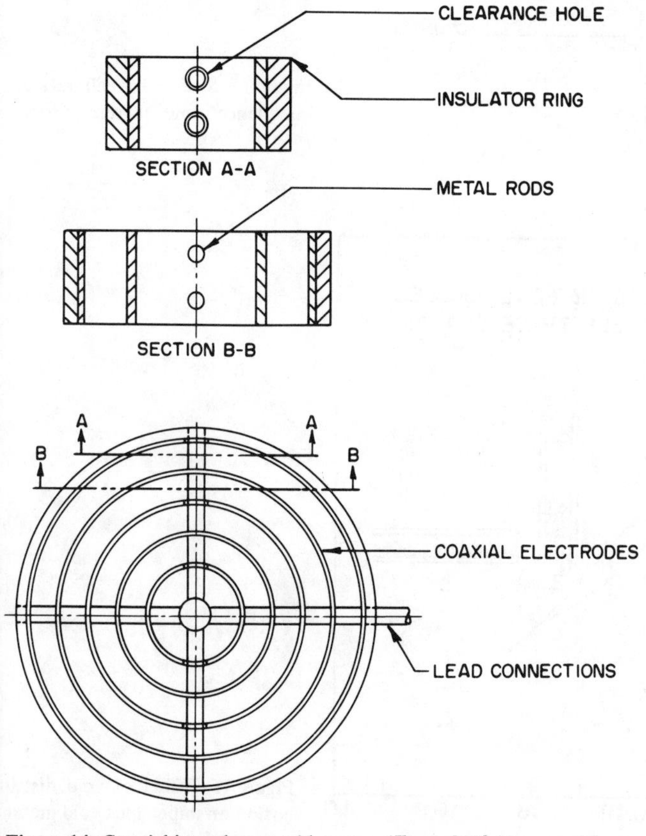

Figure 14 Coaxial impedance void meter. *(From [60].)*

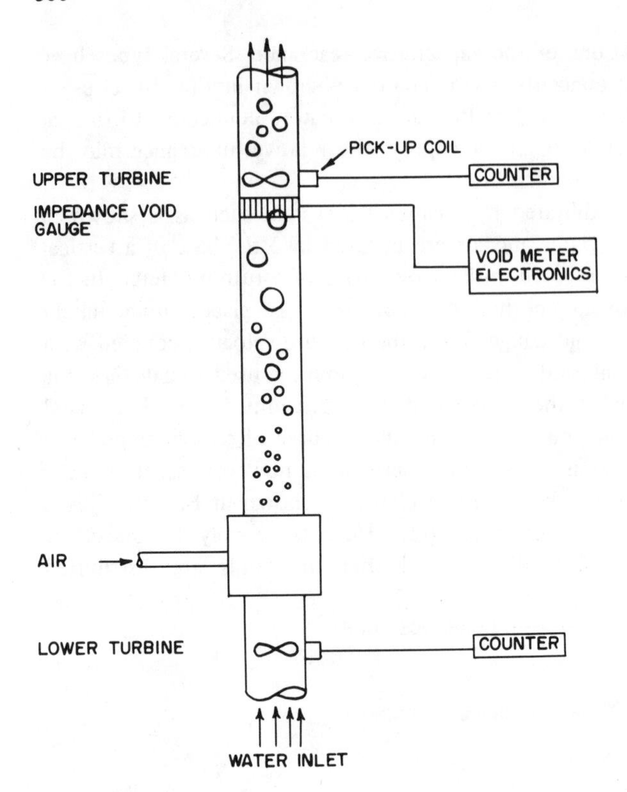

Figure 15 System for calibrating impedance void meter. *(From [60].)*

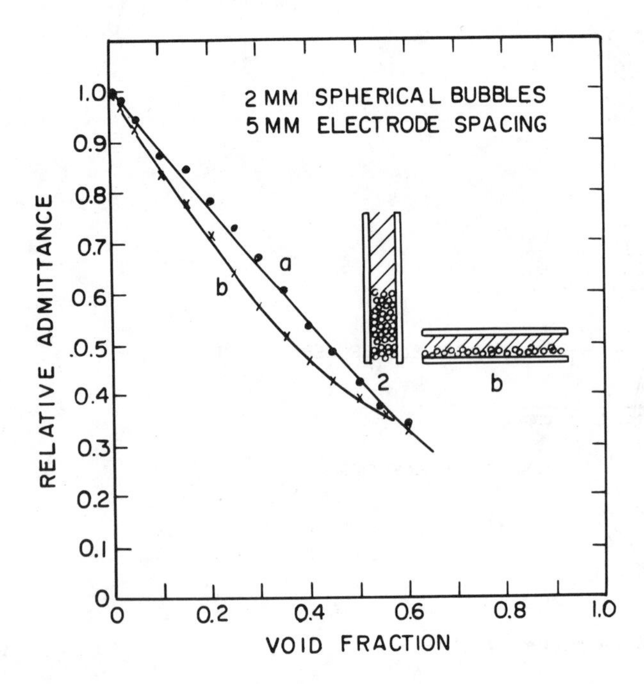

Figure 16 Effect of void distribution on impedance void meter output. *(From [60].)*

horizontally or vertically with respect to the parallel plates of the void meter, and on the relative dielectric constants for the two phases.

Similar devices have been developed and used by a number of other investigators [63 to 67]. Marked effects due to void distributions at constant void fraction are also confirmed by Cimorelli and Premoli [66]. In general, these sensors are bulky, and the methodology is unsuitable for adaptation to small-volume measurements for use in small geometries.

3.3 Hot-Film Anemometer

Hot-wire and hot-film anemometers have been widely used in gases. More recently, miniature cylindrical probes have been used for accurate measurement of low velocities (for instance, <0.5 m/s in water [181]; <0.25 m/s in mercury [182]; <0.6 m/s [183]; and <0.05 m/s [184]). The larger and more sturdy wedge and conical probes have enjoyed greater success at velocities up to about 5 m/s (<3.7 m/s [185]; <5 m/s [186]; < 4 m/s [187, 188]). It has been found that hot-wire or hot-film anemometry can be used in two-component two-phase flow or in one-component two-phase flow with phase change. In the first case, an air-water flow, for example, it is possible to measure the local void fraction, the local liquid volume flux or instantaneous velocity, and the turbulence intensity of the liquid phase in conjunction with the arrival frequency of bubbles or droplets. In the second case, a steam-water flow, for example, it has been so far impossible to obtain consistent results on calibrated liquid velocity measurements.

The hot-film anemometer was tried as a two-phase flow indicator by Hsu et al. [68], Jones [70], and later by Jones and Zuber [69]. The study by Hsu et al. indicated that this instrument could be useful in measuring all the parameters measurable by the previously described instruments and, additionally, provide both temperature and phase velocity measurements. However, this instrument has not received wide acceptance to date because of its extreme fragility and short lifetime.

Figure 17 includes a close-up view of the probe, showing that it consists of a small-diameter glass by cylinder covered with a platinum coating and connected on either end to a copper lead wire. In operation, the resistance of the probe is set by means of an electric current at a value corresponding to a desired probe temperature. If the probe resistance begins to change from the preset value due to a change in temperature, the control unit responds with a change in current to maintain a constant probe resistance. This setup is shown schematically in Fig. 17, where the probe is one arm of a four-arm bridge. The output of the transducer is measured as a voltage drop across a calibrated resistor by an oscillograph.

Typical output traces for this instrument in different flow regimes are shown in Fig. 18, where it is possible to compare the traces from bubbly flow, slug flow, and mist flow with low droplet concentration. The different flow regimes are readily discernible, one from the other. The only conditions that might cause confusion are bubbly flow with large bubble concentration as opposed to mist flow with very rich droplet concentration. As seen in Fig. 18, these two traces could look quite similar; however, one distinguishing feature would be the relative point on the scale at

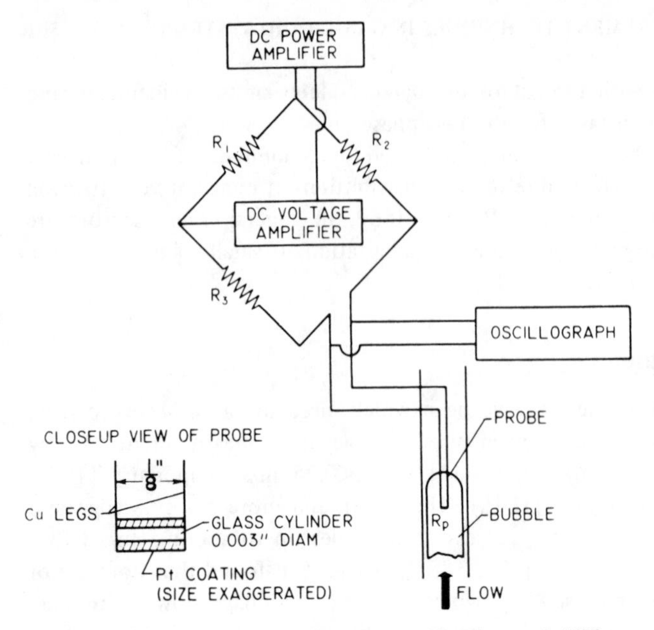

Figure 17 Hot-film anemometer system used by Hsu et al. [68].

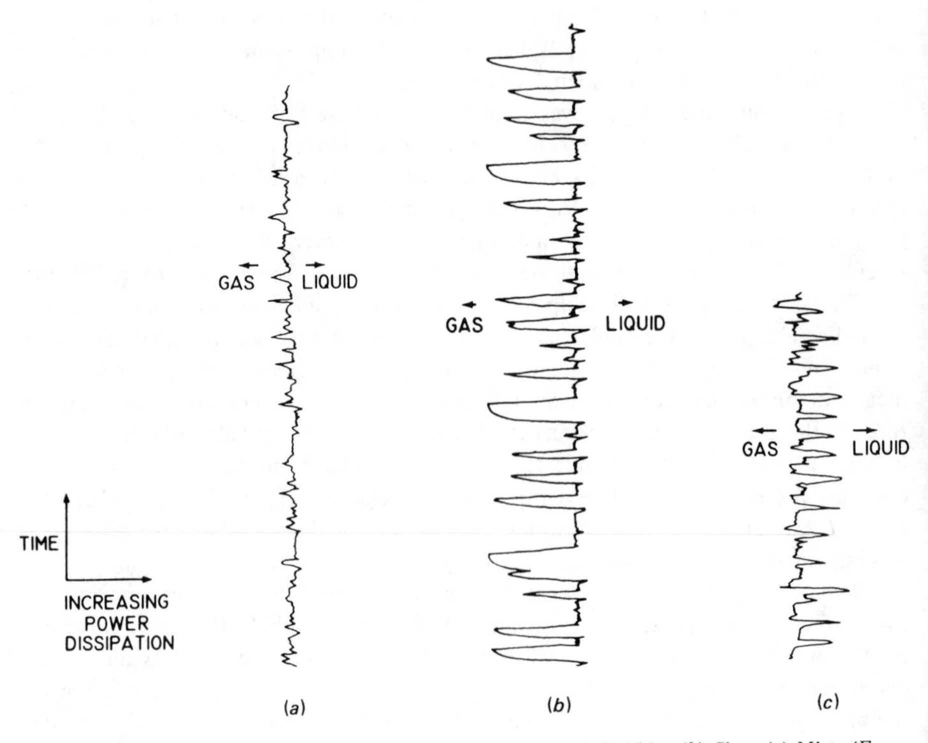

Figure 18 Typical output from the hot-film anemometer. (*a*) Bubbly. (*b*) Slug. (*c*) Mist. *(From [68].)*

which the trace occurs. In bubbly flow, there would be a larger power dissipation on the probe than in mist flow, owing to the smaller void fraction.

Figure 19 shows the typical trace of a bubble as it passes, the probe. From bottom to top in this figure, as the bubble approaches the probe, the vanguard consists of locally increased velocity as shown by increasing power dissipation. The entrance of the probe into the void is shown by a sharp decrease in power dissipation, followed almost immediately by an increase as it enters the wake. The same effect occurs again with another bubble and then, as the probe exits from the effect of the wake, the power dissipation decreases, corresponding to the decreasing local velocities. In addition, since there is apparently a relatively rapid response of the transducer when entering a void, this instrument was expected to give good quantitative information on local vapor fractions.

3.3.1 Measurements in two-component two-phase flow without phase change. Hot-wire anemometry has been used for measuring the concentration flux and the diameter histogram of liquid particles moving in a gas stream. Goldschmidt and Eskinazi [189, 190] measured the arrival frequency of liquid droplets, 1.6 to 3.3 μm in diameter, with a constant-temperature anemometer and a cylindrical probe, 4.5 μm in diameter. When the impaction frequency of the droplets is different from

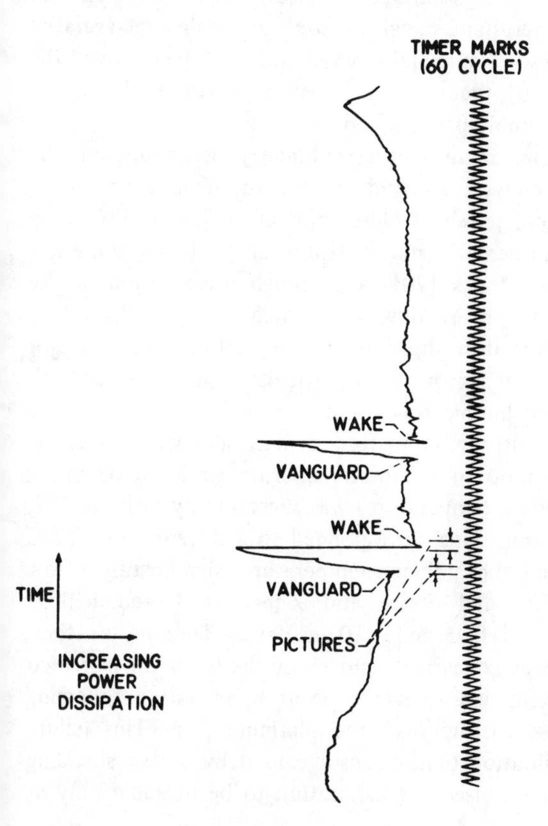

Figure 19 Wake and vanguard effects due to bubble passage by a hot-film anemometer. *(From [68].)*

the energetic frequency range of the turbulent gas stream, the signal fluctuations due to impacts can be distinguished from the fluctuations due to turbulence. In experiments by Goldschmidt and Eskinazi, results showed that the ratio of the impaction frequency to the maximum impaction frequency was insensitive to the threshold amplitude of a discriminator used to produce a binary chain of pulses due to droplet impactions. This fact has also been observed by Lackme [191] for void-fraction measurements with a resistive probe. Ginsberg [192] used the same technique to study liquid-droplet transport in turbulent pipe flow. Goldschmidt [193] determined that the measured impaction rate is lower than the true value but proportional, and should thus be calibrated against another technique.

In determining droplet-diameter histograms, Goldschmidt and Householder [72, 194] theoretically found a linear relationship between particle diameter and cooling-signal peak value, which was verified experimentally for droplet diameters lower than 200 μm. Bragg and Tevaarverk [73], however, contradicted these results and concluded that the hot wire was unsuitable for this purpose. This conflict has yet to be resolved.

Time-averaged gas velocities and gas turbulent intensities were measured by Hetsroni et al. [204] in low-concentration mist flow. The spikes due to the impingement of the liquid droplets on the hot wire were eliminated with the help of an amplitude discriminator and a somewhat simpler electronic circuit than that of Goldschmidt and Eskinazi. The resultant signal was used to obtain time-averaged gas velocity and turbulent intensities. Chuang and Goldschmidt [195] employed the hot wire as a bubble size sampler by theoretically investigating the nature of the signal due to the traverse of an air bubble past the sensor.

Despite several difficulties arising in droplet granulometry determination, the hot wire has successfully been employed for studying the turbulent diffusion of small particles suspended in turbulent jets by Goldschmidt et al. [196]. Following the studies done by Goldschmidt in aerosols and by Hsu et al. [68] in steam-water flow, and the preliminary work of Jones [70], a thorough investigation of the hot-film anemometry technique in two-phase flow was carried out by Delhaye [77, 78] using a conical constant-temperature hot-film probe that has three major advantages over the cylindrical hot-film sensor: small particulate matter carried with the fluid does not attach to the tip, bubble trajectories are less disturbed, and the relatively massive geometry is less susceptible to flow damage at higher velocities. The maximum overheat resistance ratio of the probe of 1.05 (ratio of operating resistance to resistance at ambient fluid temperature) was suggested by Delhaye [77, 78] to avoid degassing on the sensor. This corresponded to a difference of 17°C between the probe temperature and the ambient temperature, significantly below saturation temperature. Jones [53] and Jones and Zuber [69] found little difference between resistance ratios of 1.05 and 1.10 insofar as degassing on their 50-μm-diameter cylindrical sensor was concerned, and chose the latter for increased sensitivity. Degassing in their system was found to occur in operation following failure of the 8000-Å-thick quartz coating over the platinum film. This failure occurred during forced resonant vibration of the sensor caused by vortex shedding at velocities over 1.5 m/s. Degassing caused the calibration to be unstable only at velocities less than ~30 cm/s.

Delhaye [74 to 77] developed some rather sophisticated anemometry techniques for measuring void fraction in large tubes with the rugged conical hot-film probe. Using a multichannel analyzer, he obtained amplitude histograms of the voltage signal produced for a constant-temperature-sensor control system in a two-phase mixture. In Fig. 20*a*, one may see that in the ideal case the anemometer voltage is either at the voltage E_l, corresponding to the presence of liquid at a given velocity, or at the voltage E_g corresponding to all gas at the given gas velocity. The multichannel analyzer periodically determines the amplitude of the signal voltage. A block of memory locations is allocated, where each location represents a different

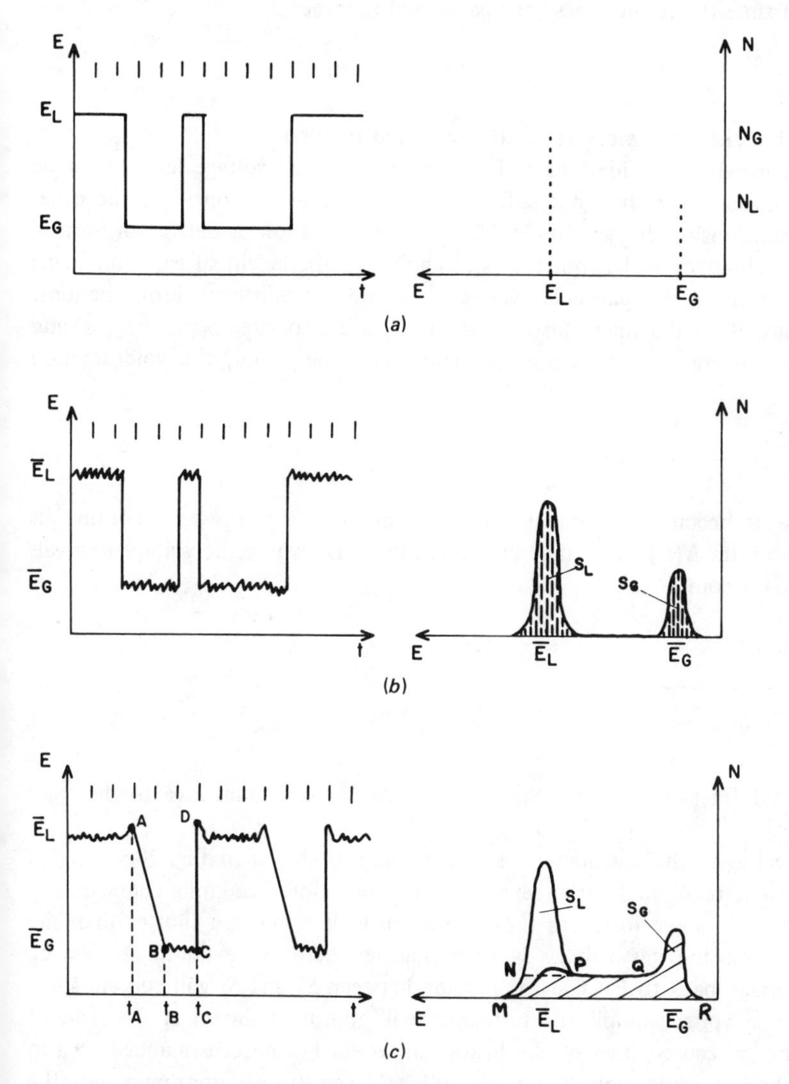

Figure 20 Delhaye's method for local void fraction measurement using multichannel analysis. (*a*) Ideal case of phase effect only. (*b*) Ideal case of phase effect with added noise. (*c*) Real case with noise (turbulence) and finite dryout and rewet times. *(From [76].)*

voltage-level range of width ΔE. The first memory location represents the voltage range E_{min} to $E_{min} + \Delta E$, while the last in the block represents $E_{max} - \Delta E$ to E_{max}. At the time the voltage level is determined by the analyzer, one count is added to the memory location representing the voltage range within which the current amplitude was found to be. In this manner, each location or channel contains a count proportional to the fraction of the total sampling time during which the voltage level was within the range assigned to that channel. For the ideal case, then, Fig. 20 shows that the histogram on the right would contain N_l counts for liquid voltage amplitude and N_g counts for gas voltage amplitude. The percentage of time the probe "sees" the gas would then be

$$t_g = \frac{N_g}{N_g + N_l} \tag{2}$$

which could be taken as a measure of the local void fraction α.

In the somewhat less ideal case (Fig. 20*b*) where each voltage level has some fluctuations associated with it but still the switching from one phase to the other occurs instantaneously, the amplitude histogram would show a cluster of vertical lines centered about E_l and another cluster about E_g, the height of each line being the number of times the particular voltage level was encountered during periodic sampling. Thus, if all the lines clustered about E_g are associated with the gas, and all the lines clustered about E_l are associated with the liquid, the void fraction would be

$$\alpha = \frac{\Sigma\, N_{gi}}{\Sigma\, N_{gi} + \Sigma\, N_{li}} \tag{3}$$

For this case it becomes convenient to begin thinking in terms of a continuous amplitude spectrum $N(E)$, such that in the limiting case where the voltage intervals vanish, Eq. (3) becomes

$$\alpha = \frac{\int_{A_g} N(E)\, dE}{\int_{A_T} N(E)\, dE} = \frac{S_g}{S_g + S_l} \tag{4}$$

where the void fraction becomes the ratio of the gas histogram area to the total histogram area.

In the real case, the anemometer signal is more as shown in Fig. 20*c*, where a finite time AB is required for the sensor to dry out before becoming characteristic of all gas (BC). In some instances the gas residence time may be shorter than the dryout time, and the gas voltage is never reached before the liquid appearance forces the voltage back to E_l. Thus, the region between S_g and S_l will contain some counts, and the voltage amplitude histogram will appear as shown at the right of Fig. 20*c*. The gas-caused area of the histogram would be the crosshatched area in the figure, having some counts for all voltage levels up to approximately the maximum for all liquid.

Delhaye calculated the void fraction locally by application of Eq. (4), where, in practice,

$$\alpha = \frac{A_{MNPQR}}{A_T} \tag{5}$$

To a first approximation, then, the local void fractions were calculated as the ratio of the hatched area to the total area, which then compared favorably with radiation absorption methods (gamma rays). Notice that trigger levels for S_g and S_l affect the result. Delhaye adjusted both to get an accurate comparison between the resulting line-averaged void fraction and an independent gamma-ray measurement but developed no specific formula governing these settings. The liquid time-averaged velocity and the liquid turbulent intensity were calculated with the nonhatched area of the amplitude histogram (Fig. 20*c*) and the calibration curve of the probe immersed in the liquid. The same method has been extensively used by Serizawa [54] for measuring the turbulent characteristics and local parameters of air-water two-phase flow in pipes.

A different processing method was proposed by Resch and Leutheusser [197] and Resch et al. [198] in a study of bubble two-phase flow in hydraulic jumps. The nonlinearized analog signal from the anemometer is digitally analyzed. A change of phase is recognized when the amplitude between two successive extremes of the signal is higher than a fluctuation threshold level ΔE. In this way the liquid mean velocities and turbulence levels are obtained, along with bubble-size histograms. ΔE was chosen to be in a plateau region of ΔE versus measured void fraction.

Jones [53] and Jones and Zuber [69] used a discriminator applied to the raw anemometer signal to obtain a binary signal representative of local void fraction but found the cutoff level needed to be adjusted, depending on the local velocity, to a point just below the minimum value for a liquid. Even though the threshold value was set at every point in the traverse, errors in average void fraction were encountered when calibrated against an x-ray measurement. These errors were found to be dependent on the liquid volume flux and the mean void fraction.

Comparisons (Fig. 21) show that significant errors exist, especially at low liquid throughout rates. By choosing a relationship between the corrected and measured void fraction as

$$\alpha_c(y) = [1 + f_c(y)]\,\alpha_m(y) + \alpha_{zc} \tag{6}$$

where

$$f_c(y) = C_1(1 - \alpha_m)\left(1 - \frac{4y^2}{s^2}\right) \tag{7}$$

and

$$\alpha_{zc} = \frac{C}{K_\alpha + j_l}(1 - \alpha_m)\left(1 - \frac{4y^2}{s^2}\right) \tag{8}$$

and averaging in y, a relationship between C_1 and the averages could be obtained as

$$C_1 = \frac{\bar{\alpha}_c - \bar{\alpha}_m - \bar{\alpha}_{zc}}{\overline{f_c \alpha_m}} \tag{9}$$

C and K_α for all data were found to be 0.0055 and 0.028 m/s, respectively, while the averages in Eq. (8) were found from the data. Good results were obtained, as shown in Fig. 22.

By counting the number of times the output of the discriminator changed from one level to another, Jones [2] also obtained local values for interface frequency. He also measured the liquid-volume flux directly by time-averaging the linearized signal equal to the liquid velocity when the sensor was in liquid, and zero when the sensor was in gas. Liquid velocities were obtained by pointwise division of the measured liquid flux by the measured void fraction. The results were somewhat questionable, however, owing to the cracking of the 8000-Å-thick quartz coating mentioned previously. No attempt was made to measure the turbulent fluctuations.

Serizawa [54] used a conical probe of much more sturdy construction and larger size, similar to that of Delhaye [77, 78]. In bubbly and slug flow in air-water mixtures he used multichannel analysis techniques to obtain the frequency spectrum of the velocity signal, including fluctuations up to ~2 m/s. Ishigai et al. [199] used an anemometer to measure liquid-film thicknesses.

3.3.2 Measurements in one-component, two-phase flow with phase change. The earliest paper on hot-wire anemometry in two-phase flow seems to have been published by Katarzhis et al. [200]. This preliminary and crude approach was

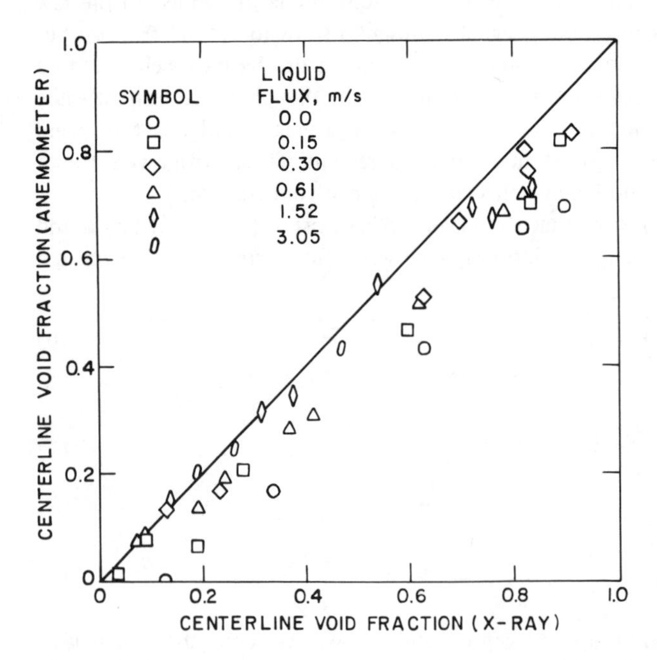

Figure 21 Uncorrected anemometer output for average void fraction. *(From [69].)*

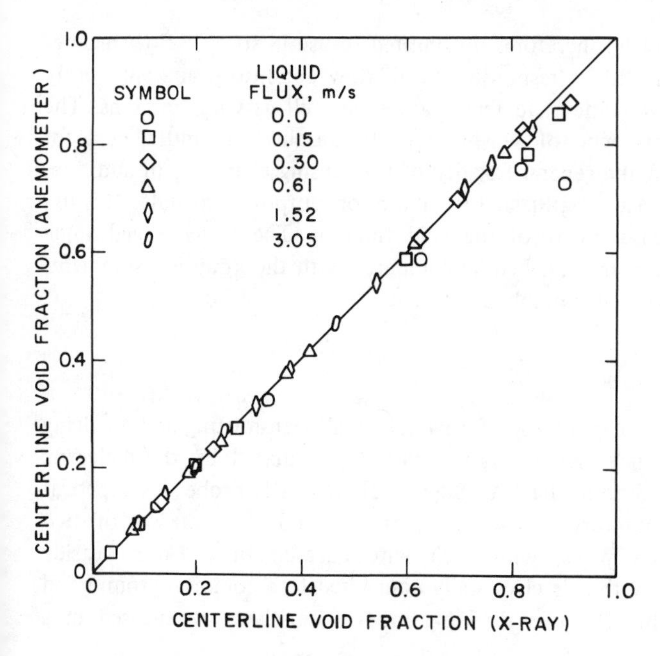

Figure 22 Corrected anemometer output for average void fraction per Eq. (6). *(From [69].)*

followed by the work of Hsu et al. [68]. These authors, by comparing the signal with high-speed movies, concluded that hot-wire anemometry was a potential tool for studying the local structure of two-phase flow, in particular for determining the flow pattern and for measuring the local void fraction. Hsu et al. [68] specified that in steam-water flow the only reference temperature is the saturation temperature. If water velocity measurements are carried out, the probe temperature must not exceed saturation temperature by more than 5°C to avoid nucleate boiling on the sensor. Conversely, if only a high sensitivity to phase change is desired, then the superheat should range between 5° and 55°C, causing nucleate boiling to occur on the probe when the liquid phase is present, and a resultant shift to forced-convective vapor heat transfer when the vapor phase is present.

The low electrical conductivity of freons enables bare wires to be used instead of hot-film probes. Shiralkar [201] used a 5-μm boiling tungsten wire with a very short active length (0.125 mm), so that the whole active zone would generally be inside a bubble or droplet. Local void fraction was determined by an amplitude discriminator with an adjustable threshold level. For void fractions lower than 0.3, the threshold was set just under the liquid level, whereas for high void fraction (0.8), it was set just above the vapor level. For void fractions ranging from 0.3 to 0.8, the threshold was set halfway between the liquid and vapor levels. The method was subsequently applied by Dix [202] and Shiralkar and Lahey [203].

3.3.3 Summary–anemometers. The major advantages of the hot-film sensor over the needle-type conductivity probe is that the sensor is self-contained, not requiring a secondary electrode, and is, therefore, not limited to use in specific flow regimes. This type of sensor appears to be responsive to all flow conditions and apparently provides information on velocities and temperatures as well as void fractions. The major drawbacks with this type of sensor seem to be the high initial cost for electronics and probes and the general fragility of the sensing elements. In addition, an independent method must be utilized for calibration purposes; to date, this has been a chordal x-ray measurement of the void fraction. The line-averaged value obtained from the anemometer may then be compared with the x-ray measurement, and parameters adjusted to yield suitable agreement.

3.4 Radio-Frequency Probe

A relatively new development is the use of separate, small transmitting and receiving antennas with radio-frequency (RF) signals amplitude-modulated by the dielectric coefficient of the surrounding media [50, 80 to 82]. The RF probe developed at Brookhaven National Laboratory, shown in Figs. 23 and 24, consists of two 0.25-mm-diameter insulated wires, with each wire encased in a 1-mm outside diameter stainless steel tube that is electrically connected to a common ground and acts as an electrical shield. The two shielding tubes themselves are encased in a

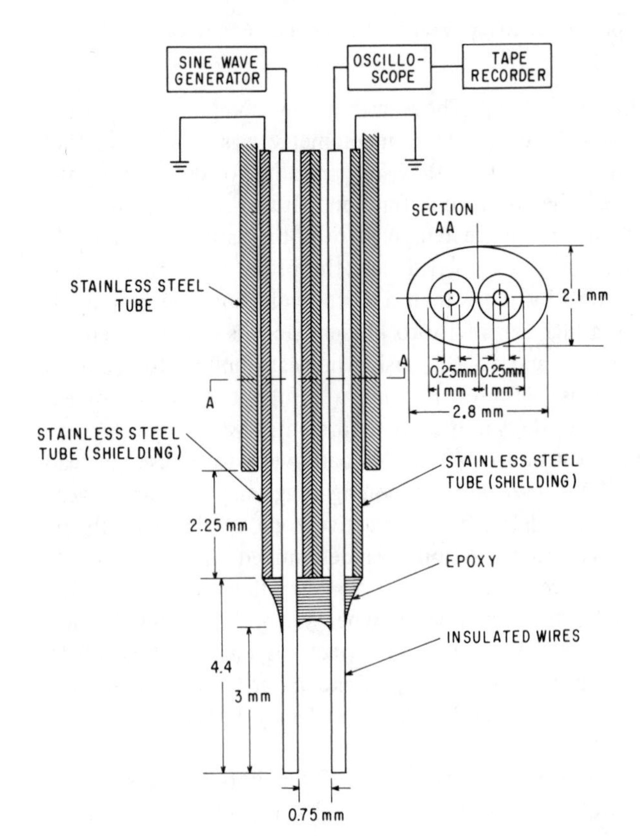

Figure 23 Schematic representation of the RF probe.

Figure 24 The RF probe.

larger stainless steel tube, which acts as a holder and provides rigidity. The sensitive part of the probe, the probe tip, is formed by extending the two insulated wires by about 3 mm from the end of the shielding tubes. To prevent water from entering into the stainless steel tubes, each of the end connections is covered with a thin layer of epoxy including the tip of the two insulated wires, which are also covered to insulate them from the surrounding medium, water or air. When a dc voltage is applied across one of the wires and the common ground, zero voltage is measured across the second wire and the ground. In operation, one of the wires is used as an emitter to which a sine wave is applied from a function generator. The second wire is used as a receiving antenna, and its output is fed directly to an oscilloscope or to a magnetic tape recorder after amplification. Similar RF probes were previously described in the literature [50, 81] but a systematic study of the response characteristics was not undertaken.

When a sine wave is applied to the transmitting antenna (input), the amplitude of the received signal (output) varies with the signal frequency of the input (from 100 to 10^7 Hz), both in air and in water (Fig. 25). Depending on the input frequency, the signal amplitude in water can be higher than the signal in air or vice versa. The RF probe seems to act as a band-pass filter. For a 500-kHz, 22.7-V peak-to-peak sine-wave input, the output voltage of the RF probe was observed to be dependent on the static immersion depth of the insulated nonshielded portion of the probe tip into the water. The output increased linearly with the immersion depth, reached a maximum, and then decreased and leveled off at the all-water signal level. An additional fact observed was that the output versus input curve as presented in Fig. 25 depends on the tube or pipe diameter in which the probe is immersed. Thus, before undertaking any application of this probe for a specific geometry, a careful signal optimization with input frequency should be performed.

The probe with a 500-kHz, 22.7-V peak-to-peak sine-wave input was also checked during the passage of bubbles with known velocities and lengths. Figure 26 presents the detailed output of the RF probe during the passage of a bubble (obtained digitally with 20-μs resolution). The output decreases from its water level to the air level with penetration into the bubble and stays almost constant during the passage of the bubble. When the water impinges on the probe tip again, the signal increases, passes through a maximum, decreases, and then levels off at the steady air level. A possible explanation for this maximum was proposed by Fortescue [83] as being due to the additional capacitance of the water surrounding the insulated unshielded wires. Grounding the water close to the tip with a separate copper wire was shown to eliminate this maximum.

In Fig. 27, bubble velocities determined with the RF probe are compared with velocities determined from the output of two light-source detectors. Two penetration time intervals were measured from the RF-probe output (Fig. 26), at t_1 and t_2. By considering a typical characteristic length of the sensitive part of the tip (3

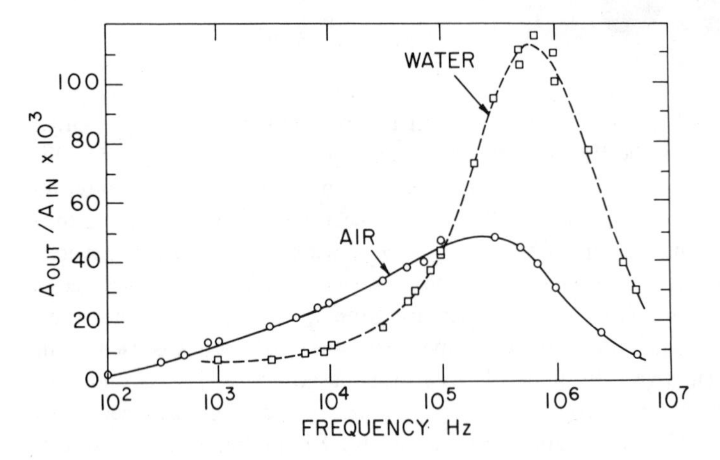

Figure 25 Ratio of RF probe output to input voltage level as a function of input sine-wave frequency, for the probe tip in air and in water.

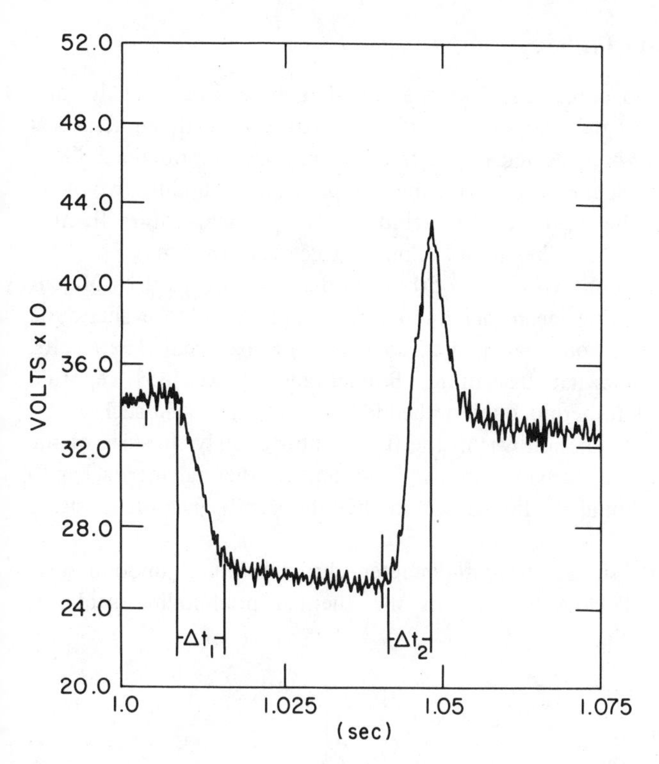

Figure 26 Expanded output of the RF probe during passage of the bubble.

and 2.75 mm), a bubble velocity was calculated. The actual dimension of the sensitive part of the tip was around 3 mm (see Fig. 24) but was difficult to determine exactly, owing to the geometry and construction. The bubble velocities determined by the two independent methods agree with each other within about ~10%. Thus, with an RF probe, the average bubble velocity can be determined from the passage time of either interface, air-water or water-air, along the insulated wires. This is true irrespective of the amplitude of the signal, which may change with fluid state, purity, or test geometry.

The probe output levels for water and air did not change with the bubble velocity in the range considered (up to 160 cm/s). Figure 28 depicts that the bubble sizes as determined by the two independent methods agree with each other with a maximum deviation of about 10%. The bubble lengths recorded by the RF probe are 10% higher at low bubble velocities, around 30 cm/s. This may be due to surface-tension effects during the penetration, which become important at these low bubble velocities.

In summary, the RF probe has a relatively simple construction, and, once it is tuned properly to the test geometry, seems to provide information on both bubble sizes and velocities. More work is, however, needed to check the response of the probe in complex two-phase pipe-flow conditions.

3.5 Microthermocouple Probes

The classical microthermocouple enables one to determine both steady and statistical characteristics of the temperature. If it is combined with an electrical phase indicator, data regarding the local void fraction may also be obtained. Both are discussed below, although the latter seems more appropriate in boiling two-phase flows. Experiments on boiling heat transfer include studies of temperature fluctuations near a heated surface with either pool boiling or forced convection.

A microthermocouple probe with wires 50 μm in diameter was used by Marcus and Dropkin [84] in measuring mean and fluctuating temperatures to evaluate the thickness of the superheated liquid layer in contact with a heated wall. The results, although timely, were somewhat inaccurate. Bonnet and Macke [85] reported results obtained with a microthermocouple imbedded in a resin block in such a way that only 20 μm of the hot junction was in the flow. Unfortunately, the size of the probe, 80 μm, produced a disturbance in the flow, and its thermal inertia led to extra vaporization of the liquid on the sensor so that the significance of the signal was not clear.

Temperature profiles using a 125-μm-diameter chromel-alumel junction were measured by Lippert and Dougall [86] in the thermal pool-boiling sublayer.

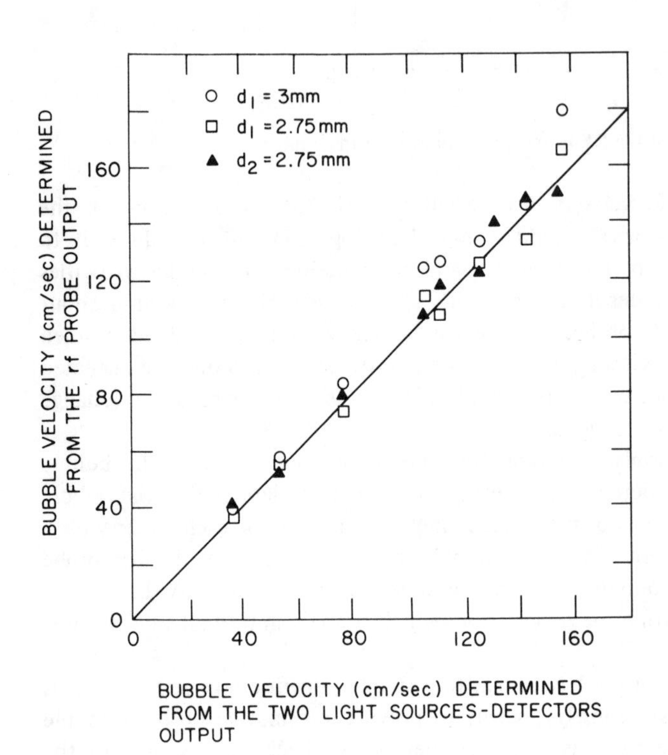

Figure 27 Comparison of bubble velocity as determined by two independent methods–RF probe and two light sources and detectors.

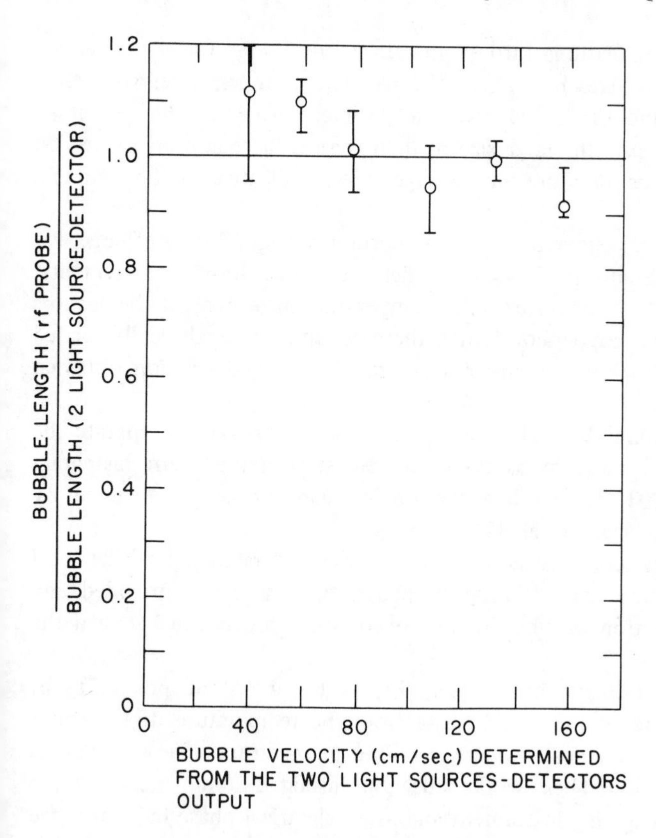

Figure 28 Ratio of bubble lengths as determined by the RF probe and the two light-source detectors as a function of bubble velocity.

According to the authors, the large-diameter-thermocouple data were shown to be reasonable by the results of tests in water, freon-113, and methyl alcohol.

The interaction between bubbles and a 25-μm-diameter microthermocouple was examined by Jacobs and Shade [87]. These authors, and Van Stralen and Sluyter [88], were primarily concerned with the thermocouple response time. Their investigations of the response time were augmented by Subbotin et al. [205], who examined the behavior of bubbles hitting different types of thermocouples.

Stefanovic et al. [89] verified the adequacy of a signal from a 40-μm-diameter thermocouple by recording the impact of a bubble on the hot junction using high-speed movies. Amplitude histograms were obtained in pool boiling and in forced-convection boiling. The authors separated steam and water temperature histograms by assuming that the predominant phase had a symmetrical distribution of temperature (Fig. 29). The identical assumption was used by Afgan et al. [90, 91]. Superheat-layer thickness measurements were conducted in saturated and subcooled nucleate boiling by Wiebe and Judd [92], employing a 75-μm chromel-constantan microthermocouple. A time-averaged temperature was determined by integrating the temperature signal.

One of the first investigations into temperature profiles in forced-convection boiling was carried out by Treschov [93]. The results appear less interesting than those obtained by Jiji and Clark [94] with a chromel-constantan thermocouple, 0.25 mm in diameter. Despite the large size of their sensor, these authors succeeded in measuring an average temperature and average values indicative of the temperature extremes.

Local subcooled boiling, characterized by important nonequilibrium effects, was studied by Walmet and Staub [95] with the help of several local measurements: pressure, void fraction, and temperature. For temperature measurement the authors used a large 0.15 × 0.2 mm copper-constantan thermocouple and analytically related the measured value to the liquid temperature through the void fraction obtained using x-rays.

In his study of flashing flow of water, Barois [96] proposed to separate the distributions of steam and water by assuming that the steam temperature histogram was symmetrical (Fig. 29), rather than the predominant phase, as assumed by Stefanovic et al. [89] and Afgan et al. [90, 91].

Although all these experiments contributed to the understanding of the local structure of two-phase flow with change of phase, they have not provided any reliable statistical information on the distribution of the temperature between the liquid and vapor phases.

The work done by Delhaye et al. [98, 99] is based on the possibility of separating the temperature of the liquid phase from the temperature of the vapor phase, and of giving the statistical properties of the temperature of each phase as well as the local void fraction. These workers used an insulated 20-μm thermocouple both as a temperature-measuring instrument and as an electrical phase indicator (see Sec. 3.1.2 by using a Kohlrausch bridge to sense the presence of a liquid conductor between the noninsulated junction and ground. The phase signal was used to route the thermocouple signal to two separate 1000-channel subgroups of a multichannel analyzer, thus providing separate histograms of liquid and vapor temperatures, as

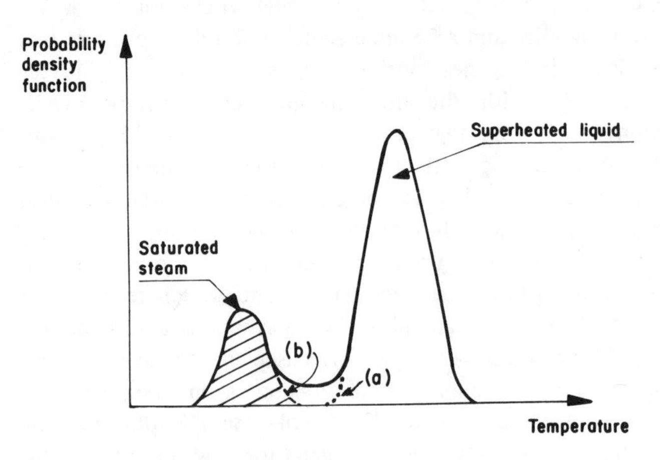

Figure 29 Separation of steam and water distributions. (*a*) According to Stefanovic et al. [90, 91]. (*b*) According to Barois [96]. *(From [98, 99].)*

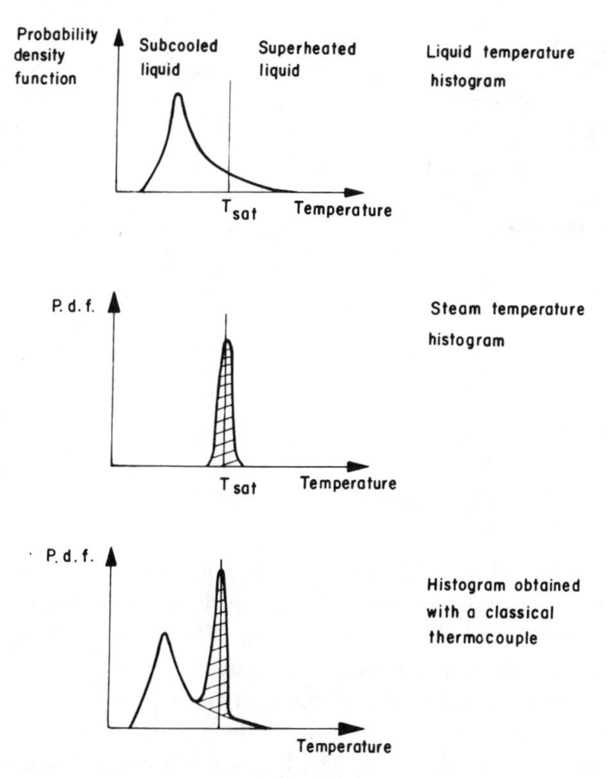

Figure 30 Subcooled boiling-temperature histograms according to Delhaye et al. [98, 99]. (*a*) Liquid temperature. (*b*) Steam temperature. (*c*) Coupled classical temperature histogram.

shown in Fig. 30 for a subcooled-boiling case. Comparison of these histograms with those in Fig. 29 clearly shows the inconsistency in the assumptions of Stefanovic et al. [89], Afgan et al. [90, 91], and Barois [96].

Van Paassen [97] did a detailed study of the microthermocouple as a droplet size sampler, showing good agreement between theory and experiment in determining droplet sizes between 3 and 1188 μm. Detection frequencies of up to 1 kHz were obtained for small droplets.

3.6 Optical Probes

An optical probe is sensitive to the change in the refractive index of the surrounding medium and is thus responsive to interfacial passages. This allows measurements of local void fraction or interface passage frequencies to be obtained, even in a nonconducting fluid. By using two sensors and a cross-correlation method, information may be obtained on a transit velocity [50]. A major advantage of such systems over others is the extremely high frequency response, limited only by photoelectric electronics and photon statistics. The next best seems to be the conductivity probe, with reported frequency response as high as 200 kHz [50].

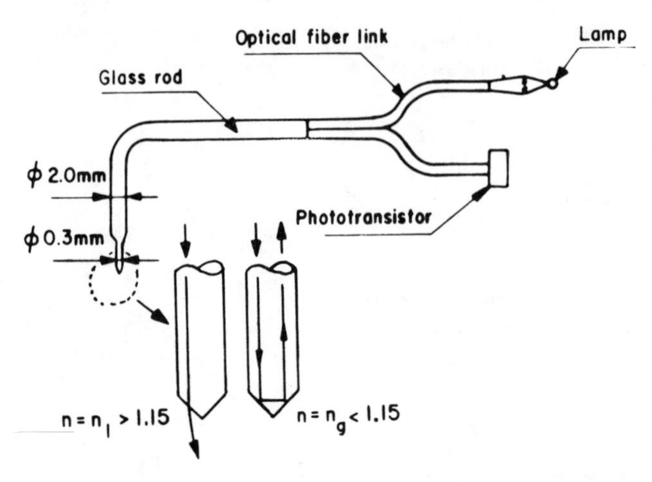

Figure 31 Typical optical probe system. *(From [100, 101].)*

3.6.1 Glass-rod system. This probe (Fig. 31) consists of a glass rod 2 mm in diameter, reduced to 0.3 mm at one end [100 to 103]. The small tip of the rod is ground and polished to the form of a right-angled guide. The light from a quartz-iodine lamp is focused on one of the branched ends of the light guide. A phototransistor is located at the other branched end of the light guide.

Light is transmitted parallel with the rod axis toward the tip of the probe. When a light beam strikes the surface at an angle of 45°, it emerges from the probe or is reflected back, depending upon the refractive indices of the surrounding material n and the probe material n_0 according to Snell's law. Thus, for a glass rod when $n_0 = 1.62$, if the incident internal angle between the light and the polished tip is 45°, light is reflected back along the rod if $n < 1.15$ and exits from the rod if $n > 1.15$. Table 1 gives possible combinations in which this probe can be successfully used.

For signal processing, all the cited authors used a discriminator to transform the actual signal into a binary signal. A trigger level is set at a value above the background level corresponding to the case where the probe is immersed in the liquid. Miller and Mitchie [100, 101] arbitrarily set the trigger level 10% of the pulse amplitude above the all-liquid level obtained by comparing the value of the local void fraction at a given point to the volume void fraction measured with quick-closing valves. Bell et al. [102] set the trigger level halfway between the

Table 1 Liquid-vapor systems in which $n_g < 1.15$ and $n_l > 1.15$ (adequate for use of a 45°-tipped optic probe rod)

System	n_g	n_l
Steam-water	1.00	1.33
Air-water	1.00	1.33
Freon-freon vapor	1.02	1.25

all-water and all-gas signal levels, while in a study of droplet jet flow, Kennedy and Collier [103] related the trigger level and the droplet time fraction with the sizes of the probe and the droplets. It should be noted that significant variations in the results can sometimes be obtained with variations in the trigger level.

3.6.2 Fiber-bundle system. The basic element of this probe is a 30-μm-diameter glass fiber that consists of a central core and an outer cladding. Several hundred such elements are tied together in a Y-shaped bundle similar in appearance to that shown in Fig. 31, with a light source and a phototransistor. The active end of this bundle is glued to a glass rod, 0.5 mm in diameter and 1 mm long, itself coated with a glass of lower refractive index (Fig. 32). The extremity of the glass rod is ground and polished. The operation of this device is similar to that of the glass-rod system. Hinata [104] obtained S-shaped curves of local void fraction versus trigger level, with a plateau corresponding to a given value of the void fraction. He used the trigger-level value corresponding to this plateau.

3.6.3 U-shaped fiber system. One of the major drawbacks of the glass-rod and fiber-bundle systems is the large dimension of the sensitive part of the probe (respectively, 0.3 and 0.5 mm). An alternative sensor configuration developed by Danel and Delhaye [105] has a distinct size advantage. This probe consists of a single coated optical fiber, 40 μm in diameter. The overall configuration is similar to that shown in Fig. 33, with a miniaturized lamp and a phototransistor chosen for its high sensitivity. The active element of the probe is obtained by bending the fiber into a U shape and protecting the entire fiber, except the U-shaped bend, inside a stainless steel tube 2 mm in diameter. The active part of the probe has a characteristic size of 0.1 mm [106], as shown in Fig. 33.

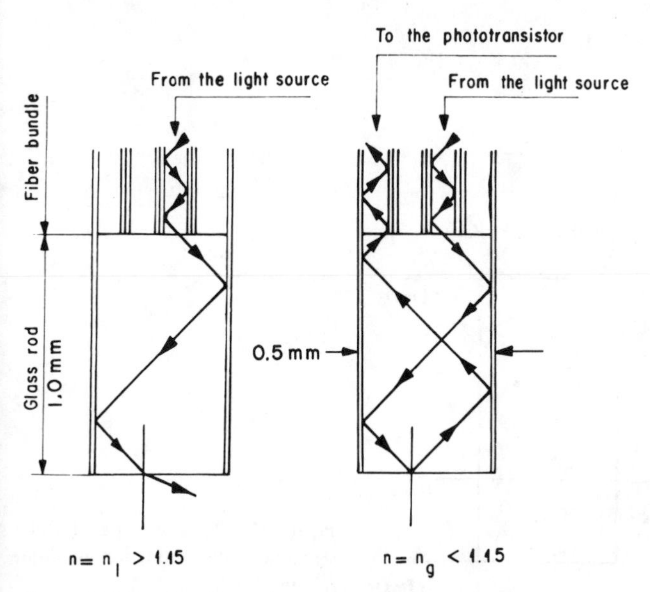

Figure 32 Fiber-bundle optical sensor. *(From [104].)*

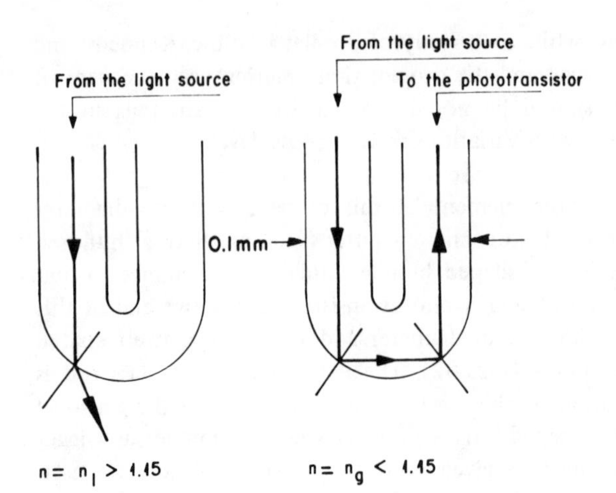

Figure 33 U-shaped fiber-optical sensor. *(From [105].)*

Signal processing for this system was taken one step further. A typical signal delivered by the probe is shown in Fig. 34. The voltage U can be divided into a static component U_0, which was reported to vary with local void fraction, and a fluctuating component u. Since U_0 corresponds to a sensitive part of the probe completely immersed in the liquid, the change in U_0 can be due to (1) the response time inherent in hydrodynamic and optoelectronic phenomena when the interface is pierced by the probe, and (2) the scattering of light by the bubbles surrounding the probe.

The fluctuating component constitutes the interesting part of the signal, while the maximum value U_{max} corresponds to the sensor completely immersed in the gas and does not depend on the local void fraction. Signal analysis is accomplished

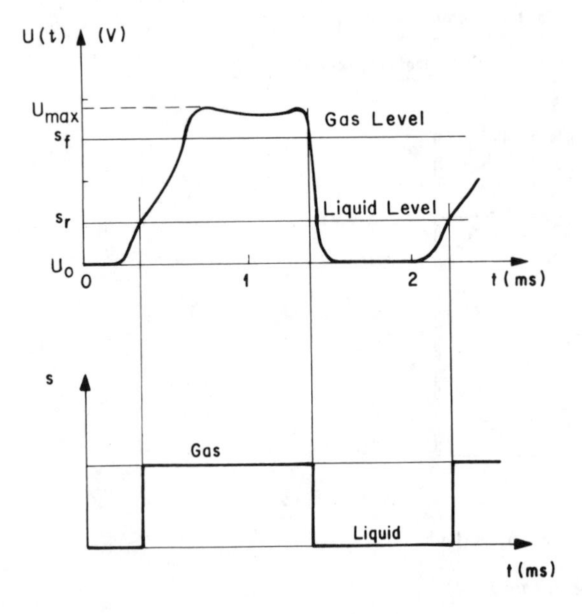

Figure 34 Typical optical-probe signal and discrimination method. *(From [106].)*

through two adjustable thresholds S_r and S_f, which enable the signal to be transformed into a square-wave signal (Fig. 34). Consequently, the local void fraction α is a function of S_r and S_f, which are adjusted and then held fixed during a traverse to obtain agreement between the profile average and a gamma-ray measurement of void fraction. Since the signal shape varies with void fraction, and since changes in S_r and S_f alter the result, local measurement of void fraction can be quite in error. This is true because of the compensating errors, even though the average is in agreement with an independent measurement.

3.6.4 Wedge-shaped fiber system. All previously described methods suffer from difficulties associated with signal processing, hydrodynamic response-time uncertainties, and fragility (especially the U-shaped fiber system). Signal processing at best seems dependent on a global reference. The variable signal levels described by Delhaye and his coworkers [50, 105, 106] can yield an accurate local result only at a specific void fraction. At other values of void fraction, corresponding to different physical locations, values are distorted due to differences between dryout and rewet response characteristics. The overall effect is a distortion in the void-fraction profiles measured in a flowing mixture.

To circumvent these difficulties, a new optical probe was developed by Abuaf et al. [107, 108]. A schematic of the probe as developed is given in Fig. 35, along with the light source and amplifier circuit diagram used in the apparatus. Two 0.125-mm fibers were inserted into a 0.5-mm outside diameter stainless steel tube. The two fibers were fused together at one end by means of a minitorch, forming a slightly enlarged hemispherical bead similar to that shown in Fig. 33. This fused end of the fibers was then pulled into the tube and cemented in place. The fibers were separated at the opposite end and encased in two pieces of stainless steel tubing (0.25 mm outside diameter). The ends of the fibers and the bifurcation were then coated with epoxy for strength. The tip of the probe containing the fused end of the fibers was ground and polished at a 45° angle to the axes of the fibers, thus forming an included angle of 90° at the finished probe tip. The resultant probe geometry is extremely rugged and capable of withstanding considerable abuse without altering its characteristics, even to the extent of being dropped on its tip.

After the free ends of the two fibers were ground flat and polished, one of them was placed in front of an incandescent light source (3 V), and the other in front of a Hewlett-Packard PIN photodiode (5082-4024). An amplifier with a design rise time of 20 μs [109] was used to enhance the output before it entered the readout device (Fig. 35). The electronic response of the system was checked by means of a light-emitting diode (LED) placed in front of the probe tip. The LED output was modulated by using a signal generator so that rectangular light pulses of different spacings and widths were emitted, simulating the passage of bubbles. The rise time of the output was thus verified to be 20 μs as specified. The amplitude of the probe output did not change with the frequency of the input signal of the LED.

The hydrodynamic response of the probe to the passage of an interface or bubble was investigated as described in [107, 108]; single bubbles could be generated and forced past the probe while independent methods were used to

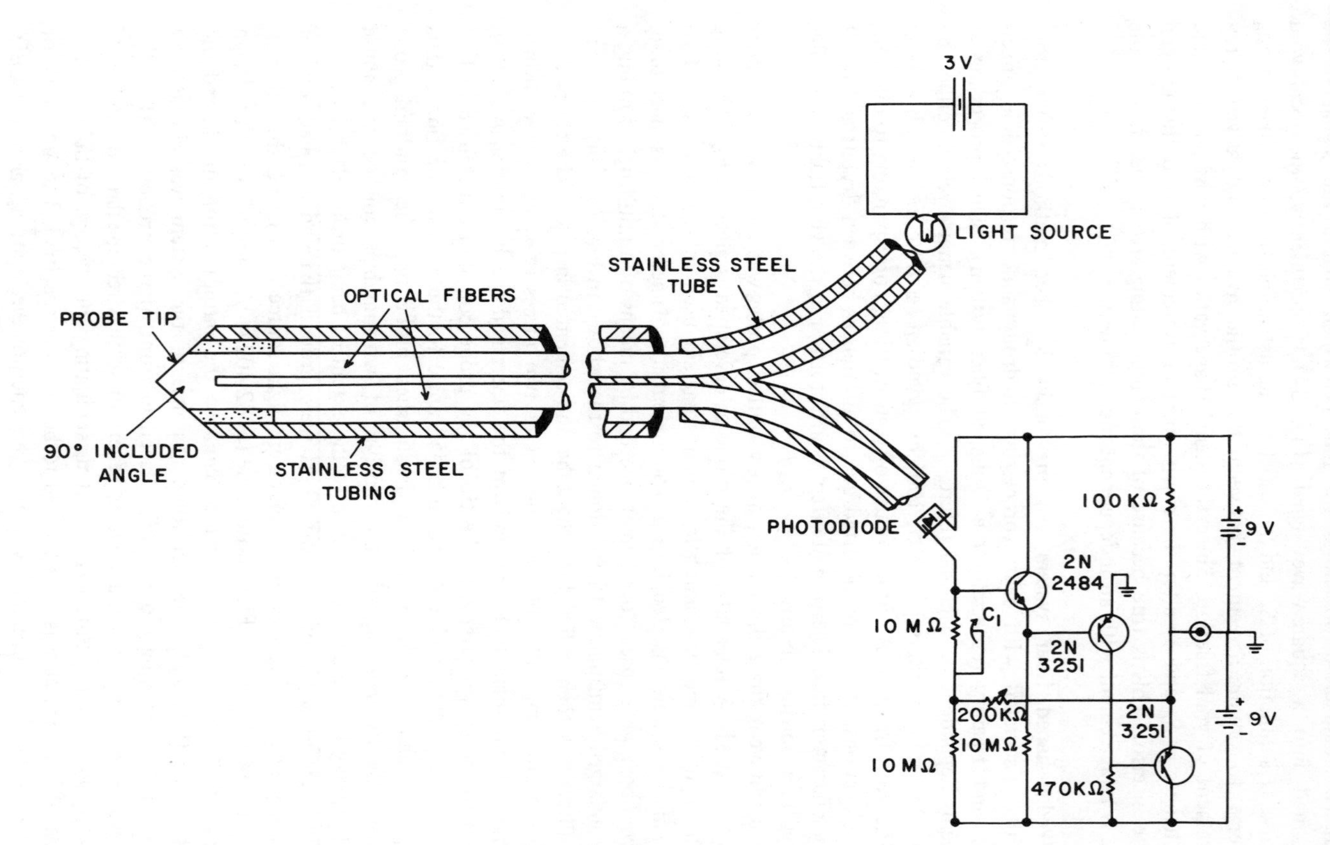

Figure 35 Schematic representation of the optical probe.

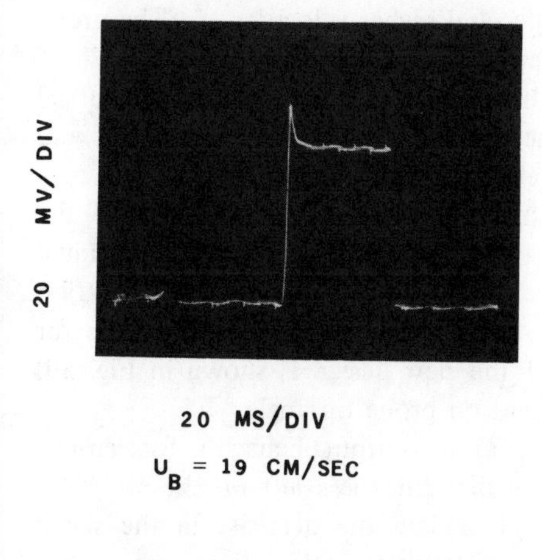

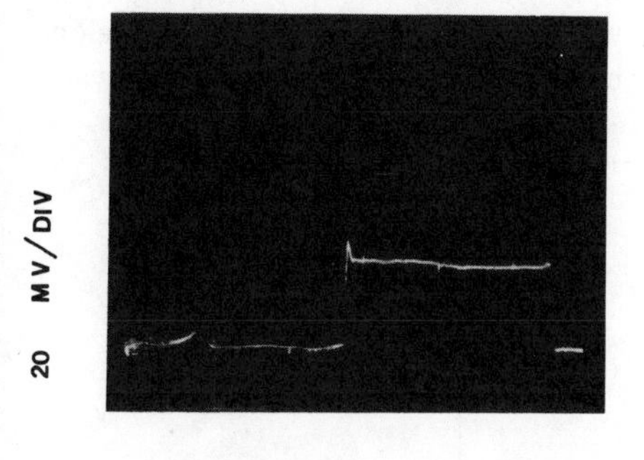

Figure 36 Typical oscillograms of the wedge-shaped optical-probe output.

determine the velocity. Typical oscillograms of the probe output during the passage of the bubble are presented in Fig. 36. The probe output is shown as a function of time for two cases where the bubble velocities were 19 and 74 cm/s, respectively. It was observed that when the tip was immersed in water the probe output was always zero, without any artificial bias, a significant improvement over previous optical probes. As the bubble hit the probe, the output was seen to increase and, after an overshoot, to level off at a steady value. At the end of the passage of the bubble, the signal dropped to its original water level of zero. The bubble penetration time was clearly observed to be larger than the time it takes the probe tip to be immersed in water (Fig. 37). Also, the signal amplitude decreased with increasing bubble velocity (Fig. 38), although both bubbles had almost the same length (void fraction). These two effects were investigated in some detail. Results obtained with two different probes are presented in Fig. 38 as a function of bubble velocity. Although the two probes had steady air-signal amplitudes of 125 and 600 mV, the ratio I/I_0 follows the same consistent pattern. A similar observation was made by

Miller and Mitchie [101] "with smaller bubbles and higher velocities. . . . The probe signal generated under these conditions . . . never reached maximum amplitude."

An important conclusion that can be drawn from Fig. 38 is that the optical probe is able to measure the local interface velocity as well as the local void fraction, after proper calibration within the velocity range observed.

A computer program was written to study the theoretical response of the output to hydrodynamic conditions at the tip of the probe by tracing individual rays of light from their source to the detector. The effect of varying liquid-film thicknesses on the probe tip was included. A comparison of geometric effects for the probe of Danel and Delhaye [105] and the new design is shown in Fig. 39, where the new geometry is seen to be dependent on probe tip angle.

As a possible explanation of the optical-probe output behavior for various bubble velocities, it was proposed that a water-film thickness left on the probe tip and increasing with the bubble velocity could explain the decrease in the signal intensity that was observed experimentally. Computer calculations were thus extended to study the signal attenuation that would be caused by a variable film thickness at the probe tip. The water layer was assumed to increase linearly along the flat face. β_W was defined as the angle between the outside face of the water

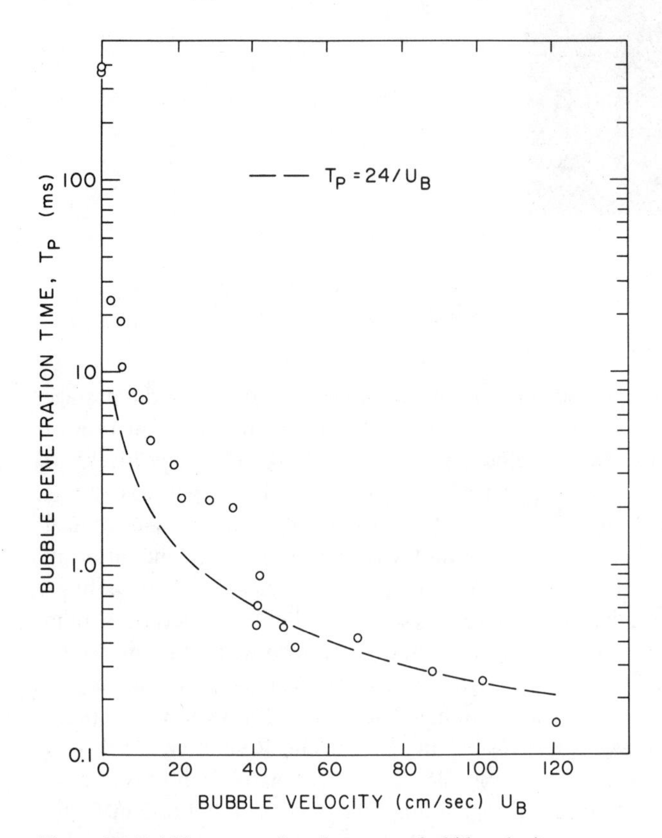

Figure 37 Bubble penetration time versus bubble velocity.

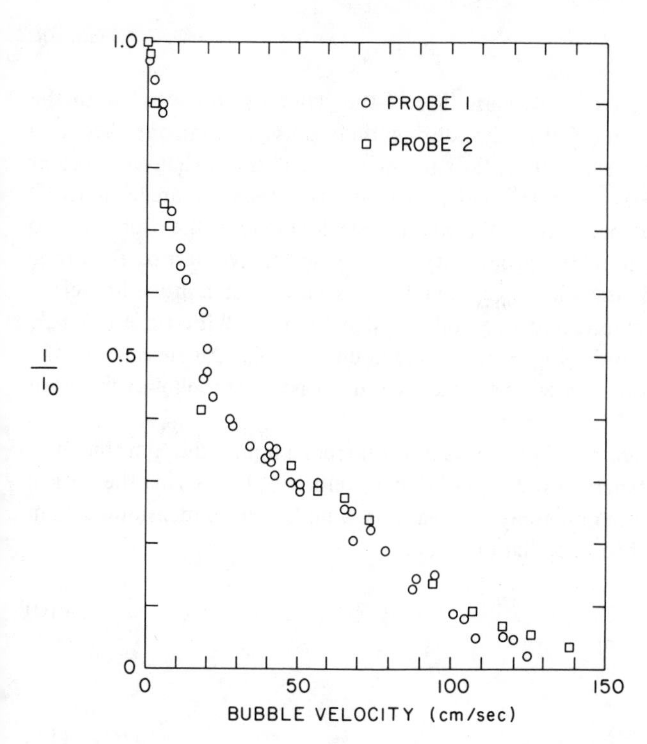

Figure 38 Probe signal I at a given bubble velocity divided by the steady air signal I_0, versus bubble velocity.

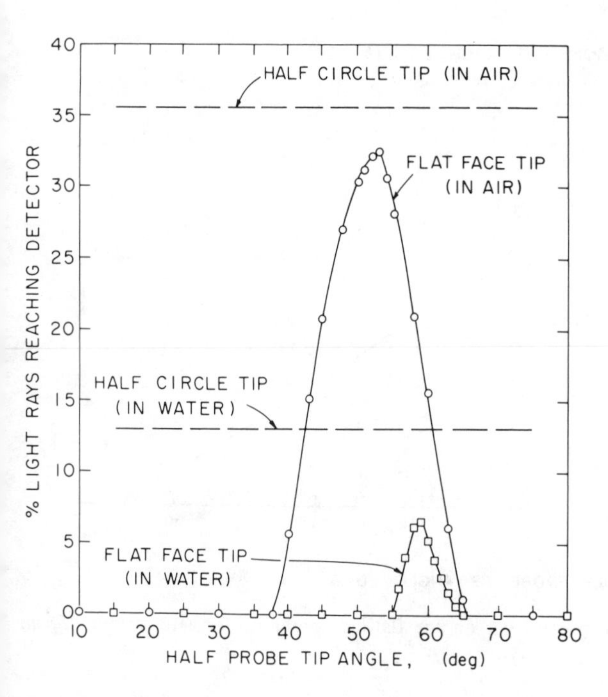

Figure 39 Percent of light rays reaching the detector as a function of the half probe-tip angle for the probe tip in air and in water. Comparison between the half-circle and the flat-face probe tips. *(From [108].)*

layer in contact with the air and the glass face of the probe tip in contact with the water layer. The attenuation of the rays during their passage from one media to another was not taken into account. Although the 52° half tip angle gave a higher signal in air when compared to the 45° half probe-tip angle, the large angle tip (52°) was found to be strongly dependent on the water-layer thickness on the probe tip. To show this strong dependence of the probe output on the water layer left on the probe tip, Fig. 40 gives the maximum angle β_{max} that can be sustained on a probe tip before the coherent light rays are refracted out and the signal is zero. Within the tip-angle range of interest, $37° < \beta_t < 65°$ (Fig. 39), the maximum angle β_{max} increases, and the sensitivity of the probe to the water-layer thickness decreases, for the lower values of tip angle.

It is predicted that when a cylinder is withdrawn from a liquid, the film thickness remaining on the cylinder increases with withdrawal velocity [110]. This theoretical prediction was checked experimentally for various liquids. The dimensionless film thickness [111] was related to a capillary number

$$N_{ca} = \frac{\mu U_w}{\sigma} \tag{10}$$

and a Goucher number

$$N_{Go} = \frac{R}{a} \tag{11}$$

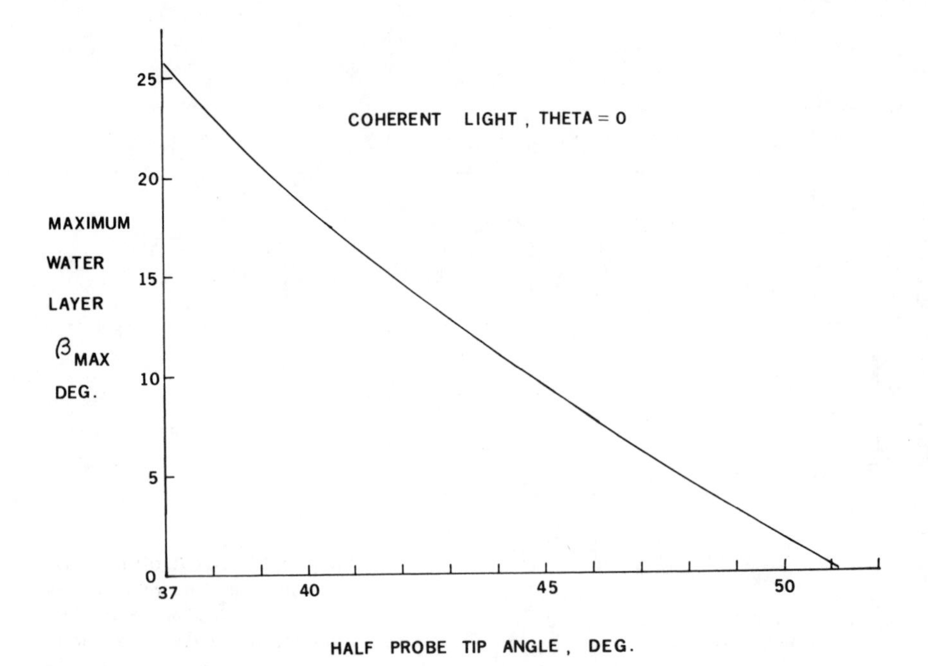

Figure 40 Maximum thickness of water layer on the flat-face probe tip for zero output. *(From [108].)*

where μ and σ are the viscosity and surface tension of the liquid, U_w is the withdrawal velocity, R is the wire radius, and a is the capillary length defined as $a = (2\sigma/\rho g)^{1/2}$. In addition, White and Tallmadge [112] observed that experiments conducted with distilled water provided film thicknesses almost twice those predicted by the theory. This fact is still unexplained. In any event, the results may be used to obtain Fig. 41, showing the increase in film thickness with increasing velocity.

The two parts of the theory explaining probe behavior thus are (1) the amplitude of the signal versus the water-film thickness and (2) the water-film thickness versus the interface passage velocity. A combination of the two will yield the predicted variation of signal amplitude with interfacial velocity, as shown in Fig. 42, in good qualitative agreement with the observed behavior. Differences may perhaps be explained by the presence of nonlinear films, slanted optical surfaces instead of surfaces colinear with the probe direction, etc.

4. IN–STREAM SENSORS WITH MECHANICAL OUTPUT

A large class of devices for taking measurements in two-phase flow have been designed for use primarily in annular flow. The single-needle conductance probe, when traversed toward a film, has been used for measuring film thickness. The dual, flush-mounted Harwell conductivity probe [27] and the ring-type CISE probes [36] were also designed to measure film thicknesses. Likewise, a number of devices with mechanical output have been designed for measuring specific properties of annular flow. The major impetus is the general idea that the critical heat-flux phenomenon is predominantly an annular-flow phenomenon caused by dryout or disruptive of the liquid film.

In all these sensors, the general principle is to remove from the flowing stream a certain portion of the fluid to determine the flow rate of either one phase or the other, usually the liquid phase. The sensor may be either stationary or designed with traversing mechanisms to determine the transverse distribution of the particular

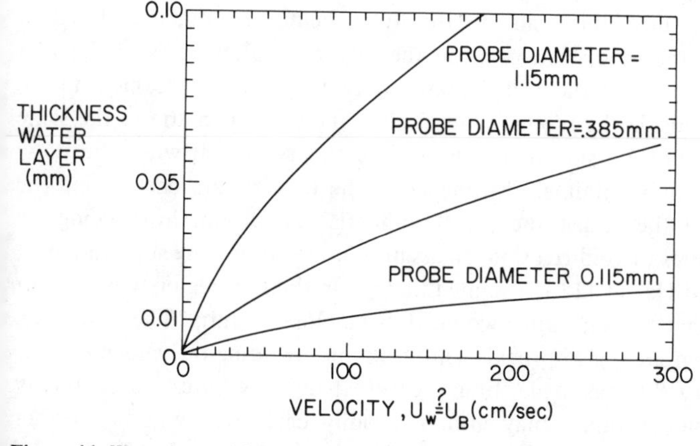

Figure 41 Water-layer thickness versus wire withdrawal velocity. *(From [111].)*

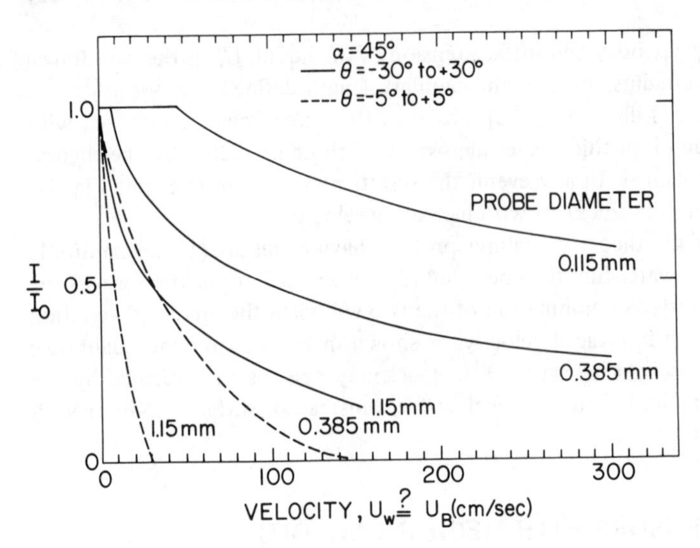

Figure 42 Crossplot of normalized probe signal I/I_0 versus wire withdrawal velocity for three probe diameters and two values of the acceptance angle. *(From [108].)*

parameter under study. In some more intricate designs, the amount of suction is determined by matching the probe-inlet static pressure with the channel static pressure to have minimum effect on the flow streamlines. The major development of these instruments may be attributed to the Atomic Energy Research Establishment (AERE) at Harwell, England, and the Centro Informazioni Studi Esperanzi (CISE) in Milan, Italy.

4.1 Wall Scoop

The wall-scoop device was developed at CISE and was reported by Cravarolo and Hassid [113] and Adorni et al. [114]. This device, shown in Fig. 43, consists of a moveable scoop built into the wall of the tube being investigated. In their 2.5-cm-diameter tube, the scoop can measure the integrated flow rates in the film region from 0.13 to 2 mm from the wall. To obtain a sample, the valve is opened until the static pressure just inside the scoop is identical to the static pressure at the same axial location on the opposite side of the tube. The theory is that when the static pressures are matched, only minimal disturbance to the flow stream occurs, and the sample is thus taken at the undisturbed, or "isokinetic" condition. In reducing the data to obtain local values of void fraction, an assumption regarding the slip ratio must be made. Cravarolo and Hassid [113] assumed that gas in the film region would occur as small bubbles and that the slip ratios would thus be close to unity. The flow rates, coupled with the known area, also provided information regarding the velocities once the assumption of unity slip was made. In any event, if the flow rates are accurately known, the phase volume fluxes may also be readily calculated without further assumptions.

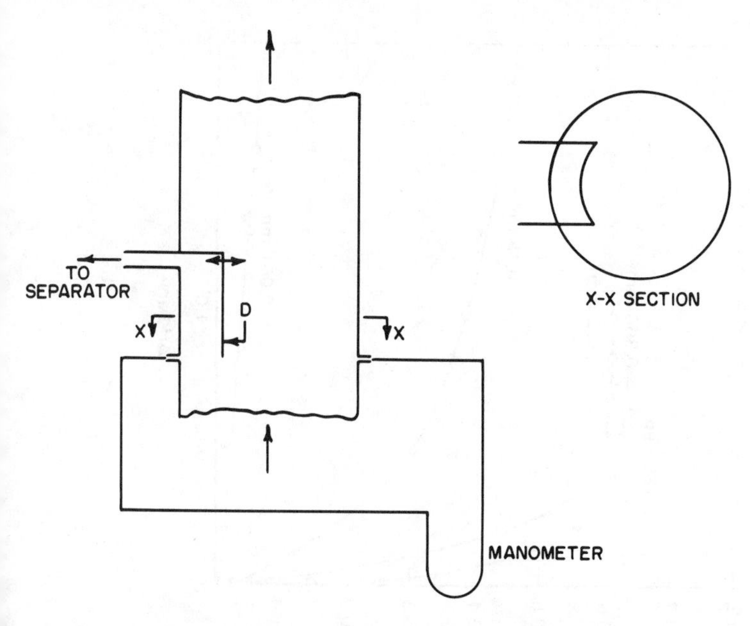

Figure 43 Schematic of wall scoop used at CISE. *(From [113].)*

One basic disadvantage of this type of instrument is that it disturbs the flow regime irrevocably. In other words, once the film is removed from the wall, there is no way of putting it back in the same distribution. Another disadvantage is its limited size, which tends to give erroneous readings for very thin or very thick films. In addition, while the method provides reasonable results in two-component flows, single-component flows are more difficult to determine accurately, owing to phase-change effects caused by heat losses and pressure drops, and under non-equilibrium conditions they become virtually impossible to measure accurately.

Truong Quang Minh and Huyghe [115] used a somewhat similar device, wherein a circumferential slit was provided at a point in a tube. The axial width of the slit could be varied as shown in Fig. 44 (left). They found that the film flow rate as measured would become almost independent of both slit width and differential pressure for widths over a certain minimum and differential pressures above some minimum. This behavior is shown in Fig. 44 (right). As discussed by Moeck [116], both this device and that designed at CISE can underestimate the film flow rate when large waves are present.

4.2 Porous Sampling Sections

Another device that has been used for annular-flow studies is the porous sampling section, fabricated sometimes from a sintered porous plate and sometimes from a fine-mesh screen [117 to 127]. The purpose of this porous sample section is to remove the film of water selectively in annular or dispersed annular flow, to measure the film flow rates.

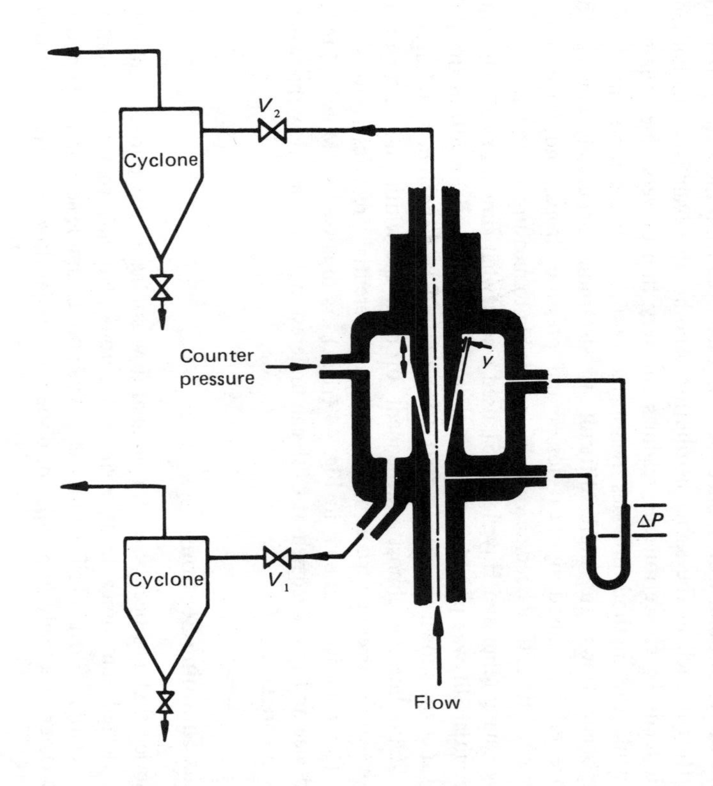

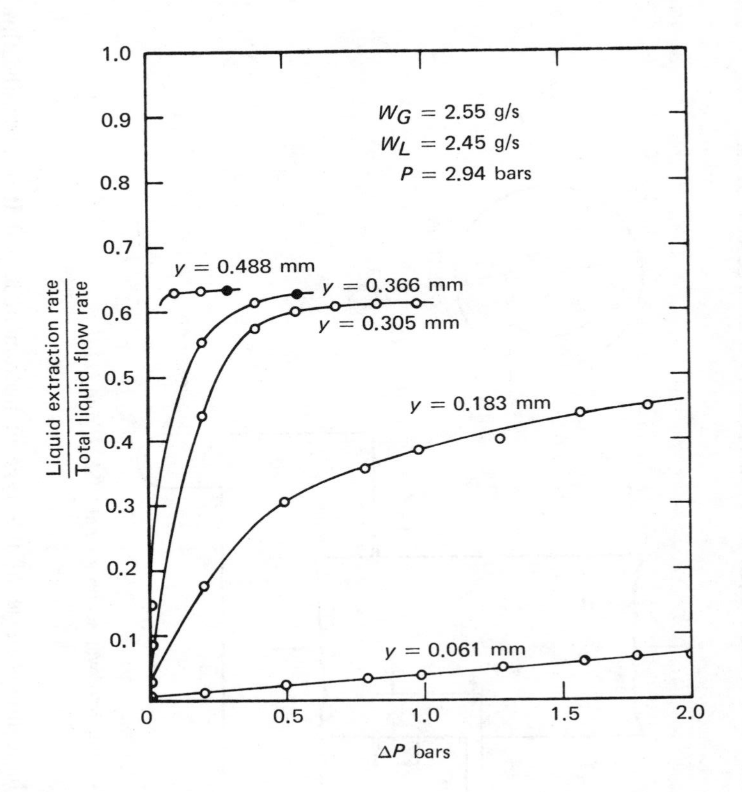

Figure 44 Film suction device used by Truong Quang Minh and Huyghe [115].

Typical of these devices is that used at General Electric's Atomic Power Equipment Department [117] in an annular test section (Fig. 45*a*). Pressure is decreased in the plenum chamber behind the porous plate by opening a sampling valve. In two-component flow the phases are usually separated and measured separately, while in single-component flow the mixture is condensed in a calorimetric device and a heat balance is applied to calculate the mass flow rates of the separate phases. In the case of two components, the liquid flow rate is usually plotted as a function of the pressure drop across the plate, and curves similar to those for the wide-gap slit sampling device shown in Fig. 45*b* are usually obtained; that is, beyond some nominally small differential pressure, the sample flow rate does not change for a wide range of differential pressures. The concept of this behavior is that if the sintered plate is sufficiently long, the entire film is removed with little pressure drop, and thereafter only additional amounts of core gas are pulled through the porous plate. If sufficient

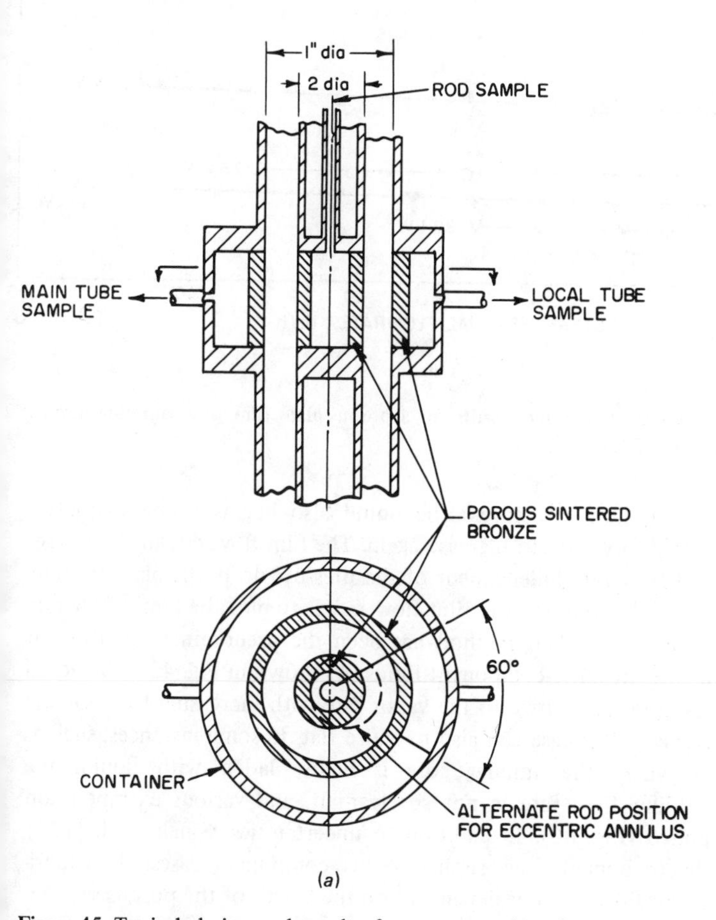

Figure 45 Typical design and results for a porous-plate film flow-sampling device. (*a*) Porous plate. *(From [117, 118].)*

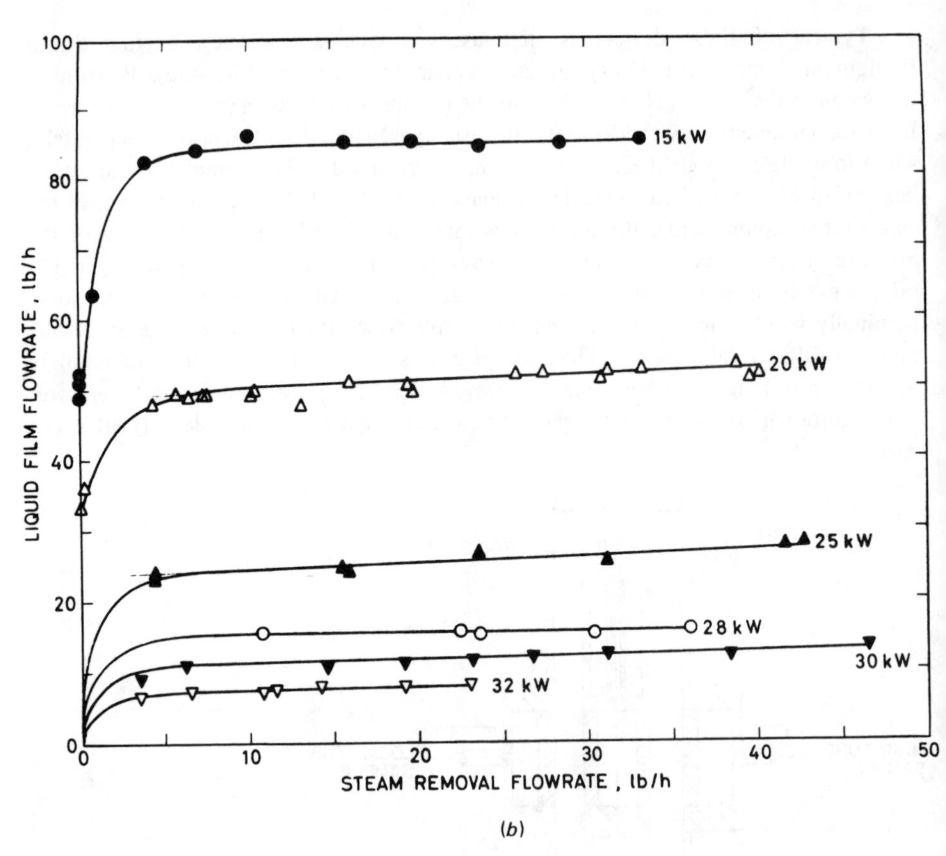

Figure 45 (*Continued*) Typical design and results for a porous-plate film flow-sampling device. (*b*) Flow data. *(From [117, 118].)*

pressure drop is applied, however, core-borne liquid also begins to be extracted, causing the liquid sample flow rate to increase again. The film flow rate in these cases is defined as the liquid flow rate independent of the pressure drop, the plateau value.

A second method of determining the film flow rate is to plot the liquid flow rate versus the gas-phase flow rate. This method has been the predominant method of determining the film flow in single-component fluids. As shown in Fig. 45*b*, the liquid removal rate becomes quite insensitive to the vapor rate with increasing total sample flow [118]. The plateau in this case can also be quite flat. In some instances, such as for low quality or where the annular core is highly laden with liquid, the determination of the film flow rate is not so clear-cut, and various extrapolation schemes must be applied (Fig. 46). In addition to uncertainties, Singh et al. [125] showed that in single-component flow, similar to two-component cases, the experimentally determined film flow rate was dependent on the length of the porous section, owing to deposition of core-borne droplets and increasingly more efficient capture of the liquid in large roll waves with longer plates.

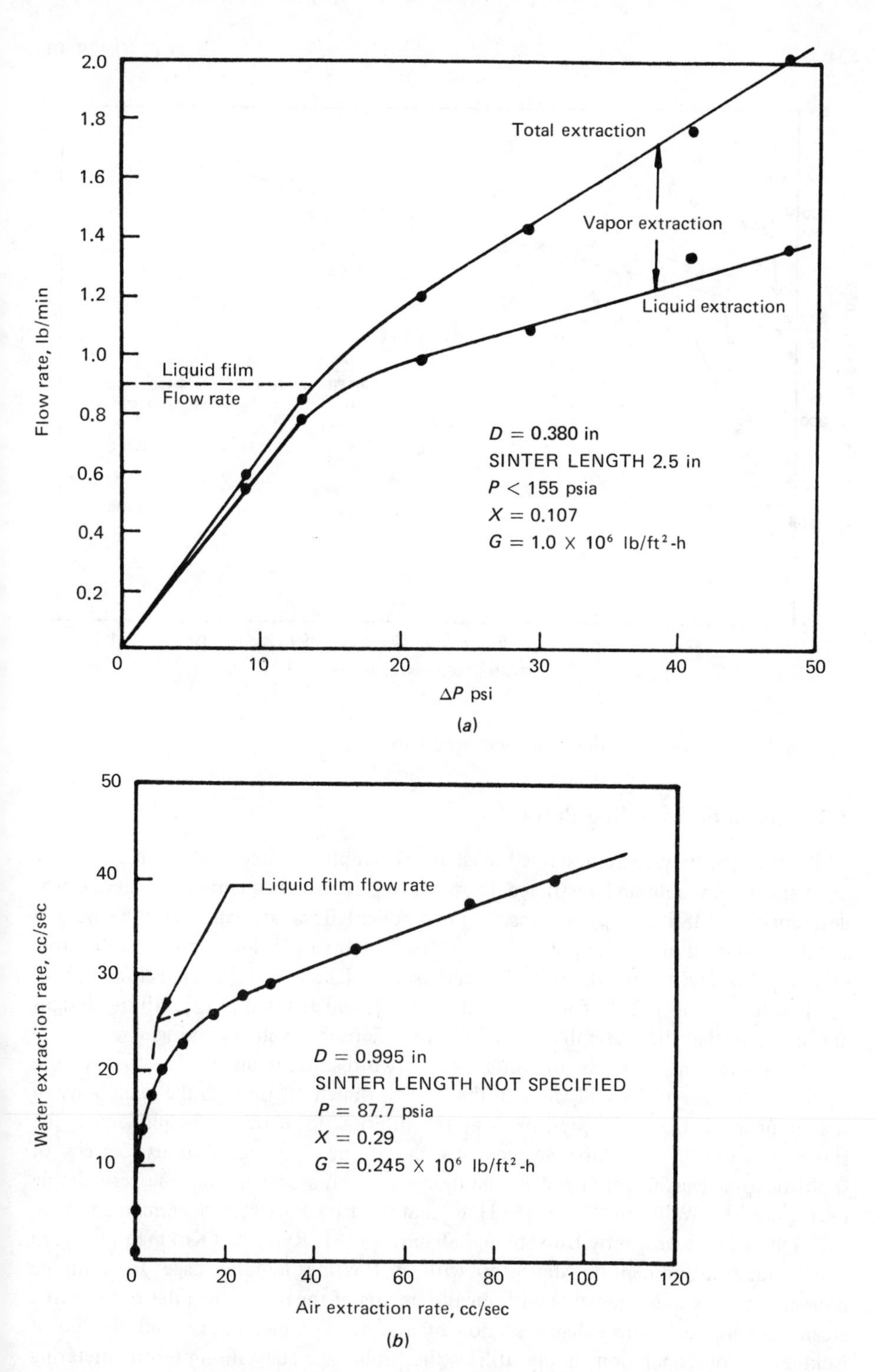

Figure 46 (*a*) Data of Staniforth et al. [127]. (*b*) Data of Schraub et al. [126].

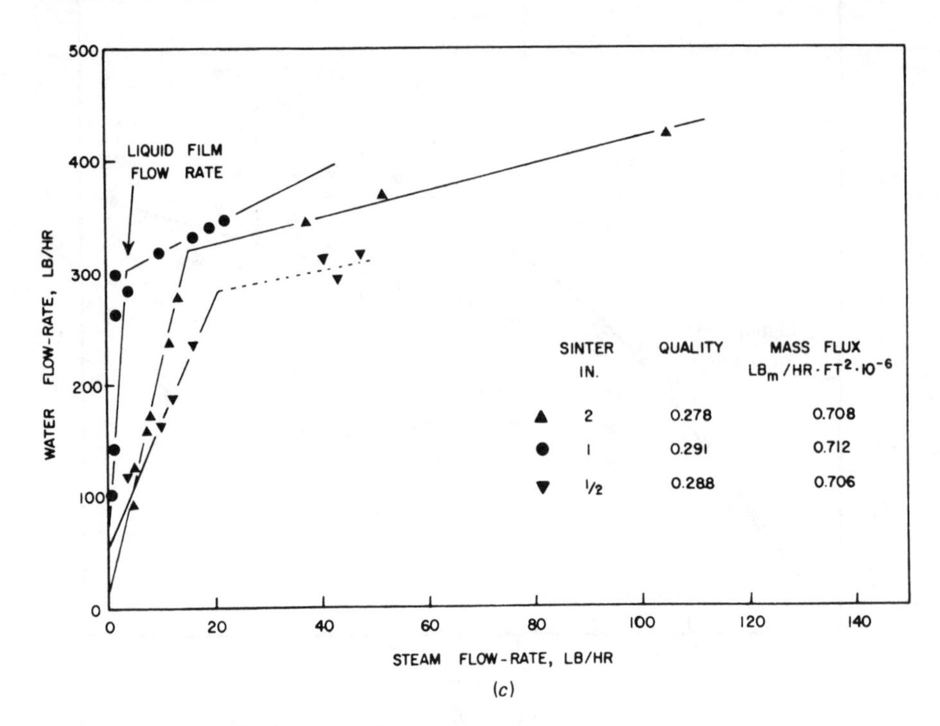

Figure 46 (*Continued*) (*c*) Various film-sampling results.

4.3 Isokinetic Sampling Probe

Where the two previously described mechanical sampling devices have been stationary or slightly moveable and restricted to measuring film-flow quantities, a device was developed by CISE [128] for measuring component flow rates and, with the aid of a suitable assumption regarding slip, the void fraction at various locations of a tube cross section. This device has subsequently been used by Lahey et al. [129], Schraub [126, 130], Adorni et al. [128], Todd and Fallon [131], and Jannsen [132]. All the designs are similar in that they resemble a small Pitot-static probe as shown in Fig. 47.

In general, this probe is used in either of two distinct manners. The first is quite similar to the use of the wall scoop in that flow is drawn off through the main body of the probe until the static pressure near the interior tip of the probe just equals the static pressure in the tube adjacent to the probe opening. Various degrees of sophistication can be employed in the design of such a probe, from the very simple device used by Wallis and Steen [133] to that used by Burick, Scheuerman and Falk [134] based on a design by Dussord and Shapiro [135]. Ryley and Kirkman [136], in fact, added a momentum deflector with a floating impulse cage to combine momentum flux measurements with measurements of the mass flow rates of saturated steam and liquid in the exhaust section of a steam turbine. The second method of isokinetic probe operation simply utilizes the probe as a stagnation-pressure metering device, where reverse purging of one phase or the other, usually the gas, is used to provide a metering reference.

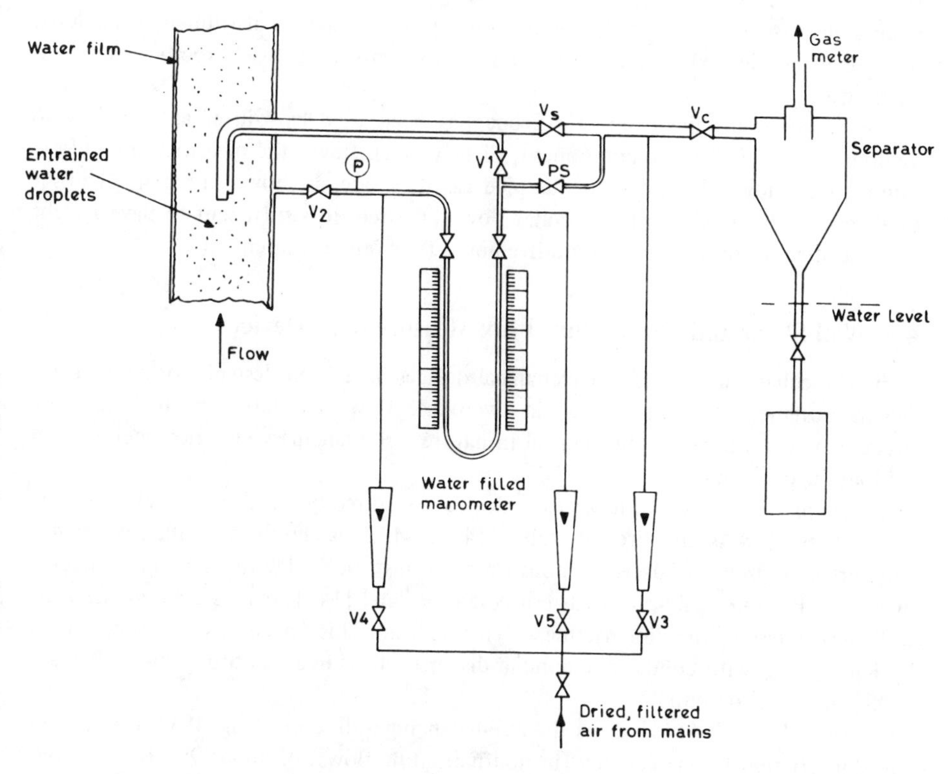

Figure 47 Isokinetic probe system designed by CISE. *(From [128].)*

A number of problems occur with the use of the isokinetic probe, not the least of which is the necessity to compensate for pressure losses between the probe tip entrance and the static-pressure port. For small probes Schraub [130] outlines a method of sampling at various conditions around the isokinetic condition and then iterating on the proper conditions by checking the integrated flow profiles against the known values. This is good for two-component systems with axisymmetric geometries and perhaps would also work in equilibrium single-component systems where this check is possible. In planar or grossly two-dimensional geometries, such a method would be extremely difficult to apply. In addition, since the corrections are flow-dependent, virtually every reading requires a different correction, and profile distortion of measured results would occur.

Another problem lies in the fact that different assumptions lead to different results for the void fraction. The CISE [128] assumed a constant slip in the fluid approaching the probe, whereas Schraub [130] assumed variable slip. Schraub [137] mentions, however, that only slightly differing results are obtained with different assumptions. In addition, Shires and Riley [138] show that if the probe is much larger than the dispersed phase, the vapor volume flow fraction is measured, whereas if the reverse is true, the vapor volume fraction α itself is measured. In spite of these problems, many workers have used this instrument with varying degrees of success, mainly because it appeared to be the only general class of instrument capable of providing the desired combination of measurements of phase velocities and void fraction.

A variation on the isokinetic probe has been used by Gill et al. [139], who demonstrated that for two-component, slightly laden flows, the measured liquid flow rate is practically independent of sampled gas flow rate, as shown in Fig. 48. Thus, if only gas-core liquid content in annular flow is desired, it is sufficient to have a slight amount of gas to obtain a representative liquid flow-rate measurement.

4.4 Wall Shear and Momentum Flux Measurement Devices

Various mechanical and electromechanical devices have been designed or adapted for the measurement of skin friction in two-phase flow. The latter are included here because they are basically mechanical in nature, using ancillary electrical methods for readout purposes only.

Perhaps the most widely used device for directly measuring wall shear in single-phase flow is the Preston tube [140], which is simply a small, right-angled total-pressure probe. In use, the mouth rests against the wall with its opening directed upstream. Preston's calibration, as corrected by Patel [141], relates the dimensionless wall shear stress to the dimensionless dynamic head. This device was used successfully by King [142] with condensing annular-dispersed flow in a 1-in tube, and by Jannsen [143] in a nine-rod bundle.

Cravarolo et al. [144] devised a null-balancing wall sensor (Fig. 49) for measuring the skin friction to an accuracy (in nonfluctuating flows) of about 2%. In slug flow, however, they were unable to balance their device. The pressure drop required to just begin to move the sleeve off the lower support, and that required to drive the sleeve off the top support, are noted. The average value provides the shear stress without the effects of friction and inertia. Similar to the Preston tube, this device can give the wall shear for a relatively small area of a tube.

Rose [145, 146] suspended his test section between two rubber connectors, which allowed it to move freely in the vertical direction (Fig. 50). Attached somewhere to the vertical section was a linear variable differential transformer (LVDT) assembly that detected small changes in the position of the test section itself. The channel was first balanced in an equilibrium position and then deflected by known forces to obtain a calibration. Then, starting at the equilibrium position, flow of two-phase mixture was passed through the tube. The deflection of the tube was then due only to the shear stress on the wall and the difference in pressure between the inlet and the outlet of the test section acting on the end areas of the tube. A similar device was designed by King [142], but, because of hardware difficulties, was never actually put into service; the method does, however, appear promising.

A device for measuring exit momentum flux was built and used by Rose, and subsequently by Andeen and Griffith [147]. This instrument, shown in Fig. 51, is simply a device to change by 90° the direction of flow at the annular channel and measure the force required to do this. With an LVDT apparatus, similar to that described for his wall shear-stress instrument, Rose measures the forces needed to

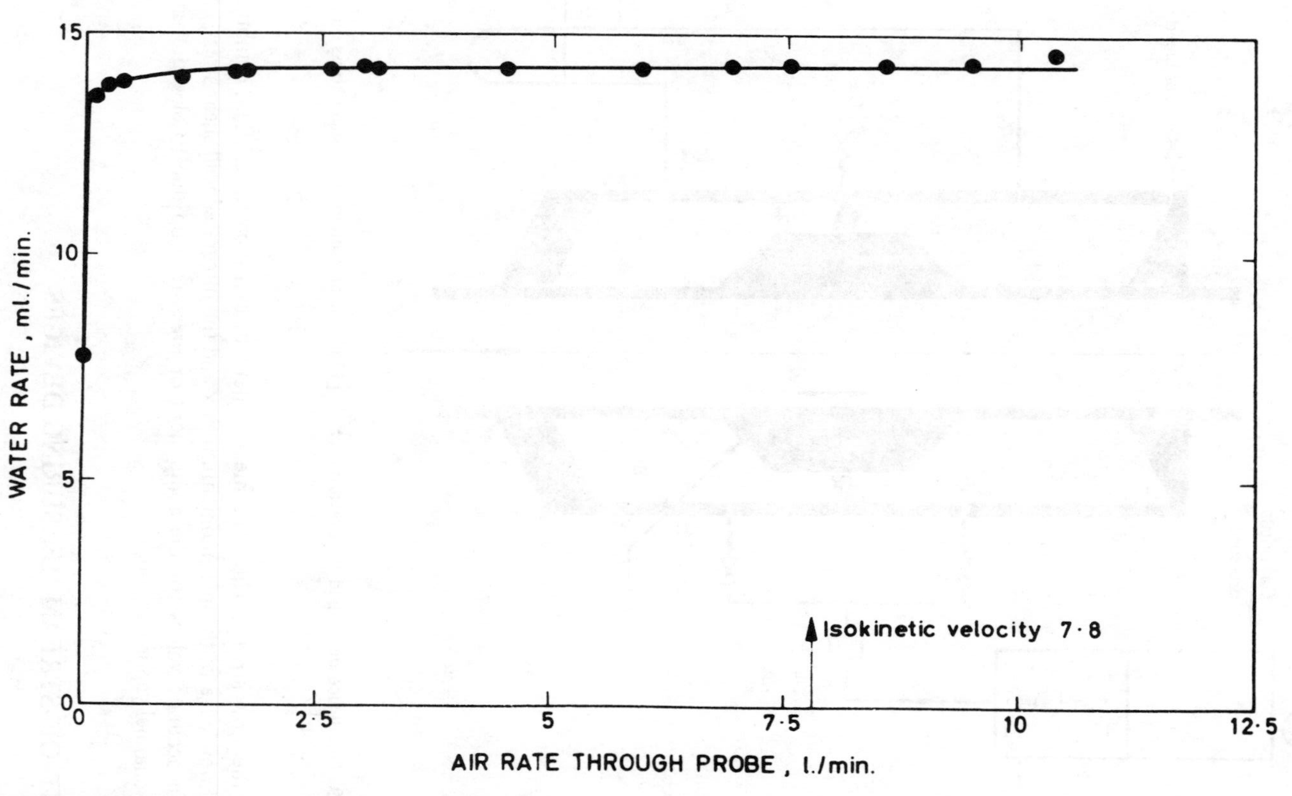

Figure 48 Typical results from an isokinetic sampling probe used in an air-water system. Data of Gill et al. [139].

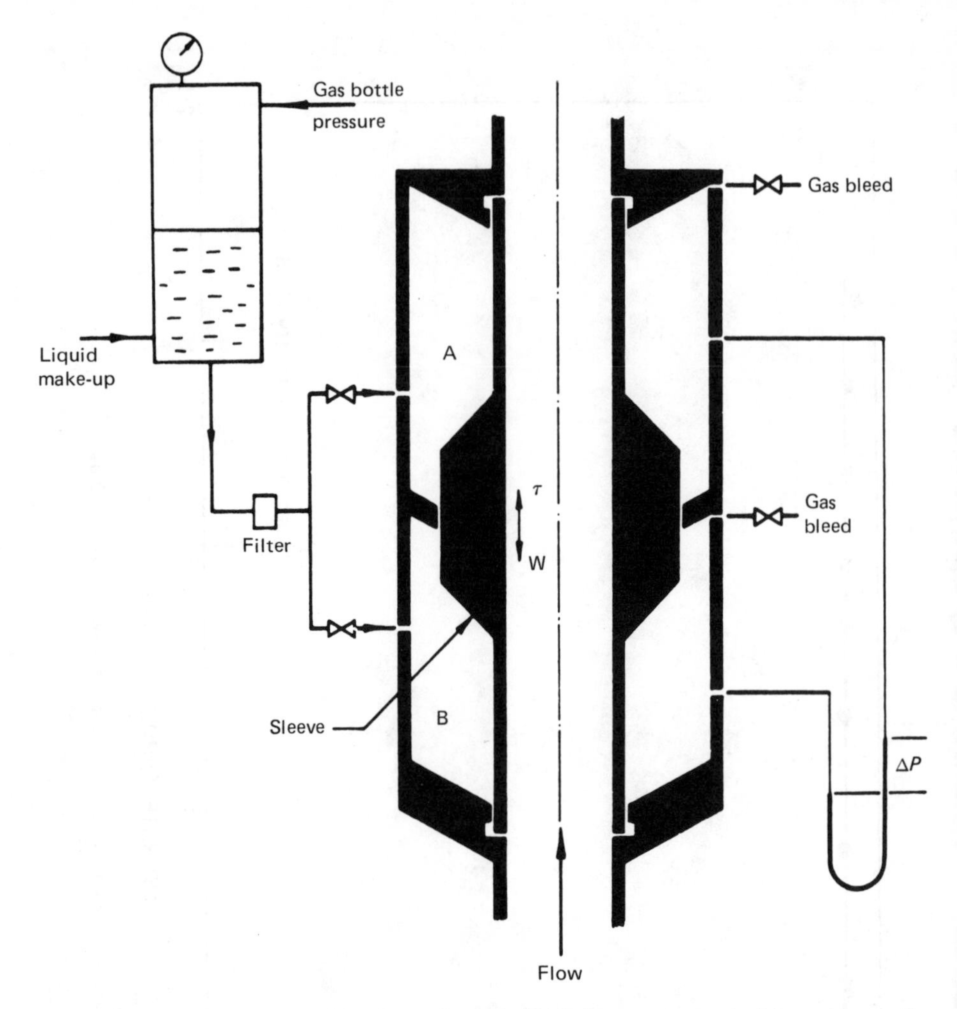

Figure 49 The device developed by Cravarolo et al. [144] for measurement of shear stress on the wall of a conduit.

deflect the flow stream. Thus, the measurement of exit momentum flux, combined with a knowledge of the inlet momentum flux and the integrated wall shear stress, can result in accurate values for the component or pressure drop due to elevation and average channel vapor fraction.

5. OUT-OF-STREAM MEASURING DEVICES

This class of instruments consists of those whose measurement ability depends on the attentuation or reflection of electromagnetic radiation or streams of atomic particles. Excellent reviews of light-photography methods are given by Arnold and Hewitt

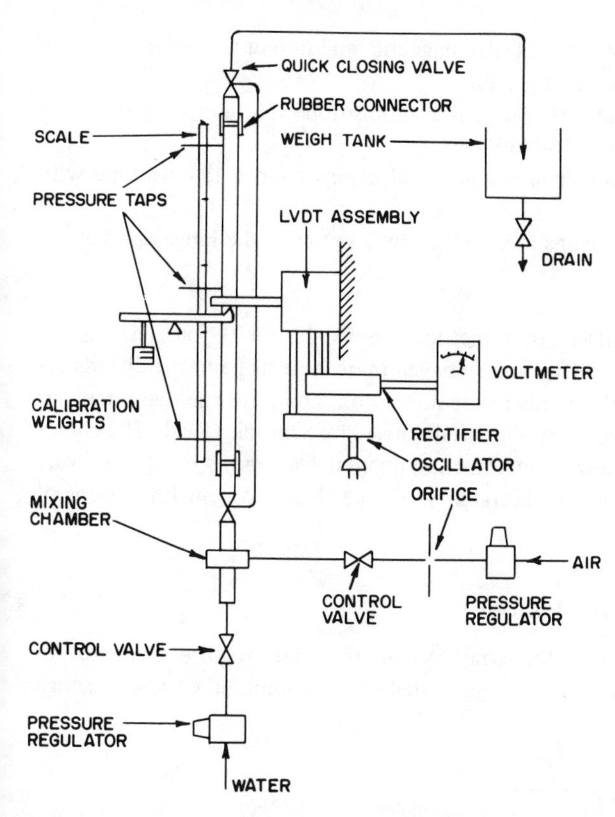

Figure 50 Diagram of the vertical-tube apparatus used by Rose [146].

[148], Cooper et al. [149], and Hsu et al. [150]. The majority of those methods are aimed at obtaining qualitative information. Indeed, visual examination of photographs has been one of the major methods of flow-pattern determination to date. On the other hand, the attenuation of invisible radiation has come to be the most widely used method of obtaining quantitative measurements of void fraction to date. Light-photography methods will not, therefore, be discussed here.

Of all the attenuation methods, the most popular involves the attenuation of the strength of a concentrated beam of photons (gamma or x-rays). This method is popular mainly because measurements may be made of space-averaged void fraction along a chord length without disturbing the fluid to a noticeable degree. This attenuation technique has generally been used to obtain average measurements of void fraction in steady systems [154 to 156] or systems with slowly varying transients (<10 Hz). Schroch [158], however, has had some success with more rapidly varying signals, and Jones [53, 157, 159] has reported a system capable of measuring transients in the millisecond range. More recently, high-intensity, high-energy systems have been developed and used for rapid transients in large pipes, owing to the increasing emphasis on large-scale nuclear safety tests [171, 172].

There are basically four types of void measurement systems using attenuation techniques:

1. X-ray systems in which a source of electromagnetic radiation in the general range of 25 to 60 keV is provided by an x-ray tube
2. Gamma-ray systems in which the source is a radioisotope that emits photons with energies usually between 40 to 100 keV
3. Gamma-ray systems that obtain a stream of electrons from a radioisotope with energies up to 10 meV
4. Neutron systems, where neutrons are supplied by a source in the range up to about 1 meV

The general problems of attenuation methods are similar for all four systems and are discussed shortly. The basic differences between these systems lie in the differences in the attenuation laws and the hardware required to accomplish the measurement. Otherwise, the overall concepts are similar, as discussed by Schrock [160]. The choice of a method usually is dependent on the experimental constraints, cost, hardware availability, etc., rather than on the desirability of a particular system for reasons of accuracy or ease of application.

5.1 X-Ray and Gamma-Ray Methods

From a source standpoint, only the x-ray system does not require a nuclear or radioisotopic source. Instead, x-rays are generated in a vacuum tube, where a high

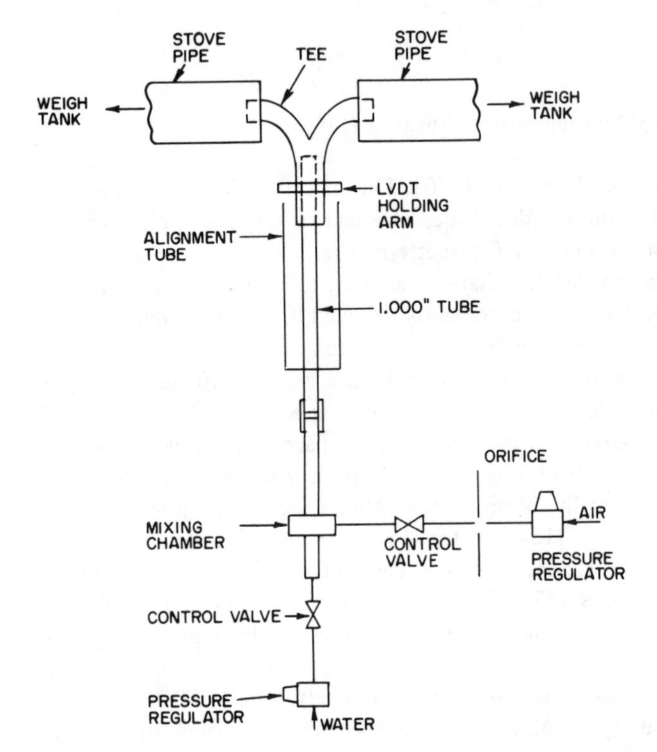

Figure 51 Schematic diagram of the exit-momentum efflux measurement system used by Rose [146] and Andeen [147].

potential is applied between a target of a specific material and a heated filament. Electrons that are "boiled" off the filament are accelerated in the potential field toward the target. The target material may be any element, but it is usually of high melting temperature and high atomic number, such as tungsten or molybdenum. Since a large amount of heat is usually generated, the target is sometimes made hollow to permit circulation of cooling water. About 99% of the electrons striking the target simply give off their energy in the form of heat upon being decelerated. The remaining 1% give off a spectrum of electromagnetic radiation, x-rays, having energies from zero up to the maximum energy of an incident electron. In addition, since some electrons knock bound electrons from atoms of the target material, other electrons fall back into these vacancies, giving off particular quanta of x-radiation characteristic of the material and the energy level vacated. This latter effect tends to produce localized maxima in the x-ray energy spectrum at these characteristic energies. Filtering of the emitted x-ray beam may be used to remove lower-energy radiation, producing a beam of nearly monoenergetic x-rays near the characteristic energy.

The advantage of using x-rays lies in the lack of bulky source holders and the repeatability and long-term stability of the source length. The disadvantages lie in the short-term unsteadiness in the source, due to the alternating portion of the applied voltage. This problem has severely limited the capabilities of transient systems such as those used by Schrock [158] and Zuber et al. [161]. Jones [157, 159], however, used special filtering techniques in the high-voltage side of his x-ray transformer supply to reduce the induced thermal ripple in the beam strength [160 to 162] to a negligible amount. In an attempt to minimize the effects of short-term unsteadiness, many users have employed dual-beam x-ray tubes, where one beam was used for the test while the other served as a reference. In theory, the ratio of the intensities of the two beams removes the original source from consideration.

A schematic of such a system is shown in Fig. 52. This is a typical dual-beam system; in this case, a tungsten target was operated at 100 kV with a tungsten filter. The result was a sharp resonant peak in the 50 to 70 keV range, corresponding to the lowest electron quantum level K_α of 58.5 keV, and the second K_β level of 68 keV. Detection and measurement of beam intensity was by means of a thallium-activated sodium iodide crystal, which absorbs x-ray wavelength radiation and gives off radiation in the visible range in its place. While the crystal has some self-absorption, the emergent intensity of the visible light is proportional to the incident x-radiation intensity. Light-piping was employed to direct the scintillated light to photomultiplier tubes that produced an electric current proportional to the incident light intensity. The output from the two photomultipliers was then differentially amplified and recorded. Additional problems in such a system include the inability to obtain well-matched photomultiplier tubes, differences in the fatigue or short-term aging characteristics of the tubes, and problems in maintaining a stable excitation voltage for the pair of detectors.

For both x-rays and gamma rays, the intensity I of a beam of original intensity I_0 is given by

$$I = I_0 e^{-\mu x} \tag{12}$$

where μ is the linear absorption coefficient, dependent on both the material and the

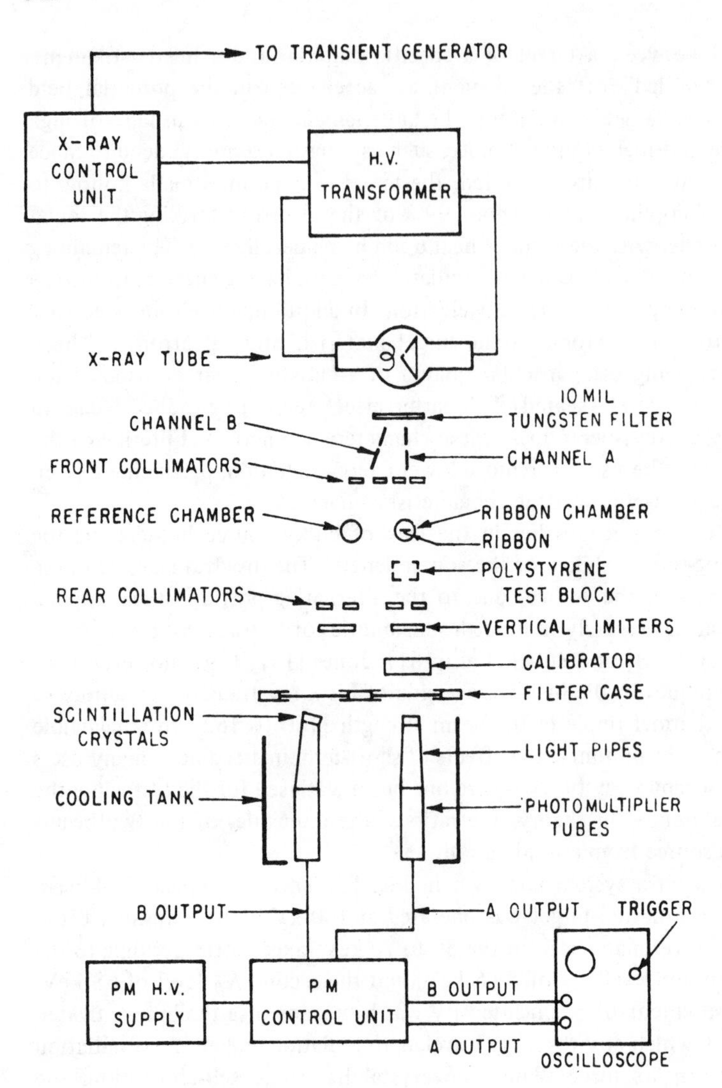

Figure 52 Schematic of x-ray densitometer system. *(From [160].)*

energy of the photons, and x is the thickness of the absorber. The major difference between the x-ray and gamma-ray systems is the source. Gamma rays are identical in nature to x-rays. Historically, however, gamma rays were considered as extremely energetic, being obtained from radioisotopes and having wavelengths much shorter than those generally associated with x-rays. In general usage, however, the term now applies to any photons originating from a nucleus, while the term x-ray is reserved for similar but extranuclear radiation.

Since gamma rays are obtained from nuclear disintegration, the source strength is time-dependent. Thus, the half-life of the source material becomes as important a consideration as the strength of the beam. Since the uncertainty (1 standard deviation) in a measurement of beam intensity is equal to the square root of the number of events measured, an intense beam from a highly radioactive source must be used to minimize observation time. The level of radioactivity, measured in curies (1 Ci = 2.22×10^{12} disintegrations/min), is dependent on the half-life (time to disintegrate 50% of the material) and the mass of the element (total number of atoms available for radioactive decay). A short-half-life material requires frequent recalibration, whereas a long-half-life material of acceptable disintegration rate must be massive and necessarily hazardous to handle. In addition, most systems have been concerned with measuring voids in water-metal or freon-metal systems. Since the linear absorption coefficient for photon attenuation decreases more rapidly with increasing energy for water than for most metals, energies less than 100 keV are desirable from a sensitivity standpoint. The three radioisotopes that have found the most use are compromises based on these factors; they are thulium-170, samarium-145, and gadolinium-153. Table 2 gives the half-life and gamma energy for these materials.

The advantages of gamma sources over x-rays are principally in cost, short-term stability, and simplicity. Since short-term stability is not a problem, dual-beam systems become unnecessary. In addition, high-voltage supplies and regulating systems are not required.

In general, x-ray and gamma-ray systems comprise the most popular void-measuring technique used to date. Extensive discussion of the errors associated with these systems was presented by Hooker and Popper [162], with exception of errors due to the fluctuations in nonlinear signals [157, 159, 163 to 166]. In [162], a one-shot method was designed to measure the voids for the entire cross section of a 0.5- × 2.175-in rectangular channel by collimating a thulium-170 source parallel to the wide plates of the channel. The authors identify the following errors:

1. Errors in the electronics system due to such characteristics as amplifier drift, photomultiplier-gain sensitivity to small changes in the supply voltage, and temperature sensitivity of the sodium-iodide crystal
2. Errors in measuring techniques due to measurement of gammas that reach the detector by some path other than through the test section, strip-chart reading errors, and errors due to calibration at nontest conditions
3. Errors due to decay of the source
4. Errors due to preferential phase distributions

Table 2 Popular isotopes for void measurements [160]

Isotope	Half-life, days	Gamma energy, keV
Samarium-145	240	39–61
Gadolinium-153	240	42–72
Thulium-170	170	52–84

For a uniform distribution of voids, Hooker and Popper concluded that the maximum absolute error in such a system is about 2.5% voids over the entire range of 0 to 100% voids. Similar conclusions have been reached in [157, 159, 160, 163] and an estimated absolute error of 1.7% voids is given for stationary measurement of local void fraction, and 2.3% voids for local measurement during a transverse void-profile scan.

The error estimates indicated above do not take into consideration errors due to void-streaming effects in preferential void distributions, nor errors due to linear averaging of a nonlinear, fluctuating quantity. Hooker and Popper discuss the former errors, associated with nonhomogeneous phase distributions. Basically these errors arise when liquid and vapor phases are separate, the limiting geometry possible being layers of liquid and gas whose planes are parallel to the beam of photons. In this case, the attenuated beam can be considered as two separate beams, one having passed through the gas and the other having passed solely through the liquid. The total intensity is, then, the sum of the individual intensities, each of which has been separately attenuated. The result is not the ideal exponential attenuation law normally employed in data reduction. Petrick and Swanson [206] verified this source of errors and showed that typical one-shot measurement techniques could be off by up to 40%. Errors of the same order or larger were also predicted in [159, 163] to arise due to the fact that two-phase flow is by nature a nonstationary phenomenon. Thus, time-averaging of a quantity that is nonlinearly related to the desired quantity, in this case void fraction, leads to significant errors when the void fraction is calculated from the average. These errors were verified by Harms and coworkers [164 to 166] in consideration of their neutron experiments.

Void-streaming errors may be significantly reduced or eliminated by reduction of the beam size with respect to the voids and by reducing the length of the measurement path, thereby reducing the magnitude of the fluctuations in intensity. Problems associated with the fluctuation source of errors may be eliminated by linearizing the signal with respect to the desired quantity, or by sampling on time scales smaller than the fluctuation periods of interest. Many workers have used fine collimation to reduce photon streaming errors [161, 156, 167 to 169]. One has used linearization to eliminate fluctuation errors [157, 159] and, to the writer's knowledge, only one group [164 to 166] has attempted to use alternative methods (i.e., short-term sampling) to circumvent this problem. In the latter case, however, sampling periods were not judged (by this writer) to be sufficiently short to eliminate all errors.

Quite recently, gamma-ray and x-ray methods have been extended to multibeam techniques in which multiple beams, spreading radially from a single source or in parallel from separate sources to individual detectors, are used. These methods have been utilized most recently in the large-scale testing programs being undertaken for nuclear safety studies. Single-source systems have been described by many workers, such as Smith [170], who used five beams in a vertical plane plus two references to characterize phase distributions in a horizontal pipe during blowdown experiments. Cut-metal windows (sometimes berillium filled) were used for high-pressure access, while otherwise standard scintillation crystal and photomultiplier instrumentation methods were used. Similarly, Yborrando [171] utilized a gamma-ray system having a single source. Such devices are useful compromises for discerning cross-sectional flow variations where flow stratification exists under transient conditions, but their usefulness is limited where a high degree of accuracy is required. Lassahan et al. [172] recently summarized the technology of x-ray and gamma-ray techniques.

All but one multibeam system to date have been static, nontraversing, systems usually with cesium-137 or cobalt-60. As such, except in high water-to-metal mass thickness ratio systems, sensitivity tends to be limited, and parasitic attenuation high. In a newly described system, however, Abuaf et al. [173] use instead a parallel five-beam, traversing system utilizing five separate thulium-170 sources. The solid-state detection and readout system has been previously described [174] utilizing cadmium-teluride crystal detectors requiring no elaborate thermal-stabilization techniques and using standard nuclear instrumentation. The 84-keV resonance is used, since it has the highest activation efficiencies of about ≡3%, requiring only 30 to 35 Ci total source activity for each effective curie at 84 keV. Also, 84-keV source strengths of 1 to 5 Ci have been reported, with activation up to 30 Ci or more expected. The major difficulty encountered was impurity activation, heretofore unencountered by other isotope researchers. It was found that any impurity having a highly efficient, high-decay-energy resonance would lead to difficulties necessitating unwieldy shielding. If impurity levels were maintained at less than 100 parts per billion, then 2.5 cm of lead would be sufficient at a 30-Ci (84-keV) activation level to keep radiation levels 1 ft away at 3 mR/h or less except in the beam. At high source strengths, however, the low-energy edge yields an extremely sensitive beam with good transient-response capabilities.

5.2 Beta-Ray Methods

In 1961, Perkins et al. [175] made the following assessment of gamma-attentuation methods:

> The γ-method is reasonably accurate for the determination of void volumes under the following conditions: (1) the voids are distributed in a homogeneous manner; (2) the test section offers a radiation path of greater than about 2500 mg/cm^2 (equivalent to one inch of water); and (3) the void fraction is greater than 25%. For smaller void fractions, smaller channels, or preferential distributions, the gamma technique may not offer sufficient accuracy.

While the methods and accuracy of beta-ray techniques have improved considerably since then, the methods they outlined are worthwhile as representative of the beta-ray method of void measurement. Specifically, beta rays are really made up of a stream of electrons emitted at high energy from a radioisotope. The basic method of attenuation for electrons is different from that for photons. Attenuation of x-rays and gamma rays is caused by photoelectric effects, where x-rays are completely absorbed by ejecting electrons from the absorbing material, and by Compton or recoil scattering, where only partial energy loss of a photon occurs with the electron ejection and a new photon of longer wavelength is emitted. In both cases the electron density of the absorbing medium is important because interaction is primarily with material electrons. Likewise, the electron density is important for the absorption of betas or free electrons. Just as when electrons decelerating in a target of an x-ray tube generate x-rays, so do they decelerate in an absorber. This loss of energy is caused by both electron interaction and nuclear interaction. The result is some x-ray production in a continuous spectrum of wavelengths, called *bremstrahlung*. Up to a certain thickness of a material, this absorption is exponential, similar to photon attenuation. However, over a certain thickness, the absorption coefficient is nearly the same for all absorbing materials. The mass per unit area required to produce a given beta absorption is nearly independent of material type and is called the *stopping distance* or *range* of the material. The thickness required to stop a beta particle is then the range divided by the material density. The range of an electron, as seen in Fig. 53, is necessarily dependent on the initial energy, since absorption is mainly by collision. Thus, for the water density of 1 g/cm^3, a 2-meV electron will be stopped by about 1 cm of water.

While beta absorption is exponential for thin absorbers, Perkins et al. [175] found that a more general law for betas is

$$\ln \frac{I}{I_0} = -2.177\,(e^{3.69Ax} - 1) \tag{13}$$

where A is proportional to the equivalent linear absorption coefficient.

For small thickness, Eq. (13) degenerates to

$$I = I_0 e^{-7.8Ax} \qquad x \simeq 0 \tag{14}$$

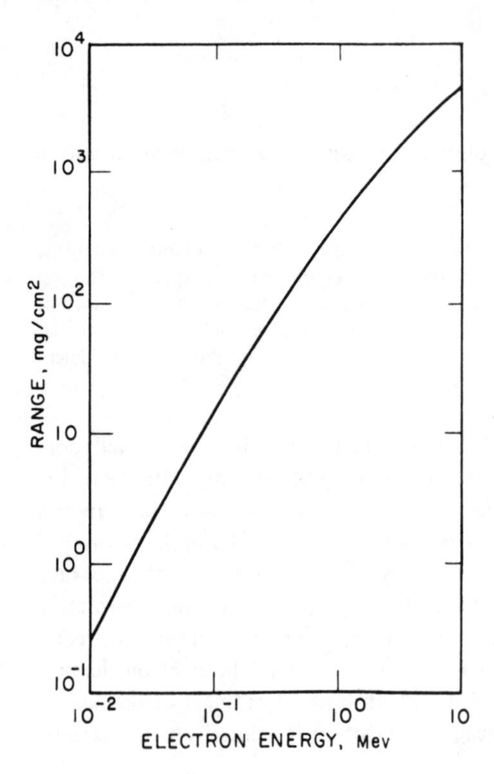

Figure 53 Range of beta particles in water. *(From [160].)*

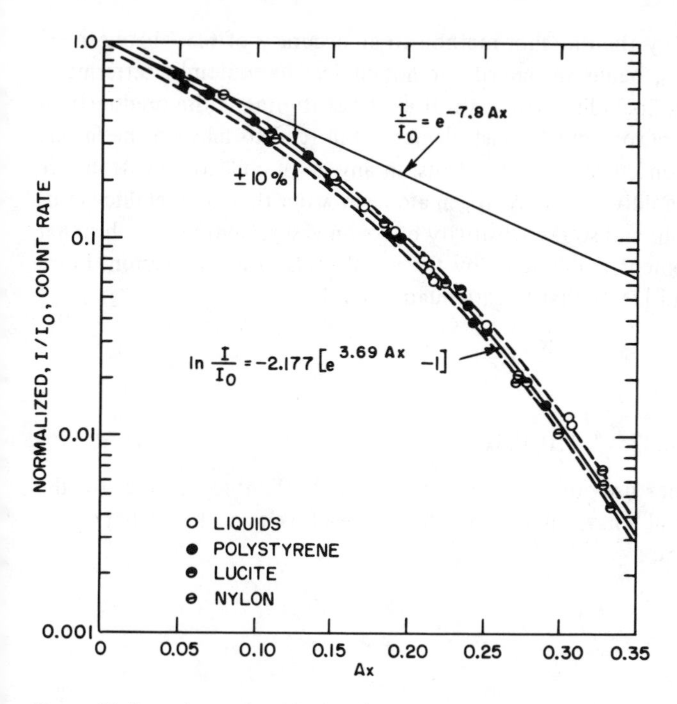

Figure 54 Experimental results for absorption of betas. *(From [175].)*

These two equations are compared in Fig. 54, where it is seen that exponential-like absorption occurs for values of Ax up to about 0.05.

There are two major advantages in using betas for void measurements where complete absorption does not occur. First, the sensitivity is greater. Perkins et al. [175], using a strontium-90 source producing 2.26-meV betas, found a sensitivity ratio between the full and empty case of 82:1, whereas for photons of about 60 keV the ratio would have been about 1.15:1. In addition, shielding is not a major obstacle, since betas are absorbed readily in most dense materials. However, a major disadvantage stems from the advantage of high sensitivity. That is, owing to the high attenuation rates, massive sources are required to obtain event rates at a test detector that would allow accurate transients to be measured. Since electrons are absorbed in the source itself (self-shielding), the use of betas becomes impractical for this application.

5.3 Neutron Methods

Neutron attenuation has been used occasionally to measure void fraction in high-pressure channels. Sha and Bonilla [176] employed an antimony-beryllium neutron source to measure voids within 3% in simulated rod-bundle geometry. Sensitivities of over 20 times that of thulium-170 gammas were obtained. More recently, Moss and Kelly [177] measured film thicknesses in a heat pipe to an accuracy of 0.006 in out of 0.125 in, while Harms and coworkers [178 to 180] measured void fractions by the one-shot method to an accuracy of 6.5% between 3% and 70% voids. Neutrons, being uncharged, are not subject to coulomb scattering as are the charged particles, including beta rays. If elastic scattering is the predominant mode of neutron deceleration, the lightest elements will tend to take on the largest percentage of the neutron current energy. Thus, in any given collision, neutrons are attenuated much more rapidly by the hydrogen atoms in water than by metallic atoms in, say, a test-section wall, and so the sensitivity of void measurement approaches that of the gamma-ray equipment, unhindered by the metal walls of a test section. For a water system it was found [180] that the attenuation law is

$$I = I_0 B(x) e^{-\mu x} \tag{15}$$

where

$$B(x) = 1 + 1.011x^2 - 0.475x^4 + 0.488x^6 \tag{16}$$

Currently, however, the use of neutrons is extremely limited, owing to the inaccessibility and general unavailability of nuclear reactors and the high cost of intense alternative sources.

NOMENCLATURE

a	capillary length
A, A_i	area, interfacial area
C, C_1	coefficients
d_1, d_2	RF probe tip dimensions
E, E_{max}	anemometer probe voltage, voltage maximum
fc	correlation function
g	gravitational acceleration
i	enthalpy
I	photon intensity
j	volume flux
K	correlation coefficient
n	index of refraction
N	count
N_{Ca}	capillary number
N_{Go}	Gocher number
q''	heat flux
R, R_p	radius, probe resistance
S	area of Probability Density Function (PDF)
t	time
$\Delta t_1, \Delta t_2$	time delays associated with d_1, d_2
U	optical probe signal, velocity
x	thickness
y	distance from wall
z	streamline coordinate

α vapor-volume or area fraction
β vapor-volume flow-rate fraction
β_w water-layer angle (optical probe)
μ viscosity, linear attenuation coefficient
Γ_v rate of vapor generation per unit volume
ξ_i interfacial perimeter in A
σ surface tension
ρ density

Subscripts

B bubble
c computed
g gas
i initial or interfacial
l liquid
m measured
max maximum
v vapor
w withdrawal

REFERENCES

1. O. C. Jones, Jr. and N. Zuber, Slug-Annular Transition with Particular Reference to Narrow Rectangular Duct, presented at the International Seminar for Momentum, Heat and Mass Transfer in Two-Phase Energy and Chemical Systems, Dubrovnik, Yugoslavia, 1978.
2. O. C. Jones, Jr. and N. Zuber, Interfacial Passage Frequency for Two-Phase, Gas-Liquid Flows in Narrow Rectangular Ducts, in *Heat and Fluid Flow in Water Reactor Safety,* pp. 5–10, London, 1977.
3. O. C. Jones, Jr. and P. Saha, Non-Equilibrium Aspects of Water Reactor Safety, in O. C. Jones and S. G. Bankoff (eds.) *Thermal and Hydraulic Aspects of Nuclear Reactor Safety,* vol. 1, *Light Water Reactors,* ASME, 1977.
4. O. C. Jones, Jr. and N. Zuber, Post-CHF Heat Transfer: A Non-Equilibrium, Relaxation Model, presented at the AIChE-ASME Heat Transfer Conference, Salt Lake City, 1977.
5. B. W. LeLourneau and A. E. Bergles (eds.), *Two-Phase Flow Instrumentation,* ASME National Heat Transfer Conference, Minneapolis, 1969.
6. G. F. Hewitt and P. C. Lovegrove, Experimental Methods in Two-Phase Flow Studies, EPRI Rep. NP-118, 1976.
7. J. M. Delhaye and O. C. Jones, A Summary of Experimental Methods for Statistical and Transient Analysis of Two-Phase Gas Liquid Flow, Argonne Rep. ANL-76-75, 1976.
8. G. F. Hewitt, *Measurements of Two-Phase Flow Parameters,* in press, Academic, New York.
9. O. C. Jones and J. M. Delhaye, Transient and Statistical Measurement Techniques for Two-Phase Flows: A Critical Review, *Int. J. Multiphase Flow,* vol. 3, pp. 89–116, 1976.
10. Y. Y. Hsu (ed.), Two Phase Flow Instrumentation Review Group Meeting, NUREG-0375, 1977.
11. J. G. Collier, *Convective Boiling and Condensation,* McGraw-Hill, New York, 1972.
12. E. S. Kordyban and T. Ranov, Experimental Study of the Mechanism of Two-Phase Slug

Flow in a Horizontal Tube, *Winter Annu. ASME Symp. on Multiphase Flow,* Philadelphia, 1963.
13. V. J. Solomon, Construction of a Two-Phase Flow Regime Transition Detector, M.S. thesis, Mechanical Engineering Department, MIT, 1962.
14. P. Griffith, The Slug-Annular Flow Regime Transition at Elevated Pressure, Argonne Rep. ANL-6796, 1963.
15. L. G. Neal and S. G. Bankoff, A High Resolution Resistivity Probe for Determination of Local Void Properties in Gas-Liquid Flow, *AIChE J.,* vol. 9, p. 490, 1963.
16. G. P. Nassos, Development of an Electrical Resistivity Probe for Void Fraction Measurements in Air-Water Flow, Argonne Rep. ANL-6738, 1963.
17. G. B. Wallis, Joint US–Euratom Research and Development Program, School of Engineering Quarterly Progress Report, Dartmouth College, July 1963.
18. E. Jannsen, Two-Phase Flow and Heat Transfer in Multirod Geometries: Eighteenth Quarterly Progress Report January 1–March 31, 1970, GEAP-10214, General Electric Co., 1970.
19. R. E. Haberstroh and P. Griffith, The Slug-Annular Two-Phase Flow Regime Transition, ASME Paper 65-HT-52, 1965.
20. H. Chevalier, C. Lakme, and J. Max, Device for the Study of Bubble Flow within a Pipe, English Patent Application 36315/65, August 1965.
21. G. C. Gardner and P. H. Neller, Phase Distributions in Flow of an Air-Water Mixture Round Bends and Past Obstructions at the Wall, Paper 12, *IMechE Conf.,* Bristol, 1969.
22. H. S. Yu and E. M. Sparrow, Experiments on Two-Component Stratified Flow in a Horizontal Duct, ASME Paper 68-HT-14, August 1968.
23. E. Jannsen, Two-Phase Flow and Heat Transfer in Multirod Geometries: Eighteenth Quarterly Progress Report January 1–March 31, 1970, GEAP-10214, General Electric Co., 1970.
24. K. Akagawa, Fluctuations of Void Ratio in Two-Phase Flow, *Bull. JSME,* vol. 7, p. 122, 1964.
25. J. F. Lafferty and F. G. Hammitt, A Conductivity Probe for Measuring Local Void Fraction in Two-Phase Flow, *Nucl. Appl.,* vol. 3, p. 317, 1967.
26. E. Jannsen, Two-Phase Flow and Heat Transfer in Multirod Geometries: Fourth and Fifth Quarterly Progress Reports, GEAP-5056, General Electric Co., January 1966.
27. J. G. Collier and G. F. Hewitt, Film Thickness Measurements, ASME Paper 64-WA/HT-41, 1964.
28. G. F. Hewitt and P. C. Lovegrove, Comparative Film Thickness and Holdup Measurements, AERE-M-1203, United Kingdom Atomic Energy Authority, 1963.
29. N. Hall-Taylor and G. F. Hewitt, The Motion and Frequency of Large Disturbance Waves in Annular Two-Phase Flow of Air-Water Mixture, AERE-R-3952, United Kingdom Atomic Energy Authority, 1962.
30. G. F. Hewitt and G. B. Wallis, Flooding and Associated Phenomena in Falling Film Flow in a Tube, AERE-R-4022, United Kingdom Atomic Energy Authority, 1963.
31. G. F. Hewitt, H. A. Kearsey, P. M. C. Lacy, and D. J. Pulling, Burnout and Film Flow in the Evaporation of Water in Tubes, AERE-R-4864, United Kingdom Atomic Energy Authority, 1965.
32. D. Butterworth, Air-Water Climbing Film Flow in an Eccentric Annulus, AERE-R-5787, United Kingdom Atomic Energy Authority, 1968.
33. E. O. Moeck, The Design, Instrumentation, and Commissioning of the Water-Air-Fog Experimental Rig (WAFER), APPE-1, Atomic Energy of Canada Ltd., 1964.
34. G. A. Wickhammer, E. O. Moeck, and I. P. L. Mac Donald, Measurement Techniques in Two-Phase Flow, AECL-2215, Atomic Energy of Canada Ltd., 1964.
35. G. F. Hewitt, R. D. King, and P. C. Lovegrove, Techniques for Liquid Film and Pressure Drop Studies in Annular Two-Phase Flow, AERE-R-3921, United Kingdom Atomic Energy Authority, 1962.
36. N. Adorni, I. Casagrande, L. Cravarolo, A. Hassid, and M. Silvestri, Experimental Data on

Two-Phase Adiabatic Flow; Liquid Film Thickness, Phase and Velocity Distribution, Pressure Drop in Vertical Gas-Liquid Flow, EURAEC-150 (CISE Rep. R35), Centro Informazioni Studi Esperanzi, Milan, Italy, 1961.
37. G. F. Hewitt and P. C. Lovegrove, Comparative Film Thickness and Holdup Measurements in Vertical Annular Flow, AERE-M-1203, United Kingdom Atomic Energy Authority, 1963.
38. Y. Iida and K. Kobayasi, Distributions of Void Fraction above a Horizontal Heating Surface in Pool Boiling, *Bull. JSME,* vol. 12, pp. 283–290, 1969.
39. Y. Iida and K. Kobayasi, An Experimental Investigation of the Mechanism of Pool Boiling Phenomena by a Probe Method, *Heat Transfer,* vol. 5, pp. 1–11, 1970.
40. K. Kobayasi, Measuring Method of Local Phase Velocities and Void Fraction in Bubbly and Slug Flows, presented at the Fifth International Heat Transfer Conference, Round Table RT-1, 1974.
41. M. Reocreux and J. C. Flamand, Etude de l'utilisation des sondes resistives dans des ecoulements diphasiques a grande vitesse, CENG, STT, Rapport interne 111, 1972.
42. H. Tawfik, S. A. Alpay, and E. Rhodes, Resistivity Probe Error Study Using a Two-Dimensional Simulation of the Electrical Field in a Two-Phase Media, *Proc. 2d Multiphase Flow and Heat Transfer Symp. Workshop,* Miami Beach, 1979.
43.–44. J. Sheppard, private communication.
45. K. Sekoguchi, H. Fukui, and Y. Sato, Flow Characteristics and Heat Transfer in Vertical Bubble Flow, *Proc. Japan-U.S. Seminar on Two-Phase Flow Dynamics,* Kobe, 1979, pp. 107–127.
46. J. Block, private communication.
47. S. W. Gouse, Jr., Void Fraction Measurement, AD-600524, United States Atomic Energy Commission, 1964.
48. L. S. Neal and S. G. Bankoff, A High Resolution Resistivity Probe for Determination of Local Void Properties in Gas-Liquid-Flow, *AIChE J.,* vol. 9, pp. 49–54, 1963.
49. K. Sekoguchi, H. Fukui, T. Matsuoka, and K. Nishikawa, Investigation into the Statistical Characteristics of Bubbles in Two-Phase Flow, *Trans. JSME,* vol. 40, pp. 2295–2310, 1974.
50. J. P. Galaup, Contribution a l'etude des methodes de mesure en ecoulement diphasique, doctoral thesis, Universite Scientifique et Medicale de Grenoble, Institute National Polytechnique de Grenoble, 1975.
51. T. Uga, Determination of Bubble Size Distribution in a BWR, *Nucl. Eng. Des.,* vol. 22, pp. 252–261, 1972.
52. N. Ibragimov, V. P. Bobkov, and N. A. Tychinskii, Investigation of the Behavior of the Gas Phase in a Turbulent Flow of a Water-Gas Mixture in Channels, *Teplofiz. Vys. Temp.,* vol. 11, pp. 1051–1061; also *High Temp.,* vol. 11, pp. 935–944, 1973.
53. O. C. Jones, Statistical Considerations in Heterogeneous, Two-Phase Flowing Systems, Ph.D. thesis, Rensselaer Polytechnic Institute, Troy, N.Y., 1973.
54. A. Serizawa, Fluid-Dynamic Characteristics of Two-Phase Flow, Ph.D. thesis, Institute of Atomic Energy, Kyoto University, 1974.
55. A. S. Telles and A. E. Dukler, Statistical Characteristics of Thin, Vertical Wavy Liquid Films, *I&EC Fundam.,* vol. 9, pp. 412–421, 1970.
56. A. E. Dukler, Characterization, Effects, and Modeling of the Wavy Gas-Liquid Interface, in G. Hetsroni, S. Sideman, and J. P. Hartnett (eds.), *Progress in Heat and Mass Transfer,* vol. 6, pp. 207–234, Pergamon, Oxford, 1972.
57. K. J. Chu and A. E. Dukler, Statistical Characteristics of Thin, Wavy Films, *AIChE J.,* vol. 20, pp. 695–706, 1974.
58. H. Lecroart and J. Lewi, Mesures locales et leur interpretation statistique pour un ecoulement diphasique a grande vitesse et taux de vide, Societe Hydrotechnique de France, Douziemes Journees de l'Hydraulique, Paris, 1972, Question IV, Rapport 7.
59. H. Lecroart and R. Porte, Electrical Probes for Study of Two-Phase Flow at High Velocity, presented at the International Symposium on Two-Phase Systems, Haifa, Israel, 1972.

60. A. J. J. Wamsteker et al., The Application of the Impedance Method for Transient Void Fraction Measurement and Comparison with the X-ray Attenuation Technique, EURAEC-1109, Atomic Energy of Canada Ltd., 1964.
61. J. Leung, The Occurrence of Critical Heat Flux during Blowdown with Flow Reversal, M.Sc. thesis, Northwestern University, Evanston, Ill., 1976.
62. L. Cimorelli and R. Evangelisti, The Application of the Capacitance Method for Void Fraction Measurement in Bulk Boiling Conditions, *Int. J. Heat Mass. Transfer,* vol. 10, p. 277, 1967.
63. I. Bencze and I. Oerbeck, Development and Application of an Instrument for Digital Measurement and Analysis of Void Using an AC Impedance Probe, KR-73, Swedish Atomic Energy Commission, Studsvik, Sweden, 1964.
64. I. Oerbeck, Impedance Void Meter, KR-32, 1962.
65. L. Cimorelli, M. DiBartolomeo, and A. Premoli, Void Fraction Measurement in a Boiling Channel Using the Impedance Method, RT-ING-(65) 7, 1965.
66. L. Cimorelli and A. Premoli, Measurement of Void Fraction with Impedance Gage Technique, *Energ. Nucl.,* vol. 13, p. 12, 1966.
67. D. S. Nielson, Void Fraction Measurements in an Out-of-Pile High-Pressure Rig MK II-A by the Impedance Bridge Method, RISO-M-894, 1969.
68. Y. Y. Hsu, F. F. Simon, and R. W. Grahm, Application of Hot Wire Anemometry for Two-Phase Flow Measurement such as Void Fraction and Slip Velocity, presented at the Two-Phase Flow Symposium at the Winter Annual ASME Metting, Philadelphia, 1963.
69. O. C. Jones and N. Zuber, Use of a Hot-Film Anemometer for Measurement of Two-Phase Void and Volume Flux Profiles in a Narrow Rectangular Channel, *AIChE Sym. Ser.,* vol. 74, pp. 191–204, 1978.
70. O. C. Jones, Preliminary Investigation of Hot Film Anemometer in Two-Phase Flow, TID-24104, General Electric Co., 1966.
71. V. Goldschmidt and S. Eskmazi, Two-Phase Turbulent Flow in a Plane Jet, ASME Paper 66-WA/APM-6, 1966.
72. V. Goldschmidt and M. K. Householder, The Hot Wire Anemometer as an Aerosol Droplet Size Sampler, *Atmos. Environ.,* vol. 3, p. 643, 1969.
73. G. M. Bragg and J. Tevaarverk, The Effect of a Liquid Droplet on a Hot Wire Anemometer Probe, Paper 2-2-19, presented at the First Symposium on Flow–Its Measurement and Control in Science and Industry, Pittsburgh, 1971.
74. J. M. Delhaye, Measurement of the Local Void Fraction Anemometer, CEA-R-3465(E), Grenoble, 1968.
75. J. M. Delhaye, Anemometre a temperature constante etalonnage des sondes a film chaude dans les liquides, CENG Rapport TT 290, 1968.
76. J. M. Delhaye, Mesure de taux de vide local en ecoulement diphasique eau-air par un anemometre a film chaud, CENG Rapport TT 79, 1967.
77. J. M. Delhaye, Theoretical and Experimental Results about Air and Water Bubble Boundary Layers, presented at the Novosibirsk Symposium, 1968.
78. J. M. Delhaye, Hot Film Anemometry in Two-Phase Flow, presented at the Symposium on Two-Phase Flow Instrumentation at the National Heat Transfer Conference, Minneapolis, 1969.
79. A. E. Bergles, J. P. Roos, and J. G. Bourne, Investigation of Boiling Flow Regimes and Critical Heat Fluxes, Final Summary Report, Dynatech Corp. Rep. NYO-3304-13, 1968.
80. N. Abuaf, A. Swoboda, and G. A. Zimmer, Reactor Safety Research Programs, Quarterly Progress Report, BNL-NUREG-50747, p. 175, Brookhaven National Laboratory, 1977.
81. N. Abuaf, T. P. Feierabend, G. A. Zimmer, and O. C. Jones, Radio Frequency (R-F) Proper for Bubble Size and Velocity Measurements, BNL-NUREG-50997, Brookhaven National Laboratory, 1979.
82. N. Abuaf, T. P. Feierabend, G. A. Zimmer, and O. C. Jones, Radio Frequency (R-F) Prober for Bubble Size and Velocity Measurements, *Rev. Sci. Instrum.,* vol. 50, pp. 1260–1263, 1979.

83. T. Fortescue, personal communication.
84. B. D. Marcus and D. Dropkin, Measured Temperature Profiles within the Superheated Boundary Layer above a Horizontal Surface in Saturated Nucleate Pool Boiling of Water, *J. Heat Transfer,* vol. 87C, pp.333–341, 1965.
85. C. Bonnet and E. Macke, Fluctuations de temperature dans la paroi chauffante et dans le liquide au cours de l'ebullition nucleee, EUR 3162f, 1966.
86. T. E. Lippert and R. S. Dougall, A Study of the Temperature Profiles Measured in the Thermal Sublayer of Water, Freon-113, and Methyl Alcohol during Pool Boiling, *J. Heat Transfer,* vol. 87C, pp. 333–341, 1965.
87. J. Jacobs and A. H. Shade, Measurement of Temperatures Associated with Bubbles in Subcooled Pool Boiling, *J. Heat Transfer,* vol. 91C, pp. 123–128, 1969.
88. S. J. D. Van Stralen and W. M. Sluyter, Local Temperature Fluctuations in Saturated Pool Boiling of Pure Liquids and Binary Mixtures, *Int. J. Heat Mass Transfer,* vol. 12, pp. 187–198, 1969.
89. N. Stefanovic, N. Afgan, V. Pislar, and L. J. Jovanovic, Experimental Investigation of the Superheated Boundary in Forced Convection Boiling, in *Heat Transfer,* vol. 5, Elsevier, Amsterdam, 1970.
90. N. Afgan, L. J. Jovanovic, M. Stefanovic, and V. Pislar, An Approach to the Analysis of Temperature Fluctuation in Two-Phase Flow, *Int. J. Heat Mass Transfer,* vol. 16, pp. 187–194, 1973.
91. N. Afgan, M. Stefanovic, L. J. Jovanovic, and V. Pislar, Determination of the Statistical Characteristics of Temperature Fluctuation in Pool Boiling, *Int. J. Heat Mass Transfer,* vol. 16, pp. 249–256, 1973.
92. J. R. Wiebe and R. L. Judd, Superheat Layer Thickness Measurements in Saturated and Subcooled Nucleate Boiling, *J. Heat Transfer,* vol. 93C, pp. 455–461, 1971.
93. G. G. Treschov, Experimental Investigation of the Mechanism of Heat Transfer with Surface Boiling of Water, *Teploenergetika,* vol. 3, pp. 44–48, 1957.
94. L. M. Jiji and J. A. Clark, Bubble Boundary Layer and Temperature Profiles for Forced Convection Boiling in Channel Flow, *J. Heat Transfer,* vol. 86C, pp. 50–58, 1964.
95. G. E. Walmet and F. W. Staub, Electrical Probes for Study of Two-Phase Flows, in B. W. LeTourneau and A. E. Bergles (eds.), *Two-Phase Flow Instrumentation,* pp. 84–101, ASME, National Heat Transfer Conference, Minneapolis, 1969.
96. G. Barois, Etude experimentale de l'autovaporisation d'un ecoulement ascendant adiabatique d'eau dans un canal de section uniforme, doctoral thesis, Faculte des Sciences de l'Universite de Grenoble, 1969.
97. C. A. A. Van Paassen, Thermal Droplet Size Measurements Using a Thermocouple, *Int. J. Heat Mass Transfer,* vol. 17, pp. 1527–1548, 1974.
98. J. M. Delhaye, R. Semeria, and J. C. Flamand, Void Fraction, Vapor and Liquid Temperatures: Local Measurements in Two-Phase Flow Using a Microthermocouple, *J. Heat Transfer,* vol. 95C, pp. 365–370, 1973.
99. J. M. Delhaye, R. Semeria, and J. C. Flamand, Mesure du taux de wide et des temperatures du liquid et de la vapeur ecoulement diphasique avec changement de phase a l'aide d'un microthermocouple, CEA-R4302, Grenoble, 1972.
100. N. Miller and R. E. Mitchie, Electrical Probes for Study of Two-Phase Flows, in B. W. LeTourneau and A. E. Bergles (eds.), *Two-Phase Flow Instrumentation,* pp. 82–88, ASME, 1969.
101. N. Miller and R. E. Mitchie, Measurement of Local Voidage in Liquid/Gas Two-Phase Flow Systems, *J. Br. Nucl. Energy Soc.,* vol. 9, pp. 94–100, 1970.
102. R. Bell, B. E. Boyce, and J. G. Collier, The Structure of a Submerged Impinging Gas Jet, *J. Br. Nucl. Energy Soc.,* vol. 11, pp. 183–193, 1972.
103. T. D. A. Kennedy and J. G. Collier, The Structure of an Impinging Gas Jet Submerged in a Liquid, in *Multi-Phase Flow Systems, ICE Symp. Ser.* 38, vol. II, Paper J4, 1974.
104. S. Hinata, A Study on the Measurement of the Local Void Fraction by the Optical Fibre Glass Probe, *Bull. JSME,* vol. 15, pp. 1228–1235, 1972.

105. F. Danel and J. M. Delhaye, Sonde optique pour mesure du taux de presence local en ecoulement diphasique, *Mesures-Regulation-Automatisme,* pp. 99–101, 1971.
106. J. M. Delhaye and J. P. Galaup, Measurement of Local Void Fraction in Freon-12 with a 0.1 mm Optical Fiber Probe, private communication.
107. N. Abuaf, O. C. Jones, and G. A. Zimmer, Response Characteristics of Optical Probes, ASME Preprint 78-WA/HT-3, 1978.
108. N. Abuaf, O. C. Jones, Jr., and G. A. Zimmer, Optical Probe for Local Void Fraction and Interface Velocity Measurements, *Rev. Sci. Instrum.,* vol. 49, pp. 1090–1094, 1978.
109. Y. Y. Hsu (ed.), *Two-Phase Flow Instrumentation Review Group Meeting,* NUREG-0375, United States Nuclear Regulation Commission, 1977.
110. D. A. White and J. A. Tallmadge, A Gravity Corrected Theory for Cylinder Withdrawal, *AIChE J.,* vol. 13, pp. 745–750, 1967.
111. J. A. Tallmadge and D. A. White, Film Properties and Design Procedures in Cylinder Withdrawal, *Inc. Eng. Chem. Process Des. Dev.,* vol. 7, pp. 503–508, 1968.
112. D. A. White and J. A. Tallmadge, A Theory of Withdrawal of Cylinders from Liquid Baths, *AIChE J.,* vol. 12, pp. 233–339, 1966.
113. L. Cravarolo and A. Hassid, Phase and Velocity Distribution in Two-Phase Adiabatic Dispersed Flow, CISE-R-98, Centro Informazioni Studi Esperanzi, Milan, Italy, 1963.
114. N. Adorni, P. Alia, L. Cravarolo, A. Hassid, and E. Pedrocchi, An Isokinetic Sampling Probe for Phase and Velocity Distribution Measurements in Two-Phase Flow near the Wall of the Conduit, CISE-R-89, Centro Informazioni Studi Esperanzi, Milan, Italy, 1963.
115. Truong Quang Minh and J. Huyghe, Measurement and Correlation of Entrainment Fraction in Two-Phase, Two-Component, Annular Dispersed Flow, CENG Rep. TT-52, Grenoble, 1965.
116. E. O. Moeck, Measurement of Liquid Film Flow and Wall Shear Stress in Two-Phase Flow, *Symp. on Two-Phase Flow Instrum.,* National Heat Transfer Conference, Minneapolis, 1969.
117. E. Jannsen, Two-Phase Flow and Heat Transfer In Multirod Geometries, Third Quarterly Progress Report April to July 1965, GEAP-4933, General Electric Company, 1965.
118. G. F. Hewitt, H. A. Kearsey, P. M. C. Lacy, and D. J. Pulling, Burnout and Film Flow in the Evaporation of Water in Tubes, AERE-4864, United Kingdom Atomic Energy Authority, 1965.
119. E. O. Moeck, Annular Dispersed Two-Phase Flow and Critical Heat Flux, AECL-3656, Atomic Energy of Canada Ltd., 1970.
120. G. F. Hewitt and G. B. Wallis, Flooding and Associated Phenomena in Falling Film Flow in a Tube, AERE-4-4022, United Kingdom Atomic Energy Authority, 1963.
121. G. F. Hewitt, H. A. Kearsey, P. M. C. Lacy, and D. J. Pulling, Burnout and Nucleation in Climbing Film Flow, AERE-R-4374, United Kingdom Atomic Energy Authority, 1963.
122. G. F. Hewitt, P. M. C. Lacy, and B. Nichols, Transitions in Film Flow in a Vertical Tube, AERE-R-4614, United Kingdom Atomic Energy Authority, 1965.
123. L. B. Cousins, W. H. Denton, and G. F. Hewitt, Liquid Mass Transfer in Annular Two-Phase Flow, AERE-R-4926, United Kingdom Atomic Energy Authority, 1965.
124. D. Butterworth, Air-Water Climbing Film Flow in an Eccentric Annulus, AERE-R-5787, United Kingdom Atomic Energy Authority, 1968.
125. K. Singh, C. C. St. Pierre, W. A. Crago, and E. O. Moeck, Liquid Film Flow Rates in Two-Phase Flow of Steam and Water at 1000 psia, *AIChE J.,* vol. 15, p. 51, 1969.
126. F. A. Schraub, R. L. Simpson, and E. Jannsen, Two-Phase Flow and Heat Transfer in Multirod Geometries: Air-Water Flow Structure Data for Round Tube, Concentric and Eccentric Annulus, and Nine-Rod Bundle, GEAP-5739, General Electric Co., 1969.
127. R. Staniforth, G. F. Stevens, and R. W. Wood, An Experimental Investigation into the Relationship between Burnout and Film Flow Rate in a Uniformly Heated Round Tube, AEEW-R-430, United Kingdom Atomic Energy Authority, 1965.
128. N. Adorni, I. Casagrande, L. Cravarolo, A. Hassid, and M. Silvestri, Experimental Data on Two-Phase Adiabatic Flow; Liquid Film Thickness, Phase and Velocity Distributions,

Pressure Drops in Vertical Gas-Liquid Flow, CISE-R-35 (EUREAC-150), Centro Informazioni Studi Esperanzi, Milan, Italy, 1961.

129. R. T. Lahey, Jr., B. S. Shiralkar, and D. W. Radcliff, Subchannel and Pressure Drop Measurements in a Nine-Rod Bundle for Diabatic and Adiabatic Conditions, GEAP-13049, General Electric Co., 1970.
130. F. A. Schraub, Isokinetic Sampling Probe Techniques Applied to Two-Component, Two-Phase Flow, ASME Paper 67-WA/FE-28, 1967; also CEAP-5287, 1966.
131. K. W. Todd and D. J. Fallon, Erosion Control in the Wet-Steam Turbine, *Proc. IME,* vol. 35, p. 180, London, 1965.
132. E. Jannsen, Two-Phase Flow and Heat Transfer in Multirod Geometries: Second Quarterly Progress Report, January or April 1965, GEAP-4863, General Electric Co., 1965.
133. G. B. Wallis and D. A. Steen, Two-Phase Flow and Boiling Heat Transfer: Quarterly Progress Report for July to September 1963, NYO-10,488, School of Engineering, Dartmouth College, 1963.
134. R. J. Burick, C. H. Scheuerman, and A. Y. Falk, Determination of Local Values of Gas and Liquid Mass Flux in Highly Loaded Two-Phase Flow, Paper 1-5-21, presented at the First Symposium on Flow–Its Measurements and Control in Science and Industry, Pittsburgh, 1971.
135. J. L. Dussord and A. H. Shapiro, A Deceleration Probe for Measuring Stagnation Pressure and Velocity of a Particle-Laden Gas Stream, *Jet. Propul.,* p. 24, January 1958.
136. D. J. Ryley and G. A. Kirkman, The Concurrent Measurement of Momentum and Stagnation Enthalpy in a High Quality Wet Steam Flow, Paper 26, I Mech E Thermodynamics and Fluid Mechanics Convention, Bristol, 1968.
137. F. A. Schraub, Isokinetic Probe and Other Two-Phase Sampling Devices: A Survey, presented at the Symposium on Two-Phase Flow Instrumentation, National Heat Transfer Conference, Minneapolis, 1969.
138. G. L. Shires and P. J. Riley, The Measurement of Radial Voidage Distribution in Two-Phase Flow by Isokinetic Sampling, AEEW-M-650, United Kingdom Atomic Energy Authority, 1966.
139. L. E. Gill, G. F. Hewitt, J. W. Hitchon, and P. M. C. Lacy, Sampling Probe Studies of the Gas Core in Annular Two-Phase Flow, pt. 1, The Effect of Length on Phase and Velocity Distributions, *Chem. Eng. Sci.,* vol. 18, p. 525, 1963.
140. J. H. Preston, Determination of Turbulent Skin Friction by Means of Pitot Tubes, *J. R. Aeronaut. Soc.,* vol. 58, p. 109, 1954.
141. V. C. Patel, Calibration of the Preston Tube and Limitations of Its Use in Pressure Gradients, *J. Fluid Mech.,* vol. 23, p. 185, 1965.
142. C. W. King, Measurement of Wall Shear Stress on a High Velocity Vapor Condensing in a Vertical Tube, Ph.D. thesis, University of Connecticut, 1970.
143. E. Jannsen, Two-Phase Flow and Heat Transfer in Multirod Geometries, Eight Quarterly Progress Report, July to October 1966, GEAP-5300, General Electric Co., 1966.
144. L. Cravarolo, A. Giorgini, A. Hassid, and E. Pedrocchi, A Device for the Measurement of Shear Stress on the Wall of a Conduit–Its Application in Mean Density Determination in Two-Phase Flow Shear Stress Data in Two-Phase Adiabatic Vertical Flow, CISE-R-82 (EURAEC-930), Centro Informazioni Studi Esperanzi, Milan, Italy, 1964.
145. S. C. Rose, Some Hydrodynamic Characteristics of Bubbly Mixtures Flowing Vertically Upwards in Tubes, Sc.D. thesis, MIT, 1964.
146. S. C. Rose and P. Griffith, Flow Properties of Bubbly Mixtures, ASME Paper 65-HT-58, 1965.
147. G. B. Andeen and P. Griffith, The Momentum Flux in Two-Phase Flow, MIT-3496-1, Massachusetts Institute of Technology, Cambridge, 1965.
148. C. R. Arnold and G. F. Hewitt, Further Developments in the Photography of Two-Phase Flow, AERE-R-5318, United Kingdom Atomic Energy Authority, 1967.
149. K. D. Cooper, G. F. Hewitt, and B. Pinchin, Photography of Two-Phase Flow, AERE-R-4301, United Kingdom Atomic Energy Authority, 1963.

150. Y. Y. Hsu, F. J. Simoneau, F. F. Simon, and R. W. Grahm, Photographic and Other Optical Techniques for Studying Two-Phase Flow, presented at the Symposium on Two-Phase Flow Instrumentation, National Heat Transfer Conference, Minneapolis, 1969.
151. R. W. Lockart and R. C. Martinelli, Proposed Correlation of Data for Iso-Thermal Two-Phase, Two-Component Flow in Pipes, *Chem. Eng. Progr.*, vol. 44, pp. 39–48, 1944.
152. C. F. Hewitt, I. King, and P. C. Lovegrove, Holdup and Pressure Drop Measurements in Two-Phase Annular Flow of Air-Water Mixtures, AERE-R-3764, United Kingdom Atomic Energy Authority, 1964.
153. L. G. Neal, Local Parameters in Cocurrent Mercury-Nitrogen Flow, ANL-6625, Argonne National Laboratory, 1963.
154. L. Cravarolo and A. Hassid, Liquid Volume Fraction in Two-Phase, Adiabatic Systems, *Energ. Nucl.*, vol. 12, p. 11, 1965.
155. M. Petrick, Two-Phase Air-Water Flow Phenomena, ANL-5787, Argonne National Laboratory, 1958.
156. R. W. Pike, B. Wilkinson, Jr., and H. C. Ward, Measurement of the Void Fraction in Two-Phase Flow by X-ray Attenuation, *AIChE J.*, vol. 11, p. 5, 1965.
157. O. C. Jones and N. Zuber, The Interrelation between Void Fraction Fluctuation and Flow Patterns in Two-Phase Flow, *Int. J. Multiphase Flow*, vol. 2, pp. 273–306, 1975.
158. V. E. Schrock and F. B. Selph, An X-ray Densitometer for Transient Steam Void Measurement, SAN-1005, University of California, Los Angeles, 1963.
159. O. C. Jones, Determination of Transient Characteristics of an X-ray Void Measurement System for Use in Studies of Two-Phase Flow, KAPL-3859, General Electric Co., 1970.
160. V. E. Schroch, Radiation Techniques in Two-Phase Flow Measurement, presented at the Symposium on Two-Phase Instrumentation, National Heat Transfer Conference, Minneapolis, 1969.
161. N. Zuber, F. W. Staub, G. Bijwaard, and P. G. Kroeger, Steady State and Transient Void Fraction in Two-Phase Flow Systems–Final Report for the Program on Two-Phase Flow Investigation, GEAP-5417, General Electric Co., 1967.
162. H. H. Hooker and G. F. Popper, A Gamma-Ray Attenuation Method for Void Fraction Determination in Experimental Boiling Heat Transfer Test Facilities, ANL-5766, Argonne National Laboratory, 1958.
163. O. C. Jones, Procedural and Calculational Errors in Void Fraction Measurements by Particle or Photon Attenuation Techniques, KAPL-3361, General Electric Co., 1967.
164. A. A. Harms and C. F. Forrest, Dynamic Effects in Radiation Diagnosis of Fluctuating Voids, *Nucl. Sci. Eng.*, vol. 46, pp. 408–413, 1971.
165. A. A. Harms and F. A. R. Laratta, The Dynamic-Bias in Radiation Interrogation of Two-Phase Flow, *Int. J. Heat Mass Transfer,* vol. 16, pp. 1459–1465, 1973.
166. W. T. Hancox, C. F. Forrest, and A. A. Harms, Void Determination in Two-Phase Systems Employing Neutron Transmission, ASME Paper 72-HT-2, 1972.
167. T. P. Bestenbreur and C. L. Spigt, Study of Mixing between Adjacent Channels in an Atmospheric Air-Water System, presented at the Two-Phase Flow Meeting, Winfrith, 1967 (see CONF-67065-6).
168. R. Martin, Measurements of the Local Void Fraction at High Pressure in a Heating Channel, *Nucl. Sci. Eng.*, vol. 48, p. 125, 1972.
169. C. L. Spigt, A. J. J. Wamsteker, and H. F. von Vlaardingen, Review of the Measuring, Recording and Analyzing Methods in Use in the Two-Phase Flow Programme of the Laboratory of Heat Transfer and Reactor Engineering at the Technological University of Eindhoven, Report WW016-R64 (EURATOM III-17, Special TR 18), 1964.
170. A. V. Smith, A Fast Response Multi-Beam X-ray Absorption Technique for Identifying Phase Distributions during Steam-Water Blowdowns, *J. Br. Nucl. Energy Soc.*, vol. 14, pp. 227–235, 1975.
171. Y. Yborrondo, Dynamic Analysis of Pressure Transducers and Two-Phase Flow Instrumentation, presented at the Third Water Reactor Safety Research Information Meeting, Washington, D.C., 1975.

172. G. D. Lassahn, A. G. Stephens, J. D. Taylor, and D. B. Wood, X-Ray and Gamma-Ray Transmission Densitometry, presented at the International Colloquium on Two-Phase Flow Instrumentation, Idaho Falls, Idaho, 1979.
173. N. Abuaf, G. A. Zimmer, and O. C. Jones, private communication.
174. G. A. Zimmer, B. J. C. Wu, W. J. Leonhardt, N. Abuaf, and O. C. Jones, Pressure and Void Distributions in a Converging-Diverging Nozzle with Non-Equilibrium Water Vapor Generation, BNL-NUREG-26003, Brookhaven National Laboratory, 1979.
175. H. C. Perkins, Jr., M. Yusuf, and G. Leppert, A Void Measurement Technique for Local Boiling, *Nucl. Sci. Eng.*, vol. 11, p. 304, 1961.
176. W. T. Sha and C. F. Bonilla, Out-of-Pile Steam-Fraction Determination by Neutron-Beam Attenuation, *Nucl. Appl.*, vol. 1, p. 69, 1965.
177. R. A. Moss and A. J. Kelly, Neutron Radiographic Study of Limiting Planar Heat Pipe Performance, *Int. J. Heat and Mass Transfer,* vol. 13, p. 491, 1970.
178. A. A. Harms and C. F. Forrest, Dynamic Effects in Radiation Diagnosis of Fluctuating Voids, *Nucl. Sci. Eng.*, vol. 46, p. 408, 1971.
179. A. A. Harms, S. Lo, and W. T. Hancox, Measurement of Time-Averaged Voids by Neutron Diagnosis, *J. Appl. Phy.*, vol. 42, p. 4080, 1971.
180. W. T. Hancox, C. F. Forrest, and A. A. Harms, Void Determination in Two-Phase Systems Employing Neutron Transmission, ASME Paper 72-HT-2, National Heat Transfer Conference, Denver, 1972.
181. H. L. Ornstein, An Investigation of Turbulent Open Channel Flow Simulating Water Desalination Flash Evaporators, Ph.D. thesis, University of Connecticut, 1970.
182. T. B. Morrow and S. J. Kline, The Evaluation and Use of Hot-Wire and Hot-Film Anemometers in Liquids, Stanford University Rep. MD-25, 1971.
183. K. Hollasch and B. Gebhart, Calibration of Constant-Temperature Hot-Wire Anemometers at Low Velocities in Water with Variable Fluid Temperature, *J. Heat Transfer,* vol. 94C, pp. 17–22, 1972.
184. J. C. Hurt and J. R. Welty, The Use of a Hot-Film Anemometer to Measure Velocities below 5 cm/sec in Mercury, *J. Heat Transfer,* vol. 95C, pp. 548–549, 1973.
185. R. S. Rosler and S. G. Bankoff, Large-Scale Turbulence Characteristics of a Submerged Water Jet, *AIChE J.,* vol. 9, pp. 672–676, 1963.
186. M. Bouvard and H. Dumas, Application de la methode du fil chaud a la mesure de la turbulence dans l'eau, *Houille Blanche,* vol. 3, pp. 257–270, 1967.
187. F. Resch, Etudes sur le fil chaud et le film chaud dans l'eau, *Houille Blanche,* vol. 2, pp. 151–161, 1969.
188. F. Resch, Etudes sur le film chaud et le film chaud dans l'eau, CEA-R3510, Grenoble, 1968.
189. V. W. Goldschmidt and S. Eskinazi, Diffusion de particules liquides dans le champ retardataire d'un jet d'air plan et turbulent, in *Les Instabilites en hydraulique et en mechanique des fluides,* pp. 291–298, Societe Hydrotechnique de France, Lille, 1964.
190. V. W. Goldschmidt and S. Eskinazi, Two-Phase Turbulent Flow in a Plane Jet, *J. Appl. Mech.,* vol. 33, pp. 735–747, 1966.
191. C. Lackme, Structure et cinematique des ecoulementes diphasiques a bulles, CEA-R3202, Grenoble, 1967.
192. T. Ginsberg, Droplet Transport in Turbulent Pipe Flow, ANL-7694, Argonne National Laboratory, 1971.
193. V. W. Goldschmidt, Measurement of Aerosol Concentrations with a Hot-Wire Anemometer, *J. Colloid Sci.,* vol. 20, pp. 617–634, 1965.
194. V. W. Goldschmidt and M. K. Householder, The Hot-Wire Anemometer as an Aerosol Droplet Size Sampler, *Atmos. Environ.,* vol. 3, pp. 643–651, 1969.
195. S. C. Chuang and V. W. Goldschmidt, The Response of a Hot-Wire Anemometer to a Bubble of Air in Water, in G. K. Patterson and J. L. Zakin (eds.), *Turbulence Measurements in Liquids,* University of Missouri, Rolla, Continuing Education Series, 1969.

196. V. W. Goldschmidt, M. K. Householder, G. Ahmadi, and S. C. Chuang, Turbulent Diffusion of Small Particles Suspended in Turbulent Jets, in G. Hetsroni, S. Sideman, and J. P. Hartnett (eds.), *Progress in Heat and Mass Transfer,* vol. 6, pp. 487–508, Pergamon, Oxford, 1972.
197. F. J. Resch and J. H. Leutheusser, Le Ressaut hydraulique: Mesures de turbulence dans la region diphasique, *Houille Blanche,* vol. 4, pp. 279–293, 1972.
198. F. J. Resch, H. J. Leutheusser, and S. Alemu, Bubbly Two-Phase Flow in Hydraulic Jump, *J. Hydraul. Div. Am. Soc. Civ. Eng.,* Proc. Paper 10297, pp. 137–149, 1974.
199. S. Ishigai, S. Nakanisi, T. Koizumi, and Z. Oyabu, Hydrodynamics and Heat Transfer of Vertical Falling Liquid Films (Part I: Classification of Flow Regimes), *Bull. JSME,* vol. 15, pp. 594–602, 1972.
200. A. K. Katarzhis, S. I. Kosterin, and B. I. Sheinin, An Electric Method of Recording the Stratification of the Steam-Water Mixture, *Izv. Akad. Nauk SSSR,* vol. 2, pp. 132–136 (A.E.R.E. Lib. Trans. 590), 1955.
201. B. S. Shiralkar, Local Void Fraction Measurements in Freon-114 with a Hot-Wire Anemometer, NEDO-13158, General Electric Co., 1970.
202. G. E. Dix, Vapor Void Fractions for Forced Convection with Subcooled Boiling at Low Flow Rates, Ph.D. thesis, University of California at Berkeley, 1971; also General Electric Co. NEDO-10491, 1971.
203. B. S. Shiralkar and R. T. Lahey, Jr., Diabatic Local Void Fraction Measurements in Freon-114 with a Hot-Wire Anemometer, ANS Trans. 15, no. 2, p. 880, American Nuclear Society, 1972.
204. G. Hetsroni, J. M. Cuttler, and M. Sokolov, Measurements of Velocity and Droplet Concentration in Two-Phase Flows, *J. Appl. Mech.,* vol. 36E, pp. 334–335, 1969.
205. V. I. Subbotin, D. N. Sorokin, and A. A. Tsiganok, Some Problems on Pool Boiling Heat Transfer, in *Heat Transfer,* vol. 5, Elsevier, Amsterdam, 1970.
206. M. Petrick and B. S. Swanson, Radiation Attenuation Methods for Measuring Density in a Two-Phase Fluid, *Rev. Sci. Instrum.,* vol. 29, pp. 1079–1085, 1958.

CHAPTER

ELEVEN

MEASUREMENT OF WALL SHEAR STRESS

Thomas J. Hanratty and Jay A. Campbell

1. INTRODUCTION

A fluid flowing past a solid boundary exerts normal and tangential stresses on it. Normal stresses or pressures are readily measured by connecting a small hole on the surface to a manometer. Consequently, numerous studies have been made of the pressure distributions on solid surfaces in contact with a flowing fluid. The measurement of the tangential or shear stresses at a surface is much more difficult. However, the extra effort needed to obtain such data is rewarding, since information about the variation of the wall shear stress on a surface is often quite useful in analyzing a flow field. The six principal methods of measuring the local wall shear stress, aside from the extrapolation of direct velocity measurements, are

The Stanton tube [1]
Direct measurement [2]
Thermal method [3]
The Preston tube [4]
The sublayer fence [5, 6]
The electrochemical technique [7, 8]

The first such measurements were reported in a historic paper by Stanton et al. [1], who wanted to determine the fluid velocity close to a boundary, to find out whether slip existed at a wall for a turbulent flow. To do this, a rectangular Pitot tube that had the wall of a pipe as one of its sides was used. The difference between the pressure measured with this Pitot tube and the static pressure was used to determine the velocity at the center of the tube. By carrying out experiments in a fully developed laminar flow, Stanton and his coworkers found that the relation between the measured pressure difference and the velocity was not given by the usual equation for a Pitot tube in a free flow. They established a calibration curve for the "effective center point" of their instrument, now called a *Stanton tube*, and used it to measure velocities in a fully developed turbulent flow. From these experiments it was shown, for the first time, that for a turbulent flow close to a wall the variation of the time-averaged velocity is given by

$$\bar{U} = \frac{\bar{\tau}_W}{\mu} y \tag{1}$$

It was established that if the effective center of the Stanton tube was located close enough to the wall, the velocity calculated from the pressure measurement could be related to the wall shear stress by using Eq. (1). The Stanton tube thereby provides an indirect method for measuring τ_W.

The local tangential force on a surface can be determined directly by allowing some portion of the surface to be movable against a restoring force. Measurements are made of the displacement of the element or of the force required to keep it in a null position. The choice of a size for the element is a balance between the need to have a sufficiently large force acting upon it and the desire to obtain local measurements of the wall shear stress. The first local measurements with such a device were made by Kempf [2], who used panels of dimensions 308 × 1010 mm to measure the drag force at several stations along the bottom of a 77-m-long pontoon. In recent years the technique has been developed sufficiently so that a movable element as small as 9 mm can be used.

Fage and Falkner [9] studied the relation between the local wall shear stress and the rate of heat transfer from small thermal elements mounted flush with the surface. In their experiments they used a nickel strip, 0.262 cm long and 26.2 cm wide, embedded 0.107 cm in an ebonite block. A heating current was passed through the nickel strip, and its temperature was determined by measuring the resistance. The current was controlled manually so that the temperature of the element was kept constant. The heating current was related to the local wall shear stress at different locations on a circular cylinder in a flow stream. Ludwieg [3] developed an instrument, using this concept, that consisted of a heated copper block that was movable and therefore that could be located at different positions over a surface. With Tillman [10], he used this instrument to establish the law of the wall for turbulent boundary layers.

Much simpler and more compact versions of this gauge were developed by Liepmann and Skinner [11], who used as the heating element a 12.7-μm platinum wire buried in a groove in the surface of bakelite, by Bellhouse and Schultz [12, 13], who used a platinum film deposited on a glass substrate, and by McCroskey and Durbin [14], who used photoetching techniques. The latter two probes have been developed into commercially available instruments that use standard hot-wire and hot-film equipment. At present, the usual mode of operation of wall film gauges is to control the average temperature of the heated element using feedback circuitry, and to determine how the electric heating current varies with wall shear stress. One of the chief difficulties in using these various heat-transfer probes is that heat is lost to the substrate as well as to the fluid, so that the effective length of the probe can be much larger than that of the heating element. Recent work by Rubesin et al. [15] with embedded heated wires and by Sandborn [16, 17] with 0.001-cm wires lying on top of the surface offer methods for greatly reducing this effect.

Preston [4] suggested what is probably the simplest method for determining wall shear stress. He measured the impact pressure on round Pitot tubes resting on a surface. The tubes had outside diameters varying from 0.74 to 3.08 mm and a ratio of inside diameter to outside diameter of 0.6. For turbulent flows the tubes were too large to be entirely in the region where Eq. (1) is valid. However, on the basis of measurements made by Ludwieg and Tillman [10], Preston argued that, for a given fluid, the velocity variation close to a wall is a universal function that depends only on the local wall shear stress and not on the geometry of the system or on the previous history of the flow field:

$$\bar{U} = \left(\frac{\bar{\tau}_W}{\rho}\right)^{1/2} f\left(\frac{y\rho(\bar{\tau}_W/\rho)^{1/2}}{\mu}\right) \tag{2}$$

In the immediate vicinity of the wall

$$f\left(\frac{y\rho(\bar{\tau}_W/\rho)^{1/2}}{\mu}\right) = \frac{y\rho(\bar{\tau}_W/\rho)^{1/2}}{\mu} \tag{3}$$

for consistency with Eq. (1).

Thus, the theoretical basis for using Preston's instrument in turbulent flows is the assumption that $\bar{U}=f(\tau_W, y)$ over a much wider range of wall distances than the viscous wall layer described by Eq. (1). If this assumption is correct, much larger Pitot tubes can be used to measure $\bar{\tau}_W$ than originally suspected. Therefore, the difficulties associated with the manufacture of a Stanton tube would not justify its choice over a conventional Pitot tube. Preston carried out experiments for fully developed turbulent flow in a pipe for which $\bar{\tau}_W$ can be calculated from pressure-drop measurements as

$$\bar{\tau}_W = \frac{d}{4}\left|\frac{dP}{dx}\right| \tag{4}$$

to establish a universal calibration curve for a round Pitot tube resting on a wall.

There has been some doubt about the accuracy of the law of the wall, and, therefore, a number of investigators have attempted to develop other devices, similar to the Stanton tube, that consist of a wall obstruction completely immersed in the viscous sublayer. One of the more interesting of these is the sublayer fence invented by Konstantinov and Dragnysh [5, 6]. The difference in pressure before and behind a sharp edge projecting through the surface, normal to the flow, is related to the wall shear stress. The advantages of this instrument over the Stanton gauge are that it gives an almost doubled pressure reading, it eliminates the necessity of a separate static-pressure tapping, and it gives readings in both forward and reversed flows.

The most recently developed instrument for measuring the velocity gradient at the wall is the electrochemical probe developed by Reiss and Hanratty [7, 8]. This instrument is the mass-transfer analog of the heated surface film. An electrochemical reaction is carried out on the surface of an electrode mounted flush with the wall. The voltage on the electrode is kept large enough so that the reaction rate is so fast that it is mass-transfer controlled, yet small enough so that no side reactions occur. Under these conditions the concentration of the reacting species is maintained at approximately zero at the electrode surface. The current flowing through the electrode circuit, which is proportional to the rate of mass transfer at the electrode surface, is then related to the velocity gradient at the surface. This instrument holds advantages over the thermal technique in that it avoids problems associated with substrate heat losses. In principle, it can be calibrated analytically and can be easily applied to situations requiring complicated sensor configurations. The mass-transfer process occurring at the electrode is characterized by large Schmidt numbers so that these probes have the desirable feature that the range of flow rates over which the concentration boundary layer is within a region where Eq. (1) describes the velocity field is quite large. These electrochemical probes have disadvantages with respect to the thermal probes in that the frequency response is not as good and in that they can be used only in liquid flows and in equipment that is compatible with the type of chemicals used.

Mitchell and Hanratty [18] showed how electrochemical probes can be used to study the time-averaged and fluctuating velocity gradient for a turbulent flow and pointed out the need to account for spatial averaging and for the time response of the concentration boundary layer. Fortuna and Hanratty [19] later improved the analysis for frequency response presented by Mitchell and Hanratty. Dimopoulos and Hanratty [20] and Son and Hanratty [21] showed how electrochemical techniques can be used for studying laminar boundary layers. Son and Hanratty [21] developed a sandwich electrode in which a pair of rectangular electrodes separated by a thin layer of insulation are oriented with their long side perpendicular to the flow. By comparing mass-transfer rates to the front and back electrodes, the direction of flow and, therefore, the separation position of a boundary layer can be determined. Tournier and Py [22] and Le Bouche and Martin [23] further developed this sandwich electrode for application to boundary layer flows. Sirkar and Hanratty [24, 25] demonstrated how a pair of rectangular electrodes in a chevron arrangement can be used to measure both components of the velocity gradient at the wall. Py and Gosse [26] suggested the use of a pair of semicircular electrodes for this purpose. However, because of the difficulty in fabricating these semicircular electrodes, Py [27] used pairs of rectangles with their long dimension parallel to the flow to measure both components of the fluctuating velocity gradient in a turbulent flow. Most of the early work with these electrochemical techniques used the ferricyanide-ferrocyanide reaction in an excess of sodium hydroxide. Py [28] suggested the use of the iodine reaction in an excess of potassium iodide. This system has been found to be more stable and to cause fewer problems with respect to electrode contamination. Consequently, its use has been favored in recent studies.

Any of the six techniques listed earlier and briefly described above can be used to measure time-averaged wall shear stresses in fully developed flows and in boundary layers with zero pressure gradient. The choice depends on the system in which they are to be used and on whether the measuring technique will interfere with the flow. For other flow situations, it appears that the heat-transfer or mass-transfer probes have the widest applicability. The possibility of measuring fluctuations in the wall shear stress as well as the time average makes them particularly attractive choices for time-varying flows. In this chapter a brief description of all six techniques will be given. However, more attention will be given the heat- and mass-transfer techniques because of their greater potential and because the theoretical problems associated with their use are particularly intricate.

2. DIRECT MEASUREMENTS

The need of aerodynamicists to measure shear stresses in high-speed flows (see [29]) has led to the development of a number of ingenious designs for compact gauges to measure directly the local wall shear stress. Winter [30] gives an excellent description of these instruments and reviews progress made in solving the following problems, which he cites to be associated with their use:

1. Provision of a transducer for measuring small forces or deflections, and the compromise between the requirement to measure local properties and the necessity of having an element of sufficient size that the force on it can be measured accurately.
2. The effect of the necessary gaps around the floating element.
3. The effects of misalignment of the floating element.
4. Forces arising from pressure gradients.
5. The effect of gravity or of acceleration if the balance is to be used in a moving vehicle.
6. Effects of temperature changes.
7. Effects of heat transfer.
8. Use with boundary layer injection or suction.
9. Effects of leaks.
10. Protection of the measuring system against transient normal forces during starting and stopping in a supersonic tunnel.

The gauge is calibrated by using a static method. A force is applied to the element by suspending weights from a thread with the instrument mounted vertically [31], or with a pulley arrangement with the instrument mounted horizontally [29]. As pointed out by Mabey and Gaudet [31], very thin threads have to be used with small gauges. In fact, these authors found it necessary to use a human hair.

Since this technique requires a portion of the wall to be movable in a direction parallel to the boundary, the sensing element must have a gap around its perimeter. Because of the presence of this gap, effects appear under flow conditions that are not taken into account in the calibration. The presence of the discontinuity at the surface alters the wall shear stress in regions close to the gap. If the element is not aligned almost perfectly with the contour of the wall, depressions or protrusions can give rise to additional forces because of flow disturbances and pressure forces acting on the protruding surfaces. If pressure variation associated with the flow is significant over the element surface, fluid circulation can occur through the gaps, and a force can exist on the head because of the difference in the fluid pressures in the upstream and downstream gaps. This latter difficulty appears to be a fundamental problem with this type of gauge, one that cannot be eliminated and may severely limit its application in flows with pressure gradients.

The use of large elements and large gaps reduces errors due to misalignment. However, too large a gap should be avoided because of the change of shear stress across the gap. Winter [30] recommends that gu^*/ν should not exceed 100, where g is the gap size.

For flows with small pressure gradients and small wall shear stress gradients, large balances may be used. Such balances are relatively easy to calibrate and require small buoyancy corrections arising from nonuniformities of the pressure in the gap around the floating plate. An example is the 368-mm-diameter gauge used by Winter and Gaudet [32] to study air flows for $16 \times 10^6 < \mathrm{Re}_x < 200 \times 10^6$ at $0.2 < \mathrm{Ma} < 2.8$. In large gradients the use of as small a gap as possible is desired so as to reduce flow through the gap. This requirement, in addition to the requirement of a small element, magnifies the alignment problem. The calibration is more difficult, and the correction for buoyancy may be uncertain.

An excellent paper describing the use of one of these small skin-friction gauges has been written by Mabey and Gaudet [31]. They describe tests performed with the Kistler gauge shown in Fig. 1. This instrument, which has a head with a diameter of 9.4 mm and which is designed to be insensitive to linear acceleration, was at one time available commercially (last sold by Sundstrand Data, Inc., Inc., Redmond, Washington). It has a feedback circuit that ensures that the floating element remains in a fixed position relative to its housing. Consequently, the gap around the element is kept uniform and can be much smaller than that required if the element is allowed to deflect, as is the case in the conventional skin-friction balance. Mabey and Gaudet mounted the balances in a smooth steel plug that could be located in holes drilled at different locations in the flat plate with which tests were made. By doing this, it was possible to align the instrument even with or slightly below the surface surrounding it before mounting in the test section.

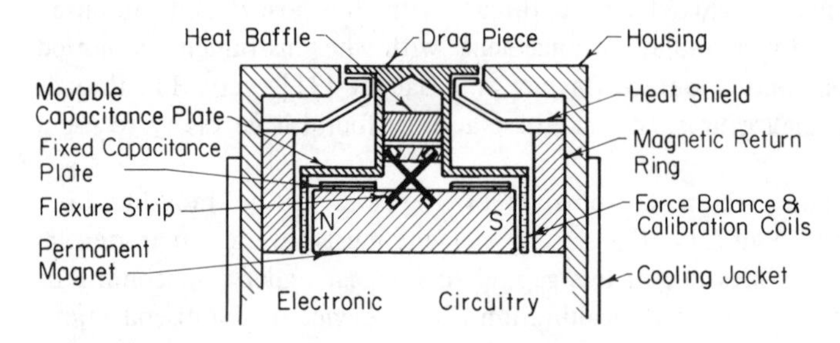

Figure 1 Construction of Kistler skin-friction gauge as presented in literature from the Kistler Instrument Co.

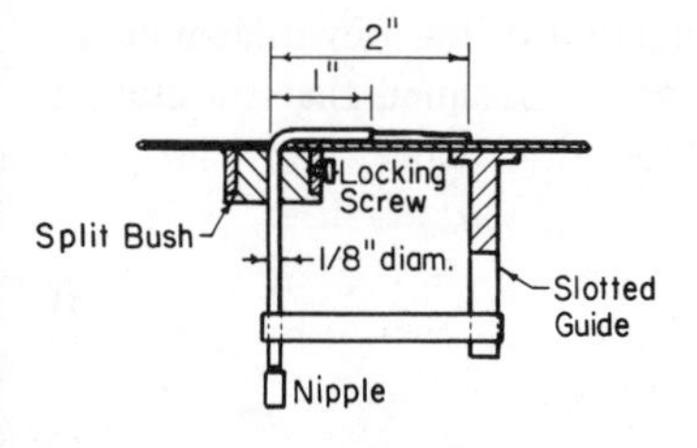

Figure 2 Construction of Preston tube as given by Preston [4].

Strain-gage balances have been used to establish the turbulent skin-friction law for a flat plate for incompressible flow [29, 33 to 35] and for compressible flows [32, 36]. They have also been used for measurements on an airplane at Mach numbers up to 4.9 [37].

3. PRESTON TUBE

The complexity of the design of the surface force gauges and the great care that must be taken in their use makes the Preston tube a particularly attractive alternative. The arrangement used by Preston is shown in Fig. 2 and is described by him as follows: "The brass stem is 1/8 in. diameter and the mouth of the pitot is 2 in. from the stem. With the exception of No. 4 pitot ($d_t = 0.1214$ in.), the front part of the tube was made of stainless steel. It was bent so that the first contact with the wall was at the mouth, and was soldered into the brass stem 1.0 in. from the mouth." In using the Preston tube, care must be taken in locating the static-pressure tap so that its reading is not being affected by blockage due to the presence of the Pitot tube at the wall. Although most Preston tubes used in tests have followed Preston's specification of a ratio of inside to outside diameter, results do not appear to be too sensitive to this parameter. For example, Patel [38] and Rechenberg [39] find no effect provided the ratio is greater than 0.2.

The principle of operation is that the instrument obeys the equations for a Pitot tube that has an effective center at $y = \frac{1}{2}K_t d_t$, where d_t is the outside tube diameter. Suppose the Preston tube lies entirely in a region close to the wall, where $U = (\tau_W/\mu)y$. Then a Reynolds number for the tube can be defined as

$$d_t^+ = \left[\frac{1}{2}\left(\frac{1}{2}K_t\right)^2 \frac{d_t^2 \tau_W}{\rho \nu^2}\right]^{1/2}$$

For large enough values of $(d_t^+)^2$ the difference of the impact pressure at the effective center of the tube from the static pressure is given by the following relation, provided $U = (\tau_W/\mu)y$:

$$\frac{\Delta P}{\tau_W} = \frac{1}{2}\left(\frac{1}{2}K_t\right)^2 \frac{d_t^2 \tau_W}{\rho \nu^2} \tag{5}$$

Young and Maas [40] have shown that for Pitot tubes in shear flow at large Reynolds numbers, $K_t = 1.30$ for a tube with a ratio of inside to outside diameter of 0.60. For very small Reynolds numbers it can be expected that viscous effects will be important, and therefore that K_t will be a function of the Reynolds number d_t^+. Equation (5) can be applied to turbulent flow if it is assumed that the flow is uniform in the spanwise direction at all instances. Then

$$\frac{\overline{\Delta P}}{\overline{\tau_w}} = \frac{1}{2}\left(\frac{1}{2} K_t\right)^2 \frac{d_t^2 \bar{\tau}_w}{\rho \nu^2} \left[1 + \frac{\overline{(\tau_w')^2}}{\bar{\tau}_w^2}\right] \tag{6}$$

where $\bar{\tau}_w$ = time-averaged wall stress

$\overline{(\tau_w')^2}$ = mean square of fluctuations in wall stress

Measurements carried out recently in a number of laboratories would indicate that $\overline{(\tau_w')^2}/\tau_w^2 \simeq 0.1$.

The most convenient system for calibrating a Preston tube is a fully developed pipe flow for which the pressure gradient is linearly related to the wall shear stress by Eq. (4). It is usually better to make the pressure-gradient measurements without the Preston tube in place, since flow blockage could affect the results. In using this calibration procedure, the pressure taps used for measuring pressure gradient should not be located in regions where the pipe walls are tapered, and tests should be carried out to ensure that the flow is symmetric [41]. Calibrations of Preston tubes in this manner in laminar flows give values of K_t close to that obtained by Young and Maas. This supports the physical interpretation that a Preston tube behaves the same as a Pitot tube located in a free shear layer.

It is not possible to apply Eqs. (5) and (6) to interpret Preston tubes in most turbulent flows, because the tube cannot be made small enough so that it samples only a region for which $\bar{U} = \bar{\tau}_w y/\mu$. The use of the Preston tube under these circumstances requires that the region of the flow that it sees be described by the law of the wall, Eq. (2). If this is true, then it can be shown by dimensional reasoning that

$$\frac{\overline{\Delta P}}{\overline{\tau_w}} = f\left(\frac{d_t^2 \bar{\tau}_w}{\rho \nu^2}\right) \tag{7}$$

The right-hand side of Eq. (7) is given by Eq. (6) for very small values of the argument. An alternative form of Eq. (7), used by Preston, is

$$\frac{\overline{\Delta P}\, d^2}{\rho \nu^2} = f\left(\frac{d_t^2 \bar{\tau}_w}{\rho \nu^2}\right) \tag{8}$$

Calibrations of Preston tubes with different diameters in fully developed turbulent pipe flows have been carried out by Preston [4], Rechenberg [39], Patel [38], and Head and Rechenberg [42]. On the basis of these measurements, Head and Ram [43] have developed the "universal calibration" given in Table 1.

The application of the Preston tube and this calibration to turbulent boundary-layer flows depends on the accuracy of the law of the wall. For turbulent flow in pipes or turbulent boundary layers with small pressure gradients, the law of the wall is usually assumed to hold provided $2d_t/d$ or d_t/δ is less than 0.1. For turbulent boundary layers with large positive or negative pressure gradients, the law of the wall holds over a smaller distance, so that d_t/δ must be an even smaller quantity if

Table 1 Calibration of Preston tube developed by Head and Ram

$\frac{\Delta P d^2}{\rho v^2}$ $\times 10^{-2}$	$\frac{\Delta P}{\tau_w}$
4.0	9.18
4.2	9.41
4.4	9.63
4.6	9.85
4.8	10.06
5.0	10.27
5.2	10.47
5.4	10.67
5.6	10.87
5.8	11.06
6.0	11.25
6.2	11.44
6.4	11.62
6.6	11.80
6.8	11.98
7.0	12.16
7.2	12.33
7.4	12.50
7.6	12.67
7.8	12.83
8.0	12.99
8.2	13.15
8.4	13.31
8.6	13.47
8.8	13.63
9.0	13.78
9.2	13.93
9.4	14.08
9.6	14.23
9.8	14.38

$\frac{\Delta P d^2}{\rho v^2}$ $\times 10^{-3}$	$\frac{\Delta P}{\tau_w}$
1.0	14.53
1.02	14.67
1.06	14.95
1.10	15.23
1.14	15.51
1.18	15.78
1.22	16.04
1.26	16.30

$\frac{\Delta P d^2}{\rho v^2}$ $\times 10^{-3}$	$\frac{\Delta P}{\tau_w}$
1.30	16.56
1.34	16.81
1.38	17.06
1.42	17.31
1.46	17.55
1.50	17.79
1.54	18.02
1.58	18.25
1.62	18.48
1.66	18.71
1.70	18.94
1.74	19.16
1.78	19.38
1.82	19.59
1.86	19.80
1.90	20.01
1.95	20.27
2.00	20.53
2.05	20.79
2.10	21.05
2.15	21.30
2.20	21.54
2.25	21.78
2.30	22.02
2.40	22.49
2.50	22.96
2.6	23.41
2.7	23.86
2.8	24.30
2.9	24.73
3.0	25.08
3.1	25.43
3.2	25.78
3.3	26.13
3.4	26.48
3.5	26.82
3.6	27.16
3.7	27.50
3.8	27.83
3.9	28.15
4.0	28.46
4.2	29.07
4.4	29.66

$\frac{\Delta P d^2}{\rho v^2}$ $\times 10^{-3}$	$\frac{\Delta P}{\tau_w}$
4.6	30.23
4.8	30.79
5.0	31.33
5.2	31.84
5.4	32.34
5.6	32.84
5.8	33.31
6.0	33.78
6.2	34.23
6.4	34.68
6.6	35.11
6.8	35.52
7.0	35.94
7.2	36.34
7.4	36.72
7.6	37.11
7.8	37.50
8.0	37.87
8.5	38.74
9.0	39.58
9.5	40.40

$\frac{\Delta P d^2}{\rho v^2}$ $\times 10^{-4}$	$\frac{\Delta P}{\tau_w}$
1.00	41.18
1.05	41.93
1.10	42.65
1.15	43.34
1.20	44.00
1.25	44.64
1.30	45.27
1.35	45.87
1.40	46.45
1.45	47.01
1.50	47.56
1.55	48.09
1.60	48.61
1.65	49.12
1.70	49.62
1.8	50.56
1.9	51.46

$\frac{\Delta P d^2}{\rho v^2}$ $\times 10^{-4}$	$\frac{\Delta P}{\tau_w}$
2.0	52.32
2.1	53.15
2.2	53.93
2.3	54.68
2.4	55.40
2.5	56.62
2.6	56.80
2.7	57.45
2.8	58.07
2.9	58.68
3.0	59.28
3.2	60.40
3.4	61.46
3.6	62.47
3.8	63.43
4.0	64.34
4.2	65.20
4.4	66.01
4.6	66.80
4.8	67.57
5.0	68.32
5.2	69.05
5.5	70.00
6.0	71.55
6.5	73.00
7.0	74.35
7.5	75.60
8.0	76.80
8.5	77.95
9.0	78.95
9.5	79.90

$\frac{\Delta P d^2}{\rho v^2}$ $\times 10^{-5}$	$\frac{\Delta P}{\tau_w}$
1.0	80.80
1.05	81.70
1.10	82.55
1.15	83.35
1.20	84.1
1.30	85.5

$\frac{\Delta P d^2}{\rho v^2}$ $\times 10^{-5}$	$\frac{\Delta P}{\tau_w}$
1.4	86.8
1.5	88.0
1.6	89.1
1.7	90.2
1.8	91.2
1.9	92.1
2.0	93.0
2.2	94.7
2.4	96.2
2.6	97.6
2.8	98.8
3.0	99.9
3.2	100.9
3.5	102.4
4.0	104.5
4.5	106.4
5.0	108.0
5.5	109.3
6.0	110.4
6.5	111.4
7.0	112.2
7.5	113.0
8.0	113.7
9.0	114.9

$\frac{\Delta P d^2}{\rho v^2}$ $\times 10^{-6}$	$\frac{\Delta P}{\tau_w}$
1.0	116.0
1.1	117.1
1.2	118.1
1.4	119.8
1.6	121.5
1.8	123.1
2.0	124.6
2.2	126.0
2.5	127.8
3.0	130.7
3.5	133.3
4.0	135.6
4.5	137.7

$\frac{\Delta P d^2}{\rho v^2}$ $\times 10^{-6}$	$\frac{\Delta P}{\tau_w}$
5.0	139.5
6.0	142.7
7.0	145.4
8.0	147.8
9.0	150.0

$\frac{\Delta P d^2}{\rho v^2}$ $\times 10^{-7}$	$\frac{\Delta P}{\tau_w}$
1.0	151.9
1.2	155.3
1.4	158.1
1.6	160.6
1.8	162.8
2.0	164.9
2.2	166.8
2.5	169.3
3.0	172.8
3.5	175.8
4.0	178.5
4.5	180.9
5.0	183.1
6.0	186.8
7.0	190.0
8.0	192.8
9.0	195.3

$\frac{\Delta P d^2}{\rho v^2}$ $\times 10^{-8}$	$\frac{\Delta P}{\tau_w}$
1.0	197.5
1.2	201.4
1.4	204.7
1.6	207.5
1.8	210.1
2.0	212.5
2.5	217.4
3.0	221.6
3.5	225.0

the universal calibration is to be used. In many situations, an example of which would be a separated region, the law of the wall is not correct and the Preston tube cannot be used if it extends beyond the region where $\bar{U} = (\bar{\tau}_w/\mu)y$.

Preston [4] demonstrated the applicability of Preston tubes to turbulent boundary layers by comparing measurements with tubes of different diameter. Head and Rechenberg [42] did essentially the same type of test by comparing measurements with Preston tubes to measurements with a Stanton tube or with a sublayer fence. Tests of Preston tubes in turbulent boundary layers on flat plates in a number of laboratories [42] have yielded results that differ by as much as 11%. This could be the type of accuracy to be expected from the Preston tube, but, more than likely, it reflects the accuracy of the methods used to evaluate the wall shear stress in these different tests. This matter has not as yet been satisfactorily resolved.

Patel [38] has presented guidelines for estimating the influence of pressure gradient on the accuracy of Preston tubes in turbulent boundary layers. For this purpose he uses the pressure gradient parameter,

$$\Delta = \frac{\nu}{\rho(u^*)^3 \; dP/dx}$$

He recommends $d_t^+ \leqslant 250$ in adverse pressure gradients with $\Delta < 0.015$, and $d_t^+ \leqslant 200$ in favorable pressure gradients with $\Delta < -0.007$, for the maximum error to be less than 6%. Operation outside the range $-0.007 < \Delta < 0.015$ is not recommended.

A number of investigators have developed methods for using Preston tubes in compressible flows [30]. The simplest of these appears to be that proposed by Bradshaw and Unsworth [44]:

$$\frac{\overline{\Delta P}}{\bar{\tau}_w} = f_i\left(\frac{d_t u^*}{\nu_w}\right) + f_c\left(\frac{d_t u^*}{\nu_w}, \frac{u^*}{a_w}\right) \tag{9}$$

where f_i is the calibration for incompressible flow, and f_c is the correction to take account of compressibility effects. They found that the following functions gave the best fit to available data for $50 < d_t u^*/\nu < 1000$:

$$f_i = 96 + 60 \log_{10} \frac{d_t u^*}{50\nu} + 23.7\left(\log_{10} \frac{d_t u^*}{50\nu}\right)^2 \tag{10}$$

$$f_c = 10^4\left(\frac{u^*}{a_w}\right)^2\left[\left(\frac{d_t u^*}{\nu_w}\right)^{0.26} - 2\right] \tag{11}$$

The Preston tube appears to offer no real advantages (and, in fact, offers disadvantages) over the force balance with respect to accurate measurement of wall shear stress. Because it presents an obstruction to the flow, it can interfere with the flow field. In three-dimensional boundary layers the direction of flow at the wall has to be known to orient the Preston tube properly. The accuracy of the Head and Ram calibration is uncertain unless the flow is an incompressible, two-dimensional, turbulent boundary layer with a moderate pressure gradient. The main reason for the choice of a Preston tube over a force balance is that it is easier to fabricate and to operate.

4. STANTON GAUGE

The Stanton gauge has been an attractive method for measuring wall shear stress because of its small size. The original design of Stanton has proved to be awkward to use, and a number of alternative arrangements have been suggested. Hool [45] proposed a surface channel that was formed on one side by the solid surface and on the other by the lower side of the tapered cutting edge of a segment of razor blade. The static pressure developed in this small enclosure was measured by a normal static-pressure hole in the solid surface. With the razor blade removed, this same static-pressure hole could be used to measure the true static pressure, and the difference between these two pressures calibrated against wall shear stress. Bradshaw and Gregory [46] applied this technique by using 0.002-in steel shim with one edge chamfered. A 0.2- × 0.1-in piece was held over a 0.04-in-square hole with cellulose adhesive. The leading edge of the shim was aligned so that it was just over the front side of the static-pressure hole. Brown [47] used a similar design, but with a 0.005-in-thick shim over a 0.020-in hole.

The values of ΔP determined with a Stanton gauge can be quite small. Consequently it is necessary to use a micromanometer with an accuracy of ±0.01 mm.

Extensive tests were carried out by East [48] with a Stanton gauge having the design shown in Fig. 3. This has the unique feature that it is held in place by a magnet that surrounds the static hole. This procedure avoids the use of an unknown thickness of glue between the blade and the surface and allows for the determina-

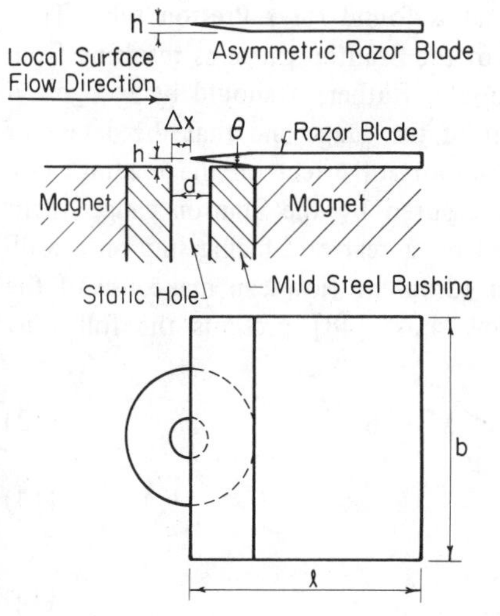

Figure 3 Design of Stanton gauge as given by East [48].

tion of the height h of the blade edge above the surface as simply the half-thickness of the blade. East recommended that the following dimensions be used:

$$\frac{d_h}{h} = 6 \qquad \frac{b}{h} = 36 \qquad \frac{l}{b} = 1 \qquad \frac{\Delta x}{h} = 0$$

where l = length of blade
b = breadth of blade
d = diameter of static hole

He used two commercially available blades with $h = 0.002$ and 0.005 in, and a fabricated blade with $h = 0.0153$ in. The blade angle was 11° to 13°. He tested the gauges against a Preston tube that used the calibration given in Table 1. The field consisted of the turbulent flow of air over a flat plate.

Calibrations of Stanton gauges have been carried out in laminar flows by Bradshaw and Gregory [46], Hool [45], and Taylor [49]. Their results can be correlated by an equation either of the form $\Delta P\, h^2 \rho/\mu^2$ versus $\tau_w h^2 \rho/\mu^2$ or, if Eq. (5) is used, of the form K_t versus $\tau_w h^2 \rho/\mu^2$ (see [50]). Taylor's experiments were carried out at extremely low Reynolds numbers, where a creeping flow approximation can be made. He showed from dimensional reasoning that in the stokesian region ($\log_{10} h^2 \tau_w \rho/\mu^2 < -0.6$)

$$\Delta P = k_s \tau_w \tag{12}$$

His experiments indicated k_s is approximately equal to 1.2. Taylor argued that at large values of the Reynolds number $(h^2 \tau_w \rho/\mu^2)^{1/2}$ the results are interpreted in the manner suggested by Eq. (5), and that K_t = const. Experiments by Hool clearly indicate that this is not the case and that K_t varies as $(h^2 \tau_w \rho/\mu^2)^{-1/5}$ at large values of the gauge Reynolds number. The effective center of the Stanton gauge is found to be less than $h/2$ at large $h^2 \tau_w \rho/\mu^2$, as has been observed by Preston [4] for flattened Pitot tubes touching a surface.

These results are quite different from what is found for a Preston tube. They suggest that the interpretation of the reading of the Stanton gauge as resulting from a stagnation flow at $y = \frac{1}{2} K_t h$ is not appropriate. Rather, it should be recognized that a recirculating flow region exists in front of the gauge and that the details of the flow in this region could be exerting an important effect on the readings (see the discussions in [46, 51]). The pressure measured by the Stanton gauge might thus be interpreted as similar to that observed on a rearward-facing step on a wall.

The calibration of a Stanton gauge in a turbulent flow can make use of the same dimensionless groups as for laminar flow. East [48] presents the following calibration for the gauge shown in Fig. 3:

$$y^* = -0.23 + 0.618x^* + 0.0165(x^*)^2 \qquad 2 < x^* < 6 \tag{13}$$

$$x^* = \log_{10} \frac{\overline{\Delta P}\, h^2 \rho}{\mu^2} \tag{14}$$

$$y^* = \log_{10} \frac{\bar{\tau}_w h^2 \rho}{\mu^2} \tag{15}$$

This calibration can be used for other Stanton gauges only if their design and dimensions closely match those used by East, who pointed out that readings can be particularly sensitive to deviations from the recommended zero value of the distance of the leading edge of the blade from the leading edge of the static hole.

In the flow range where the gauge covers distances from the wall where $U = (\tau_W/\mu)y$, the calibration curve is quite different for laminar and turbulent flows [51], and it cannot be explained by the type of reasoning presented in Eq. (6). This is not too surprising, since the circulating zone in front of the gauge for a turbulent flow would be expected to be quite different from that for a steady flow. Calibrations obtained in steady laminar flows, therefore, should not be used in turbulent flows.

In tests comparing the performance of Stanton gauges and heated film gauges in laminar boundary layers, Brown [47] has found that Stanton gauges give inaccurate readings in large unfavorable pressure gradients. He interpreted these results by suggesting that the region of the flow field "seen" by a Stanton gauge is many times its height and that this effect becomes greater as the wall shear stress decreases in regions of unfavorable pressure gradient. The inaccurate readings result because they are being influenced by regions of the field where the velocity is not varying linearly with distance from the wall. The velocity field "seen" by the gauge under test conditions was, therefore, not the same as that under calibration conditions. Brown also points out that variations in the calibrations of different investigators for laminar flows can be caused by the same effect.

These results of Brown would indicate that Stanton gauges cannot be regarded as reliable devices in boundary layers with large pressure gradients. It is possible that this conclusion is only applicable to laminar boundary layers. Consequently, there is a need for tests in turbulent boundary layers similar to those carried out by Brown. Experiments by Head and Rechenberg [42] for turbulent flows with moderately favorable pressure gradients that exist in the entry region of a pipe have yielded encouraging results.

Studies with Stanton gauges in compressible flows have been carried out by Winter and Gaudet [32] and by Gadd et al. [50]. However, no generally accepted method of correcting for compressibility effects has been developed.

5. SUBLAYER FENCE

Because of the larger pressure readings with a sublayer fence, the height of the fence can be somewhat smaller than that of the Stanton gauge. A sketch of the fence used by Rechenberg [39] is shown in Fig. 4. The design presented by Konstantinov and Dragnysh [6] is more elaborate but has the advantage that the height of the fence can be varied with a screw mechanism.

The flow over a fence consists of a recirculation zone both in front of and in back of the fence. Observations regarding the influence of a two-dimensional obstacle on a wall made by Brown [47] with reference to Stanton gauges and studies by Good and Joubert [52] indicate the region of disturbed flow is much

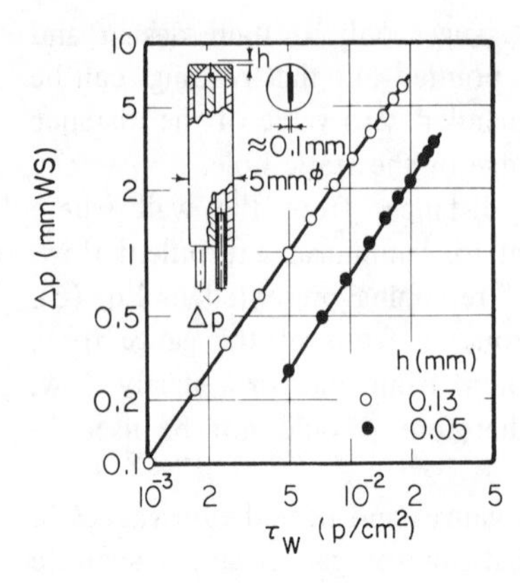

Figure 4 Assembly and calibration curve for sublayer fence presented by Rechenberg [39].

larger than the height of the fence. Consequently, the height would have to be much smaller than the thickness of the viscous sublayer for it to be influenced in a turbulent flow only by the region where $\bar{U} = \tau_W y/\mu$. As has already been pointed out for the Stanton gauge, the advantage of the sublayer fence might not be as great as originally anticipated.

Head and Rechenberg [42] calibrated a sublayer fence and a Preston tube in a turbulent flow with zero pressure gradient, and compared their readings in a flow with unfavorable pressure gradients. Agreement was noted in moderately unfavorable pressure gradients. However, in strongly unfavorable pressure gradients, the two instruments indicated different values of the wall shear stress. It is quite possible that both instruments were giving erroneous readings. Until more data are available, the sublayer fence is of unknown accuracy in flows with strong pressure gradients. Achenbach [53, 54] has used the sublayer fence in studies of flow around a cylinder and a sphere.

6. ANALYSIS OF HEAT- OR MASS-TRANSFER PROBES

6.1 Design Equation for a Two-Dimensional Mass-Transfer Probe

The difficulties cited above in the use of floating heads, Preston tubes, Stanton gauges, and sublayer fences to measure wall shear stress have stimulated considerable research in recent years on the application of flush-mounted thermal- or mass-transfer (electrochemical) probes. These have the advantages that they can be used in a wide variety of flows, that they do not interfere with the flow, that they offer the possibility of measuring time-varying flows, and, in the case of electrochemical probes, that calibration is not necessary.

The principle of operation is illustrated in Fig. 5 for the case of a two-dimensional wall element aligned with its long side perpendicular to the direction of mean flow. The fluid at the surface of the wall element is controlled at a concentration C_W or temperature T_W, which is different from that in the bulk fluid. The rate of mass or heat transfer between the fluid and the wall element is then measured. If the element is small enough in the flow direction, the concentration or thermal boundary layer over the element will be so thin that it lies within a region for which the velocity is given by

$$U = Sy \tag{16}$$

where S is the magnitude of the velocity gradient at the wall. A calibration is established between the measured mass-transfer or heat-transfer rate and the velocity gradient S. If the viscosity of the fluid is known, then the wall shear stress can be evaluated from the measured S, since $\tau_W = \mu S$.

An analytical expression for the calibration can be derived, provided the following conditions are satisfied:

1. The scalar boundary layer is within the region where $U = Sy$.
2. The flow is homogeneous over the surface of the element.
3. δ_C is small enough compared to the width W of the electrode so that diffusion in the spanwise direction can be neglected.
4. Forced convection is large enough so that diffusion in the x direction can be neglected.
5. Natural convection is small compared to forced convection.
6. The scalar boundary layer is small enough so that the influence of turbulent transport in the y direction can be neglected.

The mass balance equation for a two-dimensional field is given as

$$\frac{\partial C}{\partial t} + Sy\frac{\partial C}{\partial x} + V\frac{\partial C}{\partial y} = D\left(\frac{\partial^2 C}{\partial y^2} + \frac{\partial^2 C}{\partial x^2} + \frac{\partial^2 C}{\partial z^2}\right) \tag{17}$$

Because of assumption 1, Sy is substituted for U. Assumptions 2, 5, and 6 allow for $V\,\partial C/\partial y$ to be neglected. Assumptions 3 and 4 allow for $D\,\partial^2 C/\partial z^2$ and $D\,\partial^2 C/\partial x^2$ to be neglected. Equation (17) can therefore be simplified to

$$\frac{\partial C}{\partial t} + Sy\frac{\partial C}{\partial x} = D\frac{\partial^2 C}{\partial y^2} \tag{18}$$

If a pseudo-steady-state assumption is made, then

$$Sy\frac{\partial C}{\partial x} = D\frac{\partial^2 C}{\partial y^2} \tag{19}$$

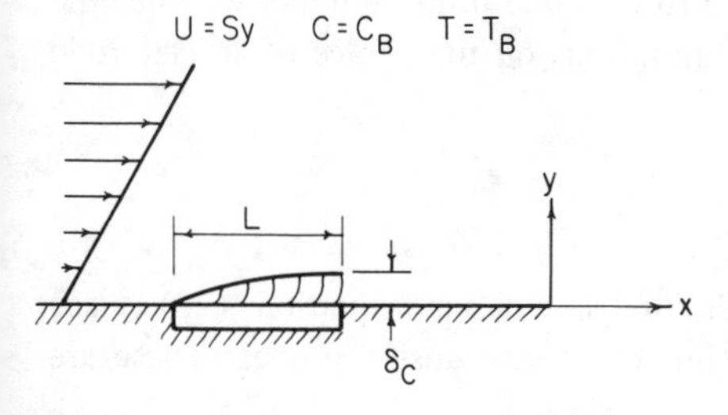

Figure 5 Description of mass- or heat-transfer technique for the case of negligible diffusion in the x direction.

This implies that the flow is steady or that it is changing slowly enough with time so that the concentration field is described by the steady-state solution to Eq. (18). Equation (19) is solved using the boundary conditions

$$C = C_W \qquad \text{for} \qquad y = 0 \tag{20}$$

$$C = C_B \qquad \text{for large } y,\ x = 0 \tag{21}$$

The solution for $C(y, x)$ is given by Mitchell and Hanratty [18]. From it, the average mass-transfer rate can be calculated, since

$$\langle N \rangle = \frac{1}{L} \int_0^L D \left(\frac{\partial C}{\partial y} \right)_{y=0} dx \tag{22}$$

Mitchell and Hanratty [18] give the following expression for the mass-transfer coefficient $K = \langle N \rangle / (C_B - C_W)$:

$$\frac{KL}{D} = 0.807 Z^{1/3} \tag{23}$$

where

$$Z = \frac{SL^2}{D} = (L^+)^2 \text{ Sc} \tag{24}$$

Equation (23) is the basic design relation for mass-transfer wall gauges. It indicates a disadvantage of this type of instrument in that the mass-transfer coefficient varies with only the cube root of S.

The use of boundary condition (20) is not quite correct. For the case of an electrochemical probe, the voltage V_0 on the electrode and not the concentration is kept constant. The rate of reaction at the electrode surface is a function of both the electrode voltage and the concentration, $R(V_0, C_W)$. Thus, the boundary condition (20) should be replaced with

$$-D \left(\frac{\partial C}{\partial y} \right)_W = C_W k_R(V_0) \qquad \text{for } y = 0 \tag{25}$$

if the rate equation is of first order. The rate constant $k_R(V_0)$ is a strong function of V_0 so that, at large voltages, $C_W \to 0$ in order that $-D(\partial C/\partial y)_W$ remain finite. Although this method maintains C_W at an approximately zero value over most of the surface, it cannot be done at $x = 0$, where the solution presented by Mitchell indicates an infinite local mass-transfer rate. It would therefore be of interest to solve Eq. (19) using boundary condition (25) to examine in more detail whether the kinetics of the surface reaction places any limitations on the choice of an electrode length.

6.2 Limitations of the Design Equation

The assumptions made in the derivation of the design equation (23) place limitations on its application to mass-transfer and heat-transfer probes. These are now explored.

First consider the condition that $U = Sy$. If δ_v is the thickness of the region over which this is valid, then it is necessary that $\delta_c/\delta_v < 1$. The thickness of the concentration boundary layer can be estimated using the concept of a Nernst diffusion layer, $\delta_c = D/K$. If K is calculated from Eq. (23), the condition $\delta_c/\delta_v < 1$ requires

$$0.807 \frac{S^{1/3}\delta_v}{L^{1/3}D^{1/3}} > 1 \tag{26}$$

In laminar flows this can usually be satisfied for mass-transfer probes, except possibly in regions where there is a large pressure gradient. Spence and Brown [55] explored the influence of pressure gradient for laminar flows by solving Eq. (19), using

$$U = Sy + \frac{y^2}{2\mu}\frac{dP}{dx} \tag{27}$$

They obtained the relation

$$\left(\frac{KL}{D}\right)^3 = (0.807)^3\ Z + \frac{25}{19 \times 9}\left(\frac{KL}{D}\right)^{-1}\left(\frac{L^3\ dP/dx}{D\mu}\right) \tag{28}$$

They found that the second term on the right side of Eq. (28) can be important in boundary layers with large unfavorable pressure gradients, and that for fully developed laminar flow in a channel the corrections are less than 2% of S provided $Z^{1/3}\ h_c/L > 28$, where h_c is the channel height.

For turbulent flows, the region where $U = Sy$ can be quite thin, and condition (26) is more restrictive. If $\delta_v = 5\nu/u^*$ is substituted in Eq. (26), the following limitation on the length of the mass-transfer surface is obtained:

$$L^+ < 64\ \mathrm{Sc} \tag{29}$$

Another limitation on the length is imposed by the requirement that turbulent transport of mass be negligible if design equation (23) is to to used. Son and Hanratty [56, 57] and Shaw and Hanratty [58] have carried out studies of turbulent mass transfer to surfaces with different electrode lengths at large Sc and found that Eq. (23) describes the results provided $L^+ < 700$. For $\mathrm{Sc} = 1$, it would be expected that turbulent transport would become important when $\delta_c > \delta_v$, that is, when the thickness of the concentration boundary layer is larger than the region where $U = Sy$. Thus, it is estimated that turbulent transport will not be important provided the length of the electrode is chosen to satisfy the more restrictive of the two conditions $L^+ < 64$ Sc and $L^+ < 700$.

Fluctuations in the wall shear stress should not be measured with mass-transfer probes with $L^+ > 700$, since measured fluctuations in the mass-transfer rate could be coupled with the turbulent velocity fluctuations associated with the turbulent transport as well as with fluctuations in S. However, if it is desired only to measure the time-averaged wall velocity gradient $\bar{S}$, then the condition that turbulent transport effects be negligible is not a limitation, provided the design equation (23) is modified so as to take into account the additional effect of turbulent mass transport described in [56 to 58].

A lower limit on the length of the mass-transfer surface is imposed by the requirement that forced convection dominate over molecular diffusion in determining the rate of mass transfer. The errors involved in neglecting diffusion in the streamwise direction have been assessed by solving the equation

$$Sy \frac{\partial C}{\partial x} = D\left(\frac{\partial^2 C}{\partial x^2} + \frac{\partial^2 C}{\partial y^2}\right) \tag{30}$$

using the boundary conditions at $y = 0$ of

$$\begin{aligned} C &= C_w \qquad \text{for } 0 < x < L \\ \frac{\partial C}{\partial y} &= 0 \qquad \text{for } x < 0,\ x > L \end{aligned} \tag{31}$$

Ling [59] presents the following numerical solution, valid for $Z \geqslant 50$:

$$\frac{KL}{D} = 0.807Z^{1/3} + 0.19Z^{-1/6} \tag{31a}$$

On the basis of this equation, it is seen that for $Z > 200$ the error will be less than 5% if the second term is ignored. This requires that

$$(L^+)^2 \text{ Sc} > 200 \tag{32}$$

Py and Gosse [26] have also carried out numerical solutions of Eq. (30) with boundary condition (31) and have found, for $Z < 50$, that the mass-transfer rate becomes insensitive to variations in S and that it is not practical to use this instrument even if the design equation is modified to take account of the effect of streamwise diffusion.

Limitations are also placed on the width of the electrode to minimize the effects of diffusion in the spanwise direction. To estimate these, it is assumed that $W/\delta_C > 10$. Using the estimate $\delta_C = L^{1/3} D^{1/3}/0.807\ S^{1/3}$ already derived, the following condition is obtained:

$$0.807Z^{1/3} \frac{W}{L} > 10 \tag{33}$$

For $W/L = 2$, this requires a lower limit of $Z = 200$; for $W/L = 10$, Z should be greater than 2.

To summarize: For surfaces with $W/L > 2$, the length of the electrode for laminar flows should be chosen so that $200 < (L^+)^2$ Sc < 64 Sc. For typical electrochemical reactions, where Sc $\simeq 1000$, this requires $0.5 < L^+ < 64{,}000$. With turbulent flows it is probably wise to use $W/L > 10$ if the surface is to respond only to flow in the x direction. For measurement of the time average of the turbulent wall shear stress, the design equation is not applicable for $L^+ > 700$, and it would be necessary to calibrate. However, it is recommended that $L^+ < 700$ if the turbulent fluctuations in the wall shear stress are to be measured.

With electrochemical mass-transfer probes it is usually quite easy to operate within these operating conditions. The chief limitation usually involves chemical considerations. At very high flow rates the mass-transfer rate becomes large compared to the reaction velocity constant, and it is therefore not possible to maintain boundary condition (20).

6.3 Nonhomogeneous Two-Dimensional Laminar Flows

Usually the mass-transfer or heat-transfer element is small enough so that the assumption of uniform flow is acceptable for two-dimensional laminar flows. However, it might not hold close to a stagnation point or a separation point, where the wall shear stress vanishes. The application of mass-transfer and heat-transfer probes under these circumstances has been considered in [20, 60 to 62].

The velocity gradient is considered to be varying over the mass-transfer surface so that, from the continuity equation, the velocity normal to the surface is given by $V = -(y^2/2)\, \partial S/\partial x$, and the steady mass-balance equation is

$$Sy \frac{\partial C}{\partial y} - \frac{y^2}{2} \frac{\partial S}{\partial x} \frac{\partial C}{\partial y} = D \frac{\partial^2 C}{\partial y^2} \tag{34}$$

The solution of the equation for the case $S = S_0 + \gamma x$ is given by

$$K = 0.807 \frac{D^{2/3}}{L^{1/3}} \left(\frac{2}{3}\right)^{2/3} \left[\frac{(S_0 + \gamma L)^{3/2} - S_0^{3/2}}{\gamma L}\right]^{2/3} \tag{35}$$

At a stagnation or separation point $S_0 = 0$, $\gamma = (\partial S/\partial x)_s$, and

$$K = 0.807(D^2\gamma)^{1/3} \left(\tfrac{2}{3}\right)^{2/3} \tag{36}$$

For $\gamma L/S_0 \ll 1$, Eq. (35) simplifies to

$$K = 0.807 \frac{D^{2/3} S_0^{1/3}}{L^{1/3}} \left(1 + \frac{1}{6} \frac{\gamma L}{S_0} + \cdots\right) \tag{37}$$

Now $\gamma L/S_0$ is approximately equal to the ratio of the electrode length to the distance from stagnation or separation. It is therefore concluded that for two-dimensional steady flows the error from nonuniform flow will be less than 10%, provided measurements are made at distances of five gauge lengths or greater from stagnation or separation. However, even if this precaution is observed, it is necessary, close to stagnation or separation points, to take into account the influence of pressure gradient and to ensure that Z is large enough for the design equation to be valid.

6.4 Frequency Response

The time response of a mass-transfer probe to a fluctuating flow field is primarily associated with the capacitance effect of the concentration boundary layer. For a slowly varying velocity field, a pseudo-steady-state assumption can be made whereby the instantaneous mass-transfer coefficient is related to the instantaneous velocity by the same equation as for steady flow:

$$\frac{K(t)L}{D} = 0.807\left(\frac{L^2}{D}\right)^{1/3} [S_x(t)]^{1/3} \tag{38}$$

Mitchell and Hanratty [18] and Sirkar and Hanratty [24] defined

$$K(t) = \bar{K}(t) + k(t) \tag{39}$$

$$S_x(t) = \bar{S}_x(t) + s_x(t) \tag{40}$$

and showed that, for sufficiently small $s_x/\bar{S}_x$, Eq. (38) can be written in the form

$$\frac{\bar{K}L}{D} = 0.807\left(\frac{L^2}{D}\right)^{1/3} \bar{S}_x^{1/3}\left(1 - \frac{1}{9}\frac{\overline{s_x^2}}{\bar{S}_x^2} + \cdots\right) \tag{41}$$

$$\frac{k}{K} = \frac{1}{3}\frac{s_x}{\bar{S}_x} - \frac{1}{9}\left(\frac{s_x^2}{\bar{S}_x^2} - \frac{\overline{s_x^2}}{\bar{S}_x^2} + \cdots\right) \tag{42}$$

Thus, to zeroth order, this analysis indicates that $\bar{K}$ is related to $\bar{S}_x$ by the same equation derived for steady flow, and that the measured mass-transfer fluctuations are related to the measured fluctuations in the velocity gradient by

$$\frac{k}{K} = \frac{1}{3}\frac{s_x}{\bar{S}_x} \tag{43}$$

It would be expected that, for large frequency fluctuations in S_x, the concentration field would not change rapidly enough so as to maintain approximately the same conditions as exist at steady state. Consequently the fluctuations in the mass-transfer coefficient predicted by the above equation could be too large. Mitchell and Hanratty [18] and Fortuna and Hanratty [19] carried out an analysis of the time response of electrochemical probes by solving Eq. (18) for the case where $s_x/\bar{S}_x$ is small. Because of this assumption the following linearized form of Eq. (18) can be used:

$$\frac{\partial c}{\partial t} + \bar{S}_x y \frac{\partial c}{\partial y} + s_x y \frac{\partial \bar{C}}{\partial x} = D\frac{\partial^2 c}{\partial y^2} \tag{44}$$

where

$$\bar{C} = \frac{C_B}{\Gamma(\frac{4}{3})}\int_0^{\eta} e^{-t^3}\, dt \tag{45}$$

$$\eta = y\left(\frac{\bar{S}_x}{9Dx}\right)^{1/3} \tag{46}$$

If the fluctuating velocity gradient is given as a harmonic function,

$$s_x = \hat{s}_R\, e^{i\omega t} \tag{47}$$

then the solution of Eq. (44) can be given as

$$\hat{c}(y, x) = [\hat{c}_R(y, x) + i\hat{c}_I(y, x)]\, e^{i\omega t} \tag{48}$$

If Eq. (48) is substituted into Eq. (44), the following equation for $\hat{c}$ is obtained:

$$i\omega\hat{c} + \bar{S}_x y \frac{\partial \hat{c}}{\partial x} + \hat{s}_R y \frac{\partial \bar{C}}{\partial x} = D \frac{\partial^2 \hat{c}}{\partial y^2} \tag{49}$$

This was solved by Fortuna and Hanratty [19] using the boundary conditions

$$\hat{c} = 0 \qquad \text{for } y = 0 \text{ and large } y \tag{50}$$

For $\omega \rightarrow 0$, the solution is the pseudo-steady-state approximation

$$\hat{c} = \frac{\hat{s}_R}{\bar{S}_x} \frac{1}{3} y \frac{\partial \bar{C}}{\partial y} \tag{51}$$

$$\frac{k}{K} = \frac{1}{3} \frac{s_x}{\bar{S}_x} \tag{52}$$

Any general time-varying s_x can be represented as a Fourier series. Because of the linearization assumption, the function describing the time-varying mass-transfer coefficient can be constructed from the solution presented by Fortuna and Hanratty [19] as the sum of a number of Fourier components having the same frequencies as the forcing function $s(t)$.

For the case of a turbulent field, the mean squared values of the fluctuating velocity gradient and the fluctuating mass-transfer coefficient can be represented in terms of spectral density functions:

$$\overline{k^2} = \frac{1}{2\pi} \int_0^\infty W_k \, d\omega \tag{53}$$

$$\overline{s_x^2} = \frac{1}{2\pi} \int_0^\infty W_{s_x} \, d\omega \tag{54}$$

Mitchell and Hanratty [18] represented the relation between W_{s_x} and W_k in the following way:

$$W_{s_x} = 9 \frac{\bar{S}_x^2}{\bar{K}^2} \frac{W_k}{A_d^2} \tag{55}$$

where $1/A_d^2 \geqslant 1$. Values of $1/A_d^2$ calculated by Fortuna and Hanratty [19] are plotted in Fig. 6 in a way suggested by McMichael [63]. From this plot, it is seen that $A_d^2 \simeq 1$ for $L^+(\omega^+)^{3/2}\,\mathrm{Sc}^{1/2} \leqslant 1$.

This result indicates that the larger the Schmidt number is, the smaller is the frequency at which the behavior of the mass-transfer probe departs from the pseudo-steady-state solution because of damping in the concentration boundary layer. Electrochemical mass-transfer probes usually operate at a Schmidt number of about 1000. Since heat-transfer probes can be used with air, with Prandtl number about 0.7, they are superior to electrochemical probes with respect to the frequency response of the concentration boundary layer.

Situations involving large-amplitude oscillations cannot be analyzed by the

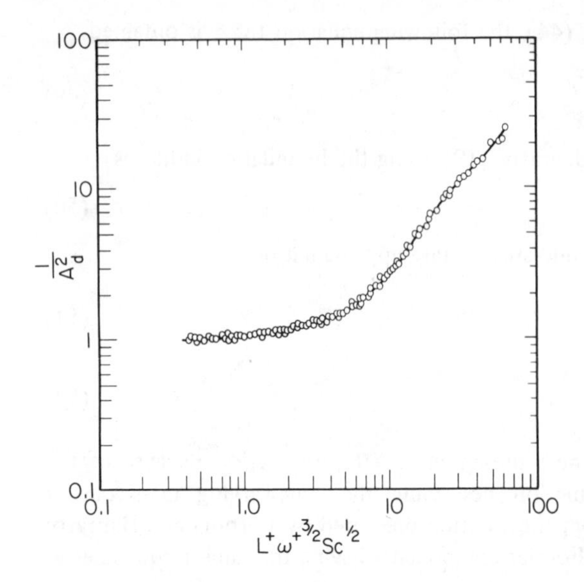

Figure 6 Correction factor for frequency response of an electrochemical shear stress probe.

methods developed by Mitchell and Hanratty [18] and by Fortuna and Hanratty [19]. For low-frequency, large-amplitude oscillations a pseudo-steady-state assumption can be made. However, the linearized equation (52) cannot be used to relate s_x to k. Instead, the full nonlinear pseudo-steady-state solution, Eq. (38), should be used. Solving Eq. (38) for S_x, one obtains the relations

$$S_x = \beta K^3 \tag{56}$$

$$\beta = \left(\frac{L}{D}\right)^3 \left(\frac{1}{0.807}\right)^3 \frac{D}{L^2} \tag{57}$$

$$\bar{S}_x = \beta \overline{K^3} \tag{58}$$

$$s_x = \beta(K^3 - \overline{K^3}) \tag{59}$$

It is seen from Eq. (56) that the measured signal must be cubed to determine S_x. This is cumbersome if analog methods are used, and can be avoided if it is possible to use the linearization assumption, Eqs. (41) and (43). No general methods are available to correct for the frequency response of the concentration boundary layer when the amplitude of the oscillation is large. In such cases it is probably best to design the test so that damping due to the concentration boundary layer will be small.

An additional complication can arise with large-amplitude oscillations for which the mean value of S_x is zero. In this case there can be periods during which the value of $Z = S_x L^2/D$ is too small for the design equation to be applicable. A discussion of the application of heat- or mass-transfer wall probes to such situations has been given by Pedley [64].

Very extensive discussions of the frequency response of electrochemical probes both for turbulent flows and for unsteady laminar boundary layers have been presented by Py [27, 65] and by Tournier and Py [22].

6.5 Turbulence Measurements

Turbulent flows constitute one of the most important types of fluctuating flows for which mass-transfer probes are used. Fully developed turbulent flows give values of $(\overline{s_x^2}/\bar{S}_x^2)^{1/2}$ of 0.30 to 0.36. From Eq. (58) the following approximate relation between $\bar{S}_x$ and the measured $\bar{K}$ is obtained using the pseudo-steady-state assumption:

$$\frac{\bar{K}L}{D} = 0.807\left(\frac{L^2}{D}\right)^{1/3} \bar{S}_x^{1/3}\left(1 - \frac{1}{9}\frac{\overline{s_x^2}}{\bar{S}_x^2}\right) \tag{60}$$

It is seen that an error of only 3 to 4% would be made in using linear theory,

$$\frac{\bar{K}L}{D} = 0.807\left(\frac{L^2}{D}\right)^{1/3} \bar{S}_x^{1/3} \tag{61}$$

to evaluate $\bar{S}_x$. Mitchell and Hanratty [18] have shown that if the pseudo-steady-state approximation is applicable and if the probability function describing s_x is gaussian, the error in calculating $(\overline{s_x^2}/\bar{S}_x^2)^{1/2}$ from $(\overline{k^2}/\bar{K}^2)^{1/2}$ with the linear relations

$$\frac{k}{\bar{K}} = \frac{1}{3}\frac{s_x}{\bar{S}_x} \qquad \frac{\overline{k^2}}{\bar{K}^2} = \frac{1}{9}\frac{\overline{s_x^2}}{\bar{S}_x^2} \tag{62}$$

is less than 3% if $(\overline{s_x^2}/\bar{S}_x^2)^{1/2}$ is less than 0.5.

These considerations would suggest that it is not necessary to use the more complicated nonlinear relations (58) and (59) for mass-transfer probes in turbulent flows to evaluate $\bar{S}_x$ and $(\overline{s_x^2}/\bar{S}_x^2)^{1/2}$. The chief difficulties in applying Eq. (52) are spatial averaging of the fluctuations over the sensor surface and the frequency response of the concentration boundary layer. Both effects will lead to an underprediction of $\overline{s_x^2}$.

Reiss and Hanratty [7, 8] discovered that turbulent velocity fluctuations close to a wall are dominated by flow structures that have small spanwise dimensions and that are greatly elongated in the flow direction. These results suggest that spatial averaging in the flow direction is not important but that, if the spanwise dimension of the probe is not small compared to the spanwise dimension of the wall eddies, the assumption of a homogeneous flow over the sensor surface will no longer be valid. Under these circumstances the following equation [18] relates the measured mean square value of the mass-transfer fluctuations to the true local value $\overline{k^2}$:

$$\overline{k_m^2} = \frac{2\overline{k^2}}{W^2}\int_0^W (W - g)R_z(g)\,dg \tag{63}$$

where R_z is the circumferential correlation coefficient for the mass-transfer fluctuations. Mitchell and Hanratty [18] developed from Eq. (63) the curve shown in Fig. 7 for $(\overline{k^2})^{1/2}/(\overline{k_m^2})^{1/2}$, where Λ_z is the circumferential scale of the mass-transfer fluctuations. Mitchell and Hanratty [18] have suggested that for turbulent pipe flow $\Lambda_z^+ \simeq 12$. Therefore, it is estimated that W^+ must be less than 8 for the error associated with spatial averaging to be less than 5%. This is quite restrictive for

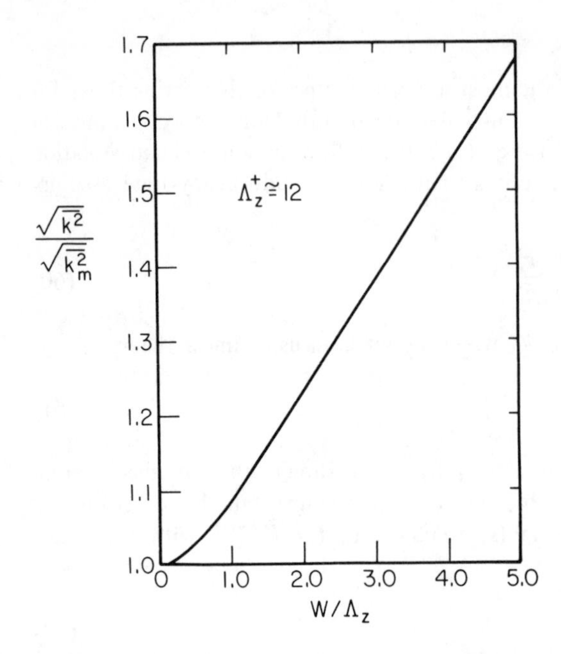

Figure 7 Nonuniform flow correction for electrochemical probes of varying widths.

rectangular electrodes, so it might be necessary to use electrodes with $W^+ > 8$ and Fig. 7 to correct the measurements.

From the analysis already presented, it was shown that damping by the concentration boundary layer would be unimportant for $L^+(\omega^+)^{3/2}\ \mathrm{Sc}^{1/2} \leqslant 1$. For pipe turbulence the median frequency of s_x is $\omega^+_{\text{median}} \simeq 2\pi\ 0.009$ [66]. This suggests that for no serious damping to occur at the median frequency, $Z = (L^+)^2\mathrm{Sc}$ should be less than 5600. For $\mathrm{Sc} = 1$, this requires $L^+ < 75$. However, since electrochemical mass-transfer probes operate at $\mathrm{Sc} = 1000$, the much more restrictive condition of $L^+ < 2.4$ is obtained for them. For such probes it will probably be necessary to correct for frequency response by using Eqs. (53), (54), and (55) and Fig. 6. This will require the measurement of the frequency spectrum of k, and not just $\overline{k^2}$.

The methods just outlined, for correcting for spatial averaging and frequency response, have been developed for the situation in which either of these effects is separately affecting the determination of $\overline{s_x^2}$. The usual case is that both effects are simultaneously operative. No well-established methods for dealing with this situation have been developed. The approach taken at present is to assume the effects are independent and to apply both corrections to the measurement.

Recent measurements of the probability distribution of s_x have indicated that it is not gaussian, as assumed in [18], but highly skewed with occasional positive pulses in s_x of the same magnitude as $\bar{S}_x$. Thus, the arguments presented earlier should probably be reexamined. Present experience would suggest that no serious errors are involved in calculating $\bar{S}_x$ and $\overline{s_x^2}$ using the linear relations (52), (61), and (62), but that it might be more appropriate to use the nonlinear relations (58) and (59) between K and S_x to evaluate the probability distribution of S_x.

7. EFFECT OF CONFIGURATION OF MASS-TRANSFER PROBE

7.1 Circular Probes

Circular mass-transfer probes offer considerable advantage over (and are much more widely used than) rectangular probes, since they are more compact and are more easily fabricated. They can be constructed simply by gluing a wire in a hole in the wall and sanding the end of the wire so it is flush with the surface.

Reiss and Hanratty [8] analyzed the performance of a circular mass-transfer surface as indicated in Fig. 8. The rate of mass transfer to the strip shown is assumed to be given by the equation derived earlier for a two-dimensional flow:

$$\frac{N_i}{C_B - C_w} = 0.807 \frac{D^{2/3} S^{1/3}}{(2R_e \sin \alpha)^{1/3}} 2R_e \sin \alpha \, dz \tag{64}$$

The total mass-transfer rate is then given as

$$K = \frac{0.807 D^{2/3} S^{1/3}}{\pi R_e^2} \int_0^{\pi} \frac{1}{2} (2R_e \sin \alpha)^{5/3} \, d\alpha \tag{65}$$

If an equivalent length for the surface is defined as

$$L_e = \left[\frac{\pi R_e^2}{\int_0^{\pi} \frac{1}{2}(2R_e \sin \alpha)^{5/3} \, d\alpha} \right]^3 = 0.81356 \, (2R_e) \tag{66}$$

then the design equation for a circular surface is obtained as

$$K = \frac{0.807 D^{2/3} S^{1/3}}{L_e^{1/3}} \tag{67}$$

A disadvantage of the circular surface compared to an approximately two-dimensional rectangular surface is that diffusion in the spanwise direction can be more important. Numerical solutions of the mass balance equations that include molecular diffusion in the flow and spanwise directions have been carried out by Py [65]. These suggest that Eq. (67) can be used provided $Z = (L_e^+)^2 \text{ Sc} > 1000$. For $\text{Sc} = 1000$, an operating range of $1 < L_e^+ < 700$ is therefore recommended for turbulent flow to ensure that the design equation holds.

In turbulent flow the instantaneous S is given by

$$S = S_x \left[1 + \left(\frac{s_z}{S_x} \right)^2 \right]^{1/2} \tag{68}$$

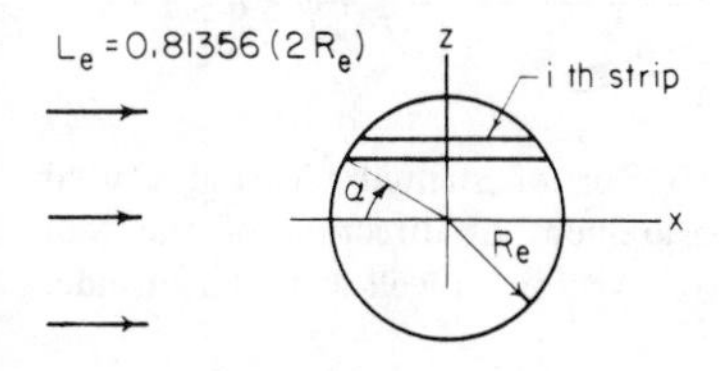

Figure 8 Evaluation of the effective length of a circular electrode.

where the x coordinate is in the direction of mean flow. The quantity s_z/S_x may be defined in terms of the angle between the instantaneous direction of the flow at wall and the x axis:

$$\tan\theta = \frac{s_z}{S_x} \tag{69}$$

Measurements by Sirkar [67] show that $\theta < 17°$ for 99.5% of the time. This indicates that the circular surface is sensitive to S_x and not to s_z; that is, the equation

$$K = 0.807\,\frac{D^{2/3}S_x^{1/3}}{L_e^{1/3}} \tag{70}$$

will be in error by less than 5% for 99.5% of the time.

7.2 Slanted Transfer Surface

Mitchell [68] was the first to consider the use of a wall transfer surface to measure the direction of the velocity gradient vector at a wall. For this purpose, he proposed the use of a slant surface of the type shown in Fig. 9 and showed that such a surface would be sensitive to the two components of the wall velocity gradient, S_x and s_z. This type of surface was tested by Sirkar and Hanratty [24]. On the basis of these tests they developed the chevron-electrode arrangement shown in Fig. 10. Karabelas and Hanratty [69] have suggested an interesting modification of the chevron arrangement that could be useful in three-dimensional boundary-layer flows.

By using the Reiss method, the slant gauge shown in Fig. 10 can be analyzed to give

$$K = 0.807D^{2/3}\left[\frac{S\sin(\phi-\theta)}{L}\right]^{1/3}\left[1+\frac{L\cot(\phi-\theta)}{5W}\right] \tag{71}$$

For $L/W \to \infty$ it is seen from the above relation that K depends only on the component of S perpendicular to the leading edge of the surface.

For the chevron arrangement shown in Fig. 10, the following results are obtained for $-\phi+\psi \leqslant \theta \leqslant \phi+\psi$, where $\tan\psi = L/W$:

$$K_1 = 0.807D^{2/3}\left[\frac{S\sin(\theta-\phi)}{L}\right]^{1/3}\left[1+\frac{L}{5W}\cot(\phi-\theta)\right] \tag{72}$$

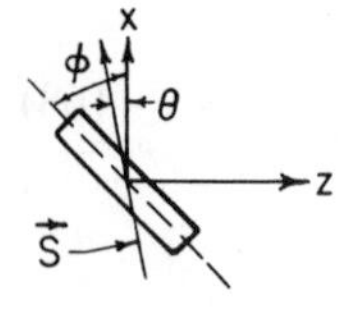

Figure 9 Single slanted electrode.

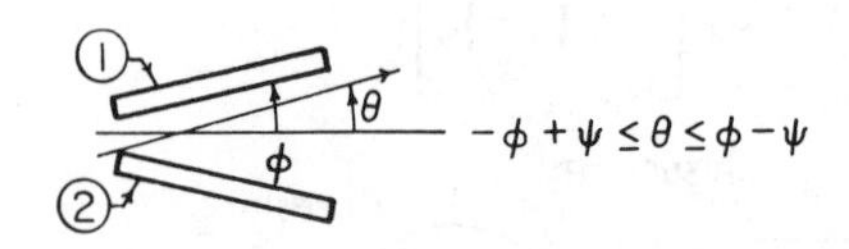

Figure 10 Pair of slanted electrodes used for measurement of direction of the wall shear-stress vector, as well as its magnitude.

$$K_2 = 0.807D^{2/3}\left[\frac{S\sin(\phi+\theta)}{L}\right]^{1/3}\left[1+\frac{L}{5W}\cot(\phi+\theta)\right] \tag{73}$$

From measurements of K_1 and K_2, the quantities S and θ can be calculated from Eqs. (72) and (73) to give the magnitude and direction of the velocity gradient at the wall. This calculation can be implemented by combining the measurements in the following way, as suggested by Tournier and Py [22]:

$$\frac{K_1 - K_2}{K_1 + K_2} = g_1(\theta) \tag{74}$$

$$K_1 + K_2 = \frac{S^{1/3}D^{2/3}}{L^{1/3}}\, g_2(\theta) \tag{75}$$

The functions $g_1(\theta)$ and $g_2(\theta)$ are evaluated from Eqs. (72) and (73) as follows:

$$g_1(\theta) = \frac{f_1 - f_2 + \tau_1(f_1f_3 - f_2f_4)}{f_1 + f_2 + \tau_1(f_1f_3 + f_2f_4)} \tag{76}$$

$$g_2(\theta) = \frac{1.5}{\Gamma(\frac{4}{3})9^{1/3}}\,[f_1 + f_2 + \tau_1(f_1f_3 + f_2f_4)] \tag{77}$$

$$f_1 = [\sin(\phi-\theta)]^{1/3} \qquad f_3 = \cot(\phi-\theta)$$

$$f_2 = [\sin(\phi+\theta)]^{1/3} \qquad f_4 = \cot(\phi+\theta)$$

$$\tau_1 = \frac{L}{5W}$$

From measurements of K_1 and K_2, the angle θ is first calculated from Eq. (74). Then, using this value of θ and the measurement of $K_1 + K_2$, the magnitude of the velocity gradient S can be evaluated from Eq. (75). One of the difficulties with this approach is that the function $g_1(\theta)$ is multivalued and therefore carries the restriction $\theta < \phi$. In the design of the chevron pair, the selection of ϕ is a compromise. Small values give greater sensitivity to s_z and give better spatial resolution in the spanwise direction, but ϕ cannot be so small that it violates the above restriction.

For turbulent flow, where K_1 and K_2 are functions of time, it would probably be necessary to use a computer to solve Eqs. (74) and (75) for instantaneous values of S and θ. An approach taken by Sirkar and Hanratty [25] can avoid this difficulty. Nonlinear terms in the fluctuating quantities are ignored. The mean flow is assumed to be in the x direction. It is assumed that $(|s_z|\cot\theta)/|\bar{S}_x + s_x| < 1$. Then Eqs. (74) and (75) can be rearranged in the following form after substituting $S_x = \bar{S}_x + s_x$, $S_z = s_z$, $S^2 = (\bar{S}_x + s_x)^2 + s_z^2$, and $\theta = \tan^{-1}[s_z/(\bar{S}_x + s_x)]$:

$$\bar{K} = 0.807D^{2/3}\left(\frac{\bar{S}_x\sin\phi}{L}\right)^{1/3}\left(1+\frac{L\cot\phi}{5W}\right) \tag{78}$$

$$\frac{K_1 - K_2}{2\bar{K}} = \frac{1}{3}\frac{s_z}{\bar{S}_x}\,\frac{\cot\phi - (2L\cot^2\phi)/5W}{1 + (L/5W)\cot\phi} \tag{79}$$

$$\frac{K_1 + K_2 - 2\bar{K}}{2\bar{K}} = \frac{1}{3}\frac{s_x}{\bar{S}_x} \tag{80}$$

Equations (79) and (80) show that s_z is directly proportional to the difference in the fluctuating signals and that s_x is directly proportional to the sum of the fluctuating signals. These operations can be easily implemented on analog circuits.

The correction of the signals from pairs of slant surfaces for frequency response was evaluated by Py [65].

7.3 Other Methods for Measuring Direction

Py and his coworkers have developed configurations for measuring the direction of the wall velocity gradient other than the chevron. Py and Gosse [26] suggested the use of two semicircular surfaces separated by insulation, as sketched in Fig. 11. The pair is arranged in such a way that the insulation lies along the direction of mean flow. For the case of negligible insulation thickness,

$$\frac{s_z}{S} = 4.20\,\frac{K_1 - K_2}{K_1 + K_2} \qquad Z > 50 \tag{81}$$

$$K_1 + K_2 = 1.73\left(\frac{D^2}{d_e}\right)^{1/3} S^{1/3} \qquad Z > 10^3 \tag{82}$$

This arrangement is quite attractive because it is more compact than the pair of slant surfaces and because the relations between the measured mass-transfer rates and the velocity field are so simple: $g_1(\theta) = \text{const}$ and $g_2(\theta) = \text{const}$. Py [65] also showed that the frequency response to s_z is better than for pairs of slant surfaces. However, this probe has not been used in experimental studies because of perceived difficulties in its fabrication and in controlling the thickness of the insulation.

The pair of rectangular surfaces [27, 70] shown in Fig. 12 is easier to fabricate than a pair of semicircles, and it can be made quite compact in the spanwise direction. The equations describing its performance are of the same form as Eqs. (74) and (75). Calculations presented by Py [27, 65] show that rectangular pairs of surfaces with large L/W ratios are not attractive choices, because the functions $g_1(\theta)$ and $g_2(\theta)$ indicate too strong a dependence on θ at small values of θ and too weak a dependence at large values of θ. Because of possible difficulties in controlling the thickness of the insulation and the dimensions of the surfaces, it might be necessary to determine the form of $g_1(\theta)$ and $g_2(\theta)$ empirically. In his studies of turbulent spanwise velocity fluctuations and of flow around slanted cylinders, Py [22, 27, 65] used pairs of rectangular probes with an L/W ratio of 2.5. These were made from

Figure 11 Twin semicircular electrodes.

Figure 12 Twin rectangular electrodes.

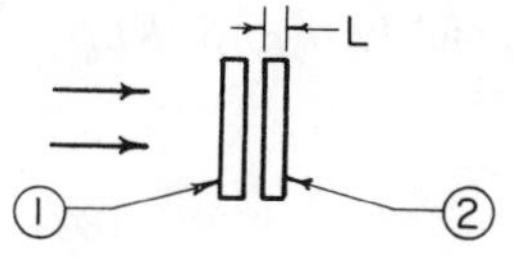

Figure 13 Sandwich electrode.

two platinum ribbons of dimensions 0.5 X 0.1 mm, separated by a strip of Mylar that was 8 μm thick.

7.4 Sandwich Elements

The use of two elements separated by a thin layer of insulation with the configuration shown in Fig. 13 was suggested by Son and Hanratty [21] to detect the direction of a two-dimensional flow. If the flow is in the forward direction, surface 2 gives a smaller value of K than surface 1, since it lies in the wake of the concentration boundary layer from 1. Work done since then has indicated that such a probe design holds other advantages as well.

Py [27, 65] has shown that

$$2L(K_1 - K_2) = g_0 \left[\frac{S_x (2L)^2}{D} \right]^{1/3} \tag{83}$$

where g_0 is a constant dependent on the insulation thickness. Thus measurements of $K_1 - K_2$ provide a means of determining S_x. Py has shown that the above equation is valid for $Z > 10$, so that a pair of surface elements appears to have an advantage over a single element in that they can be used at smaller Z. Py also demonstrated that the use of a composite signal consisting of two-thirds of $K_1 + K_2$ and one-third of $K_1 - K_2$ improves the frequency response over that obtained with a single element by almost an order of magnitude.

These elements have been applied extensively to the study of flow around a cylinder [22, 23, 62, 71]. For this purpose Lebouche [62] derived the following equation for the case of negligible insulation thickness, analogous to Eq. (35) already presented for single elements:

$$K_1 - K_2 = 0.420 \left(\frac{D^2 S_x}{2L} \right)^{1/3} \left(1 - \frac{0.803L \; \partial S_x / \partial x}{S_x} \right) \tag{84}$$

8. HEAT-TRANSFER PROBES

8.1 Analysis

The analysis presented in Sec. 6 for mass-transfer wall probes is also applicable to the thermal element shown in Fig. 14. The results already derived are directly

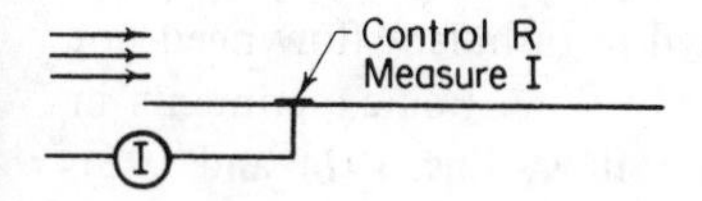

Figure 14 Wall heat-transfer probe.

transferable if $\rho C_p T$ is substituted for C, $\alpha_T = k_T/\rho C_p$ for D, and hL/k_T for KL/D. The theoretical design equation (23) can therefore be written as

$$\frac{q}{\Delta T} = 0.807 \frac{C_P^{1/3} k_T^{2/3}}{L^{1/3} \mu^{1/3}} (\rho \tau_w)^{1/3} \tag{85}$$

Since the heat loss from the thermal element is related to the heating current I and the resistance by the relation $q = I^2 R/A_e$, Eq. (85) can be written in the form

$$\frac{I^2 R}{\Delta T} = A(\rho \tau_w)^{1/3} \tag{86}$$

where $A = 0.807 A_e C_p^{1/3} k_T^{2/3}/L^{1/3}\mu^{1/3}$ is a weak function of temperature.

Experiments [3, 13] give the following result:

$$\frac{I^2 R}{\Delta T} = A(\rho \tau_w)^{1/3} + B \tag{87}$$

The term B, which represents the heat loss to the substrate, can often be larger than $A(\rho\tau_w)^{1/3}$, the heat loss to the fluid. The effective length of the thermal element calculated from experimentally determined values of A is found to be many times greater than the actual length of the heating element. This happens because heat is being transferred to the fluid both from the heating element and from the substrate.

This large heat loss to the substrate greatly complicates the use of wall heat-transfer gauges. The sensitivity of the instrument to changes in τ_w is diminished with respect to a mass-transfer probe, particularly at small τ_w. Since the heat loss to the substrate and the effective length of the thermal element are not predictable, the thermal gauge must be calibrated to determine A and B.

From the mass transfer analysis it is found that $SL_{\text{eff}}^2/\alpha_T = (L_{\text{eff}}^+)^2$ Pr should be greater than 200 for forced convection to dominate, and that L_{eff} should be less than $(0.807)^3\ S\delta_v^3/\alpha_T$ to ensure that the thermal boundary layer is within a region where the velocity is varying linearly with distance from the wall. For a turbulent fluid this latter restriction leads to the condition $L_{\text{eff}}^+ < 64$ Pr. Because of losses from the heating element, it is often quite difficult to satisfy the above condition for gas flows. It is then necessary to calibrate the instrument in a turbulent flow using the law-of-the-wall assumptions discussed in Sec. 3. Under these circumstances it can no longer be expected that the heat loss will vary with $\bar{\tau}_w^{1/3}$, so that the design equation becomes

$$\frac{(\bar{I})^2 R}{\Delta T} = A_t(\rho \bar{\tau}_w)^{1/n} + B_t \tag{88}$$

where A_t, B_t, and n must be determined empirically, and $\bar{\tau}_w$ and $\bar{I}$ are the time-averaged wall shear stress and current.

In circumstances where $L_{\text{eff}}^+ < 64$ Pr, the exponent $1/n$ equals $1/3$. Even for these cases it is desirable to calibrate the gauge in turbulent flow. For reasons given in the next section, the constants A_t and B_t determined in turbulent flow need not be equal to the constants A and B determined in steady flow. For experiments in which the calibration has been carried out in laminar flow, Eqs. (91) and (92), developed in the next section, should be used when the instrument is to be applied to a turbulent flow for $L^+_{\mathrm{eff}} < 64$ Pr.

8.2 Use in Turbulent Flows

If $L^+_{\mathrm{eff}} < 64$ Pr, then $\delta_c < \delta_v$ and it is possible to use thermal wall probes to measure fluctuations in τ_W. If the resistance of the heating element is held constant and the current or the voltage drop $E = IR$ is measured, Eq. (87) can be rearranged as follows for an incompressible fluid:

$$\tau_W^{1/3} = A^\dagger E^2 + B^\dagger \tag{89}$$

$$A^\dagger = \frac{1}{A\,\Delta T\,R\rho^{1/3}} \tag{90}$$

$$B^\dagger = -\frac{B}{A\rho^{1/3}} \tag{91}$$

The simplifications that result, in the case of mass-transfer probes, from using linearized equations may not be realized for heat-transfer probes. Sandborn [16, 17], who was the first to point out this difficulty, indicates that errors of the order of 10% in $\bar{\tau}_W$ and of the order of 50% in $[\overline{(\tau'_W)^2}]^{1/2}$ may be experienced if linearization techniques are used. Consequently $\bar{\tau}_W$ and τ'_W should be evaluated with the relations

$$\bar{\tau}_W = \overline{(A^\dagger E^2 + B^\dagger)^3} \tag{92}$$

$$\tau'_W = (A^\dagger E^2 + B^\dagger)^3 - \overline{(A^\dagger E^2 + B^\dagger)^3} \tag{93}$$

Because these equations involve evaluating the sixth power of E, it is not convenient to use analog methods. It is necessary to digitize the measured function $E(t)$ and use a computer to evaluate $\bar{\tau}_W$ and $\tau'_W(t)$ from Eqs. (92) and (93).

Quite often it is necessary to use a turbulent flow for which $\bar{\tau}_W$ is known (such as fully developed flow in a pipe) to calibrate the wall heat-transfer probe, rather than a laminar flow. This introduces additional problems if it is desired to use the probe to measure τ'_W. Expansion of Eq. (92) with $E = \bar{E} + e$ gives the following relation:

$$\bar{\tau}_W = (A^\dagger)^3\,[C1] + 3(A^\dagger)^2 B^\dagger\,[C2] + 3A^\dagger(B^\dagger)^2\,[C3] + (B^\dagger)^3 \tag{94}$$

where
$$\begin{aligned}[C1] &= \bar{E}^6 + 15\bar{E}^4\overline{e^2} + 20\bar{E}^3\overline{e^3} + 15\bar{E}^2\overline{e^4} + 6\bar{E}\overline{e^5} + \overline{e^6}\\ [C2] &= \bar{E}^4 + 6\bar{E}^2\overline{e^2} + 4\bar{E}\overline{e^3} + \overline{e^4}\\ [C3] &= \bar{E}^2 + \overline{e^2}\end{aligned} \tag{95}$$

It is the usual practice to fit an equation of the form

$$\bar{\tau}_W^{1/3} = A_t^\dagger \bar{E}^2 + B_t^\dagger \tag{96}$$

to the calibration measurements of $\bar{\tau}_W$ versus $\bar{E}$. From Eq. (94) it is seen that $\bar{\tau}_W$ is a function of $\overline{e^2}$, $\overline{e^3}$, $\overline{e^4}$, $\overline{e^5}$, and $\overline{e^6}$, in addition to $\bar{E}$. Consequently, if the moments of the fluctuating voltage cannot be ignored, then $A_t^\dagger$ and $B_t^\dagger$ will not be equal to the constants $A^\dagger$ and $B^\dagger$ that would be determined in a steady flow.

Sandborn [16, 17] has discussed two methods for determining $A^\dagger$ and $B^\dagger$ from calibration measurements in turbulent flow. In the first of these, $A_t^\dagger$ and $B_t^\dagger$ are determined from the best fit of Eq. (96) to measurements of $\bar{E}$ versus $\bar{\tau}_W$. It is initially assumed that $A^\dagger \simeq A_t^\dagger$ and $B^\dagger \simeq B_t^\dagger$. The calibration measurements are digitized, and τ_W is calculated for each of the calibration points by using Eq. (89). This gives a distribution function for τ_W. If the value of $\bar{\tau}_W$ calculated from this distribution function is in error, then it is evident that the above assumption is not correct. New values of $A^\dagger$ and $B^\dagger$ are assumed, and the calculation is repeated until the values of $\bar{\tau}_W$ obtained from the distribution function for τ_W agree with the measured values of $\bar{\tau}_W$.

A second approach is to use Eq. (94) directly. Values of $\bar{E}$ and of the moments of e are determined at each of the values of $\bar{\tau}_W$ at which the calibration was conducted. The constants $A^\dagger$ and $B^\dagger$ are determined by a least-squares fit of Eq. (93) relating the determined values of [C1], [C2], and [C3] to $\bar{\tau}_W$. Sandborn has suggested that a very close approximation to the correct values of $A^\dagger$ and $B^\dagger$ can be obtained by ignoring the much more difficultly determined $\overline{e^3}$, $\overline{e^4}$, $\overline{e^5}$, and $\overline{e^6}$.

Since the design equation for the heat-transfer wall probes cannot be linearized when they are used to determine turbulent fluctuations, it does not appear feasible to devise a method for correcting for frequency response. The best practice at present would be to design the experiment so that the pseudo-steady-state approximation is acceptable. The results presented in Sec. 6.4 can be used to investigate the influence of the concentration boundary layer on the frequency response. However, at present there is little theoretical guidance on how to estimate the influence of the substrate.

8.3 Compressible Flows

Because the constant A is a weak function of temperature, Eq. (89) or the following modification may be used to determine the quantity $\rho\tau_W$ in compressible flows:

$$(\rho\tau_W)^{1/3} = \frac{1}{A\,\Delta T\,R}\,E^2 - \frac{B}{A} \tag{97}$$

Experiments by Owen and Bellhouse [72], Diaconis [73], Bellhouse and Schultz [12], and Liepmann and Skinner [11] have demonstrated that probes calibrated under subsonic conditions can be used to carry out measurements in supersonic flows.

9. EXPERIMENTAL PROCEDURES FOR MASS-TRANSFER PROBES

9.1 The Electrochemical Cell

The mass-transfer probe used in shear-stress measurements is part of an electrochemical cell. A voltage applied to the cell drives a reaction at the probe or test

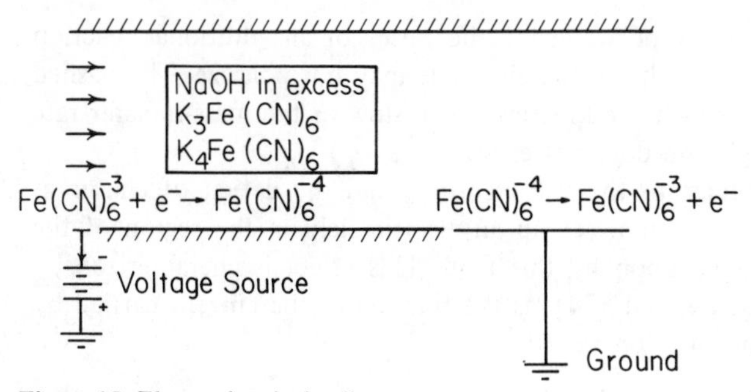

Figure 15 Electrochemical cell.

electrode. The reverse reaction occurs at the counterelectrode. The current produced by the reaction can be related to the molar flux $\langle N \rangle$ at the probe by using Faraday's law,

$$I = n_e F A_e \langle N \rangle \tag{98}$$

where F = Faraday's constant
A_e = probe area
n_e = number of electrons involved in the stoichiometric equation

In Fig. 15 the test electrode is operated cathodically by applying a negative voltage. The ferrocyanide produced at the cathode is oxidized at the anode or counter-electrode to ferricyanide. The current produced by this transfer of electrons flows through the circuit made by the solution and ground in a clockwise direction. The applied voltage is controlled so that the reaction at the test electrode is diffusion controlled, i.e., polarized.

Figure 16 is a plot of the cell current I as a function of applied voltage V_0 for two different systems. Position B, the plateau, represents an operating condition under which the test electrode is polarized. Since an increase in applied voltage results in the same current, the reaction rate is independent of the kinetics. The concentration of the reacting species at the wall is approximately zero because the reaction rate is large. The portion of the solid curve represented by A indicates operating conditions where the wall concentration is not zero. The kinetics cannot keep pace with the mass transfer, and the reaction is said to be *kinetically limited*. Operating conditions in portion C of the curve are typically due to additional

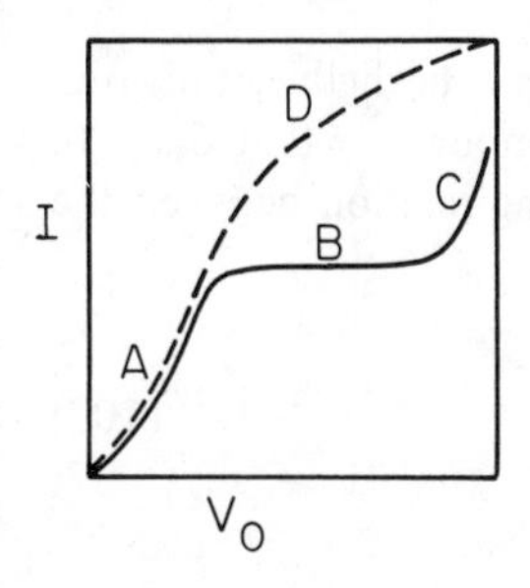

Figure 16 Electrode polarization. A, Kinetically controlled; B, mass-transfer controlled; C, side reactions; and D, above critical K_{limit}. *(From [74].)*

reactions, such as the hydrolysis of water. If the onset of an additional reaction occurs at a low enough applied voltage, the electrode may not polarize. The dashed line D represents a system in which the kinetics are so slow or the mass-transfer rate so great that a polarization plateau does not exist.

The application of a voltage to the test electrode and the deficit of charge at the electrode, caused by the reaction, set up an electric field in the region of the probe. Ions migrate when acted upon by this field. This effect is characterized by the transference number T_R, defined [74] as the fraction of the current carried by an ion in the absence of concentration variations:

$$\langle N \rangle = \frac{I}{A_e n_e F} (1 - T_R) \tag{99}$$

It is necessary to minimize the effect of migration so that the transfer of the reacting species to the test electrode is controlled by molecular diffusion. This is accomplished by adding an excess of neutral electrolyte so as to have a small transference number. In the system shown in Fig. 15, the use of a 2-M solution of sodium hydroxide results in a transference number of about 0.001 [75].

The test electrode is commonly operated near the center of the plateau region. Doubling the concentration of the reacting species doubles the driving force for mass transfer and therefore should double the current characterizing the plateau. An increase in the flow rate increases the mass-transfer coefficient and, therefore, the plateau current. However, if the flow rate is too large, the mass-transfer rate can be too rapid to polarize the electrode. The limiting flow rate for polarization can be increased by decreasing the concentration of the reacting species in the electrolyte.

With no reacting species present, a small current (tenths of a microampere) may appear at voltages below that at which water hydrolyzes. Ranz [76] attributes this to capacity and double-layer effects (see [74]). When operating with low concentrations or low currents, one should determine this residual current and subtract it from the measured current.

In the cell shown in Fig. 15 the voltage is expressed with respect to ground and not with respect to a reference voltage (as is common in electrochemistry). With a large counterelectrode, it may be assumed that the reference voltage is constant. Because of this choice of reference electrode, there could be some variation of the polarization voltage with system design if there is an appreciable IR drop through the solution. Therefore it is necessary to determine the polarization curve to define the operating voltage.

If conditions are maintained so that migration effects are negligible, only the test reaction is occurring, and the rate of reaction is large enough so that $C_w \simeq 0$, then Eqs. (98) and (23) can be combined to give the following relation between the cell current and the velocity gradient:

$$I = 0.807 \frac{D n_e F A_e C_B}{L} Z^{1/3} \tag{100}$$

9.2 The Electrolyte

The selection of an electrolyte is based on the following requirements:

1. It must react electrochemically at voltages where other reactions do not occur.
2. It must have a high reaction-rate constant.
3. It must contain nonreacting ions that eliminate migration effects.
4. It should be easy to use in a flow loop (nontoxic, nonflammable, easy to store, etc.).
5. It must not produce negative effects at the electrodes or on the system (such as probe poisoning or corrosion).

In principle there should be numerous electrolytic reactions that could be used [77]. The reduction of oxygen has been applied by Lin and coworkers [78], Ranz [76], and Reiss [79]. Lin et al. [78] also reduced quinone in a strongly buffered solution. The reduction of cuprous ion is mentioned by Mizushina [80]. Reiss [79] and Lin et al. [78] studied the oxidation of ferrocyanide. The most popular reaction seems to be the reduction of ferricyanide [76, 78, 81 to 84]:

$$Fe(CN)_6^{3-} + e^- \rightarrow Fe(CN)_6^{4-} \tag{101}$$

Concentrations of 0.01 M to 0.1 M are commonly used. The reduction of triiodide, which was first suggested by Py [28], is now used extensively at the University of Illinois [85 to 87]:

$$I_3^- + 2e^- \rightarrow 3I^- \tag{102}$$

Sodium hydroxide (0.5 to 2 M) is commonly used in the ferricyanide system as a neutral electrolyte to make migration effects negligible. It does not interfere with the reaction, and it is a good conductor. Potassium iodide is used in the triiodide system in 0.02 to 0.5 M concentrations [85].

Although the ferricyanide system has been widely used, there are many problems associated with it. Ferricyanide decomposes slowly with light to form hydrogencyanide [75]:

$$Fe(CN)_6^{4-} + 2H_2O \xrightarrow{\text{light}} Fe(CN)_5H_2O^{3-} + OH^- + HCN \tag{103}$$

The cyanide ions poison the electrodes [84]. Ferricyanide is also known to decompose with oxygen and light, but not as rapidly as ferrocyanide [84]. Alkaline solutions may produce not only cyanide but also ferric hydroxide, which could foul the electrodes [88]:

$$2Fe(CN)_6^{3-} + 6OH^- \rightleftharpoons 2Fe(OH)_3 + 12CN^- \tag{104}$$

Jenkins and Gay [84] also mention the possibility of an Fe^{++}/O_2 cell being set up when dissolved oxygen is present. Oxygen may also decompose the ferrocyanide ion to iron oxide, so that surfaces become coated with an oxide film [84]. Most of these problems can be overcome by frequently making up fresh solutions and operating within an opaque system under an atmosphere of nitrogen [80]. This limits the use of the system in air-liquid flow systems, although it has been tried [81] with very short exposure times.

The fouling of the electrode surface necessitates frequent cleaning. This is done normally with mild soap and a soft rag, followed by an alcohol or carbon tetrachloride rinse [86]. Mizushina [80] suggests buffing with soft paper to remove the oxide coat. Jenkins and Gay [84] and Mizushina [80] advocate operating the test electrode as a cathode in a 5% NaOH solution at 10 to 20 mA/cm^2 (cathodic cleaning).

An additional disadvantage of the ferricyanide system is that glycerols and sucrose cannot be used to increase the range of solution viscosities because they react with the ferricyanide ion. This is not the case with the iodine system (see, for example, [87, 89]). The iodine reaction has also been used successfully in the presence of drag-reducing agents [87, 89]. Shaw [85] has added sodium hydroxide to vary the viscosity of the ferricyanide system, but the use of high caustic concentrations increases the dangers associated with the experiment.

Problems common to both the ferricyanide and triiodide systems are the reduction of dissolved oxygen [68], the hydrolysis of water at high applied voltages, the high corrosivity, and the need for enclosure. Enclosure of the triiodide solution is necessary to minimize the loss of iodine by vaporization and to avoid corrosion caused by iodine vapors. Periodic addition of iodine is necessary to make up for vapor losses. The high vapor pressure of iodine also limits its use in gas-liquid flow studies.

Because of the corrosiveness of the electrolytes, special care has to be taken in designing the experiment. Polyvinyl chloride, acrylic resin, and stainless steel are the usual construction materials for the flow loop.

9.3 The Counterelectrode

In both the ferricyanide and triiodide systems, the counterelectrode is the anode. It has a much larger area than the test electrode, to ensure that the anodic reaction is not limiting. It is normally a portion of the flow loop, such as a section of stainless steel pipe, and it is usually located downstream of the test electrode to diminish interference effects. Stainless steel and nickel are suitable for the construction of an anode for the ferricyanide system. Stainless steel is the preferred material in the triiodide system.

9.4 The Test Electrode

Platinum is the material used for the construction of the test electrode for the triiodide system. It has also been used in the ferricyanide system [67, 85, 90], but Jenkins and Gay [84] claim that platinum cathodes produce chemical polarization. The use of nickel cathodes is more popular in the ferricyanide system.

The geometry of the test electrode is determined by the constraints outlined in Sec. 6. Figure 17 depicts a circular test electrode. A hole is drilled through a section of acrylic resin pipe perpendicular to the surface at which the stress is to be

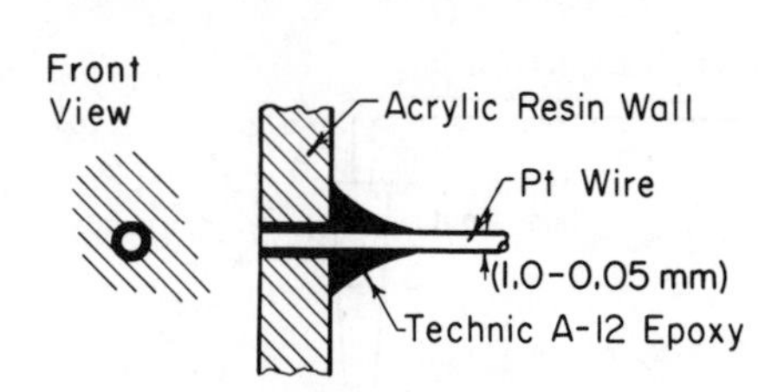

Figure 17 Circular electrode.

measured. The hole is filled with epoxy, and a platinum wire is inserted through the hole so it protrudes slightly on the solution side. After the epoxy has hardened, the probe surface is sanded with progressively finer grades of emery cloth until it is flush with the wall. The probe is then cleaned as discussed in Sec. 9.2. The manufacture of the slant or chevron electrode is described in [67, 90]; the sandwich electrode in [91]; the split-circle electrode in [87]; the split rectangular electrode in [27]; and multiple slant electrodes in [86].

The determination of the area of the test electrode is critical, especially when it is very small. The sanding of small probes tends to smear the area of the electrode exposed to the fluid. Contamination may also alter the effective area. Son [91], in his studies of flow around a cylinder, measured electrode dimensions with a calibrated microscope. Mitchell [68] assumed the electrode had the same area as the edge of the sheet used in its manufacture and reported results for shear stress in pipe flow that varied only a few percent from the accepted values. The electrodes used were rectangular, 0.0075 to 0.052 cm long, and 0.05 to 0.155 cm wide.

The probe area can be determined by calibration against a larger probe of known area. If the flow geometry is one in which the flow field is known (e.g., pipe flow, couette flow) the probe may also be calibrated by using empirical or theoretical correlations for the velocity gradient at the wall.

9.5 The Flow Loop

The chemical complications associated with the use of the electrochemical technique require that the measurements be taken under controlled conditions. At present, the use of electrochemical techniques is limited to experimental laboratory studies, mainly because of the lack of suitable naturally occurring electrolytes for field studies. The fluid properties (D, ν, C_B) must be held constant to ensure precise results. This is accomplished by recirculation and temperature control. The loop shown in Fig. 18 is similar to one used by Reiss [92], Chorn [87], and Mitchell [68]. The system is completely enclosed, and clear pipe is painted black for the ferricyanide studies. Most elements are of plastic or stainless steel to resist corrosion. A filter is included to remove any particles that might foul the electrodes.

9.6 Measurement of Fluid Properties

Precise measurement of fluid properties is essential if Eq. (100) is to be used and a calibration avoided. Equation (100) requires the values of the diffusion coefficient

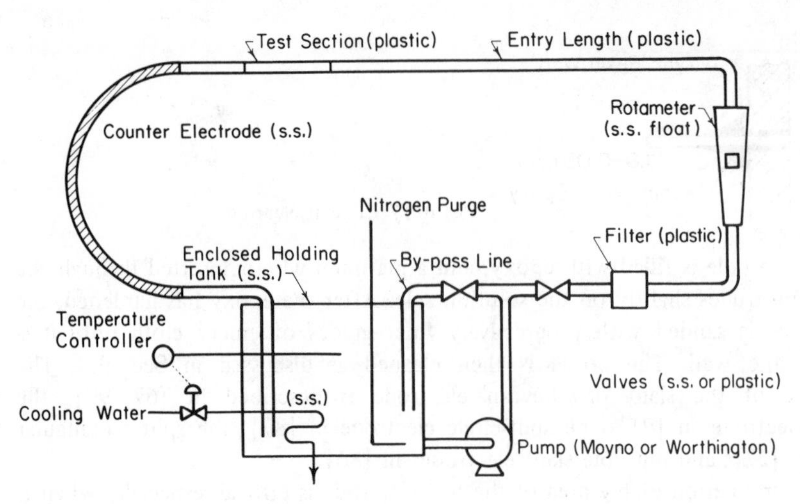

Figure 18 Flow loop of 1-in-diameter pipe.

and the bulk concentration of the reacting species. If the diffusion coefficient is to be determined from correlations in the literature, the fluid viscosity, density, and temperature must be known. An alternative procedure would be to calibrate the probe with the electrolyte in a system where the velocity gradient is known.

The literature contains much information on the diffusion coefficients of the ferricyanide and triiodide ions. Chin [93] determined the diffusivities of the ferricyanide and triiodide ions by studying mass transfer with a rotating disk. Bazan and Arvia [94] also used a rotating disk for ferricyanide, as did Newson and Riddiford [95] for triiodide. Lin et al. [78] used the Nernst equation for ions at infinite dilution,

$$D = \frac{GT\,\lambda_i^0}{z_i F^2} \tag{105}$$

where G = universal gas constant

z_i = ionic charge on species i

along with tables of the equivalent conductance λ_i^0 to calculate the diffusion coefficient for ferricyanide. The capillary method of Anderson and Saddington [96] was used by Eisenberg et al. [82, 97] for ferricyanide. Shaw [98] and Fortuna [90] used a couette-flow apparatus to study triiodide and ferricyanide. Bazan and and Arvia [94] showed that the diffusion coefficient is only very weakly dependent on the concentration of the diffusing entities. Fortuna [90] showed that the addition of drag-reducing agents does not significantly affect the diffusion coefficient.

In most cases the results of these studies can be represented as

$$D \,\alpha\, \frac{T}{\mu} \tag{106}$$

The following correlations are fits of the data and apply only to the range of conditions cited:

$$D_{\text{ferricyanide}} = 2.5 \times 10^{-10} \frac{T}{\mu} \qquad \text{cm}^2 \cdot \text{P/(s} \cdot \text{K)} \tag{107}$$

with

$$\begin{aligned} T &= 10 \text{ to } 40°\text{C} \\ C_{\text{NaOH}} &= 0.5 \text{ to } 2.1\ M \\ C_{\text{Ferri}} &= 0.38 \text{ to } 300 \times 10^{-3}\ M \end{aligned} \tag{108}$$

$$D_{\text{triiodide}} = 7.4 \times 10^{-8} \left(\frac{\rho}{\mu}\right)^{1.054}$$

with μ/ρ and D in centimeters squared per second and

$$\begin{aligned} T &= 25°\text{C} \\ C_{\text{KI}} &= 0.1 \text{ to } 0.5\ M \\ C_{\text{I}_3} &= 0.12 \text{ to } 3.7 \times 10^{-3}\ M \\ C_{\text{sucrose}} &= 0 \text{ to } 1.58\ M \end{aligned}$$

With electrolytes or conditions other than those cited above, the diffusion coefficient should be measured using one of the methods described in the references.

The iodometric titration to a clear end point discussed by Kolthoff and Furman [88] is used to obtain both the ferricyanide and the triiodide concentrations. They react with thiosulfate ion to produce tetrathionate

$$I_3^- + 2S_2O_3^{--} \rightarrow 3I^- + S_4O_6^{--} \tag{109}$$

$$2\text{Fe(CN)}_6^{3-} + 2S_2O_3^{--} \rightarrow 2\text{Fe(CN)}_6^{4-} + S_4O_6^{--} \tag{110}$$

in the pH range 4.5 to 9.5 [99]. Kolthoff and Furman [88] suggested the use of a starch indicator when titrating concentrations of less than 5×10^{-4} M. Titration of triiodide in an Erlenmeyer flask is recommended because of volatilization and air oxidation of iodide,

$$4I^- + O_2 + 4H^+ \rightarrow 2I_2 + 2H_2O \tag{111}$$

which is enhanced by sunlight and acidity. Ferricyanide can also be titrated with isonicotinic acid hydrazide with potentiometric indication [85, 100, 101].

9.7 Instrumentation

The circuit shown in Fig. 19 is used to apply a constant voltage to the electrochemical cell and to convert the current of reaction to a voltage. The Analog Devices 118A operational amplifier sets the voltage at V_1 to a value that is independent of the current. Adjustment of the helipot (variable resistor) selects the voltage V_2. The output of the amplifier is

$$V_A = A_A(V_1 - V_2) \tag{112}$$

where $A_A = 10^4$ to 10^8 [102], and

$$V_A = V_1 + IR_f \tag{113}$$

Combining these two equations yields

$$V_A \frac{1 - A_A}{A_A} = -IR_f - V_2 \tag{114}$$

Because A_A is so large,

$$V_A = IR_f + V_2 \tag{115}$$

The output of the amplifier equals the voltage set at V_2 plus the voltage drop through the feedback resistor R_f. By combining Eqs. (113) and (115), it is seen that V_1 equals V_2 and is independent of the current. The value of V_2 and V_1 can be measured at V_A when no current is flowing. A subtraction circuit given by Hanratty [103] and Eckelman [104] can be used to remove the portion of the output voltage attributable to the applied voltage. This subtraction may also be done after the output signal has been time-averaged.

Generally an applied voltage of between −0.3 and −0.5 V is needed to polarize an electrode in the ferricyanide or triiodine systems. This voltage is not sensitive to changes in concentration and probe area. The magnitude of the feedback resistor determines the resolution of the current in the output signal. Since cell currents vary from 100 mA to less than a microampere, resistances from 10 Ω to 2 million Ω are used. The 500-Ω resistor shown in Fig. 19 protects the operational amplifier by limiting the current entering it. The 10-pF capacitor corrects for the nonlinear frequency response of the amplifier [105].

10. EXPERIMENTAL PROCEDURES FOR HEATED-ELEMENT PROBES

10.1 Principles of Operation

The heat-transfer analog of the mass-transfer probe is operated on principles similar to those for a hot wire (see [106] and Chap. 4 of this book). The resistance of the heated element is temperature-sensitive. This allows fixing of the probe temperature by fixing the probe resistance (analogous to fixing the wall concentration at zero through polarization). The power needed to keep the probe at that temperature, as the fluid cools it, is related to the shear stress by Eq. (85).

10.2 Instrumentation

The simple feedback-controlled wheatstone bridge shown in Fig. 20 can be used to operate the probe. When the bridge is balanced,

$$\frac{R_{\text{var}}}{R_1} = \frac{R_c + R_{ps} + R}{R_{\text{int}}}$$

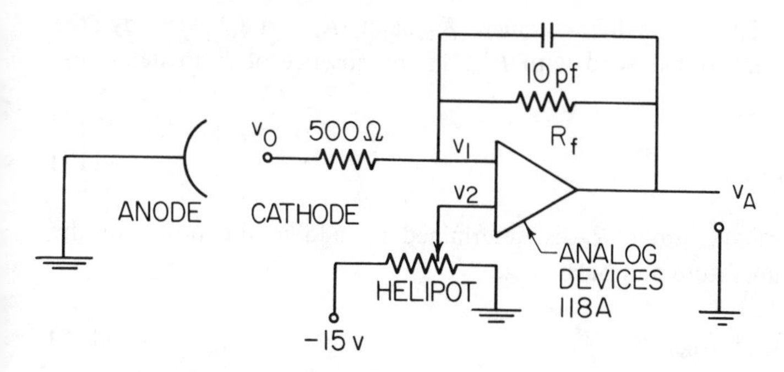

Figure 19 Basic instrumentation for electrochemical probe.

and the output V_f of the feedback amplifier is zero. When the probe is cooled, its resistance drops, causing a bridge imbalance. This is sensed by the amplifier, which signals the current source to increase its output to reheat the probe. This return to equilibrium occurs almost instantaneously, so an increase in shear stress is seen as an increase in current or bridge voltage. This type of instrumentation is sold commercially as a constant-temperature (constant-resistance) anemometer.

10.3 Calibration

Equation (87) has been found to represent calibration data for heated-film probes. The influence of fluid density ρ is usually omitted in noncompressible flows. The constants A and B are weak functions of ΔT over a small range of ΔT values. Bellhouse and Schultz [13] assumed A and B varied linearly with ΔT over large ranges to represent their measurements with

$$\frac{I^2 R}{\Delta T} = C + D\,\Delta T + (E + F\,\Delta T)\,(\rho\tau)_w^{1/3} \tag{116}$$

where

$$B = C + D\,\Delta T \qquad \text{and} \qquad A = E + F\,\Delta T$$

Resistances can be determined by balancing the anemometer bridge with the

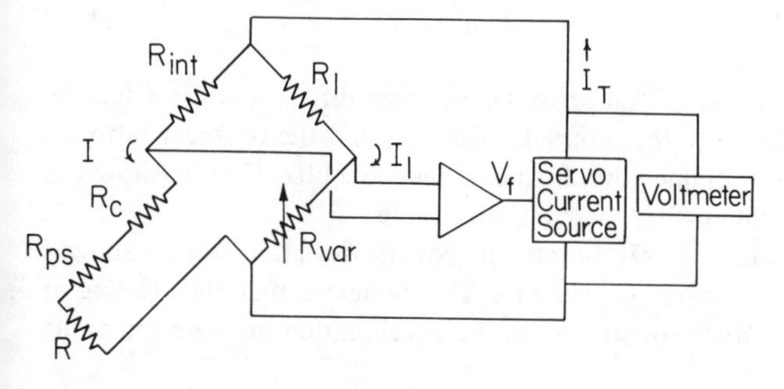

Figure 20 Simplified circuitry for heat-transfer probe.

heating current off. The internal resistance R_1 and R_{int} are given by the manufacturer. The bridge is balanced with R_{var} in the absence of R to determine $R_c + R_{ps}$:

$$R_c + R_{ps} = \frac{R_{int}}{R_1} R_{var} \tag{117}$$

The cold resistance of the probe R_0 is determined by adding the probe to the bridge at the fluid temperature and balancing:

$$R_0 = \frac{R_{int}}{R_1} R_{var} - (R_c + R_{ps}) \tag{118}$$

Its value is commonly 5 to 20 Ω [107].

In the operational mode a heating current is passed through the probe, and R_{var} is selected so that the probe resistance needed to balance the bridge is larger than R_0 by a factor of H. The selection of the overheat ratio $H = R/R_0$ sets the temperature difference and determines the sensitivity of the probe. Common overheat ratios for hot films are 1.5 in airflow and 1.1 in water flow, which correspond approximately to temperature differences of 250 and 50°C, respectively.

The temperature difference may be determined if the temperature-resistance relationship of the probe is known. Over a small temperature range the relation

$$\Delta T = \frac{H - 1}{\alpha_R} \tag{119}$$

is valid, where α_R is the temperature coefficient of resistance. The coefficient of resistance is not only a function of the material of the heated element but of impurities, defects in the lattice, oxidation of the surface, and stresses on the element [106, 108]. The coefficient should be determined for each probe, or it could be included in the constants A and B:

$$\frac{I^2 R}{H - 1} = \frac{B}{\alpha_R} + \frac{A}{\alpha_R} (\rho\tau)_w^{1/3} \tag{120}$$

The output of the anemometer is more commonly measured as a voltage $V_T = I(R + R_{int} + R_c + R_{ps}) = IR_T$. The calibration equation may now be written as

$$\frac{V_T^2 R_0 H}{R_T^2 (H - 1)} = B + A(\rho\tau)_w^{1/3} \tag{121}$$

This calibration is valid over a small range of temperature differences for a specific probe and can be used when only the internal resistances and the overheat ratio are known. A calibration curve obtained by Bellhouse and Schultz [13] is shown in Fig. 21 for a painted-on platinum film of small dimensions.

Blackwelder and Eckelmann [109] point out that the deviation from a straight line at low flows is due to natural convection. This indicates that the reading at zero flow is not an appropriate point to use in a calibration to determine the constants A and B.

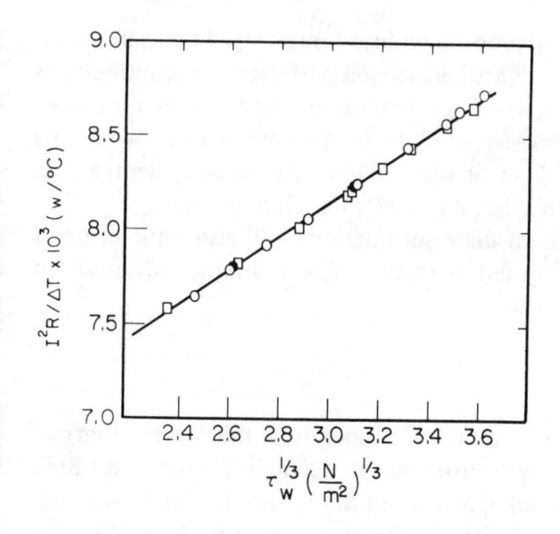

Figure 21 Calibration curve of flush-mounted thin film in a turbulent-flow annular tunnel. *(From [13].)*

10.4 Insertion of the Probe in the Wall

The positioning of the probe relative to the wall is one of the greatest sources of error in using heated films. Pessoni [110] found that the displacement of a commercial probe of the type shown in Fig. 22 from the test-section surface by 0.1 mm in a 2.5-cm air tunnel resulted in 30 to 40% deviations in the calibrations. Pessoni suggests either calibrating the probe *in situ* or calibrating it in a plug of larger area and transferring the entire plug to the test section. He used a collimated light source directed on the probe at a very shallow angle to position it flush with the surface. Simons et al. [111] also decided to mount their probes in a plug when they dis-

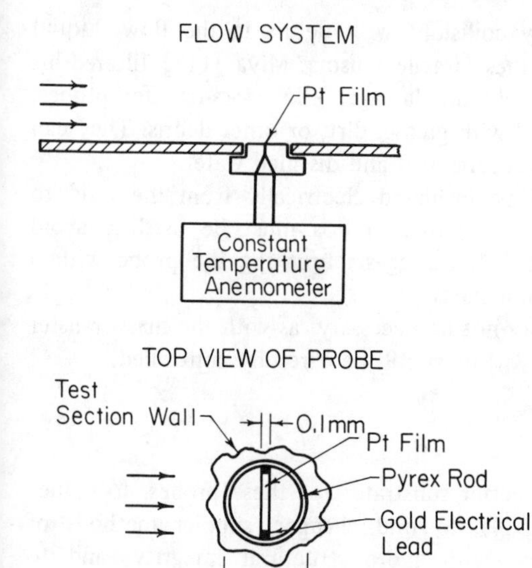

Figure 22 Hot-film probe.

covered that even the slight jarring of the probe upon insertion changed the calibration.

When commercial hot films are mounted in curved surfaces, a discontinuity is unavoidable. This problem may be overcome by the use of flexible glue on probes [112]. Wylie and Alonso [113] suggest that these contour effects would be negligible in pipes with diameters of 2 in or more. They cite an agreement within 2% between Foster's [114] results with a 2-in pipe and a flat plate.

The thermal conductivity of the immediate surroundings will also cause errors if it changes from the calibration site to the test section. This is another advantage of mounting the sensor in a plug.

10.5 Basic Probe Design

The element in most commercial flush-mounted probes is a thin layer (film) of platinum that has been deposited by sputtering or vacuum evaporation. Both techniques require very sophisticated equipment. Sputtering involves the bombardment of a source of platinum with positive ions under a moderate vacuum. Platinum atoms are sputtered (knocked off) the source and deposited on the substrate. Vacuum evaporation utilizes strong heat and vacuum to vaporize the platinum. The platinum vapor is then deposited on the substrate. Substrates are made of a good thermal insulator, usually quartz. Two high-conductivity leads run from the film and through the substrate to the anemometer. The element is protected by a quartz film of about 0.5 μm thickness [115]. Thicker films (2 μm) are used for liquid flows to electrically insulate the heated element from the fluid. This has the disadvantage, however, of decreasing the probe frequency response [116].

10.6 Difficulties Unique to Heated-Element Probes

Hot-film probes are easily damaged by collisions with debris in the flow. Liquid flows, especially, put large stresses on these fragile sensors. Miya [117] filtered his solutions to eliminate particles that could do damage. Like mass-transfer probes, heat-transfer probes may become fouled with grease, dirt, or other debris. They can be cleaned with successive rinses of 10% acetic acid and distilled water.

In liquid flows the probe should be insulated electrically from the fluid to prevent hydrolysis at the element. Low overheat ratios must be used to avoid boiling or degassing of the fluid. TSI [118] suggests brushing the probe with a camel's hair brush when bubble formation starts.

Although recycling of the fluid may not be necessary, as with the mass-transfer probe the fluid properties (T, ρ, μ, C_p, k_p) must still be carefully controlled.

10.7 New Developments

Recently work has been done to find better substrates for these probes, to reduce heat loss to the substrate (reduce L_{eff}), to find cheaper, simpler methods of construction, to develop new designs with more structural integrity, and to manufacture probes whose calibrations are reproducible.

Rubesin et al. [15] have investigated the use of a styrene copolymer substrate with a thermal conductivity one-tenth that of pyrex. The melting point, however, was too low for baking on the platinum paint used by Bellhouse and Schultz [13]. Sputtering was also unsuccessful for film sizes of less than 0.2 mm. A probe with a platinum-rhodium wire laid in a groove in the styrene (after [11]) was tried, but the waviness it produced on the surface was too great. In the final design a 25-μm platinum wire was laid on the surface and buried in a thin layer of epoxy, which was hand wiped to expose the wire. This type of probe is easier to fabricate, more reproducible (within 2%), and less likely to be destroyed. In addition, it loses less heat to the substrate than the heated-film probe because of the lower conductivity of its substrate.

A device that has recently become quite popular [112, 119, 120] is the "glue-on" probe (Fig. 23) first proposed by McCroskey and Durbin [14]. These are thin polyimide foils onto which a metal film has been sputtered or electrodeposited. The probes are inexpensive and are able to take the shape of curved surfaces. McCroskey and Durbin also claim that printed-circuit sensors are reproducible and suited to the manufacture of multielement devices. Possible disadvantages are the permanence of the probe after it is glued in place and the use of wire leads, which may cause flow disturbances. These were overcome by Coney and Simmers [112] by gluing the probe to a plastic plug and leading the wires down the side of the plug to the outside of the test section.

11. APPLICATION OF MASS-TRANSFER PROBES

11.1 Turbulent Flow in a Pipe

A system with a 1-in-diameter pipe, similar to the one shown in Fig. 18, was used by Reiss [92] to measure wall shear stress at Reynolds numbers from 3000 to 35,000. A 14-gauge (0.164-cm) nickel wire was used as the test electrode. The area of the electrode used in the calculations was based on the average wire diameter, measured with a vernier scale on a microscope. The effective length ($0.82d_e$) was 0.133 cm. The reacting ion was 0.1-*M* ferricyanide. The diffusion coefficient was determined from the equation given in [97], which is identical to Eq. (107). The remainder of the solution properties, given in Table 2, were determined using standard laboratory procedures. The polarization curves (Fig. 24) show that the electrode is polarized by an applied voltage of −0.3 to −0.5 V over a Reynolds-number range of about 4000 to 30,000. An operating voltage of −0.4 V was selected by Reiss for all flow rates measured. A feedback resistor of only 100 Ω

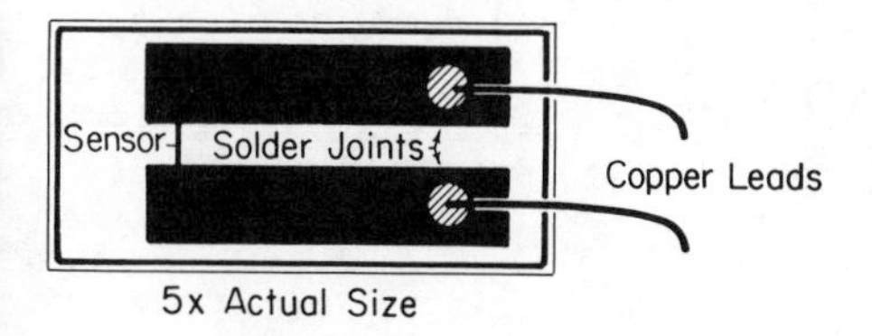

Figure 23 Glue-on probe as marketed by DISA Electronics.

Table 2 **Summary of the experiment by Reiss [92]**

Electrolyte properties		Flow conditions		Results	
$C_{\text{ferricyanide}}$	0.104 M	T	25°C	$\bar{I}$	1.96×10^{-3} A
$C_{\text{ferrocyanide}}$	0.0116 M	d_e	0.1636 cm	$\bar{K}$	9.31×10^{-3} cm/s
C_{NaOH}	2.168 M	L_{eff}	0.133 cm	$\bar{K}L/D$	239
ρ	1.094 g/cm³	A_e	0.02102 cm²	Z	2.61×10^7
μ	0.0144 g/(cm·s)	Re	34,200	$\bar{S}$	7629 s^{-1}
ν	0.01316 cm²/s	$\bar{U}_B$	176.9 cm/s	u^*	10.02 cm/s
D	5.17×10^{-6} cm²/s	V_0	−0.4 V	$(\bar{k^2})^{1/2}/\bar{K}$	0.0363
Sc	2550	R_f	100 Ω	L^+	98
		$\bar{S}_{x\,\text{Blasius}}$	6950 s^{-1}	64 Sc	160,000
				14 $Sc^{-1/2}$	0.28
				$44\,(L/W)^{3/2}\,Sc^{-1/2}$	0.87
				W^+	120
				$L^+(\omega^+)^{3/2}\,Sc^{1/2}$	66

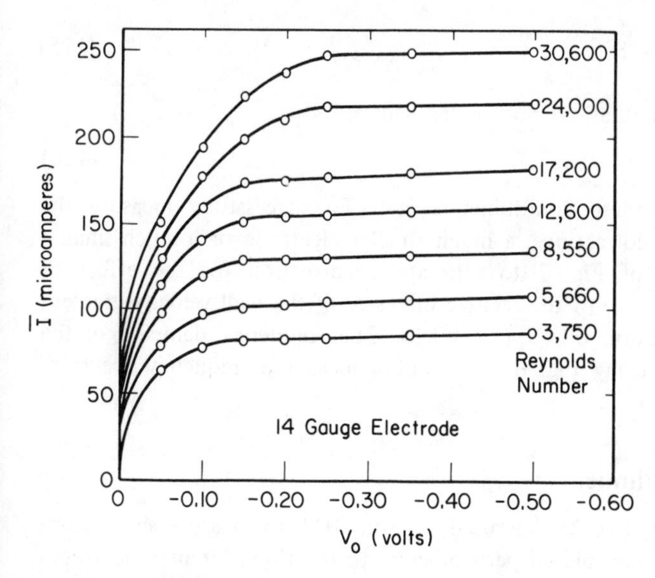

Figure 24 Polarization curves for a 14-gauge nickel electrode in a 1-in pipe flow system. *(From [92].)*

was used because of the relatively large current resulting from a high flow rate, high ferricyanide concentration, and large electrode. The flow conditions and results of a run at a Reynolds number of 34,200 are also given in Table 2.

The effective length of the electrode (made dimensionless with wall parameters) is 98 at this Reynolds number. The constraints on electrode length discussed in Sec. 6 can be represented by

$$14\ \mathrm{Sc}^{-1/2} < L^+ < 64\ \mathrm{Sc} \tag{122}$$

provided turbulent transport is negligible:

$$L^+ < 700 \tag{123}$$

and spanwise diffusion is negligible:

$$L^+ > 44\left(\frac{L}{W}\right)^{3/2}\ \mathrm{Sc}^{-1/2} \tag{124}$$

The value of the dimensionless length is within these constraints (see Table 2). The value of $\bar{S}_X$ calculated from the data using Eq. (70) is within 10% of the value calculated using the Blasius equation.

The calculated value of $\sqrt{\overline{s_x^2}}/\bar{S}_x = 3\sqrt{\overline{k^2}}/\bar{K} = 0.1089$ differs greatly from the accepted value of 0.30 to 0.36. An averaging of the fluctuations should be expected because the size of the probe is on the order of the scale of the fluctuating flow. The constraints on probe size for the measurement of the shear-stress fluctuations are that the flow be uniform over the probe surface,

$$\frac{W^+}{\Lambda_z^+} < 0.67 \qquad \text{or} \qquad W^+ < 8 \tag{125}$$

(see Fig. 7) and that the fluctuations not be damped,

$$L^+(\omega^+)^{3/2}\ \mathrm{Sc}^{1/2} \leqslant 1 \tag{126}$$

(see Fig. 6). Neither of these conditions is met. To successfully measure the fluctuations without using corrections, a much smaller electrode or a much smaller flow is needed. The use of Fig. 7 to estimate a correction for the effect of nonuniform flow yields a value of the relative intensity of the wall velocity-gradient fluctuations in the x direction of $\sqrt{\overline{s_x^2}}/\overline{S}_x = 0.23$. The frequency damping of the signal could be estimated using Fig. 6 if the entire measured frequency spectrum were available.

11.2 Flow around a Cylinder

The water tunnel shown in Fig. 25 was used by Son [91] to measure shear stress around a 1-in-diameter cylinder placed perpendicular to the flow. Circular electrodes of diameter 0.038 to 0.051 cm were used to measure average shear stresses. A sandwich electrode made of two strips of platinum sheet, 2.54 × 0.0127 cm, separated by a thickness of cellophane tape (~0.005 cm), was used to measure the direction of flow and the location of flow separation. The cylinder was rotated to vary the angular location of the test electrode. The cylinder could be removed from the test section, and the proper areas accurately measured with a microscope. Dye was also injected through holes in the cylinder walls to qualitatively measure flow direction. The measurements were all taken with respect to the front stagnation point, which was located at the position of minimum mass transfer.

Values of $|S|$, calculated from the measurements using Eq. (23), are presented in Fig. 26 for a Reynolds number of 20,000 and a circular probe of diameter 0.051 cm. In region A the wall shear stress had a periodic oscillation and a maximum at an angle of about 50°. Position S indicates the separation point detected by the

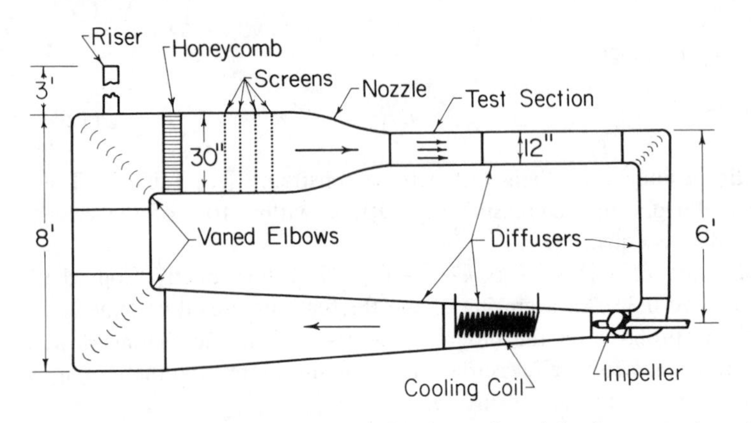

Figure 25 Water tunnel used by Son [91] in studies of the flow around a cylinder.

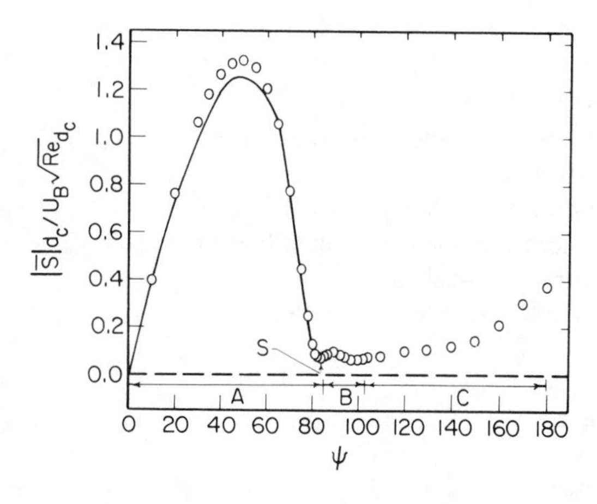

Figure 26 Wall velocity-gradient distribution around a 1-in cylinder at Re = 20,000. The solid line is the boundary-layer solution at Re = 19,000. *(From [91].)*

sandwich electrodes. Dye studies showed that a reversed flow existed close to the wall in region B. The electrode measurements indicated a unsteady periodic flow characterized by small, negative, wall shear stresses. Both the dye studies and the electrode measurements showed a highly chaotic flow in region *C*. Flow reversals are common in region *C* so the flow fluctuations may be greater than the time-averaged flow. The design equation is not applicable in this region, so the calculated values of $|S|$ have no meaning.

As discussed in Sec. 6.3, the calculations of *S* by Son and Hanratty might be in error close to the front stagnation point or the separation point, where S_x is close to zero and the wall shear stress is varying rapidly with changes in angular location. Equation (37) indicates that the use of Eq. (34) to calculate *S* introduces an error of only about −7.5% at a distance of 5° from these points.

Son and Hanratty found that measurements with the circular electrodes and with a rectangular electrode 0.038 cm long and 2.54 cm wide gave the same results in region *A*. Further support for the accuracy of the measurements is given in Fig. 26, where the solid line represents a boundary-layer calculation carried out using the Blasius series [121] and the pressure measurements of Hiemenz [122] at Re = 19,000. The agreement between the measurements and theory is quite good. In fact Son and Hanratty argued that the slight disagreement near the maximum is more likely due to errors in determining the pressure gradient rather than to errors in measuring the wall shear stress.

NOMENCLATURE

a	velocity of sound
A	parameter used to characterize heat-transfer probe in Eq. (86)
$A^\dagger$	parameter defined in Eq. (89)
$A_t^\dagger$	parameter $A^\dagger$ evaluated for a turbulent flow in Eq. (96)
A_t	calibration constant for a thermal probe in turbulent flow in Eq. (88)
A_A	amplification factor of operational amplifier

A_d	damping factor
A_e	area of electrode
b	breadth of Stanton blade
B	parameter used to characterize heat-transfer probe in Eq. (87)
$B^\dagger$	parameter defined by Eq. (91)
$B_t^\dagger$	parameter $B^\dagger$ evaluated for turbulent flow in Eq. (96)
B_t	calibration constant for a thermal probe in turbulent flow in Eq. (88)
c	fluctuating part of concentration equal, $C - \bar{C}$
$\hat{c}$	complex amplitude of a simple harmonic fluctuation in c
$\hat{c}_R$	real part of $\hat{c}$
$\hat{c}_I$	imaginary part of $\hat{c}$
C	concentration of diffusing species
C_p	heat capacity of fluid
d	pipe diameter, hole diameter
d_t	outside diameter of Pitot tube or Preston tube
d_e	electrode diameter
d_c	diameter of cylinder
D	diffusion coefficient
e	fluctuating part of E
E	voltage drop through heat-transfer element
f_c	correction to Preston-tube calibration to account for compressibility effects
f_i	calibration function for a Preston tube for incompressible flow
F	Faraday's constant
g	gap size
g_1	see Eq. (74)
g_2	Eq. (75)
g_0	constant in Eq. (83)
G	universal gas constant, 1.9872 cal/(K·mol)
h	heat-transfer coefficient, height of Stanton blade or of sublayer fence
h_c	channel height
H	overheat ratio, R/R_0
I	electric current produced by reaction in electrochemical probe, current used to heat the heat-transfer probe
I_1	electric current in balancing side of anemometer bridge
I_T	total current in bridge circuit, $I + I_1$
k	fluctuating part of mass-transfer coefficient
k_R	reaction-rate constant
k_s	constant defined in Eq. (12)
k_T	thermal conductivity of the fluid, thermal conductivity of the substrate
$\overline{k_m^2}$	measured mean square value of the mass-transfer fluctuations in a turbulent flow for a probe of finite size
K	mass-transfer coefficient, $\langle N \rangle/(C_B - C_w)$

K_t ratio of effective diameter of Preston tube to outside diameter
l length of Stanton blade
L length of mass-transfer or heat-transfer element
L_e equivalent length of a circular electrode surface, $0.81356d_e$
L_{eff} effective length of heated element
Ma Mach number
n exponent in Eq. (88)
n_e number of electrons transferred in stoichiometric reaction
N local rate of mass transfer per unit area
N_i local rate of mass transfer per unit area on ith strip
dP/dx pressure gradient
ΔP difference of impact pressure and static pressure for a Preston tube or Stanton blade
Pr Prandtl number
q rate of heat transfer per unit area
R resistance of heat-transfer probe at its operating temperature
R_0 resistance of heat-transfer probe at the bulk fluid temperature
R_1 internal resistance in anemometer bridge
R_{int} internal resistance in anemometer bridge
R_{var} variable resistance in anemometer bridge
R_c resistance of cable to heat-transfer probe
R_{ps} resistance of probe support for heat-transfer probe
R_T $R + R_{\text{int}} + R_c + R_{ps}$
R_f feedback resistance
R_z circumferential spatial correlation of the mass-transfer fluctuations
R_e radius of circular electrode
Re Reynolds number based on pipe diameter or cylinder diameter
s_x fluctuating part of wall velocity gradient in x direction
s_z fluctuating part of wall velocity gradient in z direction
$\hat{s}_R$ amplitude of a simple harmonic fluctuation in s_x assumed to be real
S magnitude of velocity gradient at wall
S_x velocity gradient at wall in x direction
S_0 velocity gradient at wall evaluated at center of wall element
Sc Schmidt number
t time
T temperature of fluid, temperature of a heated element
T_R transference number
ΔT temperature difference, $T_w - T_B$
u^* friction velocity $(\bar{\tau}_w/\rho)^{1/2}$
U streamwise velocity of fluid
V normal component of fluid velocity
V_0 voltage applied to electrode in electrochemical cell
V_1, V_2 voltage inputs to operational amplifier in instrumentation of mass-transfer probe

V_A	output of operational amplifier in instrumentation of mass-transfer probe
V_f	output of amplifier on anemometer bridge
V_T	voltage drop across anemometer bridge
W	width of mass-transfer or heat-transfer element
W_k	spectral-density function of k
W_{sk}	spectral-density function of s_x
x	distance in streamwise direction
x^*	see Eq. (14)
y	distance in the direction normal to wall
y^*	see Eq. (15)
z	distance in transverse direction
z_i	charge number of ionic species i
Z	see Eq. (24)
α	angular measurement
α_R	temperature coefficient of resistance
α_T	thermal diffusivity
β	see Eq. (57)
γ	change of velocity gradient in flow direction
δ	thickness of velocity boundary layer
δ_c	thickness of concentration boundary layer or thermal boundary layer
δ_v	thickness of region where $U = Sy$
Δ	pressure-gradient parameter
η	dimensionless similarity variable defined in Eq. (46)
θ	angle between instantaneous direction of flow at wall and x axis
λ_i^0	equivalent conductance of species i at infinite dilution
Λ_z	circumferential scale of mass-transfer fluctuations
μ	fluid viscosity
ν	fluid kinematic viscosity
ρ	fluid density
τ_1	dimensionless parameter for slant surfaces, $L/5W$
τ_w	wall shear stress
τ_w'	fluctuation in wall shear stress, $\tau_w - \bar{\tau}_w$
ϕ	angle that a slant electrode makes with x axis
ψ	angular distance around a cylinder measured from stagnation point; $\tan^{-1}(L/W)$ for mass-transfer probes
ω	frequency in radians per unit time

Superscripts

+	term made dimensionless using wall parameters u^* and ν
—	time average

Subscripts

w	fluid or flow property evaluated at conditions that prevail at wall

B fluid or flow property evaluated at bulk conditions
s at a stagnation or separation point

Other symbols

$\langle\ \rangle$ spatial average

REFERENCES

1. T. E. Stanton, D. Marshall, and C. W. Bryant, On the Condition at the Boundary of a Fluid in Turbulent Motion, *Proc. R. Soc. London Ser. A*, vol. 97, pp. 413–434, 1920.
2. G. Kempf, Neue Ergebnisse der Widerstandforschung, *Werft Reederei Hafen*, vol. 11, pp. 234–239, 1929.
3. H. Ludwieg, Ein Gerat zur messung der Wandschubspannung turbulenter Reibungschichten, *Ing. Arch.*, vol. 17, pp. 207–218, 1949 (Instrument for Measuring the Wall Shearing Stress of Turbulent Boundary Layers, NACA 7M 1284, 1950).
4. J. H. Preston, The Determination of Turbulent Skin Friction by Means of Pitot Tubes, *J. R. Aero. Soc.*, vol. 58, pp. 109–121, 1953.
5. N. I. Konstantinov, Comparative Investigation of the Friction Stress on the Surface of a Body, *Energomashinostroenie,* vol. 176, pp. 201–213, 1955 (translated 1961 DS1R RTS 1500).
6. N. I. Konstantinov and G. L. Dragnysh, The Measurement of Friction Stress on a Surface, *Energomashinostroenie,* vol. 176, pp. 191–200, 1955 (translated 1960, DSIR RTS 1499).
7. L. P. Reiss and T. J. Hanratty, Measurement of Instantaneous Rates of Mass Transfer to a Small Sink on a Wall, *AIChE J.*, vol. 8, pp. 245–247, 1962.
8. L. P. Reiss and T. J. Hanratty, An Experimental Study of the Unsteady Nature of the Viscous Sublayer, *AIChE J.*, vol. 9, pp. 154–160, 1963.
9. A. Fage and V. M. Falkner, On the Relation between Heat Transfer and Surface Friction for Laminar Flow, *Aero. Res. Counc. London,* R&M No. 1408, 1931.
10. H. Ludwieg and W. Tillmann, Investigation of the Wall Shearing Stress in Turbulent Boundary Layers, NACA TM 1285, 1950.
11. H. W. Liepmann and G. T. Skinner, Shearing-Stress Measurements by Use of a Heated Element, NACA TN 3269, 1954.
12. B. J. Bellhouse and D. L. Schultz, The Measurement of Skin Friction in Supersonic Flow by Means of Heated Thin Film Gauges, *Aero. Res. Counc. London,* R&M, p. 940, 1965.
13. B. J. Bellhouse and D. L. Schultz, Determination of Mean and Dynamic Skin Friction, Separation and Transition in Low-Speed Flow with a Thin-Film Heated Element, *J. Fluid Mech.*, vol. 24, pp. 379–400, 1966.
14. W. J. McCroskey and E. J. Durbin, Flow Angle and Shear Stress Measurements Using Heated Films and Wires, *J. Basic Eng.*, vol. 94, pp. 46–52, 1972.
15. M. W. Rubesin, A. F. Okuno, and G. G. Mateer, and A. Brosh, A Hot-Wire Surface Gage for Skin Friction and Separation Detection Measurements, NASA TM X-62, p. 465, 1975.
16. V. A. Sandborn, Surface Shear Stress Fluctuations in Turbulent Boundary Layers, *Second Symp. on Turbulent Shear Flows,* London, 1979.
17. V. A. Sandborn, Evaluation of the Time Dependent Surface Shear Stress in Turbulent Flows, ASME Preprint 79-WA/FE-17, 1979.
18. J. E. Mitchell and T. J. Hanratty, A Study of Turbulence at a Wall Using an Electrochemical Wall-Stress Meter, *J. Fluid Mech.*, vol. 26, pp. 199–221, 1966.
19. G. Fortuna and T. J. Hanratty, Frequency Response of the Boundary Layer on Wall Transfer Probes, *Int. J. Heat Mass Transfer,* vol. 14, pp. 1499–1507, 1971.
20. H. G. Dimopoulos and T. J. Hanratty, Velocity Gradients at the Wall for Flow around a

Cylinder for Reynolds Numbers between 60 and 360, *J. Fluid Mech.*, vol. 33, pp. 303–319, 1968.
21. J. S. Son and T. J. Hanratty, Velocity Gradients at the Wall for Flow around a Cylinder at Reynolds Numbers from 5×10^3 to 10^5, *J. Fluid Mech.*, vol. 35, pp. 353–368, 1969.
22. C. Tournier and B. Py, The Behaviour of Naturally Oscillating Three-Dimensional Flow around a Cylinder, *J. Fluid Mech.*, vol. 85, p. 161, 1978.
23. M. LeBouche and M. Martin, Convection forcee autour du cylindre; sensibilitie aux pulsations de l'ecoulement externe. *Int. J. Heat Mass Transfer*, vol. 18, pp. 1161–1175, 1975.
24. K. K. Sirkar and T. J. Hanratty, Limiting Behavior of the Transverse Turbulent Velocity Fluctuations Close to a Wall, *Ind. Eng. Chem. Fund.*, vol. 8, pp. 189–192, 1969.
25. K. K. Sirkar and T. J. Hanratty, The Limiting Behavior of the Turbulent Transverse Velocity Component Close to a Wall, *J. Fluid Mech.*, vol. 44, pp. 605–614, 1970.
26. B. Py and J. Gosse, Sur la Realisation d'une sonde en paroi sensible a la vitesse et a la direction de l'ecoulement, *C. R. Acad. Sci.*, vol. 269A, pp. 401–403, 1969.
27. B. Py, Etude tridimensionnelle de la sous-couche visqueuse dans une veine rectangulaire par des mesures de transfert de matiere en paroi, *Int. J. Heat Mass Transfer*, vol. 15, pp. 129–144, 1972.
28. B. Py, Sur l'Interet de la reduction de l'iode dans l'etude polarographique des ecoulements, *C. R. Acad. Sci.*, vol. 270A, pp. 202–205, 1970.
29. S. Dhawan, Direct Measurements of Skin Friction, NACA TN 2567, 1953.
30. K. G. Winter, An Outline of the Techniques Available for the Measurement of Skin Friction, *Prog. Aerosp. Sci.*, vol. 18, pp. 1–57, 1977.
31. D. G. Mabey and L. Gaudet, Some Performance of Small Skin Friction Balances at Supersonic Speeds, *J. Aircr.*, vol. 12, pp. 819–825, 1975.
32. K. G. Winter and L. Gaudet, Turbulent Boundary-Layer Studies at High Reynolds Numbers between 0.2 and 2.9. *Aero. Res. Counc. London*, R&M 3712, 1973.
33. G. Kempf, Weitere Reibungsergebnisse an ebenen glatten und rauhen Flachen, *Hydromech. Probl. Schiffsantreibs*, vol. 1, pp. 74–82, 1932.
34. F. Schultz-Grunow, Neues Reibungswiderstandsgestz fur glatte Platten, *Luftfahrforschung*, vol. 17, pp. 239–246, 1940 (New Friction Law for Smooth Plates, NASA TM 986, 1941).
35. D. W. Smith and J. H. Walker, Skin Friction Measurements in Incompressible Flow, NASA TR R26, 1959.
36. D. E. Coles, Measurement of Turbulent Friction on a Smooth Flat Plate in Supersonic Flow, *J. Aeronaut. Sci.*, vol. 21, pp. 433–448, 1954.
37. D. J. Garringer and E. J. Saltzman, Flight Demonstrations of a Skin-Friction Gage to a Local Mach Number of 4.9, NASA TN D-3820, 1967.
38. V. C. Patel, Calibration of the Preston Tube and Limitation on Its Use in Pressure Gradients, *J. Fluid Mech.*, vol. 23, pp. 185–208, 1965.
39. I. Rechenberg, Messung der turbulenten Wandschubspannung, *Z. Flugwiss.*, vol. 11, pp. 429–438, 1963.
40. A. D. Young and J. N. Mass, The Behavior of a Pitot Tube in a Transverse Total-Pressure Gradient, *Aero. Res. Counc. London*, R&M No. 1770, 1936.
41. K. C. Brown and P. N. Joubert, Measurement of Friction in Turbulent Boundary Layers, *J. Fluid Mech.*, vol. 35, pp. 737–757, 1967.
42. M. R. Head and I. Rechenberg, The Preston Tube as a Means of Measuring Skin Friction, *J. Fluid Mech.*, vol. 14, pp. 1–17, 1962.
43. M. R. Head and V. V. Ram, Simplified Presentation of Preston Tube Calibration, *Aeronaut. Q.*, vol. 22, pp. 295–300, 1971.
44. P. Bradshaw and D. Unsworth, A Note on Preston Tube Calibrations in Compressible Flow, *AIAA J.*, vol. 12, no. 9, pp. 1293–1294, 1974.
45. J. H. Hool, Measurement of Skin Friction Using Surface Tubes, *Aircr. Eng.*, vol. 28, p. 52, 1956.
46. B. A. Bradshaw and M. A. Gregory, The Determination of Local Turbulent Skin Friction from Observation in the Viscous Sub-layer. *Aero. Res. Coun. London*, R&M, p. 3202, 1961.

47. G. L. Brown, Theory and Application of Heated Films for Skin Friction Measurement, *Proc. of 1967 Heat Transfer and Fluid Mech. Inst.*, Stanford University Press, pp. 361-381, 1967.
48. L. F. East, Measurement of Skin Friction at Low Subsonic Speeds by the Razor-Blade Technique, *Aero. Res. Counc. London,* R&M 3525, 1966.
49. G. I. Taylor, Measurement with a Half-Pitot Tube, *Proc. R. Soc. London Ser. A,* vol. 166, p. 476, 1938.
50. G. E. Gadd, W. E. Cope, and J. L. Attridge, Heat Transfer and Skin-Friction Measurements at a Mach Number of 2.44 for a Turbulent Boundary Layer on a Flat Surface and in Regions of Separated Flow, *Aero. Res. Counc. London,* R&M 3148, 1959.
51. G. E. Gadd, A Note on the Theory of the Stanton Tube, *Aero. Res. Counc. London,* R&M No. 3147, 1958.
52. M. C. Good and P. N. Joubert, The Form Drag of Two-Dimensional Bluff-Plates Immersed in Turbulent Boundary Layers, *J. Fluid Mech.,* vol. 31, pp. 547-582, 1968.
53. E. Achenbach, Distribution of Local Pressure and Skin Friction around a Circular Cylinder in Cross-Flow up to $Re = 5 \times 10^6$, *J. Fluid Mech.,* vol. 34, pp. 625-639, 1968.
54. E. Achenbach, Experiments on Flow Past Spheres at Very High Reynolds Numbers, *J. Fluid Mech.,* vol. 54, pp. 565-575, 1972.
55. D. A. Spence and G. L. Brown, Heat Transfer to a Quadratic Shear Profile, *J. Fluid Mech.,* vol. 33, pp. 753-773, 1968.
56. J. S. Son, Limiting Relation for the Eddy Diffusivity Close to a Wall, M.S. thesis, Chemical Engineering Department, University of Illinois, Urbana, 1965.
57. J. S. Son and T. J. Hanratty, Limiting Relation for the Eddy Diffusivity Close to a Wall, *AIChE J.,* vol. 13, pp. 689-696, 1967.
58. D. A. Shaw and T. J. Hanratty, Turbulent Mass Transfer Rates to a Wall for Large Schmidt Numbers, *AIChE J.,* vol. 23, pp. 28-37, 1977.
59. S. C. Ling, Heat Transfer from a Small Isothermal Spanwise Strip on an Insulated Boundary, *J. Heat Transfer,* vol. C85, pp. 230-236, 1963.
60. K. R. Jolls, Flow Patterns in a Packed Bed, Ph.D. thesis, Chemical Engineering Department, University of Illinois, Urbana, 1966.
61. M. LeBouche, Transfert de matiere en regime de couche limite bidimensionnelle et a nombre de Schmidt grand, *C. R. Acad. Sci.,* vol. 270A, pp. 1757-1760, 1970.
62. M. LeBouche, Sur la Mesure polarographique de gradient parietal de vitesse dams les zones d'arret amont ou de decollement du cylinder, *C. R. Acad. Sci.,* vol. 276A, pp. 1245-1248, 1973.
63. W. J. McMichael, private communication, 1972.
64. T. J. Pedley, Hot Film in Reversing Flow, *J. Fluid Mech.,* vol. 78, pp. 513-584, 1976.
65. B. Py, Les Proprietes generales des transducteurs electrochimiques scindes, *Euromech. 90 Proc.,* 1977.
66. T. J. Hanratty, L. G. Chorn, and D. T. Hatziavramidis, Turbulent Fluctuations in the Viscous Wall Region for Newtonian and Drag Reducing Fluids, *Phys. Fluids Supp.,* pp. S112-S119, 1977.
67. K. K. Sirkar, Turbulence in the Immediate Vicinity of a Wall and Fully Developed Mass Transfer at High Schmidt Numbers, Ph.D. thesis, Chemical Engineering Department, University of Illinois, Urbana, 1969.
68. J. E. Mitchell, Investigation of Wall Turbulence Using a Diffusion-Controlled Electrode, Ph.D. thesis, Chemical Engineering Department, University of Illinois, Urbana, 1965.
69. A. J. Karabelas and T. J. Hanratty, Determination of the Direction of Surface Velocity Gradients in Three-Dimensional Boundary Layers, *J. Fluid Mech.,* vol. 34, pp. 159-162, 1968.
70. P. Duhamel and B. Py, Caractere intermittent de la sous-couche visqueuse, *AAAE 9e Colloque D'Aerodynamique Appliquee*, Paris, 1972.
71. C. Tournier and B. Py, Analyse et reconstitution spatiotemporelle de la composante circonferentielle de vitesse instantanee a proximite immediate de la paroi d'ur cylindre, *C. R. Acad. Sci.,* vol. 276A, pp. 403-406, 1973.

72. F. K. Owen and B. J. Bellhouse, Skin Friction Measurements at Supersonic Speeds, *AIAA J.*, vol. 8, pp. 1358–1360, 1970.
73. N. S. Diaconis, The Calculation of Wall Shearing Stresses from Heat-Transfer Measurements in Compressible Flow, *J. Aeronaut. Sci.*, vol. 21, pp. 201–202 (errata, p. 499), 1954.
74. J. S. Newman, *Electrochemical Systems*, pp. 1–25, Prentice-Hall, Englewood Cliffs, N.J., 1973.
75. T. J. Hanratty, Study of Turbulence Close to a Solid Wall, *Phys. Fluids Supp.*, pp. 5126–5133, 1967.
76. W. E. Ranz, Electrolytic Methods for Measuring Water Velocities, *AIChE J.*, vol. 4, pp. 338–342, 1958.
77. I. M. Kolthoff and J. J. Lingane, *Polarography*, Interscience, New York, 1952.
78. C. S. Lin, E. B. Denton, H. S. Gaskill, and G. L. Putnam, Diffusion-Controlled Electrode Reactions, *Ind. Eng. Chem.*, vol. 43, pp. 2136, 2143, 1951.
79. L. P. Reiss, Turbulent Mass Transfer to Small Sections of a Pipe Wall, M.S. thesis, Chemical Engineering Department, University of Illinois, Urbana, 1960.
80. T. Mizushina, The Electrochemical Method in Transport Phenomena, *Adv. Heat Transfer*, vol. 1, pp. 87–159, 1971.
81. A. M. Sutey and J. G. Knudsen, Mass Transfer at the Solid-Liquid Interface for Climbing Film Flow in an Annular Duct, *AIChE J.*, vol. 15, p. 719, 1969.
82. M. Eisenberg, C. W. Tobias, and C. R. Wilke, Mass Transfer at Rotating Cylinders, *Chem. Eng. Prog. Symp. Ser.*, vol. 51, no. 16, pp. 1–16, 1955.
83. E. A. Vallis, M. A. Patrick, and A. A. Wragg, Radial Distribution of Wall Fluxes in the Wall Jet Region of a Flat Plate Held Normal to an Axisymmetric, Turbulent, Impinging Jet, *Euromech. 90 Proc.*, Session IIIc, pp. 1–14, 1977.
84. J. D. Jenkins and B. Gay, Experience in the Use of the Ferri-Ferrocyanide Redox Couple for the Determination of Transfer Coefficients in Models in Shell and Tube Heat Exchangers, *Euromech. 90 Proc.*, Session IIIa, pp. 1–20, 1977.
85. D. A. Shaw, Mechanism of Turbulent Mass Transfer to a Pipe Wall at High Schmidt Number, Ph.D. thesis, Chemical Engineering Department, University of Illinois, Urbana, 1976.
86. J. G. A. Hogenes, Identification of the Dominant Flow Structure in the Viscous Wall Region of a Turbulent Flow, Ph.D. thesis, Mechanical Engineering Department, University of Illinois, Urbana, 1979.
87. L. G. Chorn, An Experimental Study of Near-Wall Turbulence Properties in Highly Drag Reduced Pipe Flows of Pseudoplastic Polymer Solutions, Ph.D. thesis, Chemical Engineering Department, University of Illinois, Urbana, 1978.
88. I. M. Kolthoff and N. H. Furman, *Volumetric Analysis*, vol. 1, p. 231, Wiley, New York, 1928.
89. G. A. McConaghy, The Effect of Drag Reducing Polymers on Turbulent Mass Transfer, Ph.D. thesis, University of Illinois, Urbana, 1974.
90. G. Fortuna, Effect of Drag-Reducing Polymers on Flow near a Wall, Ph.D. thesis, Chemical Engineering Department, University of Illinois, Urbana, 1971.
91. J. S. Son, Experimental and Computational Studies of Flow around a Circular Cylinder, Ph.D. thesis, Chemical Engineering Department, University of Illinois, Urbana, 1968.
92. L. P. Reiss, Investigation of Turbulence near a Pipe Wall Using a Diffusion Controlled Electrolytic Reaction on a Circular Electrode, Ph.D. thesis, Chemical Engineering Department, University of Illinois, Urbana, 1962.
93. D. T. Chin, An Experimental Study of Mass Transfer on a Rotating Spherical Electrode, *J. Electrochem. Soc.*, vol. 118, p. 1764, 1971.
94. J. C. Bazan and A. J. Arvia, The Diffusion of Ferro- and Ferricyanide Ions in Aqueous Solutions of Sodium Hydroxide, *Electrochm. Acta*, vol. 10, p. 1025, 1965.
95. J. D. Newson and A. C. Riddiford, Limiting Currents for the Reduction of the Tri-iodide Ion at a Rotating Platinum Disk Cathode, *J. Electrochem. Soc.*, vol. 108, p. 695, 1961.
96. J. S. Anderson and K. Saddington, The Use of Radioactive Isotopes in the Study of the Diffusion of Ions in Solution, *J. Chem. Soc.*, pp. S381–S386, 1949.

97. M. Eisenberg, C. W. Tobias, and C. R. Wilke, Selected Physical Properties of Ternary Electrolytes Employed in Ionic Mass Transfer Studies, *J. Electrochem. Soc.*, vol. 103, pp. 413–416, 1956.
98. D. A. Shaw, Measurement of Frequency Spectra of Turbulent Mass Transfer Fluctuations at a Pipe Wall, M.S. thesis, Chemical Engineering Department, University of Illinois, Urbana, 1973.
99. G. Nickless (ed.), *Inorganic Sulfur Chemistry*, pp. 200–239, Elsevier, New York, 1968.
100. J. Vulterin and J. Zyka, Investigation of Some Hydrazine Derivatives as Reductimetric Titrants, *Talanta*, vol. 10, pp. 891–898, 1963.
101. R. E. Hicks and N. Pagotto, CSIR Rep. CENG M-024, Pretoria, South Africa, 1974.
102. H. V. Malmstadt, C. G. Enke, and S. F. Crouch, *Electronic Measurements for Scientists*, pp. 94–97, Benjamin, Reading, Mass., 1974.
103. T. J. Hanratty, The Use of Electrochemical Techniques to Study Flow Fields and Mass Transfer Rates, in Z. Zaric (ed.), *Heat and Mass Transfer in Flows with Separated Regions*, pp. 139–159, Pergamon, New York, 1972.
104. L. D. Eckelman, The Structure of Wall Turbulence and Its Relation to Eddy Transport, Ph.D. thesis, Chemical Engineering Department, University of Illinois, Urbana, 1971.
105. A. Saldeen, personal communication, 1980.
106. V. A. Sandborn, *Resistance Temperature Transducers*, pp. 361–365, Metrology, Fort Collins, Colo., 1972.
107. *DISA Probe Manual*, DISA Electronics, Denmark, 1976.
108. M. W. Rubesin, A. F. Okuno, L. L. Levy, and J. B. McDevitt, and H. L. Seegmiller, An Experimental and Computational Investigation of the Flow Field about a Transonic Airfoil in Supercritical Flow with Turbulent Boundary Layer Separation, NASA TM X-73, p. 157, 1976.
109. R. F. Blackwelder and H. Eckelmann, Influence of the Convection Velocity on the Calibration of Heated Surface Elements, *Euromech. 90 Proc.*, Session I.a.1, pp. 1–10, 1977.
110. D. H. Pessoni, An Experimental Investigation into the Effects of Wall Heat Flux on the Turbulence Structure of Developing Boundary Layers at Moderately High Reynolds Numbers, Ph.D. thesis, Mechanical and Industrial Engineering Department, University of Illinois, Urbana, 1974.
111. D. B. Simons, R. M. Li, and F. D. Schall, Spatial and Temporal Distribution of Boundary Layer Shear Stress in Open Channel Flows, NSF Final Rep., Civil Engineering Department, Colorado State University, 1979.
112. J. E. Coney and D. A. Simmers, The Determination of Shear Stress in Fully Developed Laminar Axial Flow and Taylor Vortex Flow, Using a Flush-Mounted Hot Film Probe, DISA Info. Bull. 24, pp. 9–14, 1979.
113. K. F. Wylie and C. V. Alonso, Some Stochastic Properties of Turbulent Tractive Forces in Open Channel Flow, *Proc. of the Fifth Biennial Symp. on Turbulence*, Rolla, Mo., p. 181, 1979.
114. G. R. Foster, Hydraulics of Flow in a Rill, Ph.D. Thesis, Purdue University, 1977.
115. O. Christensen, New Trends in Hot-Film Probe Manufacturing, DISA Inf. Bull. 9, pp. 30–36, 1970.
116. K.E.W., Coated Hot Film Probes for Measurement in Liquids, DISA Inf. No. 3, p. 28, 1966.
117. M. Miya, Properties of Roll Waves, Ph.D. Thesis, Chemical Engineering Department, University of Illinois, Urbana, 1970.
118. *TSI Operating Manual for Model 1050 and 1050A Anemometer Modules*, pp. 2–8, Thermo Systems Inc., St. Paul, Minn., 1973.
119. A. Bertelrud, Measurement of Shear Stress in a Three-Dimensional Boundary Layer by Means of Heated Films, *Euromech. 90 Proc.*, Session I.b.2, pp. 1–14, 1977.
120. H. P. Kreplin and H. U. Meier, Application of Heated Element Techniques for the Measurement of the Wall Shear Stress in Three-Dimensional Boundary Layers, *Euromech. 90 Proc.*, Session I.B.3, pp. 1–21, 1977.
121. H. Schlichting, *Boundary Layer Theory*, 6th ed., pp. 154–162, McGraw-Hill, New York, 1960.
122. K. Hiemenz, Die Grenzschicht an einem in den Gleichformigen Flussiqkeitsstrom Eingetauchten Geraden Kreiszylinder, *Dingl. Polytechn. J.*, vol. 326, pp. 321–324, 1911.

INDEX

Page numbers in italics indicate figures or tables; *n* after a page number indicates a footnote.